Two-component Systems in Bacteria

Edited by

Roy Gross and Dagmar Beier

Biocenter
University of Würzburg
Würzburg
Germany

Caister Academic Press

Caister Academic Press
Norfolk, UK

www.caister.com

British Library Cataloguing-in-Publication Data
A catalogue record for this book is available from the British Library

ISBN: 978-1-908230-08-9

Cover design adapted from Figure 4.5

Printed and bound in Great Britain

Contents

Contributors

Nicole Acosta
Department of Biological Sciences
University of Alberta
Edmonton, AB
Canada

acostaam@ualberta.ca

Rudy Antoine
Institut Pasteur de Lille
Center for Infection and Immunity of Lille
Université Lille Nord de France
CNRS UMR 8204, F-59021
INSERM U1019, F-59019
Lille
France

rudy.antoine@pasteur-lille.fr

Carl Bauer
Department of Molecular and Cellular Biochemistry
Indiana University
Bloomington, IN
USA

bauer@indiana.edu

Dagmar Beier
Theodor-Boveri-Institut für Biowissenschaften
Lehrstuhl für Mikrobiologie
University of Würzburg
Würzburg
Germany

d.beier@biozentrum.uni-wuerzburg.de

Patricia Casino
Departamento de Genómica y Proteómica
Instituto de Biomedicina de Valencia (IBV-CSIC) and CIBERER
Valencia
Spain

pcasino@ibv.csic.es

David J. Clarke
Department of Microbiology and Alimentary Pharmabiotic Centre
University College Cork
Cork
Ireland

david.clarke@ucc.ie

Vladimira Dragnea
Department of Molecular and Cellular Biochemistry
Indiana University
Bloomington, IN
USA

vdragnea@indiana.edu

Jenny Draper
Department of Biomolecular Engineering
University of California at Santa Cruz
Santa Cruz, CA
USA

learithe@gmail.com

Elian Dupré
Institut Pasteur de Lille
Center for Infection and Immunity of Lille
Université Lille Nord de France
CNRS UMR 8204, F-59021
INSERM U1019, F-59019
Lille
France

elian.dupre@ibl.fr

Yoko Eguchi
Department of Bioscience
Graduate School of Agriculture
Kinki University
Nara
Japan

nc_eguti@nara.kindai.ac.jp

Roy Gross
Biocenter
University of Würzburg
Würzburg
Germany

roy@biozentrum.uni-wuerzburg.de

J. Mitchell Guss
School of Molecular Bioscience
The University of Sydney
Sydney
Australia

mitchell.guss@sydney.edu.au

Ralf Heermann
Center for Integrated Protein Science Munich
Ludwig-Maximilians-Universität München
Biozentrum
Bereich Mikrobiologie
Munich
Germany

heerman@lmu.de

Julien Herrou
Department of Biochemistry and Molecular Biology
The University of Chicago
Chicago, IL
USA

jherrou@uchicago.edu

Stuart Huntley
Department of Ecophysiology
Max Planck Institute for Terrestrial Microbiology
Marburg
Germany

huntley@mpi_marburg.mpg.de

Eiji Ishii
Department of Bioscience
Graduate School of Agriculture
Kinki University
Nara
Japan

ag1410322@gmail.com

Françoise Jacob-Dubuisson
Institut Pasteur de Lille
Center for Infection and Immunity of Lille
Université Lille Nord de France
CNRS UMR 8204, F-59021
INSERM U1019, F-59019
Lille
France

francoise.jacob@ibl.fr

David A. Jacques
School of Molecular Bioscience
The University of Sydney
Sydney
Australia

david.jacques@sydney.edu.au

Kirsten Jung
Center for Integrated Protein Science Munich
Ludwig-Maximilians-Universität München
Biozentrum
Bereich Mikrobiologie
Munich
Germany

jung@lmu.de

Daniela Keilberg
Department of Ecophysiology
Max Planck Institute for Terrestrial Microbiology
Marburg
Germany

daniela.keilberg@mpi-marburg.mpg.de

Paphavee Lertsethtakarn
Department of Microbiology and Environmental Toxicology
University of California at Santa Cruz
Santa Cruz, CA
USA

plertset@hotmail.com

Marisa López-Redondo
Departamento de Genómica y Proteómica
Instituto de Biomedicina de Valencia (IBV-CSIC) and CIBERER
Valencia
Spain
and
Electron Microscopy of Cells and Macromolecular Complexes
Skirball Institute
NYU Medical Center
New York
USA

maria.lopezredondo@nyumc.org

Alberto Marina
Departamento de Genómica y Proteómica
Instituto de Biomedicina de Valencia (IBV-CSIC) and CIBERER
Valencia
Spain

amarina@ibv.csic.es

Juan-Francisco Martín
Departamento de Biología Molecular
Universidad de León
Instituto de Biotecnología de León (INBIOTEC)
León
Spain

jf.martin@unileon.es

Thorsten Mascher
Department of Biology I
Microbiology
Ludwig-Maximilians-University Munich
Munich
Germany

mascher@bio.lmu.de

Alexander J. Ninfa
Department of Biological Chemistry
University of Michigan Medical School
Ann Arbor, MI
USA

aninfa@umich.edu

Karen M. Ottemann
Department of Microbiology and Environmental Toxicology
University of California at Santa Cruz
Santa Cruz, CA
USA

Ottemann@ucsc.edu

Tracy Raivio
Department of Biological Sciences
University of Alberta
Edmonton, AB
Canada

traivio@ualberta.ca

Valerie A. Ray
Department of Microbiology and Immunology
Loyola University Chicago
Chicago, IL
USA

vray@lumc.edu

Antonio Rodríguez-García
Departamento de Biología Molecular
Universidad de León
Instituto de Biotecnología de León (INBIOTEC)
León
Spain

arodg@unileon.es

Kathleen R. Ryan
Department of Plant & Microbial Biology
University of California
Berkeley, CA
USA

krr@berkeley.edu

Karen Schrecke
Department of Biology I
Microbiology
Ludwig-Maximilians-University Munich
Munich
Germany

karen.schrecke@campus.lmu.de

Stephen C. Smith
Department of Plant & Microbial Biology
University of California
Berkeley, CA
USA

scsmith@berkeley.edu

Lotte Søgaard-Andersen
Department of Ecophysiology
Max Planck Institute for Terrestrial Microbiology
Marburg
Germany

sogaard@mpi-marburg.mpg.de

Alberto Sola-Landa
Instituto de Biotecnología de León (INBIOTEC)
León
Spain

alberto.sola@unileon.es

Anna Staroń
Department of Biology I
Microbiology
Ludwig-Maximilians-University Munich
Munich
Germany

anna.staron@bio.lmu.de

Hendrik Szurmant
The Scripps Research Institute
Department of Molecular and Experimental Medicine
La Jolla, CA
USA

szurmant@scripps.edu

Jill Trewhella
School of Molecular Bioscience
The University of Sydney
Sydney
Australia

jill.trewhella@sydney.edu.au

Ryutaro Utsumi
Department of Bioscience
Graduate School of Agriculture
Kinki University
Nara
Japan

utsumi@nara.kindai.ac.jp

Juan-Jesus Vicente
Department of Plant & Microbial Biology
University of California
Berkeley, CA
USA

jjvicente@berkeley.edu

Karen L. Visick
Department of Microbiology and Immunology
Loyola University Chicago
Chicago, IL
USA

kvisick@lumc.edu

Stefanie Vogt
Department of Biological Sciences
University of Alberta
Edmonton, AB
Canada

svogt@ualberta.ca

Junshu Wang
Department of Biological Sciences
University of Alberta
Edmonton, AB
Canada

junshu@ualberta.ca

David E. Whitworth
Institute of Biological, Environmental and Rural Sciences
Aberystwyth University
Aberystwyth
UK

dew@aber.ac.uk

René Wintjens
Service de Chimie Générale (CP206/04)
Faculté de Pharmacie
Université Libre de Bruxelles
Brussels
Belgium

rene.wintjens@ulb.ac.be

Julia Wong
Department of Biological Sciences
University of Alberta
Edmonton, AB
Canada

jlw7@ualberta.ca

Jiang Wu
Department of Molecular and Cellular Biochemistry
Indiana University
Bloomington, IN
USA

jw8@indiana.edu

Current books of interest

Preface

Two-component (TCS) systems are the predominant signal transduction devices of Eubacteria, but in a limited number they are also present in Archaea and some Eukaryotes. Most TCS have a relatively simple modular architecture which generally requires at least four domains (signal input, transmitter, receiver, and signal output domains) distributed among two proteins, the sensory histidine kinase and the response regulator. During evolution, their modular architecture has enabled the employment of these systems in many different cellular processes, which need to be regulated either by extracellular of intracellular stimulus changes, by simply exchanging signal input and output modules flanking the conserved signal transducing transmitter and receiver modules. In consequence, these systems are able to respond to an extremely broad spectrum of diverse stimuli and control a vast array of cellular functions. For instance, some TCS are involved in the regulation of very basic metabolic processes such as the energy metabolism, but also in important basic features of bacterial life such as cell division and differentiation. Moreover, TCS are also engaged in special requirements of bacteria with a dedicated life style such as pathogens or symbionts which need to interact with host species and to control virulence or symbiosis factors. Highlighting the enormous impact of TCS on the biology of Eubacteria, genome sequencing revealed the presence of TCS in virtually all of them. Some environmental bacteria with quite large genomes may encode hundreds of TCS proteins. On the other hand, bacteria living in niches believed to be quite constant regarding the physicochemical conditions such as obligate intracellular pathogens or symbionts encode a strongly reduced number of TCS or, in rare cases, they are lacking them entirely.

Many basic questions about the molecular function of TCS including structure–function relationships of the signalling domains and signal transduction fidelity have been characterized since their discovery in the late 80s of the last century. Nevertheless, many basic features remain to be investigated. For instance, a major issue still concerns the identification of the stimuli perceived by many TCS and the way how these stimuli are actually sensed. Moreover, the interconnection of different TCS by cross-talk or via so-called connector proteins which allow the establishment of complex regulatory networks is a challenging task for future research.

Since their discovery thousands of articles were published about many aspects of TCS and, accordingly, it is impossible to handle these systems in a single book in a comprehensive way. Therefore, several novel aspects of signal transduction mechanisms by TCS and of their biological function in different bacteria have been selected. The book starts with a chapter illustrating bioinformatics approaches to describe and characterize TCS. Further issues discussed involve novel insights in signal perception mechanisms by TCS and the role and mode of action of atypical and essential TCS. Finally, several examples of TCS engaged in cellular differentiation processes and bacterial pathogenesis and symbiosis are described. In conclusion, the book represents a compendium of current TCS research and hopefully may help to identify important future research activities.

Finally, we would like to thank all authors for their great collaboration during preparation of this book.

Roy Gross and Dagmar Beier

Classification and Organization of Two-component Systems

1

David E. Whitworth

Abstract

Two-component systems (TCSs) are common signal transduction pathways found abundantly in most phyla except animals. The basic TCS pathway involves two multidomain proteins. The first is a histidine protein kinase (HPK), whose autokinase activity is dependent upon an environmental stimulus. The second is a response regulator (RR), onto which a phosphoryl group is transferred from the phosphorylated HPK, and which mediates phosphorylation-dependent effects within the cell. TCS proteins can be readily identified from genomic sequences, however, approaches for identifying and classifying TCS proteins are non-trivial because TCSs are multidomain, multigene pathways, which exhibit considerable heterogeneity in their gene and domain organization.

This chapter provides an overview of the domain architectures and genetic organization of TCSs, including the common TCS variants known as phosphorelays and chemosensory systems. Strategies for the identification of TCS genes are described, alongside systems for categorizing TCS proteins that exploit the evolutionary relationships between TCS proteins to provide biologically relevant insights. The development of such classification systems are reviewed in addition to the current state of the art, to help the reader navigate through the extensive literature on this topic.

Introduction

Two-component systems (TCSs) are relatively simple signalling pathways found predominantly in prokaryotes. First identified as a distinct class of signalling pathway 25 years ago (Nixon *et al.*, 1986), tens of thousands of TCSs have now been identified in sequenced bacterial genomes and metagenomes (Ulrich and Zhulin, 2007; Barakat *et al.*, 2009). The average eubacterial genome encodes 50–60 TCS proteins (5–15 in Archaebacteria and Eukaryotes), but some organisms possess more than 250 (Ashby, 2006; Whitworth, 2008; Wuichet *et al.*, 2010). TCSs regulate a wide variety of physiological processes in response to changing environmental parameters (for diverse examples see Comeau *et al.*, 1985; Wheeler and Shapiro, 1999; Takai *et al.*, 2006), including those experienced during the infection cycle by pathogenic organisms (Ewann *et al.*, 2002; Beier and Gross, 2006, 2008; Liang *et al.*, 2006; Sivaneson *et al.*, 2011). Molecular studies undertaken in many laboratories across the world have uncovered the molecular mechanisms of signalling through TCS pathways, and the structures of many TCS proteins have now been deduced (Robinson *et al.*, 2000; Stock *et al.*, 2000; West and Stock, 2001; Gao and Stock, 2009; Bourret, 2010; Stewart, 2010).

The wealth of available structural and mechanistic information has provided profound insights into the structure–function relationship of TCSs, and general principles regarding TCS signal transduction are emerging. The 'defining' feature of a typical TCS is the regulated transfer of a phosphoryl group from His to Asp amino acid residue side-chains, typically found on two separate proteins. The first protein component of a TCS is a histidine protein kinase (HPK), which acts as a signal receptor. Under stimulatory conditions HPKs autophosphorylate on a conserved His residue, with the terminal phosphoryl group of ATP being the donated phosphoryl group (Stock *et al.*, 1989). The second of the two protein components of TCSs are known as response regulators (RRs) which function as effector molecules. RRs contain several conserved Asp residues, one of which can be phosphorylated, and the effector activities of the RR are regulated by its phosphorylation state (Stock *et al.*, 1989; Parkinson and Kofoid, 1992). When a HPK has been stimulated and autophosphorylated, it binds to its partner RR and the phosphoryl group is then transferred directly from the His residue of the HPK, to the Asp residue of the RR, propagating the initial signal (Fig. 1.1). This 'direct' phosphotransfer is the most obvious difference between TCSs and the majority of eukaryotic signalling pathways, in which phosphorylation (of a Ser/Thr/Tyr residue) allosterically activates a kinase to phosphorylate other target proteins, which can often include further protein kinases (Johnson, 2011). In such systems, the activating phosphoryl group remains on the activated kinase, and is not transferred to a further substrate protein.

The signal transducing activities of TCS proteins reside within conserved protein domains, which makes databases of domain architecture such as Pfam (Finn *et al.*, 2010) very useful when identifying TCSs. The autokinase and phosphotransfer activities of HPKs are found to lie within a conserved structural unit known as a 'transmitter' domain (Kofoid and Parkinson, 1988). Transmitter domains are actually the combination of two domains – an N-terminal HisKA (Pfam entry PF00512) domain, and a C-terminal HATPase (Pfam entry PF02518) domain. HPKs are active as dimers and dimerization is also mediated by the transmitter unit, through the HisKA domain. The dimeric HisKA domain forms a four-helix bundle (Fig. 1.2) which contains the conserved HPK His residue, while the HATPase domain is monomeric, globular and has ATPase activity (Figs. 1.1 and 1.2). RRs possess a 'receiver' domain (Pfam entry PF00072; Kofoid and Parkinson, 1988) which contains the conserved Asp residue (Fig. 1.3), and which interacts with the phosphorylated

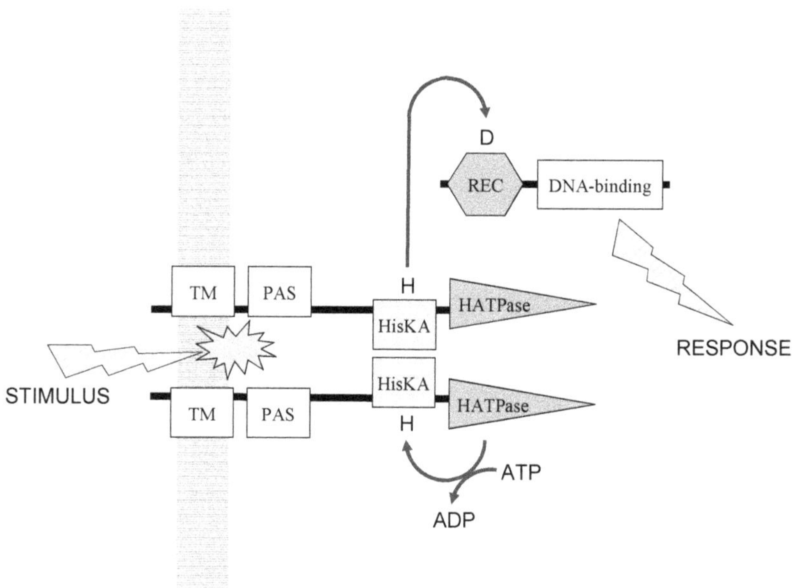

Figure 1.1 Two-component system (TCS) signalling pathways. In a typical TCS an environmental parameter is perceived by sensory domains (e.g. PAS domains) in the N-terminal region of a histidine protein kinase (HPK). HPKs generally sense stimuli across membranes, and TM denotes a transmembrane region. Stimulus perception leads to activation of the HPK HATPase domain, which leads to phosphorylation of a conserved His (H) residue in the HPK HisKA domain (HisKA domains also mediate HPK dimerization). The phosphorylated HisKA domain then interacts with the receiver domain (REC) of the partner response regulator (RR), leading to phosphotransfer onto a conserved Asp (D) residue of the receiver domain. Phosphorylation of the RR affects the activity of a C-terminal 'effector' domain (e.g. a DNA-binding domain), which results in a response to the original stimulus. Arrows denote the passage of phosphoryl groups during signal transduction.

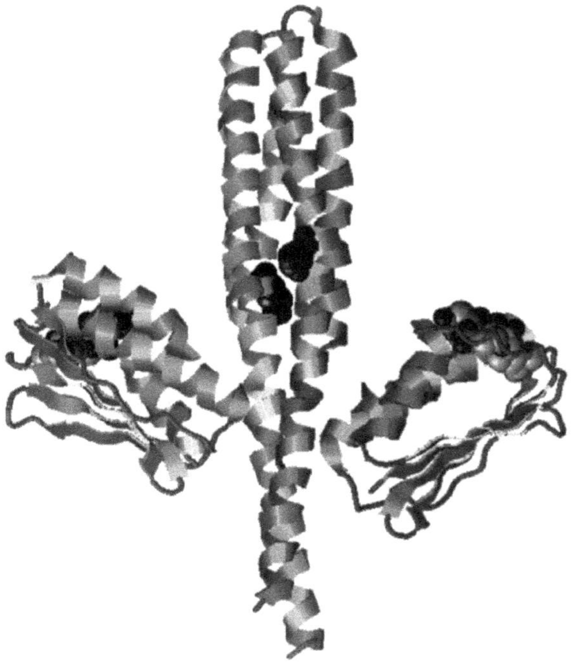

Figure 1.2 Structure of a TCS transmitter domain. The transmitter domain dimer of DesK. Globular HATPase domains (complexed with the ATP analogue AMP-PCP) are found either side of a central four-helix bundle containing a phosphorylatable His residue (dark space-filled side chains). Image created from PDB file 3GIE by Stuart Thomas.

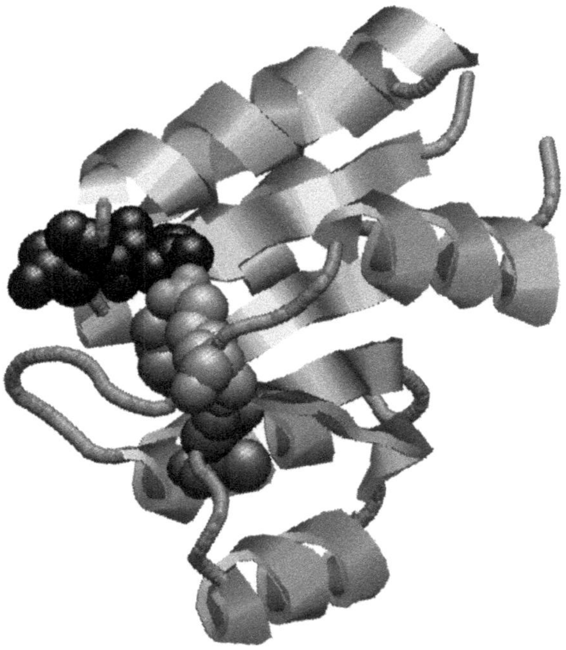

Figure 1.3 Structure of a RR receiver domain. Spo0F is a typical receiver domain, composed of five alternating sets of alpha-helices and beta-strands. A catalytic pocket is formed by conserved residues including the phosphorylatable Asp residue (space-filled side chains). Image created from PDB file 1NAT by Heather Currinn.

HisKA domain to enable HPK-RR phosphotransfer (Fig. 1.1).

While the majority of TCSs conform to the paradigm presented above, numerous systems differ significantly in their mechanisms of signal transduction, and/or their structural/genetic organization. For instance, HPKs contain a plethora of sensory domains (Mascher *et al.*, 2006; Krell *et al.*, 2010), and RRs can be found with a wide range of different effector domains (Galperin *et al.*, 2001, 2010). Many TCSs contain more than one transmitter and/or receiver domain, and how transmitter and receiver units are distributed between different proteins can vary enormously (Williams and Whitworth, 2010). Therefore, classification of TCSs is non-trivial, and a variety of approaches have emerged for identifying and categorizing TCS proteins.

Analysis of TCSs

In the earliest days of TCS research, proteins had to be identified as TCS proteins experimentally. A variety of experimental approaches were employed to assess autophosphorylation, phosphatase and phosphotransfer activities of purified proteins. Methods often followed the transfer of radioactive ^{32}P between proteins, ATP and/or water, and such approaches are still commonly used to study aspects of TCS signal transduction (Da Re *et al.*, 1999a; Skerker *et al.*, 2005; Yamamoto *et al.*, 2005; Laub *et al.*, 2007; Wegener-Feldbrügge *et al.*, 2009; Scharf, 2010). Phosphorylated amino acid residues differ in their chemical/physical stability, allowing experimental analysis of whether phosphorylation had occurred on His, Asp or other amino acid residues. Therefore by using a few standard experiments it was also feasible to define proteins under study as HPKs or RRs (Hess *et al.*, 1991; Scharf, 2010).

With the realization that HPK and RR activities resided in conserved domains, it became possible to infer HPK/RR activities from protein/gene sequences alone, allowing experimenters to focus their efforts onto proteins already highly likely to be true TCS proteins. The advent of whole genome sequences then revolutionized

TCS research, by making it possible to computationally deduce entire genomic complements of TCS genes (for example, see Mizuno *et al.*, 1996; Mizuno, 1997; Fabret *et al.*, 1999; Hutchings *et al.*, 2004; Whitworth and Cock, 2008b), rather than relying on error-prone and time-consuming experimental methods such as Southern blotting. In turn this enabled research to be undertaken into system-wide features of TCS biology, such as cross-talk (the ability of one TCS to influence signal transduction through a second TCS) and the evolution of TCSs (Skerker *et al.*, 2005; Gao *et al.*, 2008; Whitworth and Cock, 2009).

Owing to the large body of knowledge regarding the structure–function relationships within TCSs, coupled with abundant genome sequence data, it is now possible to infer various properties of TCS proteins merely from their gene sequences (for an example see Alves and Savageau, 2003). In addition, analysis of multiple TCS gene sequences can provide information regarding their evolution, signalling mechanisms and physiological roles (Weigt *et al.*, 2009; Whitworth and Cock, 2009; Lunt *et al.*, 2010; Ashenberg *et al.*, 2011). However, the ability to perform such analyses relies on consistent and appropriate systems for the classification of TCS genes/proteins. A variety of categorization systems have been used in studies of TCSs, however an over-arching classification scheme has yet to emerge.

A related issue to that of consistent classification regards the representation of TCSs in on-line biological databases. Complete genome sequences (and their encoded TCSs) are becoming available at an ever-accelerating rate (Fig. 1.4; see also Whitworth, 2008). To be of maximal benefit to the research community, the annotation of TCSs in databases should be consistent, and provide as much pertinent information as possible. Currently, the annotation of TCS genes/proteins in databases is often idiosyncratic, misleading or downright wrong. There is therefore a profound need for the adoption of a universal categorization system based on defined criteria which can be applied across all TCS – a real challenge given the diversity of TCSs found in nature.

There have been many recent reviews on various facets of TCS biology, including the structure–function relationship, regulatory mechanisms, evolution, genomics, etc. of TCSs (Whitworth and Cock, 2009; Bourret, 2010; Gao and Stock, 2010; Goulian, 2010; Krell *et al.*, 2010; Stewart, 2010; Wuichet and Zhulin, 2010, Whitworth, 2011). Focusing to some extent on

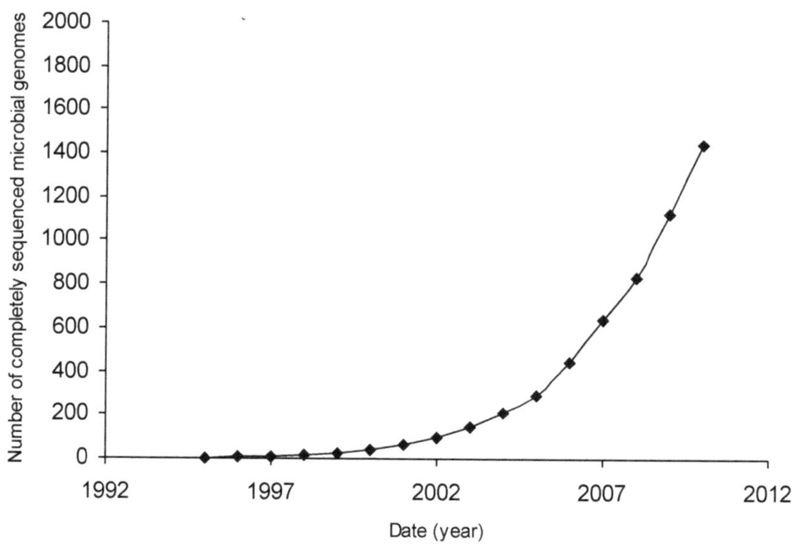

Figure 1.4 An exponential increase in genomic data. The number of completely sequenced prokaryotic genome sequences is plotted against time (release dates of completed genomes taken from http://www.ncbi.nlm.nih.gov/genomes/lproks.cgi). Every three years the number of completed genomes doubles, and the 2000th prokaryotic genome sequence will be completed in 2012. If the current acceleration in sequencing rate continues, we could expect the millionth prokaryotic genome to become available as early as 2037 (with around 50 million TCS gene sequences).

eubacterial TCSs, this chapter provides an overview of the domain architectures of TCS proteins, describes the genetic organization of TCSs, and compares classification systems that have been applied to TCSs.

Common TCS architectures and variants

Typical TCSs are encoded as two genes, usually adjacent in the genome, with one gene encoding a HPK and the other encoding a RR (Williams and Whitworth, 2010). However, in a large proportion of cases (15%; Cock and Whitworth, 2007b), the genes encoding a simple TCS (a single HPK and RR gene pair), appear to have fused together resulting in one gene encoding a protein with both a transmitter domain and a receiver domain. Such proteins are typically described as hybrid kinases and can function as entire TCSs in isolation (Fig. 1.5). Examples include HsfB and EspA of *Myxococcus xanthus* (Whitworth and Cock, 2008a). EspA has a FHA,PAS,PAS,transmitter,receiver domain architecture and regulates the timing of development (Cho and Zusman, 1999), while in HsfB the receiver domain precedes the transmitter domain, and the HsfB hybrid kinase gene is encoded alongside that of a partner response regulator, HsfA, which mediates the response to heat shock (Ueki and Inouye, 2002).

Another common modification of the basic TCS scheme is the involvement of further signalling domains, expanding the TCS into a larger pathway known as a phosphorelay (Fig. 1.5). Phosphorelays are the dominant form of TCS signalling in many eukaryotes (Santos and Shiozaki, 2001; Mizuno, 2005; Wuichet *et al.*, 2010). The phosphorelay starts with a HPK and ends with a RR, but rather than phosphotransfer occurring directly between the two proteins, the phosphoryl group is transferred firstly from the HPK onto a receiver domain (phosphorylated on an Asp

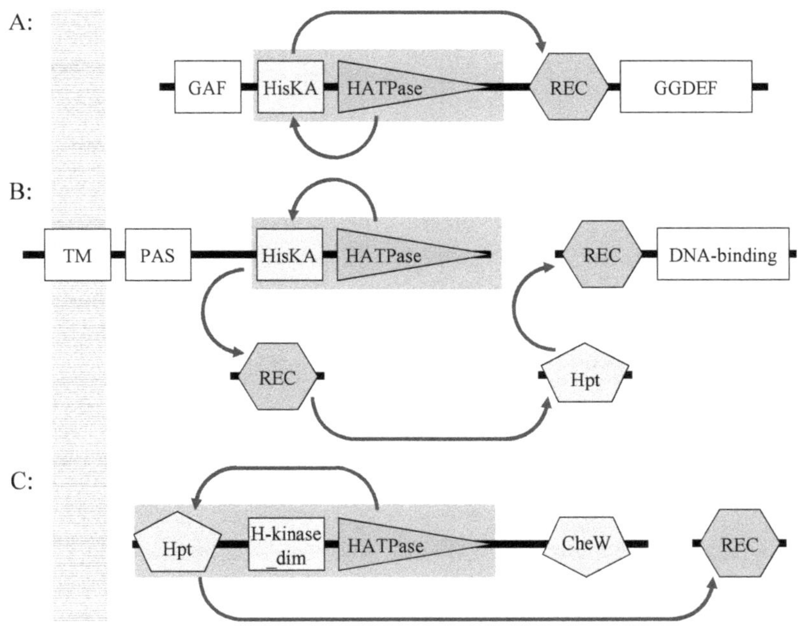

Figure 1.5 Hybrid kinases, phosphorelays and chemosensory TCSs. HPKs are shown as monomers for simplicity, with transmitter domains enclosed in grey boxes, and arrows denoting the passage of phosphoryl groups during signal transduction. (A) Some TCS are comprised fused HPK and RR proteins, containing both transmitter (paired HisKA and HATPase domains) and receiver domains. Such systems are typically cytoplasmic rather than transmembrane, and are referred to as hybrid kinases. (B) In phosphorelays, phosphoryl groups are transferred successively between transmitter, receiver, Hpt and receiver domains, sometimes encoded by as many as four separate proteins. (C) Chemosensory HPKs possess a relatively unusual variant of the typical transmitter domain, and no sensory domains. Instead they have a CheW domain for interacting with a separate membrane-bound receptor protein. Chemosensory systems are also usually cytoplasmic, and chemosensory RRs typically comprise just a receiver domain, without an associated effector domain.

residue), then onto a phosphotransfer domain (on a His residue), and then finally onto the RR (Appleby *et al.*, 1996). Phosphotransfer domains include isolated HisKA domains, normally found associated with the HATPase domain within a transmitter, and another four-helix bundle domain known as a Hpt (Pfam entry PF01627) domain. The four domains comprising a phosphorelay (transmitter, receiver 1, HisKA/Hpt, receiver 2) can be found on separate proteins. A good example of this is the sporulation phosphorelay of *Bacillus subtilis* (Perego, 1998), in which HPKs contribute to phosphorylation of the RR Spo0A (receiver2) through a relay involving Spo0F (a receiver1) and Spo0B (a Hpt/HisKA domain analogue). The phosphorylation states of Spo0F and Spo0A are regulated not only by kinases, but also by specific phospho-aspartyl phosphatases, resulting in an extremely sophisticated signalling network (Perego, 1998). However, two or more of the phosphorelay TCS domains are often located on single polypeptides. For instance, the TorRS, ArcBA and EvgSA phosphorelays of *Escherichia coli* are encoded by just two proteins (Mizuno, 1997). The first protein includes transmitter, receiver and Hpt domains (TorS, ArcB and EvgS), while the last receiver domain is found on a separate RR protein (TorR, ArcA and EvgA).

A final common departure from the typical TCS scheme worthy of mentioning here are chemosensory TCSs, often described as CheA/CheY systems after the prototypical chemosensory system of *E. coli* (Simon *et al.*, 1989). These TCSs typically regulate chemotactic behaviour, but are increasingly being found to also regulate diverse physiological processes including gene regulation (Kirby, 2009). Chemosensory TCSs possess 'atypical' transmitter domains, where a vestigial HisKA domain (H-kinase_dim) is present upstream of a HATPase domain, but only serves a role in dimerization, and is not phosphorylated by the HATPase domain. Instead a Hpt domain is found further upstream which is phosphorylated by the HATPase domain (Fig. 1.5) and which then transfers the phosphoryl group to a partner RR (Mourey *et al.*, 2001). These RRs are also unusual in that they have no obvious output domain, consisting solely of a receiver domain. As with standard TCSs and phosphorelays, both transmitter and receiver domains can often be found together on a single polypeptide as a hybrid (CheAY) kinase.

Thus there are several common variations of the typical TCS, with hybrid kinases, phosphorelays and chemosensory systems all adding complexity to the basic scheme.

Identification of TCS genes

The computational identification of TCS, phosphorelay or chemosensory proteins is most straightforwardly performed by looking for the presence of particular signalling domains (transmitter, receiver, HisKA and/or Hpt domains) in the protein sequence. Originally this was achieved by looking for sequence similarity with previously characterized TCS proteins/genes using programs such as BLAST (Altschul *et al.*, 1990). However, most TCS proteins contain multiple domains including sensory (input) and/or effector (output) domains in addition to domains that define the protein as part of a TCS. Input and output domains are usually also found in non-TCS proteins, so care must be taken that the region of sequence similarity with a TCS protein includes TCS-defining domains.

A simpler approach is to screen the protein/gene product of interest for the presence of a TCS-defining domain, made possible by databases of domain 'sequence profiles', such as Pfam (Finn *et al.*, 2010). A number of web-servers that allow users to input a query sequence, and return a domain architecture for the query protein are available, and a selection of these are described in the 'Web resources' section at the end of this chapter. Such resources enable users to identify conserved domains within a protein of interest with ease, however, when using domain profiles to identify and categorize TCS proteins there are still many potential pitfalls which researchers must be aware of.

Domain analysis of TCS proteins

Multiple sequence profile 'models' exist for several TCS domains, for a variety of reasons. Firstly, there has been duplication of profiles due to coverage of domains by more than one library of domain profiles. For instance receiver domains are represented in the Pfam (PF00072 – response_reg domain), and SMART (smart00448 – REC domain) databases (Letunic *et al.*, 2009; Finn *et al.*, 2010), and the two models can each miss some receiver domains identified by the other profile. Multiple profiles can also arise because of the definition of discrete subfamilies within a domain type as more domain sequences became available. For instance in Pfam there are multiple models for HisKA domains, some of which recognize only a relatively specific subset of HisKA domains: HisKA (PF00512), HisKA_2 (PF07568), HisKA_3 (PF07730), and HWE_HK (PF07536). These four profiles share some degree of sequence similarity, but if used in isolation, define overlapping subsets of HPKs. Thus if searching protein sequences for potential TCS proteins, it is best to use all available profiles to reduce false-negative results. The situation is even worse for HATPase/transmitter domains and for RRs, for which several tens of profiles are available (mainly multidomain profiles). Thankfully, several internet servers, including the CD-Search tool provided by the National Center for Biotechnology Information, combine domain profiles from multiple domain profile libraries for users to query (see 'Web resources' section at end of chapter).

While HisKA, Hpt and receiver domains appear to be unique to TCS proteins, HATPase domains can also be found in a wide range of proteins, including topoisomerases and chaperones. To eliminate false-positive results from such proteins, it is normal to exclude from further analysis any proteins that give HATPase hits, but do not contain HisKA or receiver domains, and/or that do contain domains characteristic of topoisomerases, chaperones, etc. Commonly excluded domains include DNA gyrase/topoisomerase IV (pfam00521, pfam00986, pfam02518), Hsp90 (pfam00183), MutL (COG0323), and RsbW anti-sigma factor (COG2172) (Cock and Whitworth, 2007a; Galperin and Nikolskaya, 2007).

When searching a protein sequence for the presence of a conserved TCS domain, it is usual to impose an e-value cut-off, and domain profile 'hits' are only considered further if the hit e-value is less than the cut-off value. The e-value for a hit is based on how well the query sequence fits the domain profile, and is dependent on how much of the query sequence matches the domain profile. Thus when searching for the presence of small domains, good 'hits' can often be above the cut-off value, purely because the region matching the profile is short, because the profile itself is short. In addition, many domain profile searches apply a low-complexity filter as default, masking regions of 'low complexity' from the query, effectively reducing further the length of any possible match. Regions within TCS domains are often defined as low-complexity, and masking them from the initial query can thus frequently result in false-negative results. It is therefore highly recommended to turn off any low-complexity filter when performing domain profile searches.

When considering TCS proteins, these factors are particular problems for the identification of HisKA domains, which are around half the size of receiver and HATPase domains, and yet vitally important in defining whether HATPase-containing proteins are HPKs. When assessing the complement of HPKs within a prokaryotic genome, it is not unusual to find a large proportion of proteins containing HATPase domains, but with no hit to a HisKA profile. The P2CS database tackles this problem by taking each HATPase-containing protein that apparently lacks a HisKA, and defining a region upstream of the HATPase domain which is most likely to contain a HisKA domain. A sequence profile is then created by making a multiple alignment of the ten residue sequences surrounding the phosphorylatable His residue of all HisKA sequences encoded by the same genome. This profile is then used to screen upstream of the HATPase domains apparently lacking partner HisKA domains (Barakat *et al.*, 2009). While this helps find HisKA domains within transmitter units, HisKA domains will still be relatively under-represented when they occur elsewhere, for instance as separate proteins or as part of RR proteins.

Another problem when identifying TCS proteins is caused by problems in the initial annotation of genes in genomes. To a surprising extent open reading frames encoding proteins are not annotated as protein-coding genes and thus are subsequently not assessed for the presence of TCS domains by most online resources. By searching a theoretical translation of all open reading frames, it is possible to recover 2% more TCS proteins than just searching predicted proteomes. The magnitude of the problem also varies significantly between genomes, with as much as 24% of the TCS protein complement not being represented in some predicted proteomes (Barakat *et al.*, 2009).

No matter how rigorous the computational approach employed is, there will inevitably be false-negative and false-positive results even when characterizing sets of TCS proteins using domain-based approaches. Manual curation is still viewed as a high-cost but high-benefit activity in database curation and genome annotation (see, for example, Baumgartner *et al.*, 2007; Hettne *et al.*, 2010), and manual input can be an important part of the characterization process when studying TCSs. The manual curation process is necessarily non-formulaic and non-prescriptive, but there are recurring themes that can guide curators. The most important of these is to look beyond the protein to the genomic neighbourhood of the encoding gene. For instance, nearly two-thirds of TCS genes are adjacent to other TCS genes, while HPK genes are often (50%) found adjacent to that of a RR (Williams and Whitworth, 2010). Thus, if a protein contains a HATPase, but with no obvious HisKA, it can be confidently described as a HPK if its gene is adjacent to that encoding a RR. Such arguments can also provide support if a TCS domain profile hit is similar to, but higher than, the cut-off being used.

And then there are a few proteins that will defy all attempts to classify them, regardless of whether automatic or manual approaches are being used. For instance MXAN_0258 from *M. xanthus* is annotated in Genbank as an ATPase domain-containing protein, and P2CS does not regard it as a TCS protein at all. Conserved domain searches (E-value cut-off of 0.01) show the presence of both HATPase and receiver domains, but no HisKA or Hpt domains. This would normally lead to the protein being classified as a hybrid kinase, however the two domain hits overlap by approximately 80 residues (about half of each domain), and both 'domains' lack key conserved residues associated with ATPase activity and phosphoryl group transfer. Presumably neither domain is functional, however the gene is upstream of that for a RR, suggesting that whatever it is doing, Mxan_0258 is somehow involved in TCS signalling.

Fortunately there are now a few publicly accessible databases aimed at biologists, which provide domain architecture analysis for tens of thousands of TCS proteins, although they do largely focus on TCSs of prokaryotes (see 'Web resources' section below). While such databases provide a rigorous, consistent and detailed analysis of large numbers of TCS proteins, if a researcher is interested in a small number of proteins, it is always worthwhile for them to spend time checking for themselves the domain architectures and genetic organization of their TCS proteins of interest.

What's in a name?

Since their discovery there have been thousands of publications regarding TCS biology. Many authors have focused on particular facets of TCS biology and they have described TCSs in ways most relevant to their facet of interest. This has led to a plethora of names and naming conventions in the TCS literature. For instance many authors make no distinction between (non-hybrid) HPKs and hybrid kinases, as they both possess histidine protein kinase activity. However, in other ways (non-hybrid) HPKs and hybrid kinases are fundamentally different: Some hybrid kinases can act as TCSs in isolation, whereas (non-hybrid) HPKs require a separate RR protein for signalling. Some authors argue that hybrid HPKs with an N-terminal receiver domain can be considered as RRs with a phosphorylation-dependent output domain (Galperin, 2006, 2010; Wuichet and Zhulin, 2010), while those with N-terminal transmitter domains are hybrid HPKs (Wuichet and Zhulin, 2010), even though more than 50% of hybrid kinases are found encoded separately from other TCS genes (Williams and Whitworth, 2010), implying that they constitute complete TCSs.

Even at the crudest level of description, TCSs and TCS proteins have been given multiple names. TCSs are commonly described as

His-Asp Phosphotransfer (HAP) pathways, two-component signal transduction (TCST) pathways, and two-component system (2CS, or 2CST) pathways. HPKs are also known as histidine kinases (HKs), sensor kinases (SKs), and even protein histidine kinases (PHKs). Luckily RRs do not suffer from such an identity crisis, with the term 'response regulator' meeting almost universal acceptance.

To distinguish between non-hybrid HPKs and hybrid HPKs, a variety of terms have been coined, including typical/atypical, orthodox/unorthodox, and classical/non-classical. These descriptors are also often used to distinguish between HPKs with the two varieties of transmitter domains – with those containing HisKA domains being classic/orthodox, and those containing Hpt domains being unorthodox/non-classical (for examples see Parkinson and Kofoid, 1992; Jourlin *et al.*, 1997; Mizuno, 1997; Kofoid and Parkinson, 1998; Barakat *et al.*, 2009). The MiST2 database also distinguishes between 'hybrid histidine kinases' with transmitter domains upstream of receivers, and 'hybrid response regulators' with receiver domains N-terminal to transmitter domains (Ulrich and Zhulin, 2010), while the P2CS database (Barakat *et al.*, 2009) defines 'unorthodox' HPKs as hybrid HPKs that also contain a Hpt domain (and are therefore likely to be part of phosphorelays).

The diversity of naming systems may seem a trivial issue, however the historical lack of a consistent nomenclature has left a legacy of problems. One notable arena where such problems occur is in genome annotation. Annotation of a genome usually involves the assignment of functional categories and descriptions to genes based on homology to previously annotated genes. Thus if a newly sequenced gene (*gene1*) is found to be homologous to a HPK, it is annotated as a HPK. However, often the newly sequenced gene will actually encode a hybrid kinase, possessing a receiver domain as well as a transmitter domain. Subsequently, another newly sequenced gene (*gene2*) will be found to be most similar to *gene1* and in turn will also be annotated as encoding a HPK. However, the product of *gene2* may possess a transmitter domain, or a receiver domain, or both domains, and thus should be annotated as a HPK, a RR or a hybrid kinase, respectively.

Examples of such misannotation are commonplace. For example, SM_b20357 is a RR from *Sinorhizobium meliloti* 1021, which contains a single receiver domain, yet it is annotated as a 'putative histidine kinase protein'. Similarly XCV0783 is a protein from *Xanthomonas campestris* pv. *vesicatoria* strain 85–10 which possesses HAMP and transmitter domains (but no receiver domain), yet is annotated as an 'EnvZ family two-component response regulator'. Currently 1.5% of all proteins now classified as RRs in the P2CS database were originally described in some way as 'sensor kinase' proteins in the annotated genome files as retrieved from Refseq/Genbank.

Categorization of TCS proteins

As the number of available sequences increased, it soon became apparent that several different input and output domains were relatively common in TCS proteins (Grebe and Stock, 1999). For instance, 63% of RRs possess one of just four DNA-binding domains (Galperin, 2006), while PAS/PAC, HAMP and GAF domains are particularly common input domains of HPKs (Whitworth and Cock, 2008b). Around 30% of HPKs have a PAS domain, 30% have a HAMP domain and about 7% possess a GAF domain (www.p2cs.org). With increasing numbers of examples, it also became clear that TCSs with the same output domains tended to have similar input domains, similar HisKA subtypes, and conserved gene order (Fabret *et al.*, 1999; Grebe and Stock, 1999). This suggested that the evolution of TCSs had occurred by duplication and divergence from a relatively small pool of ancestral TCS 'families' (Koretke *et al.*, 2000; Whitworth and Cock, 2009). The ability to allocate TCS proteins to families is important as RRs of one family tend to interact with HPKs of the same family (although of course there are notable exceptions), which provides insights into the specificity of interactions between HPK and RRs (Bjilsma and Groisman, 2003; Laub and Goulian, 2007), which can in turn guide experimenters (see later).

RR families

Categorizing RRs into families is a straightforward task, as most RRs consist of a sole receiver domain with a single output domain (Ashby, 2004; Galperin, 2006, 2010). For instance a RR containing an N-terminal receiver domain and a C-terminal Trans_reg_C domain would be classified as an OmpR family member. Families are named after prominent (well-studied) members, although several families have multiple names in the literature. For example, the OmpR family is also described as the PhoB family, while the NarL family (RRs containing a GerE DNA-binding domain, PF00196) is also known as the FixJ or GerE family (Grebe and Stock, 1999; Koretke *et al.*, 2000; Galperin, 2006; Whitworth and Cock, 2008b).

Nevertheless, there are many cases where RRs cannot be easily placed into a family using such an approach. A large group of RRs fall into this category as they lack any obvious output domain. The genes for such RRs are often found in operons with chemosensory HPKs (CheA homologues) and such proteins are thus often classified as CheY family members. However, many RRs lacking output domains nonetheless have significant regions N-terminal or C-terminal to the receiver domain, and are not encoded in the vicinity of chemosensory HPKs – presumably such regions do represent domains, but domains that have not yet been identified as such bioinformatically. The P2CS database does not classify such RRs as CheY family members, instead they are 'unclassified'. Other unclassified proteins include those with multiple receiver domains, those with multiple output domains and some with very 'rare' output domains (too few for a RR family to have been defined yet).

HPK families

HPKs families are more troublesome to define than those of RRs, as family membership does not correlate strongly with the presence of particular input domains. Fabret *et al.* (1999) studied the sequences around the phosphorylated histidine of the HPKs of *B. subtilis* and grouped the HPKs into six 'classes'. Most HPKs of *B. subtilis* are encoded adjacent to RR genes, and it was found that gene order (whether HPK>RR>or RR>HPK>) and the family of RR, was conserved for particular classes of HPKs. For instance all Class II HPKs are encoded upstream of NarL family RRs (Fabret *et al.*, 1999). Nevertheless within each class/family of HPK there is considerable diversity in the nature of the input domain (Fabret *et al.*, 1999). A similar pangenomic analysis by Grebe and Stock (1999) took HPKs from multiple genomes and divided them into families of related sequences using a phylogenetic approach. HPKs were divided into 11 families. Again it was shown that the individual families of HPKs were associated with particular families of RRs (Grebe and Stock, 1999), and this classification scheme continues to be widely used (for example see Shi *et al.*, 2008).

Phylogenetic approaches have remained the preferred method for classifying HPK families; however using different datasets and cut-offs, the number of families identified can vary considerably, for instance, Kim and Forst (2001) identified only five HPK families. In a related approach, sequence alignments of HPKs from a single family can be used to create family-specific hidden Markov models (HMMs) to assign HPKs to predefined families (for an example, see Hutchings *et al.*, 2004).

Phylogenetic/sequence-based approaches

In addition to defining families of HPKs, phylogenetic approaches have also been applied to classify RRs based on sequence similarities between their receiver domains, however phylogenetic approaches have many limitations. Phylogenetic methods rely on multiple sequence alignments (and can therefore only be used on receiver or transmitter domains extracted from whole protein sequences), manual curation of alignments is often necessary prior to the creation of dendrograms, while the computing power necessary to undertake phylogenetic analysis increases exponentially with the number of sequences considered. A comparison of phylogenetic and domain-based classification schemes also found evidence of significant incongruence between the two approaches (Grebe and Stock, 1999). This was proposed to be due to domain shuffling during evolution, which had been proposed by an earlier study (Pao and Saier, 1995). Similarly, a degree of incongruence between phylogenetic classification of transmitter and their partner RRs is often seen (Koretke *et al.*, 2000; Chen *et al.*, 2004; Alm *et al.*, 2006; Whitworth and Cock, 2009). For instance the HPKs encoded adjacent to Pho family RRs will be mostly Pho family members, but some individuals might be Ntr family HPKs. This suggests recombination has occurred between TCSs of different families during evolution, and raises questions as to which classification approach (phylogenetic vs. domain architecture) is more appropriate for the research question at hand (Pao and Saier, 1995; Alm *et al.*, 2006; Qian *et al.*, 2008; Whitworth and Cock, 2009).

As described above, RRs lacking output domains are often classified as CheY family members owing to the absence of any obvious output domains. However, many such proteins are not encoded in the vicinity of CheA homologues, and do not appear to be components of chemosensory systems (Jenal and Galperin, 2009). Such proteins are described as single-domain response regulators (SD-RRs), or standalone receiver domain proteins. Phylogenetic approaches and gene co-localization arguments can often distinguish between CheY homologues and other SD-RRs. Generally, when performing phylogenetic analyses on receiver domain sequences, clades tend to be composed exclusively of proteins from a single RR family, as would be expected given their common evolutionary origin. CheY homologues (those encoded adjacent to a CheA homologue) tend to cluster together as a clade in dendrograms (Koretke *et al.*, 2000, Whitworth and Cock, 2008b), while non-CheY SD-RRs can be found occurring in most RR clades, suggesting evolution through loss of the RR output domain (Whitworth and Cock, 2008b, 2009).

Another sequence-based categorization approach related to phylogenetic analysis involves using whole protein sequences to define 'clusters' of putative orthologues. This approach has been implemented in the P2CS database (www.p2cs.org) and allows sets of orthologous TCS proteins from all completed genomes to be accessed directly (Barakat *et al.*, 2011). A comparable approach has been applied to the TCS of cyanobacteria, with orthology groups defined by bidirectional best hits between pairs of genomes (Ashby and Houmard, 2006).

Functional and mechanistic annotation

Assigning TCS proteins to families is a worthwhile activity; however, it has led to much confusion in genome annotations. For instance, RRs of different families share sequence similarity through their receiver domains, and owing to domain shuffling the receiver domains of two RRs from different domain-based families can be particularly similar. Similarity-based annotation can therefore misassign family membership. A related consideration relates to functional annotation, which attempts to assign biological roles to gene products. Sequence similarity is used as a proxy for orthology to map the functions of experimentally characterized genes onto putative orthologues in sequenced genomes. However, it is worth noting that organisms with multiple members of a TCS family generally regulate different processes. For example, the *E. coli* PhoB/OmpR family of TCSs regulate diverse processes, including potassium homeostasis (KdpE), copper tolerance (CusR) and trimethylamine N-oxide respiration (TorR), in addition to phosphatase expression (PhoB) and osmoregulation (OmpR) (Yagil *et al.*, 1975; Comeau *et al.*, 1985; Walderhaug *et al.*, 1992; Jourlin *et al.*, 1997; Munson *et al.*, 2000). However, all PhoB/OmpR family members in a genome are often ascribed the same role, even though only a subset of such proteins will truly have such a role. For example, *Clostridium botulinum B* str. Eklund 17B possesses 28 OmpR family RRs, of which seven are annotated as regulating phosphatase expression, and 11 are annotated as being VanR, which regulates vancomycin resistance. This is a general problem of functional annotation when considering regulatory proteins, where a basic architecture has been duplicated, and copies have diverged/adapted to use a conserved mechanism to regulate subtly different behaviours, for instance binding to different DNA sequences and thereby regulating expression of different sets of genes. Similarly, the assignment of HPK function based on the presence/absence of particular input domains is also flawed as most input domains are known to respond to diverse stimuli (Krell *et al.*, 2010).

However, it is increasingly possible to gain some limited functional information from sequence data. For instance, the presence of transmembrane (TM) domains suggests a transmembrane localization of HPK and the sensation of a stimulus which is either extra-compartmental, or membranous. Conversely, HPKs lacking TMs helices are likely to reside in the cytoplasm, sensing a stimulus internal to the cell (Galperin, 2005). While the presence of a particular input/output domain cannot generally be used to suggest physiological role *per se*, it can sometimes suggest what type of stimulus is being sensed, or the mechanism of response elicitation. So for RRs with DNA-binding output domains it is clear that altered transcription is the consequence of altered TCS activity, even though the particular genes being regulated might be obscure. Similarly, RRs with GGDEF domains will likely be producing cyclic-di-GMP, even though the physiological consequences of increased c-di-GMP production may not be clear.

Gene organization

The majority of prokaryotic TCS domains (62%) are encoded in pairs at loci within genomes (both as hybrid genes and as paired TCS genes). Another 32% of transmitter and receiver domains are found encoded individually in a single gene without any neighbouring TCS genes (Williams and Whitworth, 2010), and such genes and their resulting protein products are usually described as 'orphans'. There is a significant phyletic component to the organization of TCS genes. Some organisms appear to favour encoding TCS proteins as orphan genes (e.g. *Xanthomonas*), while others (e.g. the *Firmicutes*), encode a large proportion of their TCS proteins as paired genes (Williams and Whitworth, 2010).

In prokaryotes there is often tight coupling between co-function and co-location in the genome, with genes involved in the same pathway being co-regulated, often transcribed together forming an operon. Thus when TCS genes are found paired in a replicon they will typically be encoding components of a single TCS pathway. Slaving expression of a HPK/RR gene pair to a single promoter ensures that the genes for both components of the pathway are expressed together, and mechanisms such as translational coupling (often manifested as overlapping start and stop codons) also guarantee that translation of both pathway components will happen together (Oppenheim and Yanofsky, 1980; Normark *et al.*, 1983). In bacteria adjacent genes are generally encoded on the same DNA strand, and in around a quarter of such cases, the two genes overlap by at least one nucleotide (Fukuda *et al.*, 2003). A large proportion of paired TCS genes are also co-directional and overlap (Cock and Whitworth, 2007b). The most common overlaps are of 1 or 4 bp (51%; Cock and Whitworth, 2007b), where the ATG start codon of the second gene overlaps with the stop codon of the preceding gene, with sequences such as TG<u>A</u>, or <u>ATG</u>A (bases within start codons underlined). Longer overlaps also exist between TCS genes, with the phase of overlap (relative reading frame) dictated by selective forces acting on doubly coding DNA sequences (Cock and Whitworth, 2010).

When an RR or HPK gene is an orphan, the partner TCS protein (and often multiple partners) will be encoded elsewhere in the replicon and it may not be obvious which TCS protein(s) is the functional partner, especially in genomes which have a large number of orphan TCS genes. Recently methods have been developed which allow the prediction of partnerships between TCS proteins from their primary structure. In 1999, Fabret *et al.* divided the HPKs and RRs of *B. subtilis* into families and noted that some amino acid residues thought to mediate the HPK–RR interaction were globally conserved, others were conserved within families, while others yet were variable within a family. They proposed that family-specific residues restricted interactions to partners of the same family, while variable residues gave within-family specificity (Fabret *et al.*, 1999). As such, it seemed likely that an understanding of the evolutionary forces acting on particular amino acids at the HPK–RR interface would provide insights into the specificity of the HPK–RR interaction (Hoch and Varughese, 2001). Current approaches use training datasets of paired transmitter and receiver domain sequences which are known to interact, inferred from paired TCS genes in genomes. In general multiple sequence alignments are generated from the paired receiver and transmitter domain sequences, which allow identification of residues which have co-varied between transmitter and receiver domains. Such co-varying residues typically represent residues which bind to one another when the receiver domain binds to its partner transmitter domain. Understanding the pattern of which amino acid residues can interact favourably with one another at different positions in the proteins, allows the prediction of which transmitter and receiver domains not in the training dataset should be able to interact with one another (Burger and van Nimwegen, 2008; Skerker *et al.*, 2008; Szurmant *et al.*, 2008; Cock, 2009; Kolesov and Mirny, 2009; Weigt *et al.*, 2009; Lunt *et al.*, 2010; Schug *et al.*, 2010). In addition to making testable predictions, such methods also provide supporting evidence for TCS partnerships suggested by usually laborious experimental approaches that map partnerships (Skerker *et al.*, 2005; Yamamoto *et al.*, 2005; Dortay *et al.*, 2008; Whitworth *et al.*, 2008). Understanding acquired through such modelling has also allowed the rational engineering of TCS specificity, providing insights into the evolution of specificity (Capra *et al.*, 2010; Siryaporn *et al.*, 2010), while similar approaches have now also been applied to chemosensory TCSs, allowing their interaction specificities to be re-wired (Bell *et al.*, 2010).

While the pairing of a HPK and a RR gene implies the gene products act together to form a TCS, there are common examples of genetic loci encoding more than two TCS domains and/or proteins (5292 currently in P2CS), and in such cases it is difficult to ascribe a structure to the TCS signalling pathway. However, many such loci (31%) encode four TCS domains, and typically these will be two transmitter/phosphotransfer domains and two receiver domains, implying the resulting proteins function as a phosphorelay (Williams and Whitworth, 2010). However, TCS loci sometimes (1002 currently in P2CS) encode more than four TCS domains, and in such cases there is usually an 'overabundance' of receiver domains (across the pangenome receiver domains are 31% more common than transmitter domains). Such phenomena suggest that receiver domains do not necessarily act solely as links on the chain of phosphotransfer between signal reception and response elicitation, but rather can have additional accessory functions within a signalling pathway. Examples of such accessory receiver domain functions are becoming increasingly commonplace. For example the RedCDEF phosphorelay of *M. xanthus* comprises three receiver domains and two transmitter domains (Higgs *et al.*, 2005), while the CbbRRS phosphorelay of *Rhodopseudomonas palustris* possesses two accessory receiver domains (Romagnoli and Tabita, 2006). In the CbbRRS system, a receiver domain within the hybrid kinase CbbSR dictates which of the two RRs, CbbRR1 or CbbRR2 becomes phosphorylated (Romagnoli and Tabita, 2007). Similarly, in the Che TCS of *Helicobacter pylori*, phosphoryl groups are transferred from CheA to either the RR CheY1, or to a receiver domain at the C-terminus of CheA (CheY2). Phosphoryl groups can also flow from CheY1 to CheA to CheY2, and it is proposed that CheY2 acts as a phosphate sink to modulate the half-life of CheY1 phosphorylation (Jiménez-Pearson *et al.*, 2005). There are also some cases where a core RR possesses additional accessory functions. For instance, DivK of *C. crescentus* is the cognate RR of the HPK DivJ, but it also acts as an allosteric activator of two HPKs, DivJ and PleC (Paul *et al.*, 2008).

Distribution of TCS domains between genes

When considering the distribution of TCS domains between genes, similar TCSs can exhibit different patterns of covalent linkages between their TCS domains. For instance in a transmitter, receiver$_1$,phosphotransfer,receiver$_2$ phosphorelay, a two-gene distribution is the most common form (TR$_1$P R$_2$), however other distributions are seen (e.g. TR$_1$ PR$_2$). We are currently unable to explain what selective forces discriminate between such variant patterns (Appleby *et al.*, 1996). However, slight headway has been made regarding the simplest TCSs – those which consist of a single transmitter domain and a single receiver domain, and which can be found as either a single hybrid kinase gene, or paired HPK and RR genes. More than 80% of all HPKs (both non-hybrid HPKs and hybrid kinases) possess transmembrane (TM) helices (Cock and Whitworth, 2007b), suggesting they sense extracellular stimuli (Galperin, 2005). In addition, more than half of all RRs possess DNA-binding output domains (www.P2CS.org). TM helices and DNA-binding domains each need to be positioned in appropriate (and separate) subcellular locations to function, which largely precludes the formation of a functional hybrid kinase by gene fusion. In addition, in order to maintain appropriate membrane insertion and/or DNA-binding activity, fusion will generally only result in a functional hybrid kinase if the HPK gene is encoded upstream of the RR. As the order of paired TCS genes correlates with the TCS family, as does the nature of the HPK input domain and whether the output domain is DNA-binding, fusion rate is highly dependent upon TCS family membership (Cock and Whitworth, 2007b).

It seems that the genomic localization of a TCS domain is related to that of its interaction partners. Thus HPKs and RRs that interact with one another tend to be co-located in the genome. However, large numbers of TCS genes are orphaned, with little clues as to which other TCS protein(s) they interact with. Nevertheless, computational models now allow the prediction of partnerships between TCS domains, and it would be really exciting to investigate the relationship between patterns of genetic and genomic organization and interaction partnerships of TCS proteins. For instance do RRs which can receive phosphoryl groups from two distantly located HPKs tend to be found as orphans, or are they co-located with one of the HPKs? And how do the partnership network and genomic co-localization co-evolve?

Related analyses are starting to develop genome metrics which capture facets of TCS organization in a number which can be easily compared between genomes (Galperin and Kolker, 2006; Galperin *et al.*, 2010). For instance the number of TCS (and other regulators) in a genome relates to its IQ, while the proportion of HPKs that sense intracellular stimuli defines an organism's introvertedness (Galperin, 2005). Other proposed metrics capture features of genetic organization such as the percentage of orphaned TCS genes (dissemination) and the proportion of TCS loci encoding complex TCS pathways (sophistication) (Williams and Whitworth, 2010). Genome metrics are useful in order to compare across genomes and identify aspects of TCS organization that differ between genomes. They also provide a description of 'normality' for any particular TCS feature, such as the proportion of TCS genes found in pairs. Normality is important as it is very easy to focus attention on genomes or TCSs which exhibit extreme properties or unusual organizations. For instance, genetic loci with particularly large numbers of TCS domains (five or more) are fascinating, but they are rare (only 2% of all TCS gene loci).

As might be expected, all TCS gene metrics exhibit a significant phyletic component, with closely related organisms tending to possess similar TCS gene/domain architectures, and distantly related organisms exhibiting very different organizations (Galperin *et al.*, 2010; Williams and Whitworth, 2010). Therefore we must be aware that our perception of what constitutes 'normal' TCS gene sets, may nevertheless be biased by which genomes have been sequenced, and by the relative lack of sequenced genomes in some phyla (Whitworth, 2008).

Summary and perspectives

TCSs are important signalling pathways found in large numbers across sequenced genomes, and they exhibit considerable diversity in their domain and gene organization. Classification schemes are needed to catalogue and represent TCSs in biological databases, however there are diverse classification schemes and naming conventions currently in use. Most classification schemes define categories of TCS proteins based on their constituent domains, which also allows their assignment to particular TCS families. Alternative approaches use phylogenetic analyses to assign family membership, based on the evolutionary heritage of TCS domains.

Despite some progress in understanding the relationship between gene organization, domain architecture and evolution, there remain many gaps in our knowledge, particularly regarding how TCS domains are divided between gene/proteins, and how the evolution of TCS organization is shaped by TCS protein partnerships. Nevertheless, current knowledge does allow limited predictions regarding the functional properties of TCS proteins, and *de novo* predictions of interaction partnerships, which rely upon consistent methods of classification/categorization.

There is currently a need for a 'harmonization' of categorization schemes and hopefully online databases will facilitate the adoption of a community-sanctioned scheme. However, such a categorization should be as complete as possible, with annotation of gene structure, domain architecture, and family assignment (domain-based, phylogenetic and cluster-based approaches) as a minimum. No contemporary resource provides such a complete description. However, the P2CS database currently just lacks family assignment based on phylogenetic analysis and the development team hope to incorporate such a feature in a future update. Partnerships between TCS domains also need to be built into databases, whether experimentally or computationally derived, and not just transmitter–receiver interactions, but also transmitter–transmitter partnerships and receiver–receiver interactions (Gao *et al.*, 2008; Goodman *et al.*, 2009). Such a feature would be an invaluable guide to experimental biologists wishing to define signalling interactions between TCS proteins.

Web resources

While TCSs are numerous and widespread, there are few web resources devoted to their biology. Nevertheless TCSs are so abundant that examples of TCSs are included in many other biological databases. The list below highlights the few databases that focus exclusively or largely on TCSs. In addition, web-servers that enable analyses which are particularly useful for studying TCS are described. The three main TCS-focused databases are P2CS, MiST2 and the Census of prokaryotic response regulators, while KEGG provides organism-specific views of the presence/absence of a subset of TCS orthologues:

P2CS

Prokaryotic two-component systems (P2CS) is a TCS-centred database aimed at biologists (Barakat *et al.*, 2009, 2011). Sets of TCSs can be accessed by organism, by metagenome, by sequence cluster, or by domain architecture, and data can be downloaded in a variety of formats. Each TCS is fully described and links are provided for querying external servers. The genome context of each TCS is also characterized and TCS loci are viewable in an interactive map or genome browser. Two particular strengths of P2CS are that it searches an ORFeome for TCS domains to find TCS genes missed during genome annotation, and it also builds and uses a sequence profile of an organism's HPK H-box motifs to identify possible phosphorylation sites of HPKs that apparently lack such motifs. At the time of writing, P2CS contains data from 1125 prokaryotic genomes and 39 metagenomes, and can be accessed at http://www.p2cs.org.

MiST2

The Microbial Signal Transduction (MiST) database (Ulrich and Zhulin, 2007) has been recently updated to MiST2 (Ulrich and Zhulin, 2010). The database includes all prokaryotic genomes (including draught genomes), classifying TCS proteins, one-component system proteins, chemotaxis proteins and ECF sigma factors. In MiST2 summary information and sequence data can be obtained for each gene/protein, TCS gene organization is classified and there is a cart feature for genes of interest to the user. MiST2 can be found at http://mistdb.com/.

Census of prokaryotic response regulators

This dataset provides a per-genome classification of RRs based on their output domains (Galperin, 2010). Individual entries are linked to servers for further analysis and general databases. The census can be accessed at http://www.ncbi.nlm.nih.gov/Complete_Genomes/RRcensus.html

KEGG

KEGG is a general database of genes, enzymes and metabolism, however it has pages devoted to TCSs. The pages provide useful guides to the topology of HPKs, their phosphotransfer to RRs, and the presence/absence of TCS orthologues in individual genomes (Nikitin *et al.*, 2004). The TCS pages of KEGG can be entered directly at http://www.genome.jp/kegg/pathway/ko/ko02020.html.

Owing to the variety of input and output domains, and the diversity of domain architectures, normal sequence similarity-based methods of gene classification are inappropriate for many TCSs (Galperin, 2010). Instead, domain architecture analysis provides a more useful and intuitive way of categorizing TCS proteins. While there are numerous websites that enable such searches, three webtools that the author routinely uses are SMART, CD-Search and CDART, while the presence of transmembrane helices (suggestive of intra- or extracellular signal perception) is provided by TMHMM.

SMART

The Simple Modular Architecture Research Tool (SMART) is provided at http://smart.embl-heidelberg.de/. SMART allows an analysis of the domain architecture of inputted protein sequences with a graphical representation of the results (including transmembrane helices), and allows the retrieval of proteins containing user–specified combinations of domains (Letunic *et al.*, 2009).

CD-Search/CDD

The Conserved Domain Database (CDD) hosted by the National Center for Biotechnology Information (NCBI) combines domain profiles from multiple sources, including SMART, Pfam, TIGRFAMs (Haft *et al.*, 2003), and COG (Tatusov *et al.*, 2003). Individual protein sequences can be characterized using CD-Search, and a batch submission option is provided to query multiple proteins simultaneously. CDD can be found at: http://www.ncbi.nlm.nih.gov/Structure/cdd/wrpsb.cgi

CDART

The Conserved Domain Architecture Retrieval Tool (CDART) is available at http://www.ncbi.nlm.nih.gov/Structure/lexington/lexington.cgi. CDART takes a protein sequence and identifies its conserved domains (Geer *et al.*, 2002). It also displays protein domain architectures that share domains with the query, with links to individual examples (results can also be filtered by domain composition).

TMHMM Server v. 2.0

This webtool based on the hidden Markov model created by Sonnhammer *et al.* (1998) predicts the presence of transmembrane helices in a protein sequence, providing an output of predicted protein topology. It is useful for defining whether HPKs are likely to be responsive to cytoplasmic or extracytoplasmic stimuli (Galperin, 2005). TMHMM Server v. 2.0 can be accessed at http://www.cbs.dtu.dk/services/TMHMM/.

Acknowledgements

Images within Figs. 1.2 and 1.3 were kindly prepared by Heather Currinn and Stuart Thomas while undergraduate students at Aberystwyth University.

References

Alm, E., Huang, K., and Arkin, A. (2006). The evolution of two-component systems in bacteria reveals different strategies for niche adaptation. PLoS Comp. Biol. 2, e143.

Altschul, S.F., Gish, W., Miller, W., Myers, E.W., and Lipman, D.J. (1990). Basic local alignment search tool. J. Mol. Biol. *215*, 403–410.

Alves, R., and Savageau, M.A. (2003). Comparative analysis of prototype two-component systems with either bifunctional or monofunctional sensors: differences in molecular structure and physiological function. Mol. Microbiol. *48*, 25–51

Appleby, J.L., Parkinson, J.S., and Bourret, R.B. (1996). Signal transduction via the multi-step phosphorelay: not necessarily a road less traveled. Cell *86*, 845–848.

Ashby, M.K. (2004). Survey of the number of two-component response regulator genes in the complete and annotated genome sequences of prokaryotes. FEMS Microbiol. Lett. *231*, 277–281.

Ashby, M.K. (2006). Distribution, structure and diversity of "bacterial" genes encoding two-component proteins in the Euryarchaeota. Archaea. 2, 11–30.

Ashby, M.K., and Houmard, J. (2006). Cyanobacterial two-component proteins: structure, diversity, distribution, and evolution. Microbiol. Mol. Biol. Rev. *70*, 472–509.

Ashenberg, O., Rozen-Gagnon, K., Laub, M.T., and Keating, A.E. (2011). Determinants of homodimerization specificity in histidine kinases. J. Mol. Biol. *413*, 222–235.

Barakat, M., Ortet, P., Jourlin-Castelli, C., Mejean, V., and Whitworth, D.E. (2009). P2CS: a two-component system resource for prokaryotic signal transduction research. BMC Genomics *10*, 315.

Barakat, M., Ortet, P., and Whitworth, D.E. (2011). P2CS: a database of prokaryotic two-component systems. Nucleic Acids Res. *39*, D771–D776.

Baumgartner, W.A., Jr., Cohen, K.B., Fox, L.M., Acquaah-Mensah, G., and Hunter, L. (2007) Manual curation is not sufficient for annotation of genomic databases. Bioinformatics *23*, i41–i48.

Beier, D., and Gross, R. (2006). Regulation of bacterial virulence by two-component systems. Curr. Opin. Microbiol. *9*, 143–152.

Beier, D., and Gross, R. (2008). The BvgS/BvgA phosphorelay system of pathogenic *Bordetellae*: structure, function and evolution. Adv. Exp. Med. Biol. *631*, 149–160.

Bell, C.H., Porter, S.L., Strawson, A., Stuart, D.I., and Armitage, J.P. (2010). Using structural information to change the phosphotransfer specificity of a two-component chemotaxis signalling complex. PLoS Biol. *8*, e1000306.

Bijlsma, J.J.E., and Groisman, E.A. (2003). Making informed decisions: regulatory interactions between two-component systems. Trends Microbiol. *11*, 359–366.

Bourret, R.B. (2010). Receiver domain structure and function in response regulator proteins. Curr. Opin. Microbiol. *13*, 142–149.

Burger, L., and van Nimwegen, E. (2008). Accurate prediction of protein–protein interactions from sequence alignments using a Bayesian method. Mol. Syst. Biol. *4*, 165.

Capra, E.J., Perchuk, B.S., Lubin, E.A., Ashenberg, O., Skerker, J.M., and Laub, M.T. (2010). Systematic dissection and trajectory-scanning mutagenesis of the molecular interface that ensures specificity of two-component signalling pathways. PLoS Genet. *6*, e1001220.

Chen, Y.T., Chang, H.Y., Lu, C.L., and Peng, H.L. (2004). Evolutionary analysis of the two-component systems in *Pseudomonas aeruginosa* PAO1. J. Mol. Evol. *59*, 725–737.

Cho, K., and Zusman, D.R. (1999). Sporulation timing in *Myxococcus xanthus* is controlled by the *espAB* locus. Mol. Microbiol. *34*, 714–725.

Cock, P.J.A. (2009). Two-component regulation: modelling, predicting and identifying protein–protein interactions and assessing signalling networks of bacteria. PhD Thesis, University of Warwick, UK.

Cock, P.J.A., and Whitworth, D.E. (2007a). Evolution of gene overlaps: relative reading frame bias in prokaryotic two-component system genes. J. Mol. Evol. *64*, 457–462.

Cock, P.J.A., and Whitworth, D.E. (2007b). Evolution of prokaryotic two-component system signalling pathways: gene fusions and fissions. Mol. Biol. Evol. *24*, 2355–2357.

Cock, P.J.A., and Whitworth, D.E. (2010). Evolution of relative reading frame bias in unidirectional prokaryotic gene overlaps. Mol. Biol. Evol. *27*, 753–756.

Comeau, D.E., Ikenaka, K., Tsung, K.L., and Inouye, M. (1985). Primary characterization of the protein products of the *Escherichia coli ompB* locus: structure and regulation of synthesis of the OmpR and EnvZ proteins. J. Bacteriol. *164*, 578–584.

Da Re, S.S., Deville-Bonne, D., Tolstykh, T., Veron, M., and Stock, J.B. (1999a). Kinetics of CheY phosphorylation by small molecule phosphodonors. FEBS Lett. *457*, 323–326.

Dortay, H., Gruhn, N., Pfeifer, A., Schwerdtner, M., Schmülling, T., and Heyl, A. (2008). Toward an interaction map of the two-component signaling pathway of *Arabidopsis thaliana*. J. Proteome Res. *7*, 3649–3660.

Ewann, F., Jackson, M., Pethe, K., Cooper, A., Mielcarek, N., Ensergueix, D., Gicquel, B., Locht, C., and Supply, P. (2002). Transient requirement of the PrrA-PrrB two-component system for early intracellular multiplication of *Mycobacterium tuberculosis*. Infect. Immun. *70*, 2256–2263.

Fabret, C., Feher, V.A., and Hoch, J.A. (1999). Two-component signal transduction in *Bacillus subtilis*: how one organism sees its world. J. Bacteriol. *181*, 1975–1983.

Finn, R.D., Mistry, J., Tate, J., Coggill, P., Heger, A., Pollington, J.E., Gavin, O.L., Gunasekaran, P., Ceric, G., Forslund, K., *et al.* (2010). The Pfam protein families database. Nucleic Acids Res. *38*, D211–D222.

Fukuda, Y., Nakayama, Y., and Tomita, M. (2003). On dynamics of overlapping genes in bacterial genomes. Gene *323*, 181–187

Galperin, M.Y. (2005). A census of membrane-bound and intracellular signal transduction proteins in bacteria: bacterial IQ, extroverts and introverts. BMC Microbiol. *5*, 35.

Galperin, M.Y. (2006). Structural classification of bacterial response regulators: diversity of output domains and domain combinations. J. Bacteriol. *188*, 4169–4182.

Galperin, M.Y. (2010). Diversity of structure and function of response regulator output domains. Curr. Opin. Microbiol. *13*, 150–159.

Galperin, M.Y., and Kolker, E. (2006). New metrics for comparative genomics. Curr. Opin. Biotechnol. *17*, 440–447.

Galperin, M.Y., and Nikolskaya, A.N. (2007). Identification of sensory and signal-transducing domains in two-component signalling systems. Methods Enzymol. *422*, 47–74.

Galperin, M.Y., Nikolskaya, A.N., and Koonin, E.V. (2001). Novel domains of the prokaryotic two-component signal transduction systems. FEMS Microbiol. Lett. *203*, 11–21.

Galperin, M.Y., Higdon, R., and Kolker, E. (2010). Interplay of heritage and habitat in the distribution of bacterial signal transduction systems. Mol. Biosyst. *6*, 721–728.

Gao, R., and Stock, A.M. (2009). Biological insights from structures of two-component proteins. Annu. Rev. Microbiol. *63*, 133–154.

Gao, R., and Stock, A.M. (2010). Molecular strategies for phosphorylation-mediated regulation of response regulator activity. Curr. Opin. Microbiol. *13*, 160–167.

Gao, R., Tao, Y., and Stock, A.M. (2008). System-level mapping of *Escherichia coli* response regulator dimerization with FRET hybrids. Mol. Microbiol. *69*, 1358–1372.

Geer, L.Y., Domrachev, M., Lipman, D.J., and Bryant, S.H. (2002). CDART: protein homology by domain architecture. Genome Res. *12*, 1619–1623.

Goodman, A.L., Merighi, M., Hyodo, M., Ventre, I., Filloux, A., and Lory, S. (2009). Direct interaction between sensor kinase proteins mediates acute and chronic disease phenotypes in a bacterial pathogen. Genes Dev. *23*, 249–259

Goulian, M. (2010). Two-component signaling circuit structure and properties. Curr. Opin. Microbiol. *13*, 184–189.

Grebe, T.W., and Stock, J.B. (1999). The histidine protein kinase superfamily. Adv. Microb. Physiol. *41*, 139–227.

Haft, D.H., Selengut, J.D., and White, O. (2003). The TIGRFAMs database of protein families. Nucleic Acids Res. *31*, 371–373.

Hess, J.F., Bourret, R.B., and Simon, M.I. (1991). Phosphorylation assays for proteins of the two-component regulatory system controlling chemotaxis in *Escherichia coli*. Methods Enzymol. *200*, 188–204.

Hettne, K.M., Williams, A.J., van Mulligen, E.M., Kleinjans, J., Tkachenko, V., and Kors, J.A. (2010). Automatic vs. manual curation of a multi-source chemical dictionary: the impact on text mining. J. Cheminform. *2*, 3.

Higgs, P.I., Cho, K., Evans, L.S., Whitworth, D.E., and Zusman, D.R. (2005). Four unusual two-component signal transduction homologs, RedC to RedF, are necessary for timely development in *Myxococcus xanthus*. J. Bacteriol. *187*, 8191–8195.

Hoch, J.A., and Varughese, K.I. (2001). Keeping signals straight in phosphorelay signal transduction. J. Bacteriol. *183*, 4941–4949.

Hutchings, M.I., Hoskisson, P.A., Chandra, G., and Buttner, M.J. (2004). Sensing and responding to diverse extracellular signals? Analysis of the sensor kinases and response regulators of *Streptomyces coelicolor* A3(2). Microbiology *150*, 2795–2806.

Jenal, U., and Galperin, M.Y. (2009). Single domain response regulators: molecular switches with emerging roles in cell organization and dynamics. Curr. Opin. Microbiol. *12*, 152–160.

Jiménez-Pearson, M.A., Delany, I., Scarlato, V., and Beier, D. (2005). Phosphate flow in the chemotactic response system of *Helicobacter pylori*. Microbiology *151*, 3299–3311.

Johnson, G.L. (2011). Defining MAPK interactomes. ACS Chem. Biol. *6*, 18–20.

Jourlin, C., Ansaldi, M., and Méjean, V. (1997). Transphosphorylation of the TorR response regulator requires the three phosphorylation sites of the TorS unorthodox sensor in *Escherichia coli*. J. Mol. Biol. *267*, 770–777.

Kim, D., and Forst, S. (2001). Genomic analysis of the histidine kinase family in bacteria and archaea. Microbiology *147*, 1197–1212.

Kirby, J.R. (2009). Chemotaxis-like regulatory systems: unique roles in diverse bacteria. Ann. Rev. Genet. *63*, 45–59.

Kofoid, E.C., and Parkinson, J.S. (1988). Transmitter and receiver modules in bacterial signalling proteins. Proc. Natl. Acad. Sci. U.S.A. *85*, 4981–4985.

Kolesov, G., and Mirny, L.A. (2009). Using evolutionary information to find specificity-determining and co-evolving residues. Methods Mol. Biol. *541*, 421–448.

Koretke, K.K., Lupas, A.N., Warren, P.V., Rosenberg, M., and Brown, J.R. (2000). Evolution of two-component signal transduction. Mol. Biol. Evol. *17*, 1956–1970.

Krell, T., Lacal, J., Busch, A., Silva-Jiménez, H., Guazzaroni, M.E., and Ramos, J.L. (2010). Bacterial sensor kinases: diversity in the recognition of environmental signals. Annu. Rev. Microbiol. *64*, 539–559.

Laub, M.T., and Goulian, M. (2007). Specificity in two-component signal transduction pathways. Annu. Rev. Genet. *41*, 121–145.

Laub, M.T., Biondi, E.G., and Skerker, J.M. (2007). Phosphotransfer profiling: systematic mapping of two-component signal transduction pathways and phosphorelays. Methods Enzymol. *423*, 531–548.

Letunic, I., Doerks, T., and Bork, P. (2009). SMART6: recent updates and new developments. Nucleic Acids Res. *37*, D229–D232.

Liang, X., Yu, C., Sun, J., Liu, H., Landwehr, C., Holmes, D., and Ji, Y. (2006). Inactivation of a two-component signal transduction system, SaeRS, eliminates adherence and attenuates virulence of *Staphylococcus aureus*. Infect. Immun. *74*, 4655–4665.

Lunt, B., Szurmant, H., Procaccini, A., Hoch, J.A., Hwa, T., and Weigt, M. (2010). Inference of direct residue contacts in two-component signalling. Methods Enzymol. *471*, 17–41.

Mascher, T., Helmann, J.D., and Unden, G. (2006). Stimulus perception in bacterial signal-transducing histidine kinases. Microbiol. Mol. Biol. Rev. *70*, 910–938.

Mizuno, T. (1997). Compilation of all genes encoding two-component phosphotransfer signal transducers in the genome of *Escherichia coli*. DNA Res. *4*, 161–168.

Mizuno, T. (2005). Two-component phosphorelay signal transduction systems in plants: from hormone responses to circadian rhythms. Biosci. Biotechnol. Biochem. *69*, 2263–2276.

Mizuno, T., Kaneko, T., and Tabata, S. (1996). Compilation of all genes encoding bacterial two-component signal transducers in the genome of the cyanobacterium, *Synechocystis* sp. strain PCC 6803. DNA Res. *3*, 407–414.

Mourey, L., Da Re, S., Pédelacq, J.D., Tolstykh, T., Faurie, C., Guillet, V., Stock, J.B., and Samama, J.P. (2001). Crystal structure of the CheA histidine phosphotransfer domain that mediates response regulator phosphorylation in bacterial chemotaxis. J. Biol. Chem. *276*, 31074–31082.

Munson, G.P., Lam, D.L., Outten, F.W., and O'Halloran, T.V. (2000). Identification of a copper-responsive two-component system on the chromosome of *Escherichia coli* K-12. J. Bacteriol. *182*, 5864–5871.

Nikitin, F., Rance, B., Itoh, M., Kanehisa, M., and Lisacek, F. (2004). Using protein motif combinations to update

KEGG pathway maps and orthologue tables. Genome Inform. *15*, 266–275.

Nixon, B.T., Ronson, C.W., and Ausubel, F.M. (1986). Two-component regulatory systems responsive to environmental stimuli share strongly conserved domains with the nitrogen assimilation regulatory genes *ntrB* and *ntrC*. Proc. Natl. Acad. Sci. U.S.A. *83*, 7850–7854.

Normark, S., Bergstrom, S., Edlund, T., Grundstrom, T., Jaurin, B., Lindberg, F.P., and Olsson, O. (1983). Overlapping genes. Annu. Rev. Genet. *17*, 499–525.

Oppenheim, D.S., and Yanofsky, C. (1980). Translational coupling during expression of the tryptophan operon of *Escherichia coli*. Genetics *95*, 785–795

Pao, G.M., and Saier, M.H., Jr. (1995). Response regulators of bacterial signal transduction systems: selective domain shuffling during evolution. J. Mol. Evol. *40*, 136–154.

Parkinson, J.S., and Kofoid, E.C. (1992). Communication modules in bacterial signalling proteins. Annu. Rev. Genet. *26*, 71–112.

Paul, R., Jaeger, T., Abel, S., Wiederkehr, I., Folcher, M., Biondi, E.G. Laub, M.T., and Jenal, U. (2008). Allosteric regulation of histidine kinases by their cognate response regulator determines cell fate. Cell *133*, 452–461.

Perego, M. (1998). Kinase-phosphatase competition regulates *Bacillus subtilis* development. Trends Microbiol. *6*, 366–370.

Qian, W., Han, Z.J., and He, C. (2008). Two-component signal transduction systems of *Xanthomonas* spp.: a lesson from genomics. Mol. Plant Microbe Interact. *21*, 151–161.

Robinson, V.L., Buckler, D.R., and Stock, A.M. (2000). A tale of two components: a novel kinase and a regulatory switch. Nat. Struct. Biol. *7*, 626–633.

Romagnoli, S., and Tabita, F.R. (2006). A novel three-protein two-component system provides a regulatory twist on an established circuit to modulate expression of the *cbbI* region of *Rhodopseudomonas palustris* CGA010. J. Bacteriol. *188*, 2780–2791.

Romagnoli, S., and Tabita, F.R. (2007). Phosphotransfer reactions of the CbbRRS three-protein two-component system from *Rhodopseudomonas palustris* CGA010 appear to be controlled by an internal molecular switch on the sensor kinase. J. Bacteriol. *189*, 325–335.

Santos, J.L., and Shiozaki, K. (2001) Fungal histidine kinases. Sci STKE *2001*, re1

Scharf, B. (2010). Summary of useful methods for two-component system research. Curr. Opin. Microbiol. *13*, 246–252.

Schug, A., Weigt, M., Hoch, J.A., Onuchic, J.N., Hwa, T., and Szurmant, H. (2010). Computational modeling of phosphotransfer complexes in two-component signalling. Methods Enzymol. *471*, 43–58.

Shi, X., Wegener-Feldbrügge, S., Huntley, S., Hamann, N., Hedderich, R., and Søgaard-Andersen, L. (2008). Bioinformatics and experimental analysis of proteins of two-component systems in *Myxococcus xanthus*. J. Bacteriol. *190*, 613–624.

Simon, M.I., Borkovich, K.A., Bourret, R.B., and Hess, J.F. (1989). Protein phosphorylation in the bacterial chemotaxis system. Biochimie *71*, 1013–1019.

Siryaporn, A., Perchuk, B.S., Laub, M.T., and Goulian, M. (2010). Evolving a robust signal transduction pathway from weak cross-talk. Mol. Syst. Biol. *6*, 452.

Sivaneson, M., Mikkelsen, H., Ventre, I., Bordi, C., and Filloux, A. (2011). Two-component regulatory systems in *Pseudomonas aeruginosa*: an intricate network mediating fimbrial and efflux pump gene expression. Mol. Microbiol. *79*, 1353–1366.

Skerker, J.M., Prasol, M.S., Perchuk, B.S., Biondi, E.G., and Laub, M.T. (2005). Two-component signal transduction pathways regulating growth and cell cycle progression in a bacterium: a system-level analysis. PLoS Biol. *3*, e334.

Skerker, J.M., Perchuk, B.S., Siryaporn, A., Lubin, E.A., Ashenberg, O., Goulian, M., and Laub, M.T. (2008). Rewiring the specificity of two-component signal transduction systems. Cell *133*, 1043–1054.

Sonnhammer, E.L.L., von Heijne, G., and Krogh, A. (1998). A hidden Markov model for predicting transmembrane helices in protein sequences. In Proceedings of Sixth International Conference on Intelligent Systems for Molecular Biology. Glasgow, J., Littlejohn, T., Major, F., Lathrop, R., Sankoff, D., and Sensen, S., eds (AAAI Press, Menlo Park, CA), pp. 175–182.

Stewart, R.C. (2010). Protein histidine kinases: assembly of active sites and their regulation in signalling pathways. Curr. Opin. Microbiol. *13*, 133–141.

Stock, A.M., Robinson, V.L., and Goudreau, P.N. (2000). Two-component signal transduction. Annu. Rev. Biochem. *69*, 183–215.

Stock, J.B., Ninfa, A.J., and Stock, A.M. (1989). Protein phosphorylation and regulation of adaptive responses in bacteria. Microbiol. Rev. *53*, 450–490.

Szurmant, H., Bobay, B.G., White, R.A., Sullivan, D.M., Thompson, R.J., Hwa, T., Hoch, J.A., and Cavanagh, J. (2008). Co-evolving motions at protein–protein interfaces of two-component signalling systems identified by covariance analysis. Biochemistry *47*, 7782–7784.

Takei, N., Nakajima, M., Oyama, T., Kito, R., Sugita, C., Sugita, M., Kondo, T., and Iwasaki, H. (2006). A KaiC-associating SasA-RpaA two-component regulatory system as a major circadian timing mediator in cyanobacteria. Proc. Natl. Acad. Sci. U.S.A. *103*, 12109–12114.

Tatusov, R.L., Fedorova, N.D., Jackson, J.D., Jacobs, A.R., Kiryutin, B., Koonin, E.V., Krylov, D.M., Mazumder, R., Mekhedov, S.L., Nikolskaya, A.N., *et al.* (2003). The COG database: an updated version includes eukaryotes. BMC Bioinformatics *4*, 41.

Ueki, T., and Inouye, S. (2002). Transcriptional activation of a heat-shock gene, *lonD*, of *Myxococcus xanthus* by a two component histidine-aspartate phosphorelay system. J. Biol. Chem. *277*, 6170–6177.

Ulrich, L.E., and Zhulin, I.B. (2007). MIST: a microbial signal transduction database. Nucleic Acids Res. *35* (Database issue), D386–D390.

Ulrich, L.E., and Zhulin, I.B. (2010). The MiST2 database: a comprehensive genomics resource on microbial signal transduction. Nucleic Acids Res. *38*, D401–D407.

Walderhaug, M.O., Polarek, J.W., Voelkner, P., Daniel, J.M., Hesse, J.E., Altendorf, K., and Epstein, W. (1992). KdpD and KdpE, proteins that control expression of the *kdpABC* operon, are members of the two-component sensor-effector class of regulators. J. Bacteriol. *174*, 2152–2159.

Wegener-Feldbrügge, S., and Søgaard-Andersen, L. (2009). The atypical hybrid histidine protein kinase RodK in *Myxococcus xanthus*: spatial proximity supersedes kinetic preference in phosphotransfer reactions. J. Bacteriol. *191*, 1765–1776.

Weigt, M., White, R.A., Szurmant, H., Hoch, J.A., and Hwa, T. (2009). Identification of direct residue contacts in protein–protein interaction by message passing. Proc. Natl. Acad. Sci. U.S.A. *106*, 67–72.

West, A.H., and Stock, A.M. (2001). Histidine kinases and response regulator proteins in two-component signalling systems. Trends Biochem. Sci. *26*, 369–376.

Wheeler, R.T., and Shapiro, L. (1999). Differential localization of two histidine kinases controlling bacterial cell differentiation. Mol. Cell *4*, 683–694.

Whitworth, D.E. (2008). Genomes and knowledge – a questionable relationship? Trends Microbiol. *16*, 512–519.

Whitworth, D.E. (2011). Two-component regulatory systems in prokaryotes. In Bacterial Regulatory Networks, Filloux, A., ed. (Horizon Scientific Press, Norwich, UK). In Press.

Whitworth, D.E., and Cock, P.J.A. (2008a). Myxobacterial two-component systems. In Myxobacteria: Multicellularity and Differentiation, Whitworth, D.E., ed. (ASM Press, Washington, DC), pp. 169–189.

Whitworth, D.E., and Cock, P.J.A. (2008b). Two-component systems of the myxobacteria: structure, diversity and evolutionary relationships. Microbiology *154*, 360–372.

Whitworth, D.E., and Cock, P.J.A. (2009). Evolution of prokaryotic two-component systems: insights from comparative genomics. Amino Acids *47*, 459–466.

Whitworth, D.E., Millard, A., Hodgson, D.A., and Hawkins, P.F. (2008). Protein–protein interactions between two-component system transmitter and receiver domains of *Myxococcus xanthus*. Proteomics *8*, 1839–1842.

Williams, R.H.N., and Whitworth, D.E. (2010). Genetic organisation of prokaryotic two-component system signalling pathways. BMC Genomics *11*, 720.

Wuichet, K., and Zhulin, I.B. (2010). Origin and diversification of a complex signal transduction system in prokaryotes. Sci. Signal. 29, ra50.

Wuichet, K., Cantwell, B.J., and Zhulin, I.B. (2010). Evolution and phyletic distribution of two-component signal transduction systems. Curr. Opin. Microbiol. *13*, 219–225.

Yagil, E., Bracha, M., and Lifshitz, Y. (1975). The regulatory nature of the *phoB* gene for alkaline phosphatase synthesis in *Escherichia coli*. Mol. Gen. Genet. *137*, 11–16.

Yamamoto, K., Hirao, K., Oshima, T., Aiba, H., Utsumi, R., and Ishihama, A. (2005) Functional characterization *in vitro* of all two-component signal transduction systems from *Escherichia coli*. J. Biol. Chem. *280*, 1448–1456.

Structural Basis of Signal Transduction and Specificity in Two-component Systems

2

Patricia Casino, Marisa López-Redondo and Alberto Marina

Abstract

Two-component systems (TCSs) constitute a signal transduction mechanism, mainly found in prokaryotes, which is relatively simple as they are basically formed by two proteins: a histidine kinase (HK) and a response regulator (RR). These two proteins are able to transmit all kinds of signals productively and elegantly by the use of phosphorelays in order to ensure cell survival. For this purpose, the HK possesses many qualities; first it is able to sense a signal, second it binds ATP, and third it autophosphorylates on a catalytic His residue. Subsequently, the RR comes into play to promote the final response, but first it needs to be phosphorylated on a catalytic Asp residue via a phosphoryl group transfer from the phosphorylated His of the HK. The phosphorylated RR (RR~P) can thus elicit many diverse responses, which generally involve binding to DNA to activate specific genes. Finally, the system is shut down by the dephosphorylation of the RR~P, a mechanism which can be performed by the RR itself, or assisted by the HK. Undoubtedly, understanding the molecular basis of specificity in the interaction between HK and RR couples as well as the complete catalytic mechanism between these two molecules in the signalling process is especially important from many viewpoints, including the medical, and recent advances in the structural and biochemical field have contributed decidedly.

Introduction

TCS are the most widespread signal transduction systems found in prokaryotes and they regulate many bacterial behaviours such as virulence, pathogenicity, motility, nodulation, nutrient uptake, secondary metabolite production, metabolic regulation, cell division, etc. (Galperin, 2004). TCSs have also been found in yeast and plants, but are absent in mammals, which makes them potential targets for antimicrobials drug design (Watanabe *et al.*, 2008). The designation TCS derives from the initial observation that the signalling mediated by these systems requires the presence of two molecules: a sensor called HK and an effector named RR. The stimulus sensed by receptor HK is transformed into a signal for its own ATP-dependent autophosphorylation on a His residue. Subsequently, the phosphorylated HK transmits the signal to a conserved Asp of the RR through a phosphoryl group transfer reaction. Finally, the activated RR~P acts as an effector protein which generates responses by either acting as a transcription factor for the activation/repression of target genes or having enzymatic or mechanistic activities. This basic and elegant scheme has further evolved: signal transduction can also be regulated by either the phosphatase activity shown by some HKs towards RRs to tune the level of phosphorylated RR (P~RR) or the presence of ancillary proteins which introduce a higher order of regulation versus the activities of the HK and/or the RR. A more complex regulation of TCS is found in the phosphorelay pathways, with two or more phosphotransfer steps between the multiple His- and Asp-containing proteins leading to not only hybrid HKs, but also to the presence of additional domains called histidine phosphotransfer domains (HPt) (Gao and Stock, 2009a). In this chapter, we will attempt to shed light on the catalytic mechanism of TCS by extracting information from new advances made in the field, mainly from a structural point of view.

Sequence and structure

Sequence features and general organization of HK and RR

TCS proteins have a modular architecture which allows the integration of many different stimuli into a diverse final response through a conserved mechanistic pathway. Both the HK and RR have conserved domains in charge of the His-Asp phosphotransfer machinery, as well as variable additional domains which confer specificity for each particular system. Nowadays, more that ~21,000 sequences of TCS have been reported (ExPASy Server; www.expasy.org). However, as soon as a significant number of these sequences became available, the presence of conserved regions or motifs, mainly in the HKs, was evidenced and which subsequent studies have revealed as functionally relevant. These motives, or 'boxes', have been key features to annotate new TCSs. Each box was named by the letter of the corresponding invariable residue contained in it. In this way, HKs present H, N, D, F and G boxes (Fig. 2.1). HKs can be roughly

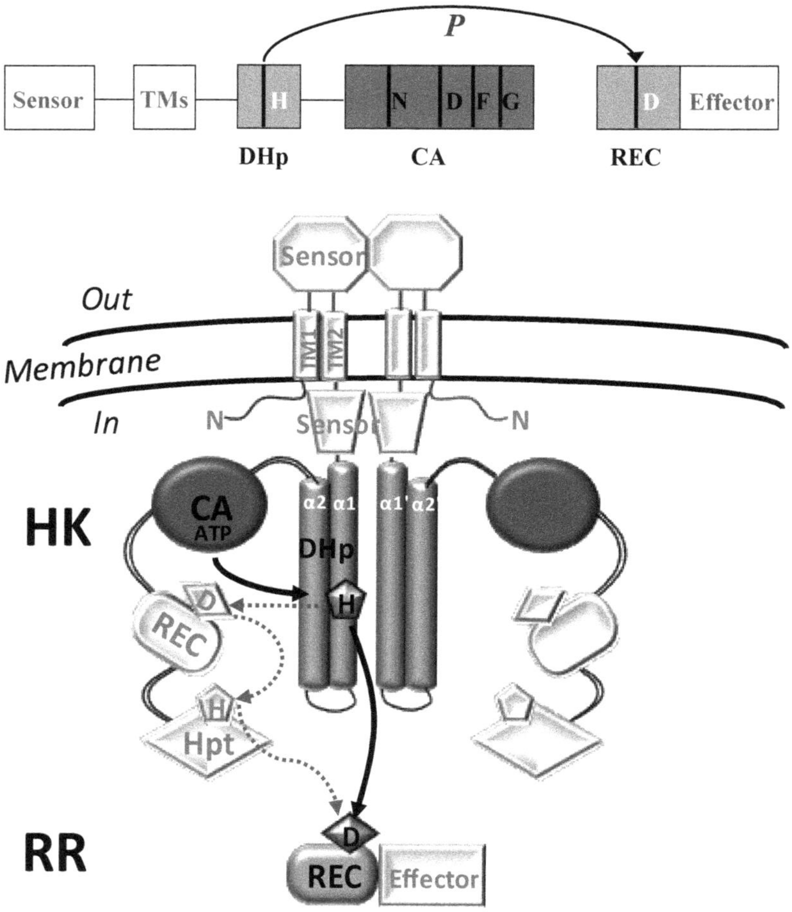

Figure 2.1 Schematic representation of the domain organization of a basic two-component system formed by a dimeric histidine kinase (HK) and a response regulator (RR). The HK contains an extracellular sensor domain connected to the cytoplasmic domains by two transmembrane helices. The cytoplasmic domains are composed of a DHp domain containing the phosphorylatable His residue in the H box and a CA domain that holds the nucleotide towards the N, D, F and G boxes. The His (H) phosphorylation is mediated by ATP followed by and subsequent phosphoryl transfer to the phosphorylatable Asp (D) in the N-terminal receiver domain (REC) of the RR (black arrow). In general, the RR contains an effector domain C-terminal to the REC in charge of eliciting the final response. Additional domains, such as Hpt and REC, are found in hybrid HK promoting additional phosphorelays (dotted grey arrows).

divided into two regions, a N-terminal sensing and a C-terminal catalytic kinase core. The largest group of HKs, the so-called orthodox, gathers dimeric membrane-embedded receptors with the N-terminal region constituting the periplasmic or extracellular sensing domain framed by two transmembrane (TM) helices and the C-terminal kinase core being localized in the cytoplasm (Fig. 2.1). Other HKs contain more TM segments, less TMs (one TM), or they even lack TMs; HKs being soluble cytoplasmic proteins belong to the last type. Additional sensing domains are present in the cytoplasmic portion of some HKs (Fig. 2.1) (Mascher *et al.*, 2006). The kinase catalytic core of the HKs is composed of two well-differentiated domains: a dimerization and histidine phosphotransfer domain (DHp) and a catalytic ATP binding domain (CA) (Fig. 2.1). The DHp domain allocates the phosphorylatable His residue and mediates HK dimerization. The CA domain binds the ATP molecule for the autophosphorylation reaction. Additional sensor or regulator domains such as PAS or HAMP, usually comprised between the TM helices and the DHp domain, are found in several HKs (Fig. 2.1). RRs typically contain two domains: the N-terminal receiver (REC) domain and the C-terminal effector domain (Fig. 2.1). The REC domain, which is extremely well conserved, contains enzymatic activity and allocates the phosphorylatable Asp, while the effector domain is diverse and renders specific responses. Therefore, RRs can be classified according to the function of their effector domain; thus 63% of RRs act as DNA binding effectors, 13% contain enzymatic activity, while the rest mediate interactions with other proteins or ligands. The remaining ~17% of RRs lack an effector domain and the REC domain mediates output signalling (Gao *et al.*, 2007).

Conserved domains of the HK

The dimerization and histidine phosphotransfer (DHp) domain

The crystal structures of the DHp domain of the HKs EnvZ (PDB: 1JOY) from *Escherichia coli* (Tomomori *et al.*, 1999), DesK (PDB: 3EHH) from *Bacillus subtilis* (Albanesi *et al.*, 2009) and HK853 (PDB: 2C2A) from *Themotoga maritima* (Marina *et al.*, 2005), revealed that it (Pfam family HisKA; http://pfam.sanger.ac.uk/) is formed by a long (> 30 Å) helical hairpin (α1–α2) which associates with the corresponding helical hairpin of the second DHp domain in an almost parallel manner (small twist of 10–20 degrees) to form a four-helix bundle in the dimeric HK (Fig. 2.2). Helices α1 and α2 of each helical hairpin are connected by an unstructured and length-variable loop (DHp loop). Helix α1 contains the H-box where the phosphorylatable His residue is allocated (Figs. 2.1 and 2.2), and this helix is connected upstream to the N-terminal sensing domains through, in some cases, the transmembrane helices. The superposition of the DHp domain of EnvZ, DesK and HK853 shows that the handedness of the DHp loop connection between α1 and α2 is not conserved (Marina *et al.*, 2005), suggesting mechanistic implications for this divergence (see below) (Casino *et al.*, 2009, 2010). The DHp domain is also connected downstream to the CA domain by a loop, variable in length, size and sequence, called the DHp-CA linker. This length variability in the DHp-CA linker could have implications for the CA domain's movement in relation to the DHp to perform the enzymatic reaction of autophosphorylation (Albanesi *et al.*, 2009; Casino *et al.*, 2009, 2010). Given the modular architecture of TCS, additional domains can be found in the signalling systems which introduce more complexity into the phosphorelay. The HPt (Pfam referred as Hpt) domains are generally found in systems where hybrid HKs are present. This domain consists in six alpha helices, of which four are arranged in a four-helix bundle folding, similarly to the DHp domain of the dimeric orthodox HKs, and it contains a phosphorylatable His residue in the middle of one of the helices forming the bundle (Fig. 2.2). The solved three dimensional structures of diverse Hpt domains found in either prokaryotes (*E. coli* ArcB, *Salmonella* CheA) (Kato *et al.*, 1998; Mourey *et al.*, 2001) or eukaryotes (*Saccharomyces cerevisiae* YPD1, maize ZmHP2) (Xu and West, 1999; Sugawara *et al.*, 2005) demonstrate that the HPt domain is a common and conserved structural motif.

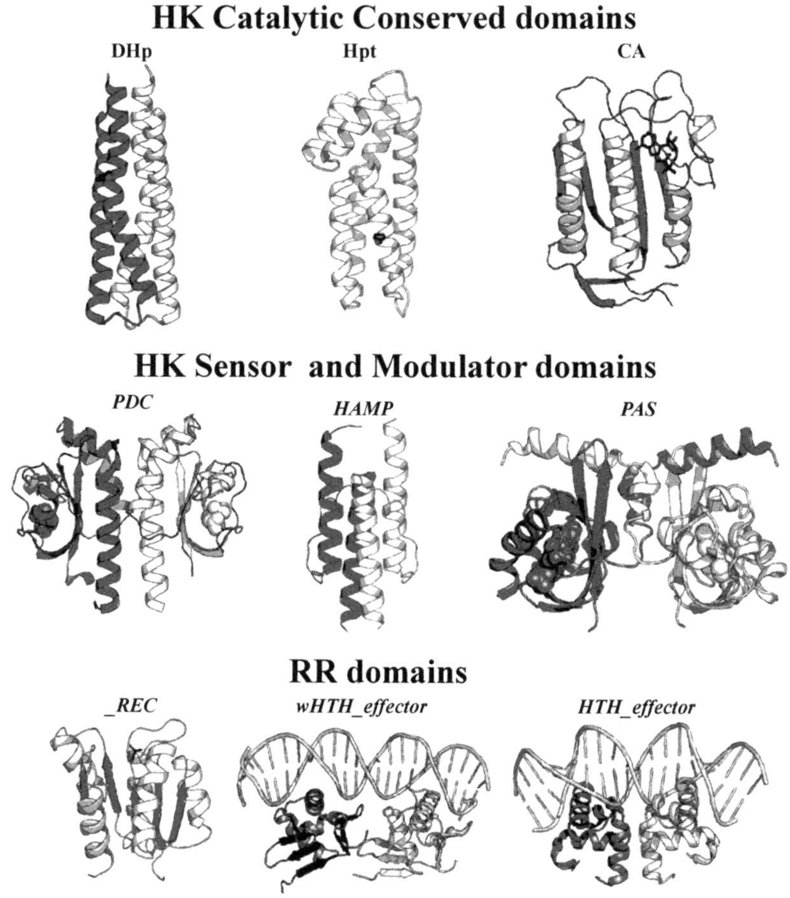

Figure 2.2 Three-dimensional structure of the basic domains found in TCS in a cartoon representation. At the top, the histidine catalytic domains comprising the DHp and CA domains from HK853 (PDB: 3DGE) and the Hpt domain form ZmHP2 (PDB: 1WN0). In the middle, some HK sensor and modulator domains are shown, the extracellular PDC-fold sensor domain found in CitA (PDB: 2J80), the cytoplasmic HAMP domain of Af1503 (PDB: 2ASW) and the cytoplasmic PAS sensor domain found in FixL (PDB: 1D06). At the bottom, the RR domains are shown by the REC domain of RR468 (PDB: 3DGE) and the effector domains of NarL bound to DNA and PhoB bound to DNA, which present a different mode of interaction with DNA. To differentiate the dimeric nature of the basic domains, subunits are coloured in grey and white. In the monomeric basic domains, the beta strands are coloured in grey while the helices are in white for clarity. The effector domain of PhoB binds to DNA as a monomer; however two monomers are shown to interact with DNA to distinguish its mechanism from NarL.

The catalytic ATP binding (CA) domain

The CA domain (Pfam HATPase_c) is a well-conserved domain formed by a 3α/5β sandwich with a characteristic left-handed βαβ connectivity (Fig. 2.2). It differs from any known ATP-binding domain of Ser/Thr/Tyr kinases, but is structurally similar to the ATPase domains of GyrB, Hsp90 and MutL, leading to the term ATPase/kinase GHKL superfamily to describe this module (Dutta and Inouye, 2000). The features describing this superfamily are the presence of four conserved motifs (I, II, III, IV) of which the first three correspond, in the HKs, to the conserved residues present in the N, D, G boxes (Fig. 2.1), which define a new ATP-binding fold known as the Bergerat fold (Dutta and Inouye, 2000). The N box contains an Asn residue which coordinates the Mg^{2+} ion required for the catalysis. The D box is responsible for ATP specificity since the side chain of the Asp residue, which names this box, interacts with the N6 atom of the ATP adenine ring. The G box contain Gly residues which act as two hinges to hold a flexible unstructured loop that is variable in length and covers the ATP-binding site called the ATP-lid. Finally, the HK-specific F box contains a Phe/Tyr which helps to support the

ATP-lid. Nucleotide binding to the CA domain induces minimal changes which are restricted to the ATP-lid. Indeed, the presence of the triphosphate nucleotide stabilizes the ATP-lid, allowing to trace this loop in the structures solved by X-ray diffraction (Bilwes *et al.*, 2001; Marina *et al.*, 2001), while in the absence of the nucleotide or the presence of the diphosphate form ATP-lid is not maintained stable unless it interacts with the RR (Bilwes *et al.*, 2001; Song *et al.*, 2004; Marina *et al.*, 2005; Casino *et al.*, 2009). Structural data have revealed that the CA domain undergoes rotational movements of the rigid body to bring the ATP closer to the His residue to either perform the autophosphorylation reaction or to stabilize the interaction with the RR (Albanesi *et al.*, 2009; Casino *et al.*, 2009).

The sensing and modulator HK domains

In conformity with the different stimuli sensed HK sensing domains are variable in sequence and structure, but can be grouped into some families. The most representative periplasmic or extracellular sensor domain families include the PAS-like domain named the PDC fold (a mixed α/β fold) found in PhoQ, DcuS, CitA (Fig. 2.2), the all-α folds found in NarX and TorS and the all-β fold (Mascher *et al.*, 2006; Cheung and Hendrickson, 2010). Some sensor domains contain double-PDC, as in the HKs type mmHK1S-ZX (Zhang and Hendrickson, 2010), or insertions which enlarge the sensor domain, as in the all-helical sensor domain of TorS (Moore and Hendrickson, 2009). Other domains, such as the CHASE 1–4 domains with a predicted α-/β fold (Anantharaman and Aravind, 2001), the Reg-prop domain which forms β-propellers, and the binding-protein (PBPb) sensing domains found in BvgS and EvgS, have been found less frequently (Mascher *et al.*, 2006).

The presence of sensing domains in the cytoplasmic portion is habitual in several HKs. These domains are inserted between the TM helices and the DHp domain, and it has been proposed that they sense internal signals or modulate the signal sensed by extracellular domains. The most characteristic ones are the PAS domain (a five-stranded anti-parallel β-sheet core flanked by α-helices) found in FixL (Miyatake *et al.*, 2000), the HAMP linker domain (two amphipathic α-helices with coiled-coil properties) and the GAF domain (a six-stranded anti-parallel β-sheet core) (Fig. 2.2) (Gao and Stock, 2009a).

RR domains

The receiver (REC) domain

The REC domain (Response_reg in Pfam) is a conserved domain with a unique fold (α_5/β_5) and a central β-sheet formed by five parallel strands surrounded by three α helices on one side and two α helices on the other side (Fig. 2.2). The catalytic centre is formed by a well conserved acidic pocket encircled by the loops connecting the end of the β strands to the beginning of the α helices named L$\beta\alpha x$ (x corresponds to the number of the β and the α element connected) (Stock *et al.*, 2000) (Figs. 2.2 and 2.3). The acidic pocket contains the phosphorylatable Asp residue, found at the beginning of the Lβα3 loop, and two other Asp residues involved in coordinating the Mg^{2+} ion that are found at the Lβα1 loop (Fig. 2.3). Additional catalytic residues are the conserved Thr/Ser residue in the Lβα4 loop, which interacts with one of the oxygen atoms of the phosphoryl group and the conserved Lys residue in the Lβα5 loop, which plays a key role in the transfer of the phosphoryl group during the different reactions in the signalling process (Fig. 2.3) (Gao and Stock, 2009b).

Several RRs or their isolated REC domains are able to autophosphorylate using small molecule phospho-donors such as phosphoramidate, acetyl phosphate, carbamyl phosphate and phosphoimidazole. They are also able to catalyse their own dephosphorylation by mediation of a water molecule, which suggests that this domain is responsible for mediating the phosphoryl transfer/phosphatase reaction from the HK. The conformational changes of the REC domain induced by phosphorylation have been extensively studied at the structural level; consequently, the mechanism of RR activation by phosphorylation, named 'two-states', is well established (Gao and Stock, 2010) (Fig. 2.3). Upon phosphorylation, the Thr/Ser residue found in the Lβα4 loop is oriented towards the active site and makes a hydrogen bond with the phosphate group, generating this movement

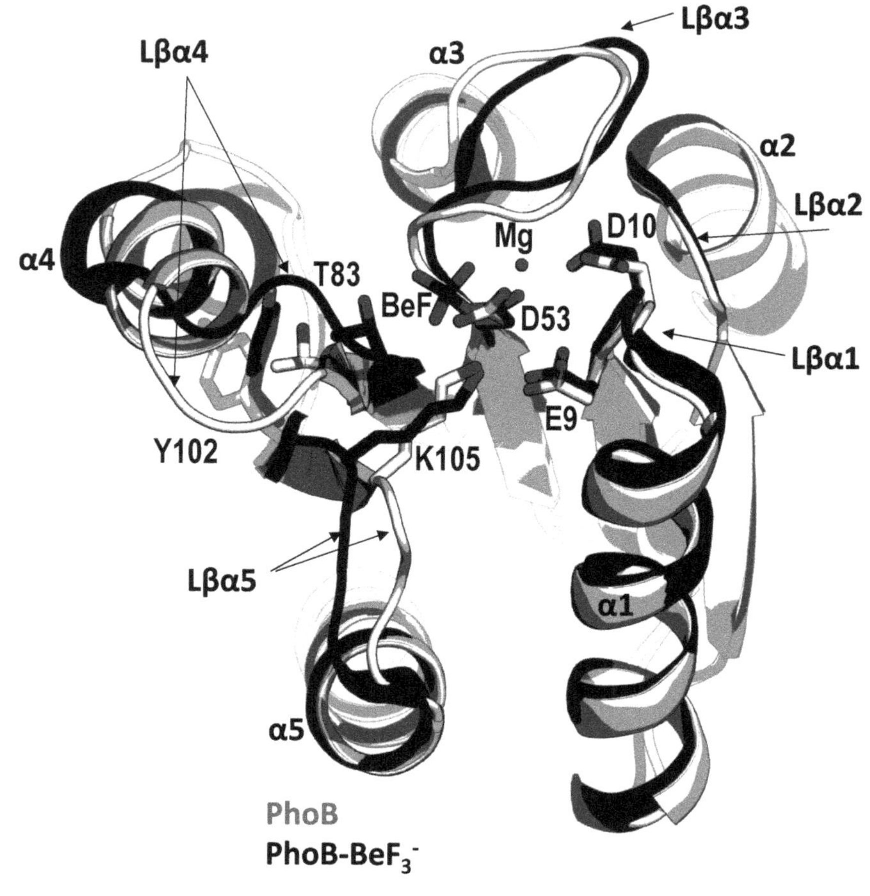

Figure 2.3 Two conformations for the active or inactive state of RRs. Cartoon representation of the RR PhoB in its active state (black, PDB: 1ZES) and in its inactive state (white, PDB:1B00). The active state refers to the phosphorylated state and is achieved when BeF_3^- is bound to PhoB. The catalytic residues involved in the active centre are shown in ball and stick, as well as the phosphoryl-mimicking BeF_3^- and the Mg^{2+} cation.

and a hydrophobic pocket that is occupied by the conserved Tyr/Phe, at the Lβα5 loop. These coordinated movements induce a conformation reorganization on the exposed α4-β5-α5 surface of the RR REC domain (Stock and Guhaniyogi, 2006). It is precisely the α4-β5-α5 surface which is generally implicated in the dimerization of the RR, habitual requirement to promote the final response. The most general form of dimerization, found in the OmpR/PhoB family (the most common RR family), is represented by the '4–5–5 dimer' where the REC domains dimerize via a 2-fold rotation axis. However, in other cases, dimerization generates an inverted 4–5–5 dimer, a 5–5 dimer and a 4–5 dimer (Gao and Stock, 2010). Therefore, the changes induced by phosphorylation on the α4-β5-α5 surface controls the RR oligomeric state and, consequently, its activity. Additionally, three-dimensional structures proving another dimerization type through a domain-swapping configuration (King-Scott *et al.*, 2007) or a dimer interface involving regions other than the α4-β5-α5 surface have been described (Gao and Stock, 2010).

The effector domain

The effector domains of the RRs are diverse: DNA-binding effector domains prevail, while RNA-binding domains, protein binding domains or domains with enzymatic activity are less abundant. The DNA-binding effector domains can be grouped into several families according to their type of folding and are named as the most representative RR of each group. The OmpR-like family, which is the most common, shows a winged helix–turn–helix (wHTH) fold (Fig. 2.2), the NarL-like family presents a helix–turn–helix (HTH) fold (Fig. 2.2), while the NtrC-like and the less frequent LytR-like, PrrA-like and AraC-type families present different variations of the HTH fold. Effector domains with enzymatic activity are variable, of which the GGDEF-type domain with diguanylate cyclase activity, the EAL domain with c-di-GMP-specific phosphodiesterase activity, the CheB-like domain showing methylesterase activity, or the protein phosphatase PP2C domain with phosphatase activity stand out (Gao *et al.*, 2007; Gao and Stock, 2009a). The protein-binding effector domains exert their function through protein–protein interactions, such as CheV-like proteins (interacting with the CheW-like domain), the Hnr-type RRs and the RRs containing the PAS, GAF, TRP or HPt domains. Finally, only one RNA-binding effector domain has been found, this being the ANTAR domain (Galperin, 2006).

Structural insights into TCS signalling

Despite the diversity of the signals to be transduced by TCSs, the basic scheme of enzymatic reactions involved in the signalling performed by these systems is conserved and implies HK autophosphorylation (kinase reaction), phosphoryl group transfer from the HK to the RR (phosphotransfer reaction) and RR dephosphorylation (phosphatase reaction) mediated by the RR or induced by the HK. Fidelity between the HK and RR pairs is a characteristic feature in TCSs; however, cross-talk between the HK and non-cognate RRs has also been described, as well as more elaborated relationships, where a HK has several cognate RRs (denominated one-to-many) (Lopez-Redondo *et al.*, 2010), or several HKs recognize a single RR (many-to-one) (Laub and Goulian, 2007). Furthermore, the interactions of the HKs and RRs with ancillary proteins, such as proteins PII, SipA or SDA, which modulate the enzymatic reactions of one or several TCSs, generate more elaborated pathways that are real signalling networks (Krell *et al.*, 2010).

HK autophosphorylation

The HKs function as constitutive dimers and perception of the signal by the sensor domain is believed to cause a conformational change that is transduced towards the kinase core to control the catalytic machinery. Thus, this portion of the enzyme acquires a conformation suitable to carry out one of the three reactions. To accomplish the autophosphorylation reaction, the binding of ATP to the CA domain is initially required, followed by the subsequent approach of the CA domain towards the DHp domain to phosphorylate the conserved His residue. This movement should align the His Nε atom which acts as an attacking nucleophile at the correct distance from the γ-phosphate of ATP, yielding the phosphorylated His (P~His) residue and ADP. The P~His bond presents chemical instability, and is especially labile at an acidic pH, although resistant to a basic pH (Attwood *et al.*, 2007). This property is a key element in the biological function of TCS as the P~N bond has a high free energy of hydrolysis, which confers an ideal feature for the phosphoryl group transfer (to RRs in the case of HKs). It is precisely owing to the high energetic P~N bond that the equilibrium of the kinase reaction is favoured towards the unphosphorylated state of the HK so that the kinase reaction only takes place owing to the high ATP/ADP intracellular ratio. This property complicates the structural determination, by crystallographic or NMR techniques, of a phosphorylated HK or a HK performing the kinase reaction. Therefore, biochemical studies have mainly provided the current limited knowledge available about the molecular basis of the kinase reaction. Recent structural data on DesK have demonstrated the dynamic nature of the DHp and CA domains to acquire the necessary conformation to allow the appropriate functional state of the protein, be it kinase, phosphotransferase or phosphatase (Albanesi *et al.*, 2009). It is amazing to note that the trapping of conformations emulating the different functional states in DesK, unphosphorylated/dephosphorylated, could also be obtained from the specific mutations of the phosphorylatable His (H188) to either Val in the case of the phosphatase state or to Glu in the case of the kinase state (Albanesi *et al.*, 2009).

cis and *trans* directionality

By considering that both the subunits in the dimeric HK are able to catalyse the kinase reaction, then two possible ways of autophosphorylation, either *cis* or *trans*, are possible (Fig. 2.4A). In the *trans* mode, the CA domain of one subunit would phosphorylate the His in the opposite subunit, while in the *cis* mode, the His would be phosphorylated by the CA domain of the same subunit. Initially, the *trans* mode was confirmed for *E. coli* NtrB and EnvZ HKs, and it was proposed that this mode was universal for all HKs (Ninfa *et al.*, 1993; Cai and Inouye, 2003). More recently, the kinase reaction in the *cis* mode has been demonstrated for HK853, PhoR and ArcB HKs (Casino *et al.*, 2009; Pena-Sandoval and Georgellis, 2010), which again leads to a debate about these mechanistic differences and their biological implications. Unfortunately, the structural information of a HK performing the autokinase reaction is not yet available, although some structural models for both modes based on the completed structural

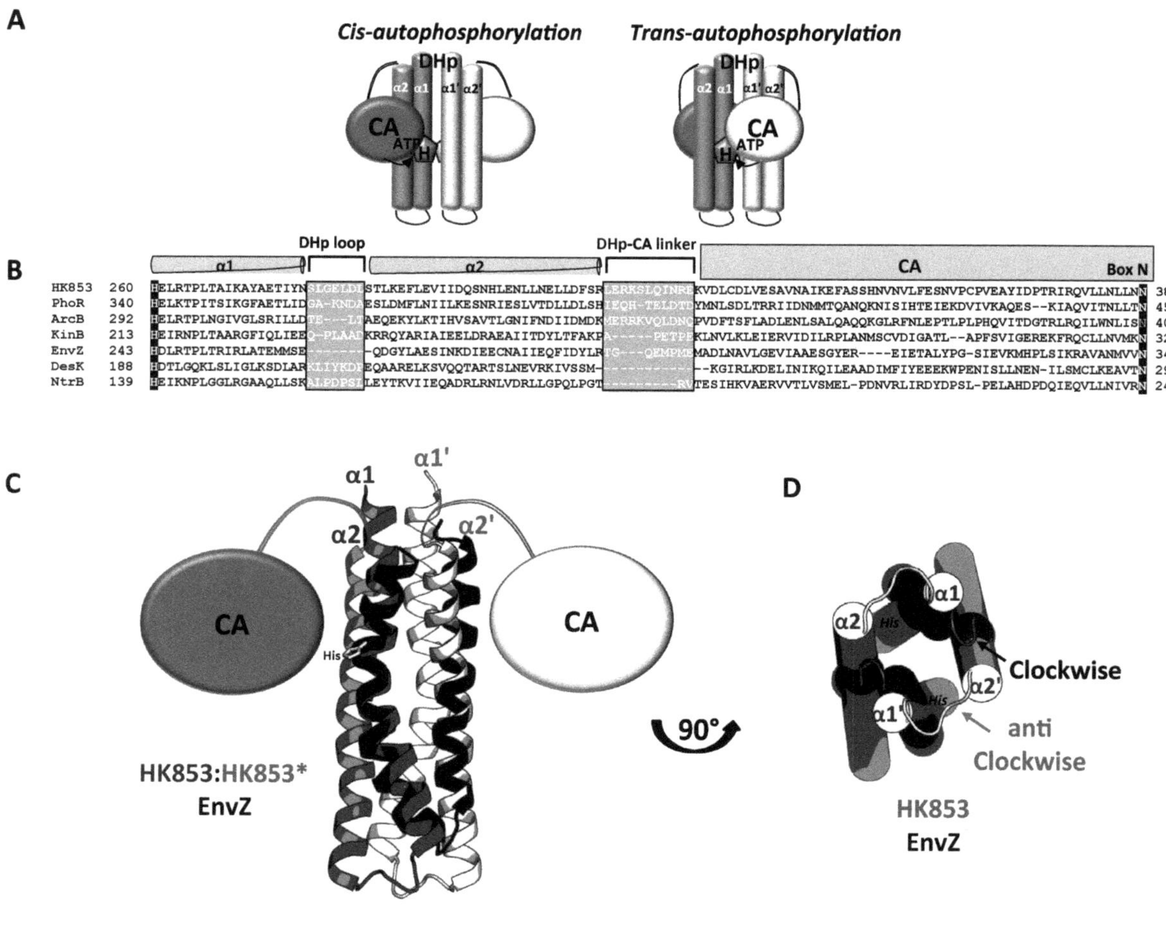

Figure 2.4 Autophosphorylation mechanism, clues for its understanding. (A) Schematic representation of the *cis*- and *trans*-autophosphorylation mechanism in a hypothetical HK. (B) Sequence and/or structure-based alignment of the HKs whose autophosphorylation mechanism is known. *trans*-autophosphorylation for *E. coli* EnvZ and NtrB and *B. subtilis* DesK and *cis*-autophosphorylation for *T. maritima* HK853, *S. aureus* PhoR and *E. coli* ArcB. *G. stearothermophilus* KinB was also introduced into the structural alignment owing to its 3-D known structure. Above the alignment, the structure features of the DHp and CA domain are drawn, and the H and N boxes are highlighted, which are crucial for autophosphorylation. (C) Superposition of HK853 with EnvZ. Cartoon representation of the dimeric DHp domain of HK853, one subunit is coloured in grey and the other in white (*). The EnvZ DHp domain is superposed to DHp HK853, but just one monomer is shown in black for clarity. The CA domain is drawn as being connected to α2 or α2*. (D) 90° rotation of (C) and representation in the cylindrical helices of the superposed DHp domains to visualize the DHp loop connecting α1 with α2 and α1* with α2*. The EnvZ cylindrical helices are coloured in black, while the HK853 cylindrical helices are depicted in white.

cytoplasmic machinery of two HKs, DesK and HK853 have been provided (Casino *et al.*, 2010; Trajtenberg *et al.*, 2010).

An initial structural explanation, which is not yet proven, for the differences of HK autophosphorylation directionality (*cis* and *trans*) was provided by Casino and co-workers, and is based on the structural comparison of the EnvZ and HK853 DHp domains (Casino *et al.*, 2009, 2010). In EnvZ, helices α1 and α2 were right-handedly connected, whereas HK853 was left-handed (Fig. 2.4C and D). Thus, the *cis* phosphorylation of HK853 would correspond to the *trans* phosphorylation in EnvZ, and, therefore, the autophosphorylation of the two proteins, although topologically non-equivalent, would appear to be structurally identical. Since the main difference between both DHp domains merely lies in the DHp loop's length (Fig. 2.4B and C), it was proposed that the size of this connector could be the structural determinant of α1–α2 connection handedness and, consequently, of autophosphorylation directionality. However, the recently reported three-dimensional structure of the HK DesK shows the same α1–α2 connection as HK853, but a *trans* mechanism has been proposed for DesK HK (Albanesi *et al.*, 2009; Trajtenberg *et al.*, 2010). So it seems that the α1–α2 connection is not the only structural determinant of autophosphorylation directionality. The DesK and HK853 structures have also provided structural evidence for the high mobility of the CA domain, which is able to adopt a different relative disposition in relation to the DHp domain (Albanesi *et al.*, 2009; Casino *et al.*, 2009). In the different HK853 and DesK structures solved, the ATP binding sites always point to opposite directions, and consequently, towards the opposite phosphoacceptor His within the dimer. It seems that these inverse orientations are imposed by the difference in length of the DHp-CA linker, suggesting that this structural element, in combination with the α1–α2 connector, could be the determinant of autophosphorylation directionality. The structure and sequence alignment of those HKs whose autophosphorylation directionality has been proven (EnvZ, HK853, DesK, NtrB, PhoR, ArcB) could support this view (Fig. 2.4B). This alignment reveals that the HKs showing a *trans* mechanism exhibit shorter connections in either the α1–α2 loop (EnvZ) or the DHp-CA linker (NtrB and DesK). Therefore, a reduction in the length of these connections could induce structural changes that would invert the autophosphorylation directionality (*cis/trans)*. In order to corroborate this hypothesis, it would be worthwhile to determine the procedure of either a HK which presents short connections in both positions, thus following the *cis* autophosphorylation pathway, or the HK chimeras receiving connectors from a HK with inverse autophosphorylation directionality. Finally, the remaining question is if this variability is of any biological relevance for the signal transduction mechanism. Since the regions proposed are strongly implicated in HK–RR recognition (see below), it is tempting to connect autophosphorylation with the phosphotransfer and phosphatase reactions, thus raising the possibility of directionality being related with the predominant *basal* activity (in the absence of a signal) of the TCS.

HK–RR phosphotransfer

The phosphotransfer reaction involves the interaction of a phosphorylated HK (HK~P) with its cognate unphosphorylated RR. The transfer of the phosphoryl group from the P~His to the catalytic Asp residue in the RR can occur through two possible nucleophilic substitution mechanism: by an associative/S_N2 mechanism or by a dissociative/S_N1 mechanism (Mildvan, 1997). In the associative mechanism, the phosphate of the phosphoryl group, the Nε atom of the His and the Oδ atom of the Asp are aligned in a transition state to form a linear pentavalent complex in such a way that the breakage of the N–P bond follows the formation of the new O–P bond. In contrast, the dissociative mechanism implies a two-step reaction; in the first, the N–P bond is broken and leads to an unstable metaphosphate intermediate which rapidly forms the bond with the Asp oxygen (Mildvan, 1997). There is very little structural and biochemical knowledge about which of these two mechanisms is used by HKs and RRs, mainly because no structural information of a HK~P interacting with its cognate RR is available. In light of this, the actual structural knowledge on the phosphotransfer reaction is derived from the structures of the phosphotransferases (HPts) Spo0B and YPD1 with their cognate Spo0F and SLN1 RRs from *B. subtilis* and *Saccharomyces cerevisiae*, respectively (Fig. 2.5A) (Varughese *et al.*, 2006; Zhao *et al.*, 2008). Although Spo0B and YPD1 HPts are not real HKs, the active centre for the phosphoryltransfer, where a α helix containing the phosphoacceptor His is clung by the RR, is structurally homologous to that observed in the HK–RR complex represented by the HK853-RR468 structure (Casino *et al.*, 2009). Thus, the deductions about the molecular basis of the phosphotransfer reaction derived from these structures can be expanded to an orthodox TCS. Both HPt–RR complexes were solved in the presence of BeF_3^-, a widely used molecule in structural studies of kinases since it mimics the phosphoryl group in the transition state (Wemmer and Kern, 2005). The BeF_3^- molecule was found in both structures to be interposed between the His and Asp phosphoacceptor residues, and to interact in line with the His Nε and the Asp Oδ atoms which are at a distance of ~ 5 Å (Fig. 2.5A). The distance, the pentavalent coordination of the BeF_3^- and the geometry of the active centre support an associative mechanism for this reaction, which has been concluded by the authors who described both structures (Varughese *et al.*, 2006; Zhao *et al.*, 2008). Furthermore, a dissociative mechanism can be ruled out since there should

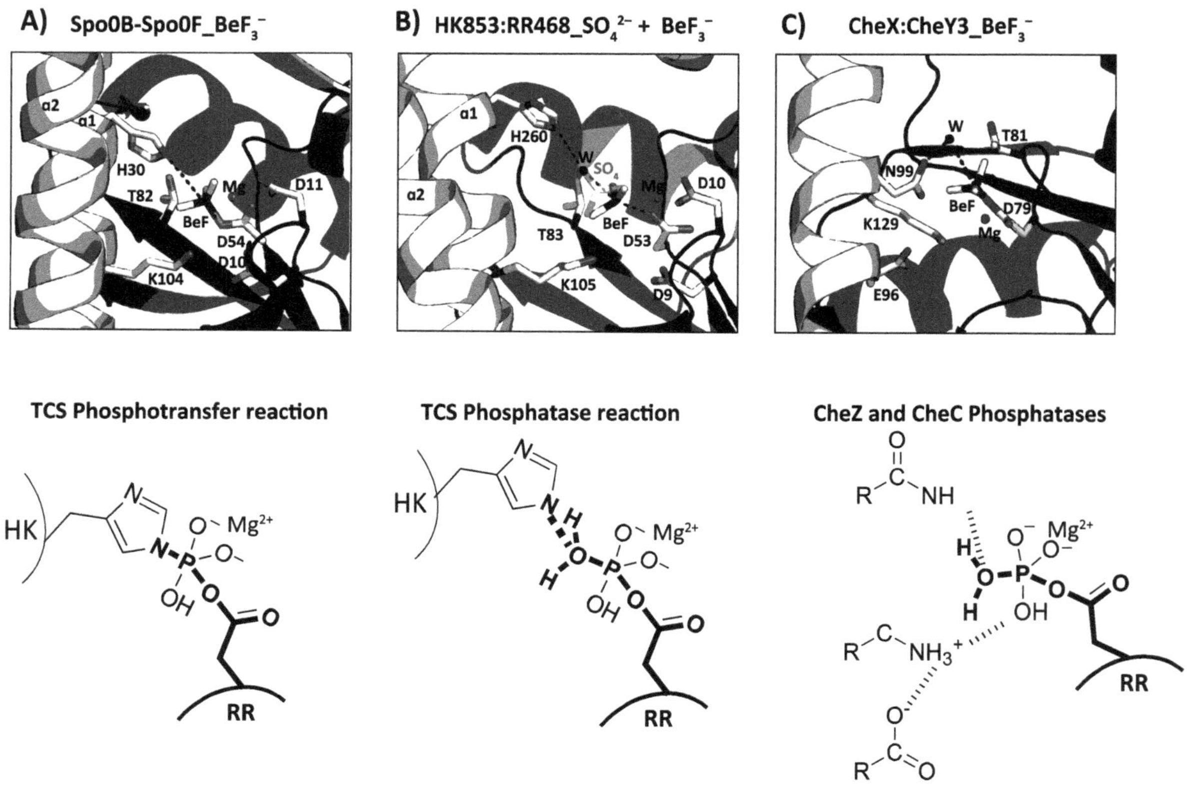

Figure 2.5 Active centres of the phosphotransfer and phosphatase reactions in TCS. (A) Active centre for the phosphotransfer reaction exemplified by the complex structure Spo0B-Spo0F formed by phosphotransferase Spo0B and RR Spo0F bound to BeF_3^-. (B) The active centre for the phosphatase reaction is shown by the complex structure HK853-RR468. The complex contains a sulfate ion at the active site that occupies almost the same position as BeF_3^- when the structure of RR468 bound to BeF_3^- is superposed with RR468 in the complex. One of the oxygens of the sulfate ion may also occupy the place of a water molecule (W), which is in charge of catalysing the phosphatase reaction as it occurs in the active site of other phosphatases like CheZ. The phosphorylatable H260 possibly polarizes the water molecule to favour the dephosphorylation of the RR. (C) The active site of phosphatase CheZ in the complex with the RR CheY3. The residue N99 in CheX polarizes the water molecule (W) to attack the BeF_3^- bound to CheY3, thus the organization of the active site is similar to the complex HK853-RR468 where the H260 could resemble the N99. The schematic reactions are drawn below.

be at least ~6.5 Å between the His Nε and the Asp Oδ, corresponding to the sum of van der Waals radii between the phosphodonor atom and the phosphorus, and between the latter and the phosphoacceptor atom, to accommodate a detached metaphosphate in the active site.

A detailed inspection of the active centres in both complexes revealed that the residues interacting with BeF_3^- are mainly provided by the RRs, except the phophodonor His and an Asn residues in the Spo0B–Spo0F complex (Fig. 2.5A). Therefore, it could be proposed that the phosphotransfer reaction is catalysed by the RR in such a way that the RR pulls the phosphoryl group off from the exposed P~His in the HK. In this transfer, the conserved Lys (K104 and K1195 in Spo0B and SLN1, respectively) emerging from the RR Lβα5 loop binds and neutralizes the negative charge of the phosphoryl group in movement (Fig. 2.5A). A second key residue in the reaction is the conserved Thr/Ser at the end of the RR β4. The relevance of both residues in this reaction is supported by the diminished or lost phosphotransferase activity of mutants at these positions (Schuster *et al.*, 2001). It is known that a divalent cation is required for the phosphotransfer reaction (usually Mg^{2+}, but also Ca^{2+} or Mn^{2+}, have been reported as suitable cations), and a Mg^{2+} cation bound to BeF_3^- is observed in both active centres which shows its typical hexa-coordination by a direct or a water-mediated interaction with the conserved negative residues (usually Asp) in the Lβα1 loop (Fig. 2.5A). This cation can promote phosphoryl transfer by neutralizing the negative charge of the transition state.

Turning off the signal

The phosphorylated form of the RR has half-lives ranging from a few seconds to several hours, or even days which, in some cases, exceed the duplication time of the harbouring microorganism. Therefore, a mechanism to dephosphorylate the stable RR~P is required to turn off the signal. In several cases, this action is accomplished by the HK which possesses phosphatase activity towards its cognate RR (Kenney, 2010). Additionally, up to four families (CheZ, CheC/CheX/FliY, Rap and Spo0E) of RR phosphatases have been described (Silversmith, 2010). Despite the structural differences between HKs and RR phosphatases, or among the four families of phosphatases, a common molecular mechanism to dephosphorylate the RR~P seems to be used for them all.

HK-mediated RR~P dephosphorylation seems to occur through a mechanism that does not correspond to the reverse phosphotransfer reaction. This conclusion was drawn several years ago when it was demonstrated that HK mutants defective in phosphotransferase activity maintain phosphatase activity (Hsing *et al.*, 1998). Indeed, exhaustive mutagenic studies on the phosphoacceptor His of the model HK EnvZ have shown that different His substitutions generate variants that present strong phosphatase activity with, obviously, impaired kinase and phosphotransferase activity (Hsing and Silhavy, 1997). For the phosphatase reaction, a mechanism has been proposed which involves the attack of a nucleophilic water molecule to P-Asp (Hsing and Silhavy, 1997). In this mechanism, the phosphoacceptor His would act as a general base by polarizing and holding in place the water molecule for the in-line attack of the phosphorous atom (Fig. 2.5B). The structure of the complex between HK853 and RR468 has been proposed as the first snapshot of this reaction (Fig. 2.5B) (Casino *et al.*, 2009). In the active centre of this complex, a sulfate ion was found between the phosphoacceptor Asp and His residues. The sulfur atom of the sulfate (interpreted as phosphorus in the phosphoryl group) is intercalated between and aligned with the His Nε and Asp Oδ atoms. One of the O atoms of sulfate (notice that sulfate has one extra oxygen in comparison with the phosphoryl group) sits on the path between sulfate S and the His Nε atoms emulating the attacking water molecule (Fig. 2.5B). The three remaining sulfate O atoms interact with the conserved Lys and the Thr/Ser catalytic residues of the RR and the Mg^{2+} cation in a similar way to what is observed in the active centre of the previously described phosphotransfer reaction (Fig. 5A,B), suggesting that these residues also assist the phosphatase reaction. Although the active centre in both reactions resembles and implicates almost the same catalytic residues, the presence of the catalytic molecule of water induces a separation of the phosphoacceptor atoms (from ~5 Å to 7.9 Å in the pass from phosphotransfer to the phosphatase reaction); consequently, a slight yet significant relocation of both proteins in each particular complex generates two non-equivalent catalytic competent HK–RR assemblies. The relative disposition of both proteins in each phosphotransfer or phosphatase complex can also be guided by the participation of an additional or a particular HK–RR interaction, as exemplified by the strict requirement of the PAS domain in several HKs to catalyse the phosphatase reaction; however, its presence does not influence the phosphotransfer reaction (Yamada *et al.*, 2009; Gutu *et al.*, 2010; Lopez-Redondo *et al.*, 2010). This structural difference would take into account the futile phosphotransfer/phosphatase cycles which could take place if both active centres and catalytic reactions were identical.

The four RR phosphatase families described seem to use a similar catalytic mechanism to desphosphorylate the RR (Casino *et al.*, 2010). Structural information of members of the three different families of phosphatases (CheZ from *E. coli*, CheX from *Borrelia burgdorferi* and RapH from *B. subtilis*) in the complex with their target RRs (CheY, CheY3 and Spo0F, respectively) is available (Zhao *et al.*, 2002; Pazy *et al.*, 2010; Parashar *et al.*, 2011). Additionally, the structure of two members (BA5174 and BA1655 from *Bacillus anthracis* (Grenha *et al.*, 2006)) of the fourth (Spo0E) phosphatase family is also known, but in their isolated form. From the structures of the complexes, it can be seen that all the phosphatases present a catalytic amide residue (Gln/Asn) which protrudes from an α helix to be accommodated in the active centre of the P~RR (Fig. 2.5C). The N atom of this amide side chain (equivalent to the His Nε atom in HKs) acts as a general base by orienting and activating the nucleophilic molecule which exerts the in-line hydrolysis of the phosphoryl group (Fig. 2.5C). In addition, a conserved acidic residue (Asp/Glu) located at the one-turn helix distance from the catalytic phosphatase amide is salt bridge bounded to the RR conserved catalytic Lys in the CheZ and CheC/CheX/FliY families (Fig. 2.5C). Thus, the phosphatase stimulates intrinsic RR phosphatase activity by positioning the nucleophile and the RR catalytic residues. The amido acid pair is characteristic of the CheZ, CheC/CheX/FliY and Spo0E families, exhibiting the DxxQ, ExxN and QelD motifs, but is missing in the Rap family since the RapH-Spo0F structure shows that RapH inserts the amide group (Gln) into the active centre of Spo0F, and that the acidic residue is absent.

Biochemical studies on the phosphatase reaction of HK NarX have revealed a sequence motif DxxxQ, conserved in the minor DHp family of HKs (HisKA_3 in Pfam) adjacent to phosphorylatable His, which is reminiscent of the CheZ phosphatase, which supports the catalytic role of Gln in this motif for the phosphatase reaction, and reveals a complex role of Asp in this motif for both autophosphorylation and phosphatase activity (Huynh *et al.*, 2010). However, the HisKA family, which is the most common type of DHp domains, contains another motif, E/DxxT/N, adjacent to the phosphorylatable His, which could be involved in phosphatase activity (Huynh *et al.*, 2010). Mutations in Glu of this motif in NtrB exhibit the kinase negative-phosphatase positive phenotype, whereas a mutation in Thr of this motif in EnvZ results in loss of phosphatase activity with a minor effect on autokinase and phosphotransfer activity. More biochemical studies in other HKs are required to clarify the role of this motif in the HK phosphatase reaction and its relationship with the corresponding phosphatase motifs. Nevertheless, phosphatase activity does not reside completely in the DHp domain as the isolated domain is much less effective in dephosphorylating the partner RR than the entire cytoplasmic DHp-CA domain (Zhu *et al.*, 2000). It is also known that the phosphatase activity of HKs can be modulated by the presence of the nucleotide (ADP) bound to the CA domain, the presence of cytoplasmic sensor domains such as PAS, as well as by residues in the ATP-lid, which are involved in the catalysis (Zhu and Inouye, 2002; Yamada *et al.*, 2009; Gutu *et al.*, 2010; Heermann and Jung, 2010; Lopez-Redondo *et al.*, 2010). The extreme complexity and differences observed in the phosphatase reaction, together with the existence of ancillary phosphatase proteins, suggest that not a unique, but several, and probably related, molecular phosphatase mechanisms should be present.

Structural insights into HK–RR recognition and specificity

Partner recognition

A single organism can have from tens to hundreds of HKs and RRs whose catalytic domains are highly conserved (Laub and Goulian, 2007). Therefore, how a HK recognizes its cognate RR to interact and perform a specific reaction is a key question in TCS signalling. Initial biochemical information about the recognition mechanism was acquired by a mutagenesis approach, performed mainly on EnvZ-OmpR TCS, which revealed 'hot spots' for complex formation that should be involved in discriminate catalytic-competent complexes for phosphotransfer versus phosphatase reactions (Hsing *et al.*, 1998). In these experiments, particular mutations impaired selectively phosphotransfer or phosphatase reactions suggesting the presence of different catalytic–competent complexes. The structure of the complex between HPt Spo0B and RR Spo0F (Zapf *et al.*, 2000) was the first structural data made available to rationalize all these biochemical observations. Although Spo0B is not a HK, it presents a HK reminiscent fold and enables the mapping of the main interactions between these proteins. RR Spo0F uses its α1 helix and Lβα loops surrounding the phosphoacceptor Asp to interact with the Spo0B four-helix bundle (equivalent to the DHp four-helix bundle of the HKs) downstream of the phosphoacceptor His (Zapf *et al.*, 2000; Hoch and Varughese, 2001). This structure led the authors to propose a structural recognition and discrimination model for the RR, where the interaction surface presents a set of partially conserved 'anchor' residues surrounded by highly variable 'recognition' residues (Hoch and Varughese, 2001). The anchor residues, conserved within families of RRs, ensure that the RR and the HK fit together in the right orientation and at a suitable distance for a productive phosphotransfer reaction. Recognition residues, highly variable between RRs, ensure fidelity by preventing productive interactions of the anchor residues between heterologous systems (Hoch and Varughese, 2001).

These initial observations were confirmed and extended to real HK–RR pairs thanks to recent biochemical and structural studies carried out by the groups of Laub and Marina, respectively. Skerker and collaborators in Laub's group, using elegant biochemical studies (Skerker *et al.*, 2008) carried out with the prototypical HK EnvZ, have revealed the counterpart of the anchor and recognition residues in the HKs, which are involved in partner binding and discrimination. In these studies, EnvZ chimeras where the segments of the DHp domain (~ 30 residues following the phosphoacceptor His and including the α1 and α2 DHp connection) are substituted for the corresponding segment of other HKs, were produced. These chimeras showed switched or 'rewired' substrate specificity by phosphorylating the cognate RR of the HK which has donated the fragment instead of OmpR, the cognate RR of EnvZ. Therefore, the specificity for the phosphotransfer reaction for HKs clearly resides in this DHp domain area (Fig. 2.6A).

Analogous rewiring experiments have also been performed subsequently in the RR with a view to changing the phosphotransfer partnering specificity with the HK. By swapping helix 1 and Lβα5 from one RR to match the corresponding residues from another RR, the RR is no longer able to be phosphorylated by its cognate HK and is phosphorylated instead by the HK partner of the second RR (Skerker *et al.*, 2008; Capra *et al.*, 2010) (Fig. 2.6A). An additional trajectory-scanning analysis performed by Laub's group on EnvZ and OmpR has demonstrated that although the specificity of a complex interaction is distributed along the interface of the complex EnvZ-OmpR, there is some hierarchy of residues which favour HK–RR interactions (Capra *et al.*, 2010). All the biochemical data were rationalized by the structure of the complex between the cytoplasmic portion of the HK HK853 and its partner RR RR468 (Casino *et al.*, 2009), which enables for the first time the visualization of the actual residues which make specific contacts between a HK and its cognate RR. This structure confirmed that the apex of the DHp domain in the HK, and helix α1 and the Lβα5 loop in the RR, were the main areas of HK–RR recognition, but it also expanded the interacting positions to previously not described residues in this and other areas (Fig. 2.6A) (Casino *et al.*, 2009). The HK–RR spatial disposition in the complex is governed by

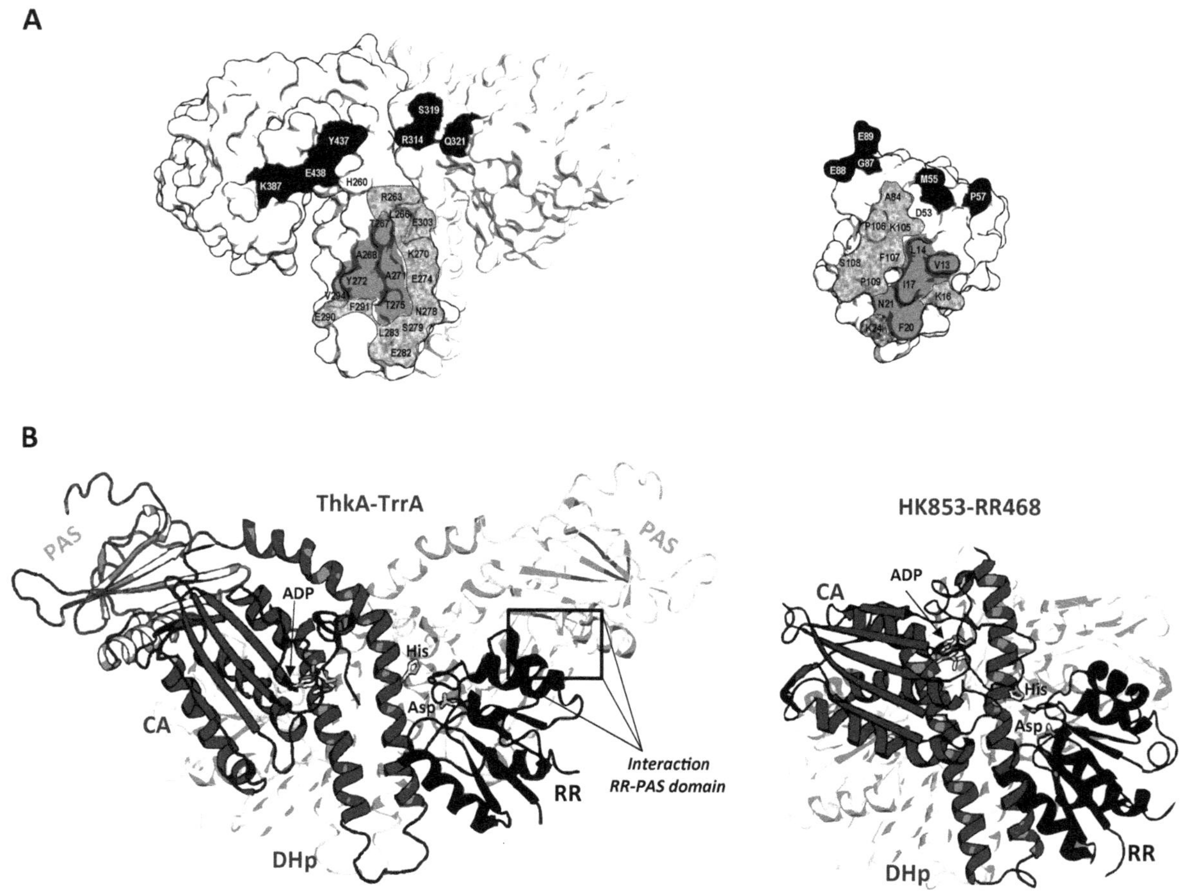

Figure 2.6 HK–RR recognition. (A) Labelled surface representation of the HK853–RR468 complex highlighting the residues involved in the specificity of recognition between one HK and its cognate RR, which are coloured in grey (these were also revealed by Laub's group in the rewiring experiments (Skerker *et al.*, 2008) are highlighted in plain grey) and the phosphorylation state recognition residues are coloured in black. Both the HK and RR are coloured in white, but in order to distinguish the individual subunits in the dimeric HK, one subunit is transparent. (B) Cartoon representation of the complex structures formed by TCS ThkA–TrrA and HK853-RR468. The ThkA–TrrA (PDB: 3A0Y) structure is solved by fitting the high-resolution structures of the isolated PAS, DHp and CA domains and the RR (TrrA bound to BeF_3^-) onto the low resolution (3.8 Å) electron density map of the ThkA–TrrA complex. The HK853-RR468 (PDB: 3DGE) structure at 2.8 Å is solved as a whole complex formed by the complete cytoplasmic domain of HK853 (DHp and CA) and the complete RR468. The dimeric representation of the HKs is represented by showing one subunit in grey and the other in white with transparency. The two RRs bound to each HK subunit are represented as one RR molecule which interacts with the grey HK subunit that is coloured in black, while the other RR molecule which interacts with the white transparent HK subunit is also white and transparent. The catalytic residues His and Asp, as well as the nucleotide ADP, are shown in sticks.

a six-helix bundle formed at the apex of the DHp domain downstream of the phosphoacceptor HK by the interaction of the four DHp helices of the dimeric HK flanked on both sides by the α1 helices of two RRs (Casino *et al.*, 2010) (Fig. 2.6B). Additionally, the structure also highlighted other minor and peripheral interacting regions (Fig. 2.6A), whose conformations are highly influenced by the phosphorylation state of each protein. It was proposed that these interactions play a key role in the discrimination of phosphotransfer versus phosphatase reactions (see below).

The structural information provided by the HK853–RR468 complex is in close agreement with a large number of previously obtained *in vivo* and *in vitro* biochemical, biophysical and genetic results which indicated the residues located in these regions as the main actors in HK–RR recognition and interaction. Furthermore, these contacts have been structurally confirmed by the low resolution structure (3.8 Å) of the ThkA–TrrA complex from *T. maritima* (Yamada *et al.*, 2009), thus extending the HK–RR contacts to the HK PAS domain (absent in HK853) (Fig. 2.6B). In addition, it seems that the molecular HK–RR interaction mechanism is exploited by ancillary proteins in order to regulate the signal transduction process as exemplified by the recognition mechanism between the RR and its phosphatase, which emulates either the HK DHp α1 helix to interact with their target RRs (see section 'Turning off the signal') or the SDA protein which binds to the corresponding DHp domain area of the KinB HK to impair the KinB–SpoF phosphotransfer (Bick *et al.*, 2009).

Phosphorylation state recognition

TCS signalling depends on a fine-tuning of the RR~P level, which is based on the balance of phosphorylation/dephosphorylation reactions in such a way that the HK and RR have to be able to discriminate between the functional state of their cognate partner to interact productively to perform the appropriate reaction, known as recognition of the partner phosphorylation state. That is, the HK~P has to be able to recognize the unphosphorylated RR to perform phosphotransfer activity and the unphosphorylated HK must be able to recognize the RR~P to perform the phosphatase activity. In addition, this mechanism should be responsible for avoiding futile RR phosphorylation/dephosphorylation cycles. Thus, upon RR phosphorylation, RR~P must dissociate before it is dephosphorylated, and, therefore, the conformation of either the HK or the RR~P (or both) at the end of RR phosphorylation must differ from that at the beginning of the reaction to ensure RR~P dissociation. Accordingly, HK~P should exhibit a low affinity for P~RR or the non-phosphorylated forms of HK and RR between them. As previously mentioned, the current scenario indicates that both the phosphotransfer and phosphatase reactions are not the reverse of each other (Gao and Stock, 2009a), which corroborates that two types of complexes should be formed. Therefore, the question is which elements are involved in the discrimination between the functional states of the HK and the RR?

The conformational changes induced by RR phosphorylation are well known and are mainly restricted to the loops connecting Lβα3 and Lβα4 (Fig. 2.3); thus, these two loops would be appropriate for monitoring the phosphorylation state of the RR. The HK853-RR468 structure shows that these RR loops interact with HK853 ATP-lid and the DHp-CA interdomain linker, respectively (Fig. 2.6A). At the same time, as it is deduced by the different HK853 and DesK structures, the ATP-lid and the DHp-CA also undergo relevant conformational changes depending on the phosphorylation state of the HK and on the presence/absence of the nucleotide. The relevance of these regions to discriminate each protein's phosphorylation situation is supported by additional data showing that several HKs and RRs of different TCSs accumulate mutations in these areas, which selectively impair the phosphotransfer or the phosphatase reaction with minor effects on the autokinase reaction (Zhu and Inouye, 2002; Depardieu *et al*, 2003; Pioszak and Ninfa, 2003). The mutational and structural information suggests that these four elements, the Lβα3 and Lβα4 loops in the RR and ATP-lid and interdomain linker in the HK, are the 'phosphorylation state recognition' determinants which guide and determine whether the complex is formed between HK~P and the RR or between the HK and RR~P, thus avoiding futile cycles. Thus, although HK–RR partner selection and recognition are imposed by the main interactions between the α1 helices of the HK DHp and RR REC domains, these peripheral interactions will allow a fine fitting between the two proteins, so they generate productive complexes which can carry out the reactions only in those cases in which both proteins have complementary phosphorylation states. At the same time, the conformational changes induced by the new state of phosphorylation of each protein after the reaction can render the complex unassembled.

Finally, in some HKs presenting a sensor PAS domain in their cytoplasmic portion, such as ThkA, WalK or NblS, it has been demonstrated that this domain is required for dephosphorylation, but not for the phosphotransfer reaction (Yamada *et al.*, 2009; Gutu *et al.*, 2010; Lopez-Redondo *et al.*, 2010). Since the ThkA–TrrA structure reveals that the PAS domain is involved in contacts with the RR (Fig. 2.6B) (Yamada *et al.*, 2009), these interactions should be relevant in the recognition of TrrA~P by ThkA and, consequently, the PAS domain could be an additional element for phosphorylation state recognition in those HKs where it is present. In the particular case of NblS, this being the first example of an orthodox HK with two partner RRs (Lopez-Redondo *et al.*, 2010), it has been demonstrated that NblS shows preference for one of the partners (SrrA) and not for the other (RpaB) in the phosphotransfer reaction, but not in the PAS-dependent phosphatase reaction where both RRs are equally recognized. A sequence analysis guided by the structures of the HK853–RR468 and ThkA–TrrA complexes shows that even the REC domains of both RRs (SrrA and RpaB) present an extremely high degree of identity (73%); four of the seven residues proposed to interact with NblS differ between both RRs (Lopez-Redondo *et al.*, 2010). These substitutions may explain the different affinity of NblS for each RR in phosphotransfer analyses. However, the residues in helix α4, the structural element that interacts with the HK PAS domain, as deduced by the ThkA–TrrA structure, are quasi-identical (only a Glu/Asp conservative change) in both RpaB and SrrA RRs (Lopez-Redondo *et al.*, 2010). This fact agrees with the similar susceptibility of the two RRs to NblS phosphatase activity and indicates the interaction between the HK PAS and RR REC domains, which are strictly required for the phosphatase reaction as the main drivers to form a dephosphorylation competent HK–RR complex.

Transmembrane signalling

The molecular mechanism involved in transducing the signal across the membrane in HKs, as well as the majority of signal perception, is one the most ulterior features awaiting to be elucidated in TCSs. Although the structural characterization of a full-length membrane placed HK is still elusive, the recently described structural contributions are decisive to provide insights into the cross-membrane signalling process. Following the 'divide and conquer' strategy, the structure of the principal HK portions (extracellular sensor, transmembrane transmitter and cytoplasmic catalytic) in the different signalling states have been determined, which allows us to put together the 'puzzle' of a complete sensor HK structure in the different transducing states (Fig. 2.7). Conformational changes induced by the signal seem to be coincident for all the portions, suggesting a limited set of 'ways' to transduce the signal. In the dimeric HK, the piston-like translation or rotation movement of helices connecting the sensor and the cytoplasmic portions via a transmembrane helix are the predominant movements associated with signal perception (Cheung and Hendrickson, 2010). Nonetheless, the possible coupling of both types of displacement has also been proposed as an integrative mechanism (Zhang and Hendrickson, 2010). These conformational changes are consistent with those observed in the transmembrane helices of transducer protein NpHtrII from *Natronomonas pharaonis* (Moukhametzianov *et al.*, 2006). In this receptor, a transmembrane helix, corresponding to the helix connecting the extra- and intracellular domains in orthodox HKs, undergoes a clockwise rotation and lateral displacement through the signal's action (Moukhametzianov *et al.*, 2006). In several HKs, additional domains, including the PAS and HAMP modules, are interposed between the end of this transmembrane helix and the DHp catalytic domain. Hulko and coworkers (2006) demonstrated that the HAMP domain forms a parallel four-helix bundle in the HK homodimer (Fig. 2.2) and proposed that the signal is relayed by altering its four-helix packing via a relative rotation of the adjacent helices, in a movement of a similar proportion to the output rotation induced by the signal on the second transmembrane helix of NpHtrII. Therefore, the signal would reach the HK catalytic portion as helical rotation and titling. Such conformational changes are observed in the structures of the DesK and HK853 DHp domains (Albanesi *et al.*, 2009; Casino *et al.*, 2009) in the different catalytic states (Fig. 2.7). These structures reveal that the DHp domains present a high plasticity switching between conformations via the rotation and bending of the four-helix bundle. The ultimate function of these different packing modes is to partly modify the contact interface

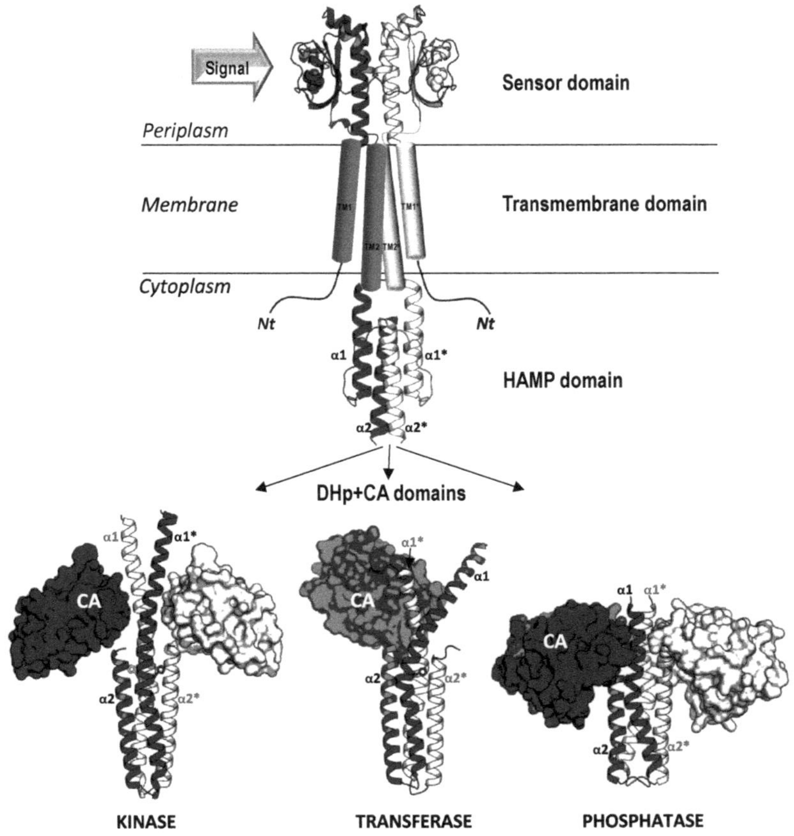

Figure 2.7 A signal transduction model in TCS. The signal reaches the sensor domain of the HK and transduces the signal towards the transmembrane helices (TM) to the cytoplasmic domains. Several cytoplasmic sensor domains, such as HAMP (PDB: 2ASW), might be found before the signal reaches the DHp domain. DHp helices' movement in response to the specific signal causes different orientations of the CA domain to drive the conformations that are competent with the autophosphorylation, phosphotransfer or phosphatase reactions. The kinase and transferase conformational states are exemplified by DesK (PDB: 3GIE and PDB: 3GIF, respectively), and the phosphatase state is shown by HK853 (PDB: 3DGE).

between the DHp and CA domains. The surface exposed by the DHp domain would permit the CA domain to come closer or to move further away from the DHp domain in rigid body movements by using the interdomain linker and other residues implied in DHp-CA contacts as a hinge. Depending on the relative movement of the DHp helices, the resulting CA disposition might either allow the alignment of the ATP molecule with the His for the kinase reaction or present a docking site for the RR in the phosphotransfer and phosphatase reactions (Fig. 2.7).

Open questions

The recent structural and biochemical advances in the characterization of TCSs have undoubtedly broadened our knowledge of these systems, but have also generated a trail of questions that need to be answered either biochemically or structurally. First of all, it is indispensable to shed light on the autophosphorylation reaction mechanism; why *cis* or *trans*? Is it relevant for signal transduction? Therefore, it would be desirable to visualize three dimensional structures of different HKs performing the autophosphorylation reaction, either in *cis* or in *trans*, and to employ chimeras to rewire HK autophosphorylation directionality to understand the molecular basis of the process, its physiological relevance and its possible relationship with the phosphotransferase or phosphatase reactions.

Other questions relate to the mechanistic differences between the phosphotransfer and the phosphatase reactions; why are they not reversible? Structures of complex HK–RR in the phosphotransfer reaction showing the intermediate state with either a metaphosphate or a pentavalent complex, and the phosphatase reaction showing the water molecule attacking the phosphoryl group bound to the RR, are necessary to obtain more in-depth information about the conformational differences and the catalytic particularities of each reaction. A subject that also needs to be considered is the action of ancillary proteins regulating the activity of HKs and RRs; thus, the structure of HK or RR in complexes with these proteins are also required to know the molecular basis of their action. Additionally, the complete cytoplasmic portion structure of hybrid HKs alone or in complex with the RR to visualize the His-Asp phosphorelay in more complex systems is required to understand this particular group of TCSs. Finally, to gain in-depth structural knowledge of the signalling mechanism of TCSs, the structure of a full-length membrane embedding HK alone and bound to its cognate RR in the presence and absence of the signal is long overdue.

Acknowledgements

This work has been supported by Spanish Ministry of Science and Innovation Grants BIO2009-10872 and BIO2010-15424. Patricia Casino is supported by CIBERER.

References

Albanesi, D., Martin, M., Trajtenberg, F., Mansilla, M.C., Haouz, A., Alzari, P.M., de Mendoza, D., and Buschiazzo, A. (2009). Structural plasticity and catalysis regulation of a thermosensor histidine kinase. Proc. Natl. Acad. Sci. U.S.A. *106,* 16185–16190.

Anantharaman, V., and Aravind, L. (2001). The CHASE domain: a predicted ligand-binding module in plant cytokinin receptors and other eukaryotic and bacterial receptors. Trends Biochem. Sci. *26,* 579–582.

Attwood, P.V., Piggott, M.J., Zu, X.L., and Besant, P.G. (2007). Focus on phosphohistidine. Amino Acids *32,* 145–156.

Bick, M.J., Lamour, V., Rajashankar, K.R., Gordiyenko, Y., Robinson, C.V., and Darst, S.A. (2009). How to switch off a histidine kinase: crystal structure of *Geobacillus stearothermophilus* KinB with the inhibitor Sda. J. Mol. Biol. *386,* 163–177.

Bilwes, A.M., Quezada, C.M., Croal, L.R., Crane, B.R., and Simon, M.I. (2001). Nucleotide binding by the histidine kinase CheA. Nat. Struct. Biol. *8,* 353–360.

Cai, S.J., and Inouye, M. (2003). Spontaneous subunit exchange and biochemical evidence for trans-autophosphorylation in a dimer of *Escherichia coli* histidine kinase (EnvZ). J. Mol. Biol. *329,* 495–503.

Capra, E.J., Perchuk, B.S., Lubin, E.A., Ashenberg, O., Skerker, J.M., and Laub, M.T. (2010). Systematic dissection and trajectory-scanning mutagenesis of the molecular interface that ensures specificity of two-component signaling pathways. PLoS Genet. *6,* e1001220.

Casino, P., Rubio, V., and Marina, A. (2009). Structural insight into partner specificity and phosphoryl transfer in two-component signal transduction. Cell *139,* 325–336.

Casino, P., Rubio, V., and Marina, A. (2010). The mechanism of signal transduction by two-component systems. Curr. Opin. Struct. Biol. *20,* 763–771.

Cheung, J., and Hendrickson, W.A. (2010). Sensor domains of two-component regulatory systems. Curr. Opin. Microbiol. *13,* 116–123.

Depardieu, F., Courvalin, P., and Msadek, T. (2003) A six amino acid deletion, partially overlapping the VanSB G2 ATP-binding motif, leads to constitutive glycopeptide resistance in VanB-type Enterococcus faecium. Mol Microbiol. *50,* 1069–1083.

Dutta, R., and Inouye, M. (2000). GHKL, an emergent ATPase/kinase superfamily. Trends Biochem. Sci. *25,* 24–28.

Galperin, M.Y. (2004). Bacterial signal transduction network in a genomic perspective. Environmen. Microbiol. *6,* 552–567.

Galperin, M.Y. (2006). Structural classification of bacterial response regulators: diversity of output domains and domain combinations. J. Bacteriol. *188,* 4169–4182.

Gao, R., and Stock, A.M. (2009a). Biological insights from structures of two-component proteins. Annu. Rev. Microbiol. *63,* 133–154.

Gao, R., and Stock, A.M. (2009b). Catalytically incompetent by design. Structure *17,* 1038–1040.

Gao, R., and Stock, A.M. (2010). Molecular strategies for phosphorylation-mediated regulation of response regulator activity. Curr. Opin. Microbiol. *13,* 160–167.

Gao, R., Mack, T.R., and Stock, A.M. (2007). Bacterial response regulators: versatile regulatory strategies from common domains. Trends Biochem. Sci. *32,* 225–234.

Grenha, R., Rzechorzek, N.J., Brannigan, J.A., de Jong, R.N., Ab, E., Diercks, T., Truffault, V., Ladds, J.C., Fogg, M.J., Bongiorni, C., *et al.* (2006). Structural characterization of Spo0E-like protein-aspartic acid phosphatases that regulate sporulation in bacilli. J. Biol. Chem. *281,* 37993–38003.

Gutu, A.D., Wayne, K.J., Sham, L.T., and Winkler, M.E. (2010) Kinetic characterization of the WalRK_{Spn} (VicRK) two-component system of *Streptococcus pneumoniae*: dependence of WalK_{Spn} (VicK) phosphatase activity on its PAS domain. J. Bacteriol. *192,* 2346–2358.

Heermann, R., and Jung, K. (2010). The complexity of the 'simple' two-component system KdpD/KdpE in *Escherichia coli.* FEMS Microbiol. Lett. *304,* 97–106.

Hoch, J.A., and Varughese, K.I. (2001). Keeping signals straight in phosphorelay signal transduction. J. Bacteriol. *183,* 4941–4949.

Hsing, W., and Silhavy, T.J. (1997). Function of conserved histidine-243 in phosphatase activity of EnvZ, the sensor for porin osmoregulation in *Escherichia coli*. J. Bacteriol. *179*, 3729–3735.

Hsing, W., Russo, F.D., Bernd, K.K., and Silhavy, T.J. (1998). Mutations that alter the kinase and phosphatase activities of the two-component sensor EnvZ. J. Bacteriol. *180*, 4538–4546.

Hulko, M., Berndt, F., Gruber, M., Linder, J.U., Truffault, V., Schultz, A., Martin, J., Schultz, J.E., Lupas, A.N., and Coles, M. (2006). The HAMP domain structure implies helix rotation in transmembrane signaling. Cell *126*, 929–940.

Huynh, T.N., Noriega, C.E., and Stewart, V. (2010). Conserved mechanism for sensor phosphatase control of two-component signaling revealed in the nitrate sensor NarX. Proc. Natl. Acad. Sci. U.S.A. *107*, 21140–21145.

Kato, M., Mizuno, T., and Hakoshima, T. (1998). Crystallization of a complex between a novel C-terminal transmitter, HPt domain, of the anaerobic sensor kinase ArcB and the chemotaxis response regulator CheY. Acta Crystallogr. D Biol. Christallogr. *54*, 140–142.

Kenney, L.J. (2010) How important is the phosphatase activity of sensor kinases? Curr. Opin. Microbiol. *13*, 168–176.

King-Scott, J., Nowak, E., Mylonas, E., Panjikar, S., Roessle, M., Svergun, D.I., and Tucker, P.A. (2007). The structure of a full-length response regulator from *Mycobacterium tuberculosis* in a stabilized three-dimensional domain-swapped, activated state. J. Biol. Chem. *282*, 37717–37729.

Krell, T., Lacal, J., Busch, A., Silva-Jimenez, H., Guazzaroni, M.E., and Ramos, J.L. (2010). Bacterial sensor kinases: diversity in the recognition of environmental signals. Ann. Rev. Microbiol. *64*, 539–559.

Laub, M.T., and Goulian, M. (2007). Specificity in two-component signal transduction pathways. Ann. Rev. Genet *41*, 121–145.

Lopez-Redondo, M.L., Moronta, F., Salinas, P., Espinosa, J., Cantos, R., Dixon, R., Marina, A., and Contreras, A. (2010). Environmental control of phosphorylation pathways in a branched two-component system. Mol. Microbiol. 78, 475–489.

Marina, A., Mott, C., Auyzenberg, A., Hendrickson, W.A., and Waldburger, C.D. (2001). Structural and mutational analysis of the PhoQ histidine kinase catalytic domain. Insight into the reaction mechanism. J. Biol. Chem. *276*, 41182–41190.

Marina, A., Waldburger, C.D., and Hendrickson, W.A. (2005). Structure of the entire cytoplasmic portion of a sensor histidine-kinase protein. Embo J. *24*, 4247–4259.

Mascher, T., Helmann, J.D., and Unden, G. (2006). Stimulus perception in bacterial signal-transducing histidine kinases. Microbiol. Mol. Biol. Rev. *70*, 910–938.

Mildvan, A.S. (1997). Mechanisms of signaling and related enzymes. Proteins 29, 401–416.

Miyatake, H., Mukai, M., Park, S.Y., Adachi, S., Tamura, K., Nakamura, H., Nakamura, K., Tsuchiya, T., Iizuka, T., and Shiro, Y. (2000). Sensory mechanism of oxygen sensor FixL from *Rhizobium meliloti*: crystallographic, mutagenesis and resonance Raman spectroscopic studies. J. Mol. Biol. *301*, 415–431.

Moore, J.O., and Hendrickson, W.A. (2009). Structural analysis of sensor domains from the TMAO-responsive histidine kinase receptor TorS. Structure *17*, 1195–1204.

Moukhametzianov, R., Klare, J.P., Efremov, R., Baeken, C., Goppner, A., Labahn, J., Engelhard, M., Buldt, G., and Gordeliy, V.I. (2006). Development of the signal in sensory rhodopsin and its transfer to the cognate transducer. Nature *440*, 115–119.

Mourey, L., Da Re, S., Pedelacq, J.D., Tolstykh, T., Faurie, C., Guillet, V., Stock, J.B., and Samama, J.P. (2001). Crystal structure of the CheA histidine phosphotransfer domain that mediates response regulator phosphorylation in bacterial chemotaxis. J. Biol. Chem. *276*, 31074–31082.

Ninfa, E.G. Atkinson, M.R., Kamberov, E.S., and Ninfa, A.J. (1993). Mechanism of autophosphorylation of *Escherichia coli* nitrogen regulator II (NRII or NtrB): trans-phosphorylation between subunits. J. Bacteriol. *175*, 7024–7032.

Parashar, V., Mirouze, N., Dubnau, D.A., and Neiditch, M.B. (2011). Structural basis of response regulator dephosphorylation by Rap phosphatases. PLoS Biol. *9*, e1000589.

Pazy, Y., Motaleb, M.A., Guarnieri, M.T., Charon, N.W., Zhao, R., and Silversmith, R.E. (2010). Identical phosphatase mechanisms achieved through distinct modes of binding phosphoprotein substrate. Proc. Natl. Acad. Sci. U.S.A. *107*, 1924–1929.

Pena-Sandoval, G.R., and Georgellis, D. (2010). The ArcB sensor kinase of *Escherichia coli* autophosphorylates by an intramolecular reaction. J. Bacteriol. *192*, 1735–1739.

Pioszak, A.A., and Ninfa, A.J. (2003). Genetic and biochemical analysis of phosphatase activity of *Escherichia coli* NRII (NtrB) and its regulation by the PII signal transduction protein. J Bacteriol. *185*, 1299–1315.

Schuster, M., Silversmith, R.E., and Bourret, R.B. (2001). Conformational coupling in the chemotaxis response regulator CheY. Proc. Natl. Acad. Sci. U.S.A. *98*, 6003–6008.

Silversmith, R.E. (2010). Auxiliary phosphatases in two-component signal transduction. Curr. Opin. Microbiol. *13*, 177–183.

Skerker, J.M., Perchuk, B.S., Siryaporn, A., Lubin, E.A., Ashenberg, O., Goulian, M., and Laub, M.T. (2008). Rewiring the specificity of two-component signal transduction systems. Cell *133*, 1043–1054.

Song, Y., Peisach, D., Pioszak, A.A., Xu, Z., and Ninfa, A.J. (2004). Crystal structure of the C-terminal domain of the two-component system transmitter protein nitrogen regulator II (NRII; NtrB), regulator of nitrogen assimilation in *Escherichia coli*. Biochemistry *43*, 6670–6678.

Stock, A.M., and Guhaniyogi, J. (2006). A new perspective on response regulator activation. J. Bacteriol. *188*, 7328–7330.

Stock, A.M., Robinson, V.L., and Goudreau, P.N. (2000). Two-component signal transduction. Annu. Rev. Biochem. *69*, 183–215.

Sugawara, H., Kawano, Y., Hatakeyama, T., Yamaya, T., Kamiya, N., and Sakakibara, H. (2005). Crystal structure of the histidine-containing phosphotransfer protein ZmHP2 from maize. Protein Sci. *14*, 202–208.

Tomomori, C., Tanaka, T., Dutta, R., Park, H., Saha, S.K., Zhu, Y., Ishima, R., Liu, D., Tong, K.I., Kurokawa, H., *et al.* (1999). Solution structure of the homodimeric core domain of *Escherichia coli* histidine kinase EnvZ. Nat. Struct. Biol. *6*, 729–734.

Trajtenberg, F., Grana, M., Ruetalo, N., Botti, H., and Buschiazzo, A. (2010). Structural and enzymatic insights into the ATP binding and autophosphorylation mechanism of a sensor histidine kinase. J. Biol. Chem. *285*, 24892–24903.

Varughese, K.I., Tsigelny, I., and Zhao, H. (2006). The crystal structure of beryllofluoride Spo0F in complex with the phosphotransferase Spo0B represents a phosphotransfer pretransition state. J. Bacteriol. *188*, 4970–4977.

Watanabe, T., Okada, A., Gotoh, Y., and Utsumi, R. (2008). Inhibitors targeting two-component signal transduction. Adv. Exp. Med. Biol. *631*, 229–236.

Wemmer, D.E., and Kern, D. (2005). Beryllofluoride binding mimics phosphorylation of aspartate in response regulators. J. Bacteriol. *187*, 8229–8230.

Xu, Q., and West, A.H. (1999). Conservation of structure and function among histidine-containing phosphotransfer (HPt) domains as revealed by the crystal structure of YPD1. J. Mol. Biol. *292*, 1039–1050.

Yamada, S., Sugimoto, H., Kobayashi, M., Ohno, A., Nakamura, H., and Shiro, Y. (2009). Structure of PAS-linked histidine kinase and the response regulator complex. Structure *17*, 1333–1344.

Zapf, J., Sen, U., Madhusudan, Hoch, J.A., and Varughese, K.I. (2000). A transient interaction between two phosphorelay proteins trapped in a crystal lattice reveals the mechanism of molecular recognition and phosphotransfer in signal transduction. Structure *8*, 851–862.

Zhang, Z., and Hendrickson, W.A.(2010). Structural characterization of the predominant family of histidine kinase sensor domains. J. Mol. Biol. *400*, 335–353

Zhao, R., Collins, E.J., Bourret, R.B., and Silversmith, R.E. (2002). Structure and catalytic mechanism of the *E. coli* chemotaxis phosphatase CheZ. Nat. Struct. Biol. *9*, 570–575.

Zhao, X., Copeland, D.M., Soares, A.S., and West, A.H. (2008). Crystal structure of a complex between the phosphorelay protein YPD1 and the response regulator domain of SLN1 bound to a phosphoryl analog. J. Mol. Biol. *375*, 1141–1151.

Zhu, Y., and Inouye, M. (2002). The role of the G2 box, a conserved motif in the histidine kinase superfamily, in modulating the function of EnvZ. Mol. Microbiol. *45*, 653–663.

Zhu, Y., Qin, L., Yoshida, T., and Inouye, M. (2000). Phosphatase activity of histidine kinase EnvZ without kinase catalytic domain. Proc. Natl. Acad. Sci. U.S.A. *97*, 7808–7813.

Redox Responding Sensor Kinases

3

Jiang Wu, Vladimira Dragnea and Carl Bauer

Abstract

Sensor kinases that respond to changes in redox have been described in several bacterial systems. This review is centred on the molecular mechanism of redox sensing in a select group of sensor kinases that have been well-defined biochemically. We cover the role of oxidized ubiquinone and redox active cysteine in controlling the kinase activity of the sensor kinases RegB and ArcB of *Rhodobacter capsulatus* and *Escherichia coli*, respectively. We also review inhibition of kinase activity that occurs upon the binding of dioxygen to haem to a PAS domain in FixL of rhizobia and upon the binding of dioxygen to haem to a GAF domain in DosT of *Mycobacterium tuberculosis*. Finally, we discuss recent evidence that the photoreceptor LovK of *Caulobacter crescentus* contains a redox active flavin that when reduced inhibits the ability of LovK to undergo a photocycle that regulates kinase activity.

Introduction

As oxygen tension decreases or increases in the environment, cells become faced with many physiological challenges. For example, many microorganisms synthesize multiple cytochrome oxidases that have different affinities for oxygen (Pitcher *et al.*, 2002; Pitcher and Watmough, 2004). Under high oxygen tension, cells synthesize a low affinity oxidase that has a high turnover rate. However under oxygen limiting conditions they often inhibit expression of the low affinity oxidase and instead express an alternative oxidase that has high affinity for oxygen at the expense of a lower turnover rate. When oxygen is very limiting, or when no oxygen is available, cells switch from respiratory growth to anaerobic growth involving substrate level phosphorylation.

Initiating redox mediated changes in growth modes are often accomplished through the use of redox and/or oxygen responding sensor kinases that phosphorylate DNA binding response regulators, which control different sets of genes. This review is centred on redox responding sensor kinases that have undergone extensive biochemical analyses. We discuss how sensor kinases RegB and ArcB both interact with ubiquinones in the ubiquinone pool. RegB and ArcB are regulators that control large global alterations in gene expression necessary for changing from respiratory to anaerobic growth modes as oxygen becomes limiting. In the case of RegB, the binding of oxidized ubiquinone that predominates under aerobic growth conditions, results in the inhibition of kinase activity. RegB also contains a universally conserved redox responding cysteine that when oxidized converts active RegB dimers into inactive RegB tetramers. We discuss how the regulation of RegB activity is independently controlled by the redox state of the ubiquinone pool and by the redox state of the redox active cysteine. In the case of ArcB we discuss evidence that oxidized ubiquinone acts as an oxidant that controls the formation of a disulfide bond that inhibits ArcB activity.

We also discuss how sensor kinases FixL and DosT utilize haem as a cofactor to bind oxygen in a way that affects kinase activity. FixL is a sensor kinase involved in regulating nitrogen fixation in response to oxygen tension. Since nitrogenase is an oxygen labile protein (Robson and Postgate,

1980), its synthesis is generally restricted to anaerobic conditions. The second haem-binding regulator is DosT that is present in Mycobacteria. DosT is a pathogenesis gene needed by these cells to respond to hypoxia conditions that occur when infecting macrophages. Interestingly, FixL uses a Per-ARNT-Sim (PAS) domain to bind haem while DosT utilizes a GAF domain to bind haem. In both cases oxygen forms an axial linkage to Fe in haem that alters the peptide backbone of these domains in a way that affects kinase activity.

Finally, we discuss recent evidence that the LovK sensor kinase from *Caulobacter crescentus* functions as a blue light absorbing photoreceptor that also senses changes in cellular redox. LovK has a PAS domain that non-covalently binds FAD as a cofactor for blue light absorption. When the flavin becomes reduced the LovK photocycle is inhibited thereby affecting the regulation of LovK activity.

The take home message is that there are many different mechanisms that sensor kinases use to regulate activity in response to changes in cellular redox conditions or in oxygen tension. While this review is centred on sensor kinases for which extensive biochemical analysis has been undertaken, it is likely that there remain many other mechanisms of redox sensing that are not yet well defined in sensor kinases.

RegB

RegB was discovered as a histidine kinase governing photosystem synthesis in response to environmental oxygen tension changes in the purple photosynthetic bacterium *Rhodobacter capsulatus* (Mosley *et al.*, 1994). In concert with its cognate response regulator RegA (Sganga and Bauer, 1992), RegB promotes a two-component signal transduction cascade to initiate photosystem synthesis under anaerobic conditions (Inoue *et al.*, 1995; Bird *et al.*, 1999). As more bacterial genomes are been sequenced, many additional RegB/RegA homologues have been found not only in photosynthetic but also in non-photosynthetic α- and γ-proteobacterial species (Elsen *et al.*, 2004; Wu and Bauer, 2007). The regulon controlled by the RegB/RegA system has also been expanded from photosynthesis genes to genes involved in electron transport (Swem *et al.*, 2001; Smart *et al.*, 2004), carbon fixation (Dubbs *et al.*, 2000; Vichivanives *et al.*, 2000), nitrogen fixation (Bauer *et al.*, 1998; Elsen *et al.*, 2000), denitrification (Laratta *et al.*, 2002), hydrogen oxidation (Elsen *et al.*, 2000), dehydrogenation (Barber and Donahue, 1998). It is now known that the RegB/RegA system serves as a highly conserved global regulatory system that provides an overlying layer of redox control on a variety of critical biological processes to delicately coordinate these processes in response to oxygen availability (Elsen *et al.*, 2004). RegB thus has a role in promoting efficient energy generation, the synthesis of substances, and in maintaining cellular redox poise.

Since its discovery, RegB has been extensively characterized in terms of kinase activity (Inoue *et al.*, 1995; Bird *et al.*, 1999; Swem *et al.*, 2003, 2006; Wu and Bauer 2010) and phosphotransfer to RegA (Du *et al.*, 1998; Bird *et al.*, 1999). The DNA-binding activity of RegA has also been well characterized (Du *et al.*, 1998; Bird *et al.*, 1999; Bowman *et al.*, 1999; Dubbs *et al.*, 2000; Swem *et al.*, 2001; Vichivanives *et al.*, 2000; Laguri *et al.*, 2003, 2006; Dangel and Tabita, 2009). However, elucidating the redox-sensing mechanism of RegB was challenging mainly because RegB is a membrane spanning sensor kinase that as a class has been notoriously hard to over-express and purify in an active form as a full-length protein that contains the membrane spanning domain. Despite challenges, efforts from several laboratories have greatly improved our understanding about signals that directly impact the kinase activity of RegB, the signal-input sites in RegB, and how RegB activity is regulated in response to several redox signals.

Because photosystem synthesis is inhibited under aerobic growth conditions, it is natural to presume that RegB kinase activity is directly inhibited by molecular oxygen. However, this hypothesis was excluded as *R. capsulatus* is capable of de-repressing photosystem synthesis under chemoautotrophic growth conditions in the presence of oxygen, hydrogen, and carbon dioxide (Madigan and Gest, 1979). Another mechanism proposed to regulate RegB was an unknown signal produced by cytochrome *cbb*$_3$ oxidase (Oh and Kaplan, 1999; Kim *et al.*, 2007). This conclusion was based on the observation that mutations of *R. sphaeroides* cytochrome cbb_3 oxidase lead to elevated aerobic expression of PrrB/PrrA (a RegB/RegA homologue from *R. sphaeroides*) regulated genes. It was therefore suggested that cytochrome cbb_3 oxidase generates an 'inhibitory' signal that represses the RegB/RegA two-component system under aerobic conditions. A later study suggested that this 'inhibitory' signal shifts PrrB activity to a phosphatase-dominant mode to silence photosystem synthesis and that the transmembrane domain of PrrB plays an important role in redox sensing (Oh and Kaplan, 2004).

Redox signalling via cysteine oxidation

In a search for a redox signal controlling RegB activity, Swem *et al.* (2003) demonstrated that a redox-active cysteine constitutes a switch that turns RegB kinase activity on and off. This suggested that one redox signal for RegB is the redox potential of the cytosol that is buffered by a number of oxidants and reductants that can react with molecular oxygen which can freely diffuse across the cell membrane. This redox-active cysteine (Cys265) is located in the cytosolic dimerization interface downstream of the H-box and is universally conserved among all known RegB homologues (Elsen *et al.*, 2004; Wu and Bauer 2007). *In vitro* analysis using truncated RegB lacking the transmembrane domain indicated that an intermolecular disulfide bond forms between RegB dimers via Cys265 under oxidizing conditions to form an inactive RegB tetramer (Swem *et al.*, 2003). Tetramer formation via disulfide bond formation in aerobically grown cells has been confirmed by *in vivo* studies. Furthermore, a Cys265 to alanine mutation resulted in increased RegB kinase activity under aerobic conditions both *in vitro* and *in vivo*. All these results lead to a proposed redox-sensing model of RegB where redox-active Cys265 functions as a signal input site responding to redox potential in the cell. Under oxidizing conditions this cysteine is thought to form an intermolecular disulfide bond that converts the active dimer into an inactive tetramer. The redox-sensing ability of Cys265 also requires a divalent metal ion, which may help to place the Cys265 to an appropriate orientation to form the disulfide bond (Swem *et al.*, 2003). Although several metal ions can fulfil this role *in vitro*, the metal used by RegB *in vivo* may be copper, because mutations in the putative copper chaperone SenC (Swem *et al.*, 2005), and putative copper transport proteins RdxI, RdxH and RdxS result in derepression of the photosystem synthesis under aerobic conditions in *R. sphaeroides* (Eraso and Kaplan, 2000; Roh and Kaplan, 2000).

Redox signalling via ubiquinone binding

Although mutation of Cys265 to alanine in full-length RegB results in elevated photosystem gene expression under aerobic growth conditions, there still remained observable redox control (Swem *et al.*, 2003). This indicated that the oxidation state of Cys265 only partially abrogates redox control *in vivo*. In search for an additional signal, Swem *et al.* (2006) demonstrated the presence of a ubiquinone-binding site in the transmembrane domain that also constituted a redox input signal. When exogenous oxidized ubiquinone-1 (an analogue of ubiquinone in *R. capsulatus*) was added to kinase reaction *in vitro*, full-length RegB kinase activity was inhibited. This was contrasted by the addition of reduced ubiquinone-1 that had no inhibitory effect. A photo affinity ubiquinone analogue [^{3}H] azidoquinone cross-linked to a GGXXNPF motif located at a short pariplasmic loop between membrane spanning helices 3 and 4 that is universally conserved among all RegB homologues (Swem *et al.*, 2006). It was also shown that cross-linking [^{3}H]-azidoquinone correlated with decreased RegB kinase activity. To confirm the role of this ubiquinone-binding site in redox sensing, a conserved residue F112 in the GGXXNPF motif was mutated to alanine and recombined into the chromosome of *R. capsulatus*. The mutant showed increased photosystem synthesis under aerobic, semi-aerobic, and anaerobic conditions.

The interaction between ubiquinone and RegB was further analysed in a study from Wu and Bauer (2010), who confirmed the interaction between RegB and ubiquinone by demonstrating that purified full-length RegB contains bound ubiquinone. The ubiquinone binding behaviour of RegB was further characterized by isothermal titration calorimetry where it was shown that both oxidized and reduced ubiquinone-0 have comparable binding constants in the ~100 μM range. Wu and Bauer (2010) also demonstrated that two universally conserved polar residues, Arg31 and Gln38, that are deeply buried in the first transmembrane helix, are involved in ubiquinone binding. Mutations of these two residues caused less bound ubiquinone content *in vitro* and increased aerobic photosystem synthesis *in vivo*.

According to these findings, a detailed model for redox sensing via ubiquinone binding has been proposed where RegB binds both reduced and oxidized ubiquinone to the identified GGXXNPF motif between helix 3 and 4 as well as to the two conserved polar residues Arg31 and Gln38 present in the first transmembrane helix (Fig. 3.1). This ubiquinone-binding site has relatively low binding affinity, which enables RegB to readily equilibrate ubiquinone bound to this region with free unbound ubiquinone present in the ubiquinone pool in the membrane (Wu and Bauer, 2010). Under anaerobic conditions the ubiquinone pool becomes more reduced (Grammel and Ghosh, 2008), so the fraction of RegB containing reduced ubiquinone would increase to allow kinase activity. This is contrasted by aerobic growth conditions where the ubiquinone pool is know to be predominantly oxidized (Grammel and Ghosh, 2008). Under this condition equilibration of RegB with the ubiquinone pool would lead to RegB containing predominantly oxidized ubiquinone that inhibits kinase activity. This equilibrium mechanism allows the kinase activity of RegB to be tuned to the redox state of the ubiquinone pool which is reflectant of the overall energy state of the cell. This non-catalytic role of ubiquinone in controlling RegB activity is distinct from that of other known ubiquinone-binding proteins, which use ubiquinone as an electron donor or acceptor.

Finally, Wu and Bauer (2010) also addressed the question whether ubiquinone binding is involved in the disulfide bond formation at Cys265. Their observation that a RegB Cys265Ser mutant still responds to the alteration of the redox state of the bound ubiquinone indicates that ubiquinone binding is an input signal that functions independently from the oxidation of Cys265. Thus, the kinase activity of RegB is controlled by two redox signals with the membrane bound ubiquinone-binding site sensing the redox state of ubiquinone pool in the membrane, while the redox state of Cys265 senses the redox potential of the cytosol.

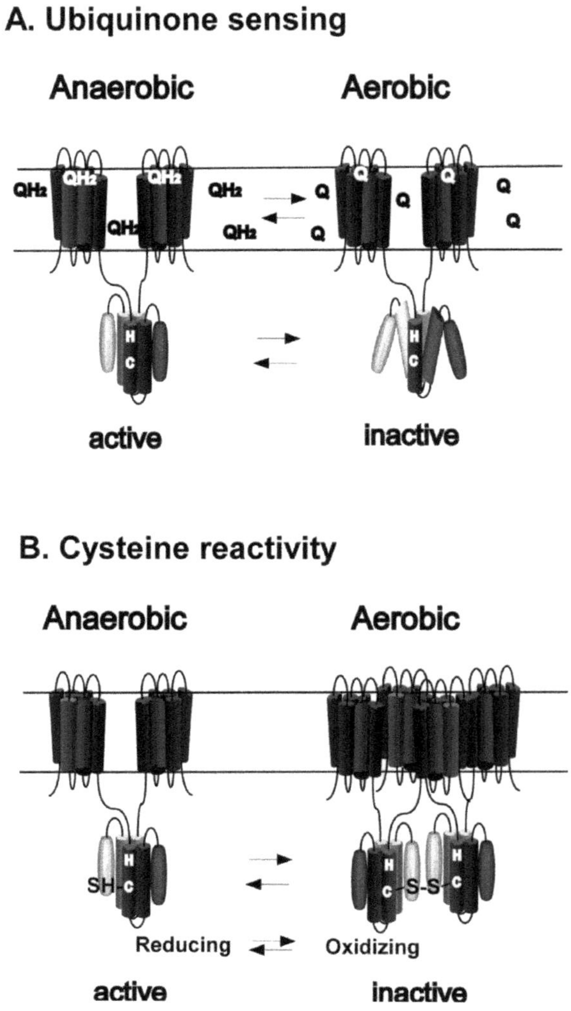

Figure 3.1 A model for RegB redox sensing. (A) Redox sensing via ubiquinone binding: Under anaerobic growth conditions, ubiquinones in the membrane are predominantly in a reduced form. Binding of reduced ubiquinone to the transmembrane domain of RegB does not affect RegB kinase activity. However, when growth conditions become aerobic, the ubiquinone pool shifts to an oxidized form. Binding of oxidized ubiquinone triggers an allosteric change that inhibits RegB kinase activity. (B) Redox sensing via disulfide bond formation: Under anaerobic growth conditions, cytosolic Cys265 is in a reduced form in the reducing cytosol allowing RegB to function as an active dimer. Under aerobic conditions, Cys265 forms an intermolecular disulfide bond converting active RegB dimer to an inactive tetramer.

Unanswered questions

Are there other oxidative modifications at C265 that regulate RegB activity? The oxidation state of cysteine has been known to function as a switch that can modulate activity of enzymes, transcription factors, and sensor proteins owing to facile redox chemistry (Antelmann and Helmann, 2011). Although highly active cysteines are commonly found to form disulfide bonds in response to a redox signal, they can also readily react with reactive oxygen species (ROS), reactive electrophilic species (RES), and reactive nitrogen species (RNS) to undergo S-thiolation, S-alkylation, or S-nitrosylation (Reddie and Carroll, 2008). For example, the redox-responsive transcriptional factor OxyR in *E.coli* has been found to differently modulate its DNA-binding activity in response to a variety of different modifications at a conserved cysteine (Kim *et al.*, 2002). Such different modifications may also occur at Cys265 in RegB, especially since Cys265 is located in a basic pocket that makes it highly reactive (Swem *et al.*, 2003). Furthermore, *in vivo* Western blot studies demonstrated that < 20% of RegB form disulfide bonds under aerobic growth conditions (Swem *et al.*, 2003), suggesting that other oxidative modifications may occur at Cys265. This hypothesis needs to be tested by the sensitive detection of these modifications *in vitro* and *in vivo*.

Is the phosphatase activity of RegB also redox controlled? Like many other histidine kinases, RegB is a bifunctional enzyme. It processes kinase activity as well as phosphatase activity on phosphorylated RegA (Bird *et al.*, 1999). While it has been demonstrated that the redox-active Cys265 and the ubiquinone-binding site both regulate RegB kinase activity, it remains unclear whether these input sites also regulate phosphatase activity. Oh *et al.* (2004) proposed that *cbb3* cytochrome *c* oxidase generates an unknown 'inhibitory' signal to stimulate RegB phosphatase activity without affecting the kinase activity under aerobic conditions, however the exact mechanism is not specified. Another possibility is that while RegB kinase activity is regulated by C265 and ubiquinone pool, the phosphatase activity may be regulated by another redox signal, thereby allowing more delicate redox regulation of the amount of RegA~P.

Finally, although it is clear that disulfide bond formation at C265 and ubiquinone binding coordinately regulate RegB activity, a clear picture of how they function at a molecular level are lacking as a molecular structure of RegB is still unavailable. Questions not yet addressed are:

1. How does the disulfide bond put RegB cytoplasmic domains (DHp dimerization and histidine phosphotransfer domain and the CA catalytic and ATP-binding domain) in an inactive conformation?
2. What residues in the ubiquinone-binding site are in contact with ubiquinone and do these contacts change when oxidized ubiquinone becomes reduced?
3. How are the DHp and CA domain structures affected by the changes in the redox state of ubiquinone bound to the transmembrane domain?

Based on the structures of DHp+CA domains from several histidine kinases, it has been proposed that relative reorientation of the helices in the DHp domain via cog wheeling and bending can control the interaction between DHp and CA domain to regulate kinase activity, phosphotransfer, and phosphatase activity (Marina *et al.*, 2005; Albanesi *et al.*, 2009; Casino *et al.*, 2009; Stewart, 2010). It remains to be seen if this mechanism holds true of RegB.

ArcB

The Arc (anoxic redox control) system was initially identified in *E.coli* as a two-component system that regulates gene expression during the transition from aerobic to anaerobic growth (Iuchi and Lin, 1988, 1989). When the growth condition becomes anaerobic, ArcB/ArcA system represses genes involved in respiratory metabolism and activates genes involved in fermentative metabolism (Brondsted and Atlung, 1994; Lynch and Lin, 1996). ArcB/A homologues also exist in other facultative anaerobes such as *Salmonella typhimurium*, *Shigella flexneri*, *Yersinia pestis*, *Erwinia carotovara*, *Photorhabdus luminescens* (McClelland *et al.*, 2001; Parkhill *et al.*, 2001; Jin *et al.*, 2002; Duchaud *et al.*, 2003), *Shewanella oneidensis* (Shroff *et al.*, 2010), and *Haemophilus influenzae* (Manukhov *et al.*, 2000). It controls between 100–150 operons in response to various respiratory growth conditions (Gunsalus and Parka, 1994; Liu and DeWulf, 2004). Genes in these operons not only contribute to TCA cycle and energy metabolism but are also involved in chromosomal replication (Lee *et al.*, 2001), stress response (Mika and Hengge, 2005; Wong *et al.*, 2007), and ageing of bacteria (Nystrom *et al.*, 1996).

The ArcB/ArcA system comprises a transmembrane sensor kinase ArcB and a response regulator ArcA (Luchi *et al.*, 1990). ArcA is a classic response regulator that consists of a N-terminal receiver domain and a C-terminal helix–turn–helix DNA binding domain. In contrast, ArcB is an atypical sensor kinase in the sense that it only has a very short periplasmic loop (16 residues) between two transmembrane helices compared with most sensor kinases that have substantial periplasmic region for signal perception. ArcB also has a complex catalytic domain architecture comprised three catalytic domains in the cytoplasmic portion: an N-terminal transmitter domain containing the His292 residue where phosphoryl transfer from ATP is initiated, a receiver domain containing an Asp576 residue that can receive phosphate from His292, and a C-terminal phophotransfer domain containing a His717 residue that can receive phosphate from Asp576 (Luchi *et al.*, 1990; Ishige *et al.*, 1994). Between these catalytic domains and the transmembrane domain, there are leucine zipper (Georgellis *et al.*, 1998) and Per-Arnt-Sim (PAS) domains (Taylor and Zhulin, 1999).

A complex phosphorelay involves more than two components

Under anaerobic growth conditions, ArcB autophosphorylates at His292 and then transphosphorylates ArcA at Asp54 via an ArcB-His292→ArcB-Asp576→ArcB-His717→ArcA-Asp54 phosphorelay (Georgellis *et al.*, 1997; Kwon *et al.*, 2000a). It was also shown that ArcA can receive a phosphoryl group at Asp54 directly from His292 in ArcB *in vitro* (Georgellis *et al.*, 1997). Phosphorylated ArcA subsequently interacts with operons to regulate gene expression. Under aerobic conditions, ArcB kinase activity is inhibited by the formation of an intermolecular disulfide bond at two cytosolic redox-active cysteines (Cys180 and Cys241) (Malpica *et al.*, 2004). Inhibition of the kinase activity is accompanied by the activation of ArcB phosphatase activity that also dephosphorylates ArcA-P via a reverse ArcA-Asp54→ArcB-His717→ArcB-Asp576→Pi phosphorelay (Georgellis *et al.*, 1998; Pena-Sandoval *et al.*, 2005). The receiver domain and the C-terminal phosphotransfer domain are required and sufficient to carry out the dephosphorylation of ArcA~P *in vivo* (Georgellis *et al.*, 1998; Pena-Sandoval *et al.*, 2005).

Interestingly, other components can join this phosphorelay to provide an additional fine-tuned redox switch. Ogino *et al.* (1998) demonstrated that SixA is a phospho-histidine phosphatase that can specifically dephosphorylate the phophoryl group at ArcB-His717 *in vitro*. SixA is required for the derepression of the *sdhCDAB* operon that codes for the succinate dehydrogenase complex involved in the TCA cycle under anaerobic respiratory conditions using nitrate as an electron acceptor (Matsubara and Mizuno, 2000). SixA allows dephosphorylation of ArcB even in the absence of oxygen, thereby allowing the expression of genes involved in anaerobic respiration that uses electron acceptors other than dioxygen.

The ArcB to ArcA phosphorelay also branches to RssB, which is an 'orphan' response regulator that mediates the proteolysis of the alternative σ-factor RpoS. Once activated by phosphorylation, RssB directly binds to RpoS and delivers it to the ClpXP protease for degradation (Becker *et al.*, 1999; Zhou *et al.*, 2001). It has been shown that ArcB can phosphorylate RssB *in vitro* and that mutations in ArcB and ArcA resulted in increased RpoS levels in growing cells (Sugiura *et al.*, 2003; Mika and Hengge, 2005). Moreover, it has been demonstrated that ArcA~P is a repressor of RpoS gene expression (Mika and Hengge, 2005). All these findings indicate that ArcB/ArcA/RssB constitute a branched 'three-component' system that regulates RpoS both, at the transcription level and by proteolysis (Mika and Hengge, 2005). Since the ArcB/ArcA system is a global regulator of the energy metabolism, this 'three-component' system apparently links the amount of RpoS, which is the master regulator of the general stress response in *E.coli*, to the cellular energy state.

Mandin and Gottesman (2010) also added a non-phosphorelay component to the ArcB/ArcA signal transduction system. They demonstrated that a small regulatory RNA called ArcZ is transcribed convergently with and overlaps the *arcB* gene. ArcZ can directly repress *arcB* transcription and can itself be repressed by ArcB/ArcA under anaerobic conditions. The same study also identified ArcZ as an activator of *rpoS* translation. Thus, it seems that the ArcB/ArcA system is not only able to use ArcZ as negative feedback loop for its own synthesis but it is also able to control ArcZ to regulate gene expression at the translational level. As an *rpoS* activator, ArcZ adds a third ArcB/ArcA-mediated level of regulation on *rpoS*, further highlighting the importance of the regulatory role of the ArcB/ArcA system in controlling the cellular stress response (Mandin and Gottesman, 2010).

Redox signalling via cysteine oxidation driven by quinone

Just like RegB, ArcB does not directly sense molecular oxygen because an ArcA-regulated operon can be derepressed by other electron acceptors under anaerobic conditions (Iuchi and Lin, 1988). Several fermentation metabolites such as D-lactate, pyruvate, and acetate were suggested to be the actual signals because they can stimulate autophosphorylation of ArcB *in vitro* (Georgellis *et al.*, 1999). However, an *in vivo* study indicated that D-lactate acts as an allosteric effector rather than a direct redox signal for ArcB (Rodriguez *et al.*, 2004).

The first evidence of a specific redox signal came from a study by Georgellis *et al.* (2001a), which indicated that ubiquinone may be a direct regulator of ArcB. The authors showed that the autophosphorylation of a truncated ArcB lacking the transmembrane domain could be inhibited by incubation with oxidized ubiquinone-0, which is a soluble analogue of ubiquinone-8, or oxidized menadione, which is a soluble analogue of menaquinone, *in vitro*. When these ubiquinone analogues were reduced by dithionite, they did not affect the autophosphorylation of the truncated ArcB. To test the signalling role of ubiquinone *in vivo*, ArcB/ArcA-regulated gene expression was analysed in an *ubiCA* mutant strain, in which a ubiquinone precursor synthesis is blocked. The results showed that the expression of an ArcA~P inducible Φ(*cydA′-lacZ*) operon fusion was increased by a factor of three under aerobic conditions compared with a wild-type strain. However, the same increase was not observed in an *ubiCA arcB* double mutant strain. Based on these data, it has been proposed that oxidized ubiquinone is the direct signal that inhibits ArcB kinase activity under aerobic conditions (Georgellis *et al.*, 2001a). A recent study showed that interference with the menaquinone synthesis pathway leads to inactivation of ArcB under anaerobic conditions, suggesting that menaquinone also participates in ArcB redox sensing *in vivo* (Bekker *et al.*, 2010). Since menaquinone has a midpoint potential that is more negative than that of ubiquinone, it was also proposed that menaquinone is responsible for the redox regulation of ArcB between 0–20% aerobiosis, and ubiquinone is responsible for redox regulation between 20–100% aerobiosis (Bekker *et al.*, 2010). The relationship between the ArcB/ArcA system and the quinone pool was also demonstrated by a study which showed that removal of genes coding for three terminal cytochrome oxidases (*cydAB, cyoABCD, and cbdAB*) and of a quinol monooxygenase gene (*ygiN*) resulted in the activation of ArcA aerobically. This activation correlated with a shift in the content of the quinone pool from ubiquinones to menaquinones (Portnoy *et al.*, 2010).

More insight into redox sensing by the ArcB/ArcA-system was provided by Malpica *et al.* (2004) who demonstrated that oxidized ubiquinone actually is used as oxidative power to drive disulfide bonds formation at two cytosolic cysteines and that redox mediated disulfide bond formation is the actual switch regulating ArcB kinase activity. This study not only verified that ArcB kinase activity can be inhibited by oxidized ubiquinonoe-0, it also showed *in vitro* that ArcB kinase activity can be inhibited by other oxidizing agents such as chloramine T or slightly increased by dithiothreitol (DTT). The cell-permeable thiol-reducing agents DTT and β-mercaptoethanol were also found to increase the expression of the ArcA~P activatable Φ (*cydA′-lacZ*) reporter gene under aerobic conditions, suggesting that cysteines may be involved in redox sensing by ArcB/ArcA *in vivo*.

In vitro mutagenesis studies have identified two conserved cysteines, Cys180 and Cys241, in the PAS domain as the redox-active cysteines (Malpica *et al.*, 2004). They demonstrated that a truncated ArcB lacking the membrane spanning domains only retains about 25% activity after incubation with oxidized ubiquinone-0. However, a similar truncated ArcB containing a Cys180Ala mutation retained >85% activity, while the truncated ArcB containing a Cys241Ala mutation retained about 35% activity and a truncated ArcB containing a Cys180Ala/Cys241Ala double mutation was not inhibited at all. After introduction of these mutations into the chromosome, the Cys180Ala, Cys241Ala, and Cys180Ala/Cys241 mutations resulted in decreased aerobic expression of the ArcA~P repressible Φ (*lldP -lacZ*) reporter to approximately 32%,84%, and 3.5% of its level, respectively, as compared with the wild-type strain. *In vitro* and *in vivo* data indicated that a mutation of either cysteine could abolish the inhibitory effect of ubiquinone to some extent, with Cys180 responsible for the major part of the activity. Thus it seems that both cysteines are required for the redox-sensing ability of ArcB, although Cys180 plays a central role.

To examine the oxidation status of the two cysteines after incubation with ubiquinone, methoxypolyethylene glycol-maleimide (MAL-PEG), a free thiol-specific probe, was used. It was found that free thiols at both cysteines were oxidized by ubiquinone *in vitro*. Western blot experiments revealed that the oxidation of Cys180 and Cys241 also resulted in dimerization of ArcB *in vivo* under aerobic conditions, presumably caused by intermolecular disulfide bond formation (Malpica *et al.*, 2004). Hence, a model of ArcB redox sensing can be proposed where two cytosolic cysteines, Cys180 and Cys241, serve as the redox switch. These two cysteines are oxidized or reduced depending on the redox state of the quinone pool (ubiquinone and menaquinone) in the membrane. Under aerobic conditions, when ubiquinone/menaquinone are oxidized, these two cysteines will use the oxidative power of ubiquinone/menaquinone to drive intermolecular disulfide bond formation to silence ArcB kinase activity.

Does the transmembrane domain of ArcB play a role in redox sensing?

Even though monitoring the redox state of the ubiquinone pool by ArcB may seem to be similar to that described above with RegB there are significant differences. Foremost is the observation that ArcB can use ubiquinone to oxidize redox-active cysteines for its redox sensing capability (Malpica *et al.*, 2004) (Fig. 3.2). This is distinctly different from RegB that utilizes allosteric binding of ubiquinone to alter activity in a way that is independent from regulation of activity observed by the oxidation of cysteine (Wu and Bauer, 2010). This also raises the question about the role of the ArcB transmembrane domain. It has been suggested that the transmembrane domain of ArcB is not involved in redox sensing and instead just serves as an anchor to keep ArcB close to its signal, the ubiquinone pool that is present in the membrane (Kwon *et al.*, 2000b). This conclusion was based on an *in vivo* mutagenesis study in which transmembrane helix 1, periplasmic loop, and transmembrane helix 2 were replaced by the corresponding segments of MalF (a subunit

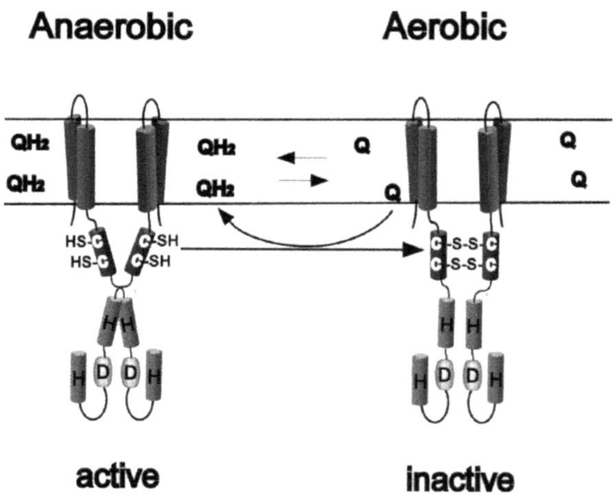

Figure 3.2 A model for ArcB redox sensing. Under anaerobic growth conditions, Cys180 and Cys241 are reduced promoting ArcB kinase activity. When growth conditions are aerobic, the ubiquinone shifts predominantly to an oxidized form. The oxidative potential of oxidized ubiquinone is used to catalyse two intermolecular disulfide bonds at Cys180 and Cys241, leading to the inhibition of ArcB kinase activity (adapted from Malpica *et al.*, 2004).

of maltose permease). The results showed that replacement of transmembrane helix 1 or helix 2 or helix 2 plus the adjacent periplasmic loop did not cause significant changes in the expression of Arc regulated reporter genes (Kwon *et al.*, 2000b). Finally, it should be pointed out that the *in vitro* data involving formation of a disulfide bond on a truncated variant of ArcB with ubiquinone analogues should be interpreted with caution as these redox-active cysteines could be non-specifically oxidized by any oxidant that has a higher redox potential than these cysteines. Thus, these soluble ubiquinone analogues could be functioning as a non-specific oxidant just like that observed by the oxidant chloramine T, which can also inhibit truncated ArcB *in vitro* (Georgellis *et al.*, 2001a; Malpica *et al.*, 2004). The only experimental evidence to support the *in vivo* specificity between the ubiquinone and the redox-active cysteines is that addition of H_2O_2 to an anaerobic culture did not affect the expression of the respective reporter gene (Malpica *et al.*, 2004). The fact that a functional ArcB homologue is present in *Haemophilus influenzae* that lacks the PAS domain containing these redox-active cysteines also suggests that ArcB homologues in other species may be able to utilize other mechanisms to sense a redox signal (Georgellis *et al.*, 2001b). Thus, although there is evidence that oxidation of redox-active cysteines and ubiquinone/menaquinone play critical roles in *E. coli* ArcB redox sensing, the function of the transmembrane domain and the relationship between the cysteines and ubiquinone/menaquinone need to be further confirmed.

To date, there are structures available of two ArcB domains. X-ray crystallographic and NMR structures of the C-terminal histidine-containing phosphotransfer domain have disclosed a five helices structure including a four-helix bundle that is thought to undergo a possible conformational change upon phosphorylation (Kato *et al.*, 1997, 1999; Ikegami *et al.*, 2001). An NMR structure of the ArcB transmembrane domain revealed a distance of 11.1 Å between the two antiparallel helices, suggesting loose helical packing. This loose packing generates flexibility of the transmembrane domain, which may be required for the signal transduction (Hendrickson, 2005). However, better understanding of the redox-sensing mechanism of ArcB will require a structure of the full-length protein.

FixL

FixL is an oxygen-sensing sensor kinase that is required for nitrogen fixation in several rhizobial species. Under anoxygenic conditions, FixL autophosphorylates and transfers its phosphate to its cognate response regulator FixJ that induces gene transcription. In *Sinorhizobium meliloti* phosphorylated FixJ induces expression of NifA that regulates nitrogen fixing genes as well as FixK that regulates a high affinity cytochrome oxidase that allows respiration under low oxygen tension (David *et al.*, 1988; de Philip *et al.*, 1990). This is contrasted by FixJ from *Bradyrhizobium japonicum* that controls expression of FixK1 which only regulates respiration but not nitrogen fixation genes (Mesa *et al.*, 2003; Sciotti *et al.*, 2003). Nitrogen fixation in this latter species is thus not under direct control of FixL–FixJ but is instead indirectly regulated as nitrogenase is an oxygen labile enzyme who's activity is dependent on reduced oxygen tension that is facilitated by respiration.

FixL proteins contain a Per-ARNT-Sim motif called a PAS domain that is known to bind haem as a cofactor for gas (O_2, NO and CO) sensing (Gilles-Gonzalez *et al.*, 1991). The PAS domain is located at the amino terminus immediately followed by a histidine kinase domain. There are some variations between species with *S. meliloti* containing a membrane spanning domain immediately preceding the haem-PAS domain, while FixL of *B. japonicum* lacks the membrane spanning domain but instead contains an additional PAS domain upstream of the haem binding PAS domain (de Philip *et al.*, 1990; Gilles-Gonzalez *et al.*, 1991).

The iron in haem binds to the PAS domain structure under anoxygenic conditions in a pentacoordinated state with four ligands to the iron provided by a nitrogen from the four haem pyrrole rings and a fifth axial ligand provided by a nitrogen from a conserved histidine (Fig. 3.3) (Miyatake *et al.*, 2000; Hao *et al.*, 2002). There are also several protein ligands to a haem propionate-7 side group with the most important being a conserved

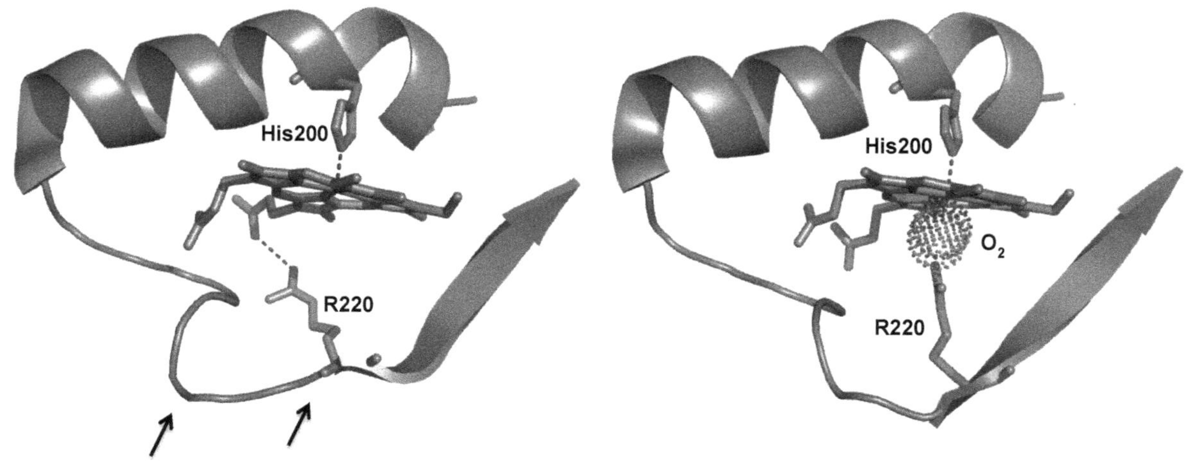

Figure 3.3 A model for dioxygen sensing by the PAS domain of FixL. Left is the structure of the FixL ferrous haem lacking oxygen (deoxy complex) (PDB: 1LXW deposited by Hao *et al*., 2002). Right: FixL containing dioxygen bound to the iron of haem as indicted by stippled balls (PDB:1DP6 deposited by Gong *et al*. 2000). Arg220 forms a hydrogen bond to a side group of the haem in the deoxygenated form, whereas in the oxygenated form Arg220 forms a hydrogen bond to the bound oxygen. This promotes an output signal as indicated by a modest change in the loop as highlighted by arrows.

arginine present in the β-G strand (Arg220 in *B. japonicum* and Arg214 in *S. meliloti*) (Miyatake *et al.*, 2000; Hao *et al.*, 2002). When oxygen is present it forms a sixth axial ligand to the haem iron as well as a hydrogen bond acceptor to Arg220 which is displaced from the haem propionate side group to form a bond to the iron bound dioxygen (Miyatake *et al.*, 2000; Hao *et al.*, 2002) (Fig. 3.3). Oxygen mediated movement of Arg220 causes a slight change in tertiary structure the most notable being a 1.6-Å movement of the FG loop that links the α-F helix to the β-G strand away from the haem centre (Gong *et al.*, 1998; Gong *et al.*, 2000; Miyatake *et al.*, 2000; Hao *et al.*, 2002; Dunham *et al.*, 2003). Presumably, movement of this loop constitutes an output signal that affects the ability of the kinase domain to phosphorylate the histidine residue. The mechanism whereby binding of dioxygen affects autophosphorylation remains unclear as there is only structural information on the haem binding PAS domain and not on the full-length FixL. However, recent time-resolved resonance Raman spectroscopy analysis on full-length FixL indicated that dynamic structural changes in the picosecond to millisecond time frame occur for both, the sensor and the kinase domain upon O_2 or CO dissociation (Hiruma *et al.*, 2007).

One interesting aspect of FixL is that the haem binding PAS domain can bind O_2, CO and NO, but only oxygen binding is an effective inhibitor of the full-length kinase (Tuckerman *et al.*, 2002; Dunham *et al.*, 2003). Time resolved X-ray crystallography structural studies with *B. japonicum* FixL showed that CO binding to iron causes movement of Leu236 and of the H- and I- β-strands which is distinctly different from movement of the α-F and β-G loops that occurs upon binding of dioxygen (Key and Moffat, 2005; Key *et al.*, 2007). Another interesting aspect of gas discrimination is a caging phenomenon that occurs to a greater extent with dioxygen than with NO (Liebl *et al.*, 2002; Kruglik *et al.*, 2007). Upon dissociation of oxygen or NO, rebinding to haem occurs remarkably effectively with oxygen (90% rebinding in 5 ps) but not nearly as efficiently with NO. It has been proposed that the Arg220 ligand to oxygen effectively recaptures displaced O_2, whereas with NO Arg220 is not providing a ligand, so that displaced NO is free to diffuse away from the iron (Liebl *et al.*, 2002).

DosS and DosT

DosS and DosT are two histidine kinases in *Mycobacterium tuberculosis* that are thought to coordinately regulate genes needed for survival upon infection of macrophages (Park *et al.*, 2003; Kendall *et al.*, 2004; Roberts *et al.*, 2004; Saini *et al.*, 2004). DosS and DosT both contain two adjacent GAF domains in their amino terminal sensor regions. GAF domains are fairly ubiquitous, and in at least one case they are known to bind cGMP as an effector (Ho *et al.*, 2000), however, in most cases the function of GAF domains is unknown. Interestingly, the first GAF domains present in DosS and DosT (called GAF-A) bind a pentacoordinated *b*-type haem with the haem able to bind O_2, NO and CO (Ioanoviciu *et al.*, 2007; Kumar *et al.*, 2007; Sousa *et al.*, 2007; Lee *et al.*, 2008). Three dimensional crystal structure of the haem binding GAF domains from DosS and DosT indicates that the haem is bound to a histidine as an axial ligand (His149 in DosS and His147 in DosT) (Podust *et al.*, 2008; Cho *et al.*, 2009). Kinase activity is inhibited upon binding of dioxygen to the ferrous form of iron in the bound haem. Kinase activity is not inhibited by binding of the other gases (Sousa *et al.*, 2007). Unlike structural changes observed in crystals of FixL that contain unbound versus bound oxygen, crystals of DosS and DosT exhibit no significant changes, whether or not dioxygen is bound to the iron in haem (Podust *et al.*, 2008; Cho *et al.*, 2009). One possibility is that crystal packing sterically interferes with movement, or alternatively, additional domains of DosS and DosT are needed to observe changes. This latter possibility is supported by analysis of the hydrogen bond network using Raman spectroscopy indicating that the second GAF domain (GAF-B) is required for the ability of O_2 to inhibit kinase activity (Yukl *et al.*, 2007). It remains unclear why the second GAF domain is needed for this inhibition.

In DosS the iron in bound haem is capable of rapidly auto-oxidizing to the ferric form, but this does not occur with DosT (Kumar *et al.*, 2007; Cho *et al.*, 2009). This has led to the proposal that DosS constitutes a redox sensor, while the related DosT could function as an O_2 sensor (Kumar *et al.*, 2007; Cho *et al.*, 2009; Kim *et al.*, 2010). This is supported by the finding that DosS is activated by the addition of a reductant such as ascorbate even when oxygen is present, while DosT does not show this response (Honaker *et al.*, 2010). This difference in reactivity to oxygen appears to reside in the observation that the GAF domain in DosT has a channel that allows dioxygen to bind the haem, while the channel in DosS is blocked by a glutamic acid (Glu87) (Cho *et al.*, 2011). Additional studies indicated that DosT has an early role in infection when oxygen is more prevalent, whereas DosS has a role when the environment becomes more anaerobic (Honaker *et al.*, 2009; Kim *et al.*, 2010).

LovK

LovK is a histidine sensor kinase from *C. crescentus* that is implicated in blue light regulation of biofilm production (Purcell *et al.*, 2007). This kinase has a PAS domain that non-covalently binds FAD and is called a flavin binding light, oxygen and voltage (LOV) domain (Purcell *et al.*, 2007). Upon light excitation of the flavin there is a photocyle initiated which involves the formation of a covalent linkage between FAD and a cysteine in the flavin binding pocket that activates the kinase (Saloman *et al.*, 2001; Crosson *et al.*, 2002; Purcell *et al.*, 2007). Shifting excited LovK to the dark allows slow reduction of the FAD–Cys bond to inhibit kinase activity (Fig. 3.3).

Although LovK has mostly been studied for its light response, a recent study by the Crossen laboratory has demonstrated that the LovK photocycle (and hence its activity) is absent when the flavin is reduced to FADH (Purcell *et al.*, 2010). Flavin bound to LovK has a midpoint potential (–260 mV) that is close to the potential of the cytoplasmic potential (~270 mV) indicating that even slight changes in cellular redox potential can alter the FAD/FADH ratio leading to alteration of LovK to undergo its photocycle. LovK also has a week affinity to the flavin indicating that the flavin binding pocket may be in equilibrium with the flavin pool. Additionally, redox active species of oxygen such as H_2O_2 also effectively oxidize the flavin pool, which would stimulate the light sensing capability of LovK. Collectively, these recent results indicate that LovK is both a redox as well as a light sensing kinase (Fig. 3.4). It is unclear if other LOV domains have similar characteristics such as loose affinity to flavin and inhibition of the photocycle by flavin reduction. However, the redox sensing characteristics of LOV domains warrant further examination, as LOV domain containing proteins are found in a number of bacterial species (Crosson *et al.*, 2003).

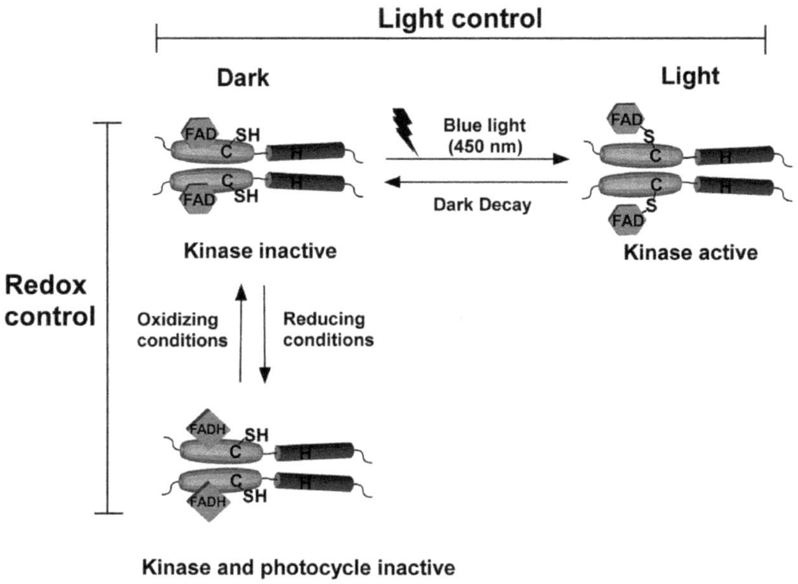

Figure 3.4 A model for redox sensing by LovK. LovK is under both light and redox control. Blue light (450 nm) absorption by a bound FAD promotes the formation of a covalent bond between FAD and a cysteine in the flavin binding pocket, resulting in increased kinase activity. Shifting excited LovK to dark breaks the FAD–Cys bond and inhibits kinase activity. Blue light activation of the kinase is abolished when the FAD cofactor is reduced to FADH by fluctuating cellular redox potential. Under such conditions, LovK kinase activity is inhibited as it is unable to undergo a photocycle.

References

Albanesi, D., Martin, M., Trajtenberg, F., Mansilla, M.C., Haouz, A., Alzari, P.M., de Mendoza, D., and Buschiazzo, A. (2009). Structural plasticity and catalysis regulation of a thermosensor histidine kinase. Proc. Natl. Acad. Sci. U.S.A. *106,* 16185–16190.

Antelmann, H., and Helmann, J.D. (2011). Thiol-based redox switches and gene regulation. Antioxid. redox signal. *14,* 1049–1063.

Barber, R.D., and Donohue, T.J. (1998). Pathways for transcriptional activation of a glutathione-dependent formaldehyde dehydrogenase gene. J. Mol. Biol. *280,* 775–784.

Bauer, E., Kaspar, T., Fischer, H.M., and Hennecke, H. (1998). Expression of the *fixR-nifA* operon in *Bradyrhizobium japonicum* depends on a new response regulator, RegR. J. Bacteriol. *180,* 3853–3863.

Becker, G., Klauck, E., and Hengge-Aronis, R. (1999). Regulation of RpoS proteolysis in *Escherichia coli*: the response regulator RssB is a recognition factor that interacts with the turnover element in RpoS. Proc. Natl. Acad. Sci. U.S.A. *96,* 6439–6444.

Bekker, M., Alexeeva, S., Laan, W., Sawers, G., Teixeira de Mattos, J., and Hellingwerf, K. (2010). The ArcBA two-component system of *Escherichia coli* is regulated by the redox state of both the ubiquinone and the menaquinone pool. J. Bacteriol. *192,* 746–754.

Bird, T.H., Du, S., and Bauer, C.E. (1999). Autophosphorylation, phosphotransfer, and DNA-binding properties of the RegB/RegA two-component regulatory system in *Rhodobacter capsulatus*. J. Biol. Chem. *274,* 16343–16348.

Bowman, W.C., Du, S., Bauer, C.E., and Kranz, R.G. (1999) *In vitro* activation and repression of photosynthesis gene transcription in *Rhodobacter capsulatus.* Mol. Microbiol. *33,* 429–437.

Brondsted, L., and Atlung, T. (1994). Anaerobic regulation of the hydrogenase 1 (*hya*) operon of *Escherichia coli*. J. Bacteriol. *176,* 5423–5428.

Casino, P., Rubio, V., and Marina, A. (2009). Structural insight into partner specificity and phosphoryl transfer in two-component signal transduction. Cell *139,* 325–336.

Cho, H.Y., Cho, H.J., Kim, Y.M., Oh, J.I., and Kang, B.S. (2009). Structural insight into the heme-based redox sensing by DosS from *Mycobacterium tuberculosis*. J. Biol. Chem. *284,* 13057–13067.

Cho, H.Y., Cho, H.J., Kim, M.H., and Kang, B.S. (2011). Blockage of the channel to heme by the E87 side chain in the GAF domain of *Mycobacterium tuberculosis* DosS confers the unique sensitivity of DosS to oxygen. FEBS Lett. *585,* 1873–1878.

Crosson, S., and Moffat, K. (2002). Photoexcited structure of a plant photoreceptor domain reveals a light-driven molecular switch. Plant Cell *14,* 1067–1075.

Crosson, S., Rajagopal, S., and Moffat, K. (2003).The LOV domain family: signaling modules coupled to diverse output domains. Biochemistry *42,* 2–10.

Dangel, A.W., and Tabita, F.R. (2009). Protein–protein interactions between CbbR and RegA (PrrA), transcriptional regulators of the *cbb* operons of *Rhodobacter sphaeroides*. Mol. Microbiol. *71,* 717–729.

David, M., Daveran, M.L., Batut, J., Dedieu, A., Domergue, O., Ghai, J., Hertig, C., Boistard, P., and Kahn, D. (1988). Cascade regulation of *nif* gene expression in *Rhizobium meliloti*. Cell *54*, 671–683.

Du, S., Bird, T.H., and Bauer, C.E. (1998). DNA binding characteristics of RegA. A constitutively active anaerobic activator of photosynthesis gene expression in *Rhodobacter capsulatus*. J. Biol. Chem. *273*, 18509–18513.

Dubbs, J.M., Bird, T.H., Bauer, C.E., and Tabita, F.R. (2000). Interaction of CbbR and RegA* transcription regulators with the *Rhodobacter sphaeroides cbbI* Promoter-operator region. J. Bacteriol. *275*, 19224–19230.

Duchaud, E., Rusniok, C., Frangeul, L., Buchrieser, C., Givaudan, A., Taourit, S., Bocs, S., Boursaux-Eude, C., Chandler, M., Charles, J.F., *et al.* (2003). The genome sequence of the entomopathogenic bacterium *Photorhabdus luminescens*. Nature Biotech. *21*, 1307–1313.

Dunham, C.M., Dioum, E.M., Tuckerman, J.R., Gonzalez, G., Scott, W.G., and Gilles-Gonzalez, M.A. (2003). A distal arginine in oxygen-sensing heme-PAS domains is essential to ligand binding, signal transduction, and structure. Biochemistry *42*, 7701–7708.

Elsen, S., Dischert, W., Colbeau, A., and Bauer, C.E. (2000). Expression of uptake hydrogenase and molybdenum nitrogenase in *Rhodobacter capsulatus* is coregulated by the RegB-RegA two-component regulatory system. J. Bacteriol. *182*, 2831–2837.

Elsen, S., Swem, L.R., Swem, D.L., and Bauer, C.E. (2004). RegB/RegA, a highly conserved redox-responding global two-component regulatory system. Microbiol. Mol. Biol. Rev. *68*, 263–279.

Eraso, J.M., and Kaplan, S. (2000). From redox flow to gene regulation: role of the PrrC protein of *Rhodobacter sphaeroides* 2.4.1. Biochemistry *39*, 2052–2062.

Georgellis, D., Lynch, A.S., and Lin, E.C. (1997). *In vitro* phosphorylation study of the arc two-component signal transduction system of *Escherichia coli*. J. Bacteriol. *179*, 5429–5435.

Georgellis, D., Kwon, O., De Wulf, P., and Lin, E.C. (1998). Signal decay through a reverse phosphorelay in the Arc two-component signal transduction system. J. Biol. Chem. *273*, 32864–32869.

Georgellis, D., Kwon, O., and Lin, E.C. (1999). Amplification of signaling activity of the arc two-component system of *Escherichia coli* by anaerobic metabolites. An *in vitro* study with different protein modules. J Biol. Chem. *274*, 35950–35954.

Georgellis, D., Kwon, O., and Lin, E.C. (2001a). Quinones as the redox signal for the *arc* two-component system of bacteria. Science *292*, 2314–2316.

Georgellis, D., Kwon, O., Lin, E.C., Wong, S.M., and Akerley, B.J. (2001b). Redox signal transduction by the ArcB sensor kinase of *Haemophilus influenzae* lacking the PAS domain. J. Bacteriol. *183*, 7206–7212.

Gilles-Gonzalez, M.A., Ditta, G.S., and Helinski, D.R. (1991). A haemoprotein with kinase activity encoded by the oxygen sensor of *Rhizobium meliloti*. Nature *350*, 170–172.

Gong, W., Hao, B., Mansy, S.S., Gonzalez, G., Gilles-Gonzalez, M.A., and Chan, M.K. (1998). Structure of a biological oxygen sensor: a new mechanism for heme-driven signal transduction. Proc. Natl. Acad. Sci. U.S.A. *95*, 15177–15182.

Gong, W., Hao, B., and Chan, M.K. (2000). New mechanistic insights from structural studies of the oxygen-sensing domain of *Bradyrhizobium japonicum* FixL. Biochemistry *39*, 3955–3962.

Grammel, H., and Ghosh, R. (2008) Redox state dynamics of ubiquinone-10 imply cooperative regulation of photosynthetic membrane expression in *Rhodospirillum rubrum*. J. Bacteriol. *190*, 4912–4921.

Gunsalus, R.P., and Park, S.J. (1994). Aerobic-anaerobic gene regulation in *Escherichia coli*: control by the ArcAB and Fnr regulons. Res. Microbiol. *145*, 437–450.

Hao, B., Isaza, C., Arndt, J., Soltis, M., and Chan, M.K. (2002). Structure-based mechanism of O2 sensing and ligand discrimination by the FixL heme domain of *Bradyrhizobium japonicum*. Biochemistry *41*, 12952–12958.

Hendrickson, W.A. (2005). Transduction of biochemical signals across cell membranes. Q. Rev. Biophys. *38*, 321–330.

Hiruma, Y., Kikuchi, A., Tanaka, A., Shiro, Y., and Mizutani, Y. (2007). Resonance Raman observation of the structural dynamics of FixL on signal transduction and ligand discrimination. Biochemistry *46*, 6086–6096.

Ho, Y.S., Burden, L.M., and Hurley, J.H. (2000). Structure of the GAF domain, a ubiquitous signaling motif and a new class of cyclic GMP receptor. EMBO J. *19*, 5288–5299.

Honaker, R.W., Leistikow, R.L., Bartek, I.L., and Voskuil, M.I. (2009). Unique roles of DosT and DosS in DosR regulon induction and *Mycobacterium tuberculosis* dormancy. Infect. Immun. *77*, 3258–3263.

Honaker, R.W., Dhiman, R.K., Narayanasamy, P., Crick, D.C., and Voskuil, M.I. (2010). DosS responds to a reduced electron transport system to induce the *Mycobacterium tuberculosis* DosR regulon. J. Bacteriol. *192*, 6447–6455.

Ikegami, T., Okada, T., Ohki, I., Hirayama, J., Mizuno, T., and Shirakawa, M. (2001). Solution structure and dynamic character of the histidine-containing phosphotransfer domain of anaerobic sensor kinase ArcB from *Escherichia coli*. Biochemistry *40*, 375–386.

Inoue, K., Kouadio, J.L., Mosley, C.S., and Bauer, C.E. (1995). Isolation and *in vitro* phosphorylation of sensory transduction components controlling anaerobic induction of light harvesting and reaction center gene expression in *Rhodobacter capsulatus*. Biochemistry *34*, 391–396.

Ioanoviciu, A., Yukl, E.T., Moënne-Loccoz, P., and de Montellano, P.R. (2007). DevS, a heme-containing two-component oxygen sensor of *Mycobacterium tuberculosis*. Biochemistry *46*, 4250–4260.

Ishige, K., Nagasawa, S., Tokishita, S., and Mizuno, T. (1994). A novel device of bacterial signal transducers. EMBO J. *13*, 5195–5202.

Iuchi, S., and Lin, E.C. (1988). *arcA* (*dye*), a global regulatory gene in *Escherichia coli* mediating repression of enzymes in aerobic pathways. Proc. Natl. Acad. Sci. U.S.A. *85*, 1888–1892.

Iuchi, S., Cameron, D.C., and Lin, E.C. (1989). A second global regulator gene (*arcB*) mediating repression of enzymes in aerobic pathways of *Escherichia coli*. J. Bacteriol. *171*, 868–873.

Iuchi, S., Matsuda, Z., Fujiwara, T., and Lin, E.C. (1990). The *arcB* gene of *Escherichia coli* encodes a sensor-regulator protein for anaerobic repression of the arc modulon. Mol. Microbiol. *4*, 715–727.

Jin, Q., Yuan, Z., Xu, J., Wang, Y., Shen, Y., Lu, W., Wang, J., Liu, H., Yang, J., Yang, F. *et al.* (2002). Genome sequence of *Shigella flexneri* 2a: insights into pathogenicity through comparison with genomes of *Escherichia coli* K12 and O157. Nucl. Acid Res. *30*, 4432–4441.

Kato, M., Mizuno, T., Shimizu, T., and Hakoshima, T. (1997). Insights into multistep phosphorelay from the crystal structure of the C-terminal HPt domain of ArcB. Cell *88*, 717–723.

Kato, M., Mizuno, T., Shimizu, T., and Hakoshima, T. (1999). Refined structure of the histidine-containing phosphotransfer (HPt) domain of the anaerobic sensor kinase ArcB from *Escherichia coli* at 1.57 A resolution. Acta Crystallo. D Biol. Crystallo. *55*, 1842–1849.

Kendall, S.L., Movahedzadeh, F., Rison, S.C., Wernisch, L., Parish, T., Duncan, K., Betts, J.C., and Stoker, N.G. (2004). The *Mycobacterium tuberculosis dosRS* two-component system is induced by multiple stresses. Tuberculosis (Edinb.) *84*, 247–255.

Key, J., and Moffat, K. (2005). Crystal structures of deoxy and CO-bound by FixLH reveal details of ligand recognition and signaling. Biochemistry *44*, 4627–4635.

Key, J., Srajer, V., Pahl, R., and Moffat, K. (2007). Time-resolved crystallographic studies of the heme domain of the oxygen sensor FixL: structural dynamics of ligand rebinding and their relation to signal transduction. Biochemistry *46*, 4706–4715.

Kim, M.J., Park, K.J., Ko, I.J., Kim, Y.M., and Oh, J.I. (2010). Different roles of DosS and DosT in the hypoxic adaptation of Mycobacteria. J. Bacteriol. *192*, 4868–4875.

Kim, S.O., Merchant, K., Nudelman, R., Beyer, W.F., Jr., Keng, T., DeAngelo, J., Hausladen, A., and Stamler, J.S. (2002). OxyR: a molecular code for redox-related signaling. Cell. *109*, 383–396.

Kim, Y.J., Ko, I.J., Lee, J.M., Kang, H.Y., Kim, Y.M., Kaplan, S., and Oh, J.I. (2007). Dominant role of the cbb_3 oxidase in regulation of photosynthesis gene expression through the PrrBA system in *Rhodobacter sphaeroides* 2.4.1. J. Bacteriol. *189*, 5617–5625.

Kruglik, S.G., Jasaitis, A., Hola, K., Yamashita, T., Liebl, U., Martin, J.L., and Vos, M.H. (2007). Subpicosecond oxygen trapping in the heme pocket of the oxygen sensor FixL observed by time-resolved resonance Raman spectroscopy. Proc. Natl. Acad. Sci. U.S.A. *104*, 7408–7413.

Kumar, A., Toledo, J.C., Patel, R.P., Lancaster, J.R., Jr., Steyn, A.J. (2007). *Mycobacterium tuberculosis* DosS is a redox sensor and DosT is a hypoxia sensor. Proc. Natl. Acad. Sci. U.S.A. *104*, 11568–11573.

Kwon, O., Georgellis, D., and Lin, E.C. (2000a). Phosphorelay as the sole physiological route of signal transmission by the arc two-component system of *Escherichia coli*. J. Bacteriol. *182*, 3858–3862.

Kwon, O., Georgellis, D., Lynch, A.S., Boyd, D., and Lin, E.C. (2000b). The ArcB sensor kinase of *Escherichia coli*: genetic exploration of the transmembrane region. J. Bacteriol. *182*, 2960–2966.

Laguri, C., Phillips-Jones, M.K., and Williamson, M.P. (2003). Solution structure and DNA binding of the effector domain from the global regulator PrrA (RegA) from *Rhodobacter sphaeroides*: insights into DNA binding specificity. Nucl. Acids Res. *31*, 6778–6787.

Laguri, C., Stenzel, R.A., Donohue, T.J., Phillips-Jones, M.K., and Williamson, M.P. (2006). Activation of the global gene regulator PrrA (RegA) from *Rhodobacter sphaeroides*. Biochemistry *45*, 7872–7881.

Laratta, W.P., Choi, P.S., Tosques, I.E., and Shapleigh, J.P. (2002). Involvement of the PrrB/PrrA two-component system in nitrite respiration in *Rhodobacter sphaeroides* 2.4.3: evidence for transcriptional regulation. J. Bacteriol. *184*, 3521–3529.

Lee, J.M., Cho, H.Y., Cho, H.J., Ko, I.J., Park, S.W., Baik, H.S., Oh, J.H., Eom, C.Y., Kim, Y.M., Kang, B.S., and Oh, J.I. (2008). O_2- and NO-sensing mechanism through the DevSR two-component system in *Mycobacterium smegmatis*. J. Bacteriol. *190*, 6795–6804.

Lee, Y.S., Han, J.S., Jeon, Y., and Hwang, D.S. (2001). The *arc* two-component signal transduction system inhibits *in vitro Escherichia coli* chromosomal initiation. J. Biol. Chem. *276*, 9917–9923.

Liebl, U., Bouzhir-Sima, L., Negrerie, M., Martin, J.L., and Vos, M.H. (2002). Ultrafast ligand rebinding in the heme domain of the oxygen sensors FixL and Dos: general regulatory implications for heme-based sensors. Proc. Natl. Acad. Sci. U.S.A. *99*, 12771–12776.

Liu, X., and De Wulf, P. (2004). Probing the ArcA-P modulon of *Escherichia coli* by whole genome transcriptional analysis and sequence recognition profiling. J. Biol. Chem. *279*, 12588–12597.

Lynch, A.S., and Lin, E.C. (1996). Transcriptional control mediated by the ArcA two-component response regulator protein of *Escherichia coli*: characterization of DNA binding at target promoters. J. Bacteriol. *178*, 6238–6249.

McClelland, M., Sanderson, K.E., Spieth, J., Clifton, S.W., Latreille, P., Courtney, L., Porwollik, S., Ali, J., Dante, M., Du, F., *et al.* (2001). Complete genome sequence of *Salmonella enterica* serovar Typhimurium LT2. Nature. *413*, 852–856.

Madigan, M.T., and Gest, H. (1979). Growth of the photosynthetic bacterium *Rhodopseudomonas capsulata* chemoautotrophically in darkness with H_2 as the energy source. J. Bacteriol. *137*, 524–530.

Malpica, R., Franco, B., Rodriguez, C., Kwon, O., and Georgellis, D. (2004). Identification of a quinone-sensitive redox switch in the ArcB sensor kinase. Proc. Natl. Acad. Sci. U.S.A. *101*, 13318–13323.

Mandin, P., and Gottesman, S. (2010). Integrating anaerobic/aerobic sensing and the general stress response through the ArcZ small RNA. EMBO J. *29*, 3094–3107.

Manukhov, I.V., Bertsova, Y.V., Trofimov, D.Y., Bogachev, A.V., and Skulachev, V.P. (2000). Analysis of HI0220 protein from *Haemophilus influenzae*, a novel structural and functional analog of ArcB protein from *Escherichia coli*. Biochemistry (Mosc.) *65*, 1321–1326.

Marina, A., Waldburger, C.D., and Hendrickson, W.A. (2005). Structure of the entire cytoplasmic portion of a sensor histidine-kinase protein. EMBO J. *24*, 4247–4259.

Matsubara, M., and Mizuno, T. (2000). The SixA phospho-histidine phosphatase modulates the ArcB phosphorelay signal transduction in *Escherichia coli*. FEBS Lett. *470*, 118–124.

Mesa, S., Bedmar, E.J., Chanfon, A., Hennecke, H., and Fischer, H.M. (2003). *Bradyrhizobium japonicum* NnrR, a denitrification regulator, expands the FixLJ-FixK2 regulatory cascade. J. Bacteriol. *185*, 3978–3982.

Mika, F., and Hengge, R. (2005). A two-component phosphotransfer network involving ArcB, ArcA, and RssB coordinates synthesis and proteolysis of sigmaS (RpoS) *in E. coli*. Genes Dev. *19*, 2770–2781.

Miyatake, H., Mukai, M., Park, S.Y., Adachi, S., Tamura, K., Nakamura, H., Nakamura, K., Tsuchiya, T., Iizuka, T., and Shiro, Y. (2000). Sensory mechanism of oxygen sensor FixL from *Rhizobium meliloti*: crystallographic, mutagenesis and resonance Raman spectroscopic studies. J. Mol. Biol. *301*, 415–431.

Mosley, C.S., Suzuki, J.Y., and Bauer, C.E. (1994). Identification and molecular genetic characterization of a sensor kinase responsible for coordinately regulating light harvesting and reaction center gene expression in response to anaerobiosis. J. Bacteriol. *176*, 7566–7573.

Nystrom, T., Larsson, C., and Gustafsson, L. (1996). Bacterial defense against aging: role of the *Escherichia coli* ArcA regulator in gene expression, readjusted energy flux and survival during stasis. EMBO J. *15*, 3219–3228.

Ogino, T., Matsubara, M., Kato, N., Nakamura, Y., and Mizuno, T. (1998). An *Escherichia coli* protein that exhibits phosphohistidine phosphatase activity towards the HPt domain of the ArcB sensor involved in the multistep His-Asp phosphorelay. Mol. Microbiol. *27*, 573–585.

Oh, J.I., and Kaplan, S. (1999). The cbb_3 terminal oxidase of *Rhodobacter sphaeroides* 2.4.1: structural and functional implications for the regulation of spectral complex formation. Biochemistry *38*, 2688–2696.

Oh, J.I., Ko, I.J., and Kaplan, S. (2004). Reconstitution of the *Rhodobacter sphaeroides* cbb_3-PrrBA signal transduction pathway *in vitro*. Biochemistry *43*, 7915–7923.

Park, H.D., Guinn, K.M., Harrell, M.I., Liao, R., Voskuil, M.I., Tompa, M., Schoolnik, G.K., and Sherman, D.R. (2003). Rv3133c/dosR is a transcription factor that mediates the hypoxic response of *Mycobacterium tuberculosis*. Mol. Microbiol. *48*, 833–843.

Parkhill, J., Wren, B.W., Thomson, N.R., Titball, R.W., Holden, M.T., Prentice, M.B., Sebaihia, M., James, K.D., Churcher, C., Mungall, K.L., *et al.* (2001). Genome sequence of *Yersinia pestis*, the causative agent of plague. Nature *413*, 523–527.

Parson, W. (1978). Quinones as secondary electron acceptors. In The Photosynthetic Bacteria, Clayton, R.K., and Sistrom, W.R., eds (Plenum Publishing Corp., New York), pp. 455–469.

Pena-Sandoval, G.R., Kwon, O., and Georgellis, D. (2005). Requirement of the receiver and phosphotransfer domains of ArcB for efficient dephosphorylation of phosphorylated ArcA *in vivo*. J. Bacteriol. *187*, 3267–3272.

de Philip, P., Batut, J., and Boistard, P. (1990). *Rhizobium meliloti* FixL is an oxygen sensor and regulates *R. meliloti nifA* and *fixK* genes differently in *Escherichia coli*. J. Bacteriol. *172*, 4255–4262.

Pitcher, R.S., and Watmough, N.J. (2004). The bacterial cytochrome cbb_3 oxidases. Biochim. Biophys. Acta *1655*, 388–399.

Pitcher, R.S., Brittain, T., and Watmough, N.J. (2002). Cytochrome cbb(3) oxidase and bacterial microaerobic metabolism. Biochem. Soc. Trans. *30*, 653–658.

Podust, L.M., Ioanoviciu, A., and Ortiz de Montellano, P.R. (2008). 2.3 A X-ray structure of the heme-bound GAF domain of sensory histidine kinase DosT of *Mycobacterium tuberculosis*. Biochemistry *47*, 12523–12531.

Portnoy, V.A., Scott, D.A., Lewis, N.E., Tarasova, Y., Osterman, A.L., and Palsson, B.O. (2010). Deletion of genes encoding cytochrome oxidases and quinol monooxygenase blocks the aerobic-anaerobic shift in *Escherichia coli* K-12 MG1655. Appl. Env. Microbiol. *76*, 6529–6540.

Purcell, E.B., Siegal-Gaskins, D., Rawling, D.C., Fiebig, A., and Crosson, S. (2007). A photosensory two-component system regulates bacterial cell attachment. Proc. Natl. Acad. Sci. U.S.A. *104*, 18241–18246.

Purcell, E.B., McDonald, C.A., Palfey, B.A., and Crosson, S. (2010). An analysis of the solution structure and signaling mechanism of LovK, a sensor histidine kinase integrating light and redox signals. Biochemistry *49*, 6761–6770.

Reddie, K.G., and Carroll, K.S. (2008). Expanding the functional diversity of proteins through cysteine oxidation. Curr. Opin. Chem. Biol. *12*, 746–754.

Roberts, D.M., Liao, R.P., Wisedchaisri, G., Hol, W.G., and Sherman, D.R. (2004). Two sensor kinases contribute to the hypoxic response of *Mycobacterium tuberculosis*. J. Biol. Chem. *279*, 23082–23087.

Robson, R.L., and Postgate, J.R. (1980) Oxygen and hydrogen in biological nitrogen fixation. Annu. Rev. Microbiol. *34*, 183–207.

Rodriguez, C., Kwon, O., and Georgellis, D. (2004). Effect of D-lactate on the physiological activity of the ArcB sensor kinase in *Escherichia coli*. J. Bacteriol. *186*, 2085–2090.

Roh, J.H., and Kaplan, S. (2000). Genetic and phenotypic analyses of the *rdx* locus of *Rhodobacter sphaeroides* 2.4.1. J. Bacteriol. *182*, 3475–3481.

Saini, D.K., Malhotra, V., and Tyagi, J.S. (2004). Cross talk between DevS sensor kinase homologue, Rv2027c, and DevR response regulator of *Mycobacterium tuberculosis*. FEBS Lett. *565*, 75–80.

Salomon, M., Eisenreich, W., Dürr, H., Schleicher, E., Knieb, E., Massey, V., Rüdiger, W., Müller, F., Bacher, A., and Richter, G. (2001). An optomechanical transducer in the blue light receptor phototropin from *Avena sativa*. Proc. Natl. Acad. Sci. U.S.A. *98*, 12357–12361.

Sciotti, M.A., Chanfon, A., Hennecke, H., and Fischer, H.M. (2003). Disparate oxygen responsiveness of two regulatory cascades that control expression of symbiotic genes in *Bradyrhizobium japonicum*. J. Bacteriol. *185*, 5639–5642.

Sganga, M.W.., and Bauer, C.E. (1992). Regulatory factors controlling photosynthetic reaction center and light-harvesting gene expression in *Rhodobacter capsulatus*. Cell *68*, 945–954.

Shroff, N.P., Charania, M.A., and Saffarini, D.A. (2010). ArcB1, a homolog of *Escherichia coli* ArcB, regulates dimethyl sulfoxide reduction in *Shewanella oneidensis* MR-1. J. Bacteriol. *192*, 3227–3230.

Smart, J.L., Willett, J.W., and Bauer, C.E. (2004). Regulation of hem gene expression in *Rhodobacter capsulatus* by redox and photosystem regulators RegA, CrtJ, FnrL, and AerR. J. Mol. Biol. *342*, 1171–1186.

Sousa, E.H., Tuckerman, J.R., Gonzalez, G., and Gilles-Gonzalez, M.A. (2007). DosT and DevS are oxygen-switched kinases in *Mycobacterium tuberculosis*. Protein Sci. *16*, 1708–1719.

Stewart, R.C. (2010). Protein histidine kinases: assembly of active sites and their regulation in signaling pathways. Curr. Opin. Microbiol. *13*, 133–141.

Sugiura, M., Aiba, H., and Mizuno, T. (2003). Identification and classification of two-component systems that affect *rpoS* expression in *Escherichia coli*. Biosci. Biotech. Biochem. *67*, 1612–1615.

Swem, D.L., Swem, L.R., Setterdahl, A., and Bauer, C.E. (2005) Involvement of SenC in assembly of cytochrome *c* oxidase in *Rhodobacter capsulatus*. J. Bacteriol. *87*, 8081–8087.

Swem, L.R., Elsen, S., Bird, T.H., Swem, D.L., Koch, H.G., Myllykallio, H., Daldal, F., and Bauer, C.E. (2001). The RegB/RegA two-component regulatory system controls synthesis of photosynthesis and respiratory electron transfer components in *Rhodobacter capsulatus*. J. Mol. Biol. *309*, 121–138.

Swem, L.R., Kraft, B.J., Swem, D.L., Setterdahl, A.T., Masuda, S., Knaff, D.B., Zaleski, J.M., and Bauer, C.E. (2003). Signal transduction by the global regulator RegB is mediated by a redox-active cysteine. EMBO J. *22*, 4699–4708.

Swem, L.R., Gong, X., Yu, C.A., and Bauer, C.E. (2006). Identification of a ubiquinone-binding site that affects autophosphorylation of the sensor kinase RegB. J. Biol. Chem. *281*, 6768–6775.

Taylor, B.L., and Zhulin, I.B. (1999). PAS domains: internal sensors of oxygen, redox potential, and light. Microbiol. Mol. Biol. Rev. *63*, 479–506.

Tuckerman, J.R., Gonzalez, G., Dioum, E.M., and Gilles-Gonzalez, M.A. (2002). Ligand and oxidation-state specific regulation of the heme-based oxygen sensor FixL from *Sinorhizobium meliloti*. Biochemistry *41*, 6170–6177.

Vichivanives, P., Bird, T.H., Bauer, C.E., and Tabita, R.F. (2000). Multiple regulators and their interactions *in vivo* and *in vitro* with the *cbb* regulons of *Rhodobacter capsulatus*. J. Mol. Biol. *300*, 1079–1099.

Wong, S.M., Alugupalli, K.R., Ram, S., and Akerley, B.J. (2007). The ArcA regulon and oxidative stress resistance in *Haemophilus influenzae*. Mol. Microbiol. *64*, 1375–1390.

Wu, J., and Bauer, C.E. (2007). RegB/RegA, a global redox-responding two-component system. In Bacterial Signal Transduction: Network and Drug Targets, Utsumi, R., ed. (Landes Bioscience Eurekah, Georgetown, Washington, DC), pp. 131–148.

Wu, J., and Bauer, C.E. (2010). RegB kinase activity is controlled in part by monitoring the ratio of oxidized to reduced ubiquinones in the ubiquinone pool. MBio. *1*, e00272–10.

Yukl, E.T., Ioanoviciu, A., de Montellano, P.R., and Moënne-Loccoz, P. (2007). Interdomain interactions within the two-component heme-based sensor DevS from *Mycobacterium tuberculosis*. Biochemistry *46*, 9728–9736.

Zhou, Y., Gottesman, S., Hoskins, J.R., Maurizi, M.R., and Wickner, S. (2001). The RssB response regulator directly targets sigma(S) for degradation by ClpXP. Genes Dev. *15*, 627–637.

BvgS of Pathogenic Bordetellae: A Paradigm for Sensor Kinases with Venus Flytrap Perception Domains

4

Françoise Jacob-Dubuisson, René Wintjens, Julien Herrou, Elian Dupré and Rudy Antoine

Abstract

The whooping cough agent *Bordetella pertussis* regulates the expression of its virulence regulon through the two-component system BvgAS. BvgA is a canonical response regulator serving as a transcriptional activator when phosphorylated. BvgS is a multidomain, hybrid sensor kinase harbouring several cytoplasmic domains that mediate a complex phospho-transfer cascade. BvgS also contains two periplasmic 'Venus flytrap' (VFT) domains in tandem. It is thus the prototype for a large family of bacterial VFT-containing sensor kinases, whose molecular mechanisms for signal perception and transduction remain to be deciphered. Ubiquitous in nature, VFT domains usually function along a 'clamshell' model, with two lobes that enclose specific ligands. Structure–function analyses of the second VFT domain of BvgS have indicated a non-canonical behaviour, whereby this domain adopts a closed conformation that gives a positive signal to the system in the absence of a bona fide ligand and can also bind negative signals that turn off the system. Sequence analyses of BvgS have shown a strong conservation of the region linking the periplasmic and cytoplasmic domains, indicating that it is essential for signal transduction. On the contrary, the linkers between VFT domains are not conserved, and thus the VFT most likely communicate with each other via their interfaces.

The Bvg regulon: role of Bvg in *Bordetella* physiology

Bordetellae are Gram-negative cocco-bacilli, which colonize the respiratory tract of a variety of hosts. *Bordetella bronchiseptica* causes chronic and sometimes asymptomatic infections in a number of mammals. This species is thought to be closely related to the ancestor of human-restricted *Bordetellae*, including *Bordetella pertussis*, the whooping cough agent, and *Bordetella parapertussis*, which causes milder human whooping cough or infections of sheep (Parkhill *et al.*, 2003; Preston *et al.*, 2004; Diavatopoulos *et al.*, 2005). These three species form the so-called *B. bronchiseptica* complex. Pathogenic *Bordetellae* have developed an infection strategy that relies on the production of a number of virulence factors including several toxins and adhesins (Gerlach *et al.*, 2001; Locht *et al.*, 2001). Expression of almost all known virulence factors is under the control of the master regulatory two-component system BvgAS (Fig. 4.1). In the Bvg regulon, genes that are positively regulated by BvgAS are called '*vags*' for virulence-activated genes. These are expressed in the so-called Bvg$^+$ phase. Over 100 genes in *B. pertussis* and more than 200 in *B. bronchiseptica* have been found to belong to this category (Cummings *et al.*, 2006). In the core regulon are found nearly all the genes coding for the known virulence factors such as toxins and adhesins as well as their assembly and secretion machineries, but also genes involved in iron acquisition, genes coding for cell envelope proteins, for folding factors, regulators etc. (Cummings *et al.*, 2006). The Bvg$^+$ phase is necessary and sufficient for initiation and development of the infectious cycle of *B. pertussis* and its relatives (Martinez de Tejada *et al.*, 1998).

Conversely, genes maximally expressed in Bvg$^-$ conditions are called '*vrgs*' for virulence-repressed genes. The set of negatively controlled genes is much more diverse between species of

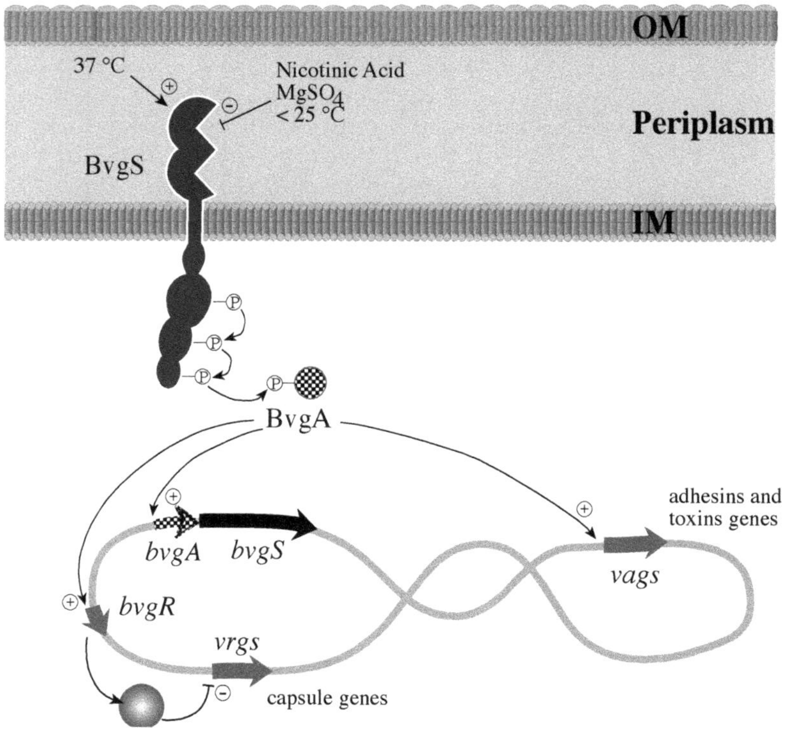

Figure 4.1 Regulation of *Bordetella* virulence by the two-component system BvgAS. Schematic representation of the bacterial chromosome with the various loci involved in virulence regulation. A single positive signal – a temperature of 37°C – has been identified so far. In the Bvg$^+$ phase, BvgS autophosphorylates and transfers the phosphoryl group to BvgA. Phosphorylated BvgA activates the transcription of the virulence-activated genes or *vags* (see text for details). *bvgR* is a *vag* whose product, BvgR, indirectly represses the virulence-repressed genes (*vrgs*). Several signals have been identified that inactivate the phosphorylation cascade in laboratory conditions, including nicotinate and sulfate ions. In the Bvg$^-$ phase, the *vags* are not transcribed. The absence of BvgR enables the *vrgs* to be expressed at a higher level. The chemical signals to which the system responds during the natural infection remain to be identified.

the *B. bronchiseptica* complex than the *vags*, most likely reflecting different capacities to adapt to life outside a mammalian host (Cotter and Miller, 1994; Martinez de Tejada *et al.*, 1998; Cummings *et al.*, 2006). Unlike *B. pertussis*, *B. bronchiseptica* is motile and able to survive for extended periods of time in the environment, and the Bvg$^-$ phase is required for *ex-vivo* adaptation to nutrient-limiting conditions (Porter and Wardlaw, 1993; Cotter and Miller, 1994). Notably chemotaxis, the production of a flagellum and the activation of metabolic and transporter genes are associated with the Bvg$^-$ phase in *B. bronchiseptica* (Cummings *et al.*, 2006). *B. pertussis* has much fewer *vrgs*, coding notably for outer membrane proteins and for the biosynthesis of a capsule (Stenson and Peppler, 1995; Antoine *et al.*, 2000; Cummings *et al.*, 2006; Neo *et al.*, 2010). Expression of the *vrgs* is minimal in the Bvg$^+$ phase, suggesting that their production is detrimental for virulence. This is corroborated by the observation that ectopic expression of the flagellum hampers infection by *B. bronchiseptica* in an animal model (Akerley *et al.*, 1995).

The role of the Bvg$^-$ phase for the *B. pertussis* biology remains poorly understood. It has been proposed to be an evolutionary remnant that has become obsolete for *B. pertussis*'s present lifestyle (Cotter and Jones, 2003). In agreement with this hypothesis, mutant bacteria blocked in the Bvg$^+$ phase because of specific mutations in BvgS ('phase-locked mutants', see below) are not defective in animal models (Cotter and Miller, 1994; Martinez de Tejada *et al.*, 1998). However, the inappropriate expression of *vrgs* was also shown to be detrimental to infection in the same models (Akerley *et al.*, 1995; Martinez de Tejada *et al.*, 1998). Nevertheless, it should be noted that such models do not fully reproduce the human infection, and thus the question of the role of the Bvg- phase for *B. pertussis* remains unresolved. It has been proposed that the Bvg^- phase might be advantageous for survival at some stage of infection, e.g. to enable the bacterium to evade the immune system (Masure, 1992).

Under laboratory conditions, BvgAS responds to environmental signals; this has been called antigenic modulation (Lacey, 1960). Thus, a temperature of 37°C activates the autophosphorylation activity of the sensor histidine-kinase BvgS. Activation of the kinase activity of BvgS triggers a complex phosphotransfer cascade ultimately resulting in the phosphorylation of the response regulator, BvgA. Phosphorylated BvgA (denoted BvgA~P) binds to the promoters of its target genes, the *vags*, serving as a transcriptional activator (Fig. 4.1). The bacterium is thus in the Bvg^+ phase. One of the targets of phosphorylated BvgA is *bvgR*, encoding a repressor of the *vrgs* (Merkel *et al.*, 1998).

Conversely, lower temperatures (< 25°C) switch off the kinase activity of BvgS, as do certain chemical signals used in the laboratory, including high millimolar concentrations of inorganic ions such as SO_4^{2-}, and organic acids such as nicotinic acid and related compounds (Lacey, 1960; Schneider and Parker, 1982; Melton and Weiss, 1993). The bacterium thus switches to the Bvg^- phase. No activating chemical signals have been identified thus far, and the agonistic or antagonistic chemical signals perceived by the bacterium in the host remain to be determined.

The target genes of BvgA have different promoter architectures. The number and spatial organization of high-affinity (primary) and low-affinity (secondary) binding sites for BvgA~P dimers determines the responsiveness of each target promoter to phosphorylated BvgA (Boucher *et al.*, 2001a,b; Jones *et al.*, 2005). Both the *fhaB-fim* operon encoding major adhesins and the *bvgAS* operon require only low concentrations of BvgA~P for activation. Therefore, they are the first to be transcribed upon a shift from the Bvg^- phase to the Bvg^+ phase and have been termed 'early *vags*' (Scarlato *et al.*, 1991; Prugnola *et al.*, 1995). In contrast, the *ptx-ptl* and *cya* operons coding for toxins and their secretion machineries are activated later, after a significant amount of BvgA~P has accumulated (Prugnola *et al.*, 1995; Steffen *et al.*, 1996; Cotter and Jones, 2003). Temporal expression of the promoters has also been observed in an animal model of infection (Veal-Carr and Stibitz, 2005). This temporal regulation is likely to be relevant in the course of infection, because adhesins mediate the first steps of colonization, while subversion of the immune functions by toxins comes into play somewhat later. In support of the importance of this regulation, it was demonstrated that inverting the temporal expression patterns of early and late genes affected *B. pertussis's* pathogenicity in an animal model of infection (Kinnear *et al.*, 2001). An additional class of genes, called intermediate, are transcribed at low concentrations of BvgA~P and repressed at high concentrations, because the occupancy of low-affinity BvgA binding sites in the gene-coding region inhibits transcription (Cotter and Miller, 1997; Deora, 2002; Williams *et al.*, 2005). This is the case of *bip*, which codes for an intimin-like protein (Deora, 2002; Cotter and Jones, 2003; Mishra and Deora, 2005). Biofilm formation also is maximal in this intermediate phase, called Bvg^i (Irie *et al.*, 2004). It has been proposed that the Bvg^i phase might play a role under conditions found in specific microenvironments in the course of infection. Alternatively, it might be required for specific steps during infection, such as aerosol transmission between hosts (Cotter and Miller, 1994, 1997; Cotter and Jones, 2003). Altogether, BvgAS operates as a rheostat rather than as an 'on–off' switch system, determining several gene expression states which might correspond to temporally or spatially defined patterns in the course of infection.

Anatomy of the phosphotransfer cascade

BvgA is a typical response regulator of the FixJ/NarL family, with a LuxR-type helix–turn–helix (HTH) DNA-binding domain (pfam: PF00196) (Galperin, 2010). In addition to its C-terminal HTH 'output' domain, BvgA also harbours a receiver domain with a conserved aspartate residue, Asp54, serving as the acceptor site for phosphorylation. BvgA forms dimers in solution via its C-terminal domain, independent of its phosphorylation status (Beier *et al.*, 1995; Perraud *et al.*, 2000). Phosphorylation of the receiver domain by BvgS activates the binding activity of BvgA's HTH domain through conformational changes in the receiver domains that are transmitted to the HTH domains. The BvgA~P dimers bind to target promoters in a cooperative manner, and an imperfect heptad repeat upstream of the −35 sequence box has been identified as BvgA~P's primary binding site (Roy and Falkow, 1991; Boucher *et al.*, 2001b). Cooperative binding to secondary sites downstream is highly dependent on binding to the primary site and leads to the formation of a stable ternary BvgA~P-DNA-RNA polymerase complex (Stibitz, 1994; Boucher *et al.*, 1997; Stibitz, 1998; Jones *et al.*, 2005).

BvgS is a 135-kDa cytoplasmic membrane protein that contains a large periplasmic moiety (see below) separated from a large cytoplasmic domain by a transmembrane segment (Fig. 4.2). It is called a 'hybrid' (or unorthodox) sensor kinase, because it harbours additional cytoplasmic domains downstream of a bona fide His-kinase domain. Thus, rather than being transferred directly from the 'transmitter' domain of BvgS, which harbours a typical dimerization-His phosphotransfer (DHp) module and a catalytic-ATP binding module (CA), to the receiver domain of BvgA, the phosphoryl group travels along a complex phosphorelay comprising an Asp-containing receiver domain and a His phosphotransfer (HPt) domain. The residues crucial for the phosphorelay have been identified as His729, Asp1023 and His1172 in the kinase, receiver and HPt domains of BvgS, respectively (Uhl and Miller, 1994, 1996a,b). Note that this numbering starts at the initiator Met. BvgS initially autophosphorylates from ATP on residue His729, and then the phosphoryl group is transferred to Asp1023 and from there to His1172 (Uhl and Miller, 1994, 1996a,b). The last two steps of phosphotransfer can occur within each monomer, in cis, or between monomers, in trans (Beier *et al.*, 1995; Uhl *et al.*, 1996a,b). The phosphorylated HPt domain is the phosphodonor for BvgA's Asp54. Deletion experiments have indicated that all three domains of BvgS (kinase, receiver and HPt) are required for BvgA phosphorylation, because a direct transfer of the phosphoryl group from His729 in the kinase domain to Asp54 in BvgA cannot occur (Uhl and Miller, 1994). This stands in contrasts with other hybrid sensor kinases, such as ArcB and VirA for instance, where the additional domains are dispensable and rather serve regulatory functions (Pazour *et al.*, 1991; Chang and Winans, 1992; Iuchi and Lin, 1992). In BvgS, the HPt domain is the essential determinant of the specificity of BvgS for BvgA. Thus, unlike the BvgS receiver domain, the HPt domain cannot be replaced by a homologous domain from another hybrid sensor kinase (Beier *et al.*, 1995; Perraud *et al.*, 1998). On the part of BvgA, specificity for BvgS is provided by BvgA's receiver domain.

The two-component sensor proteins dimerize via the DHp module of their 'transmitter', i.e. histidine kinase domain, and both *cis* and *trans* phosphorylation exist, depending on the system (Swanson *et al.*, 1993; Stock, 1996; Marina *et al.*, 2005; Casino *et al.*, 2009). The cytoplasmic portion of BvgS, and more specifically the transmitter domain, was shown to homodimerise *in vitro* independently of phosphorylation (Beier *et al.*, 1995; Dupré, unpublished). *In vivo* complementation experiments using partial merodiploids of *bvgS* carrying different mutations of the conserved motifs indicated that phosphorylation can take place *in trans*, corroborating the idea that the cytoplasmic domain dimerizes (Beier *et al.*, 1995). Dimerization is also supported by *in vitro* experiments showing that phosphotransfer can occur between BvgS domains produced separately (Perraud *et al.*, 1998, 2000).

No specific accessory protein with a phosphatase activity towards BvgS~P or BvgA~P has been identified thus far. However, BvgS itself appears to be subject to auto-dephosphorylation (Uhl and Miller, 1996a). Recent structural and biochemical works have indicated that dephosphorylation in classical two-component systems involves the transmitter domain acting on the phospho-response regulator (Yamada *et al.*, 2009; Casino *et al.*, 2010; Huynh *et al.*, 2010). According to this scenario, in the phosphotransfer reaction the Asp residue of the receiver domain attacks the His~P bond, while in the dephosphorylation reaction specific residues of the kinase activate a water molecule to attack the Asp~P bond (Huynh *et al.*,

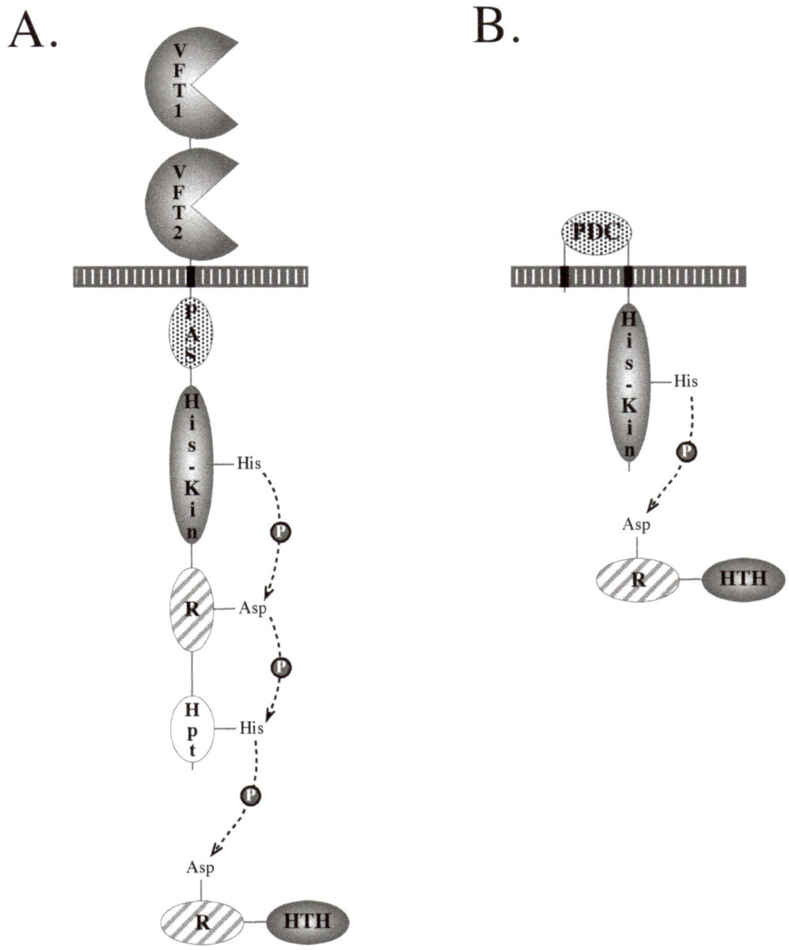

Figure 4.2 Anatomy of BvgAS and comparison with 'classical' two-component systems. BvgS harbours three putative perception domains, two Venus flytrap (VFT) periplasmic domains in tandem and a cytoplasmic Per-ARNT-Sim (PAS) domain. Unlike classical sensor kinases of two-component systems it most likely contains a single transmembrane domain (see text). The cytoplasmic portion harbours several domains that participate in the phosphorylation cascade. The histidine-kinase domain (denoted His-Kin) is actually composed of a dimerization His-containing domain and a catalytic domain, like classical two-component sensor kinases. Following autophosphorylation of the kinase domain on its conserved His residue, the phosphoryl group is first transferred to an Asp-containing receiver domain (R) and then to a histidine phosphotransfer domain (HPt). The latter domain finally phosphorylates the receiver domain of BvgA. The complexity of the phosphorylation cascade makes BvgS a hybrid sensor kinase. Note that in some hybrid systems the additional modules, receiver and HPt, form separate proteins. The receiver domain of BvgS is homologous to receiver domains found in response regulators, hence the use of the same symbols for both. B. Classically, two-component sensor kinases are composed of a single perception domain called PDC for the names of three model kinases (PhoP, DcuS and CitA). PDC domains have a fold very similar to that of the cytoplasmic PAS domain. The PDC domain is flanked by two transmembrane segments. The cytoplasmic portion of classical sensor kinases is composed of a kinase domain similar to that of BvgS, and the phosphoryl group is passed directly from the His of the kinase to the Asp of the response regulator's receiver domain. Activating chemical signals have been identified for several classical two-component kinases.

2010). Subtle movements of the His-containing dimerization domain of the kinase (DHp) brought about by signalling from upstream domains would determine the relative positions of the catalytic, DHp and receiver domains relative to each other and thus which activity – transferase, kinase or phosphatase – predominates in a given situation (Casino *et al.*, 2010). However, the scenario might not apply to complex phosphorelays such as that of BvgS, where a 'reverse signal decay' has been proposed (Huynh and Stewart, 2011). Thus, the Asp residue of the receiver domain appears to play an essential part for the dephosphorylation both of the His-P of the transmitter and of His-P of the HPt domain. After transfer of the phosphoryl group to the receiver from either the transmitter or the HPt domains, the phosphorylated receiver rapidly undergoes hydrolysis (Uhl and Miller, 1996a,b). In addition, both the receiver domain of BvgS and the receiver domain of the response regulator BvgA have been shown to be able to autophosphorylate from acetyl-phosphate *in vitro*, unlike the HPt domain, thus bringing additional sophistication to the system (Perraud *et al.*, 2000).

The complexity of the phosphorylation cascade probably contributes to endow the BvgAS system with a rheostat-like behaviour rather than an on/off mode of action. Altogether, the rheostat-like response of BvgAS to a range of environmental conditions – temperature, modulator concentrations, etc. – is thus most likely provided by several factors, including the presence of several perception domains in BvgS (see below), a sophisticated phosphorelay and the architecture of the target promoters.

Perception domains of BvgS

BvgS harbours three domains potentially involved in signal sensing, and although the physiological signals to which it responds in its natural environments remain unclear, genetic evidences indeed indicate that BvgS, or more specifically these domains are directly involved in perception. For instance, the majority of sequence differences between the BvgS proteins of pathogenic bordetellae of the *B. bronchiseptica* complex are located in the two periplasmic domains, and the sensitivity to modulation is affected by the shuffling of these domains between species (Martinez de Tejada *et al.*, 1996; Herrou *et al.*, 2009). In addition, various large deletions or small insertions in the periplasmic domains were found to inactivate BvgS (Miller *et al.*, 1992). Conversely, a number of spontaneous point mutations in the first portion of the cytoplasmic domain – that corresponds to a PAS domain and flanking sequences (see below) – render the system unresponsive to the negative modulators nicotinate and sulfate ions (Miller *et al.*, 1992; Goyard *et al.*, 1994; Manetti *et al.*, 1994). These mutations are dominant over the inactivating mutations located in the periplasmic domains described above (Miller *et al.*, 1992). More recently, we also identified a point mutation in the second periplasmic domain that abolishes BvgS activity, and other mutations in the same domain that modify its responsiveness to modulation (Herrou *et al.*, 2009; Herrou *et al.*, 2010).

BvgS harbours a cytoplasmic domain located below the transmembrane segment, which has all the features of a PAS domain (Periodic circadian protein- Aryl hydrocarbon receptor nuclear translocator (ARNT)- Single minded protein) (Fig. 4.2). PAS domains are found in many proteins in the three kingdoms of life, from eubacteria and archaea to higher eukaryotes, and they are involved in the sensing of light, redox potentials, diatomic gases, xenobiotics or other metabolic conditions (Ponting and Aravind, 1997; Zhulin *et al.*, 1997; Taylor *et al.*, 1999; Moglich *et al.*, 2009). PAS domains are mostly cytoplasmic, and some of them require a cofactor – e.g. haem, flavine adenine nucleotide or another chromophore – for activity (Gilles-Gonzalez *et al.*, 1994; Taylor *et al.*, 1999). In prokaryotes, PAS domains are frequently found in sensor kinases of two-component systems. In one of these, up to ten PAS domains in tandem have been predicted (Microbial Signal transduction Mist2 database; Ulrich and Uhlin, 2010). The exact function of the BvgS PAS domain is not known, and there is no evidence as yet for the presence of a cofactor. Nevertheless, the mutations mentioned above clearly point to its pivotal role for signal perception and/or transduction.

In the periplasm, BvgS harbours two homologous domains in tandem that belong to the superfamily of Venus flytrap proteins (Quiocho and Ledvina, 1996). In bacteria, VFT domains are notorious as periplasmic solute-binding proteins (PBP, also known as extracytoplasmic solute receptors or ESR) that are involved in the import of small molecules via various systems, including ATP-binding cassette transporters (ABC), tripartite ATP-independent porters (TRAP), or tripartite tricarboxylic acid transporters (TTT) (Kelly and Thomas, 2001; Winnen *et al.*, 2003; Locher, 2004). VFT modules are also found in transcriptional regulators such as LacR (Friedman *et al.*, 1995) and PurR (Schumacher *et al.*, 1994) and in a number of eukaryotic receptors (Felder *et al.*, 1999) (see below). These domains are composed of two globular α/β lobes that enclose a solute-binding cavity (Quiocho and Ledvina, 1996). The two lobes articulate around a hinge, enabling typical VFT domains to switch between open and closed conformations by clamshell motions. Solute binding in the cavity tends to stabilize the closed conformation (Oh *et al.*, 1993, 1994; Bjorkman and Mowbray, 1998; Trakhanov *et al.*, 2005). The closest PBP homologues of the two BvgS VFTs are amino acid-binding PBPs such as the glutamine-binding protein of *E. coli* (Herrou *et al.*, 2009).

The fusion between one or several periplasmic VFT domains and a cytoplamic His-kinase moiety, as found in BvgS, represents an evolutionary solution for sensory transduction purposes that couples the ligand-binding capacity of VFT domains to an enzymatic activity that can be regulated. In these systems, VFT domains of sensor kinases are most likely involved in physical interactions with small molecules, similar to the other types of periplasmic sensor domains found in 'classical' His-kinases such as CitA of *Klebsiella pneumoniae* (Sevvana *et al.*, 2008), DcuS (Etzkorn *et al.*, 2008) and others (Zhang and Hendrickson, 2010). An alternative solution to sensory transduction making use of VFT domains is their recruitment as separate sensor modules in other types of two-component systems. Thus, in the LuxPQ system of *Vibrio harveyi* involved in quorum sensing, the periplasmic VFT protein LuxP interacts with the periplasmic domain of the sensor kinase LuxQ, which by itself perceives no signal. The binding by LuxP of the auto-inducer molecule AI-2 triggers a rearrangement of the PQ complex that leads to the asymmetric dimerization of two LuxPQ units. This conformational change is transmitted to the cytoplasmic moiety of LuxQ to inactivate its kinase (Neiditch *et al.*, 2005, 2006).

BvgS is a model for a large family of VFT-containing sensor kinases

One of the most original and least investigated features of BvgS compared with classical sensor kinases such as EnvZ or PhoP is the presence of tandem periplasmic VFT domains for signal perception. A handful of other sensor kinases have been described that share these characteristics, including EvgS of *Escherichia coli*, KvgS of *Klebsiella pneumoniae*, AstS *of Photorhabdus luminescens*, VieS of *Vibrio cholerae*, and RocS1 of *Pseudomonas aeruginosa* (Masuda and Church, 2002; Nishino and Yamaguchi, 2002; Eguchi *et al.*, 2003; Lai *et al.*, 2003; Derzelle *et al.*, 2004; Kulasekara *et al.*, 2005; Martinez-Wilson *et al.*, 2008).

In silico analyses of all the bacterial genomes sequenced up to February 2011 have revealed that such sensor kinases are very abundant in the eubacterial world. Thus, out of 1708 bacterial genomes whose sequences were fully available at that date, 419 putative VFT-containing sensor kinases were found in 272 genomes, including α, β, γ, δ and ε Proteobacteria, Archaea and Gram-positive bacteria. The largest phylogenetic group by far was that of γ Proteobacteria, but this largely reflects the fact that numerous *Escherichia coli* and *Pseudomonas* genomes have been sequenced. Some bacterial genomes harbour several genes coding for VFT-containing sensor kinases.

Among all of these, VFT-containing sensor kinases of *Pseudomonas* are those most closely related to BvgS. The closest homologue of BvgS identified thus far is PFLU1739 of *Pseudomonas fluorescens* SBW25. Its domain organization is identical to that of BvgS, and the periplasmic moieties of the two proteins share 41% sequence identity.

The cytoplasmic portions of all of those putative sensor kinases are quite varied; they contain different types of domains in addition to a bona fide His-kinase domain, including PAS, GAF (cGMP-specific phosphodiesterases, adenylyl cyclases and FhlA), receiver and HKsen domains (Fig. 4.3). HKsen is a newly identified type of domain found associated with histidine-kinases, for which no structure is yet available (domain

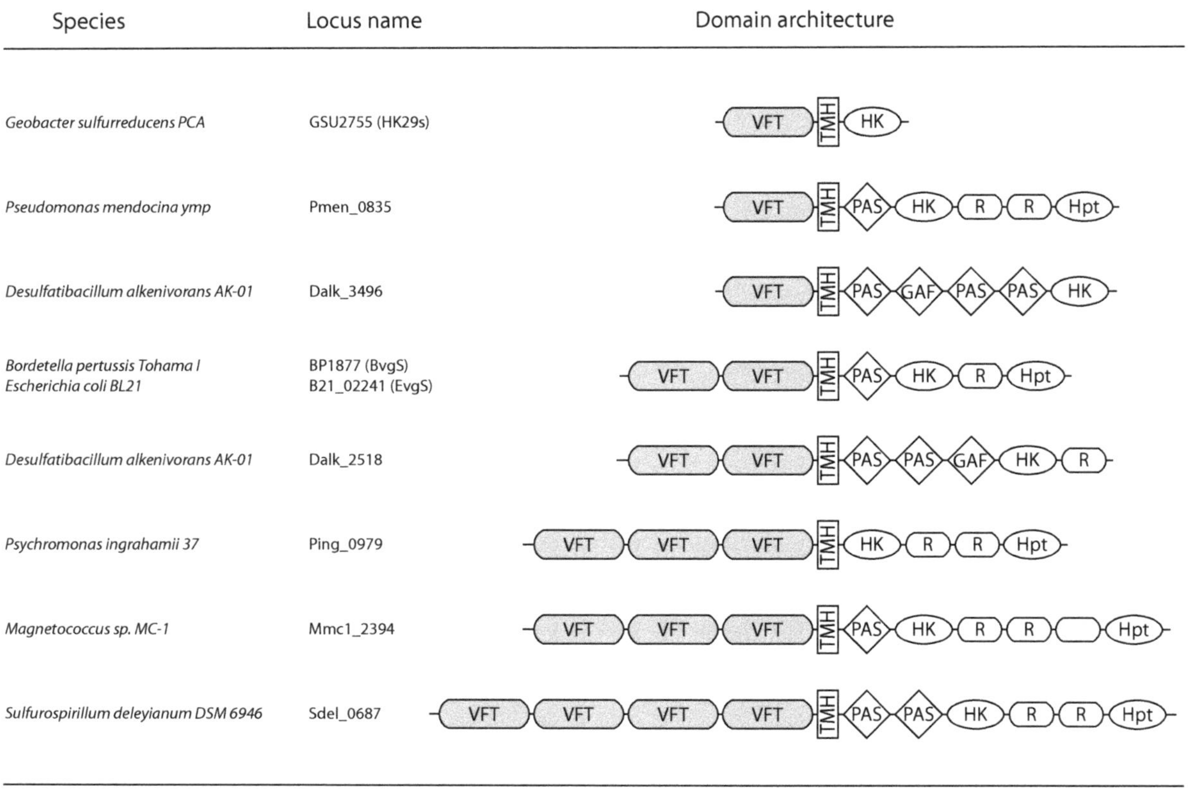

Figure 4.3 Diversity of domain composition in Venus flytrap domain-harbouring sensor kinases. Selected examples are shown. In the superfamily, 30% of the sensor kinases have a single VFT domain, and only 4% have three VFT domains or more. Note that most of these proteins are solely predicted based on genomic sequences. The various types of cytoplasmic domains are denoted by distinct symbols, with the open oval representing a non-characterized domain. TMH represents a transmembrane helix and R a receiver domain.

Hist-Kin-Sens, Pfam: PF10114). Thus, both orthodox and hybrid (or unorthodox) kinases have been coupled with VFT domains in the course of evolution to generate the class of VFT-containing sensor kinases.

VFT domains have been classified based on the topology of the polypeptide chain and notably on the structure of the hinge between the two domains of the clamshell (Quiocho and Ledvina, 1996). Three classes of VFT domains exist, with either three short β-strand segments connecting the two lobes (class 1), two segments (class 2) or a long α helix (class 3) (Ho *et al.*, 2007). The VFT domains of BvgS belong to the class 2 PBPs (see below). Most VFT domains found in His-kinases belong to the Pfam: PF00497 family, which harbours amino acid-binding PBPs.

A large proportion of all predicted VFT-containing sensor kinases, including all 23 found in Gram-positive bacteria and the two proteins found in Archaea comprise a single VFT domain. Remarkably, however, many other sensor kinases have two VFT domains in tandem, similar to BvgS, and a few of them have up to four such domains in a row (Fig. 4.3). Such architecture with several periplasmic VFT domains in tandem potentially enables the bacterium to perceive several different environmental signals, both positive and negative, and to integrate them to produce the appropriate response. Thus, a given combination of signals would determine the level of activation of the kinase downstream. This architecture of several VFT in tandem also raises the question of the mechanisms through which the signals perceived by the most N-terminal VFT domain are transmitted through the following VFT(s) to finally reach the membrane and the kinase domain underneath. In other words, the manner by which the VFT domains communicate between them will be critical to understand the molecular mechanisms of this class of sensor kinases.

The presence of several VFT domains in

tandem in many putative or confirmed sensor kinases casts doubt on whether they really adopt a topology similar to that of the classical sensor kinase model (Fig. 4.2), with two transmembrane segments flanking the periplasmic sensor domain – called extracytoplasmic in the case of Gram-positive bacteria. BvgS and its homologues have most often been represented in this manner. However, one early article on the BvgAS system reported that the periplasmic domain of BvgS is not N-terminally anchored (Stibitz and Yang, 1991). Thus, a translational fusion linking the N-terminal sequence of BvgS to alkaline phosphatase indicated that the chimeric protein was soluble rather than membrane associated, arguing in favour of a cleaved signal peptide instead of an N-terminal transmembrane segment. This is consistent with the N-terminal sequence of BvgS, which has all the features of a signal peptide, including an 'Ala-Gln-Ala' signal peptidase cleavage site.

Analyses of the N-terminal sequences of the two-VFT-containing sensor kinases have revealed that in the overwhelming majority of cases, a signal peptide rather than an uncleaved transmembrane segment is predicted. Similarly, putative sensor kinases harbouring three or four VFTs appear to have a cleavable N-terminal signal peptide. However, the situation is less-clear-cut for sensor kinases harbouring a single VFT domain. In many of them, no signal peptide cleavage site is predicted, indicating that the N-terminal sequence might correspond to a bona fide transmembrane segment. In particular, this appears to be the case of all the VFT-containing sensor kinases predicted in Gram-positive bacteria (our unpublished observations). Whether these single-VFT-containing histidine kinases really adopt a topology distinct from that of their larger homologues remains to be addressed.

Functions of other VFT-containing sensor kinases

Only a few VFT-containing sensor kinases have been investigated so far, which are described below. Efforts have been made in particular to define the conditions of expression and the functions of the two-component – or sometimes three-component – systems of which they are part in their respective bacteria. In contrast, the signals to which they respond remain largely unknown.

EvgS was found in a large number of *E. coli* isolates, and the sequences of *evgS* from approximately 70 isolates are currently available. Its architecture is identical to that of BvgS (Utsumi *et al.*, 1994). At least 37 genes were found to be dependent on EvgAS, including stationary phase acid resistance genes and multidrug efflux pumps (Eguchi *et al.*, 2003, 2004; Masuda and Church, 2002). No information is currently available on molecules to which EvgS responds, other than it is sensitive to the same modulators as BvgS (Utsumi *et al.*, 1994).

Another enterobacterial species, *Klebsiella pneumoniae*, produces a BvgS/EvgS homologue called KvgS (Lai *et al.*, 2003). A *kvgS* mutant was attenuated in a mouse model of lung infection, indicating a role of the KvgAS two-component system for the virulence of this opportunistic pathogen (Lawlor *et al.*, 2005). The *kvgS* gene appears to be up-regulated by iron limitation and oxidative stress, conditions that the bacterium may encounter in its hosts (Lai *et al.*, 2003; Lin and Peng, 2006).

Borrelia burgdorferi produces a three-VFT-containing sensor kinase called HpK1. Its cognate response regulator Rrp1 is composed of a phospho-receiver module followed by a GGDEF domain, with c-di-GMP synthase activity (Rogers *et al.*, 2009). The *rrp1-hpk1* operon is induced upon tick feeding and HpK1 is necessary for survival of *Borrelia* in tick (Caimano *et al.*, 2011). The HpK1/Rrp1 two-component system modulates the expression of a large set of genes involved in central metabolic processes and virulence mechanisms, including immune evasion, that are required for adapting to changes in nutriment availability and for survival in ticks (Rodgers *et al.*, 2009).

Photorhabdus luminescens is involved in symbiotic interactions with a nematode and in pathogenic interaction with insect larvae (Joyce *et al.*, 2006). Its complex life cycle includes a unidirectional phenotypic variation, which is controlled by the two-component system AstRS. ArtRS retards the onset of phenotypic variation, both by positively regulating the expression of *upsA*, coding for a universal stress protein, and by facilitating adaptation to stationary phase (Derzelle *et al.*, 2004). ArtSR thus regulates the expression of a large number of genes involved in transport, energy metabolism etc. (Derzelle *et al.*, 2004).

In *Yersinia enterocolitica*, YsrRS is required for the expression of *ysaE*, coding for a transcriptional activator of secreted T3SS effector genes (Walker and Miller, 2004). The regulation by YsrR requires NaCl. Interestingly, the VFT-containing YsrS sensor kinase harbours a receiver domain following its kinase domain, but no HPt module. A third component has been identified, the HPt-containing protein YsrT that provides the missing link between YsrS and YsrR (Walker *et al.*, 2010).

In *Burkholderia cenocepacia*, CblRST controls the expression of cable pili, which facilitate the binding of the bacterium to respiratory epithelial cells and to mucin, and which may also mediate *Burkholderia cenocepacia* cell–cell interactions (Tomich and Mohr, 2004a,b). In this three-component system, CblS and CblT are VFT-containing sensor kinases and CblR is the response regulator. The cytoplasmic moiety of CblS comprises His-kinase, receiver and HPt domains, while that of CblT contains only a His-kinase and a receiver but no HPt domain. Both proteins are required for cable pilus expression. The model is that CblS and CblT are physically associated and cross-phosphorylate. CblT might also act as a phosphatase, because its overexpression without that of CblS leads to a block of *cblA* (coding for the major pili subunit) expression. Expression of the *cbl* operon is modulated by environmental cues including pH, temperature and osmolarity.

Pseudomonas aeruginosa also contains complex signal transduction systems that enable its adaptation to a variety of environments. Some of these systems in particular are involved in the regulation of the complex phenotypic switch between planktonic and sessile lifestyles (Kuchma *et al.*, 2005; Kulasekara *et al.*, 2005). Thus, the two-component system RocS1A1 positively regulates the expression of *cup* fimbrial gene clusters, involved in biofilm formation. RocS1 is a BvgS homologue. In addition to its cognate response regulator RocA1, RocS1 interacts with and phosphorylates another receiver-containing protein called RocR, with an EAL domain, implicated in c-di-GMP hydrolysis (Kuchma *et al.*, 2005; Kulasekara *et al.*, 2005). RocR antagonizes the positive transcriptional activity of RocA1. RocS1 is also able to phosphorylate the response regulator RocA2, which represses the genes coding for the MexAB-OprM multidrug efflux pump (Sivaneson *et al.*, 2011). Another sensor kinase devoid of VFT domains, RocS2, also phosphorylates both RocA1 and RocA2.

Finally, the VieABS system of *Vibrio cholerae* positively regulates the expression of *toxT*, coding itself for a transcriptional activator of the *ctx* genes (Tischler *et al.*, 2002). The response regulator VieA has no HTH domain but harbours a phosphodiesterase EAL domain that hydrolyses c-di-GMP. The activity of VieA promotes mobility and virulence gene expression while repressing biofilm genes. VieS regulates these phenotypes indirectly, by controlling the phosphorylation status of VieA (Martinez-Wilson *et al.*, 2008). The environmental signals to which the system responds are unknown. As for VieB, its function remains to be determined.

Altogether, it appears that VFT-containing sensor kinases serve in a number of different processes, and the systems of which they are part are quite diverse. The signals to which they respond, and the conditions for their expression will therefore most likely be specific to each system.

Evolution of BvgAS in pathogenic bordetellae: insights into the molecular mechanisms of BvgS

B. pertussis is included in the 'bronchiseptica' complex, which comprises also the closely related species *B. bronchiseptica* and *B. parapertussis* (Parkhill *et al.*, 2003; Diavatopoulos *et al.*, 2005). In addition, the *Bordetella* genus encompasses several more recently identified species. They include *Bordetella avium* and *Bordetella hinzii*, both of which essentially cause avian infections (Arp and Cheville, 1984; Vandamme *et al.*, 1995), *Bordetella trematum* and *Bordetella holmesii*, which have been isolated from rare human infections (Weyant *et al.*, 1995; Vandamme *et al.*, 1996) and *Bordetella petrii*, a non-pathogenic species isolated from an anaerobic bioreactor (von Wintzingerode *et al.*, 2001).

While BvgS is very conserved in the 'bronchiseptica complex' (see below), it varies significantly more within the *Bordetella* genus (Gerlach *et al.*, 2004). The global architecture of BvgS is conserved between the *B. bronchiseptica* complex and the other *Bordetella* species except for *B. petrii*, but the sequences diverge significantly (Gerlach *et al.*, 2004). For *B. petrii*, genomic sequencing has revealed the presence of two copies of *bvgS* (Gross *et al.*, 2008). The two proteins, called BvgS1 and BvgS2, do not harbour large periplasmic VFT domains like that of *B. pertussis*'s BvgS or a cytoplasmic HPt domain. The cytoplasmic portion of BvgS1 contains three distinct functional modules, i.e. PAS, transmitter and receiver domains, while that of BvgS2 contains only the last two domains but no PAS. A separate gene codes for a single putative HPt protein. This large divergence most likely reflects major functional adaptations between obligate pathogens such as the members of the 'bronchiseptica complex' and *B. petrii*, an essentially environmental species.

To gain insight into the molecular mechanisms of signal perception and transduction by BvgS, we conducted a large sequence analysis of the *bvgAS* genes in a number of *Bordetella* isolates. We chose to restrict our analysis to the 'bronchiseptica complex' (Diavatopoulos *et al.*, 2005). Sequence analyses of *bvgS* and *bvgA* from of 82 isolates of the 'bronchiseptica complex', including *B. pertussis*, *B. parapertussis* and *B. bronchiseptica* strains of various geographical origins, hosts and times of isolation were carried out (Herrou *et al.*, 2009). With a genetic diversity lower than that of housekeeping genes *bvgA* is almost invariant, indicating a strong functional pressure on this gene in the 'bronchiseptica complex' (Herrou *et al.*, 2009). In contrast, the genetic diversity of *bvgS* is much higher, with 6.6% of single nucleotide polymorphisms in the entire sequence. Of note, this diversity is not homogeneously spread throughout the sequence (see below). Regarding the protein sequences deduced from the gene sequences, two molecular solutions have evolved for BvgS that correspond to what we have called the 'pertussis lineage' and the 'bronchiseptica lineage'. The 'pertussis lineage' contains all *B. pertussis* isolates as well as the *B. bronchiseptica* strains from so-called complex IV, isolated mostly from human infections (Diavatopoulos *et al.*, 2005). The 'bronchiseptica lineage' contains all other *B. bronchiseptica* strains as well as all the *B. parapertussis* isolates, irrespective of their human or sheep origin. The genetic diversity of *bvgS* is extremely reduced among *B. pertussis* isolates, indicating that BvgS may be close to equilibrium in that species or that it has undergone recently a purifying selection (Herrou *et al.*, 2009).

Analyses of the amino acid substitutions in our set of isolates indicated that selective pressure limits the range of possible substitutions at the variable positions in BvgS (Fig. 4.4) (Herrou *et al.*, 2009). Thus, only two different residues are found at any variable position, and most of these substitutions are conservative. Interestingly, the level of variation differs significantly between domains, likely pointing to greater functional pressure on some portions of the protein than others. With 14.5% variable positions, VFT1 is the most variable domain between the two lineages, although most substitutions are conservative in nature. As for VFT2, although it differs by 7.4% between the two lineages it is almost invariant in each of them, suggesting that the VFT2 domain is under strong stabilization pressure in each of the two lineages. Mapping the substitutions on structural models of VFT1 and VFT2 revealed that the substitutions are overwhelmingly localized on the external surfaces of the two domains, leaving invariant their potential ligand-binding cavities. This strongly indicates different selective pressures on the cavities and on the protein external surfaces. Purifying selective pressure on each of the VFT1 and VFT2 ligand-binding cavities, which are very different from each other (see below), argues that they have specific binding activities that are important for BvgS function.

The PAS domain is very well conserved, indicating that it is also under strong stabilizing selection and emphasizing its importance for BvgS function. Although the role and the potential ligands of the PAS domain are currently unknown, point mutations have been described in this domain that totally inactivate BvgS (Beier *et al.*, 1996; Dupré, unpublished data), in agreement

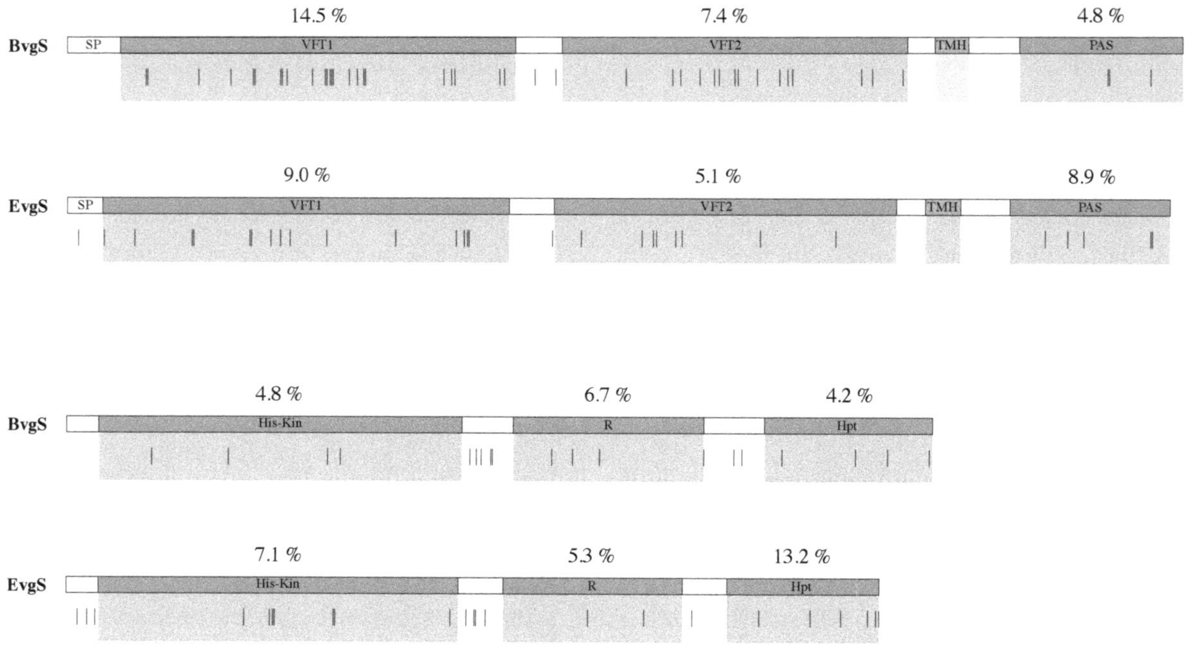

Figure 4.4 Schematic representation of BvgS and EvgS variability at the amino acid level. The various domains of the two proteins are drawn to scale and boxed. The numbers above the bars show the total percentages of variable positions in each domain. Each small vertical line represents the position of a substitution. In most cases a single substitution is found at any given position. Note that for readability only the substitutions found in at least 5% of the sequences are displayed, leaving out the positions with rare substitutions. SP represents the signal peptide, TMH the transmembrane helix, R the receiver domain, HPt the histidine phosphotransfer domain. The white boxes represent the linkers between the various domains (see text).

with a crucial role for this domain. Similarly, few mostly conservative substitutions occur in the histidine kinase domain, pointing to strong functional constraints. Finally, the receiver domain is more variable (6.7% variable positions) than the HPt domain (4.2%). Altogether, thus, the different domains have evolved at different rates.

Because 70 *evgS* gene sequences are available from various *E. coli* isolates, we also aligned the predicted amino acid sequences of these variants (Fig. 4.4). The periplasmic domain of EvgS appears to be somewhat more conserved than BvgS, with both VFT1 and VFT2 harbouring fewer variable positions than those of BvgS. The other domains appear to have evolved rather similarly to their counterparts of BvgS, except for the HPt domain of EvgS, which appears more variable than that of BvgS; the reason for this remains unclear. In most variable positions throughout the sequence, only two different residues were found, with mostly conservative substitutions, although up to three different residues were present at some sequence positions. Strikingly, as for BvgS the transmembrane region is virtually invariant, emphasizing its essential function. The conservation of the linker regions in BvgS and EvgS is quite interesting with respect to putative mechanisms of signal transmission, and it is analysed below.

Cytoplasmic and periplasmic linker regions of BvgS and EvgS: conservation and function

Our sequence analysis of the regions that join the successive domains of BvgS to one another has provided information on the functional constraints that operate on these linkers (Herrou *et al.*, 2009). Thus, with 20% variable positions, the linker between the His-kinase domain and the receiver domain has accumulated a disproportionate number of mutations relative to the rest of the protein. A similar situation prevails for the corresponding linker of EvgS. The linker between the receiver and HPt domain is also not well conserved within each of the two families of proteins. These observations argue that these two linkers do not convey a mechanical signal between domains; therefore, their main function is most likely to keep the domains physically associated. The His-kinase, receiver and HPt domains participate in the phosphotransfer reactions by direct protein–protein contacts, which most likely do not involve these linkers. In support of this model, phosphotransfer was shown to take place between fragments of the cytoplasmic moiety of BvgS expressed as separate proteins *in vitro* (Perraud *et al.*, 1998, 2000). To substantiate our hypothesis that domain–domain interactions, rather than the linkers between the domains involved in the phoshorylation cascade are important for signal transduction, we mapped the substitutions found in the BvgS receiver and HPt domains on structural models of these domains obtained by homology modelling using published structures of HPt and of kinase domains, alone or in complex with their cognate receiver partners (Kato *et al.*, 1997; Xu *et al.*, 2003; Rogov *et al.*, 2004; Marina *et al.*, 2005; Zhao *et al.*, 2008; Casino *et al.*, 2009; Yamada *et al.*, 2009; Bell *et al.*, 2010). This structural modelling analysis indicated that the predicted interface between the kinase and receiver domains is conserved in BvgS. Similarly, the putative interface of HPt with BvgA also appears to be devoid of substitutions.

The variability of the linkers joining the phosphorelay domains contrasts with the extreme conservation of some other linkers in BvgS (Fig. 4.4). Thus, the predicted transmembrane helix that joins the periplasmic and cytoplasmic moieties of BvgS is totally invariant. This transmembrane helix is prolonged by a juxtamembrane linker strongly predicted to be α-helical and leading to the PAS domain. This linker is also totally invariant, arguing that this region is crucial for transmission. Similarly, the PAS domain is followed by another α helix totally conserved, and predicted to continue into the dimerization domain (Dhp) of the His-kinase module.

A rather similar situation occurs in EvgS, with the transmembrane segment and the linkers that precede and follow it being totally invariant (Fig. 4.4). The observation made for BvgS does not appear to hold true for the linker between the PAS and kinase domains of EvgS, which harbours three variable positions. However, these substitutions are quite conservative.

For the eukaryotic or prokaryotic systems in which signal transduction across the membrane has been characterized, transmembrane helices were shown to be key elements to mechanically transfer signals between the periplasm and the cytoplasm (Call and Chou, 2010; Goldberg *et al.*, 2010). This essential function most likely accounts for the extreme degree of conservation of this region within each of the BvgS and EvgS protein families. Similarly, cytoplasmic helices such as HAMP domains or the so-called signalling (S-) helix are crucial elements in many types of signalling proteins (Chervitz and Falke, 1995; Anantharaman *et al.*, 2006; Hulko *et al.*, 2006; Taylor, 2007; Doebber *et al.*, 2008; Swain *et al.*, 2009; Dunin-Horkawicz and Lupas, 2010; Stewart *et al.*, 2010).

In support of the idea that the PAS domain and its flanking helices are pivotal elements for signal transduction in BvgS, earlier work indicated that these regions are the main sites of point mutations in the *bvgS* gene that turn the system in a permanently 'on' state, i.e. that make it unresponsive to modulating chemical agents such as nicotinate or sulfate ions (Miller *et al.*, 1992; Goyard *et al.*, 1994; Manetti *et al.*, 1994). Along the same lines, we identified a unique point mutation at codon 705, which corresponds to a position in the dimerization helix of the His-kinase domain. In a single isolate, namely the commonly used *B. pertussis* Tohama I isolate, a Lys residue replaces the Glu residue found in all the other isolates of the 'bronchiseptica complex'. Interestingly, this single substitution renders Tohama I less sensitive to negative modulation relative to the other isolates (Herrou *et al.*, 2009). The fact that so many substitutions in the PAS domain, in the helices upstream and downstream of it and in the kinase dimerization helix could be found that render the BvgAS system insensitive to negative modulation suggests that in the wild-type situation – i.e. in the absence of such a mutation -, these regions are involved in the transmission of negative signals provided by the modulators that dampen kinase activity. The concept of cytoplasmic PAS domains involved solely in signal transduction rather than in sensing has recently emerged, and it is a possibility to be considered in the case of BvgS (Etzkorn *et al.*, 2008; Slavny *et al.*, 2010). However, this hypothesis does not seem to be in good agreement with the observation that some members of the superfamily of VFT-containing sensor kinases are devoid of a PAS domain. In those cases, the presence of a PAS domain is dispensable for signal transmission between the periplasm and the kinase.

Mechanical signal transduction via α helices most likely involves one of several types of motions – shearing, piston-like or rotation-, any of which necessarily implies the formation of a bundle composed of two helices or more (Gordeliy *et al.*, 2002; Hulko *et al.*, 2006; Taylor, 2007; Moore and Hendrickson, 2009; Zhou *et al.*, 2009; Maslennikov *et al.*, 2010). Because it is likely that BvgS and its homologues possess a single rather than two transmembrane helices (see above), the most likely possibility is that this helix dimerizes in the membrane. One can thus speculate that the helical segment immediately underneath the membrane also dimerizes. This would be consistent with the observation that PAS domains themselves often form dimers (Etzkorn *et al.*, 2008; Sevvana *et al.*, 2008; Moglich *et al.*, 2009; Slavny *et al.*, 2010).

The VFT1 and VFT2 domains of BvgS and EvgS are separated by 28- and 27-residue-long spacers, respectively. The sequences of these linkers are almost invariant in BvgS and EvgS. Finally, the linker between the end of VFT2 and the membrane segment in both BvgS and EvgS is predicted to form a loop followed by an α helix, itself in the continuity of the transmembrane helix. This region is totally invariant between all isolates in each of the two protein families. Therefore, it is most likely crucial for the flow of information from the VFT domains to the membrane domain.

Analysis of the junctions between VFTs in superfamily

To enlarge our analysis of the linker between VFT1 and VFT2, we aligned the sequences of 131 putative sensor kinases harbouring two or more VFT domains in tandem. For the analysis of the linker between the last VFT and the membrane, a larger set of 237 sequences was used that included also proteins with a single periplasmic VFT. These sets of predicted proteins were much more diverse than the *Bordetella* or the *E. coli* sets with respect to linker sequences, and hardly any strict sequence conservation was found except for a Trp residue located at the very carboxyl extremity of VFT1 and immediately preceding the VFT1-VFT2 linker. The carboxy-terminus of VFT2 is characterized by the presence of a similarly located Trp residue in a majority of proteins in the superfamily. Because of the nature of the Trp side chain and the localization of each of these two conserved residues at the junction between two distinct domains, they might play specific roles in signal transmission. Note however that a group of VFT-kinases from various *Vibrionaceae* does not harbour these conserved Trp residues.

In addition to the conservation of sequence motifs, we also looked at the lengths of the linkers between VFT1 and VFT2, and between VFT2 and the membrane segment. Strikingly, for the first of these two regions there is a large distribution of sizes in the superfamily. Thus, between the carboxy-terminal extremity of VFT1 and the amino-terminal extremity of VFT2, the linker length varies from 5 to 42 residues. BvgS belongs to the most populated class of sequences, with a linker length of 27 residues. The lack of length conservation argues against the possibility of mechanical transmission of information between VFT1 and VFT2 via this linker. This is corroborated by the absence of a predicted α helix that would be sufficiently rigid for transmission of a conformational change. Instead, the linker between two VFTs is predicted to form a loop. This observation implies that signal transduction between VFT1 and VFT2 should occur through interfaces.

The length of the linker between the last VFT domain and the transmembrane segment is markedly more conserved, varying between 10 and 19 residues. Most proteins (> 60%) have linker lengths comprised between 12 and 16 residues; BvgS has 15, which is by far the most frequent length. The linker between the end of the VFT fold and the transmembrane helix is predicted to form a loop, followed by an α helix that directly precedes the predicted transmembrane helix. This more pronounced length conservation indicates that stronger functional constraints are likely to operate on this linker.

Molecular mechanisms of signal perception and transduction: the paradigm of eukaryotic receptors

Interestingly, VFT domains serving as perception modules in sensor proteins have long been recognized in eukaryotic systems (Fig. 4.5). Indeed, many receptors have made use of this type of domains in the eukaryotic world, including ionotropic receptors in neurons (iGluR), G protein-coupled receptors (metabotropic glutamate, GABA-B and pheromone receptors), natriuretic peptide/guanylate cyclase receptors and a newly identified class of 'VKR' kinases in invertebrates (O'Hara *et al.*, 1993; Laube *et al.*, 1997; Felder *et al.*, 1999; Kristensen *et al.*, 2006; Ahier *et al.*, 2009). Because some of these systems have been quite well characterized, we have chosen to describe here in a somewhat simplified manner some of their basic features to serve as a basis for functional and structural comparisons with our bacterial system.

The ionotropic glutamate receptors iGluR family includes NMDA (*N*-methyl-D-aspartate), AMPA (α-amino-3-hydroxy-5-methyl-4-isoxazole propionic acid) and kainate receptors, which mediate most excitatory neurotransmission in the central nervous system (Laube *et al.*, 1997; Kristensen *et al.*, 2006). NMDA receptors have been extensively studied for the last decade and will be used in this review as model eukaryotic systems. NMDA and similar receptors are actually dimers of dimers (Furukawa *et al.*, 2005; Sobolevsky *et al.*,

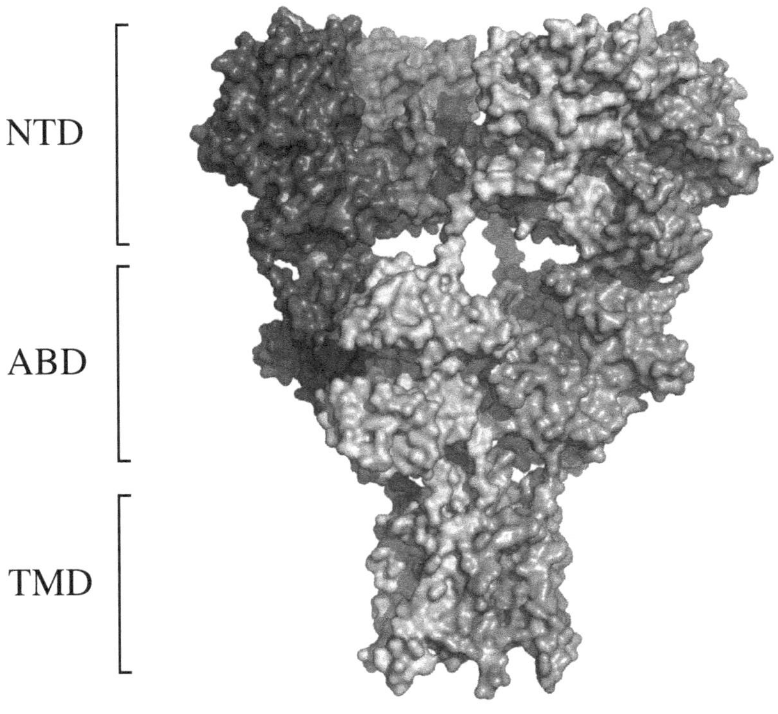

Figure 4.5 Space-filling model of an AMPA receptor (GluA2, pdb 3KG2). NTD, ADB and TMD represent the N-terminal domain (also known as ATD for amino-terminal domain), the agonist binding domain (aka LBD for ligand-binding domain) and the transmembrane domain (TMD), respectively. The NTD and ABD are both Venus flytrap domains (see text). GluA2 forms a homotetramer, and the four monomers are shown in different colours. The cavity of the ABD is empty and the ionic channel is closed in the structure presented here. The linker between the NTD and ABD (best seen in the yellow monomer) participates in signal transduction. Note that the structure displayed here is the only full-length structure available for this type of receptor. However, the system described in the text is that of the NMDA receptors, because more functional details are available for NMDA than for AMPA receptors. A colour version of this figure is located in the plate section at the back of the book.

2009), but this additional level of complexity will not be further discussed here given that dimers – which are heterodimers in the case of NMDAs – appear to be the functional units (Furukawa *et al.*, 2005) (Fig. 4.5). Each monomer (or protomer) harbours an extracellular domain composed of two VFT modules in tandem joined by a linker to an ion-permeable membrane channel (Fig. 4.5). In the resting, inactive conformation the VFTs are open and the ion channel is closed. The C-terminal, membrane-proximal VFT domain is the agonist-binding domain (ABD), and the N-terminal domain (NTD) serves for receptor trafficking and assembly and is the binding site for allosteric ligands. The ABDs form back-to-back dimers through the apposition of their first lobes. Agonist (e.g. glutamate) binding in the ABDs induces closure of their cavity, which increases the separation of their second lobes within the dimer (Furukawa *et al.*, 2005). This movement is thought in turn to increase the separation of the linkers between the ABDs and the membrane domains. This has been proposed to pull apart the transmembrane helices involved in the formation of the pore, causing the opening of the ionic channel. These receptors also have antagonists. Their binding to the ABD maintains the clamshell in an open conformation, which keeps the linkers loose and the channel closed (Kristensen *et al.*, 2006).

The first domain, NTD, modulates receptor activity by controlling channel gating. At least two types of allosteric ligands have been identified. Zn^{2+} ions are negative allosteric modulators. They bind to the cavity of the NTD and induce its closure (Paoletti *et al.*, 2000). This would exert tension on the short linker connecting the NTD and ABD, causing a rearrangement of the ABD interface characterized by a rupture of the interactions between their first lobes in the dimer. The rearrangement would release the tension of the linkers between the ABDs and the membrane domains, resulting in channel closure even though the agonist remains bound to the ABD (Gielen *et al.*, 2008, 2009). Positive allosteric modulation has also been described. Thus, polyamines bind at the interface between NTDs in the dimer of a subset of NMDA receptors, and by shielding negative charges they alleviate electrostatic repulsion and allow the close apposition of the two domains. This in turn stabilizes the NTD in an open conformation and enhances NMDA receptor activity (Mony *et al.*, 2011).

The activation/inactivation mechanisms in these eukaryotic receptor systems are thus based on a few basic principles. Firstly, the binding of agonists and antagonists in the VFT cavities occurs along the 'clamshell motions' model. Secondly, the interfaces between VFT are perturbed upon cavity closure. Thirdly, tension and relaxation of the linkers between the two extracellular VFT domains and between the second VFT domain and the transmembrane segment activate or inactivate the system. And fourthly, the binding of allosteric activators or inhibitors to the N-terminal VFT domain – in the cavity or at the interdomain interface depending on their mode of action – controls the system by modifying the interfaces and hence the linkers. Therefore, the ABD dimer interface has a key role in the transduction pathway to the channel gates (Gielen *et al.*, 2008). Similarly, the linkers between the extracellular domain and the membrane segment are central elements of the gating mechanism. Their tension and relaxation convert ligand-induced structural changes into movements of the transmembrane domains.

Which of these features are likely to be relevant to the molecular mechanisms of VFT-containing bacterial sensor kinases? Can the eukaryotic receptors contribute to our understanding of BvgS and its homologues? The VFT cavities will most likely prove to be functionally important in prokaryotic sensor kinases as they are in eukaryotic receptors. This hypothesis is corroborated by their conservation in BvgS, and it is also supported by our preliminary work on VFT2 (see below). As described above, a specific function for the linker between VFT1 and VFT2 is more contentious, especially in the view of its markedly variable length and lack of sequence conservation among putative bacterial sensor kinases harbouring several periplasmic VFTs. This variable length argues against the idea that the linker could function in a similar manner in all proteins, by transmitting signals through its tension and relaxation. As for the linker between VFT2 and the membrane domain, our *in silico* analyses have indicated that it is always predicted to adopt the same type of structure in all proteins. In addition, although no sequence conservation of this region is apparent its length is better conserved, arguing that this linker may play a role in the transduction mechanism. In all cases, it is predicted to form a loop after the VFT fold, followed by an α helix that is being continued by the transmembrane – most likely α helical – segment. How this region participates mechanically in signal transmission remains to be addressed.

The interfaces between the VFT domains of BvgS will most likely also prove to be crucial for transduction. In addition to interfaces between VFT1 and VFT2, as discussed above, it is conceivable that interfaces between identical domains (VFT1-VFT1 and VFT2-VFT2) exist and participate in the mechanism, as for the NTD-NTD or ABD–ABD interfaces in eukaryotic receptor dimers. We attempted to probe such potential interactions between the purified domains *in vitro*, but we could not gather clear evidence that the periplasmic domain of BvgS dimerizes or tetramerizes in solution (Herrou and Dupré, unpublished). It may however be difficult to draw any definitive conclusion on the oligomeric state of the periplasmic moiety of BvgS based solely on soluble recombinant proteins, because the propensity of membrane proteins to auto-associate is greatly reduced when expressed in soluble form (Grasberger *et al.*, 1986). Similarly, in solution the agonist-binding (ABD) domains of NMDA receptors did not appear to oligomerize to an appreciable extent (Furukawa *et al.*, 2005), indicating that the subsequent membrane domain stabilizes dimer formation. Likewise, the perception domains of the classical bacterial two-component sensor kinases often appear monomeric *in vitro* when devoid of their flanking transmembrane segments, although they are most likely dimeric *in vivo* (Waldburger and Sauer, 1996; Kaspar *et al.*, 1999; Pappalardo *et al.*, 2003; Cheung and Hendrickson, 2010; Zhang and Hendrickson, 2010). In agreement with these observations, both physiological and non-physiological dimeric forms of these sensor domains have been crystallized (Reinelt *et al.*, 2003; Cheung and Hendrickson, 2010). Given that the transmitter (kinase) domain of BvgS was shown to dimerize (Beier *et al.*, 1995) and that the transmembrane helix most likely does so as well, it is expected that the periplasmic domain also assembles into a higher-order organization, probably dimeric. Of note, such arrangements appear to prevail in other types of VFT that are not involved in transport systems, such as VFT modules of proteins with transcriptional regulation activities. For instance, PurR form dimers and LacR tetramers (Schumacher *et al.*, 1994; Friedman *et al.*, 1995).

Whatever the role of VFT–VFT interfaces, the topology and mechanistic model proposed for the eukaryotic receptors described above, with ligand-induced clamshell motions of VFTs eventually regulating the position of transmembrane helices appears too linear to be effective in the case of bacterial sensor kinases with three or four VFT domains in tandem. In such complex systems, it is hard to imagine how the distal VFT domains would be able to influence the membrane helix at such a long distance simply by modifying the tension/relaxation of successive linkers, without a more intricate communication system between 'layers' of VFTs.

Another mechanistic model than the one described above for iGlu receptors has been identified for the natriuretic peptide receptor (NPR), with inter–domain interfaces having been recruited for the binding of true agonists (He *et al.*, 2005). The extracellular domain of this hormone receptor contains one VFT domain, and the crystal structure of a complex between the NPR-C domain and the hormone has revealed that the cyclic peptide is bound at the dimeric interface between the two VFT domains rather than in their cavities. Comparison with the unliganded receptor has shown that dimerization is mediated by the back-to-back apposition of the first lobes of the two monomers. Ligand binding brings the second lobes of the VFTs closer together in such a way that both lobes make contacts with the peptide. This also triggers the opening of the VFT cavities. The closure of the inter-dimer gap is associated with receptor activation. From this, it appears that nature has made use of VFT domains in various manners, resulting in different switching mechanisms. Thus, the molecular solution that has evolved for VFT-containing bacterial sensor kinases might yet be different from the known paradigms. This is suggested by our structure–function analysis of BvgS's VFT2.

BvgS: A new paradigm for perception and transduction?

Because thus far only negative – modulating – signals have been identified for BvgAS, the prevailing model is that the system is active 'by default'. BvgS would then be active without the need for positive signals. The high concentrations – mM or more – that are required for negative modulation of BvgS activity by sulfate or nicotinate ions suggest that they are not the bona fide molecules sensed by the system. This model proposes that those modulators mimic natural, negative signals (antagonists) perceived by BvgS in its environment, which have not yet been identified. An alternative would be that these molecules are allosteric ligands rather than antagonists. Note that for most two-component systems, even the classical ones, the nature of the signals remains unknown and thus we cannot tell whether a 'negative signals only' model would be the exception rather than the rule in the two-component world.

In order to probe this model, we set up *in vitro* assays to identify the ligands – positive or negative – for each of the two domains using recombinant VFT1 and VFT2 proteins. Our initial hypothesis was that because the system is active under laboratory conditions, one or several molecules present in the culture medium might be the real agonists of BvgS. Therefore, BvgS might trap these molecules and become activated. According to that hypothesis, the system would be able to respond to positive and negative signals – agonists and antagonists – in opposite directions, i.e. by turning on or off its kinase activity. In particular, we tested amino acids, because the closest homologues of VFT1 and VFT2 are amino-acid-binding PBPs such as glutamine- and histidine-binding proteins, and because the metabolism of *B. pertussis* relies heavily on amino acids as energy and carbon sources, making them good candidates to be perceived by BvgS (Thalen *et al.*, 1999). We also tested the idea that small metabolites produced by *B. pertussis* might be agonists, because their production activating BvgS would also explain what is observed under laboratory growth conditions. Fluorescence spectroscopy has been widely used to analyse interactions between protein such as PBPs and their ligands (e.g. Miller *et al.*, 1983). However, in the case of BvgS, no modification of VFT1 or VFT2's fluorescence spectra – quenching, blue shift or red shift – was detected when using either the known negative modulators, the components of the growth medium or the metabolites reported to be produced by *B. pertussis* in the course of growth under laboratory conditions (Herrou *et al.*, 2010). Another technique called the 'thermal shift assay' is based on the increase of the denaturation temperature of a protein when it is bound to a ligand relative to its denaturation temperature without a ligand (Pantoliano *et al.*, 2001; Giuliani *et al.*, 2008). For VFT1 the denaturation temperature was rather low, and no small molecule was found to increase it. Conversely, for VFT2, the melting temperature was significantly increased by the addition of nicotinate and related aromatic molecules at high millimolar concentrations. There was a good correlation between the ability of various compounds to stabilize VFT2 and their modulating activity *in vivo* (Herrou *et al.*, 2010). This was the first indication that nicotinic acid might be perceived by VFT2. In contrast, no stabilizing effect was detected with sulfate ions.

In follow-up studies, we succeeded in obtaining the three-dimensional structure of VFT2 by X ray crystallography. This domain adopts a bona fide VFT fold, with two /β lobes forming a clamshell with a cleft between them, and connected by a hinge formed of two β strands. Interestingly, the two lobes were closed around the cavity, which contained several small molecules from the crystallization solution that we thus considered to be 'fortuitous', rather than authentic, ligands (Herrou *et al.*, 2010). Indeed, each of these molecules of acetate and glycerol was bound to essentially a single lobe of VFT2 and mostly via van der Waals or water-mediated hydrogen bonds, arguing that they were not bona fide ligands.

An intriguing feature of VFT2 is the highly positive electrostatic potential of its cavity, which is quite unusual among PBPs. Of note, charged surfaces were reported to be the binding sites for modulators in eukaryotic receptors (see above). Altering this electrostatic potential of VFT2 by site-directed mutagenesis in the cavity yielded a functional BvgS that became however unresponsive to modulating agents. Unlike its wild-type counterpart, this modified VFT2 protein was not stabilized by nicotinic acid (Herrou *et al.*, 2010). The X ray structure of this VFT2 variant revealed that the cavity was also in a closed conformation and contained a different set of small molecules, most likely from the crystallization solution as with the wild-type VFT2 crystal. The variant protein appeared to be stabilized in closed conformation by new, direct interactions between the two lobes caused by the introduced substitutions.

Additional substitutions were introduced in the VFT2 cavity, notably by targeting residues involved in ligand binding in the closest PBP homologues of VFT2 (Herrou *et al.*, 2010). Briefly, these allowed us to exclude the hypothesis that VFT2 binds amino acids like its PBP counterparts, showing that VFT2 has functionally diverged from that protein group. Very few replacements targeting residues involved in the network of interactions identified in the cavity affected BvgS activity, arguing that it has no high-affinity agonists. Conversely, the replacement of a phenylalanine residue in the cavity decreased the activity of BvgS and rendered it highly sensitive to modulation. The corresponding VFT2 protein was markedly less stable than its wild-type counterpart, but its stability was strongly increased by the aromatic modulators. The replacement of another phenylalanine residue nearby in the cavity abolished BvgS activity, indicating that the conformation of this region is critical.

From all these observations we infer that VFT2 represents a new paradigm for the Venus flytrap mechanism. This VFT domain is closed in the absence of bona fide agonist ligands. Interestingly, another structure of a VFT domain of a sensor kinase homologous to BvgS, HK29 of *Geobacter sulfurreducens* also displays a closed cavity, containing only water molecules (Cheung *et al.*, 2009). Thus, both VFT2 and HK29 crystallized in closed conformations in the absence of ligands. This is most likely not a coincidence, given that PBPs usually crystallize more easily in closed conformation after they have trapped a bona fide ligand. The VFT2 model that we propose is in contrast to that for the eukaryotic iGlu receptors, which are 'off' by default and respond specifically to the presence of agonists to activate the system. The closed conformation of VFT2 represents the active form, as shown by the structure of the fully active BvgS variant that did not respond to modulation. Therefore, we propose that VFT2 has no high-affinity agonists. Instead, the positive potential of the cavity is dissipated by the binding of several non-specific ligands from the milieu, thus stabilizing the closed conformation. This model is consistent with the effect of various organic and inorganic ions on Bvg activity (see below). VFT2 has also evolved to respond to negatively charged aromatic modulators. It is possible that the positive potential of the cavity facilitates their binding. Indeed, the presence of a carboxylate group appears to be a major requirement for nicotinate analogues to be effective negative modulators of BvgS (Melton and Weiss, 1993). We propose that these aromatic modulators are low-affinity antagonists of VFT2 that compete with the non-specific ligands for binding in the cavity and perturb its conformation, turning it into an inactive state. This conformational change interrupts the positive signalling that prevails by default. Unfortunately, our attempts to crystallize VFT2 with nicotinate ions were unsuccessful.

Whether the binding of low-affinity antagonists opens the VFT or keeps it closed but distorted is a matter of debate. In the case of HK29, the hinge between the two lobes appears to lack flexibility, suggesting that the cavity might not open readily (Cheung *et al.*, 2009). Other types of structural perturbations rather than cavity opening are possible. Nicotinate probably binds in the vicinity of the Phe-containing loop, and this loop itself is close to the carboxy terminus of the VFT2 domain. Thus, its displacement upon nicotinate binding might trigger some conformational change that could be easily relayed to the membrane.

Therefore, our current model is that BvgS has evolved to be active unless specific, negative chemical signals are present in its environment. This 'on state' by default might actually be the most rational type of regulation for a strictly human pathogen that does not survive for long periods of time in the external environment and that absolutely requires the Bvg^+ phase for colonization of new hosts (Weiss *et al.*, 1983). Nevertheless, the bacterium has maintained the ability to respond to negative chemical signals rather than becoming in a permanently activated state, even though such modulation-unresponsive mutants occur readily in laboratory conditions (see above). Therefore, it is probable that the Bvg^- phase and/or more likely the Bvg^i phase remains useful to *B. pertussis* at some stage of its life cycle, parts of which are as yet unclear. For instance, the duration of the infection suggests that the bacterium might persist in intracellular niches, as previously shown *in vitro* (Lamberti *et al.*, 2010). In such microenvironments, specific molecules might be present as natural antagonists of BvgS to dampen expression of the virulence factors. For instance, it has been reported that the Bvg regulon is down-regulated upon bacterial entry into macrophages (Masure, 1992).

What is the role of VFT1?

To date, little information has been gathered concerning VFT1. No ligand has been identified and no high-resolution structure is available. At least one linker insertion was reported to abolish BvgS activity (Miller *et al.*, 1992), arguing that VFT1 is not dispensable. Because the two domains are homologous to each other and to the *E. coli* glutamine-BP, a structural model of VFT1 was constructed (Herrou *et al.*, 2009), and we then started to probe the VFT1 cavity by site-directed mutagenesis as for VFT2. Targeting of residues involved in ligand binding in homologous PBPs allowed us to generate distinct phenotypes (Herrou and Dupré, unpublished). For instance, some of the substitutions made BvgS insensitive to modulators. Because we have established that nicotinate and related modulators bind to VFT2, the loss of responsiveness to this modulator caused by specific substitutions in VFT1 argues in favour of inter-domain communication. Presumably, the modified VFT1 adopts a conformation that impinges on VFT2's ability to bind the modulating molecule. We also restricted lobe mobility in VFT1, which resulted in the complete inactivation of BvgS. Based on this piece of information we speculate that in its natural conformation VFT1 provides a positive message to the system, which is on by default. Modulators such as nicotinate and its analogues are perceived by VFT2, which turns the system off, overriding the positive message originating from VFT1. VFT1 might perceive either positive or negative signal(s) on its own, to influence the state of activity of VFT2. The observation that its cavity has been conserved in the evolution of *Bordetellae* argues in favour of this idea.

Where do sulfate ions act?

We did not obtain any indication that sulfate ions bind to VFT2 similar to nicotinate. However, the positive potential of the VFT2 cavity coupled to our observations that various small molecules of the medium are found inside the cavity, suggests that sulfate might well do so if present at a sufficiently high concentration in the environment. Possibly, the nature and size of sulfate may explain why we did not observe a stabilizing effect of recombinant VFT2 by this ion in the thermal shift assay. It has been shown that at high concentrations Cl^- ions counter the modulating effect of nicotinate (Melton and Weiss, 1993). Thus it is possible that mono-charged ions are favoured over doubly charged ones. This would explain the high concentrations of sulfate required for modulation *in vivo*, several times higher than for modulation by nicotinate.

In any case, it is possible that sulfate ions act at more than one level to down-regulate the activity of BvgS. In support of this idea, we have obtained BvgS variants by site-directed mutagenesis in which the effects of the two classes of modulators were uncoupled. Using certain combinations of point substitutions, we found proteins that respond to sulfate but not to nicotinate. As BvgS may remain sensitive to sulfate even when VFT2 can no longer respond to nicotinate, sulfate may have several sites of action. Furthermore, sulfate appears to act with relatively little specificity as high concentrations are required for *in vivo* effects. A remarkable early paper reported opposing effects of various salts and inorganic ions on BvgS activity (Lacey, 1960). Cations with a single charge (Na^+, K^+, NH_4^+, Li^+) were shown to have little effect, while those with a double positive charge were essentially anti-virulence. Anions with a single charge had generally a pro-virulence effect (Cl^-, Br^-, NO_3^-), in contrast with those carrying a double charge such as sulfate. Small mono- or di-carboxylic organic anions had little or no effect, while that of longer organic anions depended on the length of their carbon chain, with longer chain compounds being the strongest modulators. The mode of action of these ions has not been deciphered. Similar 'salt effects' have also been described in eukaryotic systems (Bowie, 2002; Paternain *et al.*, 2003; Plested and Mayer, 2007; Plested *et al.*, 2008). Altogether, Lacey's data indicate that screening of charged surfaces by ions affects BvgS activity. Potentially important surfaces of BvgS in addition to the VFT cavities include VFT–VFT interfaces, as well as interfaces between cytoplasmic elements potentially involved in transmission (see above).

Conclusions and perspectives

Although BvgA/S has thus far been one of the most thoroughly characterized two-component systems among those with VFT-harbouring sensor kinases, this chapter reveals how little we actually know on the molecular mechanisms of these systems. VFT domains are ubiquitous in all kingdoms of life, and their plasticity offers several possible modes of action. Indeed, our first investigations into BvgS have indicated that the second VFT domain has evolved to function in a non-canonical manner. As for the role and the mode of action of the first VFT domain of BvgS, they remain to be deciphered. The presence of up to four VFT domains in tandem in some BvgS homologues strongly argues in favour of novel modes of signal transduction waiting to be discovered. Comparing and contrasting with the analogous eukaryotic receptors will undoubtedly remain a fruitful avenue to progress in the deciphering of signal perception and transduction by the extracytoplasmic domains of VFT-containing sensor kinases. Other major issues on BvgS and homologous sensors regard the function of the PAS domain – perception, transduction or both – and the mode of signal transmission across the membrane. BvgAS responds also to temperature, being active at 37°C and inactive at < 25°C. How this very physiological signal is perceived remains an important question. It will be essential to carry out studies on more than one system to be able to discern which features are conserved and which are more specific. Thus, while it is most likely that sensor kinases harbouring several periplasmic VFT domains in tandem have a signal peptide, the question of the presence of a signal peptide versus a transmembrane segment remains open for those proteins with a single VFT domain. The study of these other systems will also shed light on the nature of signals and their mode of perception. The Holy Grail with these systems is a mechanistic model for signal transduction by these fascinating proteins.

References

Ahier, A., Rondard, P., Gouignard, N., Khayath, N., Huang, S., Trolet, J., Donoghue, D.J., Gauthier, M., Pin, J.P., and Dissous, C. (2009). A new family of receptor tyrosine kinases with a venus flytrap binding domain in insects and other invertebrates activated by aminoacids. PLoS One *4*, e5651.

Akerley, B.J., Cotter, P.A., and Miller, J.F. (1995). Ectopic expression of the flagellar regulon alters development of the *Bordetella*-host interaction. Cell *80*, 611–620.

Anantharaman, V., Balaji, S., and Aravind, L. (2006). The signaling helix: a common functional theme in diverse signaling proteins. Biol. Direct. *1*, 25.

Antoine, R., Alonso, S., Raze, D., Coutte, L., Lesjean, S., Willery, E., Locht, C., and Jacob-Dubuisson, F. (2000). New virulence-activated and virulence-repressed genes identified by systematic gene inactivation and generation of transcriptional fusions in *Bordetella pertussis*. J. Bacteriol. *182*, 5902–5905.

Arp, L.H., and Cheville, N.F. (1984). Tracheal lesions in young turkeys infected with *Bordetella avium*. Am. J. Vet. Res. *45*, 2196–2200.

Beier, D., Schwarz, B., Fuchs, T.M., and Gross, R. (1995). *In vivo* characterization of the hybrid BvgS two-component sensor protein of *Bordetella pertussis*. J. Mol. Biol. *248*, 596–610.

Beier, D., Deppisch, H., and Gross, R. (1996). Conserved sequence motifs in the unorthodox BvgS two-component sensor protein of *Bordetella pertussis*. Mol. Gen. Genet. *252*, 169–176.

Bell, C.H., Porter, S.L., Strawson, A., Stuart, D.I., and Armitage, J.P. (2010). Using structural information to change the phosphotransfer specificity of a two-component chemotaxis signalling complex. PLoS Biol. *8*, e1000306.

Bjorkman, A.J., and Mowbray, S.L. (1998). Multiple open forms of ribose-binding protein trace the path of its conformational change. J. Mol. Biol. *279*, 651–664.

Boucher, P.E., Murakami, K., Ishihama, A., and Stibitz, S. (1997). Nature of DNA binding and RNA polymerase interaction of the *Bordetella pertussis* BvgA transcriptional activator at the *fha* promoter. J. Bacteriol. *179*, 1755–1763.

Boucher, P.E., Yang, M.S., Schmidt, D.M., and Stibitz, S. (2001a). Genetic and biochemical analyses of BvgA interaction with the secondary binding region of the *fha* promoter of *Bordetella pertussis*. J. Bacteriol. *183*, 536–544.

Boucher, P.E., Yang, M.S., and Stibitz, S. (2001b). Mutational analysis of the high-affinity BvgA binding site in the *fha* promoter of *Bordetella pertussis*. Mol. Microbiol. *40*, 991–999.

Bowie, D. (2002). External anions and cations distinguish between AMPA and kainate receptor gating mechanisms. J. Physiol. *539*, 725–733.

Caimano, M.J., Kenedy, M.R., Kairu, T., Desrosiers, D.C., Harman, M., Dunham-Ems, S., Akins, D.R., Pal, U., and Radolf, J.D. (2011). The hybrid histidine kinase Hk1 is part of a two-component system that is essential for survival of *Borrelia burgdorferi* in feeding *Ixodes scapularis* ticks. Infect. Immun. *79*, 3117–3130.

Call, M.E., and Chou, J.J. (2010). A view into the blind spot: solution NMR provides new insights into signal transduction across the lipid bilayer. Structure *18*, 1559–1569.

Casino, P., Rubio, V., and Marina, A. (2009). Structural insight into partner specificity and phosphoryl transfer in two-component signal transduction. Cell *139*, 325–336.

Casino, P., Rubio, V., and Marina, A. (2010). The mechanism of signal transduction by two-component systems. Curr. Opin. Struct. Biol. *20*, 763–771.

Chang, C.H., and Winans, S.C. (1992). Functional roles assigned to the periplasmic, linker, and receiver domains of the *Agrobacterium tumefaciens* VirA protein. J. Bacteriol. *174*, 7033–7039.

Chervitz, S.A., and Falke, J.J. (1995). Lock on/off disulfides identify the transmembrane signaling helix of the aspartate receptor. J. Biol. Chem. *270*, 24043–24053.

Cheung, J., and Hendrickson, W.A. (2010). Sensor domains of two-component regulatory systems. Curr. Opin. Microbiol. *13*, 116–123.

Cheung, J., Le-Khac, M., and Hendrickson, W.A. (2009). Crystal structure of a histidine kinase sensor domain with similarity to periplasmic binding proteins. Proteins *77*, 235–241.

Cotter, P.A., and Jones, A.M. (2003). Phosphorelay control of virulence gene expression in *Bordetella*. Trends Microbiol. *11*, 367–373.

Cotter, P.A., and Miller, J.F. (1994). BvgAS-mediated signal transduction: analysis of phase-locked regulatory mutants of *Bordetella bronchiseptica* in a rabbit model. Infect. Immun. *62*, 3381–3390.

Cotter, P.A., and Miller, J.F. (1997). A mutation in the *Bordetella bronchiseptica bvgS* gene results in reduced virulence and increased resistance to starvation, and identifies a new class of Bvg-regulated antigens. Mol. Microbiol. *24*, 671–685.

Cummings, C.A., Bootsma, H.J., Relman, D.A., and Miller, J.F. (2006). Species- and strain-specific control of a complex, flexible regulon by *Bordetella* BvgAS. J. Bacteriol. *188*, 1775–1785.

Deora, R. (2002). Differential regulation of the *Bordetella bipA* gene: distinct roles for different BvgA binding sites. J. Bacteriol. *184*, 6942–6951.

Derzelle, S., Ngo, S., Turlin, E., Duchaud, E., Namane, A., Kunst, F., Danchin, A., Bertin, P., and Charles, J.F. (2004). AstR–AstS, a new two-component signal transduction system, mediates swarming, adaptation to stationary phase and phenotypic variation in *Photorhabdus luminescens*. Microbiology *150*, 897–910.

Diavatopoulos, D.A., Cummings, C.A., Schouls, L.M., Brinig, M.M., Relman, D.A., and Mooi, F.R. (2005). *Bordetella pertussis*, the causative agent of whooping cough, evolved from a distinct, human-associated lineage of *B. bronchiseptica*. PLoS Pathog. *1*, e45.

Doebber, M., Bordignon, E., Klare, J.P., Holterhues, J., Martell, S., Mennes, N., Li, L., Engelhard, M., and Steinhoff, H.J. (2008). Salt-driven equilibrium between two conformations in the HAMP domain from *Natronomonas pharaonis*: the language of signal transfer? J. Biol. Chem. *283*, 28691–28701.

Dunin-Horkawicz, S., and Lupas, A.N. (2010). Comprehensive analysis of HAMP domains: implications for transmembrane signal transduction. J. Mol. Biol. *397*, 1156–1174.

Eguchi, Y., Oshima, T., Mori, H., Aono, R., Yamamoto, K., Ishihama, A., and Utsumi, R. (2003). Transcriptional regulation of drug efflux genes by EvgAS, a two-component system in *Escherichia coli*. Microbiology *149*, 2819–2828.

Eguchi, Y., Okada, T., Minagawa, S., Oshima, T., Mori, H., Yamamoto, K., Ishihama, A., and Utsumi, R. (2004). Signal transduction cascade between EvgA/EvgS and PhoP/PhoQ two-component systems of *Escherichia coli*. J. Bacteriol. *186*, 3006–3014.

Etzkorn, M., Kneuper, H., Dunnwald, P., Vijayan, V., Kramer, J., Griesinger, C., Becker, S., Unden, G., and Baldus, M. (2008). Plasticity of the PAS domain and a potential role for signal transduction in the histidine kinase DcuS. Nat. Struct. Mol. Biol. *15*, 1031–1039.

Felder, C.B., Graul, R.C., Lee, A.Y., Merkle, H.P., and Sadee, W. (1999). The Venus flytrap of periplasmic binding proteins: an ancient protein module present in multiple drug receptors. AAPS PharmSci. *1*, E2.

Friedman, A.M., Fischmann, T.O., and Steitz, T.A. (1995). Crystal structure of *lac* repressor core tetramer and its implications for DNA looping. Science *268*, 1721–1727.

Furukawa, H., Singh, S.K., Mancusso, R., and Gouaux, E. (2005). Subunit arrangement and function in NMDA receptors. Nature *438*, 185–192.

Galperin, M.Y. (2010). Diversity of structure and function of response regulator output domains. Curr. Opin. Microbiol. *13*, 150–159.

Gerlach, G., von Wintzingerode, F., Middendorf, B., and Gross, R. (2001). Evolutionary trends in the genus *Bordetella*. Microbes Infect. *3*, 61–72.

Gerlach, G., Janzen, S., Beier, D., and Gross, R. (2004). Functional characterization of the BvgAS two-component system of *Bordetella holmesii*. Microbiology *150*, 3715–3729.

Gielen, M., Le Goff, A., Stroebel, D., Johnson, J.W., Neyton, J., and Paoletti, P. (2008). Structural rearrangements of NR1/NR2A NMDA receptors during allosteric inhibition. Neuron *57*, 80–93.

Gielen, M., Siegler Retchless, B., Mony, L., Johnson, J.W., and Paoletti, P. (2009). Mechanism of differential control of NMDA receptor activity by NR2 subunits. Nature *459*, 703–707.

Gilles-Gonzalez, M.A., Gonzalez, G., Perutz, M.F., Kiger, L., Marden, M.C., and Poyart, C. (1994). Heme-based sensors, exemplified by the kinase FixL, are a new class of heme protein with distinctive ligand binding and autoxidation. Biochemistry *33*, 8067–8073.

Giuliani, S.E., Frank, A.M., and Collart, F.R. (2008). Functional assignment of solute-binding proteins of ABC transporters using a fluorescence-based thermal shift assay. Biochemistry *47*, 13974–13984.

Goldberg, S.D., Clinthorne, G.D., Goulian, M., and DeGrado, W.F. (2010). Transmembrane polar interactions are required for signaling in the *Escherichia coli* sensor kinase PhoQ. Proc. Natl. Acad. Sci. U.S.A. *107*, 8141–8146.

Gordeliy, V.I., Labahn, J., Moukhametzianov, R., Efremov, R., Granzin, J., Schlesinger, R., Buldt, G., Savopol, T., Scheidig, A.J., Klare, J.P., *et al.* (2002). Molecular basis of transmembrane signalling by sensory rhodopsin II–transducer complex. Nature *419*, 484–487.

Goyard, S., Bellalou, J., Mireau, H., and Ullmann, A. (1994). Mutations in the *Bordetella pertussis bvgS* gene that confer altered expression of the *fhaB* gene in *Escherichia coli*. J. Bacteriol. *176*, 5163–5166.

Grasberger, B., Minton, A.P., DeLisi, C., and Metzger, H. (1986). Interaction between proteins localized in membranes. Proc. Natl. Acad. Sci. U.S.A. *83*, 6258–6262.

Gross, R., Guzman, C.A., Sebaihia, M., dos Santos, V.A., Pieper, D.H., Koebnik, R., Lechner, M., Bartels, D., Buhrmester, J., Choudhuri, J.V., *et al.* (2008). The missing link: *Bordetella petrii* is endowed with both the metabolic versatility of environmental bacteria and virulence traits of pathogenic Bordetellae. BMC Genomics *9*, 449.

He, X.L., Dukkipati, A., Wang, X., and Garcia, K.C. (2005). A new paradigm for hormone recognition and allosteric receptor activation revealed from structural studies of NPR-C. Peptides *26*, 1035–1043.

Herrou, J., Debrie, A.S., Willery, E., Renauld-Mongenie, G., Locht, C., Mooi, F., Jacob-Dubuisson, F., and Antoine, R. (2009). Molecular evolution of the two-component system BvgAS involved in virulence regulation in *Bordetella*. PLoS One *4*, e6996.

Herrou, J., Bompard, C., Wintjens, R., Dupre, E., Willery, E., Villeret, V., Locht, C., Antoine, R., and Jacob-Dubuisson, F. (2010). Periplasmic domain of the sensor kinase BvgS reveals a new paradigm for the Venus flytrap mechanism. Proc. Natl. Acad. Sci. U.S.A. *107*, 17351–17355.

Ho, W.W., Li, H., Eakanunkul, S., Tong, Y., Wilks, A., Guo, M., and Poulos, T.L. (2007). Holo- and apo-bound structures of bacterial periplasmic heme-binding proteins. J. Biol. Chem. *282*, 35796–35802.

Hulko, M., Berndt, F., Gruber, M., Linder, J.U., Truffault, V., Schultz, A., Martin, J., Schultz, J.E., Lupas, A.N., and Coles, M. (2006). The HAMP domain structure implies helix rotation in transmembrane signaling. Cell *126*, 929–940.

Huynh, T.N., and Stewart, V. (2011) Negative control in two-component signal transduction by transmitter phosphatase activity. Mol. Microbiol. *82*, 275–286.

Huynh, T.N., Noriega, C.E., and Stewart, V. (2010). Conserved mechanism for sensor phosphatase control of two-component signaling revealed in the nitrate sensor NarX. Proc. Natl. Acad. Sci. U.S.A. *107*, 21140–21145.

Irie, Y., Mattoo, S., and Yuk, M.H. (2004). The Bvg virulence control system regulates biofilm formation in *Bordetella bronchiseptica*. J. Bacteriol. *186*, 5692–5698.

Iuchi, S., and Lin, E.C. (1992). Mutational analysis of signal transduction by ArcB, a membrane sensor protein responsible for anaerobic repression of operons involved in the central aerobic pathways in *Escherichia coli*. J. Bacteriol. *174*, 3972–3980.

Jones, A.M., Boucher, P.E., Williams, C.L., Stibitz, S., and Cotter, P.A. (2005). Role of BvgA phosphorylation and DNA binding affinity in control of Bvg-mediated phenotypic phase transition in *Bordetella pertussis*. Mol. Microbiol. *58*, 700–713.

Joyce, S.A., Watson, R.J., and Clarke, D.J. (2006). The regulation of pathogenicity and mutualism in *Photorhabdus*. Curr. Opin. Microbiol. *9*, 127–132.

Kaspar, S., Perozzo, R., Reinelt, S., Meyer, M., Pfister, K., Scapozza, L., and Bott, M. (1999). The periplasmic domain of the histidine autokinase CitA functions as a highly specific citrate receptor. Mol. Microbiol. *33*, 858–872.

Kato, M., Mizuno, T., Shimizu, T., and Hakoshima, T. (1997). Insights into multistep phosphorelay from the crystal structure of the C-terminal HPt domain of ArcB. Cell *88*, 717–723.

Kelly, D.J., and Thomas, G.H. (2001). The tripartite ATP-independent periplasmic (TRAP) transporters of bacteria and archaea. FEMS Microbiol. Rev. *25*, 405–424.

Kinnear, S.M., Marques, R.R., and Carbonetti, N.H. (2001). Differential regulation of Bvg-activated virulence factors plays a role in *Bordetella pertussis* pathogenicity. Infect. Immun. *69*, 1983–1993.

Kristensen, A.S., Geballe, M.T., Snyder, J.P., and Traynelis, S.F. (2006). Glutamate receptors: variation in structure–function coupling. Trends Pharmacol. Sci. *27*, 65–69.

Kuchma, S.L., Connolly, J.P., and O'Toole, G.A. (2005). A three-component regulatory system regulates biofilm maturation and type III secretion in *Pseudomonas aeruginosa*. J. Bacteriol. *187*, 1441–1454.

Kulasekara, H.D., Ventre, I., Kulasekara, B.R., Lazdunski, A., Filloux, A., and Lory, S. (2005). A novel two-component system controls the expression of *Pseudomonas aeruginosa* fimbrial *cup* genes. Mol. Microbiol. *55*, 368–380.

Lacey, B.W. (1960). Antigenic modulation of *Bordetella pertussis*. J. Hyg. Camb. *58*, 57–93.

Lai, Y.C., Lin, G.T., Yang, S.L., Chang, H.Y., and Peng, H.L. (2003). Identification and characterization of KvgAS, a two-component system in *Klebsiella pneumoniae* CG43. FEMS Microbiol. Lett. *218*, 121–126.

Lamberti, Y.A., Hayes, J.A., Perez Vidakovics, M.L., Harvill, E.T., and Rodriguez, M.E. (2010). Intracellular trafficking of *Bordetella pertussis* in human macrophages. Infect. Immun. *78*, 907–913.

Laube, B., Hirai, H., Sturgess, M., Betz, H., and Kuhse, J. (1997). Molecular determinants of agonist discrimination by NMDA receptor subunits: analysis of the glutamate binding site on the NR2B subunit. Neuron *18*, 493–503.

Lawlor, M.S., Hsu, J., Rick, P.D., and Miller, V.L. (2005). Identification of *Klebsiella pneumoniae* virulence determinants using an intranasal infection model. Mol. Microbiol. *58*, 1054–1073.

Lin, C.T., and Peng, H.L. (2006). Regulation of the homologous two-component systems KvgAS and KvhAS in *Klebsiella pneumoniae* CG43. J. Biochem. *140*, 639–648.

Locher, K.P. (2004). Structure and mechanism of ABC transporters. Curr. Opin. Struct. Biol. *14*, 426–431.

Locht, C., Antoine, R., and Jacob-Dubuisson, F. (2001). *Bordetella pertussis*, molecular pathogenesis under multiple aspects. Curr. Opin. Microbiol. *4*, 82–89.

Manetti, R., Arico, B., Rappuoli, R., and Scarlato, V. (1994). Mutations in the linker region of BvgS abolish response to environmental signals for the regulation of the virulence factors in *Bordetella pertussis*. Gene *150*, 123–127.

Marina, A., Waldburger, C.D., and Hendrickson, W.A. (2005). Structure of the entire cytoplasmic portion of a sensor histidine-kinase protein. EMBO J. *24*, 4247–4259.

Martinez de Tejada, G., Miller, J.F., and Cotter, P.A. (1996). Comparative analysis of the virulence control systems of *Bordetella pertussis* and *Bordetella bronchiseptica*. Mol. Microbiol. *22*, 895–908.

Martinez de Tejada, G., Cotter, P.A., Heininger, U., Camilli, A., Akerley, B.J., Mekalanos, J.J., and Miller, J.F. (1998). Neither the Bvg$^-$ phase nor the *vrg6* locus of *Bordetella pertussis* is required for respiratory infection in mice. Infect. Immun. *66*, 2762–2768.

Martinez-Wilson, H.F., Tamayo, R., Tischler, A.D., Lazinski, D.W., and Camilli, A. (2008). The *Vibrio cholerae* hybrid sensor kinase VieS contributes to motility and biofilm regulation by altering the cyclic diguanylate level. J. Bacteriol. *190*, 6439–6447.

Maslennikov, I., Klammt, C., Hwang, E., Kefala, G., Okamura, M., Esquivies, L., Mors, K., Glaubitz, C., Kwiatkowski, W., Jeon, Y.H., *et al.* (2010). Membrane domain structures of three classes of histidine kinase receptors by cell-free expression and rapid NMR analysis. Proc. Natl. Acad. Sci. U.S.A. *107*, 10902–10907.

Masuda, N., and Church, G.M. (2002). *Escherichia coli* gene expression responsive to levels of the response regulator EvgA. J. Bacteriol. *184*, 6225–6234.

Masure, H.R. (1992). Modulation of adenylate cyclase toxin production as *Bordetella pertussis* enters human macrophages. Proc. Natl. Acad. Sci. U.S.A. *89*, 6521–6525.

Melton, A.R., and Weiss, A.A. (1993). Characterization of environmental regulators of *Bordetella pertussis*. Infect. Immun. *61*, 807–815.

Merkel, T.J., Barros, C., and Stibitz, S. (1998). Characterization of the *bvgR* locus of *Bordetella pertussis*. J. Bacteriol. *180*, 1682–1690.

Miller, D.M., 3rd, Olson, J.S., Pflugrath, J.W., and Quiocho, F.A. (1983). Rates of ligand binding to periplasmic proteins involved in bacterial transport and chemotaxis. J. Biol. Chem. *258*, 13665–13672.

Miller, J.F., Johnson, S.A., Black, W.J., Beattie, D.T., Mekalanos, J.J., and Falkow, S. (1992). Constitutive sensory transduction mutations in the *Bordetella pertussis bvgS* gene. J. Bacteriol. *174*, 970–979.

Mishra, M., and Deora, R. (2005). Mode of action of the *Bordetella* BvgA protein: transcriptional activation and repression of the *Bordetella bronchiseptica bipA* promoter. J. Bacteriol. *187*, 6290–6299.

Moglich, A., Ayers, R.A., and Moffat, K. (2009). Structure and signaling mechanism of Per-ARNT-Sim domains. Structure *17*, 1282–1294.

Mony, L., Zhu, S., Carvalho, S., and Paoletti, P. (2011). Molecular basis of positive allosteric modulation of GluN2B NMDA receptors by polyamines. EMBO J. *30*, 3134–3146.

Moore, J.O., and Hendrickson, W.A. (2009). Structural analysis of sensor domains from the TMAO-responsive histidine kinase receptor TorS. Structure *17*, 1195–1204.

Neiditch, M.B., Federle, M.J., Miller, S.T., Bassler, B.L., and Hughson, F.M. (2005). Regulation of LuxPQ receptor activity by the quorum-sensing signal autoinducer-2. Mol. Cell. *18*, 507–518.

Neiditch, M.B., Federle, M.J., Pompeani, A.J., Kelly, R.C., Swem, D.L., Jeffrey, P.D., Bassler, B.L., and Hughson, F.M. (2006). Ligand-induced asymmetry in histidine sensor kinase complex regulates quorum sensing. Cell *126*, 1095–1108.

Neo, Y., Li, R., Howe, J., Hoo, R., Pant, A., Ho, S., and Alonso, S. (2010). Evidence for an intact polysaccharide capsule in *Bordetella pertussis*. Microbes Infect. *12*, 238–245.

Nishino, K., and Yamaguchi, A. (2002). EvgA of the two-component signal transduction system modulates production of the *yhiUV* multidrug transporter in *Escherichia coli*. J. Bacteriol. *184*, 2319–2323.

O'Hara, P.J., Sheppard, P.O., Thogersen, H., Venezia, D., Haldeman, B.A., McGrane, V., Houamed, K.M., Thomsen, C., Gilbert, T.L., and Mulvihill, E.R. (1993). The ligand-binding domain in metabotropic glutamate receptors is related to bacterial periplasmic binding proteins. Neuron *11*, 41–52.

Oh, B.H., Pandit, J., Kang, C.H., Nikaido, K., Gokcen, S., Ames, G.F., and Kim, S.H. (1993). Three-dimensional structures of the periplasmic lysine/arginine/ornithine-binding protein with and without a ligand. J. Biol. Chem. *268*, 11348–11355.

Oh, B.H., Kang, C.H., De Bondt, H., Kim, S.H., Nikaido, K., Joshi, A.K., and Ames, G.F. (1994). The bacterial periplasmic histidine-binding protein. Structure–function analysis of the ligand-binding site and comparison with related proteins. J. Biol. Chem. *269*, 4135–4143.

Pantoliano, M.W., Petrella, E.C., Kwasnoski, J.D., Lobanov, V.S., Myslik, J., Graf, E., Carver, T., Asel, E., Springer, B.A., Lane, P., *et al.* (2001). High-density miniaturized thermal shift assays as a general strategy for drug discovery. J. Biomol. Screen *6*, 429–440.

Paoletti, P., Perin-Dureau, F., Fayyazuddin, A., Le Goff, A., Callebaut, I., and Neyton, J. (2000). Molecular organization of a zinc binding N-terminal modulatory domain in a NMDA receptor subunit. Neuron *28*, 911–925.

Pappalardo, L., Janausch, I.G., Vijayan, V., Zientz, E., Junker, J., Peti, W., Zweckstetter, M., Unden, G., and Griesinger, C. (2003). The NMR structure of the sensory domain of the membranous two-component fumarate sensor (histidine protein kinase) DcuS of *Escherichia coli*. J. Biol. Chem. *278*, 39185–39188.

Parkhill, J., Sebaihia, M., Preston, A., Murphy, L.D., Thomson, N., Harris, D.E., Holden, M.T., Churcher, C.M., Bentley, S.D., Mungall, K.L., *et al.* (2003). Comparative analysis of the genome sequences of *Bordetella pertussis*, *Bordetella parapertussis* and *Bordetella bronchiseptica*. Nat. Genet. *35*, 32–40.

Paternain, A.V., Cohen, A., Stern-Bach, Y., and Lerma, J. (2003). A role for extracellular Na^+ in the channel gating of native and recombinant kainate receptors. J. Neurosci. *23*, 8641–8648.

Pazour, G.J., Ta, C.N., and Das, A. (1991). Mutants of *Agrobacterium tumefaciens* with elevated *vir* gene expression. Proc. Natl. Acad. Sci. U.S.A. *88*, 6941–6945.

Perraud, A.L., Kimmel, B., Weiss, V., and Gross, R. (1998). Specificity of the BvgAS and EvgAS phosphorelay is mediated by the C- terminal HPt domains of the sensor proteins. Mol. Microbiol. *27*, 875–887.

Perraud, A.L., Rippe, K., Bantscheff, M., Glocker, M., Lucassen, M., Jung, K., Sebald, W., Weiss, V., and Gross, R. (2000). Dimerization of signalling modules of the EvgAS and BvgAS phosphorelay systems. Biochim. Biophys. Acta *1478*, 341–354.

Plested, A.J., and Mayer, M.L. (2007). Structure and mechanism of kainate receptor modulation by anions. Neuron *53*, 829–841.

Plested, A.J., Vijayan, R., Biggin, P.C., and Mayer, M.L. (2008). Molecular basis of kainate receptor modulation by sodium. Neuron *58*, 720–735.

Ponting, C.P., and Aravind, L. (1997). PAS: a multifunctional domain family comes to light. Curr. Biol. *7*, R674–R677.

Porter, J.F., and Wardlaw, A.C. (1993). Long-term survival of *Bordetella bronchiseptica* in lakewater and in buffered saline without added nutrients. FEMS Microbiol. Lett. *110*, 33–36.

Preston, A., Parkhill, J., and Maskell, D.J. (2004). The Bordetellae: lessons from genomics. Nat. Rev. Microbiol. *2*, 379–390.

Prugnola, A., Arico, B., Manetti, R., Rappuoli, R., and Scarlato, V. (1995). Response of the *bvg* regulon of *Bordetella pertussis* to different temperatures and short-term temperature shifts. Microbiology *141*, 2529–2534.

Quiocho, F.A., and Ledvina, P.S. (1996). Atomic structure and specificity of bacterial periplasmic receptors for active transport and chemotaxis: variation of common themes. Mol. Microbiol. *20*, 17–25.

Reinelt, S., Hofmann, E., Gerharz, T., Bott, M., and Madden, D.R. (2003). The structure of the periplasmic ligand-binding domain of the sensor kinase CitA reveals the first extracellular PAS domain. J. Biol. Chem. *278*, 39189–39196.

Rogers, E.A., Terekhova, D., Zhang, H.M., Hovis, K.M., Schwartz, I., and Marconi, R.T. (2009). Rrp1, a cyclic-di-GMP-producing response regulator, is an important regulator of *Borrelia burgdorferi* core cellular functions. Mol. Microbiol. *71*, 1551–1573.

Rogov, V.V., Bernhard, F., Lohr, F., and Dotsch, V. (2004). Solution structure of the *Escherichia coli* YojN histidine-phosphotransferase domain and its interaction with cognate phosphoryl receiver domains. J. Mol. Biol. *343*, 1035–1048.

Roy, C.R., and Falkow, S. (1991). Identification of *Bordetella pertussis* regulatory sequences required for transcriptional activation of the *fhaB* gene and autoregulation of the *bvgAS* operon. J. Bacteriol. *173*, 2385–2392.

Scarlato, V., Arico, B., Prugnola, A., and Rappuoli, R. (1991). Sequential activation and environmental regulation of virulence genes in *Bordetella pertussis*. EMBO J. *10*, 3971–3975.

Schneider, D.R., and Parker, C.D. (1982). Effect of pyridines on phenotypic properties of *Bordetella pertussis*. Infect. Immun. *38*, 548–553.

Schumacher, M.A., Choi, K.Y., Zalkin, H., and Brennan, R.G. (1994). Crystal structure of LacI member, PurR, bound to DNA: minor groove binding by alpha helices. Science *266*, 763–770.

Sevvana, M., Vijayan, V., Zweckstetter, M., Reinelt, S., Madden, D.R., Herbst-Irmer, R., Sheldrick, G.M., Bott, M., Griesinger, C., and Becker, S. (2008). A ligand-induced switch in the periplasmic domain of sensor histidine kinase CitA. J. Mol. Biol. *377*, 512–523.

Sivaneson, M., Mikkelsen, H., Ventre, I., Bordi, C., and Filloux, A. (2011). Two-component regulatory systems in *Pseudomonas aeruginosa*: an intricate network mediating fimbrial and efflux pump gene expression. Mol. Microbiol. 79, 1353–1366.

Slavny, P., Little, R., Salinas, P., Clarke, T.A., and Dixon, R. (2010). Quaternary structure changes in a second Per-Arnt-Sim domain mediate intramolecular redox signal relay in the NifL regulatory protein. Mol. Microbiol. *75*, 61–75.

Sobolevsky, A.I., Rosconi, M.P., and Gouaux, E. (2009). X-ray structure, symmetry and mechanism of an AMPA-subtype glutamate receptor. Nature *462*, 745–756.

Steffen, P., Goyard, S., and Ullmann, A. (1996). Phosphorylated BvgA is sufficient for transcriptional activation of virulence-regulated genes in *Bordetella pertussis*. EMBO J. *15*, 102–109.

Stenson, T.H., and Peppler, M.S. (1995). Identification of two *bvg*-repressed surface proteins of *Bordetella pertussis*. Infect. Immun. *63*, 3780–3789.

Stewart, V., and Chen, L.L. (2010) The S helix mediates signal transmission as a HAMP domain coiled-coil extension in the NarX nitrate sensor from *Escherichia coli* K-12. J. Bacteriol. *192*, 734–745.

Stibitz, S. (1994). Mutations in the *bvgA* gene of *Bordetella pertussis* that differentially affect regulation of virulence determinants. J. Bacteriol. *176*, 5615–5621.

Stibitz, S. (1998). Mutations affecting the alpha subunit of *Bordetella pertussis* RNA polymerase suppress growth inhibition conferred by short C-terminal deletions of the response regulator BvgA. J. Bacteriol. *180*, 2484–2492.

Stibitz, S., and Yang, M.S. (1991). Subcellular localization and immunological detection of proteins encoded by the *vir* locus of *Bordetella pertussis*. J. Bacteriol. *173*, 4288–4296.

Stock, J. (1996). Signaling across membranes: a one and a two and a... Science *274*, 370–371.

Swain, K.E., Gonzalez, M.A., and Falke, J.J. (2009) Engineered socket study of signaling through a four-helix bundle: evidence for a yin–yang mechanism in the kinase control module of the aspartate receptor. Biochemistry *48*, 9266–9277.

Swanson, R.V., Bourret, R.B., and Simon, M.I. (1993). Intermolecular complementation of the kinase activity of CheA. Mol. Microbiol. *8*, 435–441.

Taylor, B.L. (2007). Aer on the inside looking out: paradigm for a PAS-HAMP role in sensing oxygen, redox and energy. Mol. Microbiol. *65*, 1415–1424.

Taylor, B.L., Zhulin, I.B., and Johnson, M.S. (1999). Aerotaxis and other energy-sensing behavior in bacteria. Annu. Rev. Microbiol. *53*, 103–128.

Thalen, M., van den, I.J., Jiskoot, W., Zomer, B., Roholl, P., de Gooijer, C., Beuvery, C., and Tramper, J. (1999). Rational medium design for *Bordetella pertussis*: basic metabolism. J. Biotechnol. *75*, 147–159.

Tischler, A.D., Lee, S.H., and Camilli, A. (2002). The *Vibrio cholerae vieSAB* locus encodes a pathway contributing to cholera toxin production. J. Bacteriol. *184*, 4104–4113.

Tomich, M., and Mohr, C.D. (2004a). Genetic characterization of a multicomponent signal transduction system controlling the expression of cable pili in *Burkholderia cenocepacia*. J. Bacteriol. *186*, 3826–3836.

Tomich, M., and Mohr, C.D. (2004b). Transcriptional and posttranscriptional control of cable pilus gene expression in *Burkholderia cenocepacia*. J. Bacteriol. *186*, 1009–1020.

Trakhanov, S., Vyas, N.K., Luecke, H., Kristensen, D.M., Ma, J., and Quiocho, F.A. (2005). Ligand-free and -bound structures of the binding protein (LivJ) of the *Escherichia coli* ABC leucine/isoleucine/valine transport system: trajectory and dynamics of the interdomain rotation and ligand specificity. Biochemistry *44*, 6597–6608.

Uhl, M.A., and Miller, J.F. (1994). Autophosphorylation and phosphotransfer in the *Bordetella pertussis* BvgAS signal transduction cascade. Proc. Natl. Acad. Sci. U.S.A. *91*, 1163–1167.

Uhl, M.A., and Miller, J.F. (1996a). Central role of the BvgS receiver as a phosphorylated intermediate in a complex two-component phosphorelay. J. Biol. Chem. *271*, 33176–33180.

Uhl, M.A., and Miller, J.F. (1996b). Integration of multiple domains in a two-component sensor protein: the *Bordetella pertussis* BvgAS phosphorelay. EMBO J. *15*, 1028–1036.

Ulrich, L.E., and Uhlin, I.B. (2010) The MiST2 database : a comprehenive genomics resource on microbial signal transduction. Nucl. Ac. Res. *38*, D401–D407.

Utsumi, R., Katayama, S., Taniguchi, M., Horie, T., Ikeda, M., Igaki, S., Nakagawa, H., Miwa, A., Tanabe, H., and Noda, M. (1994). Newly identified genes involved in the signal transduction of *Escherichia coli* K-12. Gene *140*, 73–77.

Vandamme, P., Hommez, J., Vancanneyt, M., Monsieurs, M., Hoste, B., Cookson, B.T., Wirsing von Koenig, C.H., Kersters, K., and Blackall, P.J. (1995). *Bordetella hinzii* sp. nov., isolated from poultry and humans. Int. J. Syst. Bacteriol. *45*, 37–45.

Vandamme, P., Heyndricks, M., Vancanneyt, M., Hoste, B., De Vos, P., Falsen, E., Kersters, K., and Hinz, K.H. (1996). *Bordetella trematum* sp. nov., isolated from wounds and ear infections in humans, and reassessment of *Alcaligenes denitrificans*. Int. J. Syst. Bacteriol. *46*, 849–858.

Veal-Carr, W.L., and Stibitz, S. (2005). Demonstration of differential virulence gene promoter activation *in vivo* in *Bordetella pertussis* using RIVET. Mol. Microbiol. *55*, 788–798.

Waldburger, C.D., and Sauer, R.T. (1996). Signal detection by the PhoQ sensor-transmitter. Characterization of the sensor domain and a response-impaired mutant that identifies ligand-binding determinants. J. Biol. Chem. *271*, 26630–26636.

Walker, K.A., and Miller, V.L. (2004). Regulation of the Ysa type III secretion system of *Yersinia enterocolitica* by YsaE/SycB and YsrS/YsrR. J. Bacteriol. *186*, 4056–4066.

Walker, K.A., Obrist, M.W., Mildiner-Earley, S., and Miller, V.L. (2010). Identification of YsrT and evidence that YsrRST constitute a unique phosphorelay system in *Yersinia enterocolitica*. J. Bacteriol. *192*, 5887–5897.

Weiss, A.A., Hewlett, E.L., Myers, G.A., and Falkow, S. (1983). Tn5-induced mutations affecting virulence factors of *Bordetella pertussis*. Infect. Immun. *42*, 33–41.

Weyant, R.S., Hollis, D.G., Weaver, R.E., Amin, M.F., Steigerwalt, A.G., O'Connor, S.P., Whitney, A.M., Daneshvar, M.I., Moss, C.W., and Brenner, D.J. (1995). *Bordetella holmesii* sp. nov., a new Gram-negative species associated with septicemia. J. Clin. Microbiol. *33*, 1–7.

Williams, C.L., Boucher, P.E., Stibitz, S., and Cotter, P.A. (2005). BvgA functions as both an activator and a repressor to control Bvg phase expression of *bipA* in *Bordetella pertussis*. Mol. Microbiol. *56*, 175–188.

Winnen, B., Hvorup, R.N., and Saier, M.H., Jr. (2003). The tripartite tricarboxylate transporter (TTT) family. Res. Microbiol. *154*, 457–465.

von Wintzingerode, F., Schattke, A., Siddiqui, R.A., Rosick, U., Göbel, U.B., and Gross, R. (2001). *Bordetella petrii* sp. nov., isolated from an anaerobic bioreactor, and emended description of the genus *Bordetella*. Int. J. Syst. Evol. Microbiol. *51*, 1257–1265.

Xu, Q., Porter, S.W., and West, A.H. (2003). The yeast YPD1/SLN1 complex: insights into molecular recognition in two-component signaling systems. Structure *11*, 1569–1581.

Yamada, S., Sugimoto, H., Kobayashi, M., Ohno, A., Nakamura, H., and Shiro, Y. (2009). Structure of PAS-linked histidine kinase and the response regulator complex. Structure *17*, 1333–1344.

Zhang, Z., and Hendrickson, W.A. (2010). Structural characterization of the predominant family of histidine kinase sensor domains. J. Mol. Biol. *400*, 335–353.

Zhao, X., Copeland, D.M., Soares, A.S., and West, A.H. (2008). Crystal structure of a complex between the phosphorelay protein YPD1 and the response regulator domain of SLN1 bound to a phosphoryl analog. J. Mol. Biol. *375*, 1141–1151.

Zhou, Q., Ames, P., and Parkinson, J.S. (2009). Mutational analyses of HAMP helices suggest a dynamic bundle model of input–output signalling in chemoreceptors. Mol. Microbiol. *73*, 801–814.

Zhulin, I.B., Taylor, B.L., and Dixon, R. (1997). PAS domain S-boxes in Archaea, Bacteria and sensors for oxygen and redox. Trends Biochem. Sci. *22*, 331–333.

Phosphatase Activity of Two-component System Transmitter Proteins

5

Alexander J. Ninfa

Abstract

Many of the two-component system transmitter proteins bring about both the phosphorylation and dephosphorylation of their cognate receiver proteins. While the mechanism of the transmitter protein kinase activity was relatively easy to discern biochemically, investigation of the phosphatase activity has proven more difficult owing to the involvement of widely dispersed portions of the transmitter protein in the activity and the corresponding requirement to study the activity using intact, full-length transmitter proteins. Here, I review the genetics, physiology, and biochemistry of the transmitter protein phosphatase activity, with special emphasis on possible mechanisms of receiver protein dephosphorylation, relationship of the transmitter phosphatase activity to the autophosphatase activity of the receiver protein, and physiological significance of the activity.

Introduction

Signal transduction via reversible covalent modification in biological systems typically employs mechanisms for both signal propagation and signal termination. This is particularly true when the time scales of the changes in input stimulation for the signalling systems are far shorter than the stability of the covalent bond that reflects a signalling state or the stability of the protein that gets modified. Thus, cyclase enzymes that produce nucleotide-derived second messengers are opposed by phosphodiesterases that destroy the second messenger, methyltransferases are opposed by demethylating enzymes, nucleotidyl transferases are opposed by enzymes that remove the nucleotidyl group, ADP ribosyl transferases are opposed by glycohydrolases, and protein kinases are opposed by the action of phosphoprotein phosphatases. In all of these cases, the chemical stability of the covalent bond that reflects a signalling state, be it an internal bond in a complex nucleotide or a covalent modification of a protein side chain, is far too great for the covalent modification to be useful in signalling rapidly changing environmental conditions, unless a mechanism for rapidly reversing the covalent modification is available. As I noted in earlier reviews (Stock *et al.*, 1989; Ninfa, 1991), the two-component signal transduction systems, in which signalling involves reversible phosphorylation and activation of the receiver (response regulator) protein, control a whole range of biological phenomena, that occur over a wide range of time scales. When this entire range of time scales is considered with reference to the bacterial growth rate and the stability of proteins and the relevant covalent bonds, it becomes clear that even relatively slow bacterial developmental pathways or adaptive responses that occur over multiple cell cycles would still in most cases be too fast for regulation by protein phosphorylation, unless there exists a mechanism for the termination of signalling by protein dephosphorylation. Thus, first principles suggest that the dephosphorylation reactions that terminate signalling play a central role in the physiological function of the two-component signal transduction systems.

A number of distinct mechanisms result in the dephosphorylation of the phosphorylated receiver proteins of two-component systems. In some cases, one or more separate phosphatase enzymes may act upon the phosphorylated receiver protein. In the regulation of development in *B. subtilis*, this arrangement allows independent control of the activity of each of the phosphatase enzymes, and thus allows complex combinatorial control of the activity of the key receiver protein involved in developmental decisions (Perego *et al.*, 1994). In the *Escherichia coli* chemotaxis system, the utilization of a separate phosphatase enzyme, CheZ, may reflect the complex and highly cooperative nature of signal propagation that employs a large and heterogeneous receptor–CheA-CheW complex. The properties of these separate phosphatases have been recently discussed (Silversmith, 2010), and they will not be reviewed here. Another mechanism leading to dephosphorylation of the receiver protein is due to their self-catalysed dephosphorylation, an activity that has been referred to as the 'autophosphatase' activity (Keener and Kustu, 1988; Fig. 5.1). The two-component system receiver proteins have widely differing autophosphatase activities, reflecting the vastly different time scales of the physiological phenomena they control (Ninfa, 1991). A very common observation concerning many phosphorylated receiver proteins is that while they contain a phospho-aspartate residue, the stability of these phosphoryl groups is significantly less than the chemical stability of phospho-aspartate residues. Furthermore, the stability of receiver protein phosphoryl groups is considerably extended upon denaturation of the proteins. Thus, they 'catalyse' their own dephosphorylation. A third mechanism for

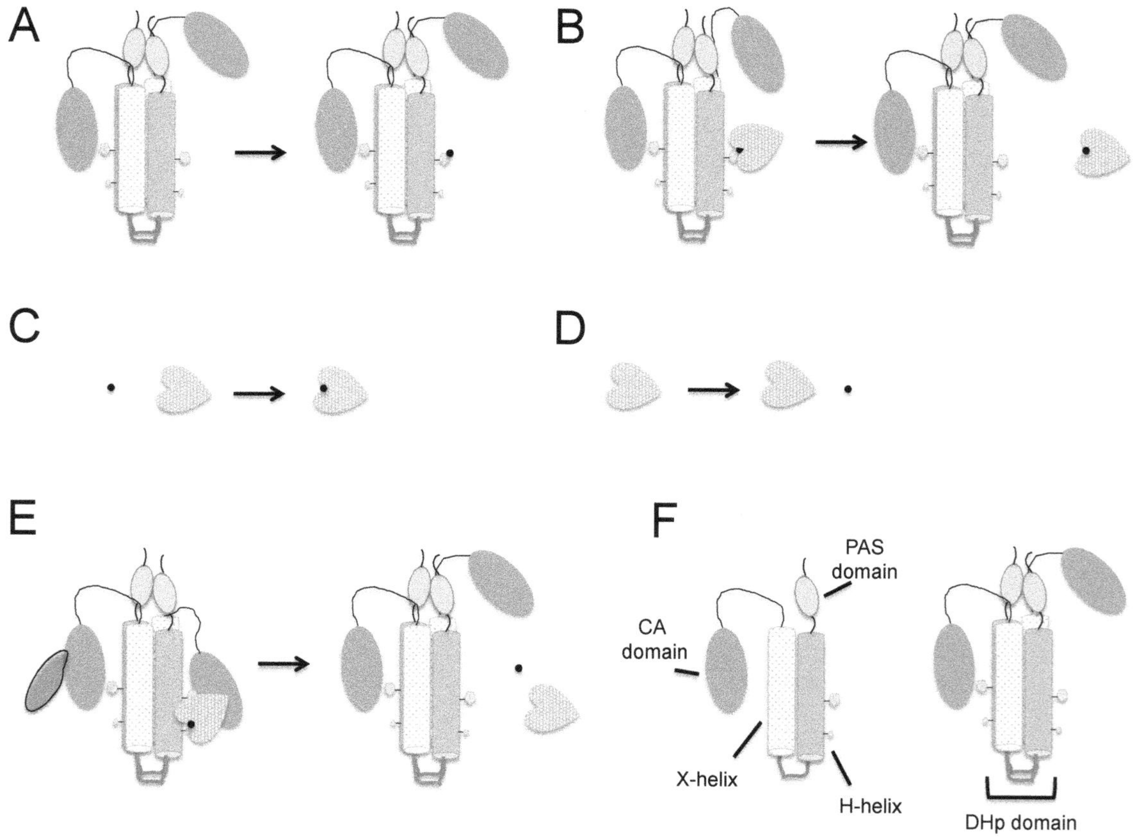

Figure 5.1 Cartoon depiction of the catalytic activities of the NRII-NRI two-component system. (A) Autophosphorylation of the NRII transmitter protein. The phosphoryl group is signified as a black dot. (B) Phosphotransfer from the autophosphorylated transmitter protein to the receiver protein. (C) Phosphorylation of the receiver protein by acetyl phosphate. (D) Autophosphatase activity of the phosphorylated receiver protein. (E) Rapid dephosphorylation of NRI~P by the complex of NRII and PII. This activity is referred to as the regulated phosphatase activity, or simply as the phosphatase activity. (F) Cartoon depiction of the transmitter protein monomer and dimer.

dephosphorylation of the phosphorylated receiver protein is dependent upon interaction with the transmitter protein (Fig. 5.1). This mechanism in some cases has been shown to result in the very rapid dephosphorylation of the phosphorylated receiver protein (Ninfa and Magasanik, 1986), and has been referred to as the 'regulated phosphatase activity' (Keener and Kustu, 1988), or more simply, as the transmitter phosphatase activity. This activity results in the hydrolysis of the phosphoryl group of the receiver without the intermediate of a phosphorylated transmitter protein, as far as we know. Finally, phosphoryl groups may be lost from receiver proteins by reversal of the phosphotransfer reaction from the transmitter, resulting in the phosphorylation of the transmitter upon its active site histidine residue (Weiss and Magasanik, 1988). This dephosphorylation mechanism then requires another step in which the transmitter is dephosphorylated to prevent re-phosphorylation of the receiver. In this chapter, I shall review the available information on the regulated phosphatase activity of two-component system transmitter proteins that results in the rapid dephopshorylation of their cognate receiver proteins. Since it has been argued that this transmitter protein phosphatase activity reflects the activation of the receiver protein autophosphatase activity, I shall also briefly consider the autophosphatase activity of the phosphorylated receiver and its role in receiver protein dephosphorylation. I will not consider reverse-phosphotransfer transfer mechanisms except to note that they have been broadly noted in certain phosphorelay type systems as well as in orthodox systems and may have an important physiological role (Georgellis *et al.*, 1998; Shi *et al.*, 1999; Freeman *et al.*, 2000; Ansaldi *et al.*, 2001). I shall start by considering information from CheY, CheZ, Spo0B, and the NRII-NRI systems and develop a working hypothesis for the mechanism of the phosphatase activity and its regulation by environmental stimuli. Next, I shall briefly review other reports examining the presence or regulation of transmitter protein phosphatase activities. I shall conclude the review with a brief consideration of the genetic and physiological data that examine the role of the transmitter protein phosphatase activity in intact cells.

Lessons from CheY

The receiver domain consists of a five-stranded β-sheet surrounded by five α-helices; the active site includes the site of modification (an aspartate at the C-terminal end of β-strand 3), two additional aspartate residues, a conserved lysine and threonine, and a Mg^{2+} ion (Stock *et al.*, 1993). The most likely mechanism for loss of the phosphoryl group from a phosphorylated receiver would be by simple hydrolysis by water or hydroxide. That is, via a nucleophilic attack by an oxygen atom from hydroxide on the phosphorus atom of the phosphoryl group. However, other mechanisms can be envisaged, such as via the formation and breakdown of a succinimide (Wolanin *et al.*, 2003). The phosphorylated form of the CheY receiver protein is dephosphorylated by two (known) routes, via a powerful autophosphatase activity and via interaction with the CheZ phosphatase, which under saturating conditions increases the dephosphorylation rate about 10-fold. Wolanin and colleagues examined the incorporation of ^{18}O into CheY as it underwent multiple cycles of dephosphorylation via both mechanisms; these investigators did not detect significant incorporation of ^{18}O into CheY (Wolanin *et al.*, 2003). This is consistent with the idea that the dephosphorylation reaction is a simple hydrolysis.

Mutations that affect the autophosphatase and CheZ-mediated dephosphorylation of CheY have been identified (Lukat *et al.*, 1991; Sanna *et al.*, 1995; Silversmith *et al.*, 2003). One of the most interesting of the mutations affects the highly conserved K109 residue that forms part of the active site. Alteration of this residue resulted in a very significant defect in the autophosphatase activity of CheY~P as well as a very significant defect in dephosphorylation by CheZ (Lukat *et al.*, 1991); in a later section I will note that the analogous mutation in the receiver domain of NRI had the same effect, eliminating both autophosphatase activity and dephosphorylation by the NRII transmitter phosphatase activity. Another group of mutations have been identified in CheY that have little effect on the CheY~P autophosphatase activity, but eliminate dephosphorylation of CheY by CheZ (Sanna *et al.*, 1995; Silversmith *et al.*, 2003). In a later section, I will note that an analogous mutation has been identified in the

NRI receiver domain, that eliminates the transmitter phosphatase action while not affecting the autophosphatase rate. These classes of mutants in CheY, and the absence of a class of mutations that eliminate the CheY autophosphatase yet have no effect on CheZ-mediated dephosphorylation, are consistent with the idea that CheZ acts by increasing the rate of the autophosphatase reaction.

Lessons from the NRII-NRI system

The NRI-NRII two-component system interacts with two additional proteins, the PII protein and the signal-transducing uridylyltransferase/uridylyl-removing enzyme (UTase/UR), to comprise a four-protein signal transduction system that regulates the initiation of transcription of nitrogen-regulated (Ntr) genes in *E. coli* (Fig. 5.2). That fact that all of these proteins were soluble simplified their purification and the biochemical characterization of their activities; this, in turn allowed detailed structure–function studies of each protein and studies of the physiological roles of their catalytic activities. Consequently, our knowledge of many aspects of both the kinase and phosphatase activities of two-component systems are most advanced for the Ntr system of *E. coli*. Furthermore, the complete reconstitution of the signal transduction system from purified components, and the analysis of how the system integrates different signals to control Ntr genes, has allowed us to have a detailed understanding of the signal processing properties of the system. In this section, I will review some of the key findings concerning the autophosphatase activity of NRI and the phosphatase activity of NRII and its regulation by PII.

NRI consists of a receiver domain, fused to an AAA^+ output domain that activates transcription. Phosphorylation of the receiver domain of NRI by NRII results in the formation of an oligomer (hexamer or heptamer) that binds to DNA enhancer sequences with high affinity, and cleaves ATP while bringing about the formation of active transcription complexes with σ^{54}-containing RNA polymerase (Ninfa *et al.*, 1987; Popham *et al.*, 1989; Porter *et al.*, 1993; Lee *et al.*, 2003). The dephosphorylation of NRI~P results in its return to the inactive state, and causes the cessation of transcription from Ntr promoters. The isolated N-terminal receiver domain of NRI appears to show the same phosphorylation and dephosphorylation properties as does the intact

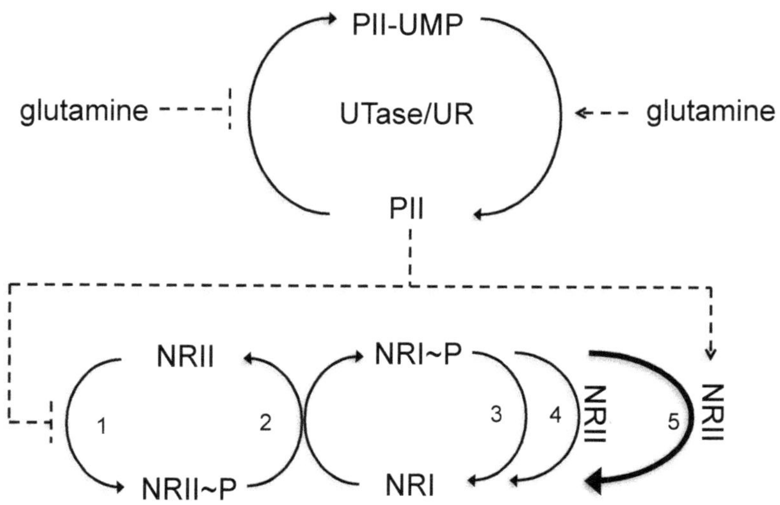

Figure 5.2 Signal transduction system for the regulation of Ntr gene transcription by the cellular glutamine concentration. The UTase/UR enzyme controls the PII uridylylation status in response to glutamine. Glutamine activates the UR activity of the UTase/UR and inhibits the UT activity of UTase/UR. Uridylylated PII subunits cannot bind to NRII. (1) NRII autophosphorylation reaction, inhibited by PII. (2) phosphotransfer reaction. (3) autophosphatase activity of NRI~P. (4) basal NRII phosphatase activity. (5) phosphatase activity of the NRII–PII complex. NRI~P is required for the activation of Ntr gene transcription.

protein (Keener and Kustu, 1988), and thus many studies were conducted using the purified N-terminal domain of NRI as a surrogate for the full-length protein. The phosphorylation and dephosphorylation of NRI are mediated by NRII, under the control of the PII protein (Figs. 5.1 and 5.2), as will be discussed below, with PII causing NRII to bring about the dephosphorylation of NRI~P. The ability of PII to interact with NRII is regulated by its reversible uridylylation and deuridylylation, catalysed by the UTase/UR, with PII uridylylation blocking its interaction with NRII (Atkinson *et al.*, 1994, Fig. 5.2). The UTase/UR is a key sensor of cellular nitrogen status; glutamine inhibits its ability to uridylylate PII and stimulates its ability to deuridylylate PII. Thus, the UTase/UR passes a signal of nitrogen sufficiency to PII (by deuridylylating it), causing PII to interact with NRII and bring about the dephosphorylation of NRI~P (Ninfa and Magasanik, 1986; Fig. 5.2).

NRII is involved in both the phosphorylation and dephosphorylation of NRI, but is essential for neither process. Like many receiver proteins, NRI can become directly phosphorylated by acetyl phosphate *in vitro*, and acetyl phosphate also seems to be able to phosphorylate NRI within intact cells (Feng *et al.*, 1992). NRI can also be phosphorylated by phosphotransfer from its cognate transmitter protein, NRII (Weiss and Magasanik, 1988). The NRI~P formed by either acetyl phosphate or NRII seemed to be identical in sharing the ability to activate *glnA* transcription *in vitro* (Feng *et al.*, 1992). In experiments with purified components, NRI has also been phosphorylated by numerous other transmitter proteins, such as CheA and EnvZ, and these transmitter proteins have been used to activate *glnA* transcription in an *in vitro* transcription assay (Ninfa *et al.*, 1988). Phosphorylation of NRI by CheA and EnvZ is thought to reflect non-physiological cross-talk resulting from the relative low specificity of the proteins and our ability to push the phosphorylation reactions *in vitro* with high protein concentrations.

NRI and its isolated receiver domain have readily discernible autophosphatase activity; the phosphorylated protein has a $t_{1/2}$ at pH 7.0 of about 4 minutes, and this stability is increased to at least several hours upon denaturation of the protein with SDS (Keener and Kustu, 1988; Weiss and Magasanik, 1988; Ninfa, 1991). As is the case with CheY (Lukat *et al.*, 1991), a mutation that altered the highly conserved lysine residue that forms part of the wall of the active site had a dramatic effect on the autophosphatase activity; in NRI the K104Q alteration essentially eliminated the autophosphatase activity (Pioszak and Ninfa, 2004). This alteration of the NRI receiver domain also eliminated the ability of the NRII–PII complex to bring about the dephosphorylation of NRI~P. By contrast, a deletion of aspartate at residue 10 of the NRI receiver domain resulted in an altered protein that displayed normal autophosphatase activity, but lacked the ability to be deuridylylated by the NRII–PII complex (Pioszak and Ninfa, 2004). These observations suggested that the NRII–PII complex brought about the rapid dephosphorylation of NRI~P by stimulating the intrinsic autophosphatase activity of NRI~P, as opposed to operating as a classical phosphatase.

The NRII protein, even in the absence of PII, displays a weak ability to bring about the dephosphorylation of NRI~P; this activity may be thought of as its 'basal' phosphatase activity (Jiang *et al.*, 2000; Fig. 5.2). This basal activity was increased many fold upon interaction of NRII with PII (Fig. 5.2). Both the basal phosphatase activity of NRII and the very robust phosphatase activity of the NRII–PII complex required the binding of an adenylylate nucleotide to NRII, but did not require the cleavage of this nucleotide. Thus, AMP-PNP acted approximately as well as did ATP to stimulate the phosphatase activity (Jiang *et al.*, 2000). Two lines of evidence indicate that the dephosphorylation of NRI~P does not proceed via a reversal of the phosphotransfer reaction by which NRII passes phosphoryl groups to NRI. First, biochemical studies have shown that phosphoryl groups were released as Pi and an NRII intermediate was not detected (Weiss and Magasanik, 1988). More importantly, the H139N altered form of NRII in which the site of NRII autophosphorylation was altered to a non-phosphorylatable residue was unable to become autophosphorylated, but displayed basal and PII-stimulated phosphatase activities (Jiang *et al.*, 2000).

Studies with truncated forms of NRII were used to investigate the requirements for the phosphatase activity. NRII consists of an N-terminal PAS domain linked to an orthodox transmitter module, which forms DHp and CA domains (Fig. 5.1). The ATP-binding CA domain of NRII is slightly atypical from the CA domains of other transmitter proteins, in that it has an additional short β-hairpin on the 'back' of the domain (on the opposite side from the cleft that binds ATP and its lid) (Song *et al.*, 2004). Although PAS domains are frequently involved in the binding of stimulatory ligands, and the N-terminal domain of NRII has been referred to as its sensory domain in the literature, in fact it is not known whether this domain plays any role in sensation. The PII protein binds in the region of the β-hairpin element on the back of the CA domain (Pioszak *et al.*, 2000). Studies with truncated proteins showed that the autophosphorylation of NRII could readily occur as long as the DHp and CA domains of NRII were present (Jiang *et al.*, 2000). Indeed, the DHp and CA domains could collaborate to produce a high level of DHp phosphorylation even when these domains were present as separate proteins (Jiang *et al.*, 2000), as in EnvZ (described in a later section). By contrast, the PII-stimulated phosphatase activity required full-length NRII. Notably, truncated forms of NRII that contained the complete DHp and CA domains, corresponding to the complete transmitter module of the protein, failed to display the PII-stimulated phosphatase activity (Jiang *et al.*, 2000). This was observed with several different constructions and with multiple protein preparations. This observation is important, because quite similar truncated forms of transmitters have been used to biochemically investigate whether other transmitter proteins have phosphatase activity. As will be noted in a later section, quite often it has been found that such truncations only display weak phosphatase activity, even though the genetic evidence for the system suggests that the transmitter protein should have phosphatase activity. Although none of the truncated versions of NRII displayed the PII-stimulated phosphatase activity, it was possible to attribute the weak basal phosphatase activity of NRII to its DHp domain (Jiang *et al.*, 2000). The result is also consistent with earlier results from the study of EnvZ, mentioned in a later section.

An important clue as to the mechanism of the NRII phosphatase activity comes from the study of NRII autophosphorylation. Although NRII is a homodimer, the two active sites of the dimer displayed different equilibrium constants for the autophosphorylation reaction, such that the autophosphorylation appears to utilize a half-of-the-sites mechanism (Jiang and Ninfa, 2000). These experiments showed that the hemi-phosphorylated form of NRII was the favoured form when NRII is incubated with ATP, and suggested that there was little flexibility in the domain interactions of NRII. (A half-of-the-sites mechanism implies limitations in the freedom of the active sites.) An hypothesis to explain the NRII apparent half-of-the-sites mechanism is depicted in Fig. 5.3 and is as follows: there is a lack of flexibility in NRII such that whenever one of the two CA domains of NRII moves closer to the central DHp domain, the other CA domain is forced to move away from the DHp domain, and vice versa. When one of the active site His139 residues within the DHp domain becomes phosphorylated, its associated CA domain moves away from it. Because of a lack of flexibility in the protein, this movement forces the other CA domain to remain in close association with its opposing histidine in the DHp domain. Presumably, the phosphoryl group from ATP may be going back and forth between the ADP and the DHp domain histidine, but since the CA domain is constrained from moving away from the hisitidine, the phosphoryl group spends most of its time on the ADP (forming ATP) and thus the histidine is mainly unphosphorylated. But, upon dephosphorylation of the other H139 of the DHp domain, its associated CA domain becomes free to move closer to it, allowing the other CA domain to move away from the DHp domain, leaving behind an autophosphorylated histidine. Such a model would explain the vastly different equilibrium constants for the phosphorylation of the two DHp histidines. Remarkably, this asymmetry in the autophosphorylation of NRII was completely lost in truncated versions of NRII (Jiang *et al.*, 2000). For example, truncated forms of NRII, unlike the wild-type

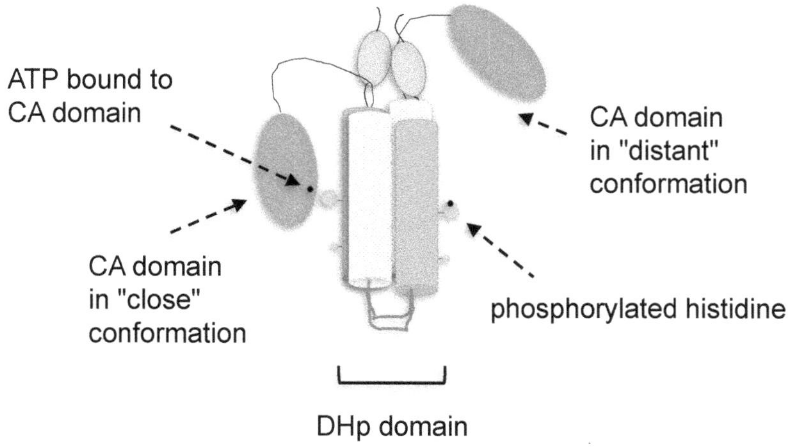

Figure 5.3 Asymmetry of NRII autophosphorylation. The two CA domains alternate between 'close' and 'distant' conformations. For a subunit to become stably autophosphorylated within the DHp domain, its opposing CA domain must be free to move into the away conformation. This strongly inhibits the simultaneous phosphorylation of the second DHpdomain active site histidine in the dimer, because its opposing CA domain is constrained to the 'close' conformation.

full-length protein, were more readily phosphorylated upon both active site histidines in the DHp domain, and both of the phosphoryl groups seem to have similar stability. Based on these data, I hypothesize that the asymmetry in NRII is essential for its phosphatase activity.

It is fairly easy to select for mutations that alter the phosphatase activity of NRII. For example, in cells lacking the UTase/UR protein, PII is always in its unmodified state and consequently the phosphatase activity of the NRII–PII complex blocks expression of *glnA* and other Ntr genes. By selecting for expression of these genes, such as by selecting for the ability to grow on a poor nitrogen source, mutations were obtained that alter either NRII or PII, and block the phosphatase activity (Atkinson and Ninfa, 1992; Kamberov *et al.*, 1994; Pioszak and Ninfa, 2003a). In another study, it was noted that *E. coli* cells contain not only PII but a second PII-like protein, GlnK, that also acts through NRII to bring about the dephosphorylation of NRI~P (van Heeswijk *et al.*, 1996). When both PII and GlnK are missing and NRII is present, the lack of the NRII phosphatase activity and the unrestrained NRII kinase activity result in an extremely severe growth defect (Atkinson and Ninfa, 1998). (This shows that the NRII phosphatase activity plays an extremely important physiological role.) This growth defect was explained as resulting from the excessive expression of an Ntr gene, *nac*, whose product caused the repression of the *serA* gene and by so doing a deficiency in the production of one carbon units (Blauwkamp and Ninfa, 2002). It was possible to isolate suppressor mutations of a strain lacking the UTase/UR that resulted in altered NRII proteins that caused a phenotype identical to that of cells lacking both PII and GlnK. Among these mutants were many that severely affected the phosphatase activity of NRII (Pioszak and Ninfa, 2003a). One unexpected conclusion from these mutant hunts was that mutations affecting the NRII phosphatase activity map throughout NRII, and were not specifically limited to the highly conserved sequences of the transmitter module. Upon characterizing many independent mutations, the mutations seemed to cluster into four regions of NRII: the N-terminal region, throughout the DHp domain, the ATP-lid of the CA domain (a stretch of amino acids that 'caps' the ATP site), and in the vicinity of the unique β-hairpin on the 'back' of the NRII CA domain (Pioszak and Ninfa, 2003a). Mutations in each location were identified that had very severe defects in the PII-activated NRII phosphatase activity.

To discern how mutations in the different

regions of NRII affect the phosphatase activity, a protein with a severe mutation in each of the four locations (N-terminus, DHp domain, ATP lid, β-hairpin) was purified and characterized. As part of this characterization, the effect of different mutations in heterodimers was investigated. Such studies were possible because techniques for the formation of NRII heterodimers *in vitro* from purified homodimeric proteins were available, having been developed during studies of the NRII autophosphorylation mechanism. Remarkably, some pairs of mutations were able to complement one another in heterodimers to produce the PII-activated phosphatase activity, while other pairs of mutants were not able to complement one another (Pioszak and Ninfa, 2003b). These observations led to another round of experimentation, in which subunits with alterations in two of the four regions were combined in heterodimers with singly altered subunits. By comparison of the complementation patterns of alterations when present in heterodimers, it was possible to deduce the basic mechanism by which the subunits of NRII collaborate in the PII-stimulated phosphatase activity (Pioszak and Ninfa, 2003b; Fig. 5.4). Specifically, PII binding in the vicinity of the β-hairpin region on one of the CA domains causes a conformational change that is transmitted through the protein, resulting in the part of the DHp from the PII-bound subunit to collaborate with the ATP lid of its opposing CA domain to bring about the dephosphorylation of NRI~P (Fig. 5.4). Notably, the mechanism is asymmetric, because the two CA domains of NRII have distinct roles in the process.

In addition to activating the phosphatase activity of NRII, the binding of PII to NRII also inhibits the NRII autophosphorylation activity (Jiang and Ninfa, 1999). Thus, it is quite possible that the two activities of NRII are reciprocally regulated, preventing a futile cycle. Unfortunately, it is difficult to prove this point of reciprocal regulation, since the inhibition of NRII autophosphorylation by PII is not complete. Although PII slows the autophosphorylation process, eventually NRII becomes autophosphorylated even in the presence of PII.

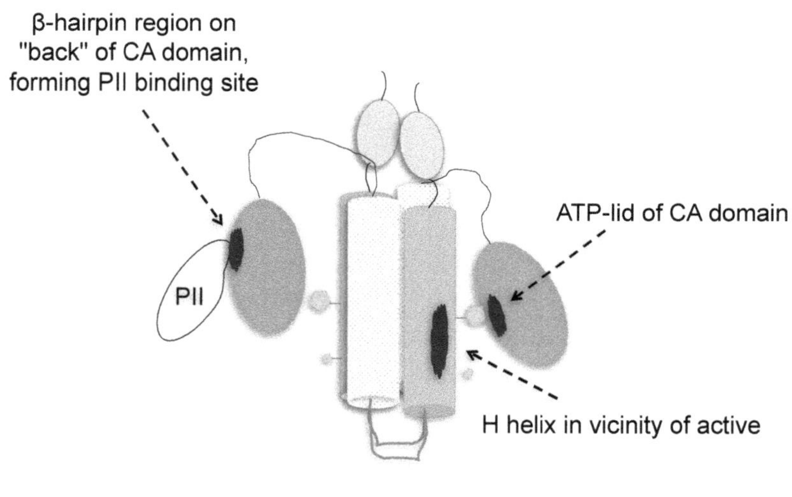

Figure 5.4 Collaboration of three different regions of NRII in the PII-activated phosphatase activity. The two CA domains of NRII play distinct roles in the phosphatase activity, as discerned from studies with heterodimeric proteins with singly and doubly mutated subunits (Pioszak and Ninfa, 2003b). Alterations on the 'back' of the CA domain in the vicinity of the β-hairpin element identify the PII-binding site. The ATP-lid of the other CA domain, and the H-box helix that opposes it are also required for the phosphatase activity. While NRII only needs one each of the three regions shown (the corresponding region on the other subunit of the dimer can be mutated), the three regions must be in the arrangement shown. For example, a functional PII site and a functional ATP lid on the same CA subunit do not allow the phosphatase activity, if the other CA domain is altered for both of these regions. Similarly, the phosphatase activity is observed with NRII heterodimers containing a single functional H-helix region, but it must be on the same subunit as the functional PII site and opposing the functional ATP-lid.

Lessons from CheZ and Spo0B

The CheZ protein plays an essential role in chemotaxis in *E. coli* and related bacteria by increasing the rate of CheY~P dephosphorylation. The atypical CheA transmitter protein of *E. coli* appears to lack phosphatase activity, and this function is instead performed by CheZ, which bears no sequence relatedness to the two-component system transmitter proteins. Interestingly, in other bacteria lacking CheZ, other types of phosphatases perform the function of dephosphorylating CheY~P (e.g. Porter *et al.*, 2008). The physiological importance of the CheY~P dephosphorylation rate is further indicated by the observations that in *E. coli* modest alterations of the CheZ level affect chemotaxis, and the activity of the protein may be controlled both by the CheY~P concentration, by its localization to a particular subcellular location, and by interactions with other members of the chemotaxis sensory apparatus (Silversmith *et al.*, 2008). Furthermore, mutations that diminish the powerful autophosphatase activity of CheY prevent chemotaxis (Lukat *et al.*, 1991).

Although the CheZ protein is not a transmitter protein, the mechanism that it uses to bring about dephosphorylation of the receiver may be highly informative of the mechanism of the phosphatase activity of the transmitter proteins. CheZ is a dimer and has two important domains, a core domain that is comprised a four-helix bundle, with each subunit contributing a helical hairpin to form the four-helix bundle, and a C-terminal domain. These two domains are connected by a flexible region of the protein, such that CheY~P is held within a pincer-like grasp (Zhao *et al.*, 2002). Of importance to our discussion is the interaction of the four-helix bundle with CheY~P, of which we may gain insight from the structure of a complex of CheZ with CheY, Mg^{2+} and the phosphoryl group analogue BeF_3 (Zhao *et al.*, 2002). In this complex, CheZ inserts two residues, Gln147 and Asp143, directly into the active site of CheY. The Asp 143 residue is on the periphery of the active site, and interacts with Asp13 of CheY, which itself helps stabilize the Mg^{2+} ion in the active site, and with Lys109. [Mutation of this Lys eliminated the autophosphatase and CheZ phosphatase activities]. The Gln147 residue of CheZ plays an even more interesting role; it is inserted directly into the centre of the active site, where it interacts with BeF_3 (the phosphoryl group analogue), the Mg^{2+} ion, and the backbone of Asp59, which itself also interacts with the Mg^{2+} ion. From its location in the structure, it has been hypothesized that the role of the Gln147 residue of CheZ is to help orient an attacking water molecule that attacks the phosphorus atom of the phosphoryl group to bring about hydrolysis. That is, the hypothesis is that CheZ Gln147 residue makes the CheY~P autophosphatase reaction much more efficient, by limiting the position of a water molecule in the active site (Zhao *et al.*, 2002). According to this hypothesis, CheZ is neither an independent phosphatase nor an activator of the CheY~P autophosphatase, but has hybrid properties; it increases the efficiency of the autophosphatase reaction by contributing residues to the active site (Zhao *et al.*, 2002). It was noted that this activity of CheZ is formally analogous to the activity of GAP proteins that increase the rate of GTP hydrolysis by GTPases by insertion of an Arg residue into the active site of the GTPase, with the Arg residue serving to stabilize the transition state (Zhao *et al.*, 2002).

The Spo0B protein is not a transmitter protein, but rather is a so-called HPt (histidine phosphotransfer protein) of the phosphorelay-type two-component system that controls development in *Bacillus subtilis*. The Spo0B protein accepts phosphoryl groups from the Spo0F receiver protein, and transfers these phosphoryl groups to the receiver domain of the Spo0A protein, which is a transcription factor that regulates development. Although Spo0B is not a transmitter protein, the study of its structure and the structure of its complex with Spo0F provided important early clues as to the nature of the transmitter–receiver interaction; Spo0B forms a four-helix bundle that interacts with the receiver in a manner very similar to the transmitter proteins (Zapf *et al.*, 2000). It has been noted that the interaction of Spo0B with Spo0F and the interaction of CheZ with CheY-Mg^{2+}-BeF_3 display remarkable similarities (Zhao *et al.*, 2002). In the Spo0B–Spo0F interaction, Spo0B inserts its His30 (the residue that gets phosphorylated) into the Spo0F active site to attack the phosphoryl group, and Asn34 of Spo0B lines the periphery of the active site, where, like the Asp143 of CheZ in the CheZ–CheY interaction, it interacts with the lysine residue that caps the active site (Lys104 of Spo0F). Since Spo0B and CheZ each interact with a receiver protein via a helical bundle, and use a very similar interaction mechanism for phosphotransfer and phosphatase activities, it has been proposed that this type of association may represent a 'general mode of interaction for the phosphotransfer reactions of receiver proteins' (Zhao *et al.*, 2002). That is, residues of the interacting proteins get shoved deep into the receiver protein active site to control whether phosphoryl groups are transferred to or from a histidine residue, or hydrolysed from the phosphorylated receiver.

A working hypothesis based upon CheY, NRII, NRI, CheZ and Spo0B

In this section, I will pull together results with the systems discussed above to speculate as to the nature of the regulated phosphatase activity and its control by environmental stimuli. I hypothesize (as before, Pioszak and Ninfa 2004) that CheZ provides a good model for the mechanism of the transmitter phosphatase, and specifically, that the transmitter protein phosphatase activity represents a stimulation of the receiver autophosphatase activity. I imagine that a portion of the DHp domain, specifically a region just slightly C-terminal to the active site His of the transmitter, interacts with the phosphorylated receiver domain and alters its conformation subtly, so as to increase the ability of water in the receiver active site to hydrolyse the phosphoryl group. In a later section, I will note that mutations in this part of several transmitter proteins can cause severe defects in the phosphatase activity, and of course we already know that the receiver interacts with this part of the transmitter, both from structural data (Yamada *et al.*, 2009) and from the simple observation of the kinase activity. Whether or not a transmitter side chain gets shoved deep into the receiver active site, as in the CheZ–CheY interaction, cannot be deduced of course, but it is not unreasonable. Environmental regulation of this activity, and the difference between the basal transmitter phosphatase activity and its maximal activity in the presence of environmental signals probably results from conformational changes in the transmitter protein that affect the ability of it to stimulate receiver dephosphorylation. For example, I imagine that it may be energetically unfavourable for the transmitter to shove a side chain into the receiver active site to speed up dephosphorylation, such that this happens vary rarely ('basal' phosphatase), but the stimulus changes the transmitter conformation subtly to facilitate the interaction. In this light, I hypothesize that the truncations of NRII that reduced the inherent tension in the protein and eliminated the asymmetry of the autophosphorylation reaction also reduced the transmitter phosphatase activity because they prevented the conformational change needed to produce the productive dephosphorylation complex. From the biochemical studies of the NRII phosphatase activity, it is clear that both the DHp domain and the ATP-lid of one of the CA domains collaborate to form the 'phosphatase active site' (Pioszak and Ninfa, 2003b; Fig. 5.4). Since these parts of the transmitter are in close association, I imagine that the ATP-lid plays a role in forcing the DHp domain into the conformation that dephosphorylates the receiver, such that when the inherent tension of the molecule is relieved upon truncation of the N-terminal domains, the binding of PII to the other CA domain cannot result in the necessary conformational change. An alternative mechanism for the transmitter phosphatase activity, consistent with the biochemical and genetic data, is that the DHp and ATP-lid of a CA domain may hold the NRI receiver domain in a pincer-like grip, analogous to the interaction of CheY with CheZ.

The hypothesis described above makes several predictions. It predicts that transmitter proteins that display the phosphatase activity will also display asymmetry in autophosphorylation, as does NRII, and that truncations (or any other mutation) that affect this asymmetry will also dramatically reduce the phosphatase activity. Because of this requirement for internal tension in the transmitter, truncations that consist solely of the transmitter module (DHp and CA domain) are not predicted to have high phosphatase activity. The hypothesis predicts that a region of the DHp domain will prove essential for the phosphatase activity, while the active site histidine is not essential. Indeed, the receiver itself probably plays a role in the transmitter phosphatase activity; since the hypothesized association between transmitter and phosphorylated receiver involves such intimate association, it is not unreasonable that the transmitter phosphatase activity is highly specific for the cognate receiver protein.

Biochemical observations of the phosphatase activity in transmitter proteins

EnvZ

An enormous effort has been made to understand the kinase and phosphatase activities of the EnvZ transmitter protein of *E. coli*, making it one of the best understood of all transmitter proteins. EnvZ responds to unknown signals to regulate the phosphorylation state of OmpR as part of the cellular adaptation to changes in medium osmolarity. The transmitter has an orthodox structure, consisting of a periplasmic domain flanked by transmembrane segments and a cytoplasmic transmitter module consisting of the DHp and CA domains. Indeed, the organization of the transmitter module into DHp and CA domains was first elucidated by the determination of the structures of these domains from EnvZ. Since the signal that regulates EnvZ, having something to do with osmolarity, is not known, it has not been possible to biochemically study the regulation of EnvZ activities. However, the EnvZ basal phosphatase activity is considerably stronger than is the comparable phosphatase activity of NRII (in the absence of PII), facilitating the investigation of this activity. Furthermore, to directly overcome the absence of knowledge concerning the stimulatory molecules, a fusion protein has been formed that links the cytoplasmic transmitter module of EnvZ to the transmembrane receptor Tar that functions as a sensory component for bacterial chemotaxis towards aspartate. This fusion protein, Taz1, brings about the aspartate-regulated expression of an OmpR target promoter, the *ompC* promoter (Utsumi *et al.*, 1989). Here, I will briefly summarize key findings concerning the phosphatase activities of EnvZ and of its derivatives, including Taz1. This subject has been reviewed (Inouye *et al.*, 2003).

Medium osmolarity controls the reciprocal regulation of *ompF* and *ompC* in *E. coli*, which encode distinct porins, with *ompF* being expressed at low osmolarity and *ompC* being expressed at high osmolarity. Expression of both promoters requires OmpR~P, and the reciprocal regulation is thought to be due to the position and affinity of OmpR~P binding sites at each promoter. Specifically, a low concentration of OmpR~P favours expression of *ompF* while a high concentration of OmpR~P favours expression of *ompC*. The ratio of the kinase and phosphatase activities of EnvZ are thought to control the level of OmpR~P and thus which of the two porin genes is expressed. In cells with an *envZ* null mutation, *ompC* expression is very low and an intermediate level of *ompF* expression is obtained, due to the phosphorylation of OmpR by acetyl phosphate. EnvZ is clearly both a positive and negative regulator of gene transcription responsible for sending a reversible signal through *ompR*, which we now know to reflect OmpR phosphorylation and dephosphorylation.

Numerous mutations in *envZ* have been identified that alter the expression of the *ompF* or *ompC* genes, and affect the kinase or phosphatase activities of EnvZ (e.g. Hsing *et al.*, 1998). Mutations throughout *envZ* can result in constitutive expression of *ompC* and loss of the EnvZ phosphatase activity. A particularly powerful mutation of this type is the *envZ11* allele, causing the alteration T247R immediately downstream from the His243 site of autophosphorylation in the DHp domain. Alteration of this residue results in a severe defect in the phosphatase activity of truncated forms of EnvZ, in the phosphatase activity of the isolated DHp domain (*in vitro* and *in vivo*), and in the phosphatase activity of a Taz1-derived fusion protein (Dutta *et al.*, 2000). It is notable that when a truncated form of EnvZ consisting of its DHp and CA domains ('EnvZc') was used, it was not possible to form a high concentration of OmpR~P *in vitro* by phosphorylation with EnvZc, since the basal phosphatase activity of EnvZc was so powerful relative to the rate of EnvZc autophosphorylation. But, when an EnvZc-type construct containing a T247 mutation was used, the powerful phosphatase activity was disabled and OmpR~P could be readily accumulated *in vitro*. Another study using the EnvZc truncated version of EnvZ examined the binding of OmpR and OmpR~P to EnvZc (Yoshida *et al.*, 2002). It was shown that both OmpR and OmpR~P bind well to EnvZc, as expected since the two forms of OmpR are substrates for the two activities of EnvZc.

A different truncated form of EnvZ consisting of the cytoplasmic transmitter module and a part of a transmembrane segment (EnvZ115) was also able to bring about the dephosphorylation of OmpR~P (Igo *et al.*, 1989). This truncated form of EnvZ is not soluble, but can be studied as a suspension of the protein in aqueous solutions. To investigate the phosphatase activity of EnvZ115, OmpR was phosphorylated using the CheA kinase, and its dephosphorylation was measured upon addition of EnvZ115 and nucleotides. Another method that was used was to phosphorylate OmpR using autophosphorylated EnvZ115 in an immune complex, and then remove the immune complex by centrifugation to obtain OmpR~P that was then used to study dephosphorylation by EnvZ115. These experiments showed that dephosphorylation of OmpR~P by EnvZ115 required adenylylate nucleotides, but apparently did not require nucleotide cleavage, since AMP-PCP worked as well as did ATP (Igo *et al.*, 1989).

Additional studies examined the roles of the different domains of EnvZ in its phosphatase activity. The isolated DHp domain was the source of the observed activity, although the phosphatase activity of this purified domain was weak and was not stimulated by nucleotides (Zhu *et al.*, 2000). Interestingly, while some mutations affecting the phosphatase activity seemed to have predictable effects on the basal phosphatase activities of various truncated versions of EnvZ, certain mutations at the C-terminal margin of the DHp domain had very significant effects on the phosphatase activity *in vivo* in the context of intact EnvZ, and had significant effects on the basal phosphatase activity of EnvZc-like proteins *in vitro*, yet had no effect on the phosphatase activity of the isolated DHp domain (Zhu *et al.*, 2000). It was hypothesized that these mutations altered the relationship of the DHp and CS domains to each other. A large number of mutations affecting the DHp domain and altering the relative kinase and phosphatase activities of EnvZ have been characterized, including mutations resulting from cysteine scanning of the domain (Inouye *et al.*, 2003). These mutations may affect the relative binding of OmpR and OmpR~P to the DHp domain as well as the efficiency of the kinase and phosphatase activities.

As noted above, although the isolated DHp domain displayed weak phosphatase activity, this activity was not stimulated by adenylylate nucleotides, unlike the situation with larger constructs such as EnvZ115 or EnvZc. This suggests that the CA domain is responsible for regulation by nucleotide, likely by binding of nucleotide to the same site used in the kinase reaction. Interestingly, addition of isolated CA domain to isolated DHp domain resulted in slight stimulation of the phosphatase activity of the DHp domain, and this was dependent on the presence of adenylated nucleotide (Zhu *et al.*, 2000).

In addition to studies using truncations such as EnvZc and EnvZ111, another study used membranes enriched in EnvZ to examine the effects of mutations on the kinase and phosphatase activities of EnvZ (Hsing *et al.*, 1998). Since the stimulus controlling EnvZ is unknown, this study essentially just examined basal kinase and phosphatase rates using the membrane preparations; a variety of mutations affecting the phosphatase activity were identified mapping in the DHp domain as well as in the N-terminal regions of EnvZ (Hsing *et al.*, 1998). These studies with intact EnvZ in membranes were entirely consistent with the genetics and physiology of the system, with the phosphatase activity of EnvZ being responsible for its negative regulation of gene expression. In another study, full-length EnvZ has been purified and reconstituted into proteoliposomes; this full-length EnvZ displayed weak phosphatase activity, that was stimulated by ADP (Jung *et al.*, 2001).

A frustrating aspect of the EnvZ-OmpR system is that a tremendous effort has been made to characterize the system biochemically, but since the regulatory effectors are unknown, only basal activities can be examined. As noted above, the use of the Taz1 fusion protein and its derivatives sidesteps this issue, since signal perception by the Tar chemoreceptor is well understood. The Taz1 protein brings about the activation of the *ompC* promoter upon its stimulation with aspartate (Utsumi *et al.*, 1989). The biochemical basis of this was studied using membranes enriched in Taz1. It was observed that aspartate seemed to have little effect on the kinase activity of Taz1, but seemed to significantly decrease the phosphatase activity of Taz1 (Jin and Inouye, 1993). Furthermore, a Taz1 derivative bearing the *envZ11* (T247R) mutation seemed to completely lack phosphatase activity and did not respond to aspartate (Jin and Inouye, 1993). This suggested that an important role of aspartate in the Taz1 system was to control the phosphatase activity. Mutational analysis of Taz1 has shown that alteration of conserved transmitter sequences eliminated the kinase activity of Taz1, and that in some cases expression of two different mutant subunits in the same cell resulted in complementation of the kinase activity and restoration of the ability to activate *ompC* expression (Yang and Inouye, 1993). In such cases, when the phosphatase activity of Taz1 was present in cells containing complementing altered Taz1 proteins, then the activation of *ompC* was regulated by aspartate. However, in some cases the complementing Taz1 subunits did not display phosphatase activity, and in these cases *ompC* expression was constitutive and not regulated by aspartate. Thus, both the kinase and phosphatase activities of Taz1 were responsible for mediating regulation of *ompC* by aspartate (Yang and Inouye, 1993).

KdpD

The KdpD–KdpE two-component system regulates the expression of the KdpFABC complex, an ATP-utilizing potassium transport system. The KdpD transmitter is thought to integrate three signals: osmolarity, potassium levels, and ATP levels (Heerman and Jung, 2010). The KdpD transmitter has an unusual arrangement, consisting of cytosolic N-terminal and C-terminal domains, with a transmembrane region in the middle. The transmitter module is in the C-terminal cytosolic part of the protein, while ATP sensing has been ascribed to the N-terminal cytosolic domain. KdpD is thought to interact with two different effector proteins, UspC and IIA^{Ntr}, with the binding of either of these proteins activating the kinase activity and decreasing the phosphatase activity (Heerman and Jung, 2010). Truncation of the N-terminal cytosolic domain gives a large increase in the phosphatase activity (Jung and Altendorf, 1998).

In one series of studies, KdpD with a C-terminal His tag was extracted from membranes with detergent, purified, and reconstituted into proteoliposomes (Jung *et al.*, 1997). The phosphatase activity of the reconstituted protein was somewhat weak, although it did require an adenylylate nucleotide (e.g. AMP-PNP). IIA^{Ntr} seemed to increase the kinase activity of KdpD many fold, while it only decreased the phosphatase activity 2- to 3-fold (Jung *et al.*, 1997). This study seems to suggest that the kinase and phosphatase were not coordinately regulated, since a much bigger effect of IIA^{Ntr} was observed on kinase activity.

In another study, membranes enriched with wild-type or mutant KdpD were used to examine the kinase and phosphatase activities of altered forms of KdpD resulting in constitutive expression of KdpFABC. Some of the mutations resulting in constitutive expression of KdpFABC resulted in altered KdpD proteins that had wild-type kinase activity, yet were severely defective in the phosphatase activity (Brandon *et al.*, 2000). Accordingly, the authors hypothesize that regulation of the KdpD phosphatase activity is the basis of the physiological control of the system. I note that in this study the phosphatase activity was ~ 1/7 as strong as the kinase activity, and the measurements were fairly crude. Nevertheless, this seems to be a promising system for biochemical studies of the phosphatase activity and its control. For further details about the KdpD–KdpE system see also Chapter 10.

FixL

The FixL–FixJ two-component system controls the expression of *fix* and *nif* genes in *Sinorhizobium meliloti* in response to oxygen availability. FixL consists of a membrane spanning region, and a cytoplasmic portion that includes a PAS domain with a bound sensory haem, followed by the transmitter module (DHp and CA domains). A truncated form consisting of the PAS domain and its associated haem and the transmitter module was purified (Lois *et al.*, 1993); this protein displayed O_2 regulation of autophosphorylation and

FixJ~P dephosphorylation. The extent of regulation varied with temperature but under optimal conditions was about 12-fold. Furthermore, a mutant was identified that appears to have a defect in the phosphatase activity (Lois *et al.*, 1993). Unfortunately, despite these very interesting initial results, the system has not been extensively studied.

PhoR

The PhoR-PhoB system mediates the cellular response to changes in phosphate availability in *E. coli*. The PhoR transmitter protein is probably regulated by an association with a phosphate transport system, as I hypothesized in an earlier review (Stock *et al.*, 1989). The transmitter has essentially no periplasmic domain, and has a cytoplasmic PAS domain fused to an orthodox transmitter module. A truncated form of PhoR consisting of the PAS-DHp-CA domains had weak phosphatase activity *in vitro*, while the isolated DHp domain had even weaker phosphatase activity *in vitro* (Carmany *et al.*, 2003).

NarXL–NarQP

The NarXL and NarPQ two-component systems form a signalling network that responds to nitrate and nitrite and controls a variety of genes including those encoding the reductases necessary for utilization of nitrate and nitrite as respiratory electron acceptors. In an early study, the NarX transmitter protein was purified from membranes and solubilized in a non-ionic detergent. The detergent-solubilized full-length NarX protein had a weak phosphatase activity that brought about the dephosphorylation of the NarL~P receiver (Walker and DeMoss, 1993).

Several groups have purified truncated forms of the NarX and NarQ transmitter proteins (Schröder *et al.*, 1994; Noriega *et al.*, 2010). Both of these transmitter proteins are comprised an N-terminal periplasmic domain flanked by transmembrane sequences, a HAMP linker region immediately C-terminal to the second transmembrane segment, and a C-terminal transmitter module (DHp and CA domains). In one study, truncated versions of these proteins were insoluble, but were soluble after an unfolding and re-folding procedure (Schröder *et al.*, 1994). It was observed that the truncated form of NarX had discernible NarL~P phosphatase activity, but that the truncated NarQ transmitter lacked this activity (Schröder *et al.*, 1994). It is curious that the phosphatase activity reported in this study apparently did not require the presence of an adenylylate nucleotide. This same study showed that acetyl phosphate could phosphorylate NarL. Another group prepared soluble truncated versions of the NarQ and NarX by fusing the cytoplasmic portions of these proteins to MBP (Noriega *et al.*, 2010). In one study with these materials, an ATP regenerating system was used to allow multiple cycles of catalysis, and the investigators examined phosphorylation and dephosphorylation of the NarL and NarP receivers by each of MBP-NarX and MBP-NarQ. It was observed that NarX displayed good phosphatase activity towards NarL~P but had minimal activity with NarP~P; conversely, NarQ displayed good NarP~P phosphatase activity but did not dephosphorylate NarL~P. In all cases, only the basal phosphatase activity was measured, which was somewhat weak. Interestingly, NarX could act as a kinase of both NarL and NarP, while NarQ acted as a kinase of NarP, but not of NarL. These results showed that the network is asymmetric, with the phosphatase activities being much greater with the 'cognate' receiver than for the 'non-cognate' receiver (Noriega *et al.*, 2010). In another study, these investigators used the MBP–NarX fusion protein (which contains only the cytoplasmic portion of NarX) to assess the phosphatase activity of wild-type and mutant proteins. They observed that within the DHp domain, the conserved glutamine residue that is the 5th reside downstream from the conserved active site His (HDxxxQ) was needed for the phosphatase activity (Huynh *et al.*, 2010). It was hypothesized that this glutamine serves to position a water molecule for nucleophilic attack, as with CheZ (Zhao *et al.*, 2002). Interestingly, this conserved glutamine is found in numerous transmitter proteins, and several other transmitter proteins have a related amino acid in this position. NRII has an asparagine at the corresponding position.

PhoQ

The PhoP–PhoQ two-component system controls the expression of virulence traits in *E. coli* and *Salmonella enterica* serovar Typhimurium in response to Mg^{2+} availability, with low Mg^{2+} causing phosphorylation of the PhoP receiver. Several studies examined the activity of PhoQ using cell membranes enriched for the protein (Castelli *et al.*, 2000; Montagne *et al.*, 2001; Regelmann *et al.*, 2002; Sanowar *et al.*, 2003). One study employed membranes enriched in PhoQ, or lacking PhoQ, or having a truncated MBP-PhoQ in the membrane, to examine the phosphorylation and dephosphorylation of purified, his-tagged PhoP (Montagne *et al.*, 2001). It was observed that the N-terminal sensory domain of the protein was required for the phosphatase activity, as was an intact active site DHp domain histidine (His 277), and that the phosphatase activity was increased by Mg^{2+}, which is sensed by the sensory domain. By contrast, the autophosphatase activity of PhoP~P required micromolar Mg^{2+}, but was not regulated by Mg^{2+} at physiological levels (Montagne *et al.*, 2001). Activities in this study were rather low, but gel electrophoresis analysis strongly supported the idea that reverse phosphotransfer had little role in the observed dephosphorylation reaction. Another study by the same investigators, also using membranes enriched in PhoQ, showed that mutations at position 48 within the N-terminal part of the protein could result in proteins that were strongly defective in the kinase activity while having high phosphatase activity; other mutations at this same position had opposite effects (Sanowar *et al.*, 2003). Based on these results, it was hypothesized that position 48 constitutes a conformational switch that allows the protein to move between kinase-dominant and phosphatase-dominant states (Sanowar *et al.*, 2003). Interestingly, when a similar study was performed with the *E. coli* protein, some of the mutations at position 48 had different effects than did the same mutations in *S. enterica* serovar Typhimurium. Apparently, the conformational switch differs somewhat between the two bacteria (Regelmann *et al*, 2002). Another group also looked at the activities of PhoQ using membranes enriched for the protein, and came to the same conclusions as above that Mg^{2+} binding to the sensory domain of the protein stimulated the phosphatase activity (Castelli *et al.*, 2000). Finally, full-length PhoQ with a C-terminal His tag has been extracted from membranes, purified in detergent, and reconstituted into proteoliposomes. Phosphatase activity was observed using the reconstituted protein, and this phosphatase activity was stimulated by ADP and by AMP-PNP (Sanowar and LeMoual, 2005). Curiously, this same group found that AMP-PNP also stimulated the dephosphorylation of PhoP~P (Sanowar and LeMoual, 2005), which raises some concern in this reviewer.

PmrB

The PmrB–PmrA two-component system responds to acidic conditions and high iron concentrations (Fe^{3+}) to activate virulence traits in *E. coli*, *Salmonella*, and *Erwinia*. The system is linked to the PhoP–PhoQ system, since PhoP~P activates the expression of PmrD, which protects PmrA~P from the phosphatase activity of PmrB. A truncated form of PmrB, PmrBc, consisting of the transmitter module has been purified from *E. coli* and *Salmonella* (Kato and Groisman, 2004; Chen *et al.*, 2011). The *Salmonella* protein was observed to have discernible phosphatase activity, while a T156R mutant form that resulted in constitutive expression of target genes *in vivo* was observed to lack phosphatase activity. Also, the phosphatase activity of PmrB could not bring about the dephosphorylation of the non-cognate XgiX~P receiver (Kato and Groisman, 2004). Interestingly, the *E. coli* truncated form of PmrB displayed about 10-fold higher phosphatase activity than did its *Salmonella* counterpart, and this is thought to be responsible for a difference in the regulatory circuitry in these bacteria (Chen *et al.*, 2011). For further information about the PmrB–PmrA system, see also Chapter 8.

AbsA1

In *Streptomyces coelicolor*, the AbsA1 (transmitter) and AbsA2 (receiver) two-component system regulates antibiotic synthesis. Specifically, phosphorylated AbsA2 appears to repress genes required for antibiotic synthesis. A his-tagged truncated version of the AbsA1 transmitter consisting of the transmitter module was examined along with his tagged and MBP tagged versions of the receiver, AbsA2 (Sheeler *et al.*, 2005). Phosphatase activity of the truncated transmitter was evident; ATP was present in the experiments but no investigation of its requirement was performed. The transmitter protein DHp domain active site histidine was not required for the phosphatase activity, as the altered H202A protein actually seemed to have slightly higher phosphatase activity. [This observation is similar to observations of the activity of *E. coli* NRII H139N]. An X-helix mutant L253A within the DHp domain seemed to have a significant defect in the phosphatase activity (Sheeler *et al.*, 2005). I note that mutation of H202 increased antibiotic production in intact cells, and mutation of L253 decreased it (Sheeler *et al.*, 2005); thus, the biochemistry and physiology are consistent with one another. For further information about AbsA1, see also Chapter 15.

MtrB

The MtrB–MtrA two-component system is involved in osmoregulation and peptidoglycan metabolism in *Corynebacterium glutamicum*. Full-length MtrB transmitter protein was hyperexpressed in *E. coli*, extracted from membranes, purified using a Strep-tag (which was placed at either the N- or C-terminal ends, with both versions reportedly acting similarly), and reconstituted into proteoliposomes (Moker *et al.*, 2007). The reconstituted protein displayed weak phosphatase activity that required ADP; it is not clear from the data presented whether the activity was due to reverse transfer or a phosphatase activity.

RegB

The RegB–RegA two-component system is a redox-responding system that controls anaerobic synthesis of the photosystem in *Rhodobacter capsulatus, R. sphaeroides, R. sulfidophilum* and *R. denitrificans*. Homologous systems are found in many bacteria including non-photosynthetic bacteria. (*Sinorhizobium meliloti, Caulobacter crescentus*). These systems have been the subject of a review (Elsen *et al.*, 2004). RegB is a somewhat atypical transmitter protein with a periplasmic domain, six transmembrane segments, and cytosolic DHp and CA domains. The full-length protein with a C-terminal His-tag as well as a truncation comprised the transmitter module were purified from *R. sphaeroides*. Curiously, the full-length purified protein was apparently not reconstituted into membrane vesicles, but simply assayed as the solubilized full-length protein (Potter *et al.*, 2002). Both versions of the transmitter protein displayed autophosphorylation activity and weak phosphatase activity; neither of these activities was regulated by oxygen and the phosphatase activity apparently did not require ATP. For further information about the RegB–RegA system, see also Chapter 3.

CpxA

The CpxR–CpxA two-component system plays a role in responding to mislocalized or misfolded proteins in the periplasmic compartment of *E coli* and related bacteria. The CpxA transmitter protein has a typical arrangement, consisting of a periplasmic domain flanked by transmembrane segments, and a cytoplasmic portion comprised the conserved transmitter module (DHp and CA domains). Experiments with intact cells showed that mutations in the periplasmic domain as well as in the transmitter module can greatly affect control. Interestingly, the T253P conversion, just downstream from the DHp active site histidine (H249) resulted in loss of negative regulation *in vivo*, analogous to the results with EnvZ (Raivio and Silhavy, 1997). A truncated form of CpxA consisting of its cytoplasmic portion fused to a cytoplasmic version of MBP has been purified, along with several mutant versions of this fusion protein (Raivio and Silhavy, 1997). The wild-type version of the fusion protein displayed phosphatase activity *in vitro*, whereas the T253P altered protein seemed to lack this activity. This is also consistent with the findings obtained with EnvZ. The phosphatase activity was activated by ATP. A 'cross-talk' method was used to conduct some of the phosphatase assays; CpxR was phosphorylated with the insoluble EnvZ115 protein, which was then removed from the suspension by centrifugation leaving behind CpxR~P for analysis of dephosphorylation rates (Raivio and Silhavy, 1997). In another study, the full-length CpxA with an added C-terminal His-tag was hyperexpressed, extracted from membranes, purified, and reconstituted into proteoliposomes (Fleischer *et al.*, 2007). This form of the transmitter protein displayed weak phosphatase activity, reducing the stability of CpxR~P from greater than one hour to about 20 min. This phosphatase activity was stimulated by ADP. For further information about the CpxR–CpxA system, see also Chapter 12.

LiaS_{Lm}

The LiaS–LiaR two-component system of *Listeria monocytogenes* responds to envelope stress and antibiotic exposure and controls the expression of genes encoding membrane proteins and a glycosyltransferase modifying enzyme. The expression of these target genes is inhibited by a third component, LiaF, which is a membrane protein that interacts with LiaS. Related systems are found in *B. subtilis, Staphylococcus aureus, Streptococcus mutans,* and *Lactococcus lactis*. In *L. monocytogenes,* deletion of the transmitter protein resulted in constitutive expression of the target genes, and this was due to phosphorylation of the receiver by acetyl phosphate (Fritsch *et al.*, 2011). Based mainly on physiological data, it was hypothesized that LiaF acts to increase the phosphatase activity of LiaS (Fritsch *et al.*, 2011). A truncated version of LiaS consisting of GST fused to the cytoplasmic part of LiaS was shown to phosphorylate the receiver, but the receiver did not accumulate. By analogy to results with EnvZ, this suggested that the transmitter protein had very strong phosphatase activity. For further information about the LiaS–LiaR system, see also Chapter 11.

DesK

The DesK–DeskR system regulates membrane fluidity in *B. subtilis* by controlling expression of the *des* gene encoding an acyl lipid desaturase. A His-tagged truncated form of the DesK transmitter comprised the transmitter module was purified, as was a GST-tagged version of the DesR receiver (Albanesi *et al.*, 2004). The truncated DesK transmitter displayed both autophosphorylation activity and (weak) phosphatase activity (Albanesi *et al.*, 2004). Specifically, the phosphorylated receiver was stable for about an hour in the absence of the transmitter but was only stable for about 15 minutes in its presence. Alteration of the active site histidine in the DHp domain did not affect the phosphatase activity *in vitro* or *in vivo*, excluding a reverse transfer mechanism. Interestingly, although the truncated DesK showed weak phosphatase activity *in vitro, in vivo* the full-length transmitter protein was required to obtain strong negative regulation. This is similar to observations with EnvZ and NRII.

VraS

The VraS–VraR two-component system of *S. aureus* responds to cell wall damage and is induced by antibiotics, regulating the expression of ~40 genes. A GST-fusion to residues 85–347 of the VraS transmitter, consisting of the transmitter module, displayed relatively weak phosphatase (Belcheva and Golemi-Kotra, 2008). For these measurements, the VraR receiver was phosphorylated with unlabelled acetyl phosphate, and its dephosphorylation by truncated VraS was examined using a mass spectrometry method (Belcheva and Golemi-Kotra, 2008).

$\text{WalRK}_{\text{Spn}}$

The WalR–WalK two-component system plays a role in cell wall maintenance and in responding to antibiotic stress in a number of Gram positive bacteria. In *Streptococcus pneumoniae* the system is also known as VicRK; the system is related to the *B. subtilis* essential YycFG system, but the structure of the external domain of the transmitter protein is quite different. The *S. pneumoniae* protein has a single transmembrane segment at its N terminus, followed by a HAMP linker, a PAS domain, and the transmitter module consisting of DHp and CA domains. A number of truncated versions of the transmitter protein were examined, that had C-terminal His-tags and N-terminal Sumo-tags (Gutu *et al.*, 2010). A version of the WalR receiver protein with an N-terminal His-tag was also purified. The phosphorylated form of WalR was very stable, and WalR~P lacking a radioactive label could be easily produced using acetyl phosphate; this allowed measurement of the WalK phosphatase activity using reverse-phase chromatography to separate WalR and WalR~P and quantify the amount of each in reaction samples (Gutu *et al.*, 2010). Deletion of the N-terminal 35 amino acids of WalK resulted in loss of the transmembrane segment and a small portion of the HAMP linker. This protein had clear phosphatase activity, that was stimulated by ATP, ADP, or ATP-γS (Gutu *et al.*, 2010). This phosphatase activity did not require the active site histidine residue of the DHp domain, as the H218A version of the protein displayed the activity. A T222R alteration (analogous to the mutation that reduced EnvZ phosphatase activity) seemed to reduce both the kinase and phosphatase activities of the protein. A smaller version of the transmitter, deleted for amino acids 1–195, consists of essentially just the DHp and CA domains; this protein displayed significantly less phosphatase activity. For further information about the WalR–WalK system, see also Chapter 7.

ThkA/TrrA

The ThkA–TrrA two-component system from *Thermatoga maritima* has provided an important structural model for the interactions of the transmitter and receiver proteins (Yamada *et al.*, 2009). The ThkA transmitter protein is characterized by (from the N-terminus) a coiled-coil domain, a GAF domain, a PAS domain, and the transmitter module consisting of DHp and CA domains. A truncation consisting of the PAS-DHp-CA portion of the protein (i.e. the NRII-like portion) was observed to have phosphatase activity in reaction mixtures that contained ATP. By contrast, a truncation that lacked the PAS domain and just contained the DHp and CA domains displayed greatly reduced phosphatase activity (Yamada *et al.*, 2009).

Physiological importance of the transmitter protein phosphatase activity

Early genetic and physiological studies of nitrogen and phosphate utilization in *E. coli*, some of which were conducted before the phosphorylation and dephosphorylation activities of the two-component systems had been discovered (Backman *et al.*, 1983; Wanner, 1983), indicated that the transmitter proteins in those systems were both positive and negative regulators of their target genes. For example, in the PhoBR system, null mutations eliminating the PhoR transmitter protein resulted in a low level constitutive expression of a target gene; by contrast null mutations in *phoB* eliminated expression of the target genes (Wanner, 1983). The explanation for this observation in the Pho system is that a separate transmitter protein brings about the phosphorylation of PhoB in cells, and PhoB is also phosphorylated directly by the small molecule phosphodonor acetyl phosphate (Wanner and Wilmes-Riesenberg, 1992). When these alternate routes for PhoB phosphorylation were blocked genetically, target gene expression was eliminated upon loss of PhoR (Wanner and Wilmes-Riesenberg, 1992). Since maintaining low expression of the Pho regulon required PhoR in wild-type cells under conditions of phosphate excess, it is obvious that PhoR is both a positive and negative regulator of Pho regulon expression.

A similar set of observations was made in studies of nitrogen assimilation control in *E. coli*. The two-component system NRII–NRI (NtrB–NtrC) mediates the control of transcription initiation of the Ntr regulon, a set of genes that enable cells to utilize poor nitrogen sources and survive stresses associated with the absence of the preferred nitrogen source, ammonia. The deletion of the NRI receiver protein, a transcriptional activator of many Ntr genes and a repressor of a few genes, resulted in the complete loss of control. By contrast, a null mutation resulting in absence of the NRII transmitter protein did not result in the complete loss of control of Ntr genes. In fact, one of these genes, *glnA* encoding glutamine synthetase, shows a remarkable level of control by the availability of ammonia, although the regulation was very slow and somewhat sloppy (Backman *et al.*, 1983). Specifically, cells lacking NRII displayed significantly higher levels of expression of *glnA* than did cells that contained NRII when growing on glucose as carbon source and ammonia as nitrogen source. Thus, the NRII transmitter protein was clearly both a positive and negative regulator of the *glnA* gene. Because of the earlier genetic studies on the Pho system, the possibility that acetyl phosphate was involved was examined, and it was observed that acetyl phosphate could directly phosphorylate NRI *in vitro*, and that the NRI~P so formed was indistinguishable in activity from that formed by phosphotransfer from NRII (Feng *et al.*, 1992). Furthermore, when both NRII and the ability of the cell to produce acetyl phosphate were eliminated by mutations, the ability to activate Ntr genes was lost, even though NRI was intact. Thus, as in the Pho system, the negative regulatory function of

NRII was responsible for blocking activation of the system by acetyl phosphate (Feng *et al*, 1992). Remarkably, those studies also showed that acetyl phosphate accumulation was controlled in part by nitrogen status in *E. coli* due to the strong phosphatase activity of NRII and its ability to act as a sink for acetyl phosphate by removing phosphoryl groups from NRI~P (Feng *et al.*, 1992). Thus, the phosphatase activity of NRII has the potential to exert a powerful influence on many cellular processes by controlling acetyl phosphate levels.

In the case of the Pho control system discussed above, multiple transmitter proteins contribute to the phosphorylation of the PhoB receiver, and the PhoR transmitter is required to counteract PhoB phosphorylation under high-phosphate conditions. A number of such cases where multiple transmitter proteins act upon a common receiver are now known, with a particularly well-characterized example being the NarXL–NarPQ system of *E. coli*, briefly summarized in a previous section. It is thought that the observed signalling in such systems is not accidental or spurious, but rather reflects a genuine physiologically important signalling network. Consequently, the term 'cross regulation' has been used to describe such phenomena (Wanner, 1992). It is notable that in these small networks, the paradigm of the *E. coli* PhoBR seems to be consistently replayed; one of the transmitter proteins displays strong phosphatase activity towards the phosphorylated receiver, while the other transmitter(s) seem to essentially only act as kinases but not as phosphatases.

In addition to the phenomenon of cross-regulation, receiver proteins can become phosphorylated by the non-physiological, spurious, phosphorylation by non-cognate transmitter proteins; this phenomenon has been termed 'cross-talk' (Ninfa *et al.*, 1988). This cross-talk occurs because of the (relatively) low specificity of the transmitter and receiver proteins, particularly with regard to the transfer of phosphoryl groups from the transmitter to the receiver. This low specificity was easily demonstrated using purified components from different two-component systems; furthermore, it was shown that overexpression of a transmitter can lead to discernible cross-talk in intact cells (e.g. Ninfa *et al.*, 1988, and several other examples provided in Ninfa, 1991). This effect of receiver protein overexpression probably reflects the occurrence of cross-talk with non-cognate transmitters. The ability of the cognate transmitter protein to suppress this cross-talk implies that it is able to mediate the dephosphorylation of the receiver protein (Stock *et al.*, 1989; Ninfa, 1991). Presumably, a low level of cross-talk occurs in cells even when receiver proteins are at their normal physiologically relevant concentrations, and the transmitter protein phosphatase activities are responsible for counteracting these spurious phosphorylation events. This role of the transmitter is most likely particularly important for the many system where the phosphorylated receiver protein activates its own expression, forming a positive feedback loop.

The correspondence of the negative regulatory function to the phosphatase activity of the transmitter proteins has been made for a number of systems, although the NRII-NRI system probably provides the clearest example, owing to our ability to assess what appears to be the fully activated phosphatase activity in this system. Mutations altering NRII that reduce the negative regulation function of the transmitter protein *in vivo* are observed to reduce the phosphatase activity *in vitro*, as noted in the previous section. Similar relationships between the negative regulatory function and the phosphatase activities of EnvZ, PhoR, and other transmitter proteins were also noted in the previous section of this review.

An important recent focus of study is the identification of cellular control systems that respond to cell wall and membrane stresses, and influence the development of antibiotic resistance. In *S. pneumoniae*, the CiaRH two-component system controls the development of resistance to a number of antibiotics. The CiaH transmitter protein is both a positive and negative regulator of the CiaR receiver protein, and a deletion of CiaH results in weak constitutive expression of the target genes of the system (Halfmann *et al.*, 2011). The physiology and genetics of the $LiaS_{Lm}$ and DesK proteins were mentioned in the previous section. The $LiaFSR_{Lm}$ three-component system controls a cell envelope stress response in *L. monocytogenes* that is inducible by antibiotics. The $LiaS_{Lm}$ transmitter is both a kinase and phosphatase for the $LiaR_{Lm}$

receiver, with the kinase and/or phosphatase activities regulated by the membrane protein $LiaF_{Lm}$ such that $LiaF_{Lm}$ blocks accumulation of the phosphorylated receiver (Fritsch *et al.*, 2011). Genetic evidence suggests that acetyl phosphate is able to phosphorylate LiaR, and that the LiaF-stimulated LiaS phosphatase activity is necessary for negative control of the system in the absence of the antibiotics that act as inducers. The DesKR system controls membrane fluidity in *B. subtilis*. In cells which contain a null mutation eliminating the transmitter DesK, overexpression of the receiver DesR lead to constitutive expression of the target genes. By contrast, in cells containing the transmitter protein, overexpression of DesR did not lead to constitutive expression of the targets. Thus, DesK is both a positive and negative regulator of the system (Albanesi *et al.*, 2004). Negative regulation by DesK required the whole protein (a truncation consisting of just the two-domain conserved transmitter module did not work), but did not require the conserved His of the DHp domain that becomes autophosphorylated as part of the kinase mechanism. This requirement for full-length DesK for its negative regulatory function is consistent with the hypothesis I presented in an earlier section.

I have already noted above that the PII protein of *E. coli* and the LiaF protein of *L. monocytogenes* are regulatory proteins that are known (PII) or thought (LiaF) to stimulate the phosphatase activity of transmitter proteins. The existence of such proteins indicates the physiological significance of the transmitter phosphatase activity. Recent work has revealed additional examples of regulatory proteins that may influence the phosphatase activity of transmitter proteins, and thus demonstrate the physiological importance of those phosphatase activities. These include MzrA, NlpE, RcsF, CpxP, TorT, B1500, and PmrD (Gerken and Misra, 2010). These proteins have a variety of cellular locations (outer membrane, periplasm, inner membrane, cytoplasm) and their common feature is that they all appear to act to increase level of receiver protein phosphorylation, possibly by inhibition of transmitter protein phosphatase activities. These auxiliary proteins can be used to connect different signalling systems, forming signalling networks comprised multiple two-component systems. Of this group of proteins, the best understood is PmrD, which, as I noted in the previous section, serves to connect the PhoP/PhoQ and PrmB/PmrA systems in *S. enterica* serovar Typhimurium (Kato and Groisman, 2004). The downstream targets of the transcriptional activator PmrA~P are controlled by signals sensed by the PhoQ transmitter protein (low Mg^{2+}) and the PmrB transmitter protein (high Fe^{3+}). The low-Mg^{2+} signal acts to promote the phosphorylation of PhoP, with PhoP~P activating the transcription of *pmrD*. The PmrD protein acts to increase the level of PmrA~P, and is necessary for the expression of the PmrA~P targets. PmrD acts to increase the cellular concentration of PmrA~P by blocking its dephosphorylation. In the absence of any other proteins, PmrA~P has a half-life of ~ 4 hours, which indicates a very weak autophosphatase activity. Interestingly, both the autophosphatase activity of PmrA~P and the PmrBc phosphatase activity were blocked by PmrD, which binds to PmrA~P (Kato and Groisman, 2004). The PmrD protein seemed to lack to ability to bind to the unphosphorylated form of PmrA, and had no significant effect on the ability of a non-cognate transmitter to autophosphorylate or transfer phosphoryl groups to PmrA (Kato and Groisman, 2004). Thus, its cellular effects seem to be due specifically to the blockage of the phosphatase activities affecting PmrA~P. These results are also consistent with the PmrB phosphatase activity representing a stimulation of the PmrA~P autophosphatase activity. Interestingly, a mutant form of PmrA has been identified, containing the R81H mutation, that appears to be resistant to dephosphorylation by PmrB *in vivo* (and PmrBc *in vitro*). This mutant protein still had autophosphatase activity that was inhibited by PmrD, but appeared to be resistant to rapid dephosphorylation by PmrBc *in vitro* (Kato and Groisman, 2004). (The properties of the PmrA-R81H mutation are similar to the properties of some of the altered forms of NRI and CheY mentioned above.) Since PmrD specifically acts as an inhibitor of PmrA~P dephosphorylation, and since mutations to constitutive expression of downstream genes were identified that specifically affected the sensitivity of PmrA~P to the

PmrB phosphatase activity, the physiological importance of the phosphatase activities of the system are evident. It has been noted that the role of PmrD in binding and protecting PmrA~P is functionally similar to the role of 14-3-3 proteins in eukaryotic cells (Kato and Groisman, 2004).

Conclusions and future trends

A common theme in signal transduction systems in higher organisms is the presence of signalling proteins that influence a physiological process indirectly, by stimulating or inhibiting the activities of other signal transduction proteins. The presence of such proteins allows the formation of complex signalling circuits that respond to multiple stimuli. For example, GAP proteins activate the cleavage of GTP by G-proteins, leading to their inactivation, and GEF proteins bring about the exchange of GDP for GTP by G-proteins, bringing about their activation. The 14-3--3 proteins bind to phosphorylated proteins and stabilize them, preventing switching. Here, I hypothesize, as before (Pioszak and Ninfa, 2004), that the transmitter protein phosphatase activity is not truly a phosphatase activity in the classical sense, but rather represents the activation of the autophosphatase activity of the phosphorylated receiver protein. This hypothesis was based upon genetic studies of CheY and NRI and other systems, and a variety of biochemical data. This hypothesis must be reconsidered if mutations are identified that eliminate the autophosphatase activity of a receiver protein, but do not alter its dephosphorylation by the cognate transmitter protein.

Based on biochemical studies of CheZ, which is not a transmitter protein but nevertheless brings about the dephosphorylation of a receiver protein, the dephosphorylation mechanism may involve an intimate interaction between transmitter and receiver proteins, in which the transmitter role is something between catalyst and activator. This intimate relationship may involve side chains from the transmitter protein helping to form the active site for dephosphorylation, and stimulating the dephosphorylation reaction by helping to position a water molecule. I hypothesized that this intimate relationship requires a conformational alteration of the DHp domain of the transmitter, and this conformation alteration in turn requires in internal tension in the transmitter that is lost upon truncation of the protein, or is unobtainable in the absence of the signals that control transmitter conformations. Hence, I hypothesize that for many transmitter proteins, the basal phosphatase activity will be significantly less than the fully activated activity. However, in other cases it may not be difficult for the transmitter protein to adopt the phosphatase conformation, and those proteins should have high basal activity, as does EnvZ. Of course, the presence or absence of basal activity is not indicative of the activity of the fully activated phosphatase activity, and thus the failure to detect activity with a truncated version of a transmitter protein in the absence of any signalling molecules is not noteworthy.

Since the transmitter protein phosphatase activity requires a distinct conformation of the transmitter, and this in turn seems to require collaboration of multiple parts of the transmitter, it has proven difficult to characterize the regulation of the activity for many classical, transmembrane transmitter proteins. In this regard, our understanding of the cytoplasmic NRII transmitter protein is informative. The transmitter protein phosphatase activity of NRII is stimulated by interaction with the PII protein. While this may at first glance seem to be a unique feature of NRII and thus not informative as to other two-component systems, the mechanism of PII action is now fairly well known, and it is clear that it serves as an allosteric activator of NRII as opposed to directly participating in the activity. Therefore, PII should be considered as playing a role similar to the stimuli that activate the phosphatase activities of the other transmitter proteins.

For many two-component system transmitter proteins, such as EnvZ and NRII, the activity of the phosphatase activity is equal to or greater than the rather weak kinase activity of the transmitter proteins. Furthermore, in many cases genetic and physiological data directly show that the negative regulatory activities of the transmitter proteins are critical to their biological function. Thus, it is important to study these phosphatase activities in order to understand the mechanisms of signal transduction in these systems.

References

Albanesi, D., Mansilla, M.C., and de Mendoza, D. (2004). The membrane fluidity sensor DesK of *Bacillus subtilis* controls the signal decay of its cognate response regulator. J. Bacteriol. *186*, 2655–2663.

Ansaldi, M., Jourlin-Castelli, C., Lepelletier, M., Theraulaz, L., and Mejean, V. (2001). Rapid dephosphorylation of the TorR response regulator by the TorS unorthodox sensor in *Escherichia coli*. J. Bacteriol. *183*, 2691–2695.

Atkinson, M.R., and Ninfa, A.J. (1992). Characterization of mutations in the *glnL* gene of *Escherichia coli* affecting nitrogen regulation. J. Bacteriol. *174*, 4538–4548.

Atkinson, M.R., and Ninfa, A.J. (1998). Role of the *glnK* gene in nitrogen-regulation of gene transcription in *Escherichia coli*. Mol. Microbiol. *29*, 431–447.

Atkinson, M.R., Kamberov, E.S., Weiss, R.L., and Ninfa, A.J. (1994). Reversible uridylylation of the *Escherichia coli* PII signal transduction protein regulates its ability to stimulate the dephosphorylation of the transcription factor nitrogen regulator I (NRI or NtrC). J. Biol. Chem. *269*, 28288–28293.

Backman, K.C., Chen, Y.M., Ueno-Nisdhio, S., and Magasanik, B. (1983). The product of *glnL* is not essential for regulation of bacterial nitrogen assimilation. J. Bacteriol. *154*, 516–519.

Belcheva, A., and Golemi-Kotra, D. (2008). A close-up view of the VraSR two-component system. J. Biol. Chem. *283*, 12354–12364.

Blauwkamp, T.A., and Ninfa, A.J. (2002). Nac-mediated repression of the *serA* promoter of *Escherichia coli*. Mol. Microbiol. *45*, 351–363.

Brandon, L., Dorus, S., Epstein, W., Altendorf, K., and Jung, K. (2000). Modulation of KdpD phosphatase implicated in the physiological expression of the Kdp ATPase of *Escherichia coli*. Mol. Microbiol. *38*, 1086–1092.

Carmany, D.O., Hollingsworth, K., and McCleary, W.R. (2003). Genetic and biochemical studies of phosphatase activity of PhoR. J. Bacteriol. *185*, 1112–1115.

Castelli, M.E., Vescovi, E.G., and Soncini, F. (2000). The phosphatase activity is the target for $Mg2^+$ regulation of the sensor protein PhoQ in *Salmonella*. J. Biol. Chem. *275*, 22948–22954.

Chen, H.D., Jewett, M.W., and Groisman, E.A. (2011). Ancestral genes can control the ability of horizontally acquired loci to confer new traits. PLoS Genet. *7*, e1002184.

Dutta, R., Yoshida, T., and Inouye, M. (2000). The critical role of the conserved Thr247 residue in the functioning of the osmosensor EnvZ, a histidine kinase/phosphatase in *Escherichia coli*. J. Biol. Chem. *275*, 38645–38653.

Elsen, S., Swem, L.R., Swem, D.L., and Bauer, C.E. (2004). RegB/RegA, a highly conserved redox-responding global two-component regulatorty system. Micrioiol. Mol. Biol. Rev. *68*, 263–279.

Feng, J., Atkinson, M.R., McCleary, W., Stock, J.B., Wanner, B.L., and Ninfa, A.J. (1992). Role of phosphorylated metabolic intermediates in the regulation of glutamine synthetase synthesis in *Escherichia coli*. J. Bacteriol. *174*, 6061–6070.

Fleischer, R., Heermann, R., Jung, K., and Hunke, S. (2007). Purification, reconstitution, and characterization of the CpxRAP envelope stress system of *Escherichia coli*. J. Biol. Chem. *282*, 8583–8593.

Freeman, J.A., Lilley, B.N., and Bassler, B.L. (2000). A genetic analytsis of the functions of LuxN: a two-component hybrid sensor kinase that regulates quorun sensing in *Vibrio harveyi*. Mol. Microbiol. *35*, 139–149.

Fritsch, F., Mauder, N., Williams, T., Weiser, J., Oberle, M., and Beier, D. (2011). The cell envelope stress response mediated by the $LiaFSR_{Lm}$ three-component system of *Listeria monocytogenes* is controlled via the phosphatase activity of the bifunctional histidine kinase $LiaS_{Lm}$. Microbiology *157*, 373–386.

Georgellis, D., Kwon, O., De Wulf, P., and Lin, E.C.C. (1998). Signal decay through a reverse phosphorelay in the Arc two-component signal transduction system. J. Biol. Cherm. *273*, 32864–32869.

Gerken, H., and Misra, R. (2010). MzrA–EnvZ interactions in the periplasm influence the EnvZ/OmpR two-component regulon. J. Bacteriol. *192*, 6371–6278.

Gutu, A.D., Wayne, K.J., Sham, L.T., and Winkler, M.E. (2010). Kinetic characterization of the $WalRK_{Spn}$ (VicRK) two-component system of *Streptococcus pneumoniae*: dependence of $WalK_{Spn}$ (VicK) phosphatase activity on its PAS domain. J. Bacteriol. *192*, 2346–2358.

Halfmann, A., Schnorpfeil, A., Müller, M., Marx, P., Günzler, U., Hakenbeck, R., and Brückner, R. (2011). Activity of the two-component regulatory system CiaRH in *Streptococcus pneumoniae* R6. J. Mol. Microbiol. Biotechnol. *20*, 96–104.

Heermann, R., and Jung, K. (2010). The complexity of the "simple" two-component systems KdpD/KdpE in *Escherichia coli*. FEMS Microbiol. Lett. *304*, 97–106.

van Heeswijk, W.C., Hoving, S., Molenaar, D., Stegman, B., Kahn, D., and Westerhoff, H.V. (1996). An alternative PII protein in the regulation of glutamine synthetase in *Escherichia coli*. Mol. Microbiol. *21*, 133–146.

Hsing, W., Russo, F.D., Bernd, K.K., and Silhavy, T.J. (1998) Mutations that alter the kinase and phosphatase activities of the two-component sensor EnvZ. J. Bacteriol *180*, 4538–4546.

Huynh, T.N., Noriega, C.E., and Stewart, V. (2010). Conserved mechanism for sensor phosphatase control of two-component signaling revealed in the nitrate sensor NarX. Proc. Natl. Acad. Sci. U.S.A. *107*, 21140–21145.

Igo, M.M., Ninfa, A.J., Stock, J.B., and Silhavy, T.J. (1989). Phosphorylation and dephosphorylation of a bacterial transcriptional activator by a transmembrane receptor. Genes Dev. *3*, 1725–1734.

Inouye, M., Dutta, R., and Zhu, Y. (2003). Regulation of porins in *Escherichia coli* by the osmosensing histidine kinase/phosphatase EnvZ. In Histidine Kinases in Signal Transduction, Inouye, M., and Dutta, R., eds (Academic Press, New York), pp. 25–46.

Jiang, P., and Ninfa, A.J. (1999). Regulation of the autophosphorylation of *Escherichia coli* NRII by the

PII signal transduction protein. J. Bacteriol. *181*, 1906–1911.

Jiang, P., and Ninfa, A.J. (2000). Asymmetry in the autophosphorylation of the two-component system transmitter protein NRII (NtrB) of *Escherichia coli*. Biochemistry *39*, 5058–5065.

Jiang, P., Atkinson, M.R., Srisawat, C., Sun, Q., and Ninfa, A.J. (2000). Functional dissection of the dimerization and enzymatic activities of *Escherichia coli* nitrogen regulator II and their regulation by the PII protein. Biochemistry *39*, 13433–13449.

Jin, T., and Inouye, M. (1993). Ligand binding to the receptor domain regulates the ratio of kinase to phosphatase activities of the signaling domain of the hybrid *Escherichia coli* transmembrane receptor, Taz1. J. Mol. Biol. *232*, 484–492.

Jung, K., and Altendorf, K. (1998). Truncation of amino acids 12–128 causes deregulation of the phosphatase activity of the sensor kinase KdpD of *Escherichia coli*. J. Biol. Chem. *273*, 17406–17410.

Jung, K., Tjaden, B., and Altendorf, K. (1997). Purification, reconstitution, and characterization of KdpD, the turgor sensor of *Escherichia coli*. J. Biol. Chem. *272*, 10847–10852.

Jung, K., Hamann, K., and Revermann, A. (2001). K^+ stimulates specifically the autokinase activity of purified and reconstituted EnvZ of *Escherichia coli*. J. Biol. Chem. *276*, 40896–40902.

Kamberov, E.S., Atkinson, M.R., Chandran, P., and Ninfa, A.J. (1994). Effect of mutations in *Escherichia coli glnL* (*ntrB*), encoding nitrogen regulator II (NRII or NtrB), on the regulated phosphatase activity involved in bacterial nitrogen assimilation. J. Biol. Chem. *269*, 28294–28299.

Kato, A., and Groisman, E.A. (2004). Connecting two-component regulatory systems by a protein that protects a response regulator from dephosphorylation by its cognate sensor. Genes Dev. *18*, 2302–2313.

Keener, J., and Kustu, S. (1988). Protein kinase and phosphoprotein phosphatase activities of nitrogen regulatory proteins NTRB and NTRC of enteric bacteria: roles of the conserved amino-terminal domain of NTRC. Proc. Natl. Acad. Sci. U.S.A. *85*, 4976–4980.

Lee, S.Y., De La Torre, A., Yan, D., Kustu, S., Nixon, B.T., and Wemmer, D.E. (2003). Regulation of the transcriptional activator NtrC1: structural studies of the regulatory and AAA^+ ATPase domains. Genes Dev. *17*, 2552–2563.

Lois, A.F., Weinstein, M., Ditta, G.S., and Helinski, D.R. (1993). Autophosphorylation and phosphatase activities of the oxygen-sensing protein FixL of *Rhizobium meliloti* are coordinately regulated by oxygen. J. Biol. Chem. *268*, 4370–4375.

Lukat, G.S., Lee, B.H., Mottonen, J.M., Stock, A.M., and Stock, J.B. (1991). Roles of the highly conserved aspartate and lysine residues in the response regulator of bacterial chemotaxis. J. Biol. Chem. *266*, 8348–8354.

Moker, N., Kramer, J., Unden, G., Kramer, R., and Morbach, S. (2007). *In vitro* analysis of the two-component system MtrB-MtrA from *Corynebacterium glutamicum*. J. Bacteriol. *189*, 3645–3649.

Montagne, M., Martel, A., and LeMoual, H. (2001). Characterization of the catalytic activities of the PhoQ histidine protein kinase of *Salmonella enterica* Serovar Typhimurium. J. Bacteriol. *183*, 1787–1791.

Ninfa, A.J. (1991). Protein phosphorylation and the regulation of cellular processes by the homologous two-component regulatory systems of bacteria. Genet. Eng. (NY) *13*, 39–72.

Ninfa, A.J., and Magasanik, B. (1986). Covalent modification of the *glnG* product, NRI, by the *glnL* product, NRII, regulates the transcription of the *glnALG* operon in *Escherichia coli*. Proc. Natl. Acad. Sci. U.S.A. *83*, 5909–5913.

Ninfa, A.J., Reitzer, L.J., and Magasanik, B. (1987). Initiation of transcription at the bacterial *glnAp2* promoter by purified *E. coli* components is facilitated by enhancers. Cell *50*, 1039–1046.

Ninfa, A.J., Ninfa, E.G. Lupas, A.N., Stock, A., Magasanik, B., and Stock, J. (1988). Cross-talk between bacterial chemotaxis signal transduction proteins and the regulators of transcription of the Ntr regulon: evidence that nitrogen assimilation and chemotaxis are controlled by a common phosphotransfer mechanism. Proc. Natl. Acad. Sci. U.S.A. *85*, 5492–5496.

Noriega, C.E., Lin, H.-Y., Williams, S.B., and Stewart, V. (2010). Asymmetric cross regulation between the nitrate-responsive NarX-NarL and NarQ-NarP two-component regulatory systems from *Escherichia coli* K-12. Mol. Microbiol. *75*, 394–412.

Perego, M., Hanstein, C., Welsh, K.M., Djavakhishvili, T., Glaser, P., and Hoch, J.A. (1994). Multiple protein-aspartate phosphatases provide a mechanism for the integration of diverse signals in the control of development in *B. subtilis*. Cell *79*, 1047–1055.

Pioszak, A.A., and Ninfa, A.J. (2003a). Genetic and biochemical analysis of the PII-dependent phosphatase activity of *Escherichia coli* NRII. J. Bacteriol. *185*, 1299–1315.

Pioszak, A.A., and Ninfa, A.J. (2003b). How the domains of NRII collaborate in the PII-activated phosphatase activity of *Escherichia coli* NRII (NtrB). Biochemistry *42*, 8885–8899.

Pioszak, A.A., and Ninfa, A.J. (2004). Mutations altering the N-terminal receiver domain of NRI (NtrC) that prevent dephosphorylation by the NRII–PII complex in *Escherichia coli*. J. Bacteriol. *186*, 5730–5740.

Pioszak, A.A., Jiang, P., and Ninfa, A.J. (2000). The *Escherichia coli* PII signal transduction protein regulates the activities of the two-component system transmitter protein NRII by direct interaction with the kinase domain of the transmitter module. Biochemistry *39*, 13450–13461.

Popham, D.L., Szeto, D., Keener, J., and Kustu, S. (1989). Function of a bacterial activator protein that binds to transcriptional enhancers. Science *243*, 629–635.

Porter, S.C., North, A.K., Wedel, A.B., and Kustu, S. (1993). Oligomerization of NtrC at the *glnA* enhancer is required for transcriptional activation. Genes Dev. *7*, 2258–2273.

Porter, S.L., Roberts, M.A.J., Manning, C.S., and Armitage, J.P. (2008). A bifunctional kinase-phosphatase in bacterial chemotaxis. Proc. Natl. Acad. Sci. U.S.A. *105*, 18531–18536.

Potter, C.A., Ward, A., Laguri, C., Williamson, M.P., Henderson, P.J.F., and Phillips-Jones, M.K. (2002). Expression, purification, and characterization of full-length histidine protein kinase RegB from *Rhodobacter sphaeroides*. J. Mol. Biol. *320*, 201–213.

Raivio, T.L., and Silhavy, T.J. (1997). Transduction of envelope stress in *Escherichia coli* by the Cpx two-component system. J. Bacteriol. *179*, 7724–7733.

Regelmann, A.G., Lesley, J.A., Mott, C., Stokes, L., and Waldburger, C.D. (2002). Mutational analysis of the *Escherichia coli* PhoQ sensor kinase: differences with the *Salmonella enterica* serovar Typhimurium PhoQ protein and in the mechanism of Mg^{2+} and Ca^{2+} sensing. J. Bacteriol. *184*, 5468–5478.

Sanna, M.G., Swanson, R.V., Bourret, R.B., and Simon, M.I. (1995). Mutations in the chemotactic response regulator, CheY, that confer resistance to the phosphatase activity of CheZ. Mol. Microbiol. *15*, 1069–1079.

Sanowar, S., and LeMoual, H. (2005). Functional reconstitution of the *Salmonella typhimurium* PhoQ histidine kinase sensor in proteoliposomes. Biochem J. *390*, 769–776.

Sanowar, S., Martel, A., and LeMoual, H. (2003). Mutational analysis of the residue at position 48 in the *Salmonella enterica* serovar Typhimurium PhoQ sensor kinase. J. Bacteriol. *185*, 1935–1941.

Schröder, I., Wolin, C.D., Cavicchioli, R., and Gunsalus, R.P. (1994). Phopshorylation and dephopshorylation of the NarQ, NarX, and NarL proteins of the nitrate-dependent two-component regulatory system of *Escherichia coli*. J. Bacteriol. *176*, 4985–4992.

Sheeler, N.L., MacMillan, S.V., and Nodwell, J.R. (2005). Biochemical activities of the *absA* two-component system of *Streptomyces coelicolor*. J. Bacteriol. *187*, 687–696.

Shi, L., Liu, W., and Hulett, F.M. (1999). Decay of activated *Bacillus subtilis* Pho response regulator, PhoP~P, involves the PhoR~P intermediate. Biochemistry *38*, 10119–10125.

Silversmith, R.E. (2010). Auxiliary phosphatases in two-component signal transduction. Curr. Opinion Microbiol. *13*, 177–183.

Silversmith, R.E., Guanga, G.P., Betts, L., Zhao, R., and Bourret, R.B. (2003). CheZ-mediated dephosphorylation of the *Escherichia coli* chemotaxis response regulator CheY: role for glutamate 89. J. Bacteriol. *185*, 1495–1502

Silversmith, R.E., Levin, M.D., Schilling, E., and Bourret, R.B. (2008). Kinetic characterization of catalysis by the chemotaxis phosphatase CheZ. J. Biol. Chem. *283*, 756–765.

Song, Y., Peisach, D., Pioszak, A., Xu, Z., and Ninfa, A.J. (2004). Crystal structure of the C-terminal domain of the two-component systems transmitter protein NRII (NtrB), regulator of nitrogen assimilation in *Escherichia coli*. Biochemistry *43*, 6670–6678.

Stock, A.M., Martinez-Hackert, E., Rasmussen, B.F., West, A.H., Stock, J.B., Ringe, D., and Petsko, G.A. (1993). Structure of the Mg^{2+}-bound form of CheY and the mechanism of phosphoryl transfer in bacterial chemotaxis. Biochemistry *32*, 13375–13380.

Stock, J.B., Ninfa, A.J., and Stock, A.M. (1989). Protein phosphorylation and the regulation of adaptive responses in bacteria. Microbiol. Rev. *53*, 450–490.

Utsumi, R., Brisette, R.E., Rampersaud, A., Forst, S., Oosawa, K., and Inouye, M. (1989). Activation of bacterial porin gene expression by a chimeric signal transducer in response to aspartate. Science *245*, 1246–1249.

Walker, M.S., and DeMoss, J.A. (1993). Phosphorylation and dephosphorylation catalyzed *in vitro* by purified components of the nitrate sensing system, NarX and NarL. J. Biol. Chem. *268*, 8391–8393.

Wanner, B.L. (1983). Overlapping and separate controls on the phosphate regulon in *Escherichia coli* K-12. J. Mol. Biol. *166*, 283–308.

Wanner, B.L. (1992) Is cross regulation by phosphorylation of two-component response regulator proteins important in bacteria? J. Bacteriol. *174*, 2053–2058.

Wanner, B.L., and Wilmes-Riesenberg, M.R. (1992). Involvement of phosphotransacetylase, acetate kinase, and acetyl phosphate synthesis in control of the phosphate regulon in *Escherichia coli*. J. Bacteriol. *174*, 2124–2130.

Weiss, V., and Magasanik, B. (1988). Phosphorylation of nitrogen regulator I (NRI) of *Escherichia coli*. Proc. Natl. Acad. Sci. U.S.A. *85*, 8919–8923.

Wolanin, P.M., Webre, D.J., and Stock, J.B. (2003). Mechanism of phosphatase activity in the chemotaxis response regulator CheY. Biochemistry *42*, 14075–14082.

Yamada, S., Sugimoto, H., Kobayashi, M., Ohno, A., Nakamura, H., and Shiro, Y. (2009). Structure of PAS-linked histidine kinase and the response regulator complex. Structure *17*, 1333–1344.

Yang, Y., and Inouye, M. (1993). Requirement of both kinase and phosphatase activites of an *Escherichia coli* receptor (Taz1) for ligand-dependent signal transduction. J. Mol. Biol. *231*, 335–342.

Yoshida, T., Cai, S.J., and Inouye, M. (2002). Interaction of EnvZ, a sensory histidine kinase, with phosphorylated OmpR, the cognate response regulator. Mol. Microbiol. *46*, 1283–1294.

Zapf, J., Sen, U., Madhusudan, Hoch, J.A., and Varughese, K.I. (2000). A transient interaction between two phosphorelay proteins trapped in a crystal lattice reveals the mechanism of molecular recognition and phosphotransfer in signal transduction. Structure *8*, 851–862.

Zhao, R., Collins, E.J., Bourret, R.B., and Silversmith, R.E. (2002). Structure and catalytic mechanism of the *E. coli* chemotaxis phosphatase CheZ. Nat. Struct. Biol. *9*, 570–575

Zhu, Y., Qin, L., Yoshida, T., and Inouye, M. (2000). Phosphatase activity of histidine kinase EnvZ without kinase catalytic domain. Proc. Natl. Acad. Sci. U.S.A. *97*, 7808–7813.

Deviations From the Rule: Orphan and Atypical Response Regulators

6

Dagmar Beier

Abstract

In a classical two-component system a response regulator is activated via phosphoryl group transfer from a cognate histidine kinase in order to elicit a cellular output response to a specific stimulus. However, it is increasingly recognized that response regulator proteins do not necessarily rely on this standard activation mode. The so-called 'orphan' response regulators lack a cognate histidine kinase as phosphorylation partner. In many cases these proteins retain their function upon substitution of the phosphate-accepting aspartic acid residue in the receiver domain suggesting phosphorylation-independent mechanisms for output response control. The concept of unorthodox signalling to and activation of a response regulator is supported by the occurrence of atypical response regulators in several bacteria. Atypical response regulators comprise degenerate receiver domains with conserved residues required for the phosphotransfer reaction and signal propagation within the response regulator protein being absent, which certainly precludes canonical receiver phosphorylation. Nevertheless, atypical response regulators are involved in various cellular functions including vegetative cell growth, differentiation, secondary metabolism and virulence.

Introduction

The response regulator (RR) is the basic control element in canonical two-component signalling which acts as a phosphorylation-dependent switch mediating the output response of a bacterium to a changing environment. In the majority of cases it comprises a receiver domain and an effector domain whose activity is controlled by the phosphorylation state of the receiver. According to their function, being predominantly DNA-binding or enzymatic, the effector domains are grouped into different subfamilies (reviewed in: Stock *et al.*, 2000; Gao *et al.*, 2007). The regulatory receiver domain of the RR consists of a five-stranded parallel β-sheet surrounded by five amphipathic α-helices. It comprises a couple of signature residues which are involved in the phosphotransfer reaction and in the propagation of long-range conformational changes accompanying receiver phosphorylation. The highly conserved phosphate-accepting aspartic acid residue located at the end of β-strand 3 and two acidic residues in the β1–α1 loop, being typically aspartate/glutamate (D/E) and aspartate (D) constitute the active site of the receiver and coordinate a Mg^{2+}-ion which is essential for the phosphotransfer reaction. These residues correspond to positions 57, 12 and 13 in the receiver consensus sequence where the numbering is according to CheY of *Escherichia coli* (Volz, 1995). A highly conserved threonine/serine (T/S) residue at the end of β-strand 4 (position 87 in the consensus sequence) and a highly conserved lysine (K) residue at the end of β-strand 5 (position 109 in the consensus sequence) as well as a moderately conserved phenylalanine/tyrosine (F/Y) residue in the middle of β-strand 5 (position 106 in the consensus sequence) are crucial for the phosphorylation-induced conformational change of the receiver domain. This conformational change affects the α4–β5–α5 face of the protein via a subtle reorientation of the conformational switch residues T/S87 and F/Y106 which are positioned between the active site and the α4–β5–α5 face. In the active conformation of the receiver T/S87 and F/Y106 are oriented towards the active site, while in the inactive conformation these residues are oriented away from the active site. Additional signature residues of the consensus receiver sequence are proline 61, glycine 65 and glycine 102 located in the β3–α3 loop, at the N-terminus of α-helix 3 and in the α4–β5 loop, respectively (reviewed in: Stock *et al.*, 2000; Gao *et al.*, 2007; Bourret *et al.*, 2010). The response regulator of a classical two-component system (TCS) is phosphorylated by a cognate histidine kinase (HK) which autophosphorylates in response to a specific stimulus. However, the thorough analysis of 755 completely sequenced bacterial genomes and 39 metagenomes for genes encoding components of TCSs revealed that in many cases there is a disparity between the number of HKs and the number of RRs (Barakat *et al.*, 2009). This observation suggests that branched signal transduction pathways where the signal input of different HKs converges on the same RR, or a single HK signals to different RRs – a situation which was termed many-to-one and one-to-many relationship, respectively, by Laub and Goulian (2007) – might be less exceptional than currently assumed. Cognate HK and RR pairs are easily recognized when the corresponding genes are part of the same operon. When a RR is not encoded in close proximity to a HK or vice versa, these two-component proteins are commonly termed 'orphan' and genome wide analysis revealed that this in fact applies to the majority of two-component signalling genes (www.p2cs.org). Nonetheless, in many cases such 'orphan' proteins will form cognate HK and RR pairs of classical or branched signalling pathways, while some two-component proteins might indeed not have bona fide partners. This issue can be addressed experimentally by phosphotransfer profiling, a systematic biochemical analysis of the ability of a histidine kinase to serve as the phosphoryl group donor for the full complement of response regulators (Laub *et al.*, 2007), provided that the genome sequence of the organism of interest is available. However, notably in bacteria with a complex lifestyle experiencing versatile environmental conditions, which tend to have large numbers of two-component proteins, phosphotransfer profiling is a demanding and time-consuming task.

RR proteins apparently lacking a cognate HK which according to a more strict interpretation are bona fide orphans, will be the focus of this chapter. Therefore, only RRs remaining definitely unpaired in the genome of the respective organism, will be discussed, though still it cannot be ruled out that the respective proteins participate in a branched signalling pathway. For the bona fide orphan RRs the question arises how signalling to the effector protein is accomplished in the absence of a cognate sensor protein. Besides a cognate HK, acetyl phosphate and other low molecular weight phosphoryl group donors can serve as the source of the phosphoryl group in RR phosphorylation (McCleary and Stock, 1994; Stock *et al.*, 1995). The cellular level of acetyl phosphate is sensitive to the metabolic state of the bacterial cell (Wolfe, 2005) and might therefore play a role in activity control of some orphan RRs. Moreover, the mere concentration of a RR in the cell might control the output response mediated by the regulator protein. A couple of orphan RRs exhibit degenerate receiver domains where conserved signature residues which are crucial for the function of the receiver domain as a phosphorylation-dependent signalling module are missing. Despite of their degenerate structure these proteins were shown to have important roles in vegetative cell growth, virulence, developmental processes and secondary metabolism (Guthrie *et al.*, 1998; Aínsa *et al.*, 1999; Yang *et al.*, 2000; Schär *et al.*, 2005; Pan *et al.*, 2009). These atypical RRs presumably do not rely on protein phosphorylation to modulate their activity as effector proteins, but rather other unorthodox mechanisms are likely to mediate their function as true regulator proteins being able to elicit a cellular response to a certain stimulus. One such new mechanism is exemplified by end-product control of the atypical RR JadR1 of *Streptomyces venezuelae* (Wang *et al.*, 2009) regulating the biosynthetic genes for the production of the antibiotic jadomycin B, and others are likely to be discovered. This review summarizes the current knowledge about atypical RRs which act as transcriptional regulators. RR proteins with strongly degenerate so-called pseudo-receiver domains are also involved in twitching motility in *Pseudomonas aeruginosa* (Huang *et al.*, 2003; Kasmierczak *et al.*, 2006) and adventurous and social motility in *Myxococcus xanthus* (Yang *et al.*, 2004; Fraser *et al.*, 2007; Mauriello *et al.*, 2009). The TCS signal transduction network controlling the complex motility system of *M. xanthus* is described in Chapter 14.

DegU of *Listeria monocytogenes*

Listeria monocytogenes harbours 15 TCS with partnered HK and RR genes and an orphan RR which belongs to the NarL/FixJ family and is homologous to the RR DegU of *Bacillus subtilis* (Glaser *et al.*, 2001). DegU of *B. subtilis* (BsDegU) is involved in the control of exoprotease production, biofilm formation, competence development and swarming motility and is part of the DegUS TCS which permits different cellular responses by adjusting the cellular levels of phosphorylated DegU in a 'rheostat'-like manner (Dahl *et al.*, 1992; Kobayashi, 2007; Verhamme *et al.*, 2007). However, BsDegU also acts as a coactivator in its unphosphorylated form (Dahl *et al.*, 1992; Hamoen *et al.*, 2000). Based on the genome sequence of *B. subtilis* Fabret *et al.* (1999) performed a classification of the TCS proteins of this organism and established that all RRs of the NarL/FixJ family pair with HKs with a group II H-box sequence. *L. monocytogenes* contains a single HK of group II which is part of the Lmo1022/Lmo1021 TCS and might be able to phosphorylate other NarL-like RRs besides Lmo1022. While *L. monocytogenes* DegU (LmDegU) which comprises a consensus receiver sequence was efficiently phosphorylated by the HK DegS of *B. subtilis* (BsDegS), no significant phosphoryl group transfer from Lmo1021 to LmDegU has been observed in *in vitro* phosphorylation assays (Gueriri *et al.*, 2008b; Kamp and Higgins, 2009; Fritsch *et al.*, 2011) arguing against the possibility of cross phosphorylation *in vivo*. Promoter mapping of the Lm*degU* gene revealed an exceptionally long 5′-untranslated region of 261 bp which might suggest post-transcriptional regulation of LmDegU expression (Gueriri *et al.*, 2008a). LmDegU acts as an autorepressor by directly binding to three distinct regions which are located downstream of the transcriptional start site (Gueriri *et al.*, 2008a). The observation that knock out mutants were non-motile on soft agar plates suggested that LmDegU plays an important role in the control of flagellar gene expression in *L. monocytogenes* which occurs only at temperatures below 37°C (Knudsen *et al.*, 2004; Williams *et al.*, 2005a; Gueriri *et al.*, 2008a). Transcriptional regulation of flagellar and chemotaxis genes in this organism relies on the repressor protein MogR which at low temperature inhibits transcription by directly binding to the promoters of the respective operons (Gründling *et al.*, 2004). The action of MogR is antagonized by the anti-repressor protein GmaR which is a bifunctional protein exhibiting also O-linked N-acetylglucosamine transferase activity towards flagellin (Shen *et al.*, 2006). GmaR is encoded within the flagellar *fliN-gmaR*-operon. LmDegU was shown to be a temperature-independent transcriptional activator of the *fliN* promoter and, therefore, is indispensable for the synthesis of the anti-repressor protein GmaR which is restricted to temperatures below 37°C by post-transcriptional expression control (Kamp and Higgins, 2009). Somewhat conflicting results have been reported regarding the role of LmDegU receiver phosphorylation in motility. Kamp and Higgins (2009) observed no effect on motility in *L. monocytogenes* mutants expressing a phosphorylation-deficient LmDegU protein (LmDegUD55N) or a truncated protein consisting only of the C-terminal output domain, while a modest decrease in motility and transcription of flagellar and chemotaxis genes was detected by others (Gueriri *et al.*, 2008b; Mauder *et al.*, 2008). Gueriri *et al.* (2008b) also reported high-level transcription of motility and chemotaxis genes in a *L. monocytogenes* mutant expressing BsDegS as well as decreased transcription of these genes in a mutant lacking *pta* and *ackA* suggesting that LmDegU phosphorylation via acetylphosphate might play a role in modulating LmDegU activity *in vivo*. LmDegU seems to be a pleiotropic regulator since a knockout mutant was impaired in the growth at elevated temperature, biofilm formation and virulence in a mouse infection model (Dons *et al.*, 2004; Williams *et al.*, 2005a; Gueriri *et al.*, 2008a). The ability to form biofilms was also reduced in

L. monocytogenes Δ*pta*Δ*ackA*, strengthening the hypothesis that LmDegU activity might be fine-tuned via receiver phosphorylation (Gueriri *et al.*, 2008b). Virulence attenuation of Lm*degU* deletion mutants of *L. monocytogenes* cannot be attributed to their motility defect, since aflagellate *L. monocytogenes* mutants were not affected in their virulence properties (Dons *et al.*, 2004; Way *et al.*, 2004). According to transcriptional profiling the transcription of 29 genes not involved in motility differed more than 2-fold in a Lm*degU* deficient mutant as compared with the *L. monocytogenes* wild-type (Williams *et al.*, 2005b) and, therefore, these genes might be responsible for the pleiotropic effect caused by the deletion of Lm*degU*.

CovR and RitR in streptococci

The CovRS two-component system, also called CsrRS, is a major regulator of virulence in both group A (GAS) and group B streptococci (GBS). In GAS CovRS either directly or indirectly, controls about 15% of the genome, thereby acting mainly as a repressor (Graham *et al.*, 2002). Several important GAS virulence genes were shown to be directly repressed by CovR binding to operator sequences located upstream and downstream of the respective transcription initiation signal (reviewed in Churchward, 2007). Interestingly, *in vivo covRS* mutant strains are selected during invasive GAS infections (Engleberg *et al.*, 2001; Sumby *et al.*, 2006; Walker *et al.*, 2007) suggesting that CovRS regulation is crucial for the transition between localized soft-tissue infection and systemic spread of GAS. A thorough biochemical analysis of the phosphorylation properties of CovS and CovR is still lacking. It is widely assumed that CovR is phosphorylated by CovS which increases the DNA binding affinity of CovR (reviewed in: Churchward, 2007). However, since CovR-repressed target genes are divided in two groups whose members are differently affected by the inactivation of *covS* (Trevino *et al.*, 2009) this common view is presumably an oversimplification. Structural similarity of CovS with EnvZ suggests that phosphatase activity might be involved in its control of the activity state of CovR (reviewed in: Churchward, 2007). Experimental evidence for this hypothesis and the role of alternative phosphoryl group donors for CovR was reported by Dalton and Scott (2004), however, the analysis of *covS* knockout mutants in various GAS strains provided conflicting data to their observations (Trevino *et al.*, 2009; Kansal *et al.*, 2010; Sugareva *et al.*, 2010). Little is known about the signals which are perceived by CovS. High concentrations of environmental Mg^{2+} repress CovR-regulated virulence genes via CovS signalling (Gryllos *et al.*, 2003, 2007). Furthermore, the human antimicrobial peptide LL-37 was shown to stimulate the expression of CovR-repressed genes in a CovS-dependent manner (Gryllos *et al.*, 2008).

While in GAS the prominent mode of action of CovR is gene repression, the CovR regulon in GBS includes both CovR-activated and -repressed genes and is quite variable in diverse GBS strains (Jiang *et al*, 2008; Lembo *et al.*, 2010). The core CovR regulon of GBS comprises about 30 genes, among them important virulence factors (Jiang *et al.*, 2008; Lembo *et al.*, 2010). Interestingly, besides the conserved active site aspartate (D53) GBS CovR contains a second phosphorylation site in the receiver domain, threonine 65 (T65), which is targeted by serine/threonine kinase 1 (Stk1) (Rajagopal *et al.*, 2006; Lin *et al.*, 2009). Stk1-mediated phosphorylation of T65 negatively affects D53 phosphorylation and impairs DNA binding of CovR suggesting that different signal input pathways converge on CovR to fine-tune virulence gene expression for optimal adaptation to different host niches (Lin *et al.*, 2009). Whether GAS CovR is also regulated by a cognate serine/threonine kinase has not been investigated so far.

Streptococcus mutans is the primary aetiological agent of dental caries. Analysis of the genome sequence of strain *S. mutans* UA159 revealed the presence of 14 two-component systems whose HK and RR genes are located adjacent to each other as well as a CovR homologue which is not partnered by a CovS HK. CovS was shown to be lacking also in other isolates of *S. mutans* demonstrating that CovR, which is also known as GcrR or TarC, is a bona fide orphan in this species (Chong *et al.*, 2008). Interestingly, a piece of 146 bp from the downstream sequence of *covR* in *S. mutans* UA159 encodes a truncated HK suggesting that a deletion event occurred in an ancestor of *S. mutans* which removed the cognate HK of CovR (Chong *et al.*, 2008). Inactivation of CovR resulted in increased acid sensitivity, an impaired acid tolerance response, aberrant biofilm architecture and negatively affected cariogenesis in germ-free rats (Idone *et al.*, 2003; Dunning *et al.*, 2008). The latter phenotypes are probably related to the fact that expression of *gtfBC*, *gtfD* and *gbpC* encoding the secreted glycosyltransferases (GTFs) GTF-I, GTF-SI and GTF-S and the glucan-binding protein C, respectively, was derepressed in the *covR* knockout mutant (Idone *et al.*, 2003; Biswas and Biswas, 2006; Biswas *et al.*, 2007). The cell-associated glycosyltransferases mediate the tenacious adherence to the tooth surface via the synthesis of water-soluble and -insoluble glucans from dietary sucrose, which are then bound by surface-associated glucan-binding proteins to promote cell-to-cell aggregation (reviewed in Banas and Vickerman, 2003). *S. mutans* CovR was shown to directly bind the upstream regions of *gtfBC* and *gbpC*, thereby covering large regions of DNA comprising the −10 and −35 boxes of the respective promoters (Biswas and Biswas, 2006; Biswas *et al.*, 2007). Furthermore, *S. mutans* CovR interacted with its own promoter resulting in modest negative autoregulation (Chong *et al.*, 2008). Negative autoregulation has also been reported for CovR from GAS (Gusa and Scott, 2005), while no autoregulation of CovR expression was observed in GBS (Lamy *et al.*, 2004). Global transcriptional profiling revealed that CovR regulates 6.5% of the genes of the *S. mutans* genome, thereby acting both as a repressor and as a transcriptional activator (Chong *et al.*, 2010; Dmitriev *et al.*, 2011). Notably, several competence related genes and genes involved in oxidative stress resistance were down regulated in the *covR* mutant suggesting a role of CovR in competence development and the oxidative stress response of *S. mutans* (Dmitriev *et al.*, 2011). Regarding the environmental signals affecting CovR expression in *S. mutans* UA159 somewhat contradictory observations have been reported. Chong *et al.* (2008) observed increased transcription of *covR* at pH 7.2 compared with pH 5.5, as well as an increase in the amount of *covR* mRNA in the presence of increasing concentrations of Mg^{2+} ions. In contrast, no difference in *covR* transcription in cultures exposed to pH 7.5 or pH 5.0 was apparent in the study of Dunning *et al.* (2008), while these authors observed an increase in *covR* transcription upon manganese depletion of the culture broth. This was attributed to manganese-dependent repression of *covR* by the metal-responsive regulator protein SloR the binding of which to the *covR* promoter could be demonstrated *in vitro*.

The mechanism of CovR-mediated signal transduction in *S. mutans* remains unclear. Since *S. mutans* CovR contains a consensus receiver sequence (Fig. 6.1) it cannot be ruled out that it participates in a branched signalling pathway. Moreover, it is well conceivable that the activity of the protein is modulated by aspartate phosphorylation via cross talk from a 'non-cognate' HK or a low molecular weight phosphoryl group donor. However, this question has not been addressed experimentally so far. *In vitro* phosphorylation with acetyl phosphate did not affect the affinity of *S. mutans* CovR for the promoter region of *gftB* (Biswas and Biswas, 2006), while *in vitro* phosphorylation increases the DNA binding capability of CovR from GAS (reviewed in: Churchward, 2007). Although a serine/threonine kinase is encoded in *S. mutans*, receiver phosphorylation by a similar mechanism as observed with Stk1 in *S. agalactiae* is little likely, since T65 is not conserved in *S. mutans* CovR (Lin *et al.*, 2009). Therefore, it might also be possible that *S. mutans* CovR is always present in an active state and its activity as a transcriptional regulator is controlled solely by the amount of CovR protein present in the cell. Interestingly, the *S. mutans covR* transcript, like the orthologous transcript in GAS, contains a non-translated leader sequence of 221 nucleotides which might be involved in post-transcriptional regulation of CovR expression (Miller *et al.*, 2001; Chong *et al.*, 2008). Furthermore, Biswas and colleagues hypothesized that post-translational regulation might be involved in the expression control of *S. mutans* CovR, since they observed up-regulation of CovR expression upon inactivation of the ClpP protease (Chong *et al.*, 2008).

The genome of *Streptococcus pneumoniae* encodes 13 HK–RR pairs and an unpaired orphan RR, called RitR (Ulijasz *et al.*, 2004), which is highly similar to CovR from GAS, GBS and *S.*

```
CovR Ssu    MAKKILIAGKERNLSHFVSMELQKKDYLVDYASTGKEAMSLAHETDFDLILMSFQLSDMS 60
RitR Spn    MGKRILLLEKERNLAHFLSLELQKEQYRVDLVEEGQKALSMALQTDYDLILLNVNLGDMM 60
CovR Smu    MAKDILIIEDEKKLAKFVSLELQHEGYEVEVTYDGRSGLERALEKDFDLILLDLMLPEMD 60
CovR Spy    MTKKILIIEDEKNLARFVSLELQHEGYEVIVEVNGREGLETALEKEFDLILLDLMLPEMD 60
            * * **:  .*::*::*:*:***:: * *     *:..:. * :.::****:.. * :*

CovR Ssu    SKELATELLAIKPATVMIVVVEPTEVSQYGEEVLSYAVSYVVKPFVISDLVEQISAIFRG 120
RitR Spn    AQDFAEKLSRTKPASVIMILDHWEDLQEELEVVQRFAVSYIYKPVFIENLVARISAIFRG 120
CovR Smu    GYEVIRRLQLEKTTYVIIITAR-DSIMDIVTGLDRGADGYIVKPFAIEELLARVRAIFRR 119
CovR Spy    GFEVTRRLQTEKTTYIMMMTAR-DSIMDVVAGLDRGADDYIVKPFAIEELLARIRAIFRR 119
            . :.  .*   *.: ::::  .   .: :     :   * .*: **. *.:*: :: ****

CovR Ssu    RDFIDNNCKQVHMHAAYRDLKVDFQNRTVTRGDELINLTRREYDLLATLMNSP-EPVSRE 179
RitR Spn    RDFIDQHCSLMKVPRTYRNLRIDVEHHTVYRGEEMIALTRREYDLLATLMGSK-KVLTRE 179
CovR Smu    QD-IEKAKHSSKNDTTYRDLKLDVHNRAAIRNGETIPLTKREFDLLNALVENINQVMTRE 178
CovR Spy    QD-IESEKKVP-SQGIYRDLVLNPQNRSVNRGDDEISLTKREYDLLNILMTNMNRVMTRE 177
            :* *:.          **:* :: .:::. *. : * **:**:***  *: .   . ::**

CovR Ssu    QLLERVWKYEAASETNVVDVYIRYLRGKLDLPHQDSYIKTVRGVGYAMRD-- 229
RitR Spn    QLLESVWKYESATETNIVDVYIRYLRSKLDVKGQKSYIKTVRGVGYTMQE-- 229
CovR Smu    ELLARVWKYDQEAETNVVDVYIRYLRGKIDVPGKESYIQTVRGMGYIIRENN 230
CovR Spy    ELLSNVWKYDEAVETNVVDVYIRYLRGKIDIPGKESYIQTVRGMGYVIREK- 228
            :**  ****:   ***:*********.*:*:   :.***:****:** :::
```

Figure 6.1 Comparison of CovR homologues of streptococci. The alignment shows the amino acid sequences of CovR of *S. pyogenes* MGAS6180 (Spy), *S. mutans* UA159 (Smu), *S. suis* 05ZYH33 (Ssu) and of RitR of *S. pneumoniae* G54 (Spn). Stars below the aligned sequences highlight single fully conserved residues, colons and dots indicate conservation between groups of amino acids with strongly and weakly similar properties, respectively. Positions corresponding to the signature residues D/E12, D13, D57, P61, G65, T/S87, Y/F106 and K109 of the receiver consensus sequence are highlighted by black shading.

mutans (48.2% identity and 60.2% similarity to CovR from *S. pyogenes*). RitR is a degenerate RR protein since its receiver domain contains an asparagine residue at the position of the phosphate-accepting aspartate. Other key residues of the receiver consensus sequence are also missing in RitR, i.e. the second acidic amino acid (D13) located at the C-terminal end of β-strand 1, which is replaced by lysine, and P61 and G65 located in the turn between β-strand 3 and α-helix 3 and at the C-terminus of α-helix 3, respectively (Fig. 6.1). In accordance with its degenerate structure RitR could not be phosphorylated *in vitro* by the low-molecular weight phosphoryl group donor phosphoramidate (Ulijasz *et al.*, 2009). Deletion of *ritR* in *S. pneumoniae* resulted in virulence attenuation in a mouse respiratory tract model of infection, while pathogenicity in a thigh model remained unaffected suggesting that RitR is involved in the control of tissue specific virulence factors (Throup *et al.*, 2000; Ulijasz *et al.*, 2004). Global transcriptional profiling revealed a role of RitR in the activation and repression of genes related to iron transport, oxidative stress resistance, DNA damage repair, sugar transport and capsule biosynthesis (Ulijasz *et al.*, 2004). As observed for CovR of GAS, RitR is *in vitro* phosphorylated by a streptococcal serine/threonine kinase, StkP of *S. pneumoniae*, however, in case of RitR a so far unidentified amino acid in the DNA-binding domain is the target of phosphorylation. In turn, PhpP, the cognate phosphatase of StkP, is able to remove the phosphoryl group from phosphorylated RitR (Ulijasz *et al.*, 2009). Interestingly, transcription of the RitR target gene *piuA* encoding a Fe carrier protein was affected in *stkP* and *phpP* deletions mutants of *S. pneumoniae* indicating *in vivo* relevance of RitR phosphorylation by StkP (Ulijasz *et al.*, 2009).

A CovR homologue being an orphan RR has also been described in *Streptococcus suis* Serotype 2 which is a zoonotic swine pathogen associated with arthritis, endocarditis, meningitis, pneumonia and septicaemia (Pan *et al.*, 2009). However, *S. suis* CovR is more similar to RitR from *S. pneumoniae* (55.0% identity, 64.2% similarity) than to

CovR from GAS and GBS (47.8% identity, 56.6% similarity to GAS CovR). Like RitR *S. suis* CovR is a degenerate RR lacking most of the conserved amino acids in the receiver domain (Fig. 6.1). Nevertheless, *S. suis* CovR turned out to be an important regulator of virulence, since deletion of *covR* increased the lethality of *S. suis* Serotype 2 in experimental infection of piglets, while complementation of the Δ*covR* mutant restored the wild-type phenotype (Pan *et al.*, 2009). The Δ*covR* mutant also exhibited major phenotypic changes, i.e. growth as elongated chains, a thicker capsule and increased haemolytic activity and showed increased resistance to phagocytosis and killing by neutrophils and monocytes. Microarray analysis demonstrated that *S. suis* CovR acts as a repressor protein which controls a regulon comprising about 190 genes including several proven or putative virulence factors (Pan *et al.*, 2009). Like the other streptococci discussed in this paragraph *S. suis* encodes a serine/threonine kinase homologue. It remains to be investigated in detail whether RR phosphorylation by serine/threonine kinases constitutes a novel and general means of modulating RR activity in response to environmental stimuli in streptococci, in particular in case of degenerate RR proteins which are not target for aspartate phosphorylation.

HP1021 and HP1043 of *Helicobacter pylori*

The human gastric pathogen *H. pylori* is well known for its general paucity of regulatory genes. Besides the chemotaxis regulatory system this organism comprises only three TCS as well as two orphan RRs which exhibit an atypical receiver sequence (Tomb *et al.*, 1997). HP1021 contains a helix–turn–helix motif at the C-terminal end of its output domain and is placed into the PrrA family of RRs (www.p2cs.org). Its receiver domain lacks the canonical phosphate-accepting aspartic acid residue which is replaced by a serine (S47) as well as the conserved proline and glycine residues corresponding to positions 61 and 65 in the consensus sequence. Interestingly, the serine residue at the canonical phosphorylation site is present in the HP1021 orthologues from all bacteria of the order Campylobacterales for which sequence information is available. C-terminally adjacent to S47 there is another serine residue in HP1021 which is not strictly conserved in the Campylobacterales bacteria. The hp1021 gene is part of an operon comprising two other genes encoding a serine protease (HP1018/HP1019) and a conserved hypothetical protein (HP1020) (Pflock *et al.*, 2005; Sharma *et al.*, 2010). Transcription from the promoter upstream of hp1018/hp1019 was strongly increased at low pH, but this transcriptional induction was unaffected by the inactivation of hp1021 arguing against an autoregulatory function of the RR (Pflock *et al.*, 2005; Sharma *et al.*, 2010). Deletion of the hp1021 gene resulted in a small colony phenotype (Beier and Frank, 2000; McDaniel *et al.*, 2001) suggesting a function of HP1021 for vegetative cell growth. Transcriptional profiling revealed that HP1021 is a transcriptional activator of the *nifS* and *nifU* genes whose products are involved in the formation of Fe–S clusters (Pflock *et al.*, 2007). Since *nifS-nifU* are considered housekeeping genes in *H. pylori*, their down-regulation in the Δhp1021 mutant might be responsible for the observed growth defect. A gene cluster encoding metabolic enzymes which enables *H. pylori* to convert acetone to acetyl-CoA was also found to be strongly regulated by HP1021 (Pflock *et al.*, 2007). Transcription of this gene cluster occurred under standard *in vitro* culture conditions and did not require the presence of acetone for its induction (Pflock *et al.*, 2007). Interestingly, a *H. pylori* mutant deficient in the gene *acxB* encoding acetone carboxylase subunit α exhibited a reduced ability to colonize mice suggesting that acetone, which is present in the stomach in millimolar concentrations, might be used as an electron donor in *H. pylori's* energy metabolism (Brahmachary *et al.*, 2008). The question whether phosphorylation of S47 might be required for the proper function of HP1021 was addressed by *in vitro* phosphorylation assays and by monitoring the growth phenotype of *H. pylori* strains producing mutated derivatives of HP1021. *In vitro* phosphorylation of serine residues located N-terminally adjacent to the canonical phosphorylation site of an RR had already been observed in mutated *E. coli* CheY and NtrC proteins with 'D57N' substitutions (Moore *et al.*, 1993; Appleby and Bourret, 1999). However, *in vitro* phosphorylation of the receiver domain of HP1021 by the *H. pylori* histidine kinases ArsS, HP0244 and HP1364 or by acetylphosphate was not observed (Beier and Frank, 2000; Schär *et al.*, 2005). Mutants expressing HP1021 proteins with S47D, S47N or S47S48NA substitutions behaved like the wild-type strain arguing against a role of serine phosphorylation in the cell growth-associated function of HP1021. Furthermore, no protein modification was detected when HP1021 isolated from protein lysates of *H. pylori* cells was analysed by mass spectrometry (Schär *et al.*, 2005). Strict transcriptional regulation of the hp1021 gene seems not to be crucial for the function of HP1021 in vegetative growth, since *H. pylori* mutants in which the hp1021 gene was put under the control of other *H. pylori* promoters (P_{cagA}, P_{pfr} and P_{fecA}, respectively) showed normal growth on blood agar plates (Müller *et al.*, 2007).

HP1043 which belongs to the OmpR-family is an atypical RR with a highly degenerated receiver sequence. Comparison of HP1043 with its orthologues from the ε-proteobacteria revealed a deletion of four amino acids at the C-terminal end of β-strand 3 resulting in the lack of the phosphate-accepting aspartic acid residue. Moreover, the highly conserved aspartic acid residue corresponding to position 13 in the receiver consensus sequence is replaced by lysine and the conserved P61 and G65 residues are also not present in HP1043. Accordingly, HP1043 could not be phosphorylated *in vitro* (Beier and Frank, 2000; Schär *et al.*, 2005). NMR spectroscopy and X-ray crystallography revealed that HP1043 forms a symmetrical dimer with two functional domains being separated by a short flexible linker (Hong *et al.*, 2004, 2007; Lee *et al.*, 2006). The dimer interface that is formed by the three secondary structure elements α4–β5–α5 of the receiver resembles the α4–β5–α5 face which is adopted by other RRs of the OmpR-family in their activated phosphorylated form. Notably, tyrosine 94 corresponding to the conformational switch residue Y106 of the receiver consensus sequence has an inward orientation (Hong *et al.*, 2007) suggesting that HP1043 functions in the absence of posttranslational modification. RR HP1043 is an essential protein (Beier and Frank, 2000; McDaniel *et al.*, 2001; Schär *et al.*, 2005). *In vitro* DNA binding experiments demonstrated that HP1043 binds to the promoter of its own gene suggesting autoregulation, and to the promoter of the *tlpB* gene encoding a methyl-accepting chemotaxis protein (Delany *et al.*, 2002). Interestingly, a mutated, dimerization-deficient HP1043 protein with a F87L substitution was still able to bind with wild-type efficiency to a 29 bp sequence motif harbouring a degenerate inverted repeat which was identified as the HP1043 binding site in the promoters of the hp1043 and *tlpB* gene (Delany *et al.*, 2002; Hong *et al.*, 2007). Other target genes which might be responsible for the essential function of HP1043 are not known so far. The identification of such target genes has not only been hampered by the essential function of HP1043, but also by the fact that overexpression of the protein could not be achieved in *H. pylori*, indicating post-transcriptional or post-translational expression control (Delany *et al.*, 2002; Müller *et al.*, 2007). Recently, a small regulatory RNA was identified which is encoded upstream of HP1043 with divergent transcriptional orientation. Interestingly, this small RNA is a post-transcriptional regulator of TlpB expression (Sharma *et al.*, 2010) suggesting a regulatory interplay between HP1043 and the small RNA in controlling TlpB expression.

HP1043 is conserved among the ε-proteobacteria and the orthologous proteins show a remarkably high degree of similarity in their C-terminal DNA-binding output domains (Fig. 6.2). This similarity averages 92.9% when all the members of the ε-proteobacteria covered by the NCBI database (www.ncbi.nlm.nih.gov) are considered, suggesting also a conserved function for the HP1043 orthologues in this highly diverse group which includes both host-associated bacteria and environmental bacteria exhibiting a chemolithoautotrophic metabolism (Campbell *et al.*, 2006). A comparable high degree of conservation is found among the Firmicutes for the essential RR YycF which is involved in the control of cell wall metabolism in these bacteria (reviewed in Dubrac *et al.*, 2008). It is tempting to speculate that the HP1043 orthologues might be essential RRs in all the ε-proteobacteria, but yet this question has not been addressed experimentally. Interestingly, in three enterohepatic *Helicobacter*

```
Hhe   MRILVIEDDPTLCKNIAEMLNERSYQTDIAENLKDGEYFISIRNYDLVLADWDLSDGCGI 60
Cje   MRILVIEDEISLNKTIIDNLNEFGYQTDSSENFKDGEYFIGIRHYDLVLANWTLPDGDGA 60
Wsu   MRILIIEDEITLNKTLAEGLNEFNYQTDVAENLKDGEYYISIRNYDLVLTDWMLPDGSGL 60
Sau   MRILIIEDEVTLNKMLAEGLKEFGYQSDVVETLKDGEYYLDIRNYDLVLMDWMLPDGNSV 60
Hpy   MRVLLIEKNSVLGGEIEKGLNVKGFMADVTESLEDGEYLMDIRNYDLVMV---SDKNAL 56
      **:*:**.:  *   : . *:  .: :*  *.::**** :.**:****: .    *  .

Hhe   DIIAQVKEKSPRTPVIIISAQTSAENEIRAFKEGADDFVSKPF-SLEVLLIRIQARLRFW 119
Cje   ELVNTIKHKSPRTSVMIMSSKTDKETEIKALKAGADDFVKKPL-DFDILLARIEARLRLG 119
Wsu   DIISQVKSKSPRTAVVILSARDDKESEIEALRGGADDFIAKPF-DFDVLVARIEARLRFG 119
Sau   DIIGDIKANTPKTVVVVISARDDNESEIEALRAGADDYIRKPF-DFDVLIARIEARLRFG 119
Hpy   SFVSRIKEKHSSIVVLVSSDNPTSEEEVHAFEQGADDYIAKPYRSIKALVARIEARLRFW 116
      .::  :* : .   *:: * .   * *:.*:. ****:: **  .:. *: **:****:

Hhe   GSSVIEIEDLVINPDEEKITYRGQEIEVKGKPFEVLTHLARHRDQIVSKEQLLDAIWEEP 179
Cje   GTNVIKIEDLVIDPDEEKITYKGQDIELKGKPFEVLTHLARHSDQIVSKEQLLDAIWEEP 179
Wsu   GTNLIEIEDLIINPDEEKITYKGQEIELKGKPFEVLTHLARHRDQIVSKEQLLDAIWEEP 179
Sau   GSNIIEIEDLTINPEEEKITYKETEIELKGKPFEVLTHLARHRDQIVSKEQLLDAIWEEP 179
Hpy   GSNVIEIGDLTISPDEEKIIYKGREVEVKGKPFEVLTHLARHRDQIVSKEQLLDAIWEEP 176
      *:.:*:* ** *.*:**** *:  ::*:************** *****************

Hhe   ELVTPNVIEVAINQIRQKMDKPLNIGTIETVRRRGYRFCYPKKPSEG 226
Cje   ELVTPNVIEVAINQIRQKMDKPLNISTIETVRRRGYRFCFPKKS--- 223
Wsu   ELVTPNVIEVAINQIRQKMDKPLNISTIETVRRRGYRFCYPKGV--- 223
Sau   ELVTPNVIEVAINQIRQKMDKPLSITTIETVRRRGYRFCFPKEV--- 223
Hpy   EMVTPNVIEVAINQIRQKMDKPLGISMVETVRRRGYRFCYPKPACEE 223
      *:*********************.*  :***********:**
```

Figure 6.2 HP1043 orthologues in the ε-proteobacteria. The amino acid sequences of RR HP1043 of *H. pylori* 26695 (Hpy) and its orthologues in *H. hepaticus* ATCC 51449 (Hhe), *Campylobacter jejuni* NCTC 11168 (Cje), *Wolinella succinogenes* DSM 1740 (Wsu) and *Sulfurimonas autotrophica* DSM 16294 (Sau) are aligned. Stars below the aligned sequences highlight single fully conserved residues, colons and dots indicate conservation between groups of amino acids with strongly and weakly similar properties, respectively. Positions corresponding to the signature residues D/E12, D13, D57, P61, G65, T/S87, Y/F106 and K109 of the receiver consensus sequence are highlighted by black shading. The structural elements of the winged helix–turn–helix motif in the output domain of HP1043 consisting of an N-terminal four-stranded β-sheet, a central three-helical bundle and a C-terminal β-hairpin are indicated according to Hong *et al.* (2007). Grey rectangles above the aligned sequences indicate α-helices, black arrows indicate β-strands.

species and seven *Campylobacter* species a consensus HP1043 orthologue is accompanied by a cognate HK encoded adjacent to the RR, while in the other members of the genera *Helicobacter* and *Campylobacter* the HP1043 orthologue is an orphan RR exhibiting either a consensus receiver sequence or a atypical receiver sequence lacking the phosphate-accepting aspartic acid residue. In almost all the other members of the order Campylobacterales, which includes many environmental bacteria with both marine and terrestrial habitats, the HP1043 orthologue is the effector protein of a TCS, suggesting degeneration of this TCS in the evolution of the Campylobacterales bacteria when environmental ancestors adapted to different hosts and host niches. However, the HP1043 orthologues of the members of the small order Nautiliales (comprising the genera *Caminibacter*, *Nautilia* and *Lebetimonas*) seem to be orphan RRs.

NblR of *Synechococcus* sp.

In non-diazotrophic cyanobacteria the atypical orphan RR NblR is involved in stress survival and an environmental adaptation process called chlorosis or bleaching. By this process, which requires the small protein NblA, cyanobacteria degrade their light-harvesting phycobilisoms when experiencing stress conditions like nutrient starvation (reviewed in Grossmann *et al.*, 1994).

Transcription of *nblA* is positively regulated by NblR and NtcA, the global regulator protein for nitrogen control (Luque *et al.*, 2001). NblR, a member of the OmpR subfamily of RRs, lacks two conserved amino acids of the receiver consensus sequence, i.e. D13 and T87, which are replaced by serine and methionine (S14 and M85), respectively, while it retains the phosphate-accepting aspartic acid residue (D57 in NblR). By replacing NblR with a mutated derivative carrying a D57A substitution it was demonstrated that phosphorylation at the canonical phosphorylation site is not required for NblR function in *Synechococcus* sp. PCC 7942, since stress survival, chlorosis and *nblA* transcription was unaffected in the mutant strain (Kato *et al.*, 2008; Ruiz *et al.*, 2008). In accordance with this observation NblR could not be phosphorylated *in vitro* with acetyl phosphate (Ruiz *et al.*, 2008). Surprisingly, gel filtration chromatography indicated that recombinant NblR is a monomer in solution and yeast two-hybrid screenings using NblR as bait provided no hint for dimer formation *in vivo* (Ruiz *et al.*, 2008). However, protein–protein interactions with nitrate reductase, the H subunit of NADH dehydrogenase and the rod-shape-determining protein MreC were observed (Kato *et al.*, 2008; Ruiz *et al.*, 2008). The biological significance of these interactions which might have a role in the phosphorylation-independent activation process of NblR remains to be investigated.

Atypical RRs of streptomycetes

Filamentous soil bacteria of the genus *Streptomyces* harbour a large number of TCS signal transduction proteins. In case of *S. coelicolor* the $MiST_2$ database (http://mistdb.com) lists 140 HK and 138 RR genes, 92 of which do not have a neighbouring two-component gene. Hence, when functional data are not available, it is impossible to predict whether a RR is a bona fide orphan lacking a cognate phosphorylation partner or whether a two-component protein is classified as orphan solely because of the unpaired genomic organization of its corresponding gene. In streptomycetes the largest number of RRs whose mode of function differs from the two-component paradigm has been described (Furuya and Hutchinson, 1996; Guthrie *et al.*, 1998; Aínsa *et al.*, 1999; Yang *et al.*, 2001; O'Connor and Nodwell, 2005). Interestingly, until now these atypical RRs, which are not encoded in the vicinity of an HK gene, were found to be exclusively involved in the control of differentiation processes and in the regulation of secondary metabolite production.

At the onset of differentiation of *S. coelicolor* a white aerial mycelium is formed from the substrate mycelium which consists of multigenomic aerial hyphae subsequently developing into chains of grey exospores. This developmental process is governed by a complex regulatory cascade (Chater, 2001). Formation of the aerial mycelium requires the RR BldM which consists of a consensus receiver domain and an output-domain of the NarL/FixJ subfamily. While *bldM* null mutants were unable to erect aerial hyphae, *S. coelicolor* mutants expressing *bldM* alleles with substitutions of the putative phosphate-accepting aspartic acid residue (D54) to asparagine (D54N) or alanine (D54A) exhibited a wild-type phenotype suggesting that BldM function is independent of receiver phosphorylation at D54 (Molle and Buttner, 2000). However, mutants with *bldM* alleles carrying other substitutions affecting the receiver phosphorylation pocket, i.e. a substitution of the conserved glycine residue at position 62 (G65 in the consensus sequence) to aspartate (G62D) or of cysteine at position 7 to tyrosine (C7Y) were defective in the differentiation of the aerial mycelium into mature spore chains (Molle and Buttner, 2000). Unfortunately, these astonishing effects of different *bldM* alleles on *S. coelicolor* development have not been further addressed experimentally.

When *S. coelicolor* is grown on rich media the formation of the aerial mycelium is dependent on the surfactant protein SapB. The precursor of mature SapB and gene products which are required for post-translational processing of the SapB precursor and SapB export are encoded in the *ram* operon whose transcription is controlled by the RR RamR (O'Connor *et al.*, 2002). RamR belongs to the NarL/FixJ subfamily of RRs and displays an atypical receiver sequence with a histidine and a leucine replacing residues D/E12 and Y/F106 of the consensus sequence, while the other key positions are retained. Substitution of residues D12 or T84 (corresponding to D13 and T/S87 in the consensus sequence) to alanine or of L102 to tyrosine had no effect on the morphogenesis of *S. coelicolor*. In contrast, aerial hyphae formation was impaired when the highly conserved amino acids D56 or K105 (D57 and K109 in the consensus sequence) were exchanged or when L102 was replaced by alanine. RamR derivatives which were impaired in their *in vivo* function were further analysed *in vitro*, demonstrating that the various receiver mutations affected either both RamR dimerization and binding to the promoter of the *ram* operon (D56A, K105A), DNA binding (D56N) or the ability to form higher order complexes on the promoter by cooperative binding (L102A) (O'Connor and Nodwell, 2005). Since RamR could not be phosphorylated *in vitro* with phosphoramidate, O'Connor and Nodwell (2005) proposed that the *in vivo* function of RamR might be independent of phosphorylation, however, based on the available data a phosphorylation-dependent mode of action cannot be ruled out.

Another atypical response regulator belonging to the NarL/FixJ subfamily, WhiI, is involved in the control of spore development of *S. coelicolor* (Aínsa *et al.*, 1999). *whiI* is considered as an early sporulation gene which is required for the formation of prespore compartments in the aerial hyphae while the late sporulation genes are essential for spore maturation (Chater, 2001). Consistently, a *whiI* null mutant exhibited white, non-sporulating aerial hyphae showing no signs of sporulation-associated DNA condensation (Aínsa *et al.*, 1999). WhiI lacks the highly conserved D/E12 residue of the receiver consensus which is replaced by alanine (A26) as well as K109 being replaced by threonine (T118) (Aínsa *et al.*, 1999). Mutagenesis of the conserved amino acids of the receiver domain of WhiI lead to astonishing observations: a *S. coelicolor* mutant expressing a *whiI* allele with a S96A substitution (corresponding to T/S87 in the consensus sequence) behaved like the wild-type, while mutants with substitutions of D27 or D69 (D13 and D57 in the consensus sequence) to alanine, a D27A-D69A double replacement or a T118K substitution were able to form spore chains, but showed varying degrees of defects in spore maturation (Tian *et al.*, 2007). According to the working model derived from these data WhiI is changing from an unmodified to a modified state in response to a signal. The unmodified form of WhiI is sufficient to bring about prespore formation, while the modified state is required for spore maturation. The different WhiI alleles would then differ in their ability to undergo post-translational modification in response to the signal. Whether this modification is phosphorylation remains to be clarified. Moreover, the different functions of WhiI in spore development might also be explained by changes in its abundance with different mutant phenotypes resulting from differences in the effective concentrations of the various allelic forms of WhiI (Tian *et al.*, 2007).

Atypical response regulators controlling the synthesis of antibiotics are usually embedded in complex regulatory cascades. *S. peucetius* produces the commercially important chemotherapeutic agents daunorubicin and doxorubicin. Daunorubicin synthesis is controlled via a regulatory cascade comprising three transcriptional regulators, DnrO, DnrN and DnrI, and is feedback-regulated by the capability of daunorubicin to compete with DnrO for DNA binding (Otten *et al.*, 1995; Otten *et al.*, 2000; Ajithkumar and Prasad, 2010). DnrN is a response regulator protein from the NarL/FixJ family the expression of which is activated by DnrO and which in turn activates DnrI being the key regulator of the daunorubicin biosynthesis and resistance genes (Madduri and Hutchinson, 1995; Jiang and Hutchinson, 2006). DnrN exhibits modest deviations from the receiver consensus sequence, i.e. D13 is replaced by a glutamic acid residue (E10 in DnrR) and the moderately conserved F/Y106 in the middle of β-strand 5 is replaced by leucine (L102). Surprisingly, mutants expressing *dnrN* alleles with substitutions of the phosphate-accepting aspartic acid residue D55 to glutamate (D55E) or asparagine (D55N) retained 53% and 35% of the wild-type level of daunorubicin synthesis, which is abolished in a *dnrN* null mutant suggesting that the function of DnrN is at least partially independent of phosphorylation (Otten *et al.*, 1995). Moreover, DnrN could not be phosphorylated by acetyl phosphate *in vitro* and it was shown that recombinant DnrN binds efficiently to the *dnrI* promoter in the absence of phosphorylation (Furuya and Hutchinson, 1996).

A related regulatory pathway controls the production of the antibiotic undecylprodigiosin in *S. coelicolor*. Expression of the pathway-specific transcriptional regulator RedD is activated by the atypical response regulator RedZ which also controls several RedD-independent biosynthetic genes involved in undecylprodigiosin production (Guthrie *et al.*, 1998; Huang *et al.*, 2001). RedZ expression is regulated by a two-component system, AbsA1-AbsA2, whose input signals are unknown so far (McKenzie and Nodwell, 2007). The NarL/FixJ subfamily member RedZ lacks most of the highly conserved amino acids of the receiver phosphorylation pocket, i.e. D/E12, D57 and K109 which are replaced by cysteine (C8), valine (V52) and alanine (A102), respectively, suggesting that RedZ does not require canonical receiver phosphorylation to act as a transcriptional regulator (Guthrie *et al.*, 1998).

S. venezuelae ISP5230 produces the antibiotics chloramphenicol and jadomycin B, the latter of which is only synthesized under nutrient limitation or in the presence of environmental stresses like elevated temperatures or toxic concentrations of ethanol. The pathway specific activator of the jadomycin B biosynthesis genes is the OmpR-type response regulator JadR1 which also acts as a negative autoregulator and as a negative regulator of chloramphenicol production (Yang *et al.*, 2000; Wang *et al.*, 2009; Xu *et al.*, 2010). In addition to autoregulation, JadR1 expression is negatively controlled by the repressor protein JadR2 (Xu *et al.*, 2010). The genes encoding *jadR1* and *jadR2* are located adjacent to each other and are transcribed in the opposite direction (Yang *et al.*, 2001). In the JadR1 receiver sequence D13 is replaced by serine (S25) and the third acidic residue usually following the highly conserved D/E12D13 pair is substituted by cysteine (C26) (Yang *et al.*, 2001). JadR1 could not be phosphorylated with phosphoramidate *in vitro*. Instead, jadomycin B was shown to bind to the receiver domain of JadR1 and to inhibit the binding of JadR1 to the promoters of *jadR1* and *jadJ* being the first gene of the jadomycin biosynthetic cluster. The latter effect was also observed with various intermediates of the jadomycin biosynthetic pathway, but was not apparent with structurally unrelated antibiotics (Wang *et al.*, 2009). Therefore, the activity of JadR1 as a transcriptional regulator is likely to be modulated by the binding of the small ligand jadomycin B which creates a negative feedback loop to shut down jadomycin B synthesis when the antibiotic is becoming abundant. Yang *et al.* (2009) corroborated their concept of feedback inhibition of pathway-specific atypical response regulator function by the finding that RedZ loses its capability to bind to the *redD* promoter *in vitro* in the presence of undecylprodigiosin. It seems quite likely that small ligand binding will turn out to be a general mechanism to modulate output function replacing the canonical receiver phosphorylation in atypical response regulators. For further information about two-component systems in streptomycetes, see also Chapter 15.

ChxR of *Chlamydia*

Chlamydiae are obligate intracellular pathogens which upon host cell infection undergo a complex developmental cycle resulting in a conversion of infectious elementary bodies to metabolically active and replicating reticulate bodies which prior to host cell lysis redifferentiate into elementary bodies (reviewed in: Beatty *et al.*, 1994). The intracellular signals governing the chlamydial developmental cycle are unknown. Chlamydiae harbour a single TCS consisting of a RR of the NtrC subfamily and its cognate HK (Koo and Stephens, 2003). Another transcription factor, named ChxR, shows some similarity to RRs of the OmpR subfamily (Koo *et al.*, 2006). This similarity is most pronounced in the C-terminal domain of ChxR, which comprises a winged helix–turn–helix motif, while the N-terminal domain is only weakly reminiscent of a receiver domain (Fig. 6.3). The phosphate-accepting aspartic acid residue is replaced by glutamic acid in ChxR of *C. trachomatis* (E49), *C. pneumoniae* and *C. psittaci*, while it is retained in strain UWE25 isolated from amoebae. With the exception of the tyrosine residue (Y90) in the β5-strand corresponding to Y106 in the consensus receiver sequence the other signature residues of the receiver domain are not present in ChxR (Koo *et al.*, 2006). Full-length ChxR and the separated receiver domain of the protein were shown to form homodimers, while the C-terminal DNA-binding domain is a monomer (Hickey

```
OmpR    MQENYKILVVDDDMRLRALLERYLTEQGFQVRSVANAEQMDRLLTRESFHLMVLDLMLPG 60
PhoB    MA--RRILVVEDEAPIREMVCFVLEQNGFQPVEAEDYDSAVNQLNEPWPDLILLDWMLPG 58
ChxR    MAGPKHVLLVSEHWDLFFQTKELLNPEEYRCTIGQQYK------QELSADLVVCEYS--- 51
        *    ::*:*.:.  :       *  : ::     : .       .   .*:: :

OmpR    EDGLSICRRLRSQSNP--MPIIMVTAKGEEVDRIVGLEIGADDYIPKPFNPRELLARIRA 118
PhoB    GSGIQFIKHLKRESMTRDIPVVMLTARGEEEDRVRGLETGADDYITKPFSPKELVARIKA 118
ChxR    ----LLPREIRSPKSLEGSFVLVLLDFFDEETSVDLLDRG-FWYLIRPITPRILKSAISL 106
             . :.::  .      ::::    :*   :  *: *   *: :*:.*: * : *

OmpR    VLRRQANELPGAPSQEEAVIAFGKFKLNLGTREMFREDEPMPLTSGEFAVLKALVSHPRE 178
PhoB    VMRRIS------PMAVEEVIEMQGLSLDPTSHRVMAGEEPLEMGPTEFKLLHFFMTHPER 172
ChxR    FLSQHS------LHSVPESIRFGPNVFYVLKLTVETPEGSVHLTPSESGILKRLLINKGQ 160
        .: : :             * :    :   .  :   : .: : . *  :*: :: :  .

OmpR    PLSRDKLMNLARGREYSAMERSIDVQISRLRRMVEEDPAHPRYIQTVWGLGYVFVPDGSK 238
PhoB    VYSREQLLNHVWGTNVYVEDRTVDVHIRRLRKALEPG-GHDRMVQTVRGTGYRFS---TR 228
ChxR    LCLRKHLLEEIKNHAKAIVARNVDVHIASLRKKLG---AYGSRIVTLRGVGYLFSDDGDK 217
           *.:*::   .       *.:**:*  **: :    .:   : *: * ** *     :

OmpR    A--------- 239
PhoB    F--------- 229
ChxR    KFSQQDTKLS 227
```

Figure 6.3 Sequence of the atypical response regulator ChxR of *Chlamydia trachomatis*. ChxR is aligned with the response regulators OmpR and PhoB of *E. coli*. Stars below the aligned sequences highlight single fully conserved residues, colons and dots indicate conservation between groups of amino acids with strongly and weakly similar properties, respectively. Secondary structure elements of the receiver domain correspond to *E. coli* PhoB (Sola *et al.*, 1999). Grey rectangles above the aligned sequences indicate α-helices, black arrows indicate β-strands. Positions corresponding to the signature residues D/E12, D13, D57, T/S87, Y/F106 and K109 of the receiver consensus sequence are highlighted by black shading.

et al., 2009, 2011a,b). Crystal structure analysis revealed unique structural characteristics of the ChxR receiver domain suggesting intra- and intermolecular interactions which are different from the phosphorylation-dependent OmpR subfamily members (Hickey *et al.*, 2011b). The molecular topology of the ChxR receiver domain is β1-α1-β2-β3-β4-α2-β5-α3 with two large random coils connecting sheets β2-β3 and β3-β4. E49 and R93 replacing D57 and K109 of the consensus receiver sequence form a salt bridge and mimic the positions of these residues in phosphorylated receiver domains, while the orientation of residues L72 and Y90 which are located in the position of the canonical conformational switch residues is similar to unphosphorylated receiver domains. In contrast to canonical OmpR subfamily members the dimer interface of the ChxR receiver domain is built up largely by hydrophobic interactions. Mutational analysis demonstrated that disruption of these hydrophobic interactions abolished both dimer formation and DNA binding *in vitro*, suggesting that dimerization is essential for stable interaction of ChxR with DNA. Substitution of G49, which is involved in an intramolecular ionic interaction, with an aspartic acid or alanine residue had no effect on dimerization or the DNA-binding properties of ChxR (Hickey *et al.*, 2011a,b). ChxR was shown to bind to its own promoter *in vivo* which contains six ChxR binding sites consisting of heptanucleotide tandem repeat sequences (Hickey *et al.*, 2011). Furthermore, ChxR activated the transcription of four chlamydial genes when genomic DNA of *C. trachomatis* was *in vitro* transcribed with *E. coli* RNA polymerase (Koo *et al.*, 2006). ChxR itself is transcribed in *C. trachomatis* 10 hours after the infection of host cells corresponding to the time of metabolic activity and replication of *Chlamydia* (Koo *et al.*, 2006).

Conclusions

Atypical and presumably some of the orphan RRs with canonical receiver sequences function in a phosphorylation-independent manner. Signal transduction by these regulators might be accomplished via interactions of their receiver domain with other proteins or small molecules. Otherwise, in some atypical RRs the receiver domain might have been retained as a mere dimerization device to ensure the DNA-binding capability of the output-domain giving rise to a constitutively active 'regulator' protein. In such cases regulatory signal input has to occur on the level of expression control of the RR itself. The characterization of more orphan and atypical RRs will doubtless reveal interesting new aspects of 'one-component' signal transduction in bacteria.

References

Aínsa, J.A., Parry, H.D., and Chater, K.F. (1999). A response regulator-like protein that functions at an intermediate stage of sporulation in *Streptomyces coelicolor* A3(2). Mol. Microbiol. *34*, 607–619.

Ajithkumar, V., and Prasad, R. (2010). Modulation of *dnrN* expression by intracellular levels of DnrO and daunorubicin in *Streptomyces peucetius*. FEMS Microbiol. Lett. *306*, 160–167.

Appleby, J.L., and Bourret, R.B. (1999). Activation of CheY mutant D57N by phosphorylation at an alternative site, Ser-56. Mol. Microbiol. *34*, 915–925.

Banas, J.A., and Vickerman, M.M. (2003). Glucan-binding proteins of the oral streptococci. Crit. Rev. Oral Biol. Med. *14*, 89–99.

Barakat, M., Ortet, P., Jourlin-Castelli, C., Ansaldi, M., Mejean, V., and Whitworth, D.E. (2009). P2CS: a two-component system resource for prokaryotic signal transduction research. BMC Genomics *10*, 315.

Beatty, W.L., Morrison, R.P., and Byrne, G.I. (1994). Persistent chlamydiae: from cell culture to a paradigm for chlamydial pathogenesis. Microbiol Rev. *58*, 686–699.

Beier, D., and Frank, R. (2000). Molecular characterization of two-component systems of *Helicobacter pylori*. J. Bacteriol. *182*, 2068–2076.

Biswas, S., and Biswas, I. (2006). Regulation of the glycosyltransferase (*gtfBC*) operon by CovR in *Streptococcus mutans*. J. Bacteriol. *188*, 988–998.

Biswas, I., Drake, L., and Biswas, S. (2007). Regulation of *gbpC* expression in *Streptococcus mutans*. J. Bacteriol. *189*, 6521–6531.

Bourret, R.B. (2010). Receiver domain structure and function in response regulator proteins. Curr. Opin. Microbiol. *13*, 142–149.

Brahmachary, P., Wang, G., Benoit, S.L., Weinberg, M.V., Maier, R.J., and Hoover, T.R. (2008). The human gastric pathogen *Helicobacter pylori* has a potential acetone carboxylase that enhances its ability to colonize mice. BMC Microbiology *8*, 14.

Campbell, B.J., Engel, A.S., Porter, M.L., and Takai, K. (2006). The versatile ε-proteobacteria: key players in sulphidic habitats. Nature Rev. Microbiol. *4*, 458–468.

Chater, K.F. (2001). Regulation of sporulation in *Streptomyces coelicolor* A3(2): a checkpoint multiplex? Curr. Opin. Microbiol. *4*, 667–673.

Chong, P., Drake, L., and Biswas, I. (2008). Modulation of CovR expression in *Streptococcus mutans* UA159. J. Bacteriol. *190*, 4478–4488.

Chong, P., Chattoraj, P., and Biswas, I. (2010). Activation of the SMU.1882 transcription by CovR in *Streptococcus mutans*. PLoS One *5*, e15528.

Churchward, G. (2007). The two faces of Janus: virulence gene regulation by CovR/S in group A streptococci. Mol. Microbiol. *64*, 34–41.

Dahl, M.K., Msadek, T., Kunst, F., and Rapoport, G. (1992). The phosphorylation state of the DegU response regulator acts as a molecular switch allowing either degradative enzyme synthesis or expression of genetic competence in *Bacillus subtilis*. J. Biol. Chem. *267*, 14509–14514.

Dalton, T.L., and Scott, J.R. (2004). CovS inactivates CovR and is required for growth under conditions of general stress in *Streptococcus pyogenes*. J. Bacteriol. *186*, 3928–3937.

Delany, I., Spohn, G., Rappuoli, R., and Scarlato, V. (2002). Growth-dependent regulation of target gene promoters for binding of the essential orphan response regulator HP1043 of *Helicobacter pylori*. J. Bacteriol *184*, 4800–4810.

Dimitrev, A., Mohapatra, S.S., Chong, P., Neely, M., Biswas, S., and Biswas, I. (2011). CovR-controlled global regulation of gene expression in *Streptococcus mutans*. PLoS One *6*, e20127.

Dons, L., Eriksson, E., Jin, Y., Rottenberg, M.E., Kristensson, K., Larsen, C.N., Bresciani, J., and Olsen, J.E. (2004). Role of flagellin and the two-component CheA/CheY system of *Listeria monocytogenes* in host cell invasion and virulence. Infect. Immun. *72*, 3237–3244.

Dubrac, S., Bisicchia, P., Devine, K.M., and Msadek, T. (2008). A matter of life and death: cell wall homeostasis and the WalKR (YycGF) essential signal transduction pathway. Mol. Microbiol. *70*, 1307–1322.

Dunning, D.W., McCall, L.W., Powell, W.F., Jr, Arscott, W.T., McConocha, E.M., McClurg, C.J., Goodman, S.D., and Spatafora, G.A. (2008). SloR modulation of the *Streptococcus mutans* acid tolerance response involves the GcrR response regulator as an essential intermediary. Microbiology *154*, 1132–1143.

Engleberg, N.C., Heath, A., Miller, A., Rivera, C., and DiRita, V.J. (2001). Spontaneous mutations in the CsrRS two-component regulatory system of *Streptococcus pyogenes* result in enhanced virulence in a murine model of skin and soft tissue infection. J. Infect. Dis. *183*, 1043–1054.

Fabret, C., Feher, V.A., and Hoch, J.A. (1999). Two-component signal transduction in *Bacillus subtilis*: how one organism sees its world. J. Bacteriol. *181*, 1975–1983.

Fraser, J.S., Merlie, J.P., Jr, Echols, N., Weisfield, S.R., Mignot, T., Wemmer, D.E., Zusman, D.R., and Alber, T. (2007). An atypical receiver domain controls the dynamic polar localization of the *Myxococcus xanthus* social motility protein FrzS. Mol. Microbiol. *65*, 319–332.

Fritsch, F., Mauder, N., Williams, T., Weiser, J., Oberle, M., and Beier, D. (2011). The cell envelope stress response mediated by the LiaFSR$_{Lm}$ three-component system of *Listeria monocytogenes* is controlled via the phosphatase activity of the bifunctional histidine kinase LiaS$_{Lm}$. Microbiology *157*, 373–386.

Furuya, K., and Hutchinson, C.R. (1996). The DnrN protein of *Streptomyces peucetius*, a pseudo-response regulator, is a DNA-binding protein involved in the regulation of daunorubicin biosynthesis. J. Bacteriol. *178*, 6310–6318.

Gao, R., Mack, T.R., and Stock, A.M. (2007). Bacterial response regulators: versatile regulatory strategies from common domains. Trends Biochem. Sci. *32*, 225–234.

Glaser, P., Frangeul, L., Buchrieser, C., Rusniok, C., Amend, A., Baquero, F., Berche, P., Bloecker, H., Brandt, P., Chakraborty, T., *et al.* (2001). Comparative genomics of *Listeria* species. Science *294*, 849–852.

Graham, M.R., Smoot, L.M., Migliaccio, C.A., Virtaneva, K., Sturdevant, D.E., Porcella, S.F., Federle, M.J., Adams, G.J., Scott, J.R., and Musser, J.M. (2002). Virulence control in group A streptococcus by a two-component gene regulators system: global expression profiling and *in vivo* infection modeling. Proc. Natl. Acad. Sci. U.S.A. *99*, 13855–13860.

Grossman, A.R., Schaefer, M.R., Chiang, G.G., and Collier, J.L. (1993). The phycobilisome, a light–harvesting complex responsive to environmental conditions. Microbiol. Rev. *57*, 725–749.

Gründling, A., Burrack, L.S., Bouwer, H.G., and Higgins, D.E. (2004). *Listeria monocytogenes* regulates flagellar motility gene expression through MogR, a transcriptional repressor required for virulence. Proc. Natl. Acad. Sci. U.S.A. *101*,12318–12323.

Gryllos, I., Levin, J.C., and Wessels, M.R. (2003). The CsrR/CsrS two-component system of group A streptococcus responds to environmental Mg^{2+}. Proc. Natl. Acad. Sci. U.S.A. *100*, 4227–4232.

Gryllos, I., Grifantini, R., Colaprico, A., Jiang, S., DeForce, E., Hakansson, A., Telford, J.L., Grandi, G., and Wessels, M.R. (2007). Mg^{2+} signalling defines the group A streptococcal CsrRS (CovRS) regulon. Mol. Microbiol. *65*, 671–683.

Gryllos, I., Tran-Winkler, H.J., Cheng, M.-F., Chung, H., Bolcome, R., 3rd, Lu, W., Lehrer, R.I., and Wessels, M.R. (2008). Induction of group A streptococcus virulence by a human antimicrobial peptide. Proc. Natl. Acad. Sci. U.S.A. *105*, 16755–16760.

Gueriri, I., Cyncynatus, C., Dubrac, S., Arana, A.T., Dussurget, O., and Msadek, T. (2008a). The DegU orphan response regulator of *Listeria monocytogenes* autorepresses its own synthesis and is required for bacterial motility, virulence and biofilm formation. Microbiology *154*, 2251–2264.

Gueriri, I., Bay, S., Dubrac, S., Cyncynatus, C., and Msadek, T. (2008b). The Pta-AckA pathway controlling acetyl phosphate levels and the phosphorylation state of the DegU orphan response regulator both play a role in regulating *Listeria monocytogenes* motility and chemotaxis. Mol. Microbiol. *70*, 1342–1357.

Gusa, A.A., and Scott, J.R. (2005). The CovR response regulator of group A streptococcus (GAS) acts directly to repress its own promoter. Mol. Microbiol. *56*, 1195–1207.

Guthrie, E.P., Flaxman, C.S., White, J., Hodgson, D.A., Bibb, M.J., and Chater, K.F. (1998). A response-regulator-like activator of antibiotic synthesis from *Streptomyces coelicolor* A3(2) with an amino-terminal domain that lacks a phosphorylation pocket. Microbiology *144*, 727–738.

Hamoen, L.W., van Werkhoven, A.F., Venema, G., and Dubnau, D. (2000). The pleiotropic response regulator DegU functions as a priming protein in competence development in *Bacillus subtilis*. Proc. Natl. Acad. Sci. U.S.A. *97*, 9246–9251.

Hickey, J.M., Hefty, P.S., and Lamb, A.L. (2009). Expression, purification, crystallization and preliminary X-ray analysis of the DNA-binding domain of a *Chlamydia trachomatis* OmpR/PhoB-subfamily response regulator homolog. Acta Cryst. *F65*, 791–794.

Hickey, J.M., Lovell, S., Batteile, K.P., Hu, L., Middaugh, C.R., and Hefty, P.S. (2011a). The atypical response regulator ChxR has structural characteristics and dimer interface interactions that are unique within the OmpR/PhoB subfamily. J. Biol. Chem. *286*, 32606–32616.

Hickey, J.M., Weldon, L., and Hefty, P.S. (2011b). The atypical OmpR/PhoB response regulator ChxR from *Chlamydia trachomatis* forms homodimers *in vivo* and binds a direct repeat of nucleotide sequences. J. Bacteriol. *193*, 389–398.

Hong, E., Jung, J.W., Shin, J., Kim, J.H., Jeon, Y.H., Yamazaki, T., and Lee, W. (2004). Backbone 1H, 13C and 15N resonance assignments of the response regulator HP1043 from *Helicobacter pylori*. J. Biomol. NMR *28*, 85–86.

Hong, E., Lee, H.M., Ko, H., Kim, D.-U., Jeon, B.-Y., Jung, J., Shin, J., Lee, S.-A., Kim, Y., Jeon, Y.H., *et al.* (2007). Structure of an atypical orphan response regulator protein supports a new phosphorylation-independent regulatory mechanism. J. Biol. Chem. *282*, 20667–20675.

Huang, B., Whitchurch, C.B., and Mattick, J.S. (2003). FimX, a multidomain protein connecting environmental signals to twitching motility in *Pseudomonas aeroginosa*. J. Bacteriol. *185*, 7068–7076.

Huang, J., Lih, J., Pan, K.H., and Cohen, S.N. (2001). Global analysis of growth phase responsive gene expression and regulation of antibiotic biosynthetic pathways in *Streptomyces coelicolor* using DNA microarrays. Genes Dev. *15*, 3183–3192.

Idone, V., Brendtro, S., Gillespie, R., Kocaj, S., Peterson, E., Rendi, M., Warren, W., Michalek, S., Krastel, K., Cvitkovitch, D., and Spatafora, G. (2003). Effect of an orphan response regulator on *Streptococcus mutans* sucrose-dependent adherence and cariogenesis. Infect. Immun. *71*, 4351–4360.

Jiang, H., and Hutchinson, C.R. (2006). Feedback regulation of doxorubicin biosynthesis in *Streptomyces peucetius*. Res. Microbiol. *157*, 666–674.

Jiang, S.M., Ishmael, N., Dunning Hotopp, J., Puliti, M., Tissi, L., Kumar, N., Cieslewicz, M.J., Tettelin, H., and Wessels, M.R. (2008). Variation in the group B streptococcus CsrRS regulon and effects on pathogenicity. J. Bacteriol. *190*, 1956–1965.

Kamp, H.D., and Higgins, D.E. (2009). Transcriptional and post-transcriptional regulation of the GmaR antirepressor governs temperature-dependent control of flagellar motility in *Listeria monocytogenes*. Mol. Microbiol. *74*, 421–435.

Kansal, R.G., Datta, V., Aziz, R.A., Abdeltawab, N.F., Rowe, S., and Kotb, M. (2010). Dissection of the molecular basis for hypervirulence of an *in vivo*-selected phenotype of the widely disseminated M1T1 strain of group A streptococcus bacteria. J. Infect. Dis. *201*, 855–865.

Kato, H., Chibazakura, T., and Yoshikawa, H. (2008). NblR is a novel one-component response regulator in the cyanobacterium *Synechococcus elongatus* PCC 7942. Biosci. Biotechnol. Biochem. *72*, 1072–1079.

Kazmierczak, B.I., Lebron, M.B., and Murray, T.S. (2006). Analysis of FimX, a phosphodiesterase that governs twitching motility in *Pseudomonas aeruginosa*. Mol. Microbiol. *60*, 1026–1043.

Knudsen, G.M., Olsen, J.E., and Dons, L. (2004). Characterization of DegU, a response regulator in *Listeria monocytogenes*, involved in regulation of motility and contributes to virulence. FEMS Microbiol. Lett. *240*, 171–179.

Kobayashi, K. (2007). Gradual activation of the response regulator DegU controls serial expression of genes for flagellum formation and biofilm formation in *Bacillus subtilis*. Mol. Microbiol. *66*, 395–409.

Koo, I.C., and Stephens, R.S. (2003). A developmentally regulated two-component signal transduction system in Chlamydia. J. Biol. Chem. *278*, 17314–17319.

Koo, I.C., Walthers, D., Hefty, P.S., Kenney, L.J., and Stephens, R.S. (2006). ChxR is a transcriptional activator in *Chlamydia*. Proc. Natl. Acad. Sci. U.S.A. *103*, 750–755.

Lamy, M.C., Zouine, M., Fert, J., Vergassola, M., Couve, E., Pellegrini, E., Glaser, P., Kunst, F., Msadek, T., Trieu-Cuot, P., *et al.* (2004). CovS/CovR of group B streptococcus: a two-component global regulatory system involved in virulence. Mol. Microbiol. *54*, 1250–1268.

Laub, M.T., and Goulian, M. (2007). Specificity in two-component signal transduction pathways. Annu. Rev. Genet. *41*, 121–145.

Laub, M.T., Biondi, E.G., and Skerker, J.M. (2007). Phosphotransfer profiling: systematic mapping of two-component signal transduction pathways and phosphorelays. Methods Enzymol. *423*, 531–548.

Lee, H.M., Hong, E., Jeon, B.Y., Kim, D.U., Byun, J.S., Lee, W., and Cho, H.S. (2006). Crystallization and preliminary X-ray crystallographic study of HP1043, a *Helicobacter pylori* orphan response regulator. Biochim. Biophys. Acta *1764*, 989–991.

Lembo, A., Gurney, M.A., Burnside, K., Banarjee, A., de los Rexes, M., Connelly, J.E., Lin, W.-J., Jewell, K.A., Vo, A., *et al.* (2010). Regulation of CovR expression in group B streptococcus impacts blood–brain barrier penetration. Mol. Microbiol. *77*, 431–443.

Lin, W.-J., Walthers, D., Connelly, J.E., Burnside, K., Jewell, K.A., Kenney, L.J., and Rajagopal, L. (2009). Threonine phosphorylation prevents promoter DNA binding of the group B streptococcus response regulator CovR. Mol. Microbiol. *71*, 1477–1495.

Luque, I., Zabulon, G., Contreras, A., and Houmard, J. (2001). Convergence of two global transcriptional regulators on nitrogen induction of the stress-acclimation gene *nblA* in the cyanobacterium *Synechococcus* sp. PCC 7942. Mol. Microbiol. *41*, 937–947.

McCleary, W.R., and Stock, J.B. (1994). Acetyl phosphate and the activation of two-component response regulators. J. Biol. Chem. *269*, 31567–31572.

McDaniel, T.K., DeWalt, K.C., Salama, N.R., and Falkow, S. (2001). New approaches for validation of lethal phenotypes and genetic reversion in *Helicobacter pylori*. Helicobacter *6*, 15–23.

McKenzie, N.L., and Nodwell, J.R. (2007). Phosphorylated AbsA2 negatively regulates antibiotic production in *Streptomyces coelicolor* through interactions with pathway-specific regulatory gene promoters. J. Bacteriol. *189*, 5284–5292.

Madduri, K., and Hutchinson, C.R. (1995). Functional characterization and transcriptional analysis of the *dnrR*$_1$ locus, which controls daunorubicin biosynthesis in *Streptomyces peucetius*. J. Bacteriol. *177*, 1208–1215.

Mauder, N., Williams, T., Fritsch, F., Kuhn, M., and Beier, D. (2008). Response regulator DegU of *Listeria monocytogenes* controls temperature-responsive flagellar gene expression in its unphosphorylated state. J. Bacteriol. *190*, 4777–4781.

Mauriello, E.M.F., Nan, B., and Zusman, D.R. (2009). AglZ regulates adventurous (A-) motility in *Myxococcus xanthus* through its interaction with the cytoplasmic receptor, FrzCD. Mol. Microbiol. *72*, 964–977.

Miller, A.A., Engleberg, N.C., and DiRita, V. (2001). Repression of virulence genes by phosphorylation-dependent oligomerization of CovR at target promoters in *S. pyogenes*. Mol. Microbiol. *40*, 976–990.

Molle, V., and Buttner, M.J. (2000). Different alleles of the response regulator gene *bldM* arrest *Streptomyces coelicolor* development at distinct stages. Mol. Microbiol. *36*, 1265–1278.

Moore, J.B., Shiau, S.-P., and Reitzer, L.J. (1993). Alterations of highly conserved residues in the regulatory domain of nitrogen regulator I (NtrC) of *Escherichia coli*. J. Bacteriol. *175*, 2692–2701.

Müller, S., Pflock, M., Schär, J., Kennard, S., and Beier, D. (2007). Regulation of expression of atypical orphan response regulators of *Helicobacter pylori*. Microbiol. Res. *162*, 1–14.

O'Connor, T.J., and Nodwell, J.R. (2005). Pivotal roles for the receiver domain in the mechanism of action of the response regulator RamR of *Streptomyces coelicolor*. J. Mol. Biol. *351*, 1030–1047.

O'Connor, Kanellis, T.J., and Nodwell, J.R. (2002). The *ramC* gene is required for morphogenesis in *Streptomyces coelicolor* and expressed in a cell type-specific manner under the direct control of RamR. Mol. Microbiol. *45*, 45–57.

Otten, S.L., Ferguson, J., and Hutchinson, C.R. (1995). Regulation of daunorubicin production in *Streptomyces peucetius* by the *dnrR*$_2$ locus. J. Bacteriol. *177*, 1216–1224.

Otten, S.L., Olano, C., and Hutchinson, C.R. (2000). The *dnrO* gene encodes a DNA-binding protein that regulates daunorubicin production in *Streptomyces peucetius* by controlling expression of the *dnrN* pseudo response regulator gene. Microbiology *146*, 1457–1468.

Pan, X., Ge, J., Li, M., Wu, B., Wang, C., Wang, J., Feng, Y., Yin, Z., Zheng, F., Cheng, G., *et al.* (2009). The orphan response regulator CovR: a globally negative modulator of virulence in *Streptococcus suis* Serotype 2. J. Bacteriol. *191*, 2601–2612.

Pflock, M., Dietz, P., Schär, J., and Beier, D. (2004). Genetic evidence for histidine kinase HP165 being an acid sensor of *Helicobacter pylori*. FEMS Microbiol. Lett. *234*, 51–61.

Pflock, M., Bathon, M., Schär, J., Müller, S., Mollenkopf, H., Meyer, T.F., and Beier, D. (2007). The orphan response regulator HP1021 of *Helicobacter pylori* regulates transcription of a gene cluster presumably involved in acetone metabolism. J. Bacteriol. *189*, 2339–2349.

Rajagopal, L., Vo, A., Silvestroni, A., and Rubens, C.A. (2006). Regulation of cytotoxin expression by converging eukaryotic-type and two-component signaling mechanisms in *Streptococcus agalactiae*. Mol. Microbiol. *62*, 941–957.

Ruiz, D., Salinas, P., Lopez-Redondo, M.L., Cayuela, M.L., Marina, A., and Contreras, A. (2008). Phosphorylation-independent activation of the atypical response regulator NblR. Microbiology *154*, 3002–3015.

Schär, J., Sickmann, A., and Beier, D. (2005). Phosphorylation-independent activity of atypical response regulators of *Helicobacter pylori*. J. Bacteriol. *187*, 3100–3109.

Sharma, C.M., Hoffmann, S., Darfeuille, F., Reignier, J., Findeiss, S., Sittka, A., Chabas, S., Reiche, K., Hackermüller, J., Reinhardt, R., *et al.* (2010) The primary transcriptome of the major human pathogen *Helicobacter pylori*. Nature *464*, 250–255.

Shen, A., Kamp, H.D., Gründling, A., and Higgins, D.E. (2006). A bifunctional O-GlcNAc transferase governs flagellar motility through anti-repression. Genes Dev. *20*, 3283–3295.

Solá, M., Gomis-Rüth, F.X., Serrano, L., Gonzáles, A., and Coll, M. (1999). Three-dimensional crystal structure of the transcription factor PhoB receiver domain. J. Mol. Biol. *285*, 675–687.

Stock, A.M., Robinson, V.L., and Goudreau, P.N. (2000). Two-component signal transduction. Annu. Rev. Biochem. *69*, 183–215.

Stock, J.B., Surette, M.G., Levit, M., and Park, P. (1995). Two-component signal transduction systems: structure–function relationships and mechanisms of catalysis. In Two-component Signal Transduction, Hoch, J.A., and Silhavy, T.J., eds (ASM Press, Washington, DC), pp. 25–51.

Sugareva, V., Arlt, R., Fiedler, T., Riani, C., Podbielski, A., and Kreikemeyer, B. (2010). Serotype- and strain-dependent contribution of the sensor kinase CovS of the CovRS two-component system to *Streptococcus pyogenes* pathogenesis. BMC Microbiology *10*, 34.

Sumby, P., Whitney, A.R., Graviss, E.A., DeLeo, F.R., and Musser, J.M. (2006). Genome-wide analysis of group A streptococci reveals a mutation that modulates global phenotype and disease specificity. PLoS Pathog. *2*, e5.

Tian, Y., Fowler, K., Findlay, K., Tan, H., and Chater, K.F. (2007). An unusual response regulator influences sporulation at early and late stages in *Streptomyces coelicolor*. J. Bacteriol. *189*, 2873–2885.

Tomb, J.-F., White, O., Kerlavage, A.R., Clayton, R., Sutton, G.G., Fleischmann, R.D., Ketchum, K.A., Klenk, H.P., Gill, S., Dougherty, B.A., *et al.* (1997). The complete genome sequence of the gastric pathogen *Helicobacter pylori*. Nature *388*, 539–547.

Trevino, J., Perez, N., Ramirez-Pena, E., Liu, Z., Shelburne, S.A., 3rd, Musser, J.M., and Sumby, P. (2009). CovS simultaneously activates and inhibits the CovR-mediated repression of distinct subsets of group A streptococcus virulence factor-encoding genes. Infect. Immun. *77*, 3141–3149.

Throup, J.P., Koretke, K.K., Bryant, A.P., Ingraham, K.A., Chalker, A.F., Ge, Y., Marra, A., Wallis, N.G., Brown, J.R., Holmes, D.J., *et al.* (2000). A genomic analysis of two-component signal transduction in *Streptococcus pneumoniae*. Mol. Microbiol. *35*, 566–576.

Ulijasz, A.T., Andes, D.R., Glasner, J.D., and Weisblum, B. (2004). Regulation of iron transport in *Streptococcus pneumoniae* by RitR, an orphan response regulator. J. Bacteriol. *186*, 8123–8136.

Ulijasz, A.T., Falk, S.P., and Weisblum, B. (2009). Phosphorylation of the RitR DNA-binding domain by a Ser-Thr phosphokinase: implications for global gene regulation in the streptococci. Mol. Microbiol. *71*, 382–390.

Verhamme, D.T., Kiley, T.B., and Stanley-Wall, N.R. (2007). DegU co-ordinates multicellular behaviour exhibited by *Bacillus subtilis*. Mol. Microbiol. *65*, 554–568.

Volz, K. (1995). Structural and functional conservation in response regulators. In Two-component Signal Transduction, Hoch, J.A., and Silhavy, T.J., eds (ASM Press, Washington, DC), pp. 53–64.

Walker, M.J., Hollands, A., Sanderson-Smith, M.-L., Cole, J.N., Kirk, J.K., Henningham, A., McArthur, J.D., Dinkla, K., Aziz, R.K., Kansal, R.G., *et al.* (2007). DNase Sda1 provides selection pressure for a switch to invasive group A streptococcal infection. Nat. Med. *13*, 981–985.

Wang, L., Tian, X., Wang, J., Yang, H., Fan, K., Xu, G., Yang, K., and Tan, H. (2009). Autoregulation of antibiotic biosynthesis by binding of the end product to an atypical response regulator. Proc. Natl. Acad. Sci. U.S.A. *106*, 8617–8622.

Way, S.S., Thompson, L.J., Lopes, J.E., Hajjar, A.M., Kollmann, T.R., Freitag, N.E., and Wilson, C.B. (2004). Characterization of flagellin expression and its role in *Listeria monocytogenes* infection and immunity. Cell. Microbiol. *6*, 235–242.

Williams, T., Bauer, S., Beier, D., and Kuhn, M. (2005a). Construction and characterization of *Listeria monocytogenes* mutants with in-frame deletions in the response regulator genes identified in the genome sequence. Infect. Immun. *73*, 3152–3159.

Williams, T., Joseph, B., Beier, D., Goebel, W., and Kuhn, M. (2005b). Response regulator DegU of *Listeria monocytogenes* regulates the expression of flagella-specific genes. FEMS Microbiol. Lett. *252*, 287–298.

Wolfe, A.J. (2005). The acetate switch. Microbiol. Mol. Biol. Rev. *69*, 12–50.

Xu, G., Wang, J., Wang, L., Tian, X., Yang, H., Fan, K., Yang, K., and Tan, H. (2010). "Pseudo" γ-butyrolactone receptors respond to antibiotic signals to coordinate antibiotic biosynthesis. J. Biol. Chem. *285*, 27440–27448.

Yang, K., Han, L., He, J.,Wang, L., and Vining, L.C. (2001). A repressor-response regulator gene pair controlling jadomycin B production in *Streptomyces venezuelae* ISP5230. Gene *279, 165–173.*

Yang, R., Bartle, S., Otto, R., Stassinopoulos, A., Rogers, M., Plamann, L., and Hartzell, P. (2004). AlgZ is a filament-forming coiled-coil protein required for adventurous gliding motility of *Myxococcus xanthus*. J. Bacteriol. *186*, 6168–6178.

Essential Two-component Systems of Gram-positive Bacteria

7

Hendrik Szurmant

Abstract

The YycF/YycG two-component signal transduction system has become an enigma since its initial discovery as the only signal transduction system essential for viability in *Bacillus subtilis*, *Staphylococcus aureus* and *Streptococcus pneumoniae*. Conserved across the low G+C Gram-positives, over a decade worth of dedicated molecular investigations have identified some of the factors that control the YycG kinase activity and the genes that are regulated by the YycF transcription factor. A picture has emerged where the YycFG system is embedded in the physiology of the cell and serves to coordinate essential physiological processes such as cell wall homeostasis and cell division. Several regulatory connections with other signalling systems have been recognized. The YycFG system has surfaced as an important player in the acquisition of antibiotic resistances in clinical isolates of significant human pathogens and has become the target of dedicated antimicrobial drug screening efforts. This chapter summarizes the current state of knowledge on this and other essential two-component signalling systems in Gram-positive bacteria.

A historical view

Two-component systems have long been known to serve as a primary means of bacteria for niche adaptation to ever changing environmental conditions. Upon completion of the first bacterial genome sequencing projects, the global importance of two-component systems to the bacteria became increasingly clear when some genomes where found to feature more than 100 such signalling systems (Galperin, 2005). It also became evident that mutations in most systems had not been identified in genetic screens to confer a desired phenotype. Hence no function had been assigned to the majority of these systems. This was true for even the most scrutinized model organisms, such as *Escherichia coli* and *Bacillus subtilis*. The lack of knowledge on most systems fuelled systematic deletion studies in an attempt to identify roles for systems of yet unknown function (Fabret and Hoch, 1998; Zahrt and Deretic, 2000).

In *B. subtilis*, a systematic deletion study revealed that a single system, the YycFG system, could not be inactivated and was thus deemed a possibly essential signalling system (Fabret and Hoch, 1998). Subsequent work with conditional mutants and with strains where the system was placed under IPTG inducible control proved the essentiality of the system in the model organism *B. subtilis* (Fabret and Hoch, 1998; Fukuchi *et al.*, 2000).

The identification of an essential signalling system was perhaps somewhat surprising. It did not conform to the common perception of a role for two-component systems in niche adaptation. Rather it became clear that the system has to be embedded in important processes regulating the life cycle of the cell. A phylogenetic analysis of a limited number of available sequenced organisms at the time suggested that orthologous systems are widespread and perhaps even ubiquitous among the low-G+C Gram-positives, also known as the phylum *Firmicutes* (Fabret and Hoch, 1998). The initial identification of this system as being essential in *B. subtilis* fuelled a flurry of studies with various different aims. First, it became imperative to identify the cellular processes that the YycFG system is regulating, with the expectation that such knowledge would lead to an increased understanding of bacterial physiology. Second, the identification of orthologous systems fuelled the question whether the system was also essential and conserved a role in other organisms, in particular important human pathogens, such as *Staphylococcus aureus* and *Streptococcus pneumoniae*. Lastly, the initial identification of these systems as essential and the lack of similar systems in the animal kingdom, suggested that the YycFG system might be a good potential drug target and approaches to develop such drugs have been described.

In their entirety these questions let to over a decade worth of extensive studies on the YycFG system, mainly focusing on the *B. subtilis*, *S. pneumoniae* and *S. aureus* orthologues. These studies have made the YycFG system one of the most scrutinized bacterial signalling systems to date and efforts are ongoing. In recent years significant advances were made that clarified our understanding of the role of the system and the reasons for its essentiality. This chapter will discuss these advances.

Nomenclature

A note on nomenclature: the names YycF for the response regulator and YycG for the histidine kinase have dominated throughout the primary literature. These names derived from the *B. subtilis* systematic annotation of genes of unknown function following completion of the genome sequence (Kunst *et al.*, 1997). Since then, the system has received a flurry of names in various different organisms. The streptococci orthologues have most commonly been referred to as VicR and VicK, because of a perceived role in virulence control initially associated with the system (Wagner *et al.*, 2002). Some orthologous systems in other Gram-positives have been automatically annotated as VicRK, for instance those in the enterococci, despite these systems being closer related to the *B. subtilis* orthologues. A second name for the streptococci orthologues that appeared briefly in the literature is MicAB, to reflect a perceived role in the microaerobiosis response (Echenique and Trombe, 2001). In the last two years the names WalR and WalK were coined to reflect the role of the system in regulating cell wall metabolism (Dubrac *et al.*, 2007, 2008). While this name is enjoying increasing popularity and might very well replace other names in the future, this chapter hereafter will continue to refer to the system as the YycFG system, since this remains the dominant name in the literature. The various names however are pointed out here to ease literature searches for the interested reader.

Phylogenetic distribution

The identification of orthologous two-component systems across diverse bacteria is not necessarily a trivial undertaking. How is one to decide whether a system is the true functional orthologue or represents just another two-component system with some sequence similarity? Essential proteins typically show an increased sequence conservation when compared with non-essential ones. This is true for the YycFG system, a fact that eases functional assignment. Secondly, orthologous systems usually conserve identical domain architectures. Given a complex multidomain architecture of the YycG kinase, this also eases identification of orthologues. As will be discussed below, the YycG kinase generally conserves its complex multidomain architecture, with the exception being the streptococci, where YycG orthologues deviate from the common full-length YycG architecture by an N-terminal truncation (Ng and Winkler, 2004; Szurmant *et al.*, 2005). As a consequence streptococci YycG orthologues lack an N-terminal transmembrane helix and an extracytoplasmic domain. Finally, genomic context can often aid in the identification of orthologous systems. Regarding the YycFG system, it has been recognized that most organisms feature at least one other gene, *yycJ* (aka *vicX*, aka *walJ*) that is associated with the *yycFG* operon (Fabret and Hoch, 1998; Ng and Winkler, 2004). A majority of *yycFG* operons (again with the exception of the streptococci) additionally feature genes for YycH and YycI, two proteins that are uniquely associated with this signalling system (Fabret and Hoch, 1998; Szurmant *et al.*, 2005). In summary, sufficient information is available to unambiguously identify true orthologues of the YycFG system from bacterial genome sequences.

When using this information to scan all available bacterial genomes for clear YycFG orthologues, it becomes apparent that the YycFG system is unique to the phylum *Firmicutes* (Fig. 7.1). Within this phylum YycFG orthologues are ubiquitous in the class *Bacilli* (found in 100% of the more than 400 currently available genomes), but are only sporadically found in the class *Clostridia* (found in 20% of more than 150 available genomes). To date, the system has been studied in molecular detail in three organisms, *B. subtilis*, *S. pneumoniae* and *S. aureus*, and essentiality has been confirmed in all three (Fabret and Hoch, 1998; Martin *et al.*, 1999; Wagner *et al.*, 2002). In addition, failed attempts to delete this system suggest that it is also essential in *Enterococcus faecalis*, *Staphylococcus epidermidis*, *Streptococcus mutans* and *Listeria monocytogenes* (Kallipolitis and Ingmer, 2001; Hancock and Perego, 2004; Senadheera *et al.*, 2005; Qin *et al.*, 2006). It appears likely that at least across the class *Bacilli* essentiality is a conserved feature of the

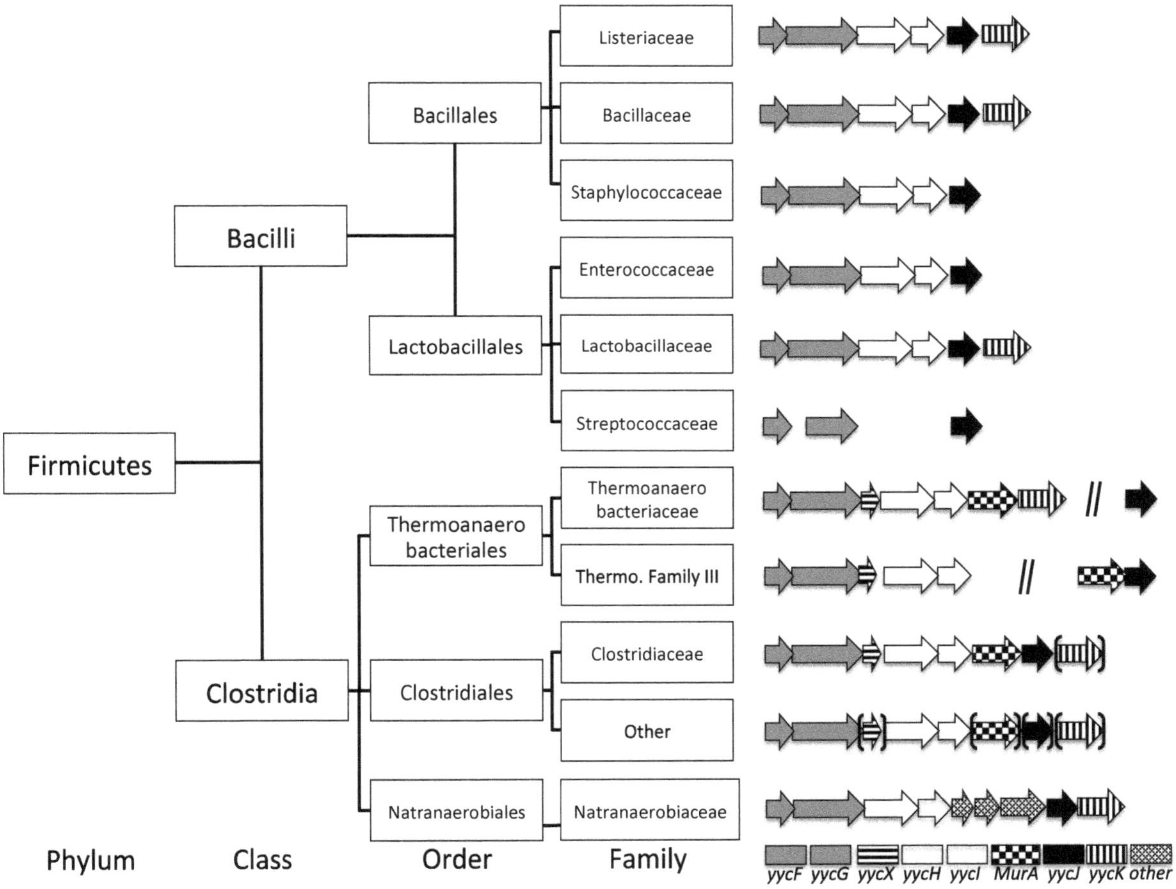

Figure 7.1 Phylogenetic distribution and operon architecture of YycFG systems. The YycFG two-component signalling system is unique to the phylum *Firmicutes*. The system is ubiquitous to all currently sequenced genomes of the *Bacilli* but only sporadically found in the *Clostridia*. The operon organization for each family of this phylum is represented by arrows. The grey arrows represent *yycF* and *yycG* genes and are drawn to gene length. With the exception of the *Streptococcaceae*, all *yyc* operons entail *yycH* and *yycI* genes (white arrows). The *yycJ* gene (black) is a member of the operon in all *Bacilli* and most *Clostridia*. Most clostridial *yyc* operons also include genes encoding YycX (horizontally striped arrow), a protein not found in the *Bacilli* and one of two copies encoding enzyme MurA (chequered arrow) that catalyses the first dedicated step in peptidoglycan biosynthesis. A gene for the htr-type serine protease YycK (vertically striped arrow) is also commonly a member of the *yyc* operon. The single *Natranaerobiaceae* genome available contains three genes of unknown function (arrows with white stars on black background). Brackets around genes indicate that not all operons within the family include the respective genes. A double forward slash indicates that the genes following are near but not part of the *yyc* operon.

YycFG system. No information is currently available whether the essentiality is also conserved across those *Clostridia* that have a genetic copy of this system.

Classification of YycFG systems

Despite the many conserved architectural features of the YycFG system and its operon, two clear subtypes, type I and type II, were observed following its initial discovery (Ng and Winkler, 2004; Szurmant *et al.*, 2005) (Figs. 7.1 and 7.2). This classification was based on differences in YycG kinase architecture and *yyc* operon structure.

The prototype for type I YycFG systems is exemplified by *B. subtilis*, where YycG is a 611 amino acid protein, and where the *yyc* operon includes *yycF, yycG, yycH, yycI, yycJ* and *yycK* genes, whose function will be discussed below. Type I YycFG systems are widespread among the orders *Bacillales* and *Lactobacillales* only excluding the family *Streptococcaceae*. The staphylococci and enterococci lack the *yycK* gene but otherwise have a type I YycFG system (Dubrac *et al.*, 2008).

The prototype for type II YycFG systems is represented by *S. pneumoniae*, where YycG is only 449 amino acids long, lacks an extracytoplasmic domain, and where the operon is restricted to *yycF, yycG* and *yycJ* genes (Wagner *et al.*, 2002). The type II YycFG system is exclusively found in the family *Streptococcaceae*. It appears that one feature of type II systems is that only the YycF response regulator but not the YycG kinase is essential (Ng *et al.*, 2003). Since the phosphorylation site on YycF, however, is essential, it is likely that this merely reflects the presence of a cross-phosphorylating kinase that is sufficiently active on YycF to replace YycG in its absence (Ng *et al.*, 2003). Alternatively small molecule phosphodonors such as acetyl phosphate can serve as phosphoryl group donors for some response regulators (Wolfe, 2010) and might explain a lack of essentiality for the kinase.

The novel observation that 20% of the *Clostridia* also feature YycFG orthologues adds some new complexity to the original classification scheme (Fig. 7.1). Regarding YycG length and architecture all clostridial YycFG systems follow type I. In addition, all of these operons feature *yycH* and *yycI* genes. Some diversity can be observed regarding the remainder of genes associated with the operon. 25 of 30 *yyc* operons of the *Clostridia* feature an additional gene not found in *Bacilli* genomes and will hereafter be referred to as *yycX*. Twenty-four of 30 operons feature genes for one or two copies encoding MurA, the enzyme that catalyses the first dedicated step in peptidoglycan biosynthesis. The presence of *yycJ* (14/30) and *yycK* (9/30) in the *yyc* operons of the *Clostridia* is less common. However, clear *yycJ* orthologues can be found across all clostridial genomes, even for those that do not feature a YycFG system.

As more and more genome sequences become available, it can be expected that novel YycFG subtypes will be identified. Currently, the *Natranaerobius thermophiles yyc* operon diverges from all other examples by harbouring several additional genes, including an enzyme annotated as glycolysis enzyme phosphoglycerate mutase as well as two genes coding for proteins of unknown functions. Such associations might be a blip in evolution, but they could also reveal new functional relations of the YycFG system with other cellular processes. In their entirety the classification schemes provide important information to distinguish potential differences in the regulatory properties of this system in the various organisms. Indeed, as will become clear below, the *B. subtilis* and *S. pneumoniae* YycG kinases are likely regulated in distinct ways reflecting their domain architecture and the presence and absence of YycHI orthologues.

Architecture and structures

Both, the YycF response regulator as well as the YycG kinase, feature some of the most common domains found for two-component systems proteins. The YycF response regulator is of the OmpR-type, featuring two domains, the response regulator receiver domain and the winged helix–turn–helix DNA-binding domain, also known as the trans_reg_c domain (Galperin, 2006) (Fig. 7.2). No structure of the full-length YycF protein is currently available. High-resolution structures are available for isolated response regulator domains of YycF from *B. subtilis* and *S. pneumoniae* (Bent *et al.*, 2004; Zhao *et al.*, 2009).

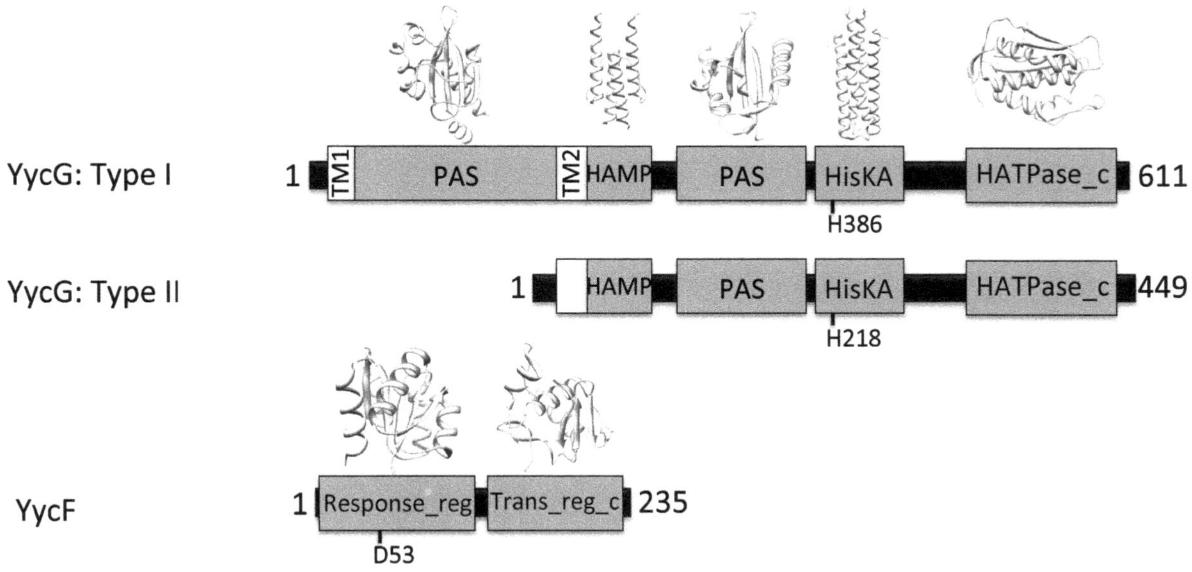

Figure 7.2 Domain architectures for YycG and YycF proteins. Two architectures have been recognized for YycG kinases. Type I is the prevalent type and includes two transmembrane helices (TM1, TM2), an extracytoplasmic PAS domain, cytoplasmic HAMP and PAS domains and the catalytic core comprising HisKA and HATPase_c domains. Type II YycG kinases are found in the *Streptococcaceae* and lack N-terminal TM1 and the extracytoplasmic domain. The amino acid lengths and the autophosphorylation site histidine are indicated for the representative YycG from *B. subtilis* (type I) and from *S. pneumoniae* (type II). Homology modelled structures of the individual domains are shown for illustration. The YycF response regulator is an OmpR-type regulator with a response_reg domain and a winged helix–turn–helix Trans_reg_c domain. The amino acid length and the aspartate site of phosphorylation is indicated for the representative YycF protein from *B. subtilis*. Structures of the individual domains from *B. subtilis* YycF are displayed for illustration.

Similarly, high-resolution structures are available for isolated DNA-binding domains of *E. faecalis*, *B. subtilis* and *S. aureus* YycF proteins (Trinh *et al.*, 2007; Okajima *et al.*, 2008). These structures have revealed no surprises and generally conform to the typical OmpR-type response regulator architecture. Perhaps these structures might in the future be useful for structure-based drug design (see below). Based on extensive microarray, gel shift and DNAase I foot printing data from *B. subtilis*, *S. aureus* and *S. pneumoniae* a cross-organismal consensus DNA binding sequence for YycF has been determined as 5′-TGTNDH-N5-BKBWRN-3′ (Dubrac *et al.*, 2008). OmpR-type response regulators typically bind six basepair direct repeats with a four to five basepair spacing, and the YycF consensus sequence follows this common pattern (Blanco *et al.*, 2002).

The YycG kinase features a complex domain-architecture (Fig. 7.2). As mentioned above two subtypes can be distinguished, that of the streptococcal YycG and that of all other YycG proteins. The former appears to be an N-terminal truncation of the full-length YycG protein. The full-length protein is ~600 amino acids large and features two transmembrane helices separated by an extracytoplasmic domain. No structure of this domain is available, but fold prediction programs unambiguously assign this sequence stretch to fold as a PAS domain (Santelli *et al.*, 2007; Chang *et al.*, 2010). PAS domains are very versatile ligand binding modules and are utilized by a large number of signalling proteins across all kingdoms of life (Moglich *et al.*, 2009). The N-terminal transmembrane helix and the extracytoplasmic PAS domain are absent in streptococcal YycG proteins indicating that this protein might respond to different signals. The remainder of the protein architecture is identical in all YycG proteins. It features a cytoplasmic HAMP domain as an immediate extension of a second transmembrane helix, followed by a cytoplasmic PAS domain and by the catalytic kinase core. HAMP domains are common structural features in transmembrane signal transduction proteins with a perceived function in transmitting a conformational change from one cellular compartment to another (Hulko *et al.*, 2006; Szurmant *et al.*, 2007c). The catalytic core comprises the HisKA domain, which harbours the site of autophosphorylation and the ATP-binding HATPase_c domain (Szurmant *et al.*, 2007c).

Currently, no crystal structures are available for any of the YycG domains. Structures can however be inferred from sequence homologues, and a homology model of the HATpase_c domain has been utilized in virtual drug screening efforts, as discussed below (Qin *et al.*, 2006). A recent report described that diffraction quality crystals of the cytoplasmic domain of YycG from *B. subtilis* have been obtained and it can thus be expected that a structure of these domains will become available in time (Zhao and Tang, 2009).

Genes associated with the *yycFG* operon

YycH and YycI

Outside of the streptococci, the *yycFG* genes are universally linked to genes coding for YycH and YycI proteins, with which they constitute an operon. YycH and YycI proteins are unique to these organisms, and proteins of homologous sequence do not exist outside of the phylum *Firmicutes*. Despite their prevalence in the *Firmicutes*, our knowledge on the molecular function of these proteins is derived solely from studies of the *B. subtilis* proteins. It is expected that the identified functions are representative for all YycH and YycI orthologues.

YycH and YycI proteins are extracytoplasmic proteins that are tethered to the cytoplasmic membrane via a single N-terminal transmembrane helix (Szurmant *et al.*, 2007b). YycH is ~460 and YycI ~280 amino acids large. The crystal structures of the extracytoplasmic portions excluding only the transmembrane helix of both proteins from *B. subtilis* have been solved and surprisingly revealed a shared fold unique to these two proteins (Szurmant *et al.*, 2006; Santelli *et al.*, 2007) (Fig. 7.3). This structural relationship was not immediately evident from sequence comparisons.

An initial indication regarding the biological function of YycH came from a transposon mutagenesis study performed on a temperature sensitive $YycF_{H215P}$ mutant strain (Szurmant *et al.*, 2005). The study was designed to screen for transposon insertions that rescued the temperature sensitive phenotype of the parental strain, with the expectation that this study might reveal genes that are functionally associated with the YycFG two-component system. Transposon insertions were repeatedly identified in the *yycH* gene. Gene disruptions of *yycH*, *yycI*, *yycJ* and *yycK* suggested that the disruption of *yycH* was solely responsible for the observed phenotype, whereas *yycI*, *yycJ* and *yycK* disruption strains did not reveal any appreciable phenotype. Subsequent studies demonstrated that YycFG protein levels remained unchanged in a *yycH* strain, but that the expression of a YycF-dependent *yocH-lacZ* reporter (see below) was drastically up-regulated in the *yycH*-disrupted strain (Szurmant *et al.*, 2005). The YycH protein was shown to face the extracytoplasmic space and with little exposure to the cytoplasm it was concluded that the most likely mode of action of YycH was the inhibition of YycG activity. Thus, *yycH* disruption should result in uncontrolled kinase activity. As a consequence of this uncontrolled kinase activity, the *yycH* disrupted strain lysed significantly faster than its parent strain suggesting that a mis-regulation of YycF-dependent genes results in a compromised cell wall (Szurmant *et al.*, 2005).

The initial disruption strain however could not be complemented by ectopic expression of the *yycH* gene, and it was subsequently found that deletion of the entire *yycH* or *yycI* genes resulted in phenotypes identical to those observed for the *yycH* disruption but not the *yycI* disruption strain (Szurmant *et al.*, 2007b). This discrepancy in phenotypes could be resolved when it became clear that ectopic expression of the respective transmembrane helices of YycH or YycI was sufficient to complement the *yycH* or the *yycI* deletion strains (Szurmant *et al.*, 2008). The original *yycI* disruption had left the YycI transmembrane helix intact. Two-hybrid analysis as well as *in vivo* cross-linking and immunoprecipitation assays demonstrated that YycG, YycH and YycI form a ternary complex of low YycG kinase activity that apparently is stabilized by transmembrane helix interaction between the three proteins (Szurmant *et al.*, 2007b, 2008; Fukushima *et al.*,

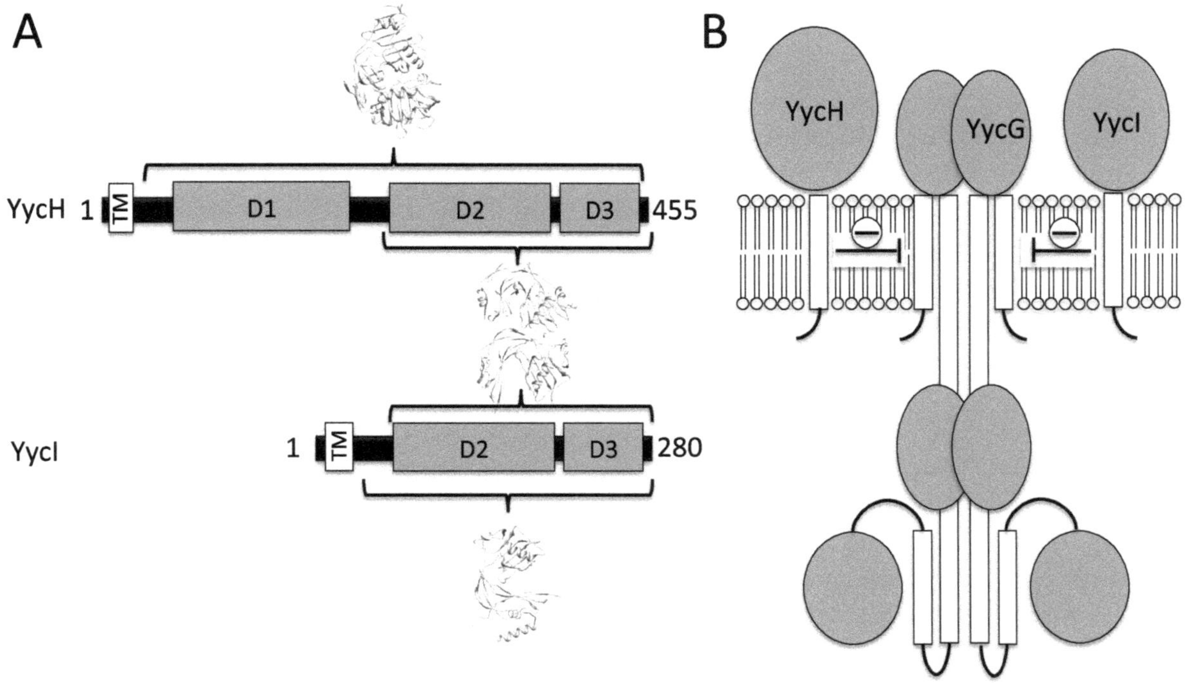

Figure 7.3 Architecture and mode of action for YycH and YycI proteins. (A) High-resolution structures for both, YycH and YycI proteins are available for *B. subtilis*. These structures have revealed a three-domain and a two-domain architecture, respectively (D1-D3). The two C-terminal domains (D2 and D3) are shared between the two proteins. Structures of the individual proteins and the shared portion are displayed for illustration. (B) YycH and YycI proteins are membrane anchored proteins and the bulk of the proteins faces the extracytoplasmic space. Both proteins interact with the YycG kinase through transmembrane helix interaction and YycG kinase activity is reduced when in complex with YycH and YycI. The entire level of activity control is absent in type II systems, which lack YycH, YycI and the TM1 helix and extracytoplasmic domain of the YycG kinase.

2011). Whereas YycH and YycI form a stable interaction even in the absence of YycG, *in vivo* interactions with YycG are only observed when both proteins are present, thereby explaining the identical phenotypes of the respective deletion strains (Fukushima *et al.*, 2011).

To date the function of the extracytoplasmic domains of YycH and YycI has remained elusive, since all phenotypes of the deletion strains could be complemented by simply expressing the corresponding transmembrane helices, and since the structures of these domains did not reveal any clues about their function. In a strain expressing a YycG construct with an internal deletion of the extracytoplasmic domain, a ternary complex was no longer observed, perhaps suggesting that the extracytoplasmic domains also contribute to the affinity of the three proteins for each other (Fukushima *et al.*, 2011).

In summary, YycH and YycI reduce the kinase activity of YycG by interacting with the transmembrane helices and perhaps to a lesser extent with the extracytoplasmic domain of the YycG kinase. This part of YycG is absent in the streptococcal orthologues and so are genes for YycH and YycI proteins, suggesting that this entire level of YycG activity control is absent in bacteria with a type II YycFG system.

YycJ

YycJ (also known as VicX or WalJ) is a ~270 amino acid single domain protein with a predicted metallo-β-lactamase fold. This fold is a particularly common enzyme fold utilized across all domains of life, with currently over 17,000 sequence examples in the databases. Proteins endowed with this fold include proteases, RNases and β-lactamases. Since YycJ conserves the metal-coordinating residues, namely two aspartate and five histidine residues, it is safe to assume that YycJ serves as some sort of hydrolytic enzyme of currently unknown specificity (Biller *et al.*, 2011; Yang *et al.*, 2011a). Numerous diverse phenotypes associated with *yycJ* deletions are described below. Since the enzymatic function of the YycJ protein is still not known and the *yycJ* deletion is associated with pleiotropic phenotypes, it still remains difficult to infer its precise molecular function.

Since *yycJ* genes are organized with the *yycFG* operon in most currently sequenced organisms, it appears obvious that the protein has to be involved in physiological processes that are associated with the YycFG signalling system. A possible involvement in the signalling process itself seems unlikely, since in *B. subtilis* YycFG-dependent expression of target genes did not change in a *yycJ* deletion strain (Szurmant *et al.*, 2007b; Biller *et al.*, 2011). However, a functional association has been demonstrated in both *B. subtilis* and *S. pneumoniae*, where *yycJ* deletion strains require a higher expression level of the *yycFG* genes for viability than a strain with a wild-type *yycJ* copy (Ng *et al.*, 2003; Biller *et al.*, 2011). This either indicates that YycF phosphorylation levels are reduced in the absence of YycJ or, alternatively, that the absence of YycJ dictates an increased requirement of the YycF-dependent enzyme activities. The latter explanation appears more likely, since changed YycF phosphorylation levels should result in altered expression of all YycF regulated genes. It thus appears probable that YycJ acts downstream of the YycFG signalling process.

There are several other phenotypes associated with the deletion of *yycJ* that give hints regarding its function. A *B. subtilis* strain with a *yycJ* deletion formed colonies with a dendritic appearance. On a sporulation plate, a heterogeneous population of sporulation deficient colonies were observed to arise at a rate of roughly 2% (Szurmant *et al.*, 2007b). Whereas the colony appearance remains mysterious, the appearance of sporulation deficient colonies was explained by the observation that a *B. anthracis yycJ* deletion strain had a mutator phenotype, i.e. it accumulated mutations at a much higher rate than a wild-type strain (Yang *et al.*, 2011a). There are two logical explanations for this phenotype. Either YycJ itself is involved in the DNA mismatch repair process (Yang *et al.*, 2011b), or, alternatively, that a substrate accumulates in the absence of YycJ that causes the mutator phenotype. To resolve this question the protein will have to be studied *in vitro* to elucidate possible enzymatic activities and functional interactions with other proteins.

A detailed study that revealed several additional phenotypes associated with YycJ was conducted in *B. subtilis* in an attempt to uncover genes involved in chromosomal segregation in dividing cells. The SpoIIIE DNA translocase typically assembles at division sites and pumps DNA unidirectional into one of the daughter cells two prevent division over unsegregated chromosomes (Wu and Errington, 1997). This protein, however, is not essential and it was argued that mutations in other genes involved in this process might result in a conditionally lethal phenotype. An association was identified with the *yycJ* gene and it was observed that strains deleted for *yycJ* frequently divide over chromosomal DNA (Biller *et al.*, 2011). Furthermore, upon blocking DNA replication wild-type strains do not divide, whereas *yycJ* strains divide frequently. This observation led to the suggestion that YycJ is involved in a process that coordinates DNA replication with cell division. Since a YycJ protein with mutated putative metal coordinating residues failed to complement the deletion strain, the YycJ enzyme activity must be required for this function. A last unexplained phenotype that was observed for both, *B. subtilis* and *S. pneumoniae yycJ* deletion strains, was an increased sensitivity to several cephalosporin antibiotics (Biller *et al.*, 2011). These compounds target the final transpeptidation step in peptidoglycan synthesis and suggest a possible link between YycJ and cell wall metabolism.

YycK

YycK (also known as YyxA) is an Htr-type serine protease featuring an N-terminal transmembrane helix, a serine protease domain and a C-terminal PDZ domain. Htr-type proteases are typically involved in degradation of mis-folded proteins and in the processing of native proteins in the extracytoplasmic space (Noone *et al.*, 2001). While YycK is commonly associated with type I *yyc* operons, this is not universally the case (Dubrac *et al.*, 2008). Deletion analysis of YycK in *B. subtilis* revealed no appreciable phenotypes and no link to YycFG signalling and function could be established. Indeed in *B. subtilis* YycK included its own SigG type promoter and is transcribed from the *yyc* promoter during the exponential growth phase and from its own promoter under conditions that favour sporulation (Fabret and Hoch, 1998). It is thus not entirely clear if a functional association between this protein and the YycFG system exists. It should be noted, however, that most genes that are regulated by YycF are targeted to the extracytoplasmic space. Perhaps YycK functions in processing some of the corresponding proteins and a partial redundancy with other Htr-type proteases (*B. subtilis* has three) masks any possible phenotypes (Noone *et al.*, 2001).

YycX

The *yycX* gene is found immediately downstream of *yycG* in most clostridial species with a *yyc* operon. *yycX* codes for a small 130 amino acids protein that features a predicted single N-terminal transmembrane helix similar to the YycH and YycI proteins. Fold assignment algorithms were unable to identify a known fold for the bulk of this protein. The *Bacilli* do not feature a homologous gene elsewhere on their chromosome suggesting that this protein functions exclusively in the *Clostridia*. A possible involvement in YycFG-dependent signalling in the *Clostridia* is awaiting experimental analysis. Its common association with *yycFG* and *yycHI* might hint to a connection with the YycGHI ternary complex.

MurA

The MurA (UDP-N-acetylglucosamine enolpyruvyl transferase) enzyme catalyses the first committed step in the biosynthesis of the peptidoglycan cell wall and is the target of the antibiotic fosfomycin. Many *Clostridia* species express one of two *murA* copies from the *yyc* operon. In addition, a *murA* gene is commonly associated with the *yycJ* gene in *Clostridia* that do not contain clear orthologues of the YycFG system. This observation is interesting since it has long been recognized that organisms of the phylum *Firmicutes* have two *murA* copies, also referred to as *murAA* and *murAB* (Du *et al.*, 2000; Kock *et al.*, 2004). This is in contrast to all other bacterial phyla, where members typically have only a single *murA* gene copy. To date it has remained unclear why the *Firmicutes* have two copies of the *murA* gene. A *B. subtilis* study demonstrated that MurAA is essential and that MurAB is unable to compensate for MurAA deletion despite these proteins being very similar in sequence and being expressed at the same time and to the same level (Kock *et al.*, 2004). A different picture emerged in *S. pneumoniae*, where the two genes are at least partially redundant with both a *murAA* and *murAB* deletion strain being viable (Du *et al.*, 2000). The observation that one MurA copy is linked to the YycFG signalling system or the YycJ protein in the *Clostridia* might suggest a functional connection between this second copy of MurA with the YycFG system in general. This possible functional connection might warrant future investigations.

Output: the YycF regulon

To rapidly gain significant insight into the role of the YycFG signalling system and to explain why this system was essential for viability it was important to define the genes whose expression was modulated by the YycF transcription regulator, also known as the YycF regulon. The typical approach to identify a regulon for a transcription factors is the measuring of global expression changes between a wild type and the transcription factor deletion strain by transcriptome analysis. The essential nature of the YycFG system dictated the design of slightly more sophisticated strategies. Approaches ranged from the generation of hybrid proteins to strains depleted for the YycFG system and in silico approaches assuming similar binding sites across different species of bacteria. These approaches were supplemented with *in vitro* gel shift and DNAase I foot printing assays, and systematic progress was made resulting in the identification of genes directly regulated by the YycFG system. To date, the regulon of YycF has been studied in some detail in three organisms, *B. subtilis*, *S. pneumoniae* and *S. aureus* (Table 7.1). The consensus that has emerged from these studies is a global theme in the regulation of genes whose products are involved in cell wall homeostasis and cell wall restructuring (Dubrac *et al.*, 2008). Perhaps somewhat surprisingly though, the regulons across the three different organism

Table 7.1 List of experimentally validated genes under direct regulation by the YycF transcription factor

Organism	Gene/ operon	Function	Regulation	Evidence	Reference
B. subtilis	*ftsAZ*	Cell division	+	GS, TR, FP	Fukuchi *et al.* (2000), Howell *et al.* (2003)
	yocH	CW hydrolase	+	MA, NA, FP, GS	Howell *et al.* (2003)
	ykvT	CW hydrolase	NK	FP	Howell *et al.* (2003)
	lytE	CW hydrolase	+	NA, GS	Bisicchia *et al.* (2007)
	ydjM	CW associated protein	+	MA, NA, GS	Bisicchia *et al.* (2007)
	yvcE (*cwlO*)	CW endopeptidase	+	MA, NA, GS	Bisicchia *et al.* (2007)
	tagAB *tagDEF*	Teichoic acid biosynthesis	–	FP	Howell *et al.* (2003)
	yoeB	Autolysin inhibitor	–	MA, NA, GS	Bisicchia *et al.* (2007)
	yjeA	Peptidoglycan deacetylase	–	MA, NA, GS	Bisicchia *et al.* (2007)
S. aureus	*ssaA*	CW amidase	+	GS, FP	Dubrac and Msadek (2004)
	isaA	CW transglycosylase	+	FP	Dubrac and Msadek (2004)
	lytM	CW endopeptidase	+	FP	Dubrac and Msadek (2004)
S. pneumoniae	*pcsB*	CW amidase	+	MA, GS, FP	Ng *et al.* (2003, 2005)
	pspA	Virulence	+	MA, GS, FP	Ng *et al.* (2005)
	lytN	CW hydrolase	+	MA, GS, FP	Ng *et al.* (2005)
	spr1875	CW binding/unknown	+	MA, GS, FP	Ng *et al.* (2005)

CW, cell wall; +, positive regulation by YycF; –, negative regulation by YycF; NK, not known; FP, DNase I foot printing; GS, gel shift mobility assay; MA, microarray; TR, transcriptional reporter; NA, Northern analysis.

have proven to be divers, possibly reflecting changing requirements on cell wall restructuring based on cell size, shape and living environments.

The YycF regulon of *Bacillus subtilis*

The first members of the *B. subtilis* YycF regulon were identified thanks to a phenotype observed for a *yycF* overexpression strain. Overexpression of *yycF* resulted in the formation of mini-cells, a phenotype typically associated with the overexpression of the gene for the cell division master regulator FtsZ (Fukuchi *et al.*, 2000). The *ftsAZ* operon is regulated by three promoters and it was determined that the proximal promoter P1 was up-regulated in the *yycF* overexpression strain. Although *ftsZ* is an essential gene, the YycF responsive promoter is not essential, suggesting that other unidentified members of the regulon were responsible for YycFG essentiality (Fukuchi *et al.*, 2000).

Since the approach that identified *ftsAZ* as members of the YycF regulon could not be extended to identify additional members, Howell *et al.* (2003) designed a clever hybrid regulator approach. They fused the YycF DNA-binding domain to the PhoP response regulator domain and argued that the hybrid regulator should respond to phosphorylation by the PhoR kinase that is known to activate under conditions of phosphate limitation. The engineered strain thus allowed them to measure global changes in expression in response to phosphate limitation by microarray analysis. These studies resulted in the identification of a single differentially regulated gene, *yocH*, coding for a putative autolysin. Subsequent gel shift and DNAase I footprinting assays on the promoter regions of *yocH* and *ftsAZ* identified a regulatory region reminiscent of an OmpR-type binding site, a direct six-basepair repeat with a five-basepair spacing. Equipped with this potential binding site, the authors scanned the genome for additional putative members of the YycF regulon. Of 11 possible candidate operons, two promoter regions were experimentally confirmed to bind YycF *in vitro*, thus identifying *ykvT* coding for a putative cell wall hydrolase, and the teichuronic acid biosynthesis *tagAB tagDEF* divergon as additional members of the YycF regulon (Howell *et al.*, 2003).

The identification of these members of the regulon failed to clarify why the YycFG system was essential. The hybrid approach likely missed a large number of possible members of the regulon since it had to be performed in a strain that still harboured a wild-type copy of *yycF*. A consensus sequence based on only two genes was also likely to miss additional members. A second transcriptome analysis utilizing a strain with an IPTG inducible copy for YycFG was undertaken to resolve this issue. This study resulted in the identification of a small number of additional members of the YycFG regulon, namely *yvcE* (also known as *cwlO*), *ydjM* and *lytE* as positively controlled genes, and *yoeB* and *yjeA* as YycF-repressed genes (Bisicchia *et al.*, 2007).

With the new members of the regulon established a consistent theme emerged for *B. subtilis*. With the exception of FtsAZ, all regulated genes code for extracytoplasmic proteins involved in cell wall metabolism. The positively regulated genes code for enzymes involved in autolysis and cell wall restructuring, the negatively regulated genes code for proteins that directly or indirectly inhibit autolysis and cell wall restructuring (Bisicchia *et al.*, 2007).

Work on multiple deletion and depletion strains demonstrated that the reason for YycFG essentiality is polygenic in nature. Two members of the YycF regulon, *lytE* and *yvcE*, could not be concurrently inactivated. Nevertheless, constitutive expression of both of these proteins did not alleviate the requirement for YycFG suggesting that the essential nature of YycFG in *B. subtilis* and possibly in other organisms with type I YycFG system is quite complex (Bisicchia *et al.*, 2007).

The YycF regulon of *Streptococcus pneumoniae*

In order to identify members of the YycFG regulon that render it essential in *S. pneumoniae* Ng and coworkers (2003) designed a strain that placed the *yycFG* operon under control of a fucose inducible promoter. Global transcriptome analysis of this strain grown with or without fucose resulted in significant changes in the expression of a large number of genes (Ng *et al.*, 2003). Phenotypes upon depletion resulted in dramatic changes in cell morphology. This information was used to scan the list of differentially expressed genes for those that could be deemed responsible for the observed phenotypes, thus identifying two prime candidates, *pcsB* and *spo0J*. In subsequent studies *spo0J* was shown to be irrelevant for the observed phenotypes. Constitutive expression of *pcsB*, however, eliminated phenotypes usually observed upon YycFG depletion. It was subsequently shown that YycF is entirely dispensable in a strain with a constitutively expressed *pcsB* gene, suggesting that unlike in *B. subtilis* the essentiality of YycF in *S. pneumoniae* is monogenic in nature (Ng *et al.*, 2003).

In a follow-up study Ng and coworkers used this information in an attempt to further define the regulon of the YycFG system of *S. pneumoniae*. A strain with a constitutively expressed *pcsB* gene was deleted for *yycF* and global expression changes were assessed by microarray analysis and compared with an isogenic strain harbouring wild-type *yycF* (Ng *et al.*, 2005). In contrast to the previous study, less drastic changes in expression were observed identifying nine potential new members of the YycF regulon. Equipped with the YycF DNA-binding motif identified for *B. subtilis* the authors showed that in addition to *pcsB*, genes for three other surface proteins with peptidoglycan binding LysM domains were directly under YycF control, namely the genes encoding the virulence factor PspA and the surface proteins Spr0096 (also known as LytN) and Spr1875 (Ng *et al.*, 2005). These studies thus support the emerging theme that YycF serves as a regulator of cell wall homeostasis in diverse organisms.

The YycF regulon of *Staphylococcus aureus*

Following the identification of a possible YycF consensus binding sequence in *B. subtilis*, Dubrac and coworkers sought to apply this knowledge to infer the YycF regulon in *S. aureus* (Dubrac and Msadek, 2004). An in silico analysis based on scanning the *S. aureus* genome for potential YycF DNA binding sites identified a set of 12 candidate genes as possible members of the YycF regulon in this organism. Some of these DNA regions were subjected to *in vitro* gel-shift and DNAse I foot printing assays, establishing genes for two surface antigens involved in virulence, SsaA and IsaA, and the peptidoglycan hydrolase LytM as members of the YycF regulon. In 2007, this effort was revisited and, by utilizing quantitative real time PCR, it became apparent that YycF positively regulates the expression of nine genes whose products are involved in cell wall degradation (Dubrac *et al.*, 2007). These studies once more demonstrate that the YycFG system conserves a role in cell wall homeostasis across the phylum *Firmicutes*.

Input: the YycG signal

In order to truly understand the role an individual signal transduction system plays in bacterial physiology, it is not sufficient to know the genes whose expression is regulated by the transcription factor. Rather knowledge on the molecular signals that regulate the activity of the sensor kinase is required to comprehend the purpose of the response. Unfortunately, the identification of an input signal is not as straightforward as the identification of the output regulon. Even for the best-studied signalling systems the activation signal has often remained elusive. Commonly, growth conditions are known that alter the activity of individual sensor kinases *in vivo*, but small molecular ligands have been identified for only a handful of signalling systems. Based on these isolated cases, the assumption is that most two-component system sensor kinases modulate their kinase activity in response to a specific small molecular ligand.

No such molecular ligand has to date been identified that modulates the activity of the YycG sensor histidine kinase. Some reports have suggested that YycG might directly sense molecular oxygen or redox potential. Such reports, however, are commonly based on the observation that the YycG kinase contains a cytoplasmic PAS domain and rely on phenotypic evidence that supports a function in redox or oxygen sensing. PAS domains were originally associated with a role in oxygen and redox response in bacteria (Zhulin *et al.*, 1997). With the accumulation of sequenced genomes it has become clear that the PAS domain is a very versatile ligand-binding domain and that it can accommodate and respond to a large variety of molecular ligands (Moglich *et al.*, 2009). In addition, PAS domains were shown to be involved in protein–protein interactions. It thus appears ill advised to suggest that a protein is directly involved in redox sensing based in large part on the observation that in contains such a domain.

In the lieu of structural information of the YycG sensing domain, which may provide direct insight in the signals perceived, several approaches have been undertaken in order to identify the input of the YycG system. As mentioned previously a transposon mutagenesis screen using a YycF$_{H215P}$ temperature sensitive strain aimed to identify mutations in genes that result in the activation of the YycFG system. This study revealed that YycH and YycI proteins serve to inactivate the YycG kinase (Szurmant *et al.*, 2007b). The molecular details of this process have been addressed and resulted in the picture that YycG kinase forms a ternary complex with YycH and YycI proteins of reduced kinase activity (Szurmant *et al.*, 2007b, 2008; Fukushima *et al.*, 2011).

An immunofluorescence based approach was undertaken for *B. subtilis* to identify the subcellular localization of the YycG kinase expecting that such information might reveal insights as to how, when and where the YycG kinase receives its activation signal (Szurmant *et al.*, 2007a; Fukushima *et al.*, 2008). Fortunately, the YycG localization pattern turned out to be insightful, although such studies with other sensor kinases in other organisms frequently were disappointing, since the proteins did not localize to any specific compartment in the cell. In contrast, the YycG kinase specifically localized to the septum in rapidly dividing cells. This septum localization was dependent on the master regulator of cell division, FtsZ. In a strain depleted for *ftsZ*, YycG no longer localized to the septum. Tested by semi-quantitative RT-PCR, depletion of either *ftsZ* or the *yycFG* operon had equivalent effects on the expression of the YycF-dependent genes, suggesting that YycG requires a properly assembled divisome for localization and activation (Fukushima *et al.*, 2008). In support of this notion, immunofluorescence studies on the YycF response regulator showed that it co-localizes with DNA in rapidly dividing cells indicating a high level of phosphorylation. In contrast, in *ftsZ* depleted cells YycF no longer exclusively localized to the DNA, suggesting a reduced level of phosphorylation.

Although these studies failed to identify a molecular ligand, they did identify the apparently essential purpose that the YycFG system serves in bacterial physiology. By combining knowledge on the activation of YycG at division sites with knowledge about the YycF regulon, a picture emerges where the YycFG system serves to coordinate cell wall restructuring with cell division (Fukushima *et al.*, 2008). Under actively dividing conditions, the YycG kinase localizes to division sites, where it is activated and induces the expression of genes whose products are required for cell wall lysis and restructuring. These activities are most certainly unwelcome under non-dividing conditions. Under these conditions, YycG no longer localizes to division sites and is not activated, resulting in the de-repression of inhibitors of the enzymes required for cell wall restructuring. The YycFG system thus represents a growth-dependent regulator of cell wall homeostasis, at least in *B. subtilis* (Fig. 7.4).

A subsequent study aimed at identifying the molecular factors that direct the YycG kinase to the septum and those that are involved in its activation (Fukushima *et al.*, 2011). This problem was addressed by two approaches: On the one hand, YycG kinase constructs were expressed in

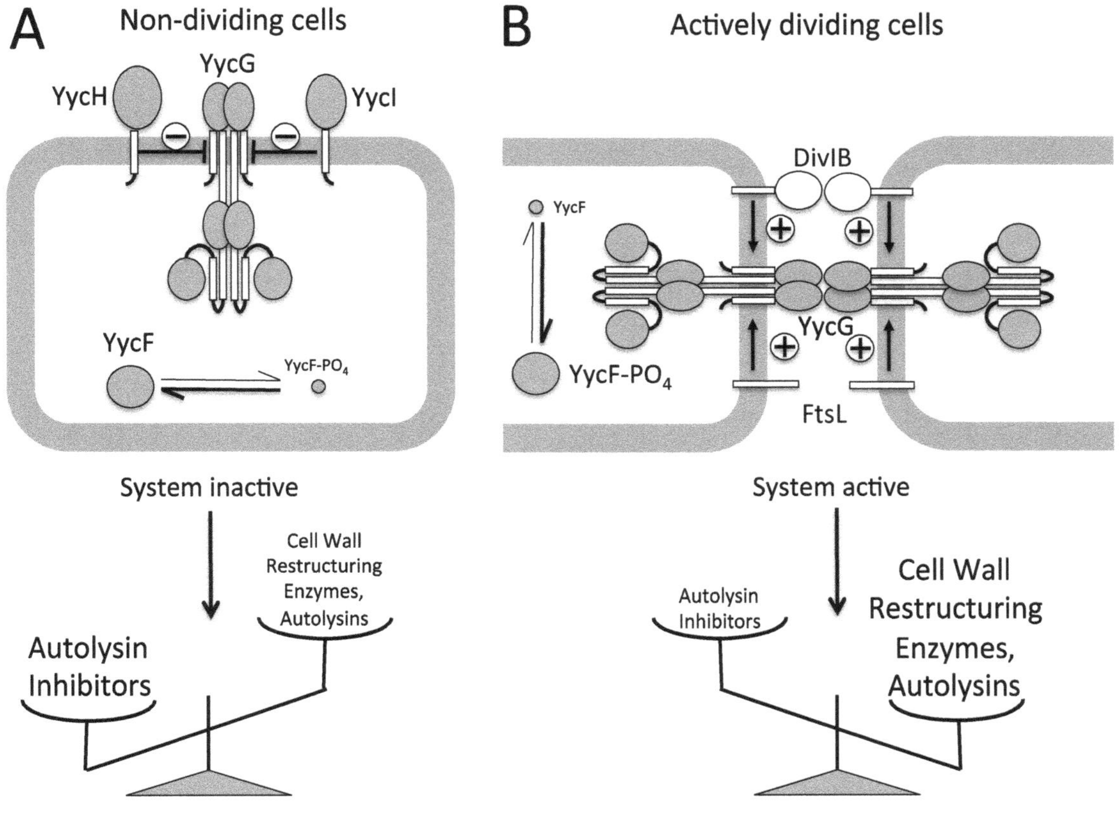

Figure 7.4 Proposed model for type I YycFG systems in coordinating cell wall restructuring with cell division. A model that ties together the current knowledge on type I YycFG systems is primarily derived from studying *B. subtilis*. In this model YycG complexes with YycH and YycI around the periphery of the cell in non-dividing cells (A). In this complexed state, YycG is inactive; the pool of phosphorylated YycF is low. As a consequence YycF fails to activate expression of autolysins and other cell wall restructuring enzymes and also no longer represses the expression of proteins that directly or indirectly inhibit the activity of the autolysins. There is reduced turnover of the cell wall. Under actively dividing conditions (B) YycG is directed to the cell division sites in an FtsZ-dependent manner. There YycH and YycI are displaced by several cell division proteins, most notably DivIB and FtsL. In this complex YycG is active; the pool of phosphorylated YycF is high. As a consequence YycF induces expression of autolysins and represses expression of their inhibitors. The cell wall rapidly turns over, a requirement for growth and cell separation.

B. subtilis that harboured deletions of individual or multiple domains. When testing activity and localization of the various constructs it became apparent that constructs deleted for any or all of the N-terminal domains down to the cytoplasmic PAS domain were constitutively active and no longer associated with the YycH and YycI proteins. Nevertheless, these YycG constructs retained the ability to localize to the septum, thus leaving only the three cytoplasmic domains, the PAS domain or the catalytic domains, as possible interaction partners with the divisome proteins. The isolated PAS domain as well as a catalytic fragment of YycG proved to be highly unstable *in vivo*. Still, experiments with the latter suggested that the PAS domain is the determinant for septum localization.

Tackling this problem from the other site, YycG localization and activation was tested in various strains that allowed for depletion of the essential cell division proteins, with the goal to identify the protein interaction partners that target YycG to the septum. Aside from FtsZ, no other protein was identified whose depletion resulted in aberrant YycG localization. Direct interaction of YycG with FtsZ could not be confirmed, neither *in vitro* with isolated proteins nor *in vivo* using a bacterial two-hybrid assay. The determinants for YycG localization thus remain unidentified. Nevertheless, some essential cell division proteins, most notably DivIB, FtsL and Pbp2B, showed strong interactions with YycG in the two-hybrid assay, and strains depleted for multiple division proteins failed to activate YycG. In conclusion, YycG localization to the septum is not sufficient for activation. Instead, YycG requires a properly assembled divisome for activation. The role of the YycH and YycI proteins in this process was also addressed by immunofluorescence. In contrast to YycG, YycH and YycI do not localize to the septum. It thus appears as though YycH and YycI serve to inactivate the fraction of YycG not localized to the septum (Fukushima *et al.*, 2011).

These studies raised the question whether the observations on YycG localization and activation are representative for all YycG systems or perhaps only for those of type I that include a full-length kinase and the *yycH* and *yycI* genes. This question was addressed by an immunofluorescence study on *S. pneumoniae* YycG. In contrast to *B. subtilis*, *S. pneumoniae* YycG did not localize to the septum, instead it localized in random patterns around the cell periphery (Wayne *et al.*, 2010). While these studies did not provide any new insights regarding the YycG activation signal in *S. pneumoniae*, they suggest that the activation signal and mechanism in type II YycG kinases might be distinct from the situation in *B. subtilis*.

Despite the lack of *S. pneumoniae* YycG localization to the septum a link to cell division for the type II YycFG system was nevertheless identified. The essential cell wall amidase PcsB, which is regulated by YycF was shown to localize to and function at the septum (Sham *et al.*, 2011). PcsB localization and activity requires the FtsEX proteins. Orthologues are widespread across bacteria, they are homologous to ABC transporters and localize to the septum in many bacteria including in *S. pneumoniae*. By regulating the activity of a YycF-controlled gene at the septum a coordination between cell wall restructuring activity with cell division might thus occur post-transcriptionally in the streptococci, in contrast to *B. subtilis* where it occurs at the level of transcription.

Given the conserved role of YycG in regulating cell wall homeostasis and restructuring, it was suggested that the YycG kinase might sense specific cell wall determinants (Dubrac *et al.*, 2008). For *B. subtilis* a detailed analysis of changes in expression of the YycF regulon in response to sublethal concentrations of several cell wall acting antibiotics suggested an involvement of the D-Ala-D-Ala moiety in the peptidoglycan precursor lipid II as a possible YycG activation signal. This hypothesis was tested in a study where the *E. faecalis* vancomycin resistance cassette was introduced into *B. subtilis* (Bisicchia *et al.*, 2011). The genes encoded by this cassette catalyse the replacement of the D-Ala-D-Ala depsipeptide with D-Ala-D-Lac. The latter has a reduced affinity for Vancomycin resulting in resistance. In exponentially growing cells no changes in the expression of the YycF regulon were observed upon introduction of the Vancomycin resistance cassette making an involvement of the D-Ala-D-Ala depsipeptide as a possible YycG signal unlikely.

Functional overlap and cross-talk with other systems

The PkrC serine/threonine kinase

The product of one of the target genes of YycF in *B. subtilis*, YocH, has been shown to stimulate its own expression in a positive feedback loop that requires the serine/threonine kinase PrkC (Shah and Dworkin, 2010). PrkC is a transmembrane spanning enzyme that directly detects extracytoplasmic muropeptides of actively growing cells and responds to the signal by activating its cytoplasmic kinase activity. A currently unidentified transcription factor has to be a subject of phosphorylation by this kinase in order to stimulate *yocH* expression. An obvious candidate would be the YycF response regulator, which is already known to regulate *yocH* expression. Consistent with this notion, the YycF response regulator conserves several serine and threonine residues at its dimerization interface. A crystal structure of *B. subtilis* YycF (Protein database ID 2ZWM) coordinated a sulfate at the dimerization interface in close proximity to T101 and perhaps might mimic a threonine-phosphorylated species of YycF. However, experimental evidence in support of the notion that YycF is the PkrC phosphorylation target has not yet been published.

The PhoPR two-component system

The PhoPR system is another two-component system common to the phylum *Firmicutes*. The PhoPR system of *B. subtilis* has long been known to be engaged in the response to a limited availability of exogenous phosphate and to regulate the expression of genes whose products are involved in scavenging of new phosphate resources (Hulett, 1996). Among the targets of the PhoPR system are the operons coding for teichuronic acid and teichoic acid biosynthesis genes (Hulett, 2002). Teichoic acid is an anionic polymer that comprises about 40% of the cell wall in exponentially growing cells (Bisicchia *et al.*, 2010). It is an essential component of the cell envelope and has a significant phosphate storage capacity. In the absence of phosphate *B. subtilis* replaces teichoic acid with the phosphate-less teichuronic acid polymer in order to scavenge the stored phosphate. The PhoPR system thus plays a role in the homeostasis of the cell envelope in correlation to the YycFG system (Bisicchia *et al.*, 2010). Interestingly, the PhoPR system is a close homologue of YycFG and both systems conserve identical domain architectures. This similarity is corroborated by the observation that several regulatory connections exist between these two systems. For instance, the PhoR kinase was shown to phosphorylate the YycF regulator *in vivo* under phosphate limiting conditions resulting in the activation of several YycF-dependent promoters (Howell *et al.*, 2006). In addition, there appears to be some overlap in the regulon controlled by these two transcription factors, since they bind to similar target sequences (Bisicchia *et al.*, 2010).

Role in antibiotic resistance

Methicillin-resistant *S. aureus* (MRSA) are an increasingly relevant threat to human health. Infections are generally treated with last line of defence antibiotics such as vancomycin and daptomycin. In recent years clinical isolates showing reduced or no susceptibility to either of these antibiotics have emerged following failed drug treatments. A large number of such isolates have now been genome sequenced to identify the genetic determinants responsible for lack of susceptibility by comparative genomics.

Vancomycin is a glycopeptide antibiotic produced by *Amycolatopsis orientalis* that exerts its function by binding to the D-Ala-D-Ala moiety of lipid II and by inhibiting peptidoglycan polymer biosynthesis and cross-linking. In contrast, daptomycin is an acidic cyclic lipopeptide antibiotic produced by *Streptomyces roseosporus* and is believed to act primarily by inserting in and depolarizing the cytoplasmic membrane. Despite these entirely different modes of action, mutations in the *yycFG* system have been linked to non-susceptibility to either antibiotic following drug treatment.

Regarding vancomycin non-susceptible *S. aureus*, also known as VISA (for vancomycin-intermediate *S. aureus*), initially two mutations had been reported to promote Vancomycin resistance. A strain was found that carried an insertion mutation in the *yyc* promoter resulting in vast overexpression of the entire *yyc* operon (Jansen *et al.*, 2007). Another isolate was described featuring a mutation in the *yycH* gene that introduced a premature stop codon (Mwangi *et al.*, 2007). Such a mutation was suspected to result in an overactive unregulated YycFG system in analogy to the function YycH plays in *B. subtilis* (Szurmant *et al.*, 2007b). Both reports thus suggested that an overactive YycFG system might have contributed to the VISA phenotype. Mutations originally described in daptomcyin-resistant isolates included point mutations in YycG (S221P and R263C) (Friedman *et al.*, 2006). Whether these mutations increased or reduced the activity of YycG or had an entirely different effect remains unknown.

A recent comprehensive report identified distinct mutations in either YycF or YycG in 10 out of 13 additional clinical isolates from patients following failed Vancomycin treatment (Howden *et al.*, 2011). Subsequent genetic analysis demonstrated that these mutations were indeed responsible for the VISA phenotype. Individual mutations were not localized to any specific domain. YycF mutants harboured point mutations in either the response regulator domain (A96T) or in the DNA-binding domain (K208R). Mutations in YycG occurred in the N-terminal transmembrane helix (L14F), the HAMP domain (G223D), the PAS domain (V268F; D337–339+I340V), the HisKA domain (N382S) or the HATPase_c domain (D469–70; A567D; T595I). Another equally comprehensive study found mutations in YycG in 24 out of 39 clinical VISA isolates, making *yycG* the most prevalent gene to feature a mutation (Shoji *et al.*, 2011). In addition, the mutations found in this study were similarly diverse to the above-described set of mutations and included point mutations and amino acid insertions dispersed across all domains of YycG.

Both studies documented phenotypes and transcriptional profiles for several of the YycG mutant strains and independently observed similar effects. The mutated strains featured a reduced autolysis activity. Microscopic analysis revealed a thicker cell wall, a phenotype that was already known to be associated with the VISA phenotype (Howden *et al.*, 2011; Shoji *et al.*, 2011). Microarray analysis demonstrated global changes in the expression of a large number of metabolic proteins, perhaps a secondary effect due to the drastic changes in cell wall metabolism.

Somewhat troubling, many of the *yycG* and *yycF* mutations studied by Howden and coworkers also promoted non-susceptibility to daptomycin, despite the fact that these strains had never been exposed to daptomycin (Howden *et al.*, 2011). This study thus conclusively demonstrated that identical mutations in the YycFG system can promote cross-resistance to the two most commonly used antibiotics against MRSA infections, despite the distinct modes of action of these two antibiotics. It remains unclear whether all the diverse mutations exert similar effects on YycFG activity. The observed phenotypes of reduced autolytic activity and thickened cell wall are more consistent with reduced YycFG activity, which is in contrast to earlier studies that suggested an overactive YycFG system to be responsible for the VISA phenotype (Howden *et al.*, 2011; Shoji *et al.*, 2011). Additional studies are clearly warranted to clarify the effects of these mutations on YycFG activity.

Drug discovery efforts

Two-component systems have long been recognized as potential broad range targets for antimicrobials. An original idea was that drugs that specifically target two-component system chemistry should show little toxicity to humans, thanks to the absence of similar chemistry in the mammalian protein complement. Before the identification of essential systems, it was suggested that such drugs should render a microbe unresponsive to its environment and thereby attenuate its virulence and ability to survive in a host organism. Unfortunately, broad range inhibitors with specificity towards two-component chemistry have not been identified. Instead, most inhibitors proved to be semi-specific protein denaturing agents and often had haemolytic activity (Stephenson *et al.*, 2000).

The discovery of the essential YycFG two-component system and its wide distribution across the phylum *Firmicutes* fuelled several targeted approaches to discover specific inhibitors for the

YycG kinase or the YycF response regulator. Such compounds can be expected to serve as specific antimicrobial drugs against important human health threats such as *S. aureus, E. faecalis* or the streptococci. Several reports outlined below have described the discovery of YycG kinase and YycF inhibitors with bactericidal activities. None of the described inhibitors has passed clinical trials and we are thus still far away from the therapeutic use of any of the reported compounds or their derivatives. Nevertheless, the ongoing efforts suggest that targeting the YycFG system remains a valuable approach and that an effective antimicrobial drug might emerge in the future.

YycG kinase inhibitors

A screen to identify natural compounds from microbial acetone extracts with specific activity towards the YycFG system was designed based on a *B. subtilis* strain carrying the temperature sensitive $YycF_{H215P}$ mutation (see above) (Watanabe *et al.*, 2003). It was argued that any compound with increased antimicrobial activity against the temperature sensitive strain over the parental wild-type strain might specifically target the YycFG system. From seven acetone extracts that followed this notion, a single compound was extracted, aranorosinol B, which was already known to exhibit bactericidal activity against some Gram-positives. *In vitro* analysis of this compound demonstrated that it inhibited YycG kinase activity at relatively high concentrations, and perhaps for that reason, initial experiments with this compound were not followed up by additional studies. Instead, utilizing this same screen along with methanol extracts from another 1368 microbes resulted in the isolation and identification of a single compound with broad antimicrobial activity against low G+C Gam-positives with a maximal inhibitory concentration below 1 μg/ml (Okada *et al.*, 2010). Consistent with the notion that this compound specifically targets the YycFG system, *B. subtilis* phenotypes observed upon exposure to this compound matched those that were observed upon depletion of the YycFG system. These phenotypes included cellular filamentation and the emergence of cells that had lost their cytoplasmic content (Ghost-cells) (Fabret and Hoch, 1998; Okada *et al.*, 2010). An *in vitro* analysis demonstrated that this compound directly binds to YycG with a K_d in the low micromolar range and that it inhibits YycG autophosphorylation with a similar IC_{50}. It remains to be seen whether this compound or its derivatives might serve as a new antimicrobial, but these initial studies look promising.

In 2006, a virtual drug screen was described that utilized a homology-modelled structure of the YycG HATPase_c domain of *S. epidermidis* as a target along with an in silico database consisting of 85,000 compounds (Qin *et al.*, 2006). 76 potential hit compounds were tested for their ability to inhibit *S. epidermidis* growth at a concentration of 200 μM. By this approach seven compounds could be singled out. These compounds were shown to inhibit bacterial growth of several Gram-positive strains, but none of them had any inhibitory effect against Gram-negatives. *In vitro* assays suggested that some of these compounds have a high affinity for the YycG kinase and inhibit its autophosphorylation activity. Some specificity could be demonstrated when these compounds were tested against another kinase, *S. epidermidis* SrrB, with no or reduced inhibitory effect against this kinase. Preliminary tests using a Vero Cell line and human erythrocytes ruled out general cytotoxic or haemolytic effects on eukaryotic cells for most of these compounds. Despite these promising results no follow up reports have surfaced that suggest that any of these compounds are being developed as antimicrobial drugs.

YycF response regulator inhibitors

Given the fact that the YycG kinase is not essential in the streptococci and that cross-phosphorylation of response regulators can be acquired relatively easily suggests that the YycF response regulator represents a better target for antimicrobials than the YycG kinase. Despite this fact the first attempt at specifically targeting the YycF protein has only recently been reported.

Gotoh *et al.* (2010) devised a clever high-throughput screen specifically targeting YycF homodimerization. In this screen the YycF response regulator domain was fused to a DNA-binding domain of the *Thermotoga maritima* transcriptional repressor IclR. IclR is known to act as a repressor in its dimeric form (Zhang *et al.*, 2002). By replacing its dimerization domain with the response regulator domain of either YycF or of several other response regulators, an *in vivo* screen was devised that extracted compounds that specifically inhibited dimerization of the YycF–IclR hybrid, but not of the other response regulator–IclR hybrids. The hybrid proteins were expressed in *E. coli* in the presence of the eGFP gene under the control of an IclR responsive promoter. It was shown that all response regulator–IclR fusion proteins inhibited eGFP expression, in contrast to a negative control strain harbouring the isolated IclR DNA-binding domain.

Upon screening an extensive chemical library the authors identified two compounds, named walrycin A and B, that inhibited the activity of YycF-IclR but not of the ArcA-IclR and OmpR–IclR hybrid proteins. Subsequent assays attempted to prove their specificity for YycF by their ability (1) to inhibit phosphorylation by YycG *in vitro*, (2) to inhibit expression of YycG responsive genes in *S. aureus* and *B. subtilis* as determined by quantitative RT-PCR, (3) to inhibit bacterial growth of both wild-type and YycF-overexpressing strains of *S. aureus* and *B. subtilis* and (4) to affect YycF homodimerization. These assays in their entity suggested that the two compounds target YycF specifically. The data indicated that rather than preventing YycF dimerization the compounds might act by promoting dimerization of a non-native and inactive conformation (Gotoh *et al.*, 2010). Again, while promising, it remains to be seen whether these compounds or their derivatives might be useful as antimicrobials or whether they prove to be toxic or ineffective.

Other essential two-component systems

The YycFG system discussed in this chapter is the most prevalent example of an essential two-component system. The first two-component proteins reported to be essential include several members of the *Caulobacter crescentus* phosphorelay that regulates cellular differentiation in this and related organisms (Wu *et al.*, 1998; Jacobs *et al.*, 1999). For more information on this pathway, see Chapter 13.

Another essential two-component system has been described in the *Actinobacteria*, the MtrAB system (Hoskisson and Hutchings, 2006; Hutchings, 2007). This system might be a functional analogue to the YycFG system of the *Firmicutes* (Winkler and Hoch, 2008). Like YycFG in *B. subtilis* MtrAB was identified as essential for *Mycobacterium tuberculosis* viability (Zahrt and Deretic, 2000). The MtrAB system is universally associated with a gene coding for the lipoprotein LpqB, making its identification simple (Hoskisson and Hutchings, 2006). Scanning all current 250+ available *Actinobacteria* genomes for this gene suggests that this two-component system is found in virtually all of them. Phenotypes associated with alterations in MtrAB expression include defects in cell division, cell wall homeostasis, reduced antibiotic resistance and attenuated virulence. Some diversity has been reported regarding MtrAB essentiality. In *M. tuberculosis* the response regulator MtrA but not the kinase MtrB is essential, much like the case for YycF and YycG in the streptococci (Zahrt and Deretic, 2000). Conversely, in *Corynebacterium glutamicum* neither MtrA nor MtrB are essential for viability, a fact that allowed for conventional identification of the MtrA regulon using an *mtrA* deleted strain (Moker *et al.*, 2004). The regulon of MtrA has since been well described in *C. glutamicum* and includes the *dnaA* gene for the essential master regulator of replication and—in analogy to the YycFG system—genes involved in cell wall metabolism and cell envelope regulation (Brocker *et al.*, 2011). It thus appears possible that most Gram-positive bacteria require a signal transduction system to integrate the restructuring of their cell wall with cellular growth.

Concluding remarks

Following a decade worth of studies on the essential YycFG system, this two-component system is now recognized as a coordinator of cell wall restructuring in diverse bacteria of the phylum *Firmicutes*. Equipped with an understanding of its role in bacterial physiology, antibiotic resistance and interconnection with other signalling systems, continued efforts are required to fully unravel the details of its regulatory role and the interconnection of physiologically important cellular processes. The groundwork has been laid to follow up on the currently emerging insights in the essential process of cell wall restructuring. Significant questions remain regarding the activation signal of the YycG kinase in the various microbes and the systems role in the development of antibiotic resistance in *S. aureus*.

Web resources

The Pfam protein family database: http://pfam.sanger.ac.uk/

The Microbial Signal Transduction database (MIST): http://mistdb.com/

The Fold and Function Assignment System (FFAS): http://ffas.ljcrf.edu/ffas-cgi/cgi/ffas.pl

References

Bent, C.J., Isaacs, N.W., Mitchell, T.J., and Riboldi-Tunnicliffe, A. (2004). Crystal structure of the response regulator 02 receiver domain, the essential YycF two-component system of *Streptococcus pneumoniae* in both complexed and native states. J. Bacteriol. *186*, 2872–2879.

Biller, S.J., Wayne, K.J., Winkler, M.E., and Burkholder, W.F. (2011). The putative hydrolase YycJ (WalJ) affects the coordination of cell division with DNA replication in *Bacillus subtilis* and may play a conserved role in cell wall metabolism. J. Bacteriol. *193*, 896–908.

Bisicchia, P., Noone, D., Lioliou, E., Howell, A., Quigley, S., Jensen, T., Jarmer, H., and Devine, K.M. (2007). The essential YycFG two-component system controls cell wall metabolism in *Bacillus subtilis*. Mol. Microbiol. *65*, 180–200.

Bisicchia, P., Lioliou, E., Noone, D., Salzberg, L.I., Botella, E., Hubner, S., and Devine, K.M. (2010). Peptidoglycan metabolism is controlled by the WalRK (YycFG) and PhoPR two-component systems in phosphate-limited *Bacillus subtilis* cells. Mol. Microbiol. 75, 972–989.

Bisicchia, P., Bui, N.K., Aldridge, C., Vollmer, W., and Devine, K.M. (2011). Acquisition of VanB-type vancomycin resistance by *Bacillus subtilis*: the impact on gene expression, cell wall composition and morphology. Mol. Microbiol. *81*, 157–178.

Blanco, A.G., Sola, M., Gomis-Ruth, F.X., and Coll, M. (2002). Tandem DNA recognition by PhoB, a two-component signal transduction transcriptional activator. Structure *10*, 701–713.

Brocker, M., Mack, C., and Bott, M. (2011). Target genes, consensus binding site, and role of phosphorylation for the response regulator MtrA of *Corynebacterium glutamicum*. J. Bacteriol. *193*, 1237–1249.

Chang, C., Tesar, C., Gu, M., Babnigg, G., Joachimiak, A., Pokkuluri, P.R., Szurmant, H., and Schiffer, M. (2010). Extracytoplasmic PAS-like domains are common in signal transduction proteins. J. Bacteriol. *192*, 1156–1159.

Du, W., Brown, J.R., Sylvester, D.R., Huang, J., Chalker, A.F., So, C.Y., Holmes, D.J., Payne, D.J., and Wallis, N.G. (2000). Two active forms of UDP-N-acetylglucosamine enolpyruvyl transferase in Gram-positive bacteria. J. Bacteriol. *182*, 4146–4152.

Dubrac, S., and Msadek, T. (2004). Identification of genes controlled by the essential YycG/YycF two-component system of *Staphylococcus aureus*. J. Bacteriol. *186*, 1175–1181.

Dubrac, S., Boneca, I.G., Poupel, O., and Msadek, T. (2007). New insights into the WalK/WalR (YycG/YycF) essential signal transduction pathway reveal a major role in controlling cell wall metabolism and biofilm formation in *Staphylococcus aureus*. J. Bacteriol. *189*, 8257–8269.

Dubrac, S., Bisicchia, P., Devine, K.M., and Msadek, T. (2008). A matter of life and death: cell wall homeostasis and the WalKR (YycGF) essential signal transduction pathway. Mol. Microbiol. *70*, 1307–1322.

Echenique, J.R., and Trombe, M.C. (2001). Competence repression under oxygen limitation through the two-component MicAB signal-transducing system in *Streptococcus pneumoniae* and involvement of the PAS domain of MicB. J. Bacteriol. *183*, 4599–4608.

Fabret, C., and Hoch, J.A. (1998). A two-component signal transduction system essential for growth of *Bacillus subtilis*: implications for anti-infective therapy. J. Bacteriol. *180*, 6375–6383.

Friedman, L., Alder, J.D., and Silverman, J.A. (2006). Genetic changes that correlate with reduced susceptibility to daptomycin in *Staphylococcus aureus*. Antimicrob. Agents Chemother. *50*, 2137–2145.

Fukuchi, K., Kasahara, Y., Asai, K., Kobayashi, K., Moriya, S., and Ogasawara, N. (2000). The essential two-component regulatory system encoded by *yycF* and *yycG* modulates expression of the *ftsAZ* operon in *Bacillus subtilis*. Microbiology *146*, 1573–1583.

Fukushima, T., Szurmant, H., Kim, E.J., Perego, M., and Hoch, J.A. (2008). A sensor histidine kinase co-ordinates cell wall architecture with cell division in *Bacillus subtilis*. Mol. Microbiol. *69*, 621–632.

Fukushima, T., Furihata, I., Emmins, R., Daniel, R.A., Hoch, J.A., and Szurmant, H. (2011). A role for the essential YycG sensor histidine kinase in sensing cell division. Mol. Microbiol. 79, 503–522.

Galperin, M.Y. (2005). A census of membrane-bound and intracellular signal transduction proteins in bacteria: bacterial IQ, extroverts and introverts. BMC Microbiol. *5*, 35.

Galperin, M.Y. (2006). Structural classification of bacterial response regulators: diversity of output domains and domain combinations. J. Bacteriol. *188*, 4169–4182.

Gotoh, Y., Doi, A., Furuta, E., Dubrac, S., Ishizaki, Y., Okada, M., Igarashi, M., Misawa, N., Yoshikawa, H., Okajima, T., *et al.* (2010). Novel antibacterial compounds specifically targeting the essential WalR response regulator. J. Antibiot. (Tokyo) *63*, 127–134.

Hancock, L.E., and Perego, M. (2004). Systematic inactivation and phenotypic characterization of two-component signal transduction systems of *Enterococcus faecalis* V583. J. Bacteriol. *186*, 7951–7958.

Hoskisson, P.A., and Hutchings, M.I. (2006). MtrAB-LpqB: a conserved three-component system in actinobacteria? Trends Microbiol. *14*, 444–449.

Howden, B.P., McEvoy, C.R., Allen, D.L., Chua, K., Gao, W., Harrison, P.F., Bell, J., Coombs, G., Bennett-Wood, V., Porter, J.L., *et al.* (2011). Evolution of multidrug resistance during *Staphylococcus aureus* infection involves mutation of the essential two component regulator WalKR. PLoS Pathog. *7*, e1002359.

Howell, A., Dubrac, S., Andersen, K.K., Noone, D., Fert, J., Msadek, T., and Devine, K. (2003). Genes controlled by the essential YycG/YycF two-component system of *Bacillus subtilis* revealed through a novel hybrid regulator approach. Mol. Microbiol. *49*, 1639–1655.

Howell, A., Dubrac, S., Noone, D., Varughese, K.I., and Devine, K. (2006). Interactions between the YycFG and PhoPR two-component systems in *Bacillus subtilis*: the PhoR kinase phosphorylates the non-cognate YycF response regulator upon phosphate limitation. Mol. Microbiol. *59*, 1199–1215.

Hulett, F.M. (1996). The signal–transduction network for Pho regulation in *Bacillus subtilis*. Mol. Microbiol. *19*, 933–939.

Hulett, F.M. (2002). The PhoP regulon. In *Bacillus subtilis* and its Closest Relatives: from Genes to Cells, Sonenshein, A.L., Hoch, J.A., and Losick, R., eds (ASM Press, Washington, DC), pp. 193–202.

Hulko, M., Berndt, F., Gruber, M., Linder, J.U., Truffault, V., Schultz, A., Martin, J., Schultz, J.E., Lupas, A.N., and Coles, M. (2006). The HAMP domain structure implies helix rotation in transmembrane signaling. Cell *126*, 929–940.

Hutchings, M.I. (2007). Unusual two-component signal transduction pathways in the actinobacteria. Adv. Appl. Microbiol. *61*, 1–26.

Jacobs, C., Domian, I.J., Maddock, J.R., and Shapiro, L. (1999). Cell cycle-dependent polar localization of an essential bacterial histidine kinase that controls DNA replication and cell division. Cell *97*, 111–120.

Jansen, A., Turck, M., Szekat, C., Nagel, M., Clever, I., and Bierbaum, G. (2007). Role of insertion elements and *yycFG* in the development of decreased susceptibility to vancomycin in *Staphylococcus aureus*. Int. J. Med. Microbiol. *297*, 205–215.

Kallipolitis, B.H., and Ingmer, H. (2001). *Listeria monocytogenes* response regulators important for stress tolerance and pathogenesis. FEMS Microbiol. Lett. *204*, 111–115.

Kock, H., Gerth, U., and Hecker, M. (2004). MurAA, catalysing the first committed step in peptidoglycan biosynthesis, is a target of Clp-dependent proteolysis in *Bacillus subtilis*. Mol. Microbiol. *51*, 1087–1102.

Kunst, F., Ogasawara, N., Moszer, I., Albertini, A.M., Alloni, G., Azevedo, V., Bertero, M.G., Bessieres, P., Bolotin, A., Borchert, S., *et al.* (1997). The complete genome sequence of the Gram-positive bacterium *Bacillus subtilis*. Nature *390*, 249–256.

Martin, P.K., Li, T., Sun, D., Biek, D.P., and Schmid, M.B. (1999). Role in cell permeability of an essential two-component system in *Staphylococcus aureus*. J. Bacteriol. *181*, 3666–3673.

Moglich, A., Ayers, R.A., and Moffat, K. (2009). Structure and signaling mechanism of Per-ARNT-Sim domains. Structure *17*, 1282–1294.

Moker, N., Brocker, M., Schaffer, S., Kramer, R., Morbach, S., and Bott, M. (2004). Deletion of the genes encoding the MtrA-MtrB two-component system of *Corynebacterium glutamicum* has a strong influence on cell morphology, antibiotics susceptibility and expression of genes involved in osmoprotection. Mol. Microbiol. *54*, 420–438.

Mwangi, M.M., Wu, S.W., Zhou, Y., Sieradzki, K., de Lencastre, H., Richardson, P., Bruce, D., Rubin, E., Myers, E., Siggia, E.D., *et al.* (2007). Tracking the *in vivo* evolution of multidrug resistance in *Staphylococcus aureus* by whole-genome sequencing. Proc. Natl. Acad. Sci. U.S.A. *104*, 9451–9456.

Ng, W.L., and Winkler, M.E. (2004). Singular structures and operon organizations of essential two-component systems in species of *Streptococcus*. Microbiology *150*, 3096–3098.

Ng, W.L., Robertson, G.T., Kazmierczak, K.M., Zhao, J., Gilmour, R., and Winkler, M.E. (2003). Constitutive expression of PcsB suppresses the requirement for the essential VicR (YycF) response regulator in *Streptococcus pneumoniae* R6. Mol. Microbiol. *50*, 1647–1663.

Ng, W.L., Tsui, H.C., and Winkler, M.E. (2005). Regulation of the *pspA* virulence factor and essential *pcsB* murein biosynthetic genes by the phosphorylated VicR (YycF) response regulator in *Streptococcus pneumoniae*. J. Bacteriol. *187*, 7444–7459.

Noone, D., Howell, A., Collery, R., and Devine, K.M. (2001). YkdA and YvtA, HtrA-like serine proteases in *Bacillus subtilis*, engage in negative autoregulation and reciprocal cross-regulation of *ykdA* and *yvtA* gene expression. J. Bacteriol. *183*, 654–663.

Okada, A., Igarashi, M., Okajima, T., Kinoshita, N., Umekita, M., Sawa, R., Inoue, K., Watanabe, T., Doi, A., Martin, A., *et al.* (2010). Walkmycin B targets WalK (YycG), a histidine kinase essential for bacterial cell growth. J. Antibiot. (Tokyo) *63*, 89–94.

Okajima, T., Doi, A., Okada, A., Gotoh, Y., Tanizawa, K., and Utsumi, R. (2008). Response regulator YycF essential for bacterial growth: X-ray crystal structure of the DNA-binding domain and its PhoB-like DNA recognition motif. FEBS Lett. *582*, 3434–3438.

Qin, Z., Zhang, J., Xu, B., Chen, L., Wu, Y., Yang, X., Shen, X., Molin, S., Danchin, A., Jiang, H., *et al.* (2006). Structure-based discovery of inhibitors of the YycG histidine kinase: new chemical leads to combat *Staphylococcus epidermidis* infections. BMC Microbiol. *6*, 96.

Santelli, E., Liddington, R.C., Mohan, M.A., Hoch, J.A., and Szurmant, H. (2007). The crystal structure of *Bacillus subtilis* YycI reveals a common fold for two members of an unusual class of sensor histidine kinase regulatory proteins. J. Bacteriol. *189*, 3290–3295.

Senadheera, M.D., Guggenheim, B., Spatafora, G.A., Huang, Y.C., Choi, J., Hung, D.C., Treglown, J.S., Goodman, S.D., Ellen, R.P., and Cvitkovitch, D.G. (2005). A VicRK signal transduction system in

Streptococcus mutans affects *gtfBCD*, *gbpB*, and *ftf* expression, biofilm formation, and genetic competence development. J. Bacteriol. *187*, 4064–4076.

Shah, I.M., and Dworkin, J. (2010). Induction and regulation of a secreted peptidoglycan hydrolase by a membrane Ser/Thr kinase that detects muropeptides. Mol. Microbiol. *75*, 1232–1243.

Sham, L.T., Barendt, S.M., Kopecky, K.E., and Winkler, M.E. (2011). Essential PcsB putative peptidoglycan hydrolase interacts with the essential FtsXSpn cell division protein in *Streptococcus pneumoniae* D39. Proc. Natl. Acad. Sci. U.S.A. *108*, E1061–E1069.

Shoji, M., Cui, L., Iizuka, R., Komoto, A., Neoh, H.M., Watanabe, Y., Hishinuma, T., and Hiramatsu, K. (2011). *walK* and *clpP* mutations confer reduced vancomycin susceptibility in *Staphylococcus aureus*. Antimicrob. Agents Chemother. *55*, 3870–3881.

Stephenson, K., Yamaguchi, Y., and Hoch, J.A. (2000). The mechanism of action of inhibitors of bacterial two-component signal transduction systems. J. Biol. Chem. *275*, 38900–38904.

Szurmant, H., Nelson, K., Kim, E.J., Perego, M., and Hoch, J.A. (2005). YycH regulates the activity of the essential YycFG two-component system in *Bacillus subtilis*. J. Bacteriol. *187*, 5419–5426.

Szurmant, H., Zhao, H., Mohan, M.A., Hoch, J.A., and Varughese, K.I. (2006). The crystal structure of YycH involved in the regulation of the essential YycFG two-component system in *Bacillus subtilis* reveals a novel tertiary structure. Protein Sci. *15*, 929–934.

Szurmant, H., Fukushima, T., and Hoch, J.A. (2007a). The essential YycFG two-component system of *Bacillus subtilis*. Methods Enzymol. *422*, 396–417.

Szurmant, H., Mohan, M.A., Imus, P.M., and Hoch, J.A. (2007b). YycH and YycI interact to regulate the essential YycFG two-component system in *Bacillus subtilis*. J. Bacteriol. *189*, 3280–3289.

Szurmant, H., White, R.A., and Hoch, J.A. (2007c). Sensor complexes regulating two-component signal transduction. Curr. Opin. Struct. Biol. *17*, 706–715.

Szurmant, H., Bu, L., Brooks, C.L., 3rd, and Hoch, J.A. (2008). An essential sensor histidine kinase controlled by transmembrane helix interactions with its auxiliary proteins. Proc. Natl. Acad. Sci. U.S.A. *105*, 5891–5896.

Trinh, C.H., Liu, Y., Phillips, S.E., and Phillips-Jones, M.K. (2007). Structure of the response regulator VicR DNA-binding domain. Acta Crystallogr. D Biol. Crystallogr. *63*, 266–269.

Wagner, C., Saizieu Ad, A., Schonfeld, H.J., Kamber, M., Lange, R., Thompson, C.J., and Page, M.G. (2002). Genetic analysis and functional characterization of the *Streptococcus pneumoniae vic* operon. Infect. Immun. *70*, 6121–6128.

Watanabe, T., Hashimoto, Y., Yamamoto, K., Hirao, K., Ishihama, A., Hino, M., and Utsumi, R. (2003). Isolation and characterization of inhibitors of the essential histidine kinase, YycG in *Bacillus subtilis* and *Staphylococcus aureus*. J. Antibiot. (Tokyo) *56*, 1045–1052.

Wayne, K.J., Sham, L.T., Tsui, H.C., Gutu, A.D., Barendt, S.M., Keen, S.K., and Winkler, M.E. (2010). Localization and cellular amounts of the WalRKJSpn (VicRKX) two-component regulatory system proteins in serotype 2 *Streptococcus pneumoniae*. J. Bacteriol. *192*, 4388–4394.

Winkler, M.E., and Hoch, J.A. (2008). Essentiality, bypass, and targeting of the YycFG (VicRK) two-component regulatory system in Gram-positive bacteria. J. Bacteriol. *190*, 2645–2648.

Wolfe, A.J. (2010). Physiologically relevant small phosphodonors link metabolism to signal transduction. Curr. Opin. Microbiol. *13*, 204–209.

Wu, J., Ohta, N., and Newton, A. (1998). An essential, multicomponent signal transduction pathway required for cell cycle regulation in *Caulobacter*. Proc. Natl. Acad. Sci. U.S.A. *95*, 1443–1448.

Wu, L.J., and Errington, J. (1997). Septal localization of the SpoIIIE chromosome partitioning protein in *Bacillus subtilis*. EMBO J. *16*, 2161–2169.

Yang, H., Sikavi, C., Tran, K., McGillivray, S.M., Nizet, V., Yung, M., Chang, A., and Miller, J.H. (2011a). Papillation in *Bacillus anthracis* colonies: a tool for finding new mutators. Mol. Microbiol. *79*, 1276–1293.

Yang, H., Yung, M., Sikavi, C., and Miller, J.H. (2011b). The role of *Bacillus anthracis* RecD2 helicase in DNA mismatch repair. DNA Repair (Amst.) *10*, 1121–1130.

Zahrt, T.C., and Deretic, V. (2000). An essential two-component signal transduction system in *Mycobacterium tuberculosis*. J. Bacteriol. *182*, 3832–3838.

Zhang, R.G., Kim, Y., Skarina, T., Beasley, S., Laskowski, R., Arrowsmith, C., Edwards, A., Joachimiak, A., and Savchenko, A. (2002). Crystal structure of *Thermotoga maritima* 0065, a member of the IclR transcriptional factor family. J. Biol. Chem. *277*, 19183–19190.

Zhao, H., and Tang, L. (2009). Crystallographic characterization of a multidomain histidine protein kinase from an essential two-component regulatory system. Acta Crystallogr. Sect. F Struct. Biol. Cryst. Commun. *65*, 346–349.

Zhao, H., Heroux, A., Sequeira, R.D., and Tang, L. (2009). Preliminary crystallographic studies of the regulatory domain of response regulator YycF from an essential two-component signal transduction system. Acta Crystallogr. Sect. F Struct. Biol. Cryst. Commun. *65*, 719–722.

Zhulin, I.B., Taylor, B.L., and Dixon, R. (1997). PAS domain S-boxes in Archaea, Bacteria and sensors for oxygen and redox. Trends Biochem. Sci. *22*, 331–333.

Molecular Mechanism of Bacterial Two-component Signal Transduction Networks via Connectors

8

Yoko Eguchi, Eiji Ishii and Ryutaro Utsumi

Abstract

Two-component system (TCS) connectors are proteins that connect different TCSs. By connecting TCSs, they integrate the signals to which each TCS responds, and enable the cell to respond flexibly to the fluctuating environment. A number of auxiliary proteins of TCSs, which bind to the components of TCSs for modification, have been reported. When the expression of such TCS auxiliary proteins is controlled by a TCS, it connects the two TCSs and is termed a TCS connector. Several TCS connectors have been identified in *Salmonella* and *Escherichia coli*. In this chapter, we describe how these TCS connectors and the TCSs they connect form a signal transduction network. TCS connectors in *Bacillus subtilis* as well as TCS auxiliary proteins as candidates of TCS connectors are also introduced. We have clarified a small portion of the predicted two-component signal transduction network as an initial description of the complicated signal network that exists in bacteria.

Introduction

Microorganisms are the most changeable and adaptable organisms. Bacteria, in particular, are found in places where animals and plants cannot survive, such as on glaciers and in hydrothermal vents located on the ocean floor. Since their appearance on earth, bacteria have evolved a variety of functions (e.g. signal transduction) to respond to environmental changes. To respond appropriately to different environmental changes, bacteria have evolved two-component signal transduction (TCS) networks, which are absent in mammals (including human beings). According to the microbial signal transduction database MiST2, 1087 sequenced bacterial genomes encode for 63,259 TCS proteins, and hence, the average genome features roughly 29 individual signalling systems (Szurmant and Hoch, 2010; Ulrich and Zhulin, 2010). TCSs are essentially ubiquitous in bacteria and are also found in archaea and some eukaryotes.

TCSs comprise a histidine kinase (HK) sensor normally residing in the inner membrane and a cognate response regulator (RR) located in the cytoplasm. The HK sensor monitors a particular environmental signal using three enzymatic activities: (i) autophosphorylation at its histidine residue using ATP; (ii) transfer of the phosphoryl group to the conserved aspartate residue of its cognate RR; and (iii) dephosphorylation of the phosphorylated cognate RR. RRs are mostly DNA-binding transcription factors whose affinities for their target promoters are modulated by phosphorylation. Consequently, alteration of the phosphorylated state of RRs results in different gene expression profiles (Gao and Stock, 2009). In addition to each TCS responding to individual environmental stresses, complex regulatory networks between the TCSs is assumed to exist, enabling bacteria to cope with a variety of environmental conditions (Utsumi, 2008). In order to analyse such TCS networks, the mRNA profiles of all 36 TCS deletion mutants of *Escherichia coli* were systematically examined (Oshima *et al.*, 2002). Overlapped target genes were found for a number of TCS pathways. The authors of this study proposed the existence of a network of functional interactions, including cross-talking and the cascade of signal transductions between different TCSs.

Fig. 8.1 shows the typical protein–protein interactions that describe bacterial signal transduction networks. They are divided into two types: cross-talk and cross-regulation. Cross-talk is a signalling system in which distinct TCSs are connected through phosphotransfer from the HK to a non-cognate RR (Laub and Goulian, 2007). Cross-phosphorylation with 25 HKs against the 34 RRs in *E. coli* was assayed *in vitro* for all combinations (Yamamoto *et al.*, 2005). Among the 34 RRs, 11 RRs were phosphorylated by more than one HK. The EnvZ, CpxA and CheA HKs were also assayed *in vitro* for cross-phosphorylation against a panel of RRs from *E. coli* (Skerker *et al.*, 2005). The results showed that EnvZ fully phosphorylated OmpR within 10 seconds, but it required as long as one hour to phosphorylate the non-cognate RR, CpxR. These results indicate that the non-cognate transfer is too slow to be relevant *in vivo* (Groban *et al.*, 2009). Given the high similarity between the sequences and structures of HKs or RRs there appears to be significant potential for cross-talk between TCSs; however, it has been shown that there is minimal cross-talk between distinct TCS pathways (Laub and Goulian, 2007). There are numerous examples of TCSs that cross-regulate one another at levels other than cross-phosphorylation (cross-talk). Understanding all of the connections between TCS pathways will be crucial in the development of a more complete understanding of how bacteria sense, adapt and respond to changes in their environment

In cross-regulation, heterodimers or connector proteins and peptides are involved in signal transductions between TCSs (Fig. 8.1). An example of HK-heterodimer formation is the orphan HK, RetS, specifically binding to another HK, GacS, in *Pseudomonas aeruginosa*. The binding of RetS inhibits the ability of GacS to autophosphorylate using ATP, and stimulates the dephosphorylation of phosphorylated GacS (Goodman *et al.*, 2009). This decreases the level of phosphorylated GacA and eliminates GacA-promoted *rsmZ* transcription. RetS exerts its regulatory action directly, and conditions/signals controlling RetS promote the formation of RetS–GacS heterodimers to the detriment of the formation of GacS-homodimers, which results in decreased levels of phosphorylated GacA (Raghavan and Groisman, 2010).

The direct interaction of proteins with TCS proteins is a recently uncovered strategy that bacteria use to integrate signals other than those detected by a given sensor. These proteins may function as negative regulators or in a positive fashion. Regulatory proteins may also target RRs to protect the phosphorylated form of RRs from dephosphorylation, or to stimulate dephosphorylation of phosphorylated RRs. TCS connectors are an emerging group of proteins that modulate the activity of HKs and RRs at the post-translational level. Since connector proteins are typically synthesized in response to signals that are different from those sensed by the cognate sensor, they often establish regulatory links between otherwise independent signal transduction pathways (Mitrophanov and Groisman, 2008). Here we will introduce connectors between bacterial TCSs and describe the critical role they play in weaving TCS networks.

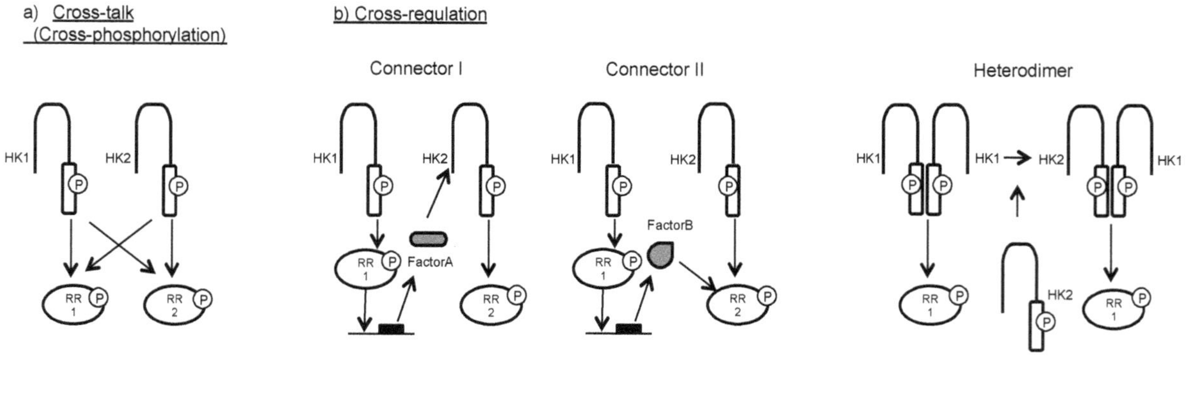

Figure 8.1 Prototype protein–protein interactions in cross-talk and cross-regulation.

Signal transduction networks in *Salmonella*

The idea of connector proteins between TCSs was first found in the *Salmonella enterica* serotype Typhimurium (*S.* Typhimurium). The virulence-related PhoQ/PhoP system is an extensively studied TCS in *Salmonella*, as well as in other pathogens such as *Pseudomonas aeruginosa*, *Yersinia pestis*, *Neisseria meningitidis* and *Erwinia carotovora*. The histidine kinase sensor PhoQ is activated at low concentrations of magnesium (Garcia Vescovi *et al.*, 1996) and promotes the phosphorylation of its cognate RR, PhoP. Increased levels of phosphorylated PhoP up-regulate or down-regulate the transcription of a set of genes, including virulence-related genes (Groisman, 2001). The PmrB/PmrA system controls resistance to the peptide antibiotic polymyxin B and to other antimicrobial compounds from human neutrophils (Roland *et al.*, 1993). This TCS independently responds to two signals: Fe^{3+}, which is sensed by the sensor PmrB (Wosten *et al.*, 2000), and low Mg^{2+}, which is sensed by the non-cognate sensor PhoQ. The latter activation by low Mg^{2+} requires both PhoQ and PhoP proteins. However, the mechanism by which the PhoQ/PhoP system transduces the low Mg^{2+} signal to the PmrB/PmrA system could not be explained by cross-talking from PhoQ to PmrA, or a classical regulatory cascade in that the PhoQ/PhoP system increased the transcription of the *pmrAB* genes (Soncini *et al.*, 1996), and had remained unknown for some time.

In 2000, PmrD, a small protein related to the polymyxin resistance, was found to mediate the signal transduction between PhoQ/PhoP and PmrB/PmrA. PmrD is composed of 85 amino acids, and its expression is controlled by the PhoQ/PhoP system. When transcription of *pmrD* is driven from a heterologous promoter, expression of the PmrA-activated genes occurs, even at repressing magnesium concentrations, and becomes independent of the *phoP* and *phoQ* genes. This PmrD protein appears to control the PmrB/PmrA system at a post-transcriptional level. Expression of PmrA-activated genes was abolished in a *pmrD* mutant experiencing low Mg^{2+}, even when the *pmrAB* genes were transcribed from a heterologous promoter. Thus, PmrD connects the PhoQ/PhoP and PmrB/PmrA systems (Kox *et al.*, 2000). Transcription of *pmrD* is not only activated at low Mg^{2+} concentrations in a PhoQ/PhoP-dependent manner, but is also subject to negative-feedback regulation by the PmrB/PmrA system in response to Fe^{3+}, thereby creating a regulatory circuit (Kato *et al.*, 2003).

Expression of PmrA-activated genes depends on the cellular level of PmrA phosphorylation, which is determined by the autophosphorylated state and the phosphatase potential of its cognate sensor PmrB. Further investigation of the mechanism of PmrB/PmrA activation by PmrD revealed that PmrD bound to the phosphorylated form of PmrA at the N-terminal domain (which harbours the putative phosphorylation site, Asp51), thus preventing both its intrinsic dephosphorylation and that promoted by PmrB. The phosphorylated form of PmrA, maintained by the binding of PmrD, promotes transcription of the PmrA-activated genes (Kato and Groisman, 2004). Since expression of PmrD is regulated by the PhoQ/PhoP system, this was the first example of a protein that connected two TCSs, and enabled a TCS to respond to the signal governing a different TCS by protecting the phosphorylated state of a response regulator (Fig. 8.2).

Activation of the PhoQ/PhoP system at low Mg^{2+} levels also accumulates the alternative sigma factor RpoS in exponentially growing *S.* Typhimurium cells. RpoS is necessary for bacterial survival in the stationary phase and under various stresses, including high osmolarity, oxidative stress, acidic pH, temperature shock and starvation for different nutrients. The production and accumulation of RpoS are tightly controlled at the levels of transcription, translation and protein stability. It has been shown that the 86-amino-acid protein IraP stabilizes RpoS by binding to the RssB protein in *E. coli*, particularly during phosphate starvation (Bougdour *et al.*, 2008). IraP is also present in *S.* Typhimurium, and its expression is directly regulated by the PhoQ/PhoP system. Thus, PhoP controls RpoS protein turnover by serving as a transcriptional activator of the *iraP* gene (Tu *et al.*, 2006), which encodes a product that enhances RpoS stability by interacting with RssB (also called MviA in *Salmonella*), the protein that normally delivers RpoS to the ClpXP protease for

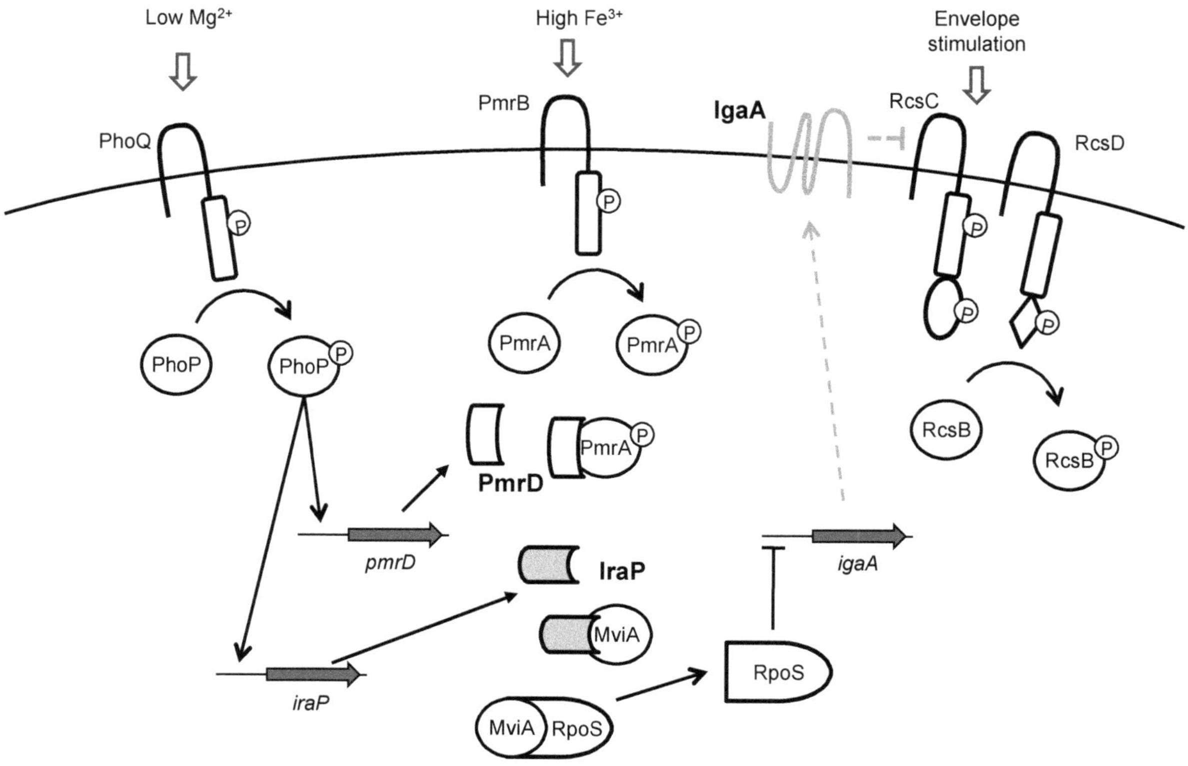

Figure 8.2 Signal transduction network via connectors in *Salmonella*. A schematic view of the signal transduction network involving three TCSs and three connectors is shown. Activation of the PhoQ/PhoP system induces the connector PmrD, which activates the PmrB/PmrA system by protecting the phosphorylated PmrA from dephosphorylation. The PhoQ/PhoP system also induces the connector IraP, which binds to the MviA RR, resulting in the accumulation of RpoS. High level of cellular RpoS suppresses the expression of the connector IgaA which activates the RcsC/RcsD/RcsB system.

degradation. Since RssB is classified as a RR, the IraP protein serves as a connector between the PhoQ/PhoP system and RssB (Fig. 8.2).

This signalling pathway has been further extended by more participants. The *igaA* is an essential gene in *S.* Typhimurium and its product, IgaA, is an inner membrane protein that represses the RcsC/RcsD/RcsB TCS system to prevent its over-activation (Cano *et al.*, 2002; Dominguez-Bernal *et al.*, 2004). The RcsC/RcsD/RcsB system regulates positively the expression of genes encoding enzymes required for synthesis of exopolysaccharide capsules, outer membrane and cell division proteins. In contrast, the *flhDC* genes that encode the flagella master regulators are repressed by this system. IgaA is required for *S.* Typhimurium virulence and this contribution is limited to ensure tight repression of the RcsC/RcsD/RcsB regulatory system when the bacteria colonize the animal host (Dominguez-Bernal *et al.*, 2004). RcsC is a hybrid histidine kinase, RcsD an intermediate phosphotransmitter, and RcsB a transcriptional regulator (Takeda *et al.*, 2001). IgaA-mediated repression of the RcsC/RcsD/RcsB system occurs at the post-translational level, and IgaA may attenuate this system by favouring the switch of phosphorylated RcsB to the dephosphorylated state. The direct target of IgaA, whether it is RcsC or other additional factors, remains to be solved (Garcia-Calderon *et al.*, 2009). In a recent study, *igaA* and the RcsC/RcsD/RcsB system have been found to be regulated by the MviA response regulator. The gene, *igaA*, is the first gene in an operon driven by a σ^{70}-dependent promoter. As aforementioned, MviA is a response regulator that participates in the degradation of RpoS. Mutation of MviA represses the expression of *igaA*, leading to the activation of the RcsC/RcsD/RcsB system (Garcia-Calderon *et al.*, 2009). Consequently, IgaA serves as another connector between MviA and the RcsC/RcsD/RcsB system.

The signal transduction network initiating from the PhoQ/PhoP system is summarized in Fig. 8.2. When the PhoQ/PhoP system is activated, the connector PmrD is induced and activates the PmrB/PmrA system. At the same time, the connector IraP is also induced and binds to the response regulator MviA, leading to cellular accumulation of RpoS. It can be presumed that the increase in the RpoS level will repress IgaA expression, leading to the activation of the RcsC/RcsD/RcsB system. In addition, the RR RcsB of the RcsC/RcsD/RcsB system is known to form heterodimers with a number of auxiliary proteins, such as RcsA, TviA, BglJ and GadE (Virlogeux *et al.*, 1996; Wehland and Bernhard, 2000; Castanie-Cornet *et al.*, 2010; Venkatesh *et al.*, 2010), to allow additional signal input by altering the activity of RcsB. One of these auxiliary proteins, TviA, is present in *S. enterica* serotype Typhi (*S.* Typhi), but not in *S.* Typhimurium. The *tviA* gene is encoded in the via*B* locus within the *Salmonella* pathogenicity island (SPI)–7 and its product, TviA controls virulence expression. Both RcsB and TviA repress expression of the flagella regulators *flhDC*, but only TviA altered *flhDC* expression in response to osmolarity. The expression of *tviA* is regulated by the EnvZ/OmpR TCS in response to osmolarity. Therefore, the RcsB protein senses osmolarity through the availability of the TviA auxiliary protein, which in turn is controlled by the osmosensitive EnvZ/OmpR system (Winter *et al.*, 2009). Hence, TviA serves as a connector between the EnvZ/OmpR and RcsC/RcsD/RcsB systems.

Besides the connector-mediated signal relays, there are a few examples of direct regulation, where a TCS directly regulates the expression of another TCS. One of these is the virulence related SsrA (also referred to as SpiR)/SsrB system regulated dually by the phosphorylated form of OmpR of the EnvZ/OmpR system. There are two promoters controlling the SsrA/SsrB system; one is upstream of *ssrA* and the other upstream of *ssrB*. The phosphorylated OmpR binds upstream of both these genes and regulates them in an uncoupled manner (Feng *et al.*, 2003). The PhoQ/PhoP system also controls the SsrA/SsrB system. In this case, PhoP binds to the *ssrB* promoter and regulates its transcription, whereas PhoP controls the SsrA sensor post-transcriptionally, involving the 5′ untranslated region of the *ssrA* message (Bijlsma and Groisman, 2005). The SsrA/SsrB system is encoded in the *Salmonella* pathogenicity island 2 (SPI-2) and genes located in this region, such as a type III secretion system and secreted effectors, are required for systemic infection in mice. The complex interplay of SsrA/SsrB, EnvZ/OmpR and PhoQ/PhoP in the regulation of the SPI-2 demonstrates the integration of multiple signals into coordinated action.

The connector proteins integrate signals from different TCSs and expand the spectrum of environments where genes can be expressed. It has been suggested that another role of the connector proteins is to affect the degree to which genes are regulated. The pathway controlled by the PhoP-dependent connector protein PmrD of *Salmonella* and the circuit in which genes are regulated directly by the transcription factor PhoP was compared by actually examining the expression levels of the genes, and also by mathematical modelling (Kato *et al.*, 2007). As a result of their study, the PmrD-mediated pathway displayed both signal amplification and persistence of expression when compared with the direct pathway.

Signal transduction networks in *Escherichia coli*

Connections between TCSs are also found and studied in *E. coli*. One of these connections initiates from the EvgS/EvgA system. The EvgS/EvgA system is the major system for conferring acid resistance to exponential-phase cells (Masuda and Church, 2002, 2003; Ma *et al.*, 2004). Since EvgS/EvgA in cells growing in LB medium is not active, a constitutively active EvgS mutant was compared with the wild-type strain using a DNA microarray in order to compare the on- and off-state of the EvgS/EvgA system (Eguchi *et al.*, 2003). Among the genes affected by EvgS activation, many of the PhoP-regulated genes were found, indicating the presence of signal transduction from the EvgS/EvgA system to the PhoQ/PhoP system (Eguchi *et al.*, 2004). This signal transduction also occurred by overexpression of the RR EvgA, under EvgS inactivating condition, which denied cross-talk from EvgS to PhoP. Expressing *phoP* and *phoQ* from a heterogeneous promoter did not activate the PhoP-regulated genes; therefore, the presence of a connector was considered (Eguchi *et al.*, 2004). A shotgun screening approach in a PhoP-reporter strain selected a small ORF, *b1500*, which activated the PhoQ/PhoP system when overexpressed. Expression of the *b1500* gene is directly regulated by the EvgS/EvgA system and *b1500* expression from a heterogeneous promoter simultaneously activated the expression of *mgtA* and other PhoP regulon genes. This activation was PhoQ/PhoP-dependent and EvgS/EvgA-independent. B1500 localizes in the inner membrane and forms a complex with the sensor PhoQ. This observation was confirmed by a bacterial two-hybrid analysis (Eguchi *et al.*, 2007). Since B1500 connected the signal transduction between EvgS/EvgA and PhoQ/PhoP systems by direct interaction with the sensor PhoQ, it was annotated as SafA (sensor-associating-factor A) (Itou *et al.*, 2009). Similar to the connector PmrD in *Salmonella*, SafA is a small protein composed of 65 amino acids. However, whereas PmrD resides in the cytoplasm and interacts with the RR, SafA is located in the inner-membrane and interacts with the sensor at the periplasm (Eguchi *et al.*, unpublished), indicating a different mode of action between the two connector proteins. Other small-sized auxiliary proteins that modify HK sensors have also been found in *E. coli*. These are MzrA (127 amino acids), whose expression is regulated by the CpxA/CpxR system and which activates the sensor EnvZ of the EnvZ/OmpR system at the periplasm (Gerken *et al.*, 2009; Gerken and Misra, 2010); and MgrB (47 amino acids), which is regulated by the PhoQ/PhoP system, and represses the activity of the sensor PhoQ also at the periplasm (Lippa and Goulian, 2009). MzrA is a SafA-like connector between CpxA/CpxR and EnvZ/OmpR systems, and MgrB is part of the feedback repression of the PhoQ/PhoP system.

Another connector found in *E. coli* is IraM (107 amino acids). Similar to IraP in *Salmonella*, IraM is induced by the activation of the PhoQ/PhoP system, and prevents RpoS degradation by binding to the response regulator RssB (Bougdour *et al.*, 2008). Thus, once the EvgS/EvgA system is activated, the induced connector SafA activates the PhoQ/PhoP system. PhoQ/PhoP then induces the connector IraM, which binds and inactivates the RR RssB. As a result, RpoS is not degraded by the ClpXP protease and accumulates within the cell (Fig. 8.3). Indeed, the transcriptome analysis of EvgS-activated exponential phase cells (Eguchi *et al.*, 2003) showed increases in the expression for ~25 genes among the 65 RpoS up-regulated genes previously reported (Loewen *et al.*, 1998). Activation of the EvgS/EvgA system initiates a transcriptional cascade of acid resistance genes encoding regulators EvgA, YdeO and the central regulator of acid resistance genes, GadE. In a recent study, the acid resistant phenotype induced by EvgS/EvgA activation was partly dependent on SafA, PhoQ, PhoP and IraM. The increased level of RpoS via SafA, PhoQ/PhoP and IraM enhances *gadE* transcription, as well as that of *gadX* (Ma *et al.*, 2002; Eguchi *et al.*, 2011). In addition, PhoQ/PhoP activation also induces expression of acid resistance genes, *gadW* and *hdeA* (Zwir *et al.*, 2005). Acid resistance studies of deletion mutants showed that deletion of *safA*, *phoQ*, *phoP* or *iraM* from an EvgS/EvgA activated strain actually reduced acid resistance of exponential phase cells. Thus, a complex network of acid resistance gene regulation initiated by the EvgS/EvgA system and integrating three TCSs with two connectors has been proposed (Fig. 8.3). EvgS/EvgA induces the EvgA-YdeO-GadE cascade of acid resistance gene regulation and a branched pathway operating via SafA-PhoQ/PhoP-IraM-RssB-RpoS enforces the acid resistance phenotype (Eguchi *et al.*, 2011).

The acid-responding RstB/RstA system is under the control of the PhoQ/PhoP system. In this case, PhoP directly binds to the promoter of the *rstAB* operon (Minagawa *et al.*, 2003) and regulates its expression. Expression of the RstA regulon *asr* (acid shock RNA) and *csgD* (a regulator of the curli operon) are both under the direct control of the PhoQ/PhoP–RstB/RstA signal relay cascade (Ogasawara *et al.*, 2007), adding another branch to the network (Fig. 8.3).

RcsB of the RcsC/RcsD/RcsB system also participates in this network. As in *Salmonella*, the RR RcsB forms heterodimers with other proteins and shows a wide range of roles in many enteric bacteria. RcsB is essential for survival during extreme acid challenge during the stationary phase and regulates transcription of acid resistance genes

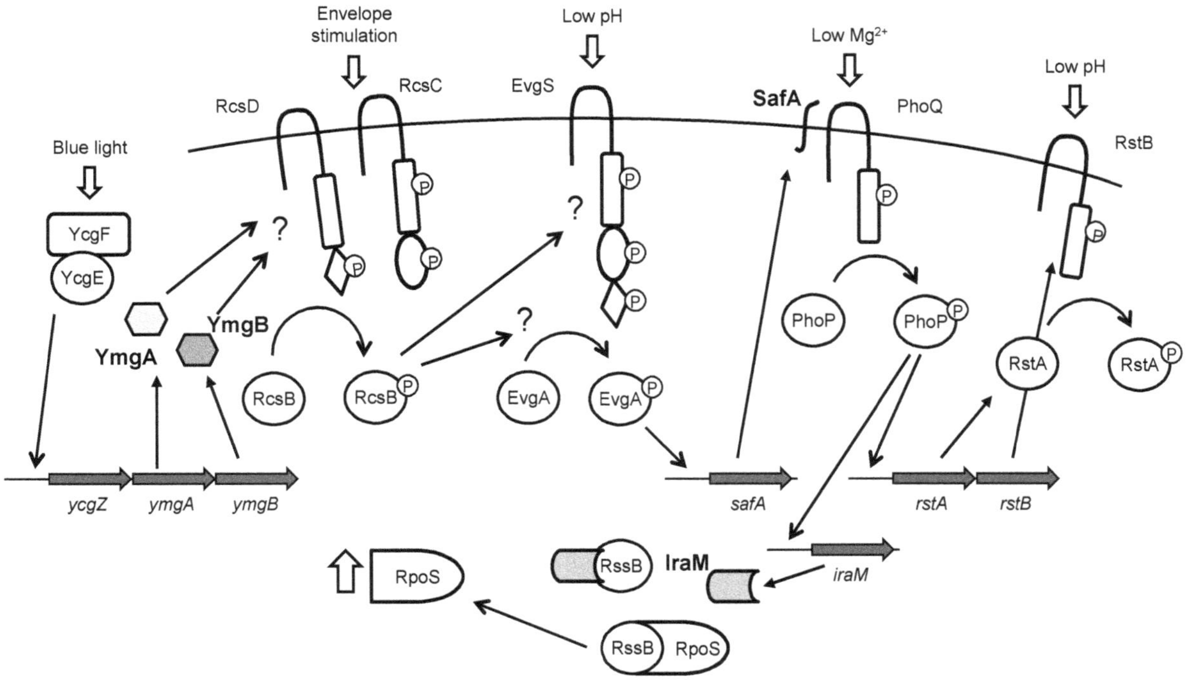

Figure 8.3 Signal transduction network via connectors in *E. coli*. A schematic view of the signal transduction to and from the EvgS/EvgA system is shown. Once the EvgS/EvgA system is activated, the induced connector SafA activates the PhoQ/PhoP system. PhoQ/PhoP then induces the connector IraM, which binds to the RssB RR, resulting in the accumulation of RpoS. Besides the connector-mediated signal transduction, the PhoQ/PhoP system directly induces the RstB/RstA system. The RcsC/RcsD/RcsB system is activated by the small proteins, YmgA and YmgB, whose expression is induced by the blue-light responding YcgF protein (not a TCS). The RcsB RR participates in signal transduction from the EvgS/EvgA system by an unknown mechanism.

by forming a heterodimer with GadE (Castanie-Cornet *et al.*, 2010). Furthermore, RcsB is also required for inducible acid resistance in the exponential phase and RcsB acts on promoters that are not GadE regulated (Johnson *et al.*, 2011). These promoters are mostly regulated by the EvgS/EvgA system, suggesting a signal transduction between RcsC/RcsD/RcsB and EvgS/EvgA systems. Studies of promoter activities and acid resistance in *ΔrcsB* strains indicate that RcsB acts downstream of EvgS. Hypotheses are that either RcsB may be required for the EvgS-dependent phosphorylation of EvgA or for the regulatory activity of EvgA (Johnson *et al.*, 2011). It is tempting to assume activation of the EvgS/EvgA system by heterodimer formation of EvgS or EvgA with a non-cognate RcsB RR.

Further upstream of the RcsC/RcsD/RcsB system are the small proteins YmgA and YmgB. Expression of these two proteins is controlled by the blue light using FAD-EAL protein YcgF (a known blue-light sensor of *E. coli*) and the MerR-like repressor YcgE. Upon blue-light irradiation, YcgF directly binds to and releases YcgE from its operator DNA, strongly inducing *ymgA*, *ymgB* and six other small proteins. The induced YmgA and YmgB activate the RcsC/RcsD/RcsB system, resulting in the production of the biofilm matrix substrate colanic acid as well as acid resistance genes and the biofilm-associated *bdm* gene and down-regulate adhesive curli fimbriae (Tschowri *et al.*, 2009). Although YcgF/YcgE is not a TCS, this blue-light sensing system is connected to the RcsC/RcsD/RcsB system by the small proteins YmgA and YmgB. These two proteins can be considered as a new type of connectors that provide additional input to a TCS pathway.

These connections of the signalling pathways found in *E. coli* are summarized in Fig. 8.3.

Other TCS connectors

The sporulation in *Bacillus subtilis* is governed by a phosphorelay unit comprised five different HKs

(KinA, KinB, KinC, KinD and KinE), a single-domain RR (Spo0F), a Hpt protein (Spo0B) and a RR (Spo0A). Spo0F receives the phosphoryl group from the five HKs, and passes it to Spo0B, which serves as a phosphate donor to Spo0A. Phosphorylated Spo0A controls the transcription of a number of genes involved in sporulation (Burbulys *et al.*, 1991; Piggot and Hilbert, 2004). This unit is under complex regulation by antikinases and phosphatases that integrate multiple signals. Several Rap and Spo0E family phosphatases interact with, and stimulate the autophosphatase activity of the phosphorylated Spo0F and Spo0A RRs, respectively. The Spo0F phosphatase activity has been demonstrated for RapA, RapB, RapE and RapH. Other members of the Rap family, RapC, RapF, RapG, RapH and RapK, associate with the DNA-binding effector domain of the ComA or DegU RRs and inhibit their abilities to regulate transcription. How these proteins bind Spo0F, ComA, or DegU for dephosphorylation or for inhibition of DNA-binding is still unclear (Silversmith, 2010). Transcription of the *rapA* and *rapE* genes is under the control of the ComP/ComA TCS for competence development (Mueller *et al.*, 1992). Hence RapA and RapE are connectors between the ComP/ComA system and the Spo0F RR. On the other hand, RapK, which suppresses the expression of genes activated by ComA, is thought to be activated indirectly by Spo0A (Auchtung *et al.*, 2006). Thus, RapK may serve as a connector from Spo0A to the ComP/ComA system.

Auxiliary proteins modifying TCS components: TCS connector candidates

We have focused on TCS connectors that link different TCSs for understanding the cross-regulation between TCSs. A different definition of a TCS connector has been introduced by Mitrophanov and Groisman (2008). Their definition is that a TCS connector 'connects' a TCS to the signal(s) controlling a different regulatory system (the signal input regulatory system need not be a TCS). Hence auxiliary proteins, which modify the components of a TCS, would most likely be termed as TCS connectors by their definition. When expression of an auxiliary protein is controlled by another TCS, then it will be a connector between TCSs, and will contribute to the TCS network. Various auxiliary proteins against HKs as well as against RRs have been identified, and are summarized in excellent reviews (Mitrophanov and Groisman, 2008; Alix and Blanc-Potard, 2009; Buelow and Raivio, 2010). Indeed many of the connectors between TCSs are included in these auxiliary proteins. These auxiliary proteins together with their targets are listed in Table 8.1. Some of these proteins may be considered as candidates of TCS connectors, because the regulation of their expression is not fully understood, and some of them may be controlled by TCSs.

The TCS auxiliary proteins differ in which component of the TCS they modify. They can be divided into HK and RR modifying proteins (Table 8.1). HK modifying proteins can further be grouped by the HK domains they target. The proteins localized in the periplasm target the sensory domain of a HK, such as the CpxP protein binding and suppressing the activity of CpxA HK in *E. coli*. Expression of CpxP is regulated by the CpxA/CpxR system, thus forming a feedback repression loop (Raivio *et al.*, 1999) (see also Chapter 12). Auxiliary proteins residing in the inner membrane also target the HK. As aforementioned, the TCS connectors, SafA and MzrA in *E. coli*, are inner membrane proteins interacting with the periplasmic region of their target HK, PhoQ and EnvZ, respectively, and activating them (Eguchi *et al.*, 2007; Gerken and Misra, 2010). The *E. coli* inner membrane protein, MgrB, also targets PhoQ at the periplasm, but suppresses PhoQ activity (Lippa and Goulian, 2009). The periplasmic domains of SafA, MzrA, and MgrB interact with their target HKs, whereas the YycH and YycI proteins in *B. subtilis* act differently. These proteins, embedded in the inner membrane by a single transmembrane helix and containing periplasmic domains, negatively regulate the essential WalK/WalR (YycG/YycF) TCS (Szurmant *et al.*, 2005, 2007). The transmembrane domains of YycH and YycI were sufficient to repress the WalK/WalR system, suggesting an interaction with the WalK HK through the transmembrane domains (Szurmant *et al.*, 2008) (see also Chapter 7). Interactions between alpha helices within biological membranes have

Table 8.1 Auxiliary proteins of TCS and their targets

Auxiliary protein	Organism	Target	Localization	Reference
Histidine kinase modification				
ChvE	*A. tumefaciens*	VirA (+)	PP	Shimoda *et al.* (1993)
CpxP	*E. coli*	CpxA(–)	PP	Raivio *et al.* (1999)
TorT	*E. coli*	TorS (–)	PP	Baraquet *et al.* (2006)
PII	*E. coli*	NtrB (–)	PP	Jiang and Ninfa (1999)
MgrB	*E. coli*	PhoQ (–)	IM	Lippa and Goulian (2009)
MzrA	*E. coli*	EnvZ (+)	IM	Gerken *et al.* (2009)
SafA	*E. coli*	PhoQ (+)	IM	Eguchi *et al.* (2007)
YycH	*B. subtilis*	WalK (–)	IM	Szurmant *et al.* (2005)
YycI	*B. subtilis*	WalK (–)	IM	Szurmant *et al.* (2007)
FeuN	*S. meliloti*	FeuQ (–)	IM	Carlyon *et al.* (2010)
LprF	*M. tuberculosis*	KdpD (+)	IM	Steyn *et al.* (2003)
LprJ	*M. tuberculosis*	KdpD (+)	IM	Steyn *et al.* (2003)
FixT	*C. crescentus*	FixL (–)	CP	Crosson *et al.* (2005)
SixA	*E. coli*	ArcB (–)	CP	Ogino *et al.* (1998)
Sda	*B. subtilis*	KinA (–)	CP	Burkholder *et al.* (2001)
KipI	*B. subtilis*	KinA (–)	CP	Wang *et al.* (1997)
Response regulator modification				
PmrD	*S.* Typhimurium	PmrA (+)	CP	Kato *et al.* (2007)
IraP	*S.* Typhimurium	RssB (–)	CP	Tu *et al.* (2006)
IraM	*E. coli*	RssB (–)	CP	Bougdour *et al.* (2008)
CheZ	*E. coli*	CheY (–)	CP	Blat and Eisenbach (1996)
RapA	*B. subtilis*	Spo0F (–)	CP	Perego *et al.* (1994)
RapB	*B. subtilis*	Spo0F (–)	CP	Perego *et al.* (1994)
RapC	*B. subtilis*	ComA (–)	CP	Lazazzera *et al.* (1999)
RapE	*B. subtilis*	Spo0F (–)	CP	Jiang *et al.* (2000)
RapK	*B. subtilis*	ComA (–)	CP	Auchtung *et al.* (2006)
RapF	*B. subtilis*	ComA (–)	CP	Bongiorni *et al.* (2005)
RapG	*B. subtilis*	DegU (–)	CP	Ogura *et al.* (2003)
RapH	*B. subtilis*	Spo0F (–), ComA (–)	CP	Smits *et al.* (2007)
YnzD	*B. subtilis*	Spo0F (–)	CP	Perego (2001)
YisI	*B. subtilis*	Spo0F (–)	CP	Perego (2001)
Spo0E	*B. subtilis*	Spo0F (–)	CP	Ohlsen *et al.* (1994)

E. coli, *Escherichia coli*; *B. subtilis*, *Bacillus subtilis*; *S. meliloti*, *Sinorhizobium meliloti*; *A. tumefaciens*, *Agrobacterium tumefaciens*; *M. tuberculosis*, *Mycobacterium tuberculosis*; *C. crescentus*, *Caulobacter crescentus*; *S.* Typhimurium, *Salmonella* enterica serotype Typhimurium; PP, periplasm; IM, inner membrane; CP, cytoplasm; (+), positive modification of the target TCS; (–), negative modification of the target TCS.

recently been highlighted along with the recognition of hydrophobic peptides (< 50 amino acids) as novel regulators within the bacterial membranes (Alix and Blanc-Potard, 2009).

In a recent study, the 83-amino-acid protein FeuN, an essential auxiliary inner membrane protein of the nitrogen-fixing symbiotic bacterium, *Sinorhizobium meliloti,* was found to negatively regulate the osmosensory TCS FeuQ/FeuP (Carlyon *et al.*, 2010). This TCS plays a critical role in symbiotic infection by positively regulating expression of the cyclic glucan exporter NdvA (Griffitts *et al.*, 2008). FeuN suppresses the lethal overactivation of the FeuQ/FeuP system by interacting directly with the FeuQ HK to attenuate FeuP phosphorylation. How FeuN expression is controlled is not yet known. Random mutagenesis studies revealed a defined region of the FeuQ protein in and around its two predicted transmembrane domains that are required for FeuN-dependent signalling modulation (Carlyon *et al.*, 2010). Whether FeuN interacts with FeuQ through the transmembrane domains or at the periplasm remains to be clarified.

Some auxiliary proteins against HKs reside in the cytoplasm and interact with the kinase domain of HKs. The *B. subtilis* HK KinA promotes the initiation of sporulation when nutrients are limiting, but sporulation can be delayed by two inhibitors of KinA, Sda or KipI (Wang *et al.*, 1997; Burkholder *et al.*, 2001). Sda and KipI bind the DHp domain of KinA and inhibit autophosphorylation, but not the phospho-transfer from KinA to the Spo0F RR. Sda expression is induced in response to perturbations in DNA replication or DNA damage and regulated by DnaA, preventing cells from sporulating until chromosome replication has been resumed or completed (Burkholder *et al.*, 2001). KipI expression is induced when cells need to use a non-preferred nitrogen source, but still have a preferred carbon source available (Wang *et al.*, 1997). For further information about auxiliary proteins which act as anti-kinases, see also Chapter 9.

Auxiliary proteins targeting RRs modify the phosphorylation state of RRs, inhibit DNA-binding of RRs, or act as anti-adaptors. As shown in Table 8.1, most of the RR modifying proteins suppress the activity of their target TCSs. In some cases, RR dephosphorylation occurs by an auxiliary phosphatase. A classical example is the CheZ phosphatase, which specifically targets the CheY RR of the chemotaxis CheA/CheY TCS in *E. coli*. CheZ controls the phosphorylation level of CheY RR. However, CheZ phosphatase activity is suppressed at low phosphorylated CheY levels, presumably to avoid intracellular phosphorylated CheY levels reaching too low a level (Silversmith *et al.*, 2008). The Rap and Spo0E families of auxiliary phosphatases regulate the sporulation in *B. subtilis*. Spo0E dephosphorylates the phosphorylated Spo0A RR in a mechanism similar to CheZ. Interestingly, members of the Rap family either function as phosphatases or as inhibitors of DNA-binding of RRs, as described in the previous section (Silversmith, 2010). Auxiliary proteins, which target RRs and activate the TCS, such as PmrD, are still few in number.

Conclusions

In this review, we have introduced the recently revealed signal transduction network by TCSs and their connectors in *S.* Typhimurium and *E. coli*. Whereas many of the TCS auxiliary proteins (Table 8.1) stimulate dephosphorylation and negatively control their target TCSs, the TCS connectors such as PmrD and SafA activate their target TCS. Another feature of these TCS connectors is their small size: PmrD (85 aa), SafA (65 aa), IraP (86 aa), IraM (107 aa) and MzrA (127 aa). In a search for small functional ORFs in *E. coli*, translation of 20 previously known or annotated small ORFs was validated, and 18 new expressed peptides were identified among the predicted proteins of 16–50 amino acids in length (Hemm *et al.*, 2008). Further characterization of these newly identified small proteins as well as other genes with unknown function may find more TCS auxiliary proteins, and perhaps more TCS connectors. Another source of new TCS connectors will be the expanding group of TCS auxiliary proteins (Table 8.1). If some of these proteins are found to be regulated by TCSs, then they will connect different TCSs and contribute to the TCS network. Although we have focused on *S.* Typhimurium and *E. coli* in this manuscript (Figs. 8.2 and 8.3), information about TCSs and their regulation is rapidly accumulating in other bacteria. Identification of novel TCS connectors and TCS networks in bacteria other than *S.* Typhimurium and *E. coli* is also important and needs to be examined. This information will deepen our understanding of the essential role of TCSs and TCS connectors in signal transduction networks that respond to environmental changes and stimuli.

Acknowledgements

The research from the Utsumi lab was partly supported by the Research and Development Program for Bio-Industry Initiatives (2006–2010) of the Bio-Oriented Technology Research Advancement Institute (BRAIN), a Grant-in-Aid for Scientific Research (A, 290248012) of the Japan Society for the Promotion of Science, Adaptable and Seamless Technology Transfer Program through target-driven RD, and 'Strategic Project to Support the Formation of Research Bases at Private Universities': Matching Fund Subsidy from MEXT (Ministry of Education, Culture, Sports, Science and Technology), 2011–2015(S1101035).

References

Alix, E., and Blanc-Potard, A.B. (2009). Hydrophobic peptides: novel regulators within bacterial membrane. Mol. Microbiol. *72*, 5–11.

Auchtung, J.M., Lee, C.A., and Grossman, A.D. (2006). Modulation of the ComA-dependent quorum response in *Bacillus subtilis* by multiple Rap proteins and Phr peptides. J. Bacteriol. *188*, 5273–5285.

Baraquet, C., Theraulaz, L., Guiral, M., Lafitte, D., Mejean, V., and Jourlin-Castelli, C. (2006). TorT, a member of a new periplasmic binding protein family, triggers induction of the Tor respiratory system upon trimethylamine N-oxide electron-acceptor binding in *Escherichia coli*. J. Biol. Chem. *281*, 38189–38199.

Bijlsma, J.J., and Groisman, E.A. (2005). The PhoP/PhoQ system controls the intramacrophage type three secretion system of *Salmonella enterica*. Mol. Microbiol. *57*, 85–96.

Blat, Y., and Eisenbach, M. (1996). Oligomerization of the phosphatase CheZ upon interaction with the phosphorylated form of CheY. The signal protein of bacterial chemotaxis. J. Biol. Chem. *271*, 1226–1231.

Bongiorni, C., Ishikawa, S., Stephenson, S., Ogasawara, N., and Perego, M. (2005). Synergistic regulation of competence development in *Bacillus subtilis* by two Rap-Phr systems. J. Bacteriol. *187*, 4353–4361.

Bougdour, A., Cunning, C., Baptiste, P.J., Elliott, T., and Gottesman, S. (2008). Multiple pathways for regulation of sigmaS (RpoS) stability in *Escherichia coli* via the action of multiple anti-adaptors. Mol. Microbiol. *68*, 298–313.

Buelow, D.R., and Raivio, T.L. (2010). Three (and more) component regulatory systems – auxiliary regulators of bacterial histidine kinases. Mol. Microbiol. *75*, 547–566.

Burbulys, D., Trach, K.A., and Hoch, J.A. (1991). Initiation of sporulation in *B. subtilis* is controlled by a multicomponent phosphorelay. Cell *64*, 545–552.

Burkholder, W.F., Kurtser, I., and Grossman, A.D. (2001). Replication initiation proteins regulate a developmental checkpoint in *Bacillus subtilis*. Cell *104*, 269–279.

Cano, D.A., Dominguez-Bernal, G., Tierrez, A., Garcia-Del Portillo, F., and Casadesus, J. (2002). Regulation of capsule synthesis and cell motility in *Salmonella enterica* by the essential gene *igaA*. Genetics *162*, 1513–1523.

Carlyon, R.E., Ryther, J.L., VanYperen, R.D., and Griffitts, J.S. (2010). FeuN, a novel modulator of two-component signalling identified in *Sinorhizobium meliloti*. Mol. Microbiol. 77, 170–182.

Castanie-Cornet, M.P., Cam, K., Bastiat, B., Cros, A., Bordes, P., and Gutierrez, C. (2010). Acid stress response in *Escherichia coli*: mechanism of regulation of *gadA* transcription by RcsB and GadE. Nucleic Acids Res. *38*, 3546–3554.

Crosson, S., McGrath, P.T., Stephens, C., McAdams, H.H., and Shapiro, L. (2005). Conserved modular design of an oxygen sensory/signaling network with species-specific output. Proc. Natl. Acad. Sci. U.S.A. *102*, 8018–8023.

Dominguez-Bernal, G., Pucciarelli, M.G., Ramos-Morales, F., Garcia-Quintanilla, M., Cano, D.A., Casadesus, J., and Garcia-del Portillo, F. (2004). Repression of the RcsC-YojN-RcsB phosphorelay by the IgaA protein is a requisite for *Salmonella* virulence. Mol. Microbiol. *53*, 1437–1449.

Eguchi, Y., Oshima, T., Mori, H., Aono, R., Yamamoto, K., Ishihama, A., and Utsumi, R. (2003). Transcriptional regulation of drug efflux genes by EvgAS, a two-component system in *Escherichia coli*. Microbiology *149*, 2819–2828.

Eguchi, Y., Okada, T., Minagawa, S., Oshima, T., Mori, H., Yamamoto, K., Ishihama, A., and Utsumi, R. (2004). Signal transduction cascade between EvgA/EvgS and PhoP/PhoQ two-component systems of *Escherichia coli*. J. Bacteriol. *186*, 3006–3014.

Eguchi, Y., Itou, J., Yamane, M., Demizu, R., Yamato, F., Okada, A., Mori, H., Kato, A., and Utsumi, R. (2007). B1500, a small membrane protein, connects the two-component systems EvgS/EvgA and PhoQ/PhoP in *Escherichia coli*. Proc. Natl. Acad. Sci. U.S.A. *104*, 18712–18717.

Eguchi, Y., Ishii, E., Hata, K., and Utsumi, R. (2011). Regulation of acid resistance by connectors of two-component signal transduction systems in *Escherichia coli*. J. Bacteriol. *193*, 1222–1228.

Feng, X., Oropeza, R., and Kenney, L.J. (2003). Dual regulation by phospho-OmpR of *ssrA/B* gene expression in *Salmonella* pathogenicity island 2. Mol. Microbiol. *48*, 1131–1143.

Gao, R., and Stock, A.M. (2009). Biological insights from structures of two-component proteins. Annu. Rev. Microbiol. *63*, 133–154.

Garcia-Calderon, C.B., Casadesus, J., and Ramos-Morales, F. (2009). Regulation of *igaA* and the Rcs system by the MviA response regulator in *Salmonella enterica*. J. Bacteriol. *191*, 2743–2752.

Garcia Vescovi, E., Soncini, F.C., and Groisman, E.A. (1996). Mg^{2+} as an extracellular signal: environmental regulation of *Salmonella* virulence. Cell *84*, 165–174.

Gerken, H., and Misra, R. (2010). MzrA–EnvZ interactions in the periplasm influence the EnvZ/OmpR two-component regulon. J. Bacteriol. *192*, 6271–6278.

Gerken, H., Charlson, E.S., Cicirelli, E.M., Kenney, L.J., and Misra, R. (2009). MzrA: a novel modulator of the EnvZ/OmpR two-component regulon. Mol. Microbiol. *72*, 1408–1422.

Goodman, A.L., Merighi, M., Hyodo, M., Ventre, I., Filloux, A., and Lory, S. (2009). Direct interaction between sensor kinase proteins mediates acute and chronic disease phenotypes in a bacterial pathogen. Genes Dev. *23*, 249–259.

Griffitts, J.S., Carlyon, R.E., Erickson, J.H., Moulton, J.L., Barnett, M.J., Toman, C.J., and Long, S.R. (2008). A *Sinorhizobium meliloti* osmosensory two-component system required for cyclic glucan export and symbiosis. Mol. Microbiol. *69*, 479–490.

Groban, E.S., Clarke, E.J., Salis, H.M., Miller, S.M., and Voigt, C.A. (2009). Kinetic buffering of cross talk between bacterial two-component sensors. J. Mol. Biol. *390*, 380–393.

Groisman, E.A. (2001). The pleiotropic two-component regulatory system PhoP–PhoQ. J. Bacteriol. *183*, 1835–1842.

Hemm, M.R., Paul, B.J., Schneider, T.D., Storz, G., and Rudd, K.E. (2008). Small membrane proteins found by comparative genomics and ribosome binding site models. Mol. Microbiol. *70*, 1487–1501.

Itou, J., Eguchi, Y., and Utsumi, R. (2009). Molecular mechanism of transcriptional cascade initiated by the EvgS/EvgA system in *Escherichia coli* K-12. Biosci. Biotechnol. Biochem. *73*, 870–878.

Jiang, M., Grau, R., and Perego, M. (2000). Differential processing of propeptide inhibitors of Rap phosphatases in *Bacillus subtilis*. J. Bacteriol. *182*, 303–310.

Jiang, P., and Ninfa, A.J. (1999). Regulation of autophosphorylation of *Escherichia coli* nitrogen regulator II by the PII signal transduction protein. J. Bacteriol. *181*, 1906–1911.

Johnson, M.D., Burton, N.A., Gutierrez, B., Painter, K., and Lund, P.A. (2011). RcsB is required for inducible acid resistance in *Escherichia coli* and acts at *gadE*-dependent and -independent promoters. J. Bacteriol. *193*, 3653–3656.

Kato, A., and Groisman, E.A. (2004). Connecting two-component regulatory systems by a protein that protects a response regulator from dephosphorylation by its cognate sensor. Genes Dev. *18*, 2302–2313.

Kato, A., Latifi, T., and Groisman, E.A. (2003). Closing the loop: the PmrA/PmrB two-component system negatively controls expression of its posttranscriptional activator PmrD. Proc. Natl. Acad. Sci. U.S.A. *100*, 4706–4711.

Kato, A., Mitrophanov, A.Y., and Groisman, E.A. (2007). A connector of two-component regulatory systems promotes signal amplification and persistence of expression. Proc. Natl. Acad. Sci. U.S.A. *104*, 12063–12068.

Kox, L.F., Wosten, M.M., and Groisman, E.A. (2000). A small protein that mediates the activation of a two-component system by another two-component system. EMBO J. *19*, 1861–1872.

Laub, M.T., and Goulian, M. (2007). Specificity in two-component signal transduction pathways. Annu. Rev. Genet. *41*, 121–145.

Lazazzera, B.A., Kurtser, I.G., McQuade, R.S., and Grossman, A.D. (1999). An autoregulatory circuit affecting peptide signaling in *Bacillus subtilis*. J. Bacteriol. *181*, 5193–5200.

Lippa, A.M., and Goulian, M. (2009). Feedback inhibition in the PhoQ/PhoP signaling system by a membrane peptide. PLoS Genet. *5*, e1000788.

Loewen, P.C., Hu, B., Strutinsky, J., and Sparling, R. (1998). Regulation in the *rpoS* regulon of *Escherichia coli*. Can. J. Microbiol. *44*, 707–717.

Ma, Z., Richard, H., Tucker, D.L., Conway, T., and Foster, J.W. (2002). Collaborative regulation of *Escherichia coli* glutamate-dependent acid resistance by two AraC-like regulators, GadX and GadW (YhiW). J. Bacteriol. *184*, 7001–7012.

Ma, Z., Masuda, N., and Foster, J.W. (2004). Characterization of EvgAS-YdeO-GadE branched regulatory circuit governing glutamate-dependent acid resistance in *Escherichia coli*. J. Bacteriol. *186*, 7378–7389.

Masuda, N., and Church, G.M. (2002). *Escherichia coli* gene expression responsive to levels of the response regulator EvgA. J. Bacteriol. *184*, 6225–6234.

Masuda, N., and Church, G.M. (2003). Regulatory network of acid resistance genes in *Escherichia coli*. Mol. Microbiol. *48*, 699–712.

Minagawa, S., Ogasawara, H., Kato, A., Yamamoto, K., Eguchi, Y., Oshima, T., Mori, H., Ishihama, A., and Utsumi, R. (2003). Identification and molecular characterization of the Mg^{2+} stimulon of *Escherichia coli*. J. Bacteriol. *185*, 3696–3702.

Mitrophanov, A.Y., and Groisman, E.A. (2008). Signal integration in bacterial two-component regulatory systems. Genes Dev. *22*, 2601–2611.

Mueller, J.P., Bukusoglu, G., and Sonenshein, A.L. (1992). Transcriptional regulation of *Bacillus subtilis* glucose starvation-inducible genes: control of *gsiA* by the ComP-ComA signal transduction system. J. Bacteriol. *174*, 4361–4373.

Ogasawara, H., Hasegawa, A., Kanda, E., Miki, T., Yamamoto, K., and Ishihama, A. (2007). Genomic SELEX search for target promoters under the control of the PhoQP-RstBA signal relay cascade. J. Bacteriol. *189*, 4791–4799.

Ogino, T., Matsubara, M., Kato, N., Nakamura, Y., and Mizuno, T. (1998). An *Escherichia coli* protein that exhibits phosphohistidine phosphatase activity towards the HPt domain of the ArcB sensor involved in the multistep His-Asp phosphorelay. Mol. Microbiol. *27*, 573–585.

Ogura, M., Shimane, K., Asai, K., Ogasawara, N., and Tanaka, T. (2003). Binding of response regulator DegU to the *aprE* promoter is inhibited by RapG, which is counteracted by extracellular PhrG in *Bacillus subtilis*. Mol. Microbiol. *49*, 1685–1697.

Ohlsen, K.L., Grimsley, J.K., and Hoch, J.A. (1994). Deactivation of the sporulation transcription factor Spo0A by the Spo0E protein phosphatase. Proc. Natl. Acad. Sci. U.S.A. *91*, 1756–1760.

Oshima, T., Aiba, H., Masuda, Y., Kanaya, S., Sugiura, M., Wanner, B.L., Mori, H., and Mizuno, T. (2002). Transcriptome analysis of all two-component regulatory system mutants of *Escherichia coli* K-12. Mol. Microbiol. *46*, 281–291.

Perego, M. (2001). A new family of aspartyl phosphate phosphatases targeting the sporulation transcription factor Spo0A of *Bacillus subtilis*. Mol. Microbiol. *42*, 133–143.

Perego, M., Hanstein, C., Welsh, K.M., Djavakhishvili, T., Glaser, P., and Hoch, J.A. (1994). Multiple protein-aspartate phosphatases provide a mechanism for the integration of diverse signals in the control of development in *B. subtilis*. Cell 79, 1047–1055.

Piggot, P.J., and Hilbert, D.W. (2004). Sporulation of *Bacillus subtilis*. Curr. Opin. Microbiol. *7*, 579–586.

Raghavan, V., and Groisman, E.A. (2010). Orphan and hybrid two-component system proteins in health and disease. Curr. Opin. Microbiol. *13*, 226–231.

Raivio, T.L., Popkin, D.L., and Silhavy, T.J. (1999). The Cpx envelope stress response is controlled by amplification and feedback inhibition. J. Bacteriol. *181*, 5263–5272.

Roland, K.L., Martin, L.E., Esther, C.R., and Spitznagel, J.K. (1993). Spontaneous *pmrA* mutants of *Salmonella typhimurium* LT2 define a new two-component regulatory system with a possible role in virulence. J. Bacteriol. *175*, 4154–4164.

Shimoda, N., Toyoda-Yamamoto, A., Aoki, S., and Machida, Y. (1993). Genetic evidence for an interaction between the VirA sensor protein and the ChvE sugar-binding protein of *Agrobacterium*. J. Biol. Chem. *268*, 26552–26558.

Silversmith, R.E. (2010). Auxiliary phosphatases in two-component signal transduction. Curr. Opin. Microbiol. *13*, 177–183.

Silversmith, R.E., Levin, M.D., Schilling, E., and Bourret, R.B. (2008). Kinetic characterization of catalysis by the chemotaxis phosphatase CheZ. Modulation of activity by the phosphorylated CheY substrate. J. Biol. Chem. *283*, 756–765.

Skerker, J.M., Prasol, M.S., Perchuk, B.S., Biondi, E.G., and Laub, M.T. (2005). Two-component signal transduction pathways regulating growth and cell cycle progression in a bacterium: a system-level analysis. PLoS Biol. *3*, e334.

Smits, W.K., Bongiorni, C., Veening, J.W., Hamoen, L.W., Kuipers, O.P., and Perego, M. (2007). Temporal separation of distinct differentiation pathways by a dual specificity Rap-Phr system in *Bacillus subtilis*. Mol. Microbiol. *65*, 103–120.

Soncini, F.C., Garcia Vescovi, E., Solomon, F., and Groisman, E.A. (1996). Molecular basis of the magnesium deprivation response in *Salmonella typhimurium*: identification of PhoP-regulated genes. J. Bacteriol. *178*, 5092–5099.

Steyn, A.J., Joseph, J., and Bloom, B.R. (2003). Interaction of the sensor module of *Mycobacterium tuberculosis* H37Rv KdpD with members of the Lpr family. Mol. Microbiol. *47*, 1075–1089.

Szurmant, H., and Hoch, J.A. (2010). Interaction fidelity in two-component signaling. Curr. Opin. Microbiol. *13*, 190–197.

Szurmant, H., Nelson, K., Kim, E.J., Perego, M., and Hoch, J.A. (2005). YycH regulates the activity of the essential YycFG two-component system in *Bacillus subtilis*. J. Bacteriol. *187*, 5419–5426.

Szurmant, H., Mohan, M.A., Imus, P.M., and Hoch, J.A. (2007). YycH and YycI interact to regulate the essential YycFG two-component system in *Bacillus subtilis*. J. Bacteriol. *189*, 3280–3289.

Szurmant, H., Bu, L., Brooks, C.L., 3rd, and Hoch, J.A. (2008). An essential sensor histidine kinase controlled by transmembrane helix interactions with its auxiliary proteins. Proc. Natl. Acad. Sci. U.S.A. *105*, 5891–5896.

Takeda, S., Fujisawa, Y., Matsubara, M., Aiba, H., and Mizuno, T. (2001). A novel feature of the multistep phosphorelay in *Escherichia coli*: a revised model of the RcsC -- > YojN -- > RcsB signalling pathway implicated in capsular synthesis and swarming behaviour. Mol. Microbiol. *40*, 440–450.

Tschowri, N., Busse, S., and Hengge, R. (2009). The BLUF-EAL protein YcgF acts as a direct anti-repressor in a blue-light response of *Escherichia coli*. Genes Dev. *23*, 522–534.

Tu, X., Latifi, T., Bougdour, A., Gottesman, S., and Groisman, E.A. (2006). The PhoP/PhoQ two-component system stabilizes the alternative sigma factor RpoS in *Salmonella enterica*. Proc. Natl. Acad. Sci. U.S.A. *103*, 13503–13508.

Ulrich, L.E., and Zhulin, I.B. (2010). The MiST2 database: a comprehensive genomics resource on microbial signal transduction. Nucleic Acids Res. *38*, D401–D407.

Utsumi, R. (2008). Bacterial signal transduction: networks and drug targets. Preface. Adv. Exp. Med. Biol. *631*, v.

Venkatesh, G.R., Kembou Koungni, F.C., Paukner, A., Stratmann, T., Blissenbach, B., and Schnetz, K. (2010). BglJ–RcsB heterodimers relieve repression of the *Escherichia coli bgl* operon by H-NS. J. Bacteriol. *192*, 6456–6464.

Virlogeux, I., Waxin, H., Ecobichon, C., Lee, J.O., and Popoff, M.Y. (1996). Characterization of the *rcsA* and *rcsB* genes from *Salmonella typhi*: *rcsB* through *tviA* is involved in regulation of Vi antigen synthesis. J. Bacteriol. *178*, 1691–1698.

Wang, L., Grau, R., Perego, M., and Hoch, J.A. (1997). A novel histidine kinase inhibitor regulating development in *Bacillus subtilis*. Genes Dev. *11*, 2569–2579.

Wehland, M., and Bernhard, F. (2000). The RcsAB box. Characterization of a new operator essential for the regulation of exopolysaccharide biosynthesis in enteric bacteria. J. Biol. Chem. *275*, 7013–7020.

Winter, S.E., Winter, M.G., Thiennimitr, P., Gerriets, V.A., Nuccio, S.P., Russmann, H., and Baumler, A.J. (2009). The TviA auxiliary protein renders the *Salmonella enterica* serotype Typhi RcsB regulon responsive to changes in osmolarity. Mol. Microbiol. *74*, 175–193.

Wosten, M.M., Kox, L.F., Chamnongpol, S., Soncini, F.C., and Groisman, E.A. (2000). A signal transduction system that responds to extracellular iron. Cell *103*, 113–125.

Yamamoto, K., Hirao, K., Oshima, T., Aiba, H., Utsumi, R., and Ishihama, A. (2005). Functional characterization *in vitro* of all two-component signal transduction systems from *Escherichia coli*. J. Biol. Chem. *280*, 1448–1456.

Zwir, I., Shin, D., Kato, A., Nishino, K., Latifi, T., Solomon, F., Hare, J.M., Huang, H., and Groisman, E.A. (2005). Dissecting the PhoP regulatory network of *Escherichia coli* and *Salmonella enterica*. Proc. Natl. Acad. Sci. U.S.A. *102*, 2862–2867.

Antikinases: Their Structures and Roles in Two-component Signalling

9

David A. Jacques, J. Mitchell Guss and Jill Trewhella

Abstract

Antikinases are protein inhibitors of the bacterial sensor histidine kinases found in two-component systems. Antikinases act by binding the highly conserved autokinase domains and thereby inhibit autophosphorylation. Antikinases offer unique opportunities to understand histidine kinase function and inhibition, and as histidine kinases are not found in mammals, knowledge of their structures and binding can potentially be exploited for the development of novel antimicrobial compounds. This chapter describes our current understanding of the three known antikinases: Sda and KipI from *Bacillus subtilis* and FixT from *Sinorhizobium meliloti*, with a focus on understanding of their detailed structures and mechanisms of action.

Introduction

Two-component systems and histidine kinases

Two-component systems are the primary mechanism by which bacteria sense and respond to environmental conditions. The spectrum of stimuli and the range of responses coordinated by two-component systems are as broad as the environmental niches to which bacteria have adapted. Despite this diversity, two-component systems are characterized by, as the name suggests, the presence of two protein components: a sensor histidine kinase responsible for stimulus perception and subsequent autophosphorylation; and a response regulator which receives a phosphoryl group from an activated histidine kinase and effects the desired cellular response that is typically initiated by altered patterns of gene expression.

As the primary environmental sensors in bacteria, the structure and function of histidine kinases have been the subject of much research in the last two decades. In general terms, histidine kinases may be characterized by two functional domains: a variable N-terminal sensor domain, and a conserved C-terminal autokinase domain.

The sensor domains are capable of recognizing diverse stimuli including amino acids, carbohydrates, divalent metals, nitrate, autoinducers, iron, oxygen, light, and temperature. The specific stimulus may be extracellular or intracellular and thus sensor domains may be either membrane-bound or cytoplasmic. With such a diversity of potential stimuli, sensor domains tend to be highly variable in their structure and frequently consist of multiple tandem domains. A thorough review of known histidine kinase sensor domain structures, their modular architecture, and their mechanisms of stimulus perception is given by Mascher and colleagues (Mascher *et al.*, 2006).

The autokinase domains (also known as the transmitter domains) carry the phosphoaccepting histidine and are responsible for the autophosphorylation reaction, as well as interaction with the cognate response regulator during the phosphotransfer reaction (Stewart, 2010). In contrast to the variable architecture of sensor domains, autokinase domains are highly conserved and are always cytoplasmic. Indeed, in genomic analyses, the presence of an autokinase domain is used to define a protein as a histidine kinase (Grebe and Stock, 1999; Kim and Forst, 2001). Autokinase domains contain two structural domains: a dimerization-histidine phosphotransfer (DHp) domain, and a catalytic ATP-binding (CA) domain (Fig. 9.1). DHp domains consist of two antiparallel α-helices that dimerize to form a four-helix bundle. The DHp domain also carries the phospho-accepting histidine, which is positioned on a solvent exposed face of the N-terminal helix. The CA domain usually catalyses the autophosphorylation reaction by binding ATP and transferring the γ-phosphate to the phospho-accepting histidine on the DHp domain of the partner molecule in the dimer; that is, the autophosphorylation occurs in *trans* (Fig. 9.1) (Krell *et al.*, 2010); although recently an example of autophosphorylation occurring in *cis* has also been described (Casino *et al.*, 2009). In the few available autokinase domain crystal structures (Marina *et al.*, 2005; Bick *et al.*, 2009) the ATP-binding site is distant from the phospho-accepting histidine. The autophosphorylation reaction therefore requires considerable relative motion of the CA and DHp domains which is believed to be facilitated by a flexible linker between them. This inherent flexibility has been used to rationalize why complete autokinase domains have proven difficult to crystallize, and hence difficult to characterize structurally.

Histidine kinases as drug targets

Protein phosphorylation as a form of signal transduction is a theme prevalent across all biology. In animals, for instance, many signals involve serine/threonine kinases or tyrosine kinases. Inhibitors of such kinases have been successfully employed for the treatment of conditions involving aberrant signal propagation, such as cancer and inflammation (Deremer *et al.*, 2008; Halin *et al.*, 2008). However, animals do not possess histidine kinases; and, as such, histidine kinase inhibitors have been postulated to have potential to guide the development of novel therapeutic antimicrobial compounds (Rowland and King, 2007; Gotoh *et al.*, 2010). As the autokinase domain is the conserved feature of all bacterial histidine kinases, inhibition of the autophosphorylation reaction potentially represents a broad-spectrum strategy to disrupt bacterial signalling pathways.

Research into the development of small-molecule histidine kinase inhibitors has met with mixed success (Matsushita and Janda, 2002). A small number of compounds are capable of targeting specific kinases. An example of this class of compound are the walkmycins which inhibit the WalK/WalR (formerly YycG/YycF) two-component system responsible for regulating cell wall metabolism in the *Firmicutes* (Watanabe et al., 2003). The development of more broad-spectrum compounds has generally focussed on competitive inhibitors for the autophosphorylation reaction. The acid lability of phosphohistidine has limited its use for the development of substrate mimics, while other compounds, such as TEP (Gilmour *et al.*, 2005) and TNP-ATP (Theodorou *et al.*, 2011) have been found to be competitive inhibitors with

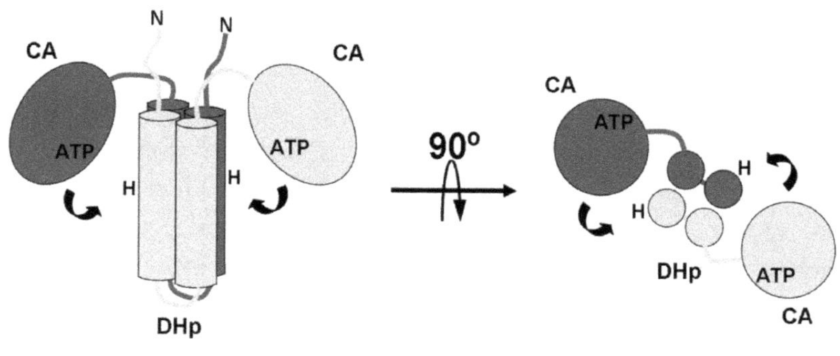

Figure 9.1 Autokinase structure and postulated mechanism. Cytoplasmic autokinases are dimers. Their DHp domains carry the phospho-accepting histidine and dimerize to form a central helical bundle with a pair of CA domains responsible for binding ATP and carrying out the autophosphorylation reaction. The CA domains in the dimer are required to undergo significant motion in order to phosphorylate the target histidine. Autophosphorylation generally, though there is at least one exception, occurs in *trans* (indicated by arrows).

respect to ATP. Recently the fluorescence of TNP-ATP has been exploited to develop a displacement assay for the high-throughput of screening of potential histidine kinase inhibitors (Guarnieri *et al.*, 2011). Despite these promising leads, no histidine kinase inhibitors are yet employed for the treatment of bacterial infection.

The discovery of the first antikinase protein, KipI from *Bacillus subtilis*, by Wang and colleagues in 1997 opened up the possibility of understanding naturally occurring histidine kinase inhibition and its possible exploitation for therapeutic benefit (Wang *et al.*, 1997). Since that time, two additional yet distinct antikinases, FixT from *Sinorhizobium meliloti* (Foussard *et al.*, 1997) and Sda from *B. subtilis* (Burkholder *et al.*, 2001), have also been described.

The sporulation phosphorelay in *B. subtilis*

To date, the most thoroughly characterized system known to employ antikinases is the sporulation phosphorelay found in *B. subtilis*. When essential nutrients are depleted in the environment, *B. subtilis* cells are capable of forming specialized 'resting' cells called endospores (or simply 'spores'). The two-component system that governs the decision to sporulate is the sporulation phosphorelay, the first 'four-component' system to be discovered (Burbulys *et al.*, 1991). Like other members of the two-component signalling family, phosphorelays are triggered by the activation of a sensor histidine kinase, with the phosphorylation of the response regulator eliciting the cellular response. However, unlike archetypal two-component systems, phosphorelays include additional phosphotransfer proteins which act as a conduit for the phosphate signal between the histidine kinase and response regulator. It is generally accepted that the expansion of the typical two-component system into multicomponent phosphorelays allows for the integration of multiple signals to prevent inappropriate triggering of complex, and metabolically expensive, physiological processes (Grebe and Stock, 1999; Buelow and Raivio, 2010).

The sporulation phosphorelay may be activated by one of five histidine kinases, named KinA, KinB, KinC, KinD, and KinE. The multiplicity of sensor kinases allows for the input of different signals in order to elicit various downstream responses (discussed below). Additionally, the genes encoding the histidine kinases are independently regulated, which may enable the cell to modulate its sensitivity to independent stimuli. In each case, once the histidine kinase is phosphorylated, the phosphoryl group is transferred to an aspartic acid residue on the first phosphotransfer protein, Spo0F. The phosphoryl group is then transferred to a histidine residue on the second phosphotransfer protein, Spo0B, and finally to an aspartic acid on the response regulator, Spo0A (Fig. 9.2).

While the specific environmental triggers for each of the five histidine kinases are not known, single knockout mutations of *kinA* or *kinB* result in a sporulation frequency that is approximately 10% of that of the wild-type (with some variation depending on the sporulation medium used) (LeDeaux *et al.*, 1995). When both *kinA* and *kinB* are knocked out, sporulation frequency is reduced by six to eight orders of magnitude. KinA and KinB are therefore known as the major sporulation kinases. The residual sporulation frequency is essentially abolished when *kinC* is knocked out as well, suggesting that it plays a minor role in sporulation. While KinC, KinD, and KinE appear to have little to do with spore formation, their respective gene knockouts interfere with biofilm development (Hamon and Lazazzera, 2001). It is now understood that Spo0A is one of the master regulators of cell differentiation, responsible for biofilm development and cannibalism (when low amounts of Spo0A~P are present); and sporulation (when high amounts of Spo0A~P are present) (Lopez and Kolter, 2009; Lopez *et al.*, 2009). As such, the phosphorelay allows for the integration of multiple signals controlling the level of Spo0A phosphorylation and hence the decision to form biofilms or sporulate (Sonenshein, 2000). Input signals may affect the outcome via regulation of the expression of phosphorelay proteins, inactivation of the phosphorelay by aspartic acid phosphatases (Rap phosphatases and Spo0E), or regulation of histidine kinase activity by stimuli and antikinases (Fig. 9.2).

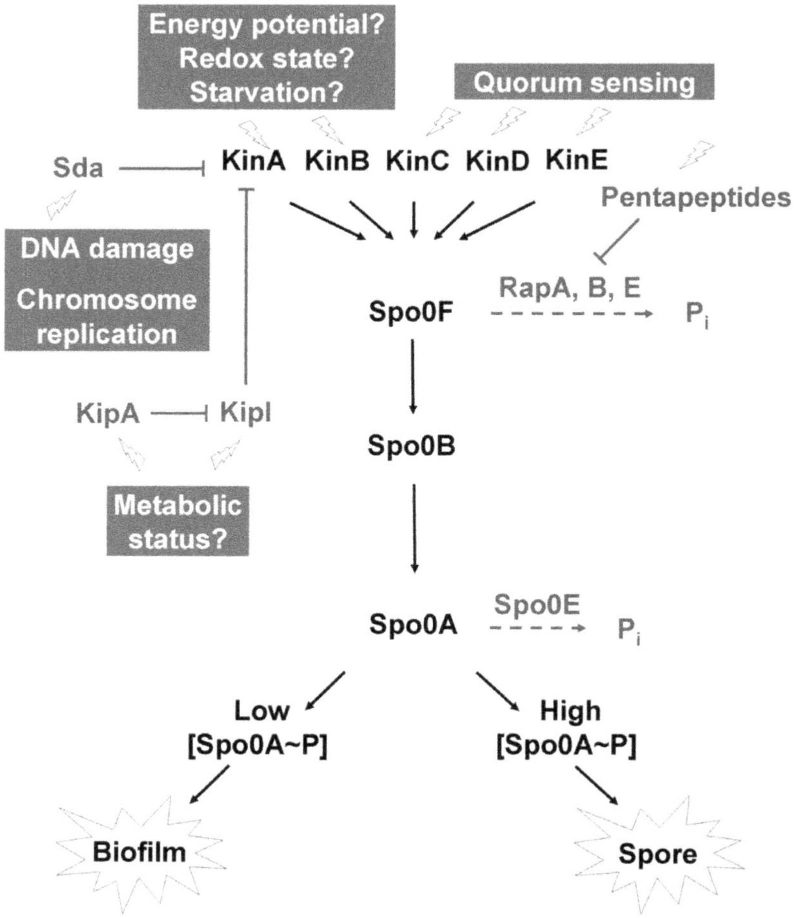

Figure 9.2 The sporulation phosphorelay in *B. subtilis*. The core phosphorelay is shown in black, regulators in grey, and stimuli in boxes. Transcriptional regulators have been omitted for clarity. Phosphate enters the phosphorelay by autophosphorylation of one of the five sensor histidine kinases (KinA-E). The phosphoryl group is transferred via Spo0F and Spo0B to Spo0A which binds DNA and affects patterns of gene expression that result in either biofilm formation or sporulation. Phosphate can be removed from the phosphorelay by the action of phosphatases (RapA, B, E, or Spo0E). The primary sporulation kinase, KinA, is inhibited by Sda (in response to DNA damage, or failure to initiate chromosome replication) or KipI (in response to an unknown metabolic signal). The antikinase activity of KipI is itself inhibited by the KipA protein. Many of the details concerning the specific stimuli that regulate the phosphorelay remain unknown.

The sporulation histidine kinase KinA

KinA (originally designated SpoIIJ (Antoniewski *et al.*, 1990) is a cytoplasmic protein and, being soluble, has become the most studied of the sporulation histidine kinases. KinA is a 606 residue protein that, like other histidine kinases, functions as a dimer (Wang *et al.*, 2001). The sensor apparatus, residues 1–382, comprises three PAS domains (labelled from the N-terminus as PAS-A, PAS-B, and PAS-C). The autokinase domain, residues 383–606, features the usual DHp and CA domains (Fig. 9.3).

Before the structures of any histidine kinase domains were known, sequence analyses identified regions of high sequence conservation within autokinase domains. One of these so-called 'homology boxes', the H-box, is the sequence flanking the conserved phospho-accepting histidine in the DHp domain (Parkinson and Kofoid, 1992). Homology boxes have been used to compare members of the histidine kinase superfamily and classify them into subfamilies (Grebe and Stock, 1999; Kim and Forst, 2001). By the definition of Grebe and Stock, KinA is a member of subfamily 4. Subfamilies 1–4 defined by Grebe

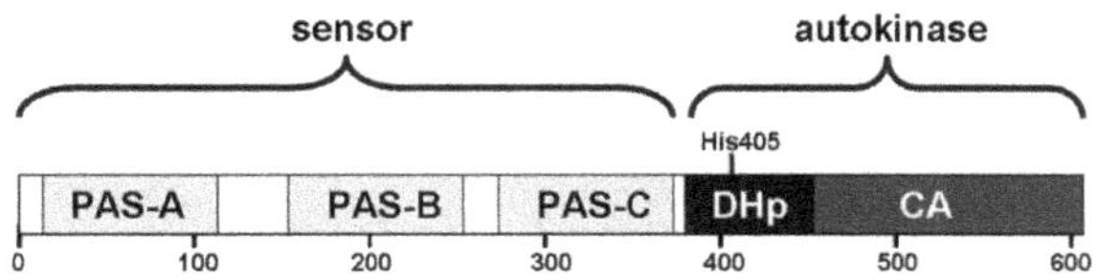

Figure 9.3 KinA domain architecture. KinA is a five domain protein with three tandem PAS domains forming the sensor, and DHp and CA domains forming the autokinase. The phospho-accepting histidine, His405, is located within the DHp domain.

and Stock, were grouped together as subfamily 1 by Kim and Forst. By either definition, the H-box is highly conserved among members of these subfamilies and contains a near invariant proline residue at position 5 after the phospho-accepting histidine. While the role of this proline is not yet fully understood, in the crystal structure of HK853 from *Thermotoga maritima* (the first structure of a complete autokinase domain) it causes a bend in the N-terminal DHp helix, resulting in a tightly packed four-helix bundle (Marina *et al.*, 2005). In contrast, the structurally related but catalytically inert, Spo0B, does not possess a proline at this site (Varughese *et al.*, 1998), and its four-helix bundle is less tightly packed (Fig. 9.4).

Despite its popularity as a model system, much about KinA remains elusive. The precise nature of kinase effector molecule or molecules is unknown,

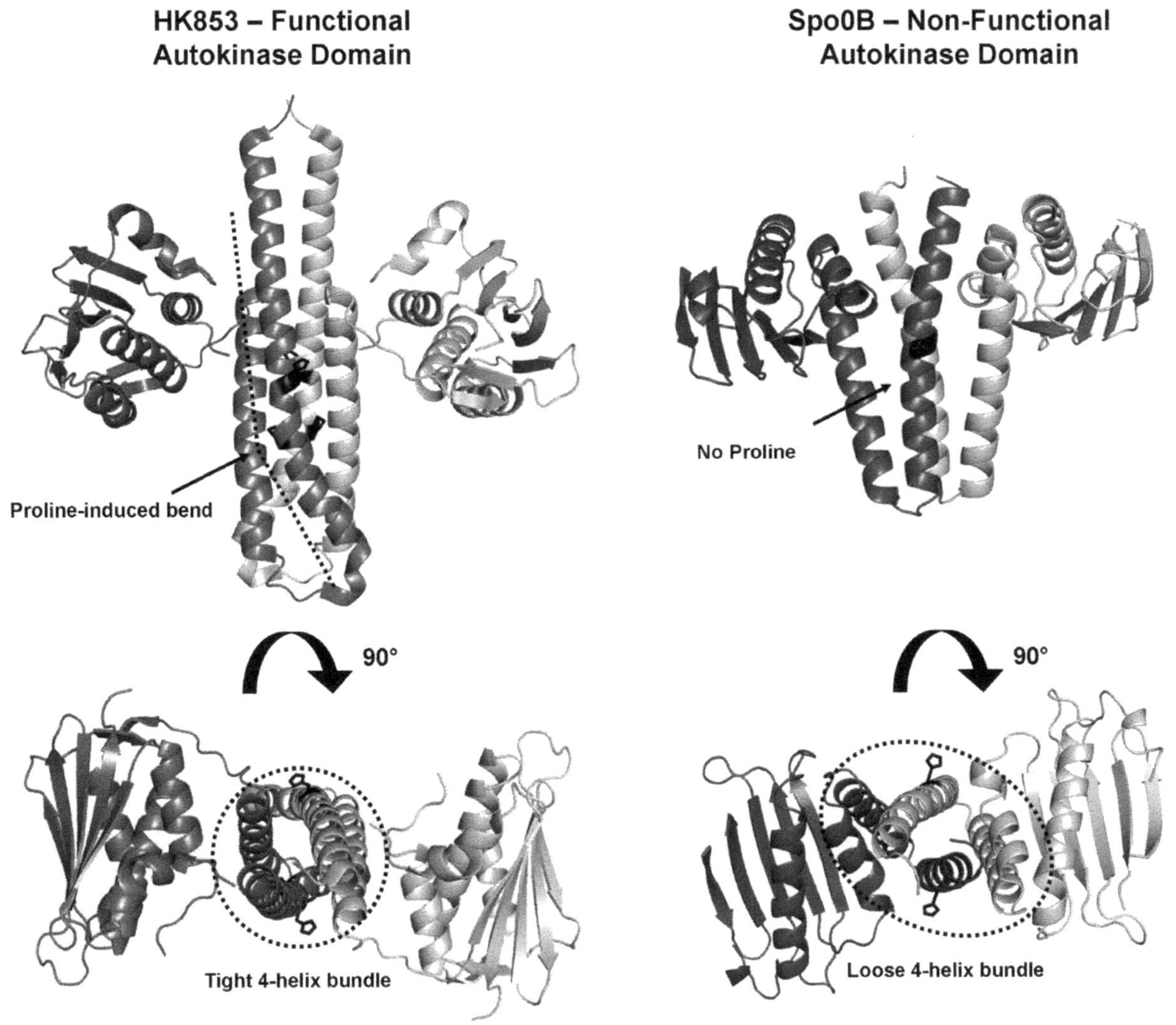

Figure 9.4 The H-box proline. The structure of HK853 from *T. maritima* shows that the conserved H-box proline occurs five residues after the phospho-accepting histidine (both residues shown as black sticks), causing a bend in the first helix of the DHp domain. The catalytically inert Spo0B does not carry a proline in this position. Consequently, the first helix is more straight, and the DHp domain is more loosely packed than that of HK853.

although the kinase is active without the addition of any putative stimuli (Perego *et al.*, 1989), raising the possibility that the sensor apparatus may respond to a stimulus by inhibiting autokinase activity. Also, very little high-resolution structural information is available, with only the structure of PAS-A having been solved by crystallography (Lee *et al.*, 2008).

The antikinase Sda

In response to defects in the initiation of DNA replication or DNA damage, *B. subtilis* cells express the small, 46-residues protein Sda, which inhibits KinA and KinB, and thereby spore formation. Mutations in the *sda* gene were originally identified as suppressors of the sporulation-incompetent phenotype of a *dnaA* strain, hence its name Sda (Burkholder *et al.*, 2001). Recently, it has also been shown that Sda is produced in bursts at the beginning of cell division where it also functions as a molecular timer ensuring that a cell enters the sporulation pathway when it contains the correct chromosome copy number and has completed DNA replication and repair (Veening *et al.*, 2009). Sda is unstable *in vivo* as it is rapidly proteolysed, consistent with its role in temporal control of sporulation regulation (Ruvolo *et al.*, 2006).

The structure of Sda has been solved by solution NMR and crystallography (Burkholder *et al.*, 2001; Jacques *et al.*, 2009) revealing a simple helix-turn helix structure for this smallest of the three known antikinases (Fig. 9.5). The KinA–Sda interaction was of particular interest as the small size of Sda might be exploited to investigate the minimal requirement for histidine kinase inhibition by a protein. Early binding studies showed that Sda binds the autokinase domain of KinA and inhibits autophosphorylation rather than preventing phosphotransfer or activating a cryptic phosphatase activity within KinA (Rowland *et al.*, 2004). Phosphorylated KinA is able to transfer the phosphate to Spo0F in the presence of bound Sda, an observation that led to the conclusion that Sda and Spo0F binding sites on KinA must be mutually exclusive (Rowland *et al.*, 2004). The co-crystal structure of Spo0F bound to Spo0B (which is structurally homologous to the autokinase domain of KinA, but is catalytically inert) revealed that Spo0F interacts with the DHp domain (Zapf *et al.*, 2000). In combination, these observations led to the proposal that Sda must bind the flexible linkers between the DHp and CA domains of KinA, with Sda either stabilizing the linker in an extended conformation or forming a steric barrier preventing ATP reaching the phospho-accepting histidine. While early mutatagenesis experiments were able to identify a cluster of aromatic residues on one face of Sda (Fig. 9.5) as responsible for mediating the interaction with KinA, the reciprocal residues on KinA remained elusive.

The structural nature of the KinA–Sda interaction was initially revealed by using small-angle solution scattering (Whitten *et al.*, 2007). The small-angle scattering of X-rays or neutrons can be used to reconstruct a low-resolution shape model of a macromolecule in solution. Using rigid-body modelling one also can test atomic models against the scattering data; e.g. models that incorporate known individual domain or subunit structures along with additional experimentally derived structural constraints. Neutron scattering can provide information concerning the shapes and

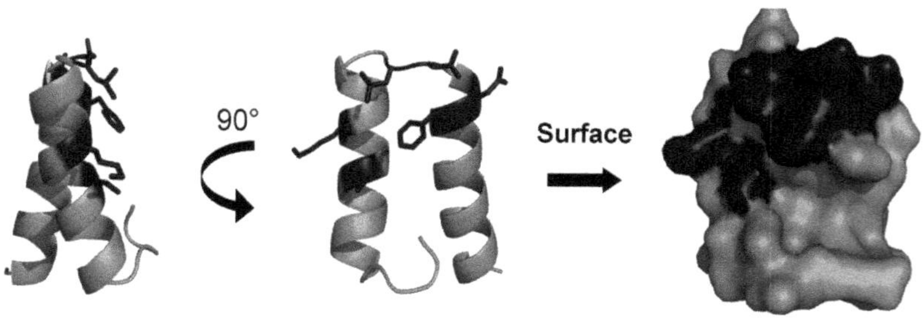

Figure 9.5 Crystal structure of Sda. Those residues identified by mutagenesis as binding to KinA (Ser12, Lys15, Leu21, Asn22, Asp24, and Phe25) are coloured black and form a near contiguous surface across one face of the molecule.

relative dispositions of individual proteins within complexes (Jacques and Trewhella, 2010). This capability arises from the different neutron scattering properties of hydrogen and deuterium which can be exploited to systematically change contributions from the deuterated and non-deuterated components to the scattering signal. The technique is referred to as contrast variation and requires the acquisition of a contrast series; a set of small-angle neutron scattering measurements from a complex formed with non-deuterated and deuterated protein components in solvents with varying D_2O contents. In the case of the KinA–Sda complex, Sda was produced in deuterated form (by expression in D_2O) and neutron scattering data for the complex were acquired in seven different D_2O-containing buffers. Using a combination of the then available NMR structure of Sda and the crystal structure of the KinA homologue, HK853 from *T. maritima* (Marina *et al.*, 2005), the KinA–Sda complex was modelled by rigid body refinement against the combined neutron contrast variation data and complementary small-angle X-ray scattering data. This model, in combination with overall shape parameters derived from the scattering data, proved that two molecules of Sda bind on opposite sides of the base of the DHp domain of KinA (Fig. 9.6). Additional confirmation as to the Sda-binding site subsequently came from the crystal structure of the KinB–Sda complex from *Geobacillus stearothermophilus* (Bick *et al.*, 2009). In this structure, only a single Sda molecule is bound to an autokinase dimer, though the authors demonstrate this to be a crystallization artefact and that two Sda molecules are required for autokinase inhibition. The location of the Sda binding site in the crystal structure is the same as that predicted by the model derived from small-angle scattering, with Sda rotated 180° – an observation beyond the resolving power of the small-angle scattering experiment. This crystal structure informed mutagenesis studies which identified KinB residues Gly224, F225, and L228 as critical for Sda inhibition. The proposed GFXXL motif, which is found C-terminal from the H-box, is conserved in all the sporulation histidine kinases, however Sda inhibition has yet to be experimentally verified for KinC, KinD and KinE. Bick and colleagues also were able to show that while HK853 bore high structural homology to their structure, it was unaffected by Sda. However, a triple mutant of HK853 in which the GFXXL motif was incorporated was inhibited by Sda, demonstrating that a histidine kinase can be engineered to become sensitive to the action of an antikinase.

Comparison of models tested against small-angle scattering data acquired from the KinA-Sda with that from KinA alone also revealed that the autokinase domain undergoes a compaction upon Sda-binding. The models suggest that Sda binding results in the CA domains collapsing towards the central DHp four-helix bundle. As solution scattering data represent a time and ensemble average, this effect might be due to a shift in the dynamic range of an inherently flexible CA domain positioning, facilitated by its flexible linkage to DHp. The binding of Sda may either cause the CA domains to be rigidly positioned, or simply constrain the range of motion, or shift the dynamic equilibrium towards the more collapsed state.

The observed Sda binding site on KinA and KinB apparently conflicts with the assumption that Sda and Spo0F can bind KinA simultaneously. The predicted binding site is supported by further mutation and protein footprinting studies, which identified specific residues at the base of the DHp domain involved in the Sda binding site (Cunningham and Burkholder, 2009). Furthermore, investigations into the kinetics of phosphotransfer between KinA and Spo0F showed that while Sda did not abolish the phosphotransfer reaction, the rate of phosphortransfer was significantly reduced (Bick *et al.*, 2009; Cunningham and Burkholder, 2009). These studies confirmed that the Sda and Spo0F binding sites were not mutually exclusive, and that the efficacy of the antikinase is not solely due to inhibition of histidine kinase autophosphorylation, but is also due to a reduction in the rate of phosphotransfer to Spo0F.

The antikinase KipI

The first histidine kinase inhibitor to be discovered was KipI (kinase interacting protein – inhibitor) (Wang *et al.*, 1997). KipI was found to inhibit the autophosphorylation reaction of KinA (but not KinB) and, like Sda, permitted phosphotransfer to Spo0F. While both KipI and Sda act as antikinases,

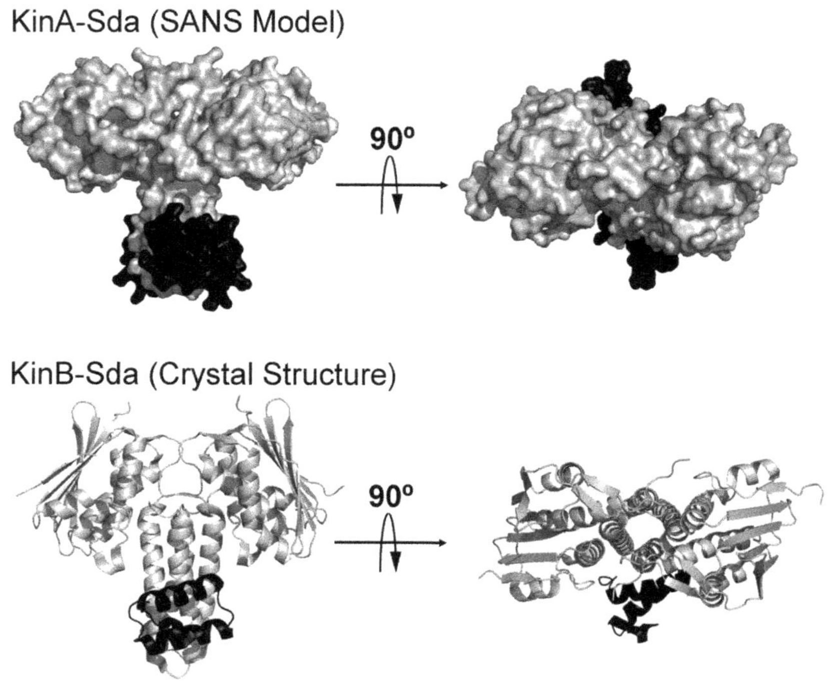

Figure 9.6 Binding of Sda to autokinase domains. Sda (black) inhibits the autokinase reaction by binding the base of the DHp domain within the histidine kinase (grey). This positioning was shown independently by small-angle scattering from the KinA–Sda complex from *B. subtilis* (top), and the co-crystal structure of Sda with KinB from *G. stearothermophilus* (bottom).

they are structurally unrelated with KipI being a two-domain protein of 240 amino acid residues homologous to the protein of unknown function PH0987 from *Pyrococcus horikoshii* (PDB entry 2PHC, 47% sequence identity, 64% similarity). The structure of an intact KipI has yet to be solved, however by using the crystal structure of PH0987 as a guide, it was possible to express and study the individual domains of KipI.

Initial small-angle X-ray scattering data showed that KipI forms an elongated dimer in solution. Further solution scattering studies revealed that the N-terminal domain (residues 1–99) is also dimeric, while the C-terminal domain (residues 100–240) is a monomer (Jacques *et al.*, 2008). The structure of the N-terminal domain was subsequently solved by NMR (PDB entry 2KWA) (Hynson *et al.*, 2010; Jacques *et al.*, 2011a) and the C-terminal domain by crystallography (PDB entry 2zp2) (Jacques *et al.*, 2008). These structures could then be used in conjunction with the scattering data to model the complete KipI dimer (Fig. 9.7). Comparison of these domains with structures of known function revealed that the N-terminal domain of KipI shares structural homology with the LuxS family, a protein responsible for the production of autoinducer 2 (AI-2), a molecule used by bacteria in inter-cellular communication during biofilm development (Lombardia *et al.*, 2006). On the other hand, the C-terminal domain is a member of the cyclophilin family. Cyclophilins are found across all species and are responsible for catalysing proline *cis/trans* isomerization (Wang *et al.*, 1997). Peptidyl-prolyl isomerases often function as chaperones, but may also mediate specific protein–protein interactions by recognizing target sequences (Huse *et al.*, 1999). In the case of the C-terminal domain of KipI, a cleft formed by a five-stranded β-sheet contained a cluster of highly conserved aromatic residues. In cyclophilins, a similar hydrophobic cleft in this position is responsible for catalysing the peptidyl-prolyl *cis/trans* isomerization, however the exact mechanism remains controversial. It has even been suggested that these enzymes need only display a hydrophobic cavity of the correct conformation to confer activity (Ikura and Ito, 2007). However, despite the structural similarity, no peptidyl-prolyl isomerase activity has been observed for KipI (Wang *et al.*, 1997).

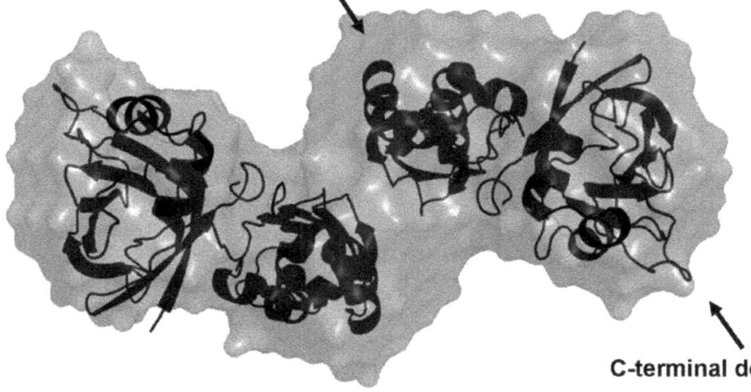

Figure 9.7 Structure of KipI. Small-angle X-ray scattering revealed KipI to be a dimer (envelope shown in grey) and that the N-terminal domain was responsible for the dimerization. Atomic models for the N-terminal and C-terminal domains were obtained by NMR and crystallography, respectively. These models (black) were used with rigid-body refinement to fit the scattering data and model the KipI dimer. The N-terminal domain shares structural homology with LuxS, a protein involved in quorum sensing, while the C-terminal domain is a member of the cyclophilin family, typically responsible for catalysing peptidyl-prolyl *cis/trans* isomerization.

In order to study the structure of the KinA–KipI complex, a strategy similar to that used for the KinA–Sda complex was employed. In this case, KipI was deuterated and the complex formed with non-deuterated KinA (autokinase domain only). Small-angle neutron scattering with contrast variation revealed that, like Sda, two molecules of KipI bind to opposite sides of the KinA DHp domain (Jacques *et al.*, 2008). Similar to the case with Sda, a significant compaction of KinA was observed when bound by KipI. Owing to the similar sizes of the N- and C-terminal domains of KipI, the rigid body modelling against the scattering data could not unambiguously assign one domain as the DHp-interacting domain using the scattering data alone. While full-length KipI is capable of binding KinA, we could not identify conditions to observe binding for either of the individual domains in isolation. This observation implies that both domains are required for the interaction, either because both interact weakly with KinA to create an avidity effect (which is inconsistent with the scattering models, which placed only one domain in contact with KinA), or that one domain modulates the affinity of the other, perhaps in response to an additional signal. Fortunately, the KinA–KipI complex itself provided a simple means of identifying which domain of KipI contacts KinA. The KinA autokinase domain has no tryptophan residues, while KipI has only one (W195) that is found in the conserved aromatic cleft on its C-terminal domain. By monitoring the fluorescence during a titration with KinA, an increase in fluorescence is observed which plateaus when KinA is in excess. This change in fluorescence is a result of the change in the solvation environment of the tryptophan upon complex formation, and demonstrates that it is the cleft in the C-terminal domain of KipI that interacts with KinA. This contact information was incorporated into the modelling of the small-angle neutron scattering data (Fig. 9.8). Additionally, Cunningham and Burkholder (2009) identified a KinA double mutant, P410L/F436S, which retains full autokinase activity but is not inhibited by the action of Sda. P410 is the conserved H-box proline in KinA and binding assays demonstrated that the P410L/F436S double mutant did not bind KipI (Jacques *et al.*, 2008). Given that the C-terminal domain of KipI bears similarity with proteins involved in proline binding, and that the tryptophan fluorescence implicated this domain with binding KinA, the absence of binding provided further evidence that it is indeed the C-terminal domain of KipI that binds the base of the DHp domain of KinA. It is interesting to note here that the P410L/F436S double mutant retained its ability to bind Sda. This result is intriguing in the light of these

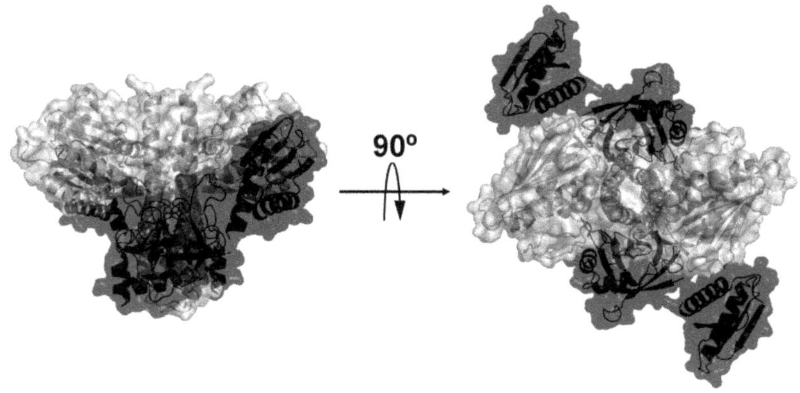

Figure 9.8 Binding of KipI to the autokinase domain of KinA. Small-angle solution scattering revealed that two molecules of KipI (black) bind the base of the DHp four-helix bundle of KinA (grey), in a location that overlaps the Sda binding site. Tryptophan fluorescence and binding studies of a mutant KinA confirm that the C-terminal domain of KipI participates in the interaction with the KinA DHp domain.

mutations abolishing the inhibitory effects of Sda. A possible explanation for this conundrum is that the P410L/F436S double mutant does not interfere with the KinA–Sda interaction, but rather is unable to transmit the inhibitory signal that normally results from Sda binding. This idea would be consistent with an allosteric model of histidine kinase inhibition involving signals transmitted via the four helical bundle formed by the DHp domains (Whitten *et al.*, 2007).

The antikinase activity of KipI is counteracted by a 337-residue binding partner, KipA (kinase interacting protein – activator, also known as an anti-antikinase) (Wang *et al.*, 1997). *In vitro* studies showed that, while KinA autokinase activity was lost in the presence of KipI, it could be restored by the addition of KipA. Similarly, while overexpression of KipI in *B. subtilis* reduced sporulation frequency, overexpression of both KipI and KipA recovered the sporulation frequency to near wild-type values. The *kipI* and *kipA* genes are the fourth and fifth, respectively, on a seven-gene operon of unknown function (referred to as the '*kip* operon', Fig. 9.9). The first gene on the *kip* operon, *ycsF*, codes for a protein that is homologous to LamB, which is responsible for the utilization of lactams as nitrogen sources in the fungus *Aspergillus nidulans* (Richardson *et al.*, 1992). The second and third genes, *ycsG* and *ycsI*, both code for membrane proteins. YcsG is similar to bacterial permeases, such as branched chain amino acid transporters, while YcsI is of unknown function. The sixth gene, *kipR*, encodes a protein involved in regulating the *kip* operon; and the seventh gene, *lipC*, encodes a phospholipase that is located in the spore coat and is involved in spore membrane development and germination (Masayama *et al.*, 2007, 2010). The *kip* operon is up-regulated under conditions of high glucose where only poor nitrogen sources (such as glutamate and allantoin) are available, suggesting that KipI and KipA are involved in responding to the metabolic status of the cell. In some strains of *B. subtilis* and other *Bacillus* species, the *kipI* and *kipA* genes are arranged such that the start codon of *kipA* occurs before the stop codon of *kipI* (Fig. 9.9). Such gene overlap in the absence of a Shine–Dalgarno sequence is consistent with translational reinitiation, a mechanism frequently used by bacteria to ensure equimolar translation from polycistronic mRNAs (Schumperli *et al.*, 1982). Searches for homologues across species reveal that KipI and KipA are found almost ubiquitously across the bacterial kingdom and are not limited to sporulating organisms. Further, genes homologous to *kipI* and *kipA* are almost always found adjacent to each other (although the order is not conserved) and are frequently fused to produce a single polypeptide. These observations strongly support the assumption that KipI and KipA are produced so that they may form a complex of equimolar stoichiometry, consistent with our observations *in vitro*.

Homology searches also revealed that KipI and

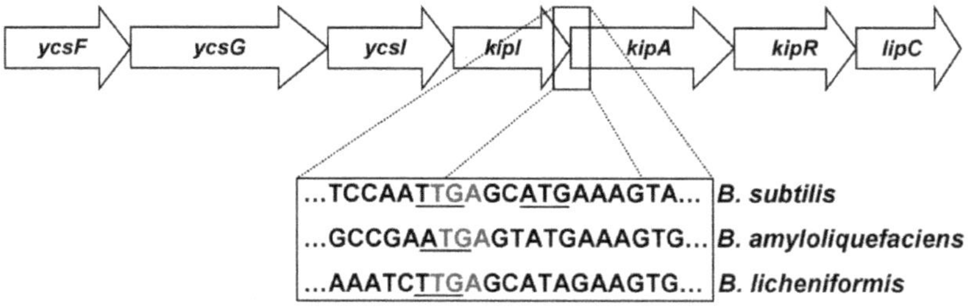

Figure 9.9 The *kip* operon. KipI and KipA are encoded on an operon of seven genes. The sequence about the KipI stop codon (grey) and the KipA start codon (underlined) is shown from three *Bacillus* species. The genes are arranged either overlapping, or within two nucleotides (two KipA start codons are underlined for *B. subtilis*, as subspecies have been annotated differently). In either case, there is no Shine–Dalgarno sequence and the stop and start codons are within ten nucleotides of each other, indicative of translational reinitiation.

KipA have been annotated as subunits 1 and 2, respectively, of allophanate hydrolase, a relatively poorly characterized enzyme involved in nitrogen metabolism. This annotation, while proven to be incorrect (Jacques *et al.*, 2011a,b), was based on studies of urea amidolyase from *Saccharomyces cerevisiae*. Urea amidolyase is a large bifunctional enzyme possessing both urea carboxylase activity (which carboxylates urea in a biotin- and ATP-dependent reaction to produce allophanate) and allophanate hydrolase activity (which hydrolyses allophanate to produce ammonia and carbon dioxide in a cofactor-independent reaction) (Whitney and Cooper, 1972). While its structure remains unknown, urea amidolyase is predicted to have several domains, which from the N-terminus include an amidase domain, a carboxylase domain, a KipA-like domain, a KipI-like domain, and a biotin ligase domain. The N-terminal amidase domain is known to be responsible for the allophanate hydrolase activity, while the remaining domains have urea carboxylase activity (Cooper *et al.*, 1980). The role of the KipI and KipA domains in the context of this enzyme is unknown, as the presence of both carboxylase and biotin ligase domains would appear to be sufficient for urea carboxylation. It is conceivable that KipI and KipA may have a regulatory role modulating the activity of urea amidolyase in response to additional signals, or may bind other proteins to form a large macromolecular assembly.

Although KipI- and KipA-homologues are highly conserved across species, their structures have been most thoroughly studied in the context of the *B. subtilis* sporulation phosphorelay. The structures of the N- and C-terminal domains of KipI have been solved by NMR and crystallography and small-angle solution scattering have provided additional information in the context of the intact KipI. No such experiments have been possible to date with KipA owing to its strong tendency to form aggregates. However, the KipI–KipA complex exhibited sufficient stability and solubility to facilitate study by small-angle neutron scattering with contrast variation. The KipI–KipA complex was shown to have a 1:1 stoichiometry and shape reconstruction from a neutron contrast series from the complex made with deuterated KipI indicated there are extensive contacts between the two components along the long dimension of the complex (Jacques *et al.*, 2011a). The models also indicated a large cleft formed between the KipI and KipA components.

The subsequent solving of the crystal structure of TTHA0988 from *Thermus thermophilus* (Jacques *et al.*, 2011a,b) (PDB entries 3OEP, 3OPF, and 3ORE) enabled more detailed analysis and interpretation of the KipI-KipA scattering data. TTHA0988 is a protein of unknown function that is homologous to a KipI–KipA fusion (Fig. 9.10). Its crystal structure revealed a four-domain protein in which domains A and B are structurally homologous to the N- and C-terminal domains of KipI as expected. Domains C and D are homologous to KipA, with domain C being a beta-sandwich domain, and domain D being of the cyclophilin family (like domain B). The structure of TTHA0988 revealed that both KipI and KipA possess C-terminal cyclophilin domains, despite having no appreciable sequence

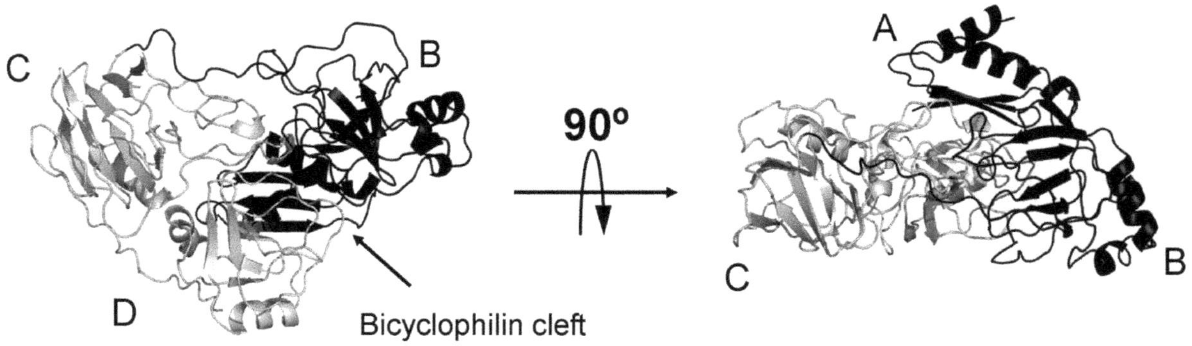

Figure 9.10 Crystal structure of TTHA0988, a structural homologue for the KipI–KipA complex. Domains A and B (black) are homologous to the N- and C-terminal domains of KipI, while domains C and D (grey) are homologous to KipA, and the overall structure is in good agreement with the shape model for KipI–KipA generated using neutron contrast variation data (not shown). From this model, it can be concluded that both KipI and KipA possess C-terminal cyclophilin domains, and that these domains (designated B and D) are positioned so that their conserved β-sheets form a large 'bicyclophilin cleft' that runs along the surface of the molecule.

homology. The structure of TTHA0988 fit the KipI-KipA small-angle scattering data well, and revealed that the cleft observed in the dummy atom model was formed between the cyclophilin domains of KipI and KipA. The juxtaposition of cyclophilin domains is interesting. The large 'bicyclophilin' cleft is the most prominent feature of both the neutron scattering-derived shape restoration model and crystal structure of TTHA0988, but it is not sufficiently wide to accommodate a four-helix bundle such as that observed in KinA suggesting that KipA does not release KipI inhibition of KinA by directly competing for KipI binding. The current model for the mechanism for the 'anti-antikinase' activity of KipA therefore remains the steric barrier model, with KipA physically preventing KipI from interacting with the DHp domain of KinA.

Nitrogen fixation in *S. meliloti*

The only other two-component system identified as being subject to antikinase regulation is found in the legume symbiont, *S. meliloti*. This aerobic, Gram-negative, nitrogen fixing bacterium may exist either in a free-living state in soil or in symbiosis with certain legume species within a specialized nitrogen-fixing organ known as a root nodule. The microoxic environment (10–30 nM dissolved O_2) of the root nodule is necessary for the expression and activity of nitrogenase, the enzyme responsible for converting atmospheric nitrogen into ammonia that is assimilated by the host plant. The adaptation to this microoxic environment is coordinated by the FixLJ two-component system.

FixL is a membrane-bound histidine kinase that detects molecular oxygen via a haem-coordinating cytoplasmic PAS domain. In the oxygen-bound state the kinase is inactive, but in response to microoxic conditions, the C-terminal autokinase domain autophosphorylates at histidine 285. The phosphoryl group is then passed to aspartate 54 on the cognate response regulator, FixJ. The FixJ~P directly activates the transcription of *nifA* and *fixK*, the gene products of which in turn activate the expression of 122 genes involved in respiration and nitrogen fixation (Bobik *et al.*, 2006) (Fig. 9.11). It was originally thought that FixK negatively regulates its own expression, however it was discovered that repression of transcription is achieved via the FixK-mediated activation of the *fixT* gene (Foussard *et al.*, 1997). *fixT* is a duplicated gene on the pSym (a megaplasmid responsible for symbiosis in *S. meliloti*) both copies of which are regulated by FixK but not by NifA. For further information about FixL, see also Chapter 3.

FixT is an antikinase that binds the autokinase domain of FixL and inhibits the autophosphorylation reaction. In this sense, the mechanism of inhibition is similar to KipI and Sda. However,

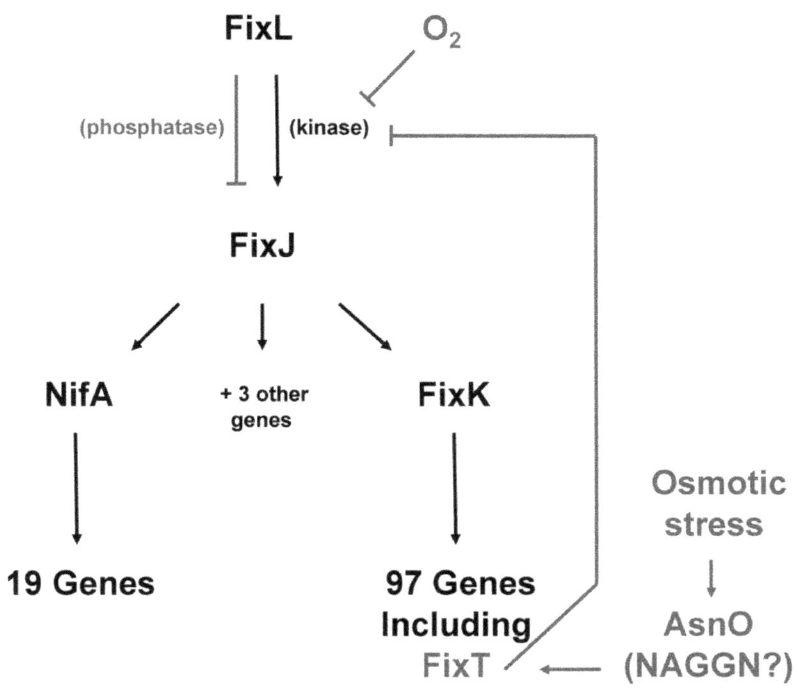

Figure 9.11 The nitrogen fixation two-component system in *S. meliloti*. Activators of the pathway are black, while inhibitors are grey. The histidine kinase, FixL, has response-regulator phosphatase and autokinase activities. The autokinase activity is inhibited by molecular oxygen, but in the microoxic environment of the root nodule the inhibition is released resulting in the phosphorylation of FixJ. FixJ~P triggers the expression of NifA and FixK, which in turn trigger the expression of 116 genes involved in nitrogen fixation as well as other functions including respiration, metabolism and transport. FixT is expressed from a FixK promoter and inhibits the autokinase activity of FixL. The kinase inhibition without concomitant phosphatase inhibition results in FixJ~P dephosphorylation and thus results in negative feedback of the system. The antikinase activity of FixT is dependent on a functional AsnO protein, which is up-regulated under osmotic stress. The requirement for AsnO suggests that its product, NAGGN, may mediate FixT activity.

while FixT alone has no phosphatase activity, the combination of FixT and FixL rapidly dephosphorylates FixJ~P resulting in the repression of *nifA* and *fixK* expression. The dephosphorylation of FixJ~P has been attributed to an intrinsic phosphatase activity of FixL that is present even under low oxygen conditions. Under such conditions the elevated level of FixJ~P is maintained by the constant ATP consumption and presence of a FixL~P pool. In the presence of FixT, autophosphorylation ceases, the FixL~P pool rapidly depletes and the FixL phosphatase removes any remaining FixJ~P.

While the structure of the haem-coordinating PAS domain of FixL has been solved (Gong *et al.*, 1998, 2000; Miyatake *et al.*, 2000), neither the autokinase domain nor FixT has been structurally characterized. Analysis of the primary sequence of FixT (115 amino acid residues) reveals that it is unrelated to Sda and KipI, and therefore may represent a third structural class of antikinase. FixT bears some homology with the receiver domains of response regulators, and it may be that the antikinase has evolved from a protein whose normal function involved DHp-binding. If so, the mechanism of inhibition would be expected to be similar to that of Sda and KipI in that it is the DHp domain that is the target for autokinase inhibition.

The antikinase activity of FixT was found to be regulated by the *asnO* gene, which is also up-regulated under microoxic conditions in a FixJ-independent manner. Mutants with the *asnO* disrupted were found to escape repression by FixT (Berges *et al.*, 2001). AsnO shares homology with glutamine-dependent asparagine synthetases; however, *asnO* inactivation did not lead to asparagine auxotrophy, suggesting that the protein is not involved in asparagine synthesis. Recently it has been shown (Sagot *et al.*, 2010) that *asnO* functions with the adjacent gene, *ngg*, to synthesize the non-ribosomal peptide *N*-acetylglutaminylglutamine amide (NAGGN). Synthesis of NAGGN is required for *S. meliloti* growth under high osmolarity and *asnO* can be induced by the addition of 0.5 M NaCl to the growth medium. The fact that *asnO* is required for FixT antikinase activity implies that FixT responds to the presence of NAGGN by inhibiting FixL. However, this hypothesis has yet to be experimentally verified.

Structural data on the proteins from this system are needed to progress our understanding of the mechanisms for inhibition, and clearly there are many of the same challenges for structural analysis as with the KinA-antikinase proteins.

Unanswered questions in regard to antikinase structures and function

As is often the case, the study of the first few antikinases to be identified appears to have raised more questions than it has answered. Further work on the role of antikinases in histidine kinase inhibition will undoubtedly provide important insights into the biology surrounding two-component systems. Some of the remaining questions, the implications for answering them, and approaches that will be important are discussed here.

The mechanism of antikinase inhibition

The question of the exact mechanism of histidine kinase inhibition by Sda and KipI remains controversial. Bick and colleagues (2009) suggested that the positioning of Sda in the KinB crystal structure prevents the CA domains from accessing the phospho-accepting histidine. That is, Sda inhibits the autokinase reaction by sterically blocking the access to the histidine. KipI is certainly capable of acting as a steric block given its size and binding location. However, the small-angle scattering data indicate that binding of either Sda or KipI is accompanied by an overall compaction of the autokinase domain of KinA involving a net movement of the CA domains towards the DHp. This result suggests that the exact mechanism of histidine kinase inhibition may be more complex and involve some sort of allosteric mechanism operating over a distance; such as a twisting of the helical bundle of the dimerization domain that transmits a signal to the sensor and/or CA domains (Whitten *et al.*, 2007). Further evidence that the steric-block mechanism may not be the whole story comes from the observation that the KinA P410L/F436S double mutant binds Sda, but retains full autokinase activity. If this bound state occurs *in vivo* then Sda alone would appear to not be sufficient to sterically prevent the CA domains from accessing the phospho-accepting histidine. The inhibition therefore may indeed involve rearrangements in the DHp helices such that the phospho-accepting histidine is positioned so that it cannot react with ATP. Another possibility is that the binding of the antikinase restricts the dynamic motion within the autokinase domain, preventing the CA domains from accessing the phospho-accepting histidine. As yet, no crystal structure has been solved for an autokinase domain in the presence and absence of an inhibitor. Until such high-resolution structures become available, the possibility that conformational changes are involved in autokinase inhibition cannot be discounted. Importantly, an understanding of how the inhibition is effected will ultimately improve our understanding of how autokinase domains function, and may even help explain how sensor domains transmit their signals to regulate autokinase activity, keeping in mind that it is the DHp domain that connects to the CA domains; inhibition and activation may well be achieved using a common architecture.

Release of antikinase inhibition

The three antikinases Sda, KipI, and FixT appear to have different mechanisms by which they are released or prevented from binding their target histidine kinase. The current thinking is that Sda is proteolysed *in situ* in order release it from its target, while the antikinase activity of KipI is antagonised by KipA. In the case of Sda, the protease responsible for antikinase disposal is not clear, but may involve ClpX (Ruvolo *et al.*, 2006), and the nature of the signals that regulate its clearance from the cell are unknown. It is known, however, that Sda is produced in bursts during replication and that it is rapidly cleared once the cell has divided. In the case of KipI it is known that by binding KipA the antikinase can no longer bind its target. It would also seem from the organization of the *kip* operon, that KipI and KipA are produced together in equimolar amounts. It would therefore seem that the regulation of KipI antikinase activity cannot take place at the level of transcription. It is conceivable that KipA initially binds KipI, but releases it in response to some additional signal. Given that expression of the *kip* operon is regulated by nitrogen and carbon metabolism, it is possible that KipA responds to some nutritional status of the cell. Similarly, as two of the *kip* operon genes code for membrane proteins, it may be that KipA responds to an extracellular stimulus. In the case of FixT, it is known that the AsnO is required for antikinase activity. While this implies that NAGGN is required for histidine kinase inhibition, such a requirement has not been demonstrated experimentally. Likewise, whether NAGGN depletion plays a role in the release of the antikinase from its target remains to be seen.

A possible role for antikinases in quorum sensing

One of the more intriguing results from studying the structure of KipI, is the homology found between its N-terminal domain and LuxS. As yet the role of the KipI N-terminal domain is not defined, but its similarity with proteins involved in quorum sensing suggests a common evolutionary ancestor. As mentioned above, it is now known that the sporulation phosphorelay plays a role in coordinating cell differentiation by modulating the amount of phosphorylated Spo0A present in the cell. It therefore is understood that the phosphorelay plays a role in quorum sensing and biofilm development, and as such it seems likely that the antikinases play some role in regulating the response of the bacterial cell to population signals. With the function of the *kip* operon so poorly defined, it is interesting to speculate on the possibility that these seven genes play a role in quorum sensing by perceiving external signals and inputting information into the phosphorelay by modulating the activity of KinA.

Antikinases in other systems

Two-component systems are the primary mechanism by which bacteria sense and respond to environmental changes. As such most bacterial species possess multiple histidine kinases that are subject to different forms of regulation. So far the most studied systems to employ the use of antikinases are the sporulation phosphorelay in *B. subtilis* and the nitrogen fixation pathway in *S. meliloti*. It seems very likely that other species utilize antikinases to regulate different processes. For example, KipI- and KipA-homologues are highly conserved across almost all bacterial species, and are certainly not restricted to sporulating organisms, but their functions are yet to be determined. The identification and characterization of antikinases across bacterial species remains one of the major challenges in the study of two-component systems, but may yield untold benefits to the study of bacterial stimulus perception, as well as the fields of medical microbiology and biotechnology.

Outlook

Modern biotechnology and bioinformatics approaches are powerful tools for identifying the likely molecular actors in complex, dynamic and highly regulated molecular networks such as the two-component systems of bacteria that are the subject of this book. Progress towards the kind of detailed mechanistic view that is essential if we are to design therapeutic drugs will always depend upon high resolution structural information. Such data are difficult to obtain for multidomain and multisubunit proteins with inherent flexibility as is the case for our KinA-antikinase molecular actors; and when that flexibility is required for function there is an imperative to understand how.

We have significantly progressed our knowledge and understanding of antikinase structure and function in the absence of a high resolution kinase-antikinase crystal structure by using the structures of individual domains and structural homologues determined by crystallography or NMR. These structural fragments have been assembled, like the pieces of a jigsaw puzzle, using a frame work provided by small-angle solution scattering and assisted by additional structural constraints from complementary methods. This approach that depends upon the judicious use of hybrid experimental data that includes data from small-angle scattering and neutron contrast variation, has been useful in studies of other multidomain kinases that are similarly recalcitrant to crystallization; specifically the serine-threonine kinases that are ubiquitous in mammals whose activation and inhibition is controlled by calcium–calmodulin (Krueger *et al.*, 2001) or cAMP (Trewhella, 2006). These studies helped define the by now classic conformational collapse of calmodulin and the calcium–calmodulin regulated kinase autoinhibition release mechanism. In the case of the cAMP-dependent protein kinase they helped define how inhibition and activation is accomplished and differentiated among protein kinase A (PKA) isoforms and how the flexible domain linkers between highly conserved regulatory domains of the different isoforms of the multisubunit protein facilitate targeting to specific locations within cells.

The application of these methods that include small-angle scattering approaches is approaching an era where it can be envisaged that automation and standardization will facilitate the non-expert being able to readily use them, e.g. work is progressing on robotic systems for sample handling at synchrotron based small-angle scattering stations (Trame *et al.*, 2004), user friendly small-angle scattering data interpretation (Petoukhov *et al.*, 2007; Whitten *et al.*, 2008), and structural modelling methods (Petoukhov and Svergun, 2007; Rambo and Tainer, 2010) are becoming widely used, and the International Union of Crystallography Journals and Small-angle Scattering Commissions and the Protein Data Bank are currently addressing the issues of requirements and standards for publication and data deposition. We thus can anticipate that, as has proven the case for the sporulation pathway, the FixLJ two-component system and its antikinase FixT, or perhaps a yet to be discovered kinase-antikinase system, may well have its secrets unveiled through application of these methods.

So far only three antikinases have been described, none of which share any sequence or structural homology. With the majority of prokaryotic genes being assigned putative or unknown functions, there is every possibility that additional antikinases will be discovered. The current studies were primarily motivated by the possibility of exploiting antikinases in the context of the development of medically relevant histidine kinase inhibitors. However, the observations made to date have raised important questions relating to the fundamental biology of two-component systems. In particular, the possible role of antikinases (or other KipI-KipA related proteins) in biofilms is especially worthy of further investigation. Questions such as this serve to highlight that we have only scratched the surface in this field, but also emphasize the rich potential for discovery in antikinase research.

References

Antoniewski, C., Savelli, B., and Stragier, P. (1990). The *spoIIJ* gene, which regulates early developmental steps in *Bacillus subtilis*, belongs to a class of environmentally responsive genes. J. Bacteriol. *172*, 86–93.

Berges, H., Checroun, C., Guiral, S., Garnerone, A.M., Boistard, P., and Batut, J. (2001). A glutamine-amidotransferase-like protein modulates FixT anti-kinase activity in *Sinorhizobium meliloti*. BMC Microbiol. *1*, 6.

Bick, M.J., Lamour, V., Rajashankar, K.R., Gordiyenko, Y., Robinson, C.V., and Darst, S.A. (2009). How to switch off a histidine kinase: crystal structure of *Geobacillus stearothermophilus* KinB with the inhibitor Sda. J. Mol. Biol. *386*,163–177.

Bobik, C., Meilhoc, E., and Batut, J. (2006). FixJ: a major regulator of the oxygen limitation response and late symbiotic functions of *Sinorhizobium meliloti*. J. Bacteriol. *188*, 4890–4902.

Buelow, D.R., and Raivio, T.L. (2010). Three (and more) component regulatory systems – auxiliary regulators of bacterial histidine kinases. Mol. Microbiol. *75*, 547–566.

Burbulys, D., Trach, K.A., and Hoch, J.A. (1991). Initiation of sporulation in *B. subtilis* is controlled by a multicomponent phosphorelay. Cell *64*, 545–552.

Burkholder, W.F., Kurtser, I., and Grossman, A.D. (2001). Replication initiation proteins regulate a developmental checkpoint in *Bacillus subtilis*. Cell *104*, 269–279.

Casino, P., Rubio, V., and Marina, A. (2009). Structural insight into partner specificity and phosphoryl transfer in two-component signal transduction. Cell *139*, 325–336.

Cooper, T.G., Lam, C., and Turoscy, V. (1980). Structural analysis of the *dur* loci in *S. cerevisiae*: two domains of a single multifunctional gene. Genetics *94*, 555–580.

Cunningham, K.A., and Burkholder, W.F. (2009). The histidine kinase inhibitor Sda binds near the site of autophosphorylation and may sterically hinder autophosphorylation and phosphotransfer to Spo0F. Mol. Microbiol. *71*, 659–677.

Deremer, D.L., Ustun, C., and Natarajan, K. (2008). Nilotinib: a second-generation tyrosine kinase inhibitor for the treatment of chronic myelogenous leukemia. Clin. Ther. *30*, 1956–1975.

Foussard, M., Garnerone, A.M., Ni, F., Soupene, E., Boistard, P., and Batut, J. (1997). Negative autoregulation of the *Rhizobium meliloti fixK* gene is indirect and requires a newly identified regulator, FixT. Mol. Microbiol. *25*, 27–37.

Gilmour, R., Foster, J.E., Sheng, Q., McClain, J.R., Riley, A., Sun, P.M., Ng, W.L., Yan, D., Nicas, T.I., Henry, K., *et al.* (2005). New class of competitive inhibitor of bacterial histidine kinases. J. Bacteriol. *187*, 8196–8200.

Gong, W., Hao, B., Mansy, S.S., Gonzalez, G., Gilles-Gonzalez, M.A., and Chan, M.K. (1998). Structure of a biological oxygen sensor: a new mechanism for heme-driven signal transduction. Proc. Natl. Acad. Sci. U.S.A. *95*, 15177–15182.

Gong, W., Hao, B., and Chan, M.K. (2000). New mechanistic insights from structural studies of the oxygen-sensing domain of *Bradyrhizobium japonicum* FixL. Biochemistry *39*, 3955–3962.

Gotoh, Y., Eguchi, Y., Watanabe, T., Okamoto, S., Doi, A., and Utsumi, R. (2010). Two-component signal transduction as potential drug targets in pathogenic bacteria. Curr. Opin. Microbiol. *13*, 232–239.

Grebe, T.W., and Stock, J.B. (1999). The histidine protein kinase superfamily. Adv. Microb. Physiol. *41*, 139–227.

Guarnieri, M.T., Blagg, B.S., and Zhao, R. (2011). A high-throughput TNP-ATP displacement assay for screening inhibitors of ATP-binding in bacterial histidine kinases. Assay Drug Dev. Technol. *9*, 174–183.

Halin, C., Fahrngruber, H., Meingassner, J.G., Bold, G., Littlewood-Evans, A., Stuetz, A., and Detmar, M. (2008). Inhibition of chronic and acute skin inflammation by treatment with a vascular endothelial growth factor receptor tyrosine kinase inhibitor. Am. J. Pathol. *173*, 265–277.

Hamon, M.A., and Lazazzera, B.A. (2001). The sporulation transcription factor Spo0A is required for biofilm development in *Bacillus subtilis*. Mol. Microbiol. *42*, 1199–1209.

Huse, M., Chen, Y.G., Massague, J., and Kuriyan, J. (1999). Crystal structure of the cytoplasmic domain of the type I TGF beta receptor in complex with FKBP12. Cell *96*, 425–436.

Hynson, R.M., Kwan, A.H., Jacques, D.A., Mackay, J.P., and Trewhella, J. (2010). ^{1}H, ^{13}C and ^{15}N backbone and side chain resonance assignments of the N-terminal domain of the histidine kinase inhibitor KipI from *Bacillus subtilis*. Biomol. NMR Assign. *4*, 167–169.

Ikura, T., and Ito, N. (2007). Requirements for peptidyl-prolyl isomerization activity: a comprehensive mutational analysis of the substrate-binding cavity of FK506-binding protein 12. Protein Sci. *16*, 2618–2625.

Jacques, D.A., and Trewhella, J. (2010). Small-angle scattering for structural biology – expanding the frontier while avoiding the pitfalls. Protein Sci. *19*, 642–657.

Jacques, D.A., Langley, D.B., Jeffries, C.M., Cunningham, K.A., Burkholder. W.F., Guss, J.M., and Trewhella, J. (2008). Histidine kinase regulation by a cyclophilin-like inhibitor. J. Mol. Biol. *384*, 422–435.

Jacques, D.A., Streamer, M., Rowland, S.L., King, G.F., Guss, J.M., Trewhella, J., and Langley, D.B. (2009). Structure of the sporulation histidine kinase inhibitor Sda from *Bacillus subtilis* and insights into its solution state. Acta Crystallogr. D Biol. Crystallogr. *65*, 574–581.

Jacques, D.A., Langley, D.B., Hynson, R.M., Whitten, A.E., Kwan, A., Guss, J.M., and Trewhella, J. (2011a). A novel structure of an antikinase and its inhibitor. J. Mol. Biol. *405*, 214–226.

Jacques, D.A., Langley, D.B., Kuramitsu, S., Yokoyama, S., Trewhella, J., and Guss, J.M. (2011b). The structure of TTHA0988 from *Thermus thermophilus*, a KipI-KipA homologue incorrectly annotated as an allophanate hydrolase. Acta Crystallogr. D Biol. Crystallogr. *67*, 105–111.

Kim, D., and Forst, S. (2001). Genomic analysis of the histidine kinase family in bacteria and archaea. Microbiology *147*, 1197–1212.

Krell, T., Lacal, J., Busch, A., Silva-Jimenez, H., Guazzaroni, M.E., and Ramos, J.L. (2010). Bacterial sensor kinases: diversity in the recognition of environmental signals. Annu. Rev. Microbiol. *64*, 539–559.

Krueger, J.K., Gallagher, S.C., Zhi, G., Geguchadze, R., Persechini, A., Stull, J.T., and Trewhella, J. (2001). Activation of myosin light chain kinase requires translocation of bound calmodulin. J. Biol. Chem. *276*, 4535–4538.

LeDeaux, J.R., Yu, N., and Grossman, A.D. (1995). Different roles for KinA, KinB, and KinC in the initiation of sporulation in *Bacillus subtilis*. J. Bacteriol. *177*, 861–863.

Lee, J., Tomchick, D.R., Brautigam, C.A., Machius, M., Kort, R., Hellingwerf, K.J., and Gardner, K.H. (2008). Changes at the KinA PAS-A dimerization interface influence histidine kinase function. Biochemistry *47*, 4051–4064.

Lombardia, E., Rovetto, A.J., Arabolaza, A.L., and Grau, R.R. (2006). A LuxS-dependent cell-to-cell language regulates social behavior and development in *Bacillus subtilis*. J. Bacteriol. *188*, 4442–4452.

Lopez, D., and Kolter, R. (2009). Extracellular signals that define distinct and coexisting cell fates in *Bacillus subtilis*. FEMS Microbiol. Rev. *34*, 134–149.

Lopez, D., Vlamakis, H., and Kolter, R. (2009). Generation of multiple cell types in *Bacillus subtilis*. FEMS Microbiol. Rev. *33*, 152–163.

Marina, A., Waldburger, C.D., and Hendrickson, W.A. (2005). Structure of the entire cytoplasmic portion of a sensor histidine-kinase protein. EMBO J. *24*, 4247–4259.

Masayama, A., Kuwana, R., Takamatsu, H., Hemmi, H., Yoshimura, T., Watabe, K., and Moriyama, R. (2007). A novel lipolytic enzyme, YcsK (LipC), located in the spore coat of *Bacillus subtilis*, is involved in spore germination. J. Bacteriol. *189*, 2369–2375.

Masayama, A., Kato, S., Terashima, T., Molgaard, A., Hemmi, H., Yoshimura, T., and Moriyama, R. (2010). *Bacillus subtilis* spore coat protein LipC is a phospholipase B. Biosci. Biotechnol. Biochem. *74*, 24–30.

Mascher, T., Helmann, J.D., and Unden, G. (2006). Stimulus perception in bacterial signal-transducing histidine kinases. Microbiol. Mol. Biol. Rev. *70*, 910–938.

Matsushita, M., and Janda, K.D. (2002). Histidine kinases as targets for new antimicrobial agents. Bioorg. Med. Chem. *10*, 855–867.

Miyatake, H., Mukai, M., Park, S.Y., Adachi, S., Tamura, K., Nakamura, H., Nakamura, K., Tsuchiya, T., Iizuka, T., and Shiro, Y. (2000). Sensory mechanism of oxygen sensor FixL from *Rhizobium meliloti*: crystallographic, mutagenesis and resonance Raman spectroscopic studies. J. Mol. Biol. *301*, 415–431.

Parkinson, J.S., and Kofoid, E.C. (1992). Communication modules in bacterial signaling proteins. Annu. Rev. Genet. *26*, 71–112.

Perego, M., Cole, S.P., Burbulys, D., Trach, K., and Hoch, J.A. (1989). Characterization of the gene for a protein kinase which phosphorylates the sporulation-regulatory proteins Spo0A and Spo0F of *Bacillus subtilis*. J. Bacteriol. *171*, 6187–6196.

Petoukhov, M.V., and Svergun, D.I. (2007). Analysis of X-ray and neutron scattering from biomacromolecular solutions. Curr. Opin. Struct. Biol. *17*, 562–571.

Petoukhov, M.V., Konarev, P.V., Kikhney, A.G.,and Svergun, D.I. (2007). ATSAS 2.1 – towards automated and web-supported small-angle scattering data analysis. J. Appl. Crystallogr. *40*, S223-S228.

Rambo, R.P., and Tainer, J.A. (2009). Bridging the solution divide: comprehensive structural analyses of dynamic RNA, DNA, and protein assemblies by small-angle X-ray scattering. Curr. Opin. Struct. Biol. *20*, 128–137.

Richardson, I.B., Katz, M.E., and Hynes, M.J. (1992). Molecular characterization of the *lam* locus and sequences involved in regulation by the AmdR protein of *Aspergillus nidulans*. Mol. Cell. Biol. *12*, 337–346.

Rowland, S.L., and King, G.F. (2007). Histidine kinases as antimicrobial targets: prospects and pitfalls. Mini Rev. Med. Chem. *7*, 1144–1154.

Rowland, S.L., Burkholder, W.F., Cunningham, K.A., Maciejewski, M.W., Grossman, A.D., and King, G.F. (2004). Structure and mechanism of action of Sda, an inhibitor of the histidine kinases that regulate initiation of sporulation in *Bacillus subtilis*. Mol. Cell. *13*, 689–701.

Ruvolo, M.V., Mach, K.E., and Burkholder, W.F. (2006). Proteolysis of the replication checkpoint protein Sda is necessary for the efficient initiation of sporulation after transient replication stress in *Bacillus subtilis*. Mol. Microbiol. *60*, 1490–1508.

Sagot, B., Gaysinski, M., Mehiri, M., Guigonis, J.M., Le Rudulier, D., and Alloing, G. (2010). Osmotically induced synthesis of the dipeptide N-acetylglutaminylglutamine amide is mediated by a new pathway conserved among bacteria. Proc. Natl. Acad. Sci. U.S.A. *107*, 12652–12657.

Schumperli, D., McKenney, K., Sobieski, D.A., and Rosenberg, M. (1982). Translational coupling at an intercistronic boundary of the *Escherichia coli* galactose operon. Cell *30*, 865–871.

Sonenshein, A.L. (2000). Control of sporulation initiation in *Bacillus subtilis*. Curr. Opin. Microbiol. *3*, 561–566.

Stewart, R.C. (2010). Protein histidine kinases: assembly of active sites and their regulation in signaling pathways. Curr. Opin. Microbiol. *13*, 133–141.

Theodorou, E.C., Theodorou, M.C., and Kyriakidis, D.A. (2011). Inhibition of the signal transduction through the AtoSC system by histidine kinase inhibitors in *Escherichia coli*. Cell Signal. *23*, 1327–1337.

Trame, C., MacDowell, A.A., Celestre, R.S., Padmore, H.A., Cambie, D., Domning, E.E., Duarte, R.M., Kelez, N., Plate, D.W., Holton, J.M., *et al.* (2004). SIBYLS – a SAXS and protein crystallography beamline at the ALS. Synchrotron Radiation Instrumentation *705*, 502–505.

Trewhella, J. (2006). Structural themes and variations in protein kinase A as seen by small-angle scattering and neutron contrast variation. Eur. Biophys. J. *35*, 585–589.

Varughese, K.I., Madhusudan, Zhou, X.Z., Whiteley, J.M., and Hoch, J.A. (1998). Formation of a novel four-helix bundle and molecular recognition sites by dimerization of a response regulator phosphotransferase. Mol. Cell *2*, 485–493.

Veening, J.W., Murray, H., and Errington, J. (2009). A mechanism for cell cycle regulation of sporulation initiation in *Bacillus subtilis*. Genes Dev. *23*, 1959–1970.

Wang, L., Grau, R., Perego, M., and Hoch, J.A. (1997). A novel histidine kinase inhibitor regulating development in *Bacillus subtilis*. Genes Dev. *11*, 2569–2579.

Wang, L., Fabret, C., Kanamaru, K., Stephenson, K., Dartois, V., Perego. M., and Hoch, J.A. (2001). Dissection of the functional and structural domains of phosphorelay histidine kinase a of *Bacillus subtilis*. J. Bacteriol. *183*, 2795–2802.

Watanabe, T., Hashimoto, Y., Yamamoto, K., Hirao, K., Ishihama, A., Hino, M., and Utsumi, R. (2003). Isolation and characterization of inhibitors of the essential histidine kinase, YycG in *Bacillus subtilis* and *Staphylococcus aureus*. J. Antibiot. (Tokyo) *56*, 1045–1052.

Whitney, P.A., and Cooper, T.F. (1972). Urea carboxylase and allophanate hydrolase. Two components of adenosine triphosphate:urea amido-lyase in *Saccharomyces cerevisiae*. J. Biol. Chem. *247*, 1349–1353.

Whitten, A.E., Jacques, D.A., Hammouda, B., Hanley, T., King, G.F., Guss, J.M., Trewhella, J., and Langley, D.B. (2007). The structure of the KinA–Sda complex suggests an allosteric mechanism of histidine kinase inhibition. J. Mol. Biol. *368*, 407–420.

Whitten, A.E., Ca, S.Z., and Trewhella, J. (2008). MULCh: modules for the analysis of small-angle neutron contrast variation data from biomolecular assemblies. J. Appl. Crystallogr. *41*, 222–226.

Zapf, J., Sen, U., Madhusudan, Hoch, J.A., and Varughese, K.I. (2000). A transient interaction between two phosphorelay proteins trapped in a crystal lattice reveals the mechanism of molecular recognition and phosphotransfer in signal transduction. Structure *8*, 851–862.

K^+ Supply, Osmotic Stress and the KdpD/KdpE Two-component System 10

Ralf Heermann and Kirsten Jung

Abstract

The KdpD/KdpE system is one of the most distributed histidine kinase/response regulator systems in bacteria. This two-component system controls expression of the *kdpFABC* operon encoding the high affinity K^+ uptake system KdpFABC. KdpD/KdpE is activated whenever the constitutively expressed K^+ uptake systems are unable to keep up with the cellular need for K^+. Typical KdpD/KdpE activating conditions comprise K^+ limitation or osmotic stress. The histidine kinase KdpD is composed of several subdomains that are important for stimulus perception, modulation of the kinase to phosphatase ratio, and signalling. The response regulator KdpE receives the phosphoryl group from KdpD and induces *kdpFABC* transcription. The three-dimensional structure of the KdpE receiver domain gives insights into the activation mechanism of this transcriptional regulator. Two accessory proteins, the universal stress protein UspC and the phosphotransferase system component IIA^{Ntr}, interact with KdpD, whereby *kdpFABC* expression is modulated under certain physiological conditions. Here we will review the nature of the stimulus of KdpD/KdpE in correlation with structural features as well as the interconnectivity of KdpD/KdpE with K^+ supply, osmotic stress response and virulence.

Introduction

Potassium (K^+), an essential element of life, is the major intracellular cation in bacteria as well as in eukaryotic cells. This fact reflects a very ancient decision of living cells to accumulate K^+ and to exclude the sodium (Na^+) ion. K^+ has four important functions in bacteria:

1 K^+ is an important osmotic solute to ensure maintenance of turgor (Epstein, 2003).
2 K^+ contributes to the regulation of the intracellular pH (Booth, 1985).
3 K^+ is important for translation by binding close to the active centre of the ribosome (Nissen *et al.*, 2000) as well as for the activation of different enzymes (Suelter, 1970).
4 K^+ is a regulator of several stress responses, e.g. chaperone Hsp70 activity (Palleros *et al.*, 1993).

Although the availability of K^+ is low in most environments (usually 0.1–10 mM), bacteria maintain a high internal K^+ concentration, which, for example was determined in *Escherichia coli* to be 0.4 M (McLaggan *et al.*, 1994). Bacteria accumulate K^+ by a number of different transport systems that vary in kinetics, energy coupling, and regulation. Importantly, the large diversity in the bacterial kingdom is not reflected in a similar diversity with respect to proteins involved in K^+ homeostasis. Only a few types of specific K^+ transport systems are found in bacteria. The major systems are the constitutively produced low affinity K^+ uptake systems Trk and Ktr, and the inducible and high affinity K^+ uptake system Kdp (Epstein, 2003). Besides these Kup, which is a single membrane protein with both modest affinity and uptake rate for K^+, is found in some bacteria (Epstein, 2003). Apart from energy-dependent K^+ uptake systems, K^+

channels like KscA exist, which allow very fast accumulation of K^+, but can only operate at sufficient external K^+ concentrations (Roux and MacKinnon, 1999).

Under physiological conditions the powerful constitutive K^+ uptake systems Trk or Ktr ensure K^+ homeostasis. Under K^+ limitation ($[K^+] < 2$ mM) or upon osmotic upshift, the inducible Kdp system scavenges K^+ from the environment with high affinity and specificity (Epstein, 1992). This K^+ uptake system is a P-type ATPase and consists of the four subunits KdpF, KdpA, KdpB and KdpC. Expression of the corresponding *kdpFABC* operon is under control of the two-component system KdpD/KdpE (Polarek *et al.*, 1992; Walderhaug *et al.*, 1992). Under *kdp*-inducing conditions, the histidine kinase KdpD autophosphorylates, transfers the phosphoryl group to the response regulator KdpE, and phosphorylated KdpE triggers *kdpFABC* transcription (Sugiura *et al.*, 1992; Voelkner *et al.*, 1993). The enzymatic activities of purified KdpD and KdpE were characterized *in vitro* (Jung *et al.*, 1997).

The KdpD/KdpE system is one of the most distributed histidine kinase/response regulator systems in bacteria. To date more than 1082 homologues are found in bacterial and archaeal species, among them are many pathogens [this information is based by searching the NCBI's Conserved Domain Architecture Retrieval Tool (CDART at http://www.ncbi.nlm.nih.gov/Structure/lexington/lexington.cgi?cmd=rps (Geer *et al.*, 2002) with the KdpD domain (pfam02702) as query].

Originally it was suggested that KdpD is a turgor sensor (Laimins *et al.*, 1981). Therefore, research on KdpD/KdpE has been focussed on the following challenging questions: What are the primary stimuli sensed by KdpD? Does KdpD sense a physical or a chemical stimulus, or both? Does KdpD/KdpE employ accessory proteins? Here we will review the nature of the stimulus for KdpD, the exceptional multidomain structure of the histidine kinase KdpD, the signalling mechanism of KdpE as well as the interconnectivity of KdpD/KdpE with K^+ supply, osmotic stress response and virulence.

The KdpD/KdpE two-component system

The histidine kinase KdpD

The histidine kinase KdpD is a membrane-anchored sensor of the cytoplasmic membrane with a modular domain organization. It consists of four transmembrane domains (TM1-TM4) and of large N- and C-terminal domains (Fig. 10.1).

Three different conserved domains were identified in the input domain of *E. coli* KdpD that are all involved in stimulus sensing (Fig. 10.1): The KdpD domain (pfam02702) that is conserved among all KdpD homologues (Heermann *et al.*, 2000, 2003a), a Usp domain (USP-OKCHK) that is similar to the universal stress protein family (Usp) (cd01987, pfam00582) (Heermann *et al.*, 2009a,b), and a GAF domain (SMART 00065) that was prototypically explored in phytochromes and cGMP-specific phosphodiesterases (McAllister-Lucas *et al.*, 1995).

The domains KdpD and Usp, located in the N-terminal part of KdpD, are well characterized. Altendorf and coworkers found that truncations within the KdpD domain resulted in functional KdpD proteins *in vitro* (Puppe *et al.*, 1996). Later on, a 'Walker-type' ATP-binding site has been identified within this region (Jung and Altendorf, 1998b). Truncated forms of KdpD lacking this site (KdpD/Δ12–228, KdpD/Δ12–395) were characterized by a deregulated phosphatase activity. A truncated KdpD derivative comprising only the domains KdpD and Usp (KdpD/1–395) induced constitutive expression of *kdpFABC in vivo* (Heermann *et al.*, 2003a). Detailed biochemical studies revealed a stabilizing function of the N-terminal domain of KdpD in complex with phospho-KdpE and the corresponding DNA-binding site (Heermann *et al.*, 2000, 2003a). Many bacteria, for example the cyanobacterium *Anabaena*, have a KdpD homologue that comprises only the N-terminal domain without the C-terminal transmitter domain and the transmembrane helices (Ballal *et al.*, 2005). Replacement of the N-terminal domain of *E. coli* KdpD with the short KdpD version of *Anabaena* resulted in a chimera that was functional in *E. coli in vivo* and *in vitro* (Ballal *et al.*, 2002). This result suggested that *Anabaena* KdpD could

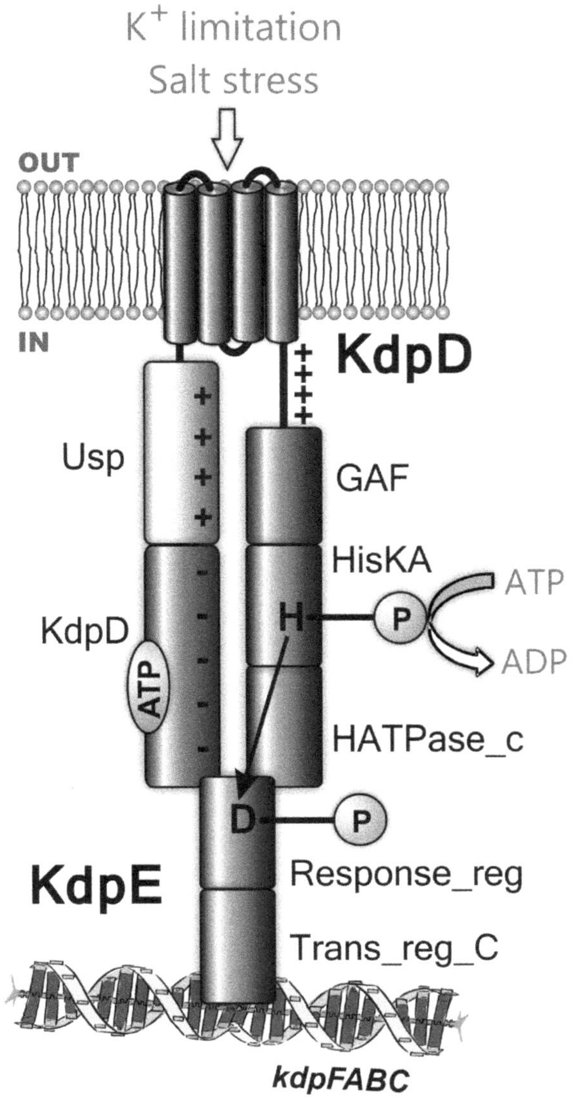

Figure 10.1 Domain organization of the KdpD/KdpE two-component system. The model of KdpD is based on hydropathy plot analysis, studies with *lacZ/phoA* fusions (Zimmann *et al.*, 1995), and use of CDART (Geer *et al.*, 2002; Heermann *et al.*, 2009b). KdpD contains the conserved domains of histidine kinases: HATPase_c (Histidine kinase-like ATPases; Histidine kinase-, DNA gyrase B-, phytochrome-like ATPases, SMART00387) and HisKA (His Kinase A phosphoacceptor domain, dimerization, and phosphoacceptor domain of histidine kinases, SMART00388). The highly conserved KdpD domain (pfam02702), the Usp domain USP-OKCHK (cd01987) and the GAF domain (SMART00065) belong to the input domain of KdpD. Positive surface charges of a domain are indicated by '+', negative by '–'. KdpE consist of an N-terminal receiver domain (Response_reg, pfam00072) fused to a C-terminal DNA-binding output domain (Trans_reg_C, cd00383).

act in a similar way as the N-terminal domain of *E. coli* KdpD.

The Usp domain within the N-terminal domain functions as binding surface for the universal stress protein UspC (see below) (Heermann *et al.*, 2009b), and it was shown to be important for internal protein dynamics allowing structural alterations within KdpD upon stimulus perception (Heermann *et al.*, 2009a). Using a domain swapping approach, the Usp domain within KdpD was replaced by KdpD-Usp domains of various bacteria and the six soluble universal stress proteins of *E. coli*, respectively. *In vivo* and *in vitro* analyses of these KdpD chimeras revealed that signalling within KdpD involves alterations of electrostatic interactions. Chimeras containing UspF or UspG not only prevented *kdpFABC* expression under salt stress, but also under K^+ limitation, although these hybrid proteins exhibited kinase and phosphatase activities *in vitro* (Heermann *et al.*, 2009a). Analysis of the wild-type KdpD-Usp tertiary structure revealed that this domain has a net positively charged surface, while both UspF and UspG are characterized by net negatively charged surfaces (Heermann *et al.*, 2009a). It is proposed that the positively charged Usp domain interacts with other positively charged residues in KdpD activating the histidine kinase by electrostatic repulsion (Fig. 10.1). Chimeras containing the negatively charged UspF or UspG remain in an inactive state owing to electrostatic attraction. A possible explanation why KdpD-UspF and KdpD-UspG are active *in vitro* but block *kdpFABC* expression *in vivo* might be that the stabilization of the phospho-KdpE/DNA complex by the N-terminal domain of KdpD is prevented in the inactive state (Heermann *et al.*, 2003a).

In the C-terminal part of the input domain, adjacent to TM4, a cluster of positively charged amino acids (Arg residues) was identified that is important for the ratio between kinase to phosphatase activities (Jung and Altendorf, 1998a). Replacement of these Arg residues (positions 503, 506, 508, 511 or 513) by Gln resulted in three groups of KdpD derivatives. Removal of the positive charge at position 511 led to the loss of phosphatase activity. On the other hand, KdpD proteins with Arg replacements at positions 503,

506 or 508 were characterized by an increase of the phosphatase activity. Since the removal of one positively charged amino acid residue was sufficient to disturb the ratio of the KdpD activities, it was proposed that electrostatic interactions within the protein affect the kinase to phosphatase equilibrium (Jung and Altendorf, 1998a).

NMR analysis of the transmembrane domains of KdpD revealed a four-helix bundle, in which the second and third helices are relatively short and loosely packed suggesting inherent flexibility (Maslennikov *et al.*, 2010). Earlier, the transmembrane domains of KdpD were thought to be essential for sensing. Truncated KdpD derivatives lacking the transmembrane domains including the Arg cluster adjacent to TM4 were characterized enzymatically inactive *in vitro* and unable to induce *kdpFABC* expression *in vivo* (Puppe *et al.*, 1996). More recent studies shed new light on the role of the transmembrane domains for KdpD sensing and signalling (Heermann *et al.*, 2003b). A truncated KdpD lacking all four transmembrane domains but retaining the Arg cluster supported *kdpFABC* expression in a K^+-dependent manner. Furthermore, truncated KdpD proteins that lack only two transmembrane domains or derivatives in which a linker peptide or two transmembrane domains of PutP, the Na^+/proline transporter of *E. coli*, replaced the missing part indicated that the transmembrane domains are not essential for sensing K^+ limitation, but important for the correct positioning of the large N- and C-terminal cytoplasmic domains to each other (Heermann *et al.*, 2003b). Although not important for sensing K^+ limitation, there are some indications that the transmembrane domains of KdpD are involved in osmosensing. A truncated KdpD lacking TM1 and TM2 was unable to sense an increase of medium osmolarity (Heermann *et al.*, 2003b). Furthermore, the systematic replacement of each single amino acid of TM1 revealed that amino acids of this transmembrane domain are involved in osmosensing but not in K^+ sensing (Stallkamp *et al.*, 2002). Mutational analysis of amino acids located within TM3, TM4, and the adjacent C-terminal hydrophilic region identified a number of KdpD derivatives that were insensitive towards the K^+ signal but sensitive towards osmotic shifts (Sugiura *et al.*, 1994).

Several investigations addressed the identification of a putative K^+-binding site in KdpD. Cells producing an N-terminal truncated, soluble KdpD (KdpD/Δ1–498) were able to respond to changes of the extracellular K^+ concentration (Rothenbücher *et al.*, 2006). Moreover, amino acid replacements located within TM4 and the adjacent region resulted in K^+-insensitive KdpD derivatives (Brandon *et al.*, 2000). It is predicted that TM4 forms a long helix that extends into the cytoplasm and contains the cluster of Arg residues (Zimmann *et al.*, 2007). Random mutagenesis of the corresponding part of the *kdpD* gene produced KdpD derivatives that caused K^+-independent *kdpFABC* expression. Therefore, it is assumed that the putative K^+-binding site is located adjacent to TM4 in the C-terminal domain. In addition, most of these KdpD derivatives exhibited altered response to osmotic stress (Zimmann *et al.*, 2007). These data indicate that this part of KdpD that includes the GAF domain is crucial for stimulus sensing and signal transmission.

The transmitter domain is typical for the protein family of histidine kinase and is located in the C-terminal part of KdpD (Fig. 10.1). It contains the typical two subdomains, HATPase_c and HisKA, that are shared by members of the histidine kinase protein family. The latter includes the autophosphorylation site His673 (Voelkner *et al.*, 1993). The tertiary structures of the HATPase_c and the HisKA domains were resolved for the histidine kinase EnvZ (Tanaka *et al.*, 1998; Tomomori *et al.*, 1999). Similar structures are predicted for KdpD as these domains are homologous to EnvZ. The HATPase_c domain encloses several highly conserved regions (G1, G2, F and N boxes), acts as ATP-binding domain, and forms an α/β sandwich fold (Tanaka *et al.*, 1998). Dimerization via the HisKA domain of EnvZ is essential for its autophosphorylation and phosphotransfer functions. The HisKA domain comprises a four-helix bundle formed by two identical helix–turn–helix subunits, revealing the molecular assembly of two active sites within the dimeric kinase (Tomomori *et al.*, 1999). KdpD functions as a homodimer (Heermann *et al.*, 1998). Co-production of two kinase inactive

KdpD derivatives KdpD/His673→Gln and KdpD/Asn788→Asp resulted in a KdpD protein that regained kinase activity *in vitro*. This result suggested that the functional state of KdpD is at least a dimer and that the kinase reaction occurs in *trans*, meaning that one subunit binds ATP in the HATPase_c domain and phosphorylates the other subunit in the HisKA domain. A similar mechanism of autophosphorylation has been demonstrated for other histidine kinases as EnvZ, NtrB and CheA (Yang and Inouye, 1991; Ninfa *et al.*, 1993; Swanson *et al.*, 1993; Wolfe and Stewart, 1993). To solve the question whether KdpD undergoes a monomer-to-dimer transition upon activation, the relative molecular masses of non-phosphorylated KdpD and phosphorylated KdpD were determined using several techniques. The molecular mass of native KdpD correlated with a KdpD dimer, and there was no difference between KdpD and phospho-KdpD. Cross-linking experiments with single Cys KdpD derivatives provided evidence for a close contact between the monomers in the transmitter domain as well as in TM1, but the Cys residues were without role for the stabilization of the dimer (Heermann *et al.*, 1998). Nevertheless, an intramolecular disulfide bridge formed between Cys852 and Cys874 was found to be important for kinase activity (Jung *et al.*, 1998).

Generally the response regulator KdpE forms a separate protein, however, in some actinomycetes such as *Saccharopolyspora erythraea* or *Streptomyces* sp. KdpD is fused to the cognate response regulator KdpE, forming a hybrid histidine kinase.

The response regulator KdpE

KdpE belongs to the OmpR/PhoB response regulator family. Like all members of this family, KdpE consists of an N-terminal receiver domain (Response_reg, pfam 00072) and a C-terminal DNA-binding output domain (Trans_reg_C, cd00383). The receiver domain contains the aspartate phosphorylation site, which is Asp52 in KdpE (Altendorf *et al.*, 1994). Replacement of Asp52 against Asn resulted in a non-phosphorylable KdpE derivative (Lucassen, 1998). In contrast, the replacement of Asp9 with Asn led to a KdpE derivative that still can be phosphorylated although with lower efficiency. This result supported the notion that Asp9 participates in the catalysis of phosphorylation and plays a similar role in the 'acid pocket' as it was already postulated for the corresponding Asp residues of other response regulators (Lukat *et al.*, 1991). DNaseI footprint analysis and gel retardation experiments demonstrated that KdpE as well as phospho-KdpE bind to a $[T_n]$-rich region upstream of the *kdpFABC* promoter (Sugiura *et al.*, 1992, 1993), whereby phospho-KdpE had a 10-fold higher affinity to the DNA than KdpE (Nakashima *et al.*, 1991; Lucassen, 1998).

The crystal structure of the receiver domain of *E. coli* KdpE was solved by X-ray crystallography in the presence and absence of the phosphoryl analogue BeF_3^- so that the phosphorylated form of the N-terminal domain of KdpE could be compared with the non-phosphorylated form (Toro-Roman *et al.*, 2005). The domain exhibits the conserved $(\beta\alpha)_5$ fold of response regulator receiver domains that consists of a central five stranded parallel β-sheet surrounded by five amphipathic helices. Glu8, Asp9, and Asp52 position an Mg^{2+} ion required for catalysis of phosphoryl transfer to Asp52. The phosphorylation site Asp52 is located in the β4–α4 loop in close distance to Ser79 and Tyr98. These amino acids are conserved in KdpE and presumably involved in the switch mechanism of activation associated with phosphorylation of Asp52 (Toro-Roman *et al.*, 2005). The activation of KdpE is analogous to other response regulators, and involves only small structural alterations within the receiver domain. The phosphoryl group is bound to one carboxylate oxygen of Asp52. Other interactions of the phosphoryl side chain include a hydrogen bond to Ser79, a salt bridge to Lys101, and contacts with the backbone nitrogen atoms of Gly54 and Ala80. Upon phosphorylation, movement of Ser79 into the active site correlates with movement of Tyr98 into an inward position where it can form a hydrogen bond with the main-chain carbonyl oxygen of Arg81, fixing and stabilizing the β4–α4 loop in an active conformation. In non-phosphorylated KdpE, Ser79 and Tyr98 are in similar active conformations, with the difference that Ser79 and the β4–α4 loop are ~1 Å farther from the active site as compared with their positions in phospho-KdpE (Toro-Roman *et al.*, 2005).

Like in other OmpR/PhoB response regulators, the dimer interface in KdpE is lined by conserved amino acids, and dimerization is hypothesized to be an important step for the activation of the response regulator's output domain (Toro-Roman *et al.*, 2005). Indeed biochemical evidence was obtained that KdpE undergoes a monomer-to-dimer transition upon phosphorylation (Lucassen, 1998).

Nature of the stimulus perceived by KdpD

The nature of the stimulus that is sensed by KdpD has been puzzling for a long time. Epstein and colleagues found that an increase of external osmolarity at a constant K^+ concentration, a condition that reduces turgor, induced *kdpFABC* expression. The induction level correlated with the magnitude of osmotic stress. Because K^+ plays a major role in maintaining turgor, it was supposed that KdpD senses a decrease in turgor or some effect thereof (Laimins *et al.*, 1981; Malli and Epstein, 1998). This model has been challenged by the finding that *kdpFABC* expression is only induced when the osmolarity is increased by a salt and not by a sugar (Gowrishankar, 1985; Sutherland *et al.*, 1986; Asha and Gowrishankar, 1993). Therefore, Mizuno and colleagues suggested that the sensing mechanisms for K^+ limitation and osmotic upshift are mechanistically different (Sugiura *et al.*, 1994). Other groups argued that the K^+ signal is related to the internal K^+ level and/or the processes of K^+ transport (Asha and Gowrishankar, 1993; Frymier *et al.*, 1997) or the external K^+ concentration (Roe *et al.*, 2000). Measurements of the cytoplasmic volume of cells exposed to different external osmolytes revealed that reduction of turgor is not the stimulus for KdpD (Hamann *et al.*, 2008). It is important to note that the level of *kdpFABC* expression is at least 10-fold higher under K^+ limitation than in response to salt stress, arguing for a specific K^+ effect on KdpD (Jung *et al.*, 2001b; Hamann *et al.*, 2008). This hypothesis is supported by the fact that extracellular Cs^+, which is taken up and significantly lowers the intracellular available K^+, induces *kdpFABC* expression (Jung *et al.*, 2001b). *In vitro* phosphorylation assays with inverted membrane vesicles (Voelkner *et al.*, 1993) or proteoliposomes (Nakashima *et al.*, 1993) demonstrated that KdpD kinase activity is stimulated by salts such as NaCl or KCl, whereby NaCl was much more effective than KCl. Another *in vitro* test system that was based on right-side-out membrane vesicles provided first evidence for an inhibitory effect of K^+ on the kinase activity when provided from the inside of the vesicles (Jung *et al.*, 2000). This finding was supported by the results obtained with the *in vitro* reconstructed signal transduction cascade, consisting of KdpD in proteoliposomes, purified KdpE, a DNA fragment comprising the KdpE-binding site, and a mixture of ATP/ADP. Using this experimental setup, an inhibitory effect of K^+ was shown. The higher the K^+ concentration, the lower the level of phosphorylated KdpE (Heermann *et al.*, 2009b; Lüttmann *et al.*, 2009). Based on these results, it was proposed that the intracellular K^+ concentration directly influences KdpD by down-regulating the autophosphorylation activity.

An increase of the ionic strength imposed by salts in the lumen of right-side-out membrane vesicles containing KdpD stimulated KdpD kinase activity (Jung *et al.*, 2000). Because of the loss of K^+ or owing to an osmotic upshift, cells lose water, a process that is associated with an increase of the concentration of all dissolved molecules and consequently an increase of the ionic strength (Record *et al.*, 1998). Thus, it is conceivable that KdpD detects alterations of the intracellular ionic strength. In general, an increase of the osmolarity imposed by salts induced *kdpFABC* expression stronger than non-polar molecules of the same osmolarity. This effect could be explained by a specific effect of salts on phospholipids or an interaction between phospholipids and KdpD. Indeed, KdpD autophosphorylation activity was found to be dependent on negatively charged phospholipids, whereby the structure of the phospholipids was of minor importance (Stallkamp *et al.*, 1999). Moreover, the lipid composition of *E. coli* changes in a K^+-dependent manner. The negatively charged phospholipid cardiolipin (net charge −2) was elevated in cells exposed to K^+ limitation (Schniederberend *et al.*, 2010).

Comparison of various KdpD sequences from different bacteria revealed that the N-terminal domain of KdpD is highly conserved and includes two motifs (Walker A and Walker B) that are very similar to the classical ATP-binding sites of ATP-requiring enzymes. By means of photoaffinity labelling with 8-azido-[α^{32}P]ATP, direct evidence was obtained for the existence of an ATP-binding site located in the N-terminal domain of KdpD (Heermann *et al.*, 2000). Truncated KdpD derivatives lacking this site were characterized by a deregulated phosphatase activity (Jung and Altendorf, 1998b). Therefore, it was proposed that binding of ATP to the N-terminal domain modulates the ratio between kinase to phosphatase activities of KdpD. Because the intracellular ATP concentration is elevated upon an osmotic upshift (Ohwada and Sagisaka, 1987), the internal ATP level is discussed as the third stimulus for KdpD.

To sum up, the current model proposes that KdpD perceives and integrates three intracellular chemical stimuli: (1) the K^+ concentration; (2) the ionic strength; and (3) the ATP concentration.

K^+ supply

The cytoplasmic K^+ pool is maintained by a set of K^+ transport systems that differ in architecture, transport mechanism and affinity for K^+. Most of these transporters or transport subunits, respectively, belong to the HKT1 K^+ symporter superfamily of both prokaryotes and eukaryotes. It is assumed that the HKT1 superfamily is evolutionarily derived from the superfamily of K^+ channels in bacteria (Durell *et al.*, 1999). The most prominent example of a bacterial K^+ channel is KcsA of *Streptomyces lividans* (Gross *et al.*, 1999). One K^+ channel subunit consists of a typical motif formed by two transmembrane domains including a P-loop that opens and closes the channel (MPM motif). Homologous motifs are found in the K^+ translocating subunits of secondary transporters. While K^+ channels are formed by four MPM motif-containing subunits, the transmembrane subunits of the HKT1 symporter proteins have four corresponding MPM motifs within a single polypeptide. These transporters putatively originate from bacterial K^+ channels by at least two gene duplications and two gene fusion events. HKT1 transporters contain a much simpler K^+ selectivity motif (only one conserved glycine residue in their P-loops filter) than K^+ channels that share a conserved Thr-Val-Gly-Tyr-Gly motif.

The multicomponent complex Trk is one of the most conserved HKT1 K^+ uptake systems in bacteria. Trk systems have moderate affinity for K^+ (around 1 mM), are generally constitutively produced, and are probably energized by H^+ transport in symport with K^+. The products of three different and unclustered genes *trkA*, *trkE*, and *trkH* are necessary for Trk activity in *E. coli* and encode a classical ABC transporter (Fig. 10.2) (Epstein and Kim, 1971; Dosch *et al.*, 1991). The homologues of *Salmonella typhimurium* are named *sapG*, *sapD* and *sapJ*, respectively, because they were previously identified as genes making the corresponding mutants sensitive towards antimicrobial peptides (Parra-Lopez *et al.*, 1994). Subsequent studies revealed that *sapD* and the other *trk* genes are required because a high K^+ transport capacity is needed for maintaining K^+ homeostasis in the presence of toxic peptides (Stumpe and Bakker, 1997). TrkA binds NADH or NAD^+ (Bossemeyer *et al.*, 1989; Schlösser *et al.*, 1993), whereas TrkE (SapD) is one of the two ATP-binding components of ABC transporters (Parra-Lopez *et al.*, 1994). The Trk system requires both the proton motive force as well as ATP for transport activity (Rhoads and Epstein, 1977). *E. coli* encodes a second transmembrane protein, named TrkG, which is very similar to TrkH and can even replace it to allow full Trk activity (Dosch *et al.*, 1991). Both belong to the HKT1 superfamily of bacterial K^+ transporters (Durell *et al.*, 1999). In contrast to TrkH, TrkG is only partially-dependent on TrkE suggesting that an ATP-binding cassette of another ABC transporter can replace TrkE/SapD (Dosch *et al.*, 1991).

Another K^+ transport system present in *E. coli* is Kup, formerly TrkD (Fig. 10.2) (Trchounian and Kobayashi, 1999). Both Trk and Kup activity probably require the proton-motive force (PMF) to energize transport under aerobic conditions (Stewart *et al.*, 1985; Zakharyan and Trchounian, 2001; Epstein, 2003). Kup activity is more important at low pH where its maximum rate exceeds that of Trk. K^+ uptake via Kup is inhibited at elevated osmolarity.

Many bacteria have another K^+ uptake system, Ktr, that is similar to Trk. Ktr consists of two

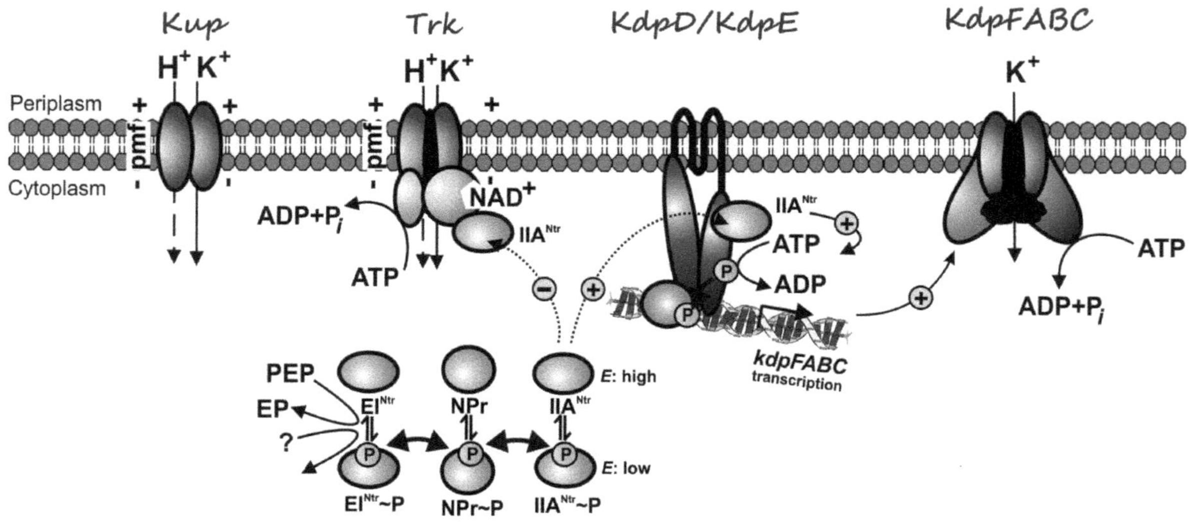

Figure 10.2 K^+ uptake in *E. coli*. *E. coli* has three different K^+ uptake systems, the constitutively produced transporters Trk and Kup, and the inducible high-affinity K^+ transporter KdpFABC. The corresponding *kdpFABC* operon is under transcriptional control of the KdpD/KdpE two-component system. Under K^+ limitation or high osmolarity, conditions when the other systems cannot keep up with the cell's requirement for K^+, expression of *kdpFABC* is induced. There is a connection between the Kdp system and Trk via the IIA^{Ntr} component of the Ntr-phosphotransferase system (Ntr-PTS; see text for details). Dephosphorylated IIA^{Ntr} binds to the energy coupling TrkA subunit and inhibits K^+ transport via Trk. Likewise dephosphorylated IIA^{Ntr} binds to the histidine kinase KdpD triggering kinase activity. As there is probable cross-talk between Ntr-PTS and the canonical carbohydrate PTS, the energy status of the cell influences K^+ transport. Fast growing cells (*E*: high) have a high demand for K^+, and prefer KdpFABC for K^+ uptake. In this way, K^+ requirements are adequately adapted to the energy status of the cell.

subunits, KtrA and KtrB, while KtrA is a distant relative of TrkA, and KtrB is distantly related to TrkH (Nakamura *et al.*, 1998). The Ktr system of *Vibrio alginolyticus* is Na^+ dependent, and it is assumed that K^+ transport is energized by the symport of Na^+ instead of H^+ (Tholema *et al.*, 1999). KtrA belongs to the KTN/RCK family of regulatory proteins and protein domains. These cytoplasmic proteins are known to regulate the activity of a number of K^+ transporters and K^+ channels upon ligand binding. KtrA contains a G-rich, ADP-moiety binding beta-alpha-beta-fold ('Rossman fold'). Binding of ATP specifically promoted complex formation of KtrA with KtrB (Kroning *et al.*, 2007). The K^+-translocating subunit KtrB is, like TrkG/TrkH a member of the HKT1 superfamily of K^+ transporters (Durell and Guy, 1999; Durell *et al.*, 1999). KtrB has an unusual long transmembrane stretch that contains several small and polar amino acids. This region is flanked by the two alpha-helices and may form a flexible gate at the cytoplasmic side of the membrane that prevents K^+ permeation to the cytoplasm (Hänelt *et al.*, 2010, 2011). *Bacillus subtilis* has two Ktr homologues, KtrAB and KtrBC. KtrAB has approximately 10-fold higher affinity for K^+ than KtrBC. Production of each system is independent of the medium K^+ concentration, but when one of both systems is inactivated at the transcriptional level, activity of the other system is enhanced. Both systems are important for growth under high osmolarity, but can only operate when moderate levels of K^+ are available (Holtmann *et al.*, 2003).

KdpFABC is a high affinity K^+ transport system and the only one whose production is regulated by environmental conditions (Fig. 10.2) (Altendorf and Epstein, 1996). Reduction in K^+ supply or salt stress rapidly increase expression of *kdpFABC*. The signal(s) caused by these changes (see above) are perceived by the histidine kinase KdpD and then transmitted to the response regulator KdpE that stimulates transcription of the *kdpFABC* operon. KdpFABC belongs to the family of P-type ATPases (Møller *et al.*, 1996), that share common structural features as well as the catalytic mechanism. KdpA is the K^+ translocating subunit that shows similarities to KcsA-like K^+-channel proteins and therefore is assumed to belong to the HKT1 superfamily of K^+ transporters (Durell *et al.*, 2000; van der Laan *et al.*, 2002). The KdpB subunit shares homologies to other P-type ATPases and represents the catalytic subunit performing ATP hydrolysis (Bramkamp and Altendorf, 2005). The KdpC subunit contains only one N-terminal transmembrane span and a large hydrophilic cytoplasmic region (Gassel and Altendorf, 2001). The soluble portion of KdpC binds ATP and is assumed to act as a catalytic chaperone in the nucleotide-binding mechanism (Ahnert *et al.*, 2006). KdpF is a small hydrophobic peptide that consists of only one transmembrane domain and is supposed to stabilize the complex (Gassel *et al.*, 1999). The complete KdpFABC complex is functional as a dimer (Heitkamp *et al.*, 2008). KdpA is in close contact with the transmembrane region of the ATP-hydrolysing KdpB subunit. The cytosolic C-terminal domain of KdpC is located close to the nucleotide binding domain of KdpB (Heitkamp *et al.*, 2009; Irzik *et al.*, 2011). In *E. coli*, K^+ homeostasis is mediated by the three K^+ uptake systems TrkG/TrkH, Kup, and KdpFABC (Fig. 10.2). The latter one is the only inducible system and only produced when the constitutive systems are not able to keep up with the cell's requirement for K^+.

Trk and Kdp are linked via a clever regulatory mechanism: a protein of a phosphotransferase system, enzyme IIA^{Ntr}, directly interacts with KdpD as well as with TrkA (Lee *et al.*, 2007; Lüttmann *et al.*, 2009). IIA^{Ntr} is a component of the Ntr-phosphotransferase system that is conserved in many proteobacteria (Deutscher *et al.*, 2006). In this system, EI^{Ntr} (encoded by *ptsP*) is phosphorylated by phosphoenolpyruvate (PEP) or another thus far unknown source, and transfers phosphoryl groups to NPr that subsequently phosphorylates protein IIA^{Ntr} (Fig. 10.2) (Rabus *et al.*, 1999; Zimmer *et al.*, 2008). The Ntr-PTS works in parallel to the phosphoenolpyruvate:carbohydrate phosphotransferase system (transport-PTS) in *E. coli*. Since no obvious phosphoryl group acceptor was identified for phospho-IIA^{Ntr}, it is speculated that this system has regulatory rather than transport function. Dephosphorylated IIA^{Ntr} binds and inhibits the NAD^+/NADH-binding Trk subunit TrkA (Lee *et al.*, 2007). In parallel, the dephosphorylated form of IIA^{Ntr} binds to KdpD, triggers kinase activity, and therefore stimulates *kdpFABC* promoter activity (Fig. 10.2). The connection between the Ntr-PTS and the Kdp system was found in a search for a potential phosphoacceptor for IIA^{Ntr}. Overproduction of IIA^{Ntr} resulted in the accumulation of a phosphorylated protein that was identified as phospho-KdpB (Lüttmann *et al.*, 2009). In parallel, a transposon clone library was screened for enhanced expression of *kdpFABC* in the presence of 5 mM K^+, a concentration that normally represses the target operon. Two independent transposon mutants were isolated out of 9000 tested, in which *kdpFABC* expression was significantly elevated. Both insertions mapped in gene *ptsP* that encodes EI^{Ntr} providing independent evidence for the interrelation between Kdp and Ntr-PTS (Lüttmann *et al.*, 2009). Two-hybrid data and biochemical analysis suggested that IIA^{Ntr} interacts with KdpD and stimulates its autokinase activity. Subsequently, the level of phospho-KdpE is increased and *kdpFABC* is induced. Since there is 'cross-talk' between the Ntr-PTS and the transport-PTS, the carbon source influenced *kdpFABC* expression. When cells utilized preferred carbohydrates like glucose, which results in dephosphorylation of the transport-PTS and also of IIA^{Ntr}, *kdpFABC* expression was enhanced. When there is strong need for K^+ in fast growing cells, then KdpFABC seems to be the preferred K^+ uptake system. In this way, bacterial K^+ requirements can be adequately adapted to the energy status of the cell.

Osmotic stress

Bacteria are frequently exposed to osmotic stress in natural habitats as well as during biotechnological applications. In nature, water availability can be low because of drought or increased salt concentrations, which leads to osmotic stress (Bremer and Krämer, 2000). During fermentations, high concentrations of both nutrients and/or products might also cause osmotic challenges. The bacterial response towards osmotic stress represents a multiphase adjustment comprising ion transport processes as well as the accumulation of compatible solutes (Fig. 10.3) (Reed and Walsby, 1985; Wood, 1999; Bremer and Krämer, 2000).

In general bacteria respond to osmotic upshifts in three overlapping phases: dehydration (loss of cell water) (phase I), adjustment of cytoplasmic solvent composition and rehydration (phase II), and cellular remodelling (phase III) (Fig. 10.3). Since the water permeability of the cytoplasmic membrane is high, a higher osmolarity outside causes a rapid loss of cellular water (phase I). Consequently, the turgor pressure is reduced. Moreover, high external osmolarity exerts pleiotropic effects on the physiology of bacteria. For example, osmotically stressed *E. coli* cells are characterized by alterations in respiration, ATP and polyamine concentration, cytoplasmic pH, and a down-regulation of central metabolic pathways combined with a decreased transcription of genes encoding ribosomal proteins (Weber and Jung, 2002). Osmotically stressed *B. subtilis* cells exhibit alterations in the composition of the cytoplasmic membrane and the cell wall, the supercoiling of plasmid DNA, defects in sporulation, iron uptake, chemotaxis and motility (Steil *et al.*, 2003).

To survive bacteria immediately counteract this dangerous situation by activating K^+ uptake systems (phase I). In both Gram-negative and Gram-positive bacteria the most rapid response to counteract osmotic upshock is a stimulation of K^+ uptake. The Trk and the Kdp system play a major role in osmostress response in *E. coli* (Epstein, 2003), while secondary active Ktr-type transporters are involved in the cellular defence against osmotic stress in *B. subtilis* and *Synechocystis* sp. PCC 6803 (Berry *et al.*, 2003; Holtmann *et al.*, 2003). Recently, first experimental evidence was provided that the K^+ channel CglK of *Corynebacterium glutamicum* is sufficient for the osmotic stress-dependent K^+ uptake, but only when these bacteria grow in an environment with sufficient K^+ (Ochrombel *et al.*, 2011). Besides its osmoregulatory role, the intracellular K^+ influences nucleoid structure, cellular energetics, cytoplasmic ionic strength as well as ion composition. K^+ glutamate accumulation may specifically counteract the effects of increased macromolecular crowding on nucleoid organization (Record *et al.*, 1998), and stimulate transcription by disengaging inhibitory polymerase interactions e.g. at the sigma(38) promoter (Lee and Gralla, 2004; Rosenthal *et al.*, 2008). Some K^+ accumulation occurs even in the presence of compatible solutes, such as glycine betaine, which are more effective than K^+ in restoring cellular hydration after an osmotic upshift. All these observations suggest that the

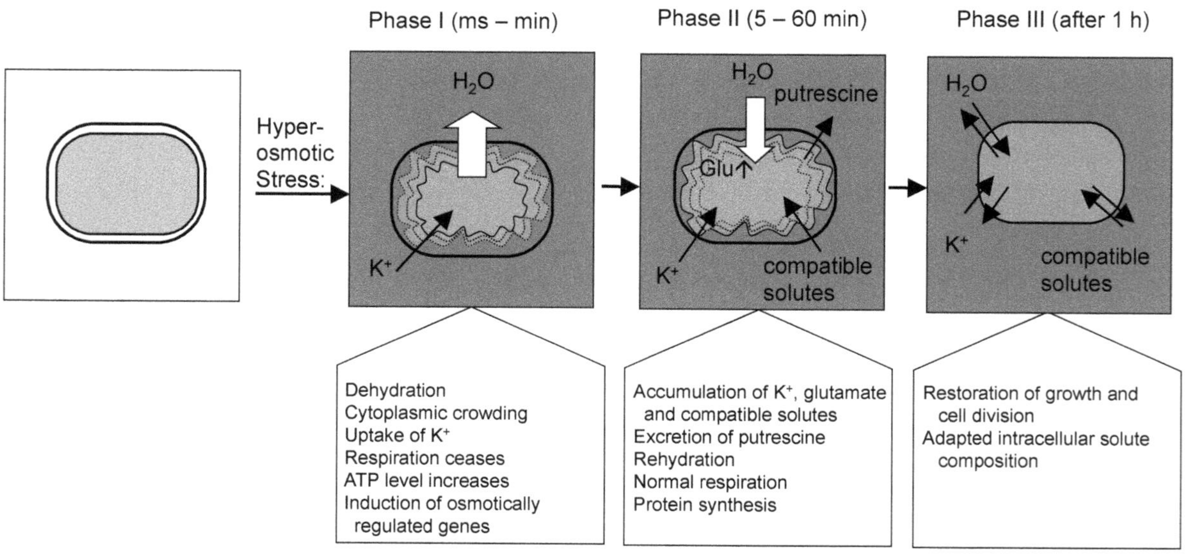

Figure 10.3 Phases of the hyperosmotic stress response and adaptation of *E. coli* K-12. Hyperosmotic stress response and adaptation is accompanied by pleiotropic structural and physiological effects as indicated. Basically three different phases of different duration can be distinguished. For details see text and the publication by Wood (1999).

adjustments in cytoplasmic K^+ level meet multiple objectives. Therefore, it is not surprising, that K^+ plays also an important role in osmosensing (see below) (Heermann and Jung, 2004; Wood, 2006).

The rapid uptake of K^+ is followed by the accumulation, either by synthesis and/or uptake, of organic low molecular weight osmolytes, termed compatible solutes owing to their compatibility with cellular processes at high internal concentrations (phase II). Besides their role as osmolytes, compatible solutes have also a stabilizing effect on proteins. Compatible solutes are restricted to a few classes including amino acids, betaines, amine derivatives, polyols, and urea which are found not only in prokaryotes but also in higher forms of life reflecting the constraints for such compounds to be compatible with macromolecules. These solutes increase the internal osmotic pressure without interfering vital cellular protein functions (Wood *et al.*, 2001).

A number of osmosensors and osmoregulated transporters has been studied in more depth. The transporters ProP of *E. coli* (Racher *et al.*, 1999), BetP of *C. glutamicum* (Rübenhagen *et al.*, 2000), and OpuA of *Lactococcus lactis* (van der Heide *et al.*, 2001), which mediate the uptake of proline and glycine betaine, respectively, clearly function both as osmosensors and osmo-regulated transporters. Furthermore, the histidine kinases KdpD and EnvZ of *E. coli* as well as MtrB of *C. glutamicum* operate as osmosensors and transmit the information via phosphorylation reactions to their cognate response regulators resulting in altered gene expression (Jung *et al.*, 2001a; Jung and Altendorf, 2002; Möker *et al.*, 2004).

A central question is how do these receptors and transporters sense alterations in the external osmolarity? Osmostress is a physico-chemical parameter that cannot be deduced to a single molecule, and as described above it causes pleiotropic effects on cell physiology. Thus, it is quite difficult to experimentally elucidate the nature of the primary stimuli osmosensors respond to. In the literature the following primary stimuli are discussed: turgor, membrane strain or shrinkage, curvature of the membrane, external or internal osmolarity, external or internal ion concentration or ionic strength, the concentration of specific cytoplasmic or periplasmic solutes, the transmembrane osmotic gradient, or macromolecular crowding (Wood, 1999; Morbach and Krämer, 2002; Sleator and Hill, 2002). It has to be emphasized that all these changes are more or less related to the extreme fast (milliseconds) water efflux along the osmotic gradient and therefore all the alterations described occur simultaneously, which makes it so difficult to discriminate the primary stimuli perceived by osmosensors *in vivo*. Nevertheless, based on *in vitro* systems with less complexity, the following physico-chemical parameters have been unravelled as primary stimuli for osmosensing proteins: the cytoplasmic K^+ and ATP levels, the composition of the cytoplasmic membrane and the ionic strength of the cytoplasm (Heermann and Jung, 2004; Wood, 2006; Krämer, 2009).

The betaine transporter BetP was designated as a K^+-specific chemosensor, because among the tested monovalent cations K^+, Rb^+ and Cs^+ activated BetP more effectively than did Na^+, NH_4^+, or choline *in vitro* (Rübenhagen *et al.*, 2001; Schiller *et al.*, 2004). Similarly, the activity of *E. coli* ProP was specifically stimulated when K^+ was supplied to solute-depleted bacteria (MacMillan *et al.*, 1999). However, whether K^+ is the specific stimulus to activate ProP is still under investigation (Culham *et al.*, 2008). It was suggested that ProP activity reflects ProP hydration (Wood, 2006). Nevertheless, expression of *proP* is K^+-dependent in *C. glutamicum* (Ochrombel *et al.*, 2011). For OpuA of *L. lactis* the impact of changing cytoplasmic ionic strength on the membrane lipids was found to be the predominant trigger for the activation of this transport protein (van der Heide *et al.*, 2001; van der Heide and Poolman, 2002; Biemans-Oldehinkel *et al.*, 2006). Therefore, Poolman and his colleagues (2002) designated OpuA as an ionic strength sensor, because diverse salts elicited this response.

The kinase activity of full-length EnvZ of *E. coli* in proteoliposomes is stimulated by monovalent cations, whereby K^+ produced the strongest effect (Jung *et al.*, 2001a). Nevertheless, it is still unclear, whether EnvZ is a true osmosensor at all (Yuan *et al.*, 2011). Although, the histidine kinase MtrB of *C. glutamicum* requires K^+ to stimulate the autokinase activity, other stimuli, such as the hydration state, are proposed to activate this protein *in vivo* (Möker *et al.*, 2007).

KdpD also responds to osmotic stress imposed by salts in *E. coli* (Sugiura *et al.*, 1994). KdpD activation occurs at a time when the intracellular K^+ concentration is already far above the normal level (Fig. 10.3, end of phase I). However, K^+ has an inhibitory effect on the autokinase activity (Jung *et al.*, 2000; Lüttmann *et al.*, 2009). Therefore, KdpD has to sense and integrate additional stimuli. Indeed, KdpD contains a regulatory ATP-binding site within the N-terminal input domain (Fig. 10.1), and a regulation of KdpD activity by ATP is conceivable because of the transient increase of the ATP concentration during osmoadaptation (Fig. 10.3) (Ohwada and Sagisaka, 1987). In addition, the kinase activity of KdpD correlates with the ionic strength and the presence of negatively charged phospholipids (Stallkamp *et al.*, 1999; Jung *et al.*, 2000). The higher the ionic strength the higher is the autokinase activity of KdpD (Voelkner *et al.*, 1993).

Recently, the universal stress protein UspC was identified that scaffolds the KdpD/KdpE signalling cascade under salt stress by interacting with the Usp domain in KdpD (Heermann *et al.*, 2009b). Scaffold proteins were discovered in eukaryotes more than 15 years ago. These proteins coordinate the physical assembly of components in a signalling pathway (Good *et al.*, 2011). UspC belongs to the Usp superfamily encompassing an ancient and conserved group of proteins that are found in bacteria, archaea, fungi, flies and plants (Kvint *et al.*, 2003). The *usp* genes encode either small Usp proteins (one Usp domain), larger versions with two Usp domains in tandem or Usp domains integrated in multidomain proteins (Siegele, 2005). *E. coli* contains six Usp proteins that are divided into two subfamilies based on sequence similarities (Gustavsson *et al.*, 2002). Three of these proteins belong to the UspA subfamily (UspA, UspC, and UspD), and two to the UspFG subfamily (UspF and UspG). Usp proteins are produced in response to different stress conditions. For example, UspA is one of the most abundant proteins in growth-arrested cells. The copy numbers of UspA, UspC, UspD, and UspE increase upon exposure of cells to starvation, moderate heat stress, oxidative stress or osmotic stress (Gustavsson *et al.*, 2002). KdpD has a Usp domain within the N-terminal input domain belonging to the UspA-subfamily (Fig. 10.1), and it was hypothesized that KdpD might form heteromers by interacting with a soluble Usp protein. It was found that UspC specifically stimulated the *in vitro* reconstructed signalling cascade (KdpD → KdpE → DNA) resulting in phosphorylation of KdpE at a K^+ concentration that would otherwise almost prevent phosphorylation (Heermann *et al.*, 2009a). In agreement, in a Δ*uspC* mutant KdpFABC production was significantly down-regulated when cells were exposed to salt stress, but unaffected under K^+ limitation. Biochemical studies revealed that UspC specifically interacts with the Usp domain in the stimulus perceiving N-terminal domain of KdpD. Even more, UspC stabilized the KdpD/phospho–KdpE/DNA complex but did not influence the enzymatic activities of KdpD. In summary, UspC can be regarded as one of the rare examples of a bacterial scaffold protein (Heermann and Jung, 2009). UspC also solved the puzzling question, how KdpD could be activated in response to osmotic stress under conditions of a high intracellular K^+ concentration.

The Kdp system and bacterial virulence

Recent studies revealed the importance of the KdpD/KdpE system in several pathogens. *Mycobacterium tuberculosis* contains only 11 histidine kinase/response regulator systems. Nevertheless, the KdpD/KdpE system was found to be responsible for the K^+-dependent *kdpFABC* expression in strain H37Rv and *M. smegmatis*. Importantly, deletion of *kdpDE* resulted in increased virulence. Mice infected with the *M. tuberculosis kdpDE* mutant died faster than those infected with wild-type bacteria (Parish *et al.*, 2003). Moreover, two lipoproteins LprF and LrpJ were identified that specifically interact with KdpD and modulate its activity (Steyn *et al.*, 2003).

The KdpD/KdpE system was found to be

crucial for the survival of the human and insect pathogen *Photorhabdus asymbiotica* in haemocytes of *Manduca sexta*. Curiously, a harmless *E. coli* strain equipped with the *P. asymbiotica* KdpD/KdpE was able to persist within hemocytes and became pathogenic towards insects by making the bacteria resistant to phagocytic killing (Vlisidou *et al.*, 2010). This finding suggested a central role for K^+ in this process and supported previous work implicating K^+ sensing in virulence of other bacteria and also in the process of killing of engulfed bacteria by neutrophils (Reeves *et al.*, 2002; Waterfield *et al.*, 2008).

In accordance, the *Yersinia pestis* KdpD/KdpE homologues promoted survival of the bacteria in human neutrophils. A *Y. pestis* Δ*kdpDE* mutant exhibited increased sensitivity towards human polymorphonuclear leucocytes (O'Loughlin *et al.*, 2010).

Listeria monocytogenes mutants lacking *kdpE* were attenuated in virulence indicating the importance of this system to encounter stressful conditions during infection (Brondsted *et al.*, 2003).

The *Staphylococcus aureus* KdpD/KdpE system couples K^+ sensing and Agr quorum sensing to program infection (Xue *et al.*, 2011). It is postulated that the KdpD/KdpE system is an important virulence regulator coordinating K^+ sensing and Agr signalling during pathogenesis in this bacterium. In agreement, a *S. aureus* Δ*kdpDE* mutant poorly survived in human blood cells compared with the wild-type (Zhao *et al.*, 2010). It has been demonstrated that autoinducer-2 (AI-2) regulates capsule polysaccharide synthase (*cap*) gene expression, which is dependent on the binding of phospho-KdpE to the *cap* promoter (Zhao *et al.*, 2010). In this way *S. aureus* might modulate its infection status by sensing external K^+ in different environments (Xue *et al.*, 2011).

Finally, KdpE was found to be involved in the expression of the LEE operon, a major virulence gene cluster of enterohaemorrhagic *E. coli* (EHEC) (Hughes *et al.*, 2009).

These examples underline the importance of the K^+-sensing KdpD/KdpE-system during colonization of pathogens in their hosts.

Outlook

The KdpD/KdpE system is one of the most conserved histidine kinase/response regulator systems. Even bacteria that acquired only a few histidine kinase/response regulator systems, e.g. *Mycobacterium*, have KdpD/KdpE homologues. Our studies on KdpD in *E. coli* underline the capability of this receptor to integrate diverse stimuli allowing optimal adaptation of the bacterium in various environments. Moreover, the link between Kdp and Ntr-PTS indicates an elegant mechanism to connect gene regulation with metabolic requirements to ensure K^+ homeostasis under various cellular conditions. It is conceivable that KdpD/KdpE in other bacterial species employ different sets of accessory proteins to fine-tune *kdpFABC* expression. In addition, bacterial genome analyses already indicated a highly variable domain composition of KdpD in various species.

In general, the structural dynamics of histidine kinases upon stimulus perception and signalling are not well understood, yet. This lack of information is mainly due to the difficulties to crystallize membrane proteins, and the concomitant unavailability of a three-dimensional structure for a full-length histidine kinase. KdpD is a multidomain histidine kinase with many interacting sites. Therefore, structural elucidation of individual soluble domains of this receptor and the subsequent reconstruction of a full-length receptor might be a promising strategy to gain insights into the structural refinement of this protein family.

In many pathogens the KdpD/KdpE system was found to be crucial during host colonization. Although the basic molecular mechanism of KdpD/KdpE functioning is known, its exact role in pathogenesis remains enigmatic. Since vertebrates lack histidine kinase/response regulator systems, KdpD/KdpE might be an interesting target for antibiotics.

References

Ahnert, F., Schmid, R., Altendorf, K., and Greie, J.C. (2006). ATP binding properties of the soluble part of the KdpC subunit from the *Escherichia coli* K^+-transporting KdpFABC P-type ATPase. Biochemistry 45, 11038–11046.

Altendorf, K., and Epstein, W. (1996). The Kdp-ATPase of *Escherichia coli*. Biomembranes 5, 403–420.

Altendorf, K., Voelkner, P., and Puppe, W. (1994). The sensor kinase KdpD and the response regulator KdpE control expression of the *kdpFABC* operon in *Escherichia coli*. Res. Microbiol. *145*, 374–381.

Asha, H., and Gowrishankar, J. (1993). Regulation of *kdp* operon expression in *Escherichia coli*: evidence against turgor as signal for transcriptional control. J. Bacteriol. *175*, 4528–4537.

Ballal, A., Heermann, R., Altendorf, K., and Jung, K. (2002). A chimeric *Anabaena/Escherichia coli* KdpD protein (Anacoli KdpD) functionally interacts with *E. coli* KdpE and activates *kdp* expression in *E. coli*. Arch. Microbiol. *178*, 141–148.

Ballal, A., Bramkamp, M., Rajaram, H., Zimmann, P., Apte, S.K., and Altendorf, K. (2005). An atypical KdpD homologue from the cyanobacterium *Anabaena* sp. strain L-31: cloning, *in vivo* expression, and interaction with *Escherichia coli* KdpD-CTD. J. Bacteriol. *187*, 4921–4927.

Berry, S., Esper, B., Karandashova, I., Teuber, M., Elanskaya, I., Rogner, M., and Hagemann, M. (2003). Potassium uptake in the unicellular cyanobacterium *Synechocystis* sp. strain PCC 6803 mainly depends on a Ktr-like system encoded by slr1509 (*ntpJ*). FEBS Lett. *548*, 53–58.

Biemans-Oldehinkel, E., Mahmood, N.A., and Poolman, B. (2006). A sensor for intracellular ionic strength. Proc. Natl. Acad. Sci. U.S.A. *103*, 10624–10629.

Booth, I.R. (1985). Regulation of cytoplasmic pH in bacteria. Microbiol. Rev. *49*, 359–378.

Bossemeyer, D., Schlösser, A., and Bakker, E.P. (1989). Specific cesium transport via the *Escherichia coli* Kup (TrkD) K^+ uptake system. J. Bacteriol. *171*, 2219–2221.

Bramkamp, M., and Altendorf, K. (2005). Single amino acid substitution in the putative transmembrane helix V in KdpB of the KdpFABC complex of *Escherichia coli* uncouples ATPase activity and ion transport. Biochemistry *44*, 8260–8266.

Brandon, L., Dorus, S., Epstein, W., Altendorf, K., and Jung, K. (2000). Modulation of KdpD phosphatase implicated in the physiological expression of the Kdp-ATPase of *Escherichia coli*. Mol. Microbiol. *38*, 1086–1092.

Bremer, E., and Krämer, R. (2000). Coping with osmotic challenges: osmoregulation through accumulation and release of compatible solutes. In Bacterial Stress Responses, Storz, G., and Hengge, R., eds (ASM Press, Washington, DC), pp. 79–97.

Brondsted, L., Kallipolitis, B.H., Ingmer, H., and Knochel, S. (2003). KdpE and a putative RsbQ homologue contribute to growth of *Listeria monocytogenes* at high osmolarity and low temperature. FEMS Microbiol. Lett. *219*, 233–239.

Culham, D.E., Romantsov, T., and Wood, J.M. (2008). Roles of K^+, H^+, H_2O, and $\Delta\psi$ in solute transport mediated by major facilitator superfamily members ProP and LacY. Biochemistry *47*, 8176–8185.

Deutscher, J., Francke, C., and Postma, P.W. (2006). How phosphotransferase system-related protein phosphorylation regulates carbohydrate metabolism in bacteria. Microbiol. Mol. Biol. Rev. *70*, 939–1031.

Dosch, D.C., Helmer, G.L., Sutton, S.H., Salvacion, F.F., and Epstein, W. (1991). Genetic analysis of potassium transport loci in *Escherichia coli*: evidence for three constitutive systems mediating uptake potassium. J. Bacteriol. *173*, 687–696.

Durell, S.R., and Guy, H.R. (1999). Structural models of the KtrB, TrkH, and Trk1,2 symporters based on the structure of the KcsA K^+ channel. Biophys. J. *77*, 789–807.

Durell, S.R., Hao, Y., Nakamura, T., Bakker, E.P., and Guy, H.R. (1999). Evolutionary relationship between K^+ channels and symporters. Biophys. J. *77*, 775–788.

Durell, S.R., Bakker, E.P., and Guy, H.R. (2000). Does the KdpA subunit from the high affinity K^+-translocating P-type KDP-ATPase have a structure similar to that of K^+ channels? Biophys. J. *78*, 188–199.

Epstein, W. (1992). Kdp, a bacterial P-type ATPase whose expression and activity are regulated by turgor pressure. Acta. Phys. Scand. Suppl. *607*, 193–199.

Epstein, W. (2003). The roles and regulation of potassium in bacteria. Prog. Nucleic Acid Res. Mol. Biol. *75*, 293–320.

Epstein, W., and Kim, B.S. (1971). Potassium transport loci in *Escherichia coli* K-12. J. Bacteriol. *108*, 639–644.

Frymier, J.S., Reed, T.D., Fletcher, S.A., and Csonka, L.N. (1997). Characterization of transcriptional regulation of the *kdp* operon of *Salmonella typhimurium*. J. Bacteriol. *179*, 3061–3063.

Gassel, M., and Altendorf, K. (2001). Analysis of KdpC of the K^+-transporting KdpFABC complex of *Escherichia coli*. Eur. J. Biochem. *268*, 1772–1781.

Gassel, M., Möllenkamp, T., Puppe, W., and Altendorf, K. (1999). The KdpF subunit is part of the K^+-translocating Kdp complex of *Escherichia coli* and is responsible for stabilization of the complex *in vitro*. J. Biol. Chem. *274*, 37901–37907.

Geer, L.Y., Domrachev, M., Lipman, D.J., and Bryant, S.H. (2002). CDART: protein homology by domain architecture. Genome Res. *12*, 1619–1623.

Good, M.C., Zalatan, J.G., and Lim, W.A. (2011). Scaffold proteins: hubs for controlling the flow of cellular information. Science *332*, 680–686.

Gowrishankar, J. (1985). Identification of osmoresponsive genes in *Escherichia coli*: evidence for participation of potassium and proline transport systems in osmoregulation. J. Bacteriol. *164*, 434–445.

Gross, A., Columbus, L., Hideg, K., Altenbach, C., and Hubbell, W.L. (1999). Structure of the KcsA potassium channel from *Streptomyces lividans*: a site-directed spin labeling study of the second transmembrane segment. Biochemistry *38*, 10324–10335.

Gustavsson, N., Diez, A., and Nyström, T. (2002). The universal stress protein paralogues of *Escherichia coli* are co-ordinately regulated and co-operate in the defence against DNA damage. Mol. Microbiol. *43*, 107–117.

Hamann, K., Zimmann, P., and Altendorf, K. (2008). Reduction of turgor is not the stimulus for the sensor kinase KdpD of *Escherichia coli*. J. Bacteriol. *190*, 2360–2367.

Hänelt, I., Wunnicke, D., Müller-Trimbusch, M., Vor der Brüggen, M., Kraus, I., Bakker, E.P., and Steinhoff, H.J. (2010). Membrane region M2C2 in subunit KtrB of the K^+ uptake system KtrAB from *Vibrio alginolyticus* forms a flexible gate controlling K^+ flux: an electron paramagnetic resonance study. J. Biol. Chem. *285*, 28210–28219.

Hänelt, I., Tholema, N., Kroning, N., Vor der Brüggen, M., Wunnicke, D., and Bakker, E.P. (2011). KtrB, a member of the superfamily of K^+ transporters. Eur. J. Cell Biol. *90*, 696–704.

Heermann, R., and Jung, K. (2004). Structural features and mechanisms for sensing high osmolarity in microorganisms. Curr. Opin. Microbiol. *7*, 168–174.

Heermann, R., and Jung, K. (2009). Stimulus perception and signaling in histidine kinases. In Bacterial Signaling, Krämer, R., and Jung, K., eds (Wiley-VCH, Weilheim, Germany), pp. 135–161.

Heermann, R., Altendorf, K., and Jung, K. (1998). The turgor sensor KdpD of *Escherichia coli* is a homodimer. Biochim. Biophys. Acta *1415*, 114–124.

Heermann, R., Altendorf, K., and Jung, K. (2000). The hydrophilic N-terminal domain complements the membrane-anchored C-terminal domain of the sensor kinase KdpD of *Escherichia coli*. J. Biol. Chem. *275*, 17080–17085.

Heermann, R., Altendorf, K., and Jung, K. (2003a). The N-terminal input domain of the sensor kinase KdpD of *Escherichia coli* stabilizes the interaction between the cognate response regulator KdpE and the corresponding DNA-binding site. J. Biol. Chem. *278*, 51277–51284.

Heermann, R., Fohrmann, A., Altendorf, K., and Jung, K. (2003b). The transmembrane domains of the sensor kinase KdpD of *Escherichia coli* are not essential for sensing K^+ limitation. Mol. Microbiol. *47*, 839–848.

Heermann, R., Lippert, M.L., and Jung, K. (2009a). Domain swapping reveals that the N-terminal domain of the sensor kinase KdpD in *Escherichia coli* is important for signaling. BMC Microbiol. *9*, 133.

Heermann, R., Weber, A., Mayer, B., Ott, M., Hauser, E., Gabriel, G., Pirch, T., and Jung, K. (2009b). The universal stress protein UspC scaffolds the KdpD/KdpE signaling cascade of *Escherichia coli* under salt stress. J. Mol. Biol. *386*, 134–148.

van der Heide, T., and Poolman, B. (2002). ABC transporters: one, two or four extracytoplasmic substrate-binding sites? EMBO Rep. *3*, 938–943.

van der Heide, T., Stuart, M.C., and Poolman, B. (2001). On the osmotic signal and osmosensing mechanism of an ABC transport system for glycine betaine. EMBO J. *20*, 7022–7032.

Heitkamp, T., Kalinowski, R., Böttcher, B., Borsch, M., Altendorf, K., and Greie, J.C. (2008). K^+-translocating KdpFABC P-type ATPase from *Escherichia coli* acts as a functional and structural dimer. Biochemistry *47*, 3564–3575.

Heitkamp, T., Böttcher, B., and Greie, J.C. (2009). Solution structure of the KdpFABC P-type ATPase from *Escherichia coli* by electron microscopic single particle analysis. J. Struct. Biol. *166*, 295–302.

Holtmann, G., Bakker, E.P., Uozumi, N., and Bremer, E. (2003). KtrAB and KtrCD: two K^+ uptake systems in *Bacillus subtilis* and their role in adaptation to hypertonicity. J. Bacteriol. *185*, 1289–1298.

Hughes, D.T., Clarke, M.B., Yamamoto, K., Rasko, D.A., and Sperandio, V. (2009). The QseC adrenergic signaling cascade in Enterohemorrhagic *E. coli* (EHEC). PLoS Pathog. *5*, e1000553.

Irzik, K., Pfrotzschner, J., Goss, T., Ahnert, F., Haupt, M., and Greie, J.C. (2011). The KdpC subunit of the *Escherichia coli* K^+-transporting KdpB P-type ATPase acts as a catalytic chaperone. FEBS J. *278*, 3041–3053.

Jung, K., and Altendorf, K. (1998a). Individual substitutions of clustered arginine residues of the sensor kinase KdpD of *Escherichia coli* modulate the ratio of kinase to phosphatase activity. J. Biol. Chem. *273*, 26415–26420.

Jung, K., and Altendorf, K. (1998b). Truncation of amino acids 12–128 causes deregulation of the phosphatase activity of the sensor kinase KdpD of *Escherichia coli*. J. Biol. Chem. *273*, 17406–17410.

Jung, K., and Altendorf, K. (2002). Towards an understanding of the molecular mechanisms of stimulus perception and signal transduction by the KdpD/KdpE system of *Escherichia coli*. J. Mol. Microbiol. Biotechnol. *4*, 223–228.

Jung, K., Tjaden, B., and Altendorf, K. (1997). Purification, reconstitution, and characterization of KdpD, the turgor sensor of *Escherichia coli*. J. Biol. Chem. *272*, 10847–10852.

Jung, K., Heermann, R., Meyer, M., and Altendorf, K. (1998). Effect of cysteine replacements on the properties of the turgor sensor KdpD of *Escherichia coli*. Biochim. Biophys. Acta *1372*, 311–322.

Jung, K., Veen, M., and Altendorf, K. (2000). K^+ and ionic strength directly influence the autophosphorylation activity of the putative turgor sensor KdpD of *Escherichia coli*. J. Biol. Chem. *275*, 40142–40147.

Jung, K., Hamann, K., and Revermann, A. (2001a). K^+ stimulates specifically the autokinase activity of purified and reconstituted EnvZ of *Escherichia coli*. J. Biol. Chem. *276*, 40896–40902.

Jung, K., Krabusch, M., and Altendorf, K. (2001b). Cs^+ induces the *kdp* operon of *Escherichia coli* by lowering the intracellular K^+ concentration. J. Bacteriol. *183*, 3800–3803.

Krämer, R. (2009). Osmosensing and osmosignaling in *Corynebacterium glutamicum*. Amino Acids *37*, 487–497.

Kroning, N., Willenborg, M., Tholema, N., Hänelt, I., Schmid, R., and Bakker, E.P. (2007). ATP binding to the KTN/RCK subunit KtrA from the K^+-uptake system KtrAB of *Vibrio alginolyticus*: its role in the formation of the KtrAB complex and its requirement *in vivo*. J. Biol. Chem. *282*, 14018–14027.

Kvint, K., Nachin, L., Diez, A., and Nyström, T. (2003). The bacterial universal stress protein: function and regulation. Curr. Opin. Microbiol. *6*, 140–145.

van der Laan, M., Gassel, M., and Altendorf, K. (2002). Characterization of amino acid substitutions in KdpA, the K^+-binding and -translocating subunit of the KdpFABC complex of *Escherichia coli*. J. Bacteriol. *184*, 5491–5494.

Laimins, L.A., Rhoads, D.B., and Epstein, W. (1981). Osmotic control of *kdp* operon expression in *Escherichia coli*. Proc. Natl. Acad. Sci. U.S.A. *78*, 464–468.

Lee, C.R., Cho, S.H., Yoon, M.J., Peterkofsky, A., and Seok, Y.J. (2007). *Escherichia coli* enzyme IIA^{Ntr} regulates the K^+ transporter TrkA. Proc. Natl. Acad. Sci. U.S.A. *104*, 4124–4129.

Lee, S.J., and Gralla, J.D. (2004). Osmo-regulation of bacterial transcription via poised RNA polymerase. Mol. Cell. *14*, 153–162.

Lucassen, M. (1998). Regulation des Kdp-Systems aus *Escherichia coli*: Biochemische Charakterisierung des Antwortregulators KdpE und Nachweis von Konformationsänderungen im Zuge der Aktivierung (Dissertation, University of Osnabrück, Germany).

Lukat, G.S., Lee, B.H., Mottonen, J.M., Stock, A.M., and Stock, J.B. (1991). Roles of the highly conserved aspartate and lysine residues in the response regulator of bacterial chemotaxis. J. Biol. Chem. *266*, 8348–8354.

Lüttmann, D., Heermann, R., Zimmer, B., Hillmann, A., Rampp, I.S., Jung, K., and Görke, B. (2009). Stimulation of the potassium sensor KdpD kinase activity by interaction with the phosphotransferase protein IIA^{Ntr} in *Escherichia coli*. Mol. Microbiol. *72*, 978–994.

McAllister-Lucas, L.M., Haik, T.L., Colbran, J.L., Sonnenburg, W.K., Seger, D., Turko, I.V., Beavo, J.A., Francis, S.H., and Corbin, J.D. (1995). An essential aspartic acid at each of two allosteric cGMP-binding sites of a cGMP-specific phosphodiesterase. J. Biol. Chem. *270*, 30671–30679.

McLaggan, D., Naprstek, J., Buurman, E.T., and Epstein, W. (1994). Interdependence of K^+ and glutamate accumulation during osmotic adaptation of *Escherichia coli*. J. Biol. Chem. *269*, 1911–1917.

MacMillan, S.V., Alexander, D.A., Culham, D.E., Kunte, H.J., Marshall, E.V., Rochon, D., and Wood, J.M. (1999). The ion coupling and organic substrate specificities of osmoregulatory transporter ProP in *Escherichia coli*. Biochim. Biophys. Acta *1420*, 30–44.

Malli, R., and Epstein, W. (1998). Expression of the Kdp ATPase is consistent with regulation by turgor pressure. J. Bacteriol. *180*, 5102–5108.

Maslennikov, I., Klammt, C., Hwang, E., Kefala, G., Okamura, M., Esquivies, L., Mörs, K., Glaubitz, C., Kwiatkowski, W., Jeon, Y.H., *et al.* (2010). Membrane domain structures of three classes of histidine kinase receptors by cell-free expression and rapid NMR analysis. Proc. Natl. Acad. Sci. U.S.A. *107*, 10902–10907.

Möker, N., Brocker, M., Schaffer, S., Krämer, R., Morbach, S., and Bott, M. (2004). Deletion of the genes encoding the MtrA-MtrB two-component system of *Corynebacterium glutamicum* has a strong influence on cell morphology, antibiotics susceptibility and expression of genes involved in osmoprotection. Mol. Microbiol. *54*, 420–438.

Möker, N., Krämer, J., Unden, G., Krämer, R., and Morbach, S. (2007). *In vitro* analysis of the two-component system MtrB-MtrA from *Corynebacterium glutamicum*. J. Bacteriol. *189*, 3645–3649.

Møller, J.V., Juul, B., and le Maire, M. (1996). Structural organization, ion transport, and energy transduction of P-type ATPases. Biochim. Biophys. Acta *1286*, 1–51.

Morbach, S., and Krämer, R. (2002). Body shaping under water stress: osmosensing and osmoregulation of solute transport in bacteria. Chem. Biochem. *3*, 384–397.

Nakamura, T., Yuda, R., Unemoto, T., and Bakker, E.P. (1998). KtrAB, a new type of bacterial K^+-uptake system from *Vibrio alginolyticus*. J. Bacteriol. *180*, 3491–3494.

Nakashima, K., Kanamaru, K., Aiba, H., and Mizuno, T. (1991). Signal transduction and osmoregulation in *Escherichia coli*. A novel type of mutation in the phosphorylation domain of the activator protein, OmpR, results in a defect in its phosphorylation-dependent DNA binding. J. Biol. Chem. *266*, 10775–10780.

Nakashima, K., Sugiura, A., and Mizuno, T. (1993). Functional reconstitution of the putative *Escherichia coli* osmosensor, KdpD, into liposomes. J. Biochem. *114*, 615–621.

Ninfa, E.G. Atkinson, M.R., Kamberov, E.S., and Ninfa, A.J. (1993). Mechanism of autophosphorylation of *Escherichia coli* nitrogen regulator II (NRII or NtrB): trans-phosphorylation between subunits. J. Bacteriol. *175*, 7024–7032.

Nissen, P., Hansen, J., Ban, N., Moore, P.B., and Steitz, T.A. (2000). The structural basis of ribosome activity in peptide bond synthesis. Science *289*, 920–930.

O'Loughlin, J.L., Spinner, J.L., Minnich, S.A., and Kobayashi, S.D. (2010). *Yersinia pestis* two-component gene regulatory systems promote survival in human neutrophils. Infect. Immun. *78*, 773–782.

Ochrombel, I., Becker, M., Krämer, R., and Marin, K. (2011). Osmotic stress response in *C. glutamicum*: impact of channel- and transporter-mediated potassium accumulation. Arch. Microbiol. *193*, 787–796.

Ohwada, T., and Sagisaka, S. (1987). An immediate and steep increase in ATP concentration in response to reduced turgor pressure in *Escherichia coli* B. Arch. Biochem. Biophys. *259*, 157–163.

Palleros, D.R., Reid, K.L., Shi, L., Welch, W.J., and Fink, A.L. (1993). ATP-induced protein–Hsp70 complex dissociation requires K^+ but not ATP hydrolysis. Nature *365*, 664–666.

Parish, T., Smith, D.A., Kendall, S., Casali, N., Bancroft, G.J., and Stoker, N.G. (2003). Deletion of two-component regulatory systems increases the virulence of *Mycobacterium tuberculosis*. Infect. Immun. *71*, 1134–1140.

Parra-Lopez, C., Lin, R., Aspedon, A., and Groisman, E.A. (1994). A *Salmonella* protein that is required for resistance to antimicrobial peptides and transport of potassium. EMBO J. *13*, 3964–3972.

Polarek, J.W., Williams, G., and Epstein, W. (1992). The products of the *kdpDE* operon are required for expression of the Kdp ATPase of *Escherichia coli*. J. Bacteriol. *174*, 2145–2151.

Puppe, W., Zimmann, P., Jung, K., Lucassen, M., and Altendorf, K. (1996). Characterization of truncated forms of the KdpD protein, the sensor kinase of the K^+-translocating Kdp system of *Escherichia coli*. J. Biol. Chem. *271*, 25027–25034.

Rabus, R., Reizer, J., Paulsen, I., and Saier, M.H., Jr. (1999). Enzyme I Ntr from *Escherichia coli*. A novel enzyme of the phosphoenolpyruvate-dependent phosphotransferase system exhibiting strict specificity for its phosphoryl acceptor, NPr. J. Biol. Chem. *274*, 26185–26191.

Racher, K.I., Voegele, R.T., Marshall, E.V., Culham, D.E., Wood, J.M., Jung, H., Bacon, M., Cairns, M.T., Ferguson, S.M., Liang, W.J., *et al.* (1999). Purification and reconstitution of an osmosensor: transporter ProP of *Escherichia coli* senses and responds to osmotic shifts. Biochemistry *38*, 1676–1684.

Record, M.T., Jr., Courtenay, E.S., Cayley, D.S., and Guttman, H.J. (1998). Responses of *E. coli* to osmotic stress: large changes in amounts of cytoplasmic solutes and water. Trends Biochem. Sci. 23, 143–148.

Reed, R.H., and Walsby, A.E. (1985). Changes in cell turgor pressure in response to increases in external NaCl concentration in the gas-vacuolate cyanobacterium *Microcystis* sp. Arch. Microbiol. *143*, 290–296.

Reeves, E.P., Lu, H., Jacobs, H.L., Messina, C.G., Bolsover, S., Gabella, G., Potma, E.O., Warley, A., Roes, J., and Segal, A.W. (2002). Killing activity of neutrophils is mediated through activation of proteases by K^+ flux. Nature *416*, 291–297.

Rhoads, D.B., and Epstein, W. (1977). Energy coupling to net K^+ transport in *Escherichia coli* K-12. J. Biol. Chem. *252*, 1394–1401.

Roe, A.J., McLaggan, D., O'Byrne, C.P., and Booth, I.R. (2000). Rapid inactivation of the *Escherichia coli* Kdp K^+ uptake system by high potassium concentrations Mol. Microbiol. 35, 1235–1243.

Rosenthal, A.Z., Kim, Y., and Gralla, J.D. (2008). Poising of *Escherichia coli* RNA polymerase and its release from the sigma 38 C-terminal tail for *osmY* transcription. J. Mol. Biol. *376*, 938–949.

Rothenbücher, M.C., Facey, S.J., Kiefer, D., Kossmann, M., and Kuhn, A. (2006). The cytoplasmic C-terminal domain of the *Escherichia coli* KdpD protein functions as a K^+ sensor. J. Bacteriol. *188*, 1950–1958.

Roux, B., and MacKinnon, R. (1999). The cavity and pore helices in the KcsA K^+ channel: electrostatic stabilization of monovalent cations. Science *285*, 100–102.

Rübenhagen, R., Ronsch, H., Jung, H., Krämer, R., and Morbach, S. (2000). Osmosensor and osmoregulator properties of the betaine carrier BetP from *Corynebacterium glutamicum* in proteoliposomes. J. Biol. Chem. *275*, 735–741.

Rübenhagen, R., Morbach, S., and Krämer, R. (2001). The osmoreactive betaine carrier BetP from *Corynebacterium glutamicum* is a sensor for cytoplasmic K^+. EMBO J. *20*, 5412–5420.

Schiller, D., Rübenhagen, R., Krämer, R., and Morbach, S. (2004). The C-terminal domain of the betaine carrier BetP of *Corynebacterium glutamicum* is directly involved in sensing K^+ as an osmotic stimulus. Biochemistry *43*, 5583–5591.

Schlösser, A., Hamann, A., Bossemeyer, D., Schneider, E., and Bakker, E.P. (1993). NAD^+ binding to the *Escherichia coli* K^+-uptake protein TrkA and sequence similarity between TrkA and domains of a family of dehydrogenases suggest a role for NAD^+ in bacterial transport. Mol. Microbiol. *9*, 533–543.

Schniederberend, M., Zimmann, P., Bogdanov, M., Dowhan, W., and Altendorf, K. (2010). Influence of K^+-dependent membrane lipid composition on the expression of the *kdpFABC* operon in *Escherichia coli*. Biochim. Biophys. Acta *1798*, 32–39.

Siegele, D.A. (2005). Universal stress proteins in *Escherichia coli*. J. Bacteriol. *187*, 6253–6254.

Sleator, R.D., and Hill, C. (2002). Bacterial osmoadaptation: the role of osmolytes in bacterial stress and virulence. FEMS Microbiol. Rev. *26*, 49–71.

Stallkamp, I., Dowhan, W., Altendorf, K., and Jung, K. (1999). Negatively charged phospholipids influence the activity of the sensor kinase KdpD of *Escherichia coli*. Arch. Microbiol. *172*, 295–302.

Stallkamp, I., Altendorf, K., and Jung, K. (2002). Amino acid replacements in transmembrane domain 1 influence osmosensing but not K^+ sensing by the sensor kinase KdpD of *Escherichia coli*. Arch. Microbiol. *178*, 525–530.

Steil, L., Hoffmann, T., Budde, I., Volker, U., and Bremer, E. (2003). Genome-wide transcriptional profiling analysis of adaptation of *Bacillus subtilis* to high salinity. J. Bacteriol. *185*, 6358–6370.

Stewart, L.M., Bakker, E.P., and Booth, I.R. (1985). Energy coupling to K^+ uptake via the Trk system in *Escherichia coli*: the role of ATP. J. Gen. Microbiol. *131*, 77–85.

Steyn, A.J.C., Joseph, J., and Bloom, B.R. (2003). Interaction of the sensor module of *Mycobacterium tuberculosis* H37Rv KdpD with members of the Lpr family. Mol. Microbiol. *47*, 1075–1089.

Stumpe, S., and Bakker, E.P. (1997). Requirement of a large K^+-uptake capacity and of extracytoplasmic protease activity for protamine resistance of *Escherichia coli*. Arch. Microbiol. *167*, 126–136.

Suelter, C.H. (1970). Enzymes activated by monovalent cations. Science *168*, 789–795.

Sugiura, A., Nakashima, K., Tanaka, K., and Mizuno, T. (1992). Clarification of the structural and functional features of the osmoregulated *kdp* operon of *Escherichia coli*. Mol. Microbiol. *6*, 1769–1776.

Sugiura, A., Nakashima, K., and Mizuno, T. (1993). Sequence-directed DNA curvature in activator-binding sequence in the *Escherichia coli kdpABC* promoter. Biosci. Biotechnol. Biochem. *57*, 356–357.

Sugiura, A., Hirokawa, K., Nakashima, K., and Mizuno, T. (1994). Signal-sensing mechanisms of the putative osmosensor KdpD in *Escherichia coli*. Mol. Microbiol. *14*, 929–938.

Sutherland, L., Cairney, J., Elmore, M.J., Booth, I.R., and Higgins, C.F. (1986). Osmotic regulation of transcription: induction of the *proU* betaine transport gene is dependent on accumulation of intracellular potassium. J. Bacteriol. *168*, 805–814.

Swanson, R.V., Bourret, R.B., and Simon, M.I. (1993). Intermolecular complementation of the kinase activity of CheA. Mol. Microbiol. *8*, 435–441.

Tanaka, T., Saha, S.K., Tomomori, C., Ishima, R., Liu, D., Tong, K.I., Park, H., Dutta, R., Qin, L., Swindells, M.B., *et al.* (1998). NMR structure of the histidine kinase domain of the *E. coli* osmosensor EnvZ. Nature *396*, 88–92.

Tholema, N., Bakker, E.P., Suzuki, A., and Nakamura, T. (1999). Change to alanine of one out of four selectivity filter glycines in KtrB causes a two orders of magnitude decrease in the affinities for both K^+ and Na^+ of the Na^+ dependent K^+ uptake system KtrAB from *Vibrio alginolyticus*. FEBS Lett. *450*, 217–220.

Tomomori, C., Tanaka, T., Dutta, R., Park, H., Saha, S.K., Zhu, Y., Ishima, R., Liu, D., Tong, K.I., Kurokawa, H., *et al.* (1999). Solution structure of the homodimeric core domain of *Escherichia coli* histidine kinase EnvZ. Nat. Struct. Biol. *6*, 729–734.

Toro-Roman, A., Wu, T., and Stock, A.M. (2005). A common dimerization interface in bacterial response regulators KdpE and TorR. Protein Sci. *14*, 3077–3088.

Trchounian, A., and Kobayashi, H. (1999). Kup is the major K^+ uptake system in *Escherichia coli* upon hyper-osmotic stress at a low pH. FEBS Lett. *447*, 144–148.

Vlisidou, I., Eleftherianos, I., Dorus, S., Yang, G., ffrench-Constant, R.H., Reynolds, S.E., and Waterfield, N.R. (2010). The KdpD/KdpE two-component system of *Photorhabdus asymbiotica* promotes bacterial survival within *M. sexta* hemocytes. J. Invertebr. Pathol. *105*, 352–362.

Voelkner, P., Puppe, W., and Altendorf, K. (1993). Characterization of the KdpD protein, the sensor kinase of the K^+-translocating Kdp system of *Escherichia coli*. Eur. J. Biochem. *217*, 1019–1026.

Walderhaug, M.O., Polarek, J.W., Voelkner, P., Daniel, J.M., Hesse, J.E., Altendorf, K., and Epstein, W. (1992). KdpD and KdpE, proteins that control expression of the *kdpABC* operon, are members of the two-component sensor-effector class of regulators. J. Bacteriol. *174*, 2152–2159.

Waterfield, N.R., Sanchez-Contreras, M., Eleftherianos, I., Dowling, A., Yang, G., Wilkinson, P., Parkhill, J., Thomson, N., Reynolds, S.E., Bode, H.B., *et al.* (2008). Rapid Virulence Annotation (RVA): identification of virulence factors using a bacterial genome library and multiple invertebrate hosts. Proc. Natl. Acad. Sci. U.S.A. *105*, 15967–15972.

Weber, A., and Jung, K. (2002). Profiling early osmostress-dependent gene expression in *Escherichia coli* using DNA macroarrays. J. Bacteriol. *184*, 5502–5507.

Wolfe, A.J., and Stewart, R.C. (1993). The short form of the CheA protein restores kinase activity and chemotactic ability to kinase-deficient mutants. Proc. Natl. Acad. Sci. U.S.A. *90*, 1518–1522.

Wood, J.M. (1999). Osmosensing by bacteria: signals and membrane-based sensors. Microbiol. Mol. Biol. Rev. *63*, 230–262.

Wood, J.M. (2006). Osmosensing by bacteria. Sci. STKE *2006*, pe43.

Wood, J.M., Bremer, E., Csonka, L.N., Kraemer, R., Poolman, B., van der, H.T., and Smith, L.T. (2001). Osmosensing and osmoregulatory compatible solute accumulation by bacteria. Comp. Biochem. Physiol. A Mol. Integr. Physiol. *130*, 437–460.

Xue, T., You, Y., Hong, D., Sun, H., and Sun, B. (2011). *Staphylococcus aureus* KdpDE two-component system couples extracellular K^+ sensing and Agr signaling to infection programming. Infect. Immun. *78*, 3506–3515.

Yang, Y., and Inouye, M. (1991). Intermolecular complementation between two defective mutant signal- transducing receptors of *Escherichia coli*. Proc. Natl. Acad. Sci. U.S.A. *88*, 11057–11061.

Yuan, J., Wei, B., Shi, M., and Gao, H. (2011). Functional assessment of EnvZ/OmpR two-component system in *Shewanella oneidensis*. PLoS One *6*, e23701.

Zakharyan, E., and Trchounian, A. (2001). K^+ influx by Kup in *Escherichia coli* is accompanied by a decrease in H^+ efflux. FEMS Microbiol. Lett. *204*, 61–64.

Zhao, L., Xue, T., Shang, F., Sun, H., and Sun, B. (2010). *Staphylococcus aureus* AI-2 quorum sensing associates with the KdpDE two-component system to regulate capsular polysaccharide synthesis and virulence. Infect. Immun. *78*, 3506–3515.

Zimmann, P., Puppe, W., and Altendorf, K. (1995). Membrane topology analysis of the sensor kinase KdpD of *Escherichia coli*. J. Biol. Chem. *270*, 28282–28288.

Zimmann, P., Steinbrügge, A., Schniederberend, M., Jung, K., and Altendorf, K. (2007). The extension of the fourth transmembrane helix of the sensor kinase KdpD of *Escherichia coli* is involved in sensing. J. Bacteriol. *189*, 7326–7334.

Zimmer, B., Hillmann, A., and Görke, B. (2008). Requirements for the phosphorylation of the *Escherichia coli* $EIIA^{Ntr}$ protein *in vivo*. FEMS Microbiol. Lett. *286*, 96–102.

Two-component Signalling in the Gram-positive Envelope Stress Response: Intramembrane-sensing Histidine Kinases and Accessory Membrane Proteins

11

Karen Schrecke[1], Anna Staroń[1] and Thorsten Mascher

Abstract

The cell envelope stress response (CESR) network monitors and maintains envelope integrity to counteract the damaging effects of cell wall antibiotics and membrane perturbating agents. Two-component systems (2CSs) involved in orchestrating CESR in Firmicutes bacteria (low G+C Gram-positive) are characterized by so-called intramembrane-sensing histidine kinases (IM-HKs). The N-terminal input domain of these proteins consists of two transmembrane helices with a very short extracellular linker of less than 20 amino acids, which is insufficient for stimulus perception. It was originally thought that these HKs sense their stimuli within the membrane interface. But subsequent studies identified accessory membrane proteins for all IM-HKs described so far. This chapter will specifically summarize the current state of knowledge on BceRS- and LiaRS-like 2CSs, which are ubiquitously distributed in Firmicutes bacteria. While BceRSAB-like systems represent antibiotic-specific detoxification modules, LiaFSR-like three-component systems mount more general CESRs. These two types of systems are genetically and functionally linked to BceAB-like ABC transporters and LiaF-like membrane-anchored regulatory proteins, respectively, which play a crucial role in sensing envelope stress and transferring the information to the cognate HKs. Accordingly, BceS- and LiaS-like IM-HKs do not function as sensor proteins, but rather as signal transfer relays between the sensor and the cognate response regulators.

Introduction

The ability of a bacterial cell to respond to changing environmental conditions necessitates the presence of signal transducing systems that respond to specific cues and initiate stimulus-specific cellular responses, usually through differential gene expression. In bacteria, three major regulatory principles are involved in transmembrane signalling: one-component systems (1CSs), two-component systems (2CSs), and alternative sigma factors of the extracytoplasmic function (ECF) family (Staroń and Mascher, 2010). In 1CSs, the sensory and regulatory domains are fused on a single polypeptide chain (Ulrich *et al.*, 2005). Because of this restrictive protein architecture, such systems predominantly respond to intracellular signals and play only a minor role in transmembrane signal transduction. In contrast, ECF sigma factors and 2CSs – which separate sensory and regulatory functions on two distinct proteins – can easily facilitate responses to intracellular, membrane-derived, and extracytoplasmic signals (Mascher *et al.*, 2006; Staroń *et al.*, 2009). Very often, such signalling systems are involved in sensing stress conditions which can potentially harm the microbial cell. In response, they mount protective countermeasures, very often through differential expression of target genes that are involved in stress-resistance (Storz and Hengge-Aronis, 2000; Storz and Hengge, 2010).

This chapter will specifically focus on 2CSs that are involved in orchestrating the cell envelope stress response (CESR) in Gram-positive bacteria

1 Karen Schrecke and Anna Staroń contributed equally to the writing of this chapter.

with a low G+C content (Firmicutes). The cell envelope of Gram-positive bacteria consists of the cytoplasmic membrane and a thick multilayered cell wall made up of peptidoglycan and teichoic acids (Foster and Popham, 2002). It differs significantly from the Gram-negative model, which contains a much thinner peptidoglycan wall lacking teichoic acids, but containing a periplasmic space surrounded by an outer membrane. Accordingly, the regulatory network orchestrating CESR also differs between Gram-positive and -negative bacteria, especially with regard to the 2CSs involved (Alba and Gross, 2004; Jordan *et al.*, 2008; MacRitchie *et al.*, 2008).

The primary 2CS of the proteobacterial CESR is the CpxAR system, which will be described in Chapter 12. In contrast, the two primary CESR 2CSs in Firmicutes bacteria – the subject of this chapter – are homologues to BceRS and LiaRS of *Bacillus subtilis*, the model organism for the low G+C Gram-positive bacteria. Another important 2CS for maintaining envelope integrity in this and other Firmicutes bacteria in the absence of stress conditions is the essential WalKR (or YycFG) system, which will be described in Chapter 7. For an overview of all signalling systems involved in mediating CESR and envelope integrity in Gram-positive bacteria (both Firmicutes and Actinobacteria), including 1CSs, ECF sigma factors, and other 2CSs, readers are referred to a comprehensive review article on the subject (Jordan *et al.*, 2008).

The envelope is a crucial structure of the bacterial cell. It gives the cell its shape, protects it against environmental threats and counteracts the high internal osmotic pressure (Höltje, 1998; Delcour *et al.*, 1999). Because of these essential functions, it is also a prime target for many antibiotics that interfere with cell wall biosynthesis (Jordan *et al.*, 2008) (see Fig. 11.1 for an overview on antibiotics relevant for this chapter). Production of antibiotics can be viewed as a means of

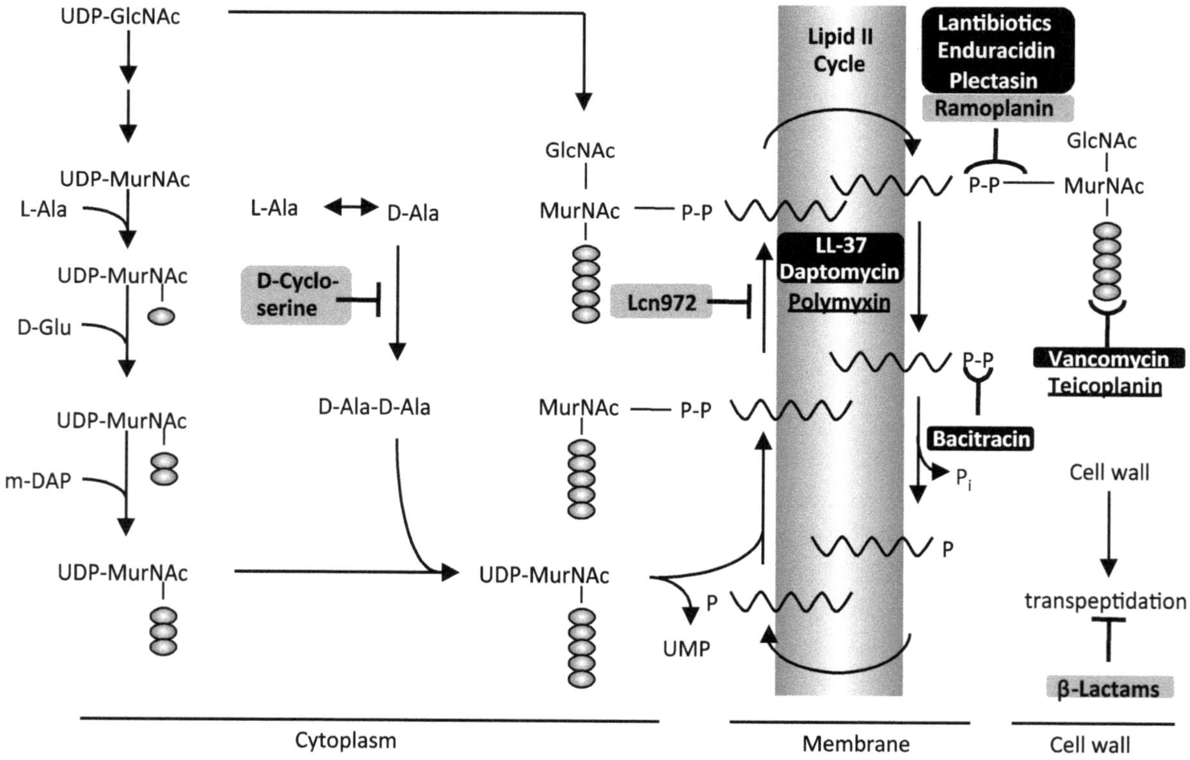

Figure 11.1 Cell wall biosynthesis of Gram-positive bacteria and its inhibition by antibiotics that induce Bce-like and Lia-like systems. Crucial steps in cell wall biosynthesis are schematically indicated, together with their cellular localization. GlcNAc, N-acetyl-glucosamine; MurNAc, N-acetyl-muramic acid. Amino acids are depicted as small grey circles, undecaprenol by the waved line. Antibiotics which induce the Lia-like systems are highlighted in grey, those inducing Bce-like systems are underlined. Compounds inducing both systems are highlighted in black. Lantibiotics is used as a collective term for actagardine, gallidermin, mersacidin, nisin, and subtilin.

interspecies competition for survival in complex and densely populated habitats, such as the soil ecosystems (D'Costa *et al.*, 2006). Not surprisingly, soil bacteria such as the Actinobacteria, but also the Firmicutes are amongst the most prolific antibiotic producers (Berdy, 2005). In order to compete, soil bacteria also need to have cunning and sensitive reflexes to respond to antibiotic threat before the cells can take irreparable damage. Hence, both production of and resistance against antibiotics are part of the microbial warfare in complex habitats.

The response of the Gram-positive model organism *B. subtilis* to antibiotics has been particularly well studied during the last 10 years, both at the level of differential gene expression and protein production, by extensively applying transcriptomics and proteomics approaches (Wecke and Mascher, 2011). In case of cell wall antibiotics, the underlying CESR regulatory network was thoroughly characterized and shown to consist of at least four ECF sigma factors and a similar number of 2CSs (Jordan *et al.*, 2008). Of the four 2CSs, three are paralogous to each other and referred to as BceRS-like 2CSs, based on the best-understood example. All of them are associated with genes encoding an ABC transporter, which are strongly induced in the presence of specific cell wall antibiotics and mediate antibiotic resistance against them (Jordan *et al.*, 2008; Staroń *et al.*, 2011). The LiaRS system is the fourth CESR 2CS of *B. subtilis*, which responds to a broader range of cell wall antibiotics (Fig. 11.1). While its physiological role is still unclear in *B. subtilis*, it seems to represent the primary general CESR system in many other Firmicutes bacteria (Jordan *et al.*, 2008). Both BceRS- and LiaRS-like 2CSs are widely distributed in this phylum and all systems investigated to date in a range of Firmicutes bacteria are involved in specific cell wall antibiotic resistances or more general CESRs (Jordan *et al.*, 2008; see Tables 11.1 and 11.2 for a detailed overview).

A unifying feature of all 2CSs involved in mediating CESR in Firmicutes bacteria is the domain architecture of their HKs. Both LiaS- and BceS-like HKs are simple, membrane-anchored proteins that harbour two transmembrane regions (TMRs) with a very short extracellular linker of fewer than 20 (usually fewer than 10) amino acids in between. Because of its small size, it was assumed that this linker cannot have a function in direct antibiotic perception. Instead, it was suggested that these kinase sense their stimuli at or from within the membrane interphase. Accordingly, these HKs were termed intramembrane-sensing (IM-) HKs (Mascher, 2006). But in the last couple of years, it became evident that both LiaS- and BceS-like HKs do not seem to function as sensor proteins themselves. Instead, they seem to act as signal transfer proteins that are wired between the primary sensor protein and their cognate response regulator (RR).

Discussing the potential mechanisms of this novel mode of stimulus perception and signal transduction will be one central aspect of this chapter. Moreover, we will present a detailed overview on the current state of knowledge on BceRS- and LiaRS-like 2CSs, respectively, with the goal of extracting general features and individual aberrations from the norm for both types of CESR 2CSs of Firmicutes bacteria. The chapter will close with a brief discussion of open questions for future studies.

BceRS-like systems

BceRS-like 2CS constitute part of antimicrobial peptide detoxification modules, which are widespread among Firmicutes bacteria (Dintner *et al.*, 2011). They are named after the paradigm example for this class of 2CS, BceRS from *B. subtilis.* A general and unique feature of these systems is their composition, as they are composed of four subunits, all of which are indispensable for sensing and transducing of the signal (Fig. 11.2A) (Bernard *et al.*, 2007; Rietkötter *et al.*, 2008). Two of these proteins constitute a 2CS, with the HK BceS and its cognate RR BceR. Another two form an ABC transporter, consisting of a nucleotide-binding domain (NBD) subunit BceA and a membrane-spanning domain (MSD) subunit BceB.

BceS-like HKs are EnvZ-like proteins that belong to a family of cell envelope stress sensors termed intramembrane-sensing HKs (Mascher, 2006). Similarly to other members of this class, they do not harbour any extracytoplasmic sensory domains. Instead, the transmembrane helices are

Table 11.1 Bce-like detoxification modules in Firmicutes bacteria

Organism	Resistance module		Inducing substrates	Phenotype of ABC mutant	Additional regulon	References
	TCS	ABC				
B. subtilis	BceRS	BceAB	Bacitracin, plectasin, actagardine, mersacidin	Sensitivity to bacitracin, actagardine, mersacidin	None	Ohki *et al.* (2003), Bernard *et al.* (2007), Rietkötter *et al.* (2008), Staroń *et al.* (2011)
B. subtilis	PsdRS/ YvcPQ	PsdAB/ YvcRS	Enduracidin, actagardine, gallidermin, nisin, subtilin	Sensitivity to enduracidin, gallidermin, nisin, subtilin	None	Rukmana *et al.* (2009), Staroń *et al.* (2011)
B. subtilis	YxdJK	YxdLM	LL-37	Not determined	*dltABCDE*, *ywaA*	Joseph *et al.* (2004), Pietiäinen *et al.* (2005)
B. licheniformis	YtsAB	YtsCD	Bacitracin	Not determined	Unknown	Wecke *et al.* (2006)
B. licheniformis	YxdJK2	YxdL2M2	Bacitracin	Not determined	Unknown	Wecke *et al.* (2006)
L. lactis	LlrG-KinG	YsaBC	Nisin	Sensitivity to nisin	Unknown	Kramer *et al.* (2006)
L. monocytogenes	VirRS	AnrAB		Defects in cell entry, adhesion, virulence, sensitivity to bacitracin, nisin and β-lactams	*dltABCD*, *mprF*	Mandin *et al.* (2005), Camejo *et al.* (2009), Collins *et al.* (2010)
S. aureus	GraRS/ ApsRS	VraFG	Indolicidin, melittin, nisin, LL-37, mersacidin, vancomycin	Increased autolytic activity, reduced infectivity, altered cell surface charge, no biofilm formation after stimulation with citrate, sensitivity to hBD3, nisin, indolicidin, LL-37, vancomycin, polimyxin B, mersacidin, Pep5, LP9, gallidermin, more sensitive to killing by human neutrophil granulocytes	*dltABCD*, *mprF*, *vraDE*[a]	Herbert *et al.* (2007), Li *et al.* (2007a), Meehl *et al.* (2007), Kraus *et al.* (2008), Sass *et al.* (2008), Shanks *et al.* (2008), Sass and Bierbaum (2009)
S. aureus	BraRS/ NsaRS/ BceRS	BraAB/ NsaAB/ BceAB	hBD3, bacitracin, nisin	Sensitivity to nisin and bacitracin	*vraDE*	Sass *et al.* (2008), Blake *et al.* (2011), Hiron *et al.* (2011), Yoshida *et al.* (2011)
S. aureus	–	VraDE	Mersacidin, hBD3, LL-37, temporin, ovispirin, dermaseptin, bacitracin, teicoplanin	Sensitivity to nisin, bacitracin, daptomycin[a], hBD3[a] and Pep5[a]	Not applicable	Sass *et al.* (2008), Pietiäinen *et al.* (2009), Hiron *et al.* (2011)
S. epidermidis	ApsRS	VraFG	hBD3, tachyplesin, nisin, LL-37, magainin, histatin, brevinin	Sensitivity to hBD3, lysis in the stationary phase, sensitivity to killing by neutrophils	*dltABCD*, *mprF*, *vraDE*	Li *et al.* (2007b), Cheung *et al.* (2010)
S. mutans	MbrCD/ BceRS	MbrAB/ BceAB	Bacitracin, defensins	Sensitivity to bacitracin and defensins	SMU.862–4, 302 and.1856c	Tsuda *et al.* (2002), Ouyang *et al.* (2010), Kitagawa *et al.* (2011)
S. pneumoniae	–	Spr0812–3/ SP0912–3	Nisin and bacitracin	Sensitivity to bacitracin, nisin, gramicidin, lincomycin	Not applicable	Becker *et al.* (2009), Majchrzykiewicz *et al.* (2010)

[a]Contradictory results from different *S. aureus* strains (see text for details).

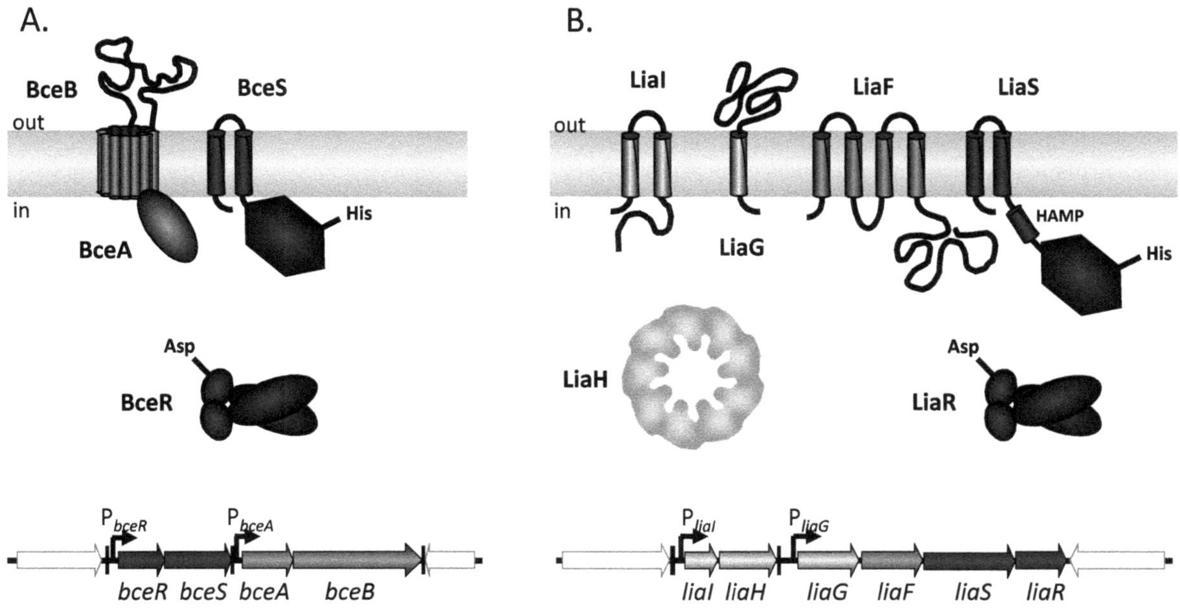

Figure 11.2 Schematic representation of the genes and proteins of the BceRSAB detoxification system (A) and the LiaFSR three-component system (B) in *B. subtilis*. Genes and proteins of the BceRSAB and LiaFSR systems are marked in dark grey (2CSs) and medium grey (subunits of the ABC transporter/LiaF). Other genes/proteins belonging to the *lia* operon are labelled in light grey. Genes flanking the *bce/lia* operon are white. Promoters are marked with bent arrows, terminators are represented by vertical bars.

connected by only a few extracellular amino acids. BceS-like proteins, in contrast to LiaS-like HKs described later, lack any additional cytoplasmic domains for intramolecular signal transfer, like HAMP or PAS domain (Aravind and Ponting, 1997; Ponting and Aravind, 1997). Therefore, they are indeed minimalistic HKs, with only a conserved DHp (dimerization histidine phosphotransfer) and catalytic ATPase domains. In response to a stimulus, they specifically phosphorylate the cognate BceR-like RR, which belongs to the OmpR family and contains a winged helix–turn–helix DNA-binding output domain.

BceAB-like ABC transporters are classified as Peptide-7-Exporter (Pep7E) in the Transport Classification Database (TCDB) (Saier *et al.*, 2009). The MSD contains ten TMRs and a large extracytoplasmic loop of approximately 200 amino acids between helices 7 and 8 (Fig. 11.2A). This domain architecture is highly conserved among all BceB-like proteins (Dintner *et al.*, 2011). Interestingly, the ABC transporter has a dual role, as it is both a resistance determinant (output) and a part of the stimulus sensing complex (input of the system). Therefore, BceAB-like ABC transporters basically regulate their own expression in response to their extracellular substrate by using a 2CS as mediator of signal transduction and gene regulation. In these unique modules, both subunits of the ABC transporter are absolutely required for stimulus sensing and subsequent activation of the kinase. Although the exact mechanism of both stimulus detection and antibiotic removal remains to be elucidated, some experimental data are slowly emerging and will be described in more detail in a later section (see 'Role of BceAB-like ABC transporters in sensing and detoxification').

Experimentally characterized Bce-like systems

A number of Bce-like systems have been characterized (Table 11.1), with the best studied example being the bacitracin resistance module BceRSAB from *B. subtilis*. BceRSAB was initially described as a part of the bacitracin stimulon in *B. subtilis* (hence the name: bacitracin efflux genes) and represent the most effective bacitracin resistance determinant in this organism (Mascher *et al.*, 2003; Ohki *et al.*, 2003). It is also responsible for conferring resistance to other peptide antibiotics, such as mersacidin and actagardine (Staroń *et al.*,

Table 11.2 LiaFSR-like three-component systems in Firmicutes bacteria

		Input			Output			
Species	Name	LiaF function	LiaS activity	Stimuli	Regulon[a]	Putative function	Phenotype	Reference
B. subtilis	LiaF-LiaSR	Negative	Bifunctional	Bacitracin, vancomycin, ramoplanin, nisin, daptomycin, rhamnolipids, detergents, organic solvents, secretion stress, alkaline shock, filamentous phage infection	*lialH*	PspA-like protein	Protection against oxidative stress, daptomycin, and rhamnolipids, Psp-like response	Petersohn *et al.* (2001), Wiegert *et al.* (2001), Mascher *et al.* (2004), Hyyryläinen *et al.* (2005), Pietiäinen *et al.* (2005), Jordan *et al.* (2006), Tam le *et al.* (2006), Hachmann *et al.* (2009), Wecke *et al.* (2009), Wolf *et al.* (2010), Wecke *et al.* (2011)
B. licheniformis	YvqF-YvqEC	Unknown	Unknown	Bacitracin, vancomycin, D-cycloserine	*yvqIH*, *yhcYZ*?	Unknown	Unknown	Wecke *et al.* (2006)
L. monocytogenes	$LiaF_{Lm}$-$LiaSR_{Lm}$	Negative	Bifunctional	Bacitracin, vancomycin, fosfomycin	$lialH_{Lm}$, $liaFSR_{Lm}$, *lmo*0181, *lmo*2484–2587, *lmo*2567, *lmo*2568,	LiaSR 2CS, ABC transporters, putative membrane and secreted proteins	Unknown	Fritsch *et al.* (2011)
S. mutans	LiaF-LiaSR	Negative	Unknown	Bacitracin, vancomycin, nisin, chlorhexidine, acid, high osmolarity, ethanol	*liaFSR-ppiB-pnpB*, *smu*753, *spxA*, *htrA*, *dagK*, *smu*1727	LiaSR 2CS, PspC-like proteins, proteins involved in membrane protein synthesis, PG biosynthesis, envelope chaperone/proteases, transcriptional regulators	Acid tolerance, biofilm formation, resistance to cell wall antibiotics	Li *et al.* (2002), Perry *et al.* (2008), Suntharalingam *et al.* (2009)

S. pneumoniae	Spr0342-HK03-RR03 (TCS03)	Unknown	Unknown	CdpD, LytA, LytC, bacitracin, nisin, tunicamycin	*lia* operon (*spr0342*-*spr*0351), *spr*0810, *spx* homologue	LiaSR 2CS, PspC-like proteins, DNA alkylation repair enzymes, choline binding proteins, proteins with stress related functions (e.g. heat shock)	Protection against self-lysis in competent cells, Psp-like response?	Haas *et al.* (2005), Eldholm *et al.* (2010)
Staph. aureus	YvqF-VraSR	Positive	Bifunctional	β-lactams, glycopeptides (vancomycin, teicoplanin), bacitracin, D-cycloserine	*orf1-yvqF-vraSR*, *pbp2*, *sgtB*, *murZ*, *fmtA*, *spsA*	VraSR 2CS, proteins associated with cell wall biosynthesis, peptidases, compounds of PTS system	β-lactam and glycopeptide resistance, maintaining integrity of cell wall	Kuroda *et al.* (2003), Boyle-Vavra *et al.* (2006), Belcheva and Golemi-Kotra, (2008), McCallum *et al.* (2011)
L. lactis	YjbB-CesSR	Unknown	Unknown	Lcn972, bacitracin, vancomycin, plantaricin C, lysozyme	*yjbB-cesS*-cesR,*llmg*2163,*spxB*	CesSR 2CS, PspC-like proteins, proteins involved in peptidoglycan synthesis/modification	Resistance against cell wall antibiotics and PG hydrolysis, Psp-like response?	Martinez *et al.* (2007), Veiga *et al.* (2007), Roces *et al.* (2009)

[a]If more than one regulated target gene is described, a selection of the most prominent regulon members is shown.

2011). Moreover, it was reported to respond to a fungal defensin, plectasin (Schneider *et al.*, 2010).

In addition, *B. subtilis* harbours two other paralogous systems named PsdRSAB and YxdJKLM–YxeA. The PsdRSAB (peptide antibiotic sensing and detoxification) module is induced by a broad range of peptide antibiotics, including enduracidin, actagardine, gallidermin, nisin, and subtilin (Staroń *et al.*, 2011). It mediates resistance against most of its inducers, but not actagardine. The third system, YxdJKLM–YxeA is the least understood of the three *B. subtilis* systems. It is induced by an antimicrobial peptide LL-37 (Pietiäinen *et al.*, 2005), and YxdJK 2CS controls expression of *yxdLM–yxeA* and *dltABCDE* operons (Joseph *et al.*, 2004). Interestingly, the expression of the YxdLM ABC transporter is induced more strongly by LL-37 in a *dlt* mutant compared with the wild type (Hyyryläinen *et al.*, 2007). Another unique feature of this system is the fact that the encoding gene locus harbours five genes instead of four. The additional gene encodes a small (115 aa) protein that is predicted to be a transmembrane protein with a DUF1093 domain. Its function is not known.

A close relative of *B. subtilis*, the bacitracin producer *B. licheniformis* also harbours three Bce-like systems. Its genes encode one homologue of BceRS, two homologues of YxdJK, but no direct homologue of PsdRS (Dintner *et al.*, 2011). YtsABCD, similarly to its direct homologue BceRSAB, was also reported to be induced by bacitracin (Wecke *et al.*, 2006). Two other systems (encoded by *yxdKLM-yxeA* and *Bli04272-yxdJK2L2M2*) are highly similar to the YxdJKLM-YxeA system of *B. subtilis*, with both systems encoding a small transmembrane protein harbouring a DUF1093 domain (in *B. licheniformis yxeA* and *Bli04272*). The second system (*Bli04272-yxdJK2L2M2*) was shown to be induced by bacitracin, while the first one (*yxdKLM-yxeA*) is not. But this lack of induction may be due to a mutation in the gene encoding the YxdJ-like RR.

Staphylococcus aureus has two complete 2CS/ABC transporter modules termed GraRS-VraFG and BraRSAB (other names used in the literature for these systems are listed in Table 11.1). Additionally, it harbours a third Pep7E-type ABC transporter, VraDE, which is not directly associated with a 2CS. GraRS was initially described as conferring resistance against glycopeptide antibiotics, specifically vancomycin (Cui *et al.*, 2005), hence the name: glycopeptide resistance associated. It is located in an operon directly upstream of genes encoding the VraFG transporter and regulates their expression (Li *et al.*, 2007a). The VraFG transporter is both necessary for sensing the peptide compounds, such as colistin or indolicidin, and for mediating resistance against them (Li *et al.*, 2007a; Meehl *et al.*, 2007). Moreover, GraRS 2CS plays an important role in virulence of *S. aureus* (Li *et al.*, 2007a; Kraus *et al.*, 2008). It also influences autolytic activity (Meehl *et al.*, 2007) and biofilm formation (Shanks *et al.*, 2008).

Expression of the second Bce-like system in *S. aureus*, BraRSAB, is induced by bacitracin and nisin (Hiron *et al.*, 2011). BraRS 2CS controls the expression of two ABC transporters, BraAB and VraDE. Interestingly, the BraAB ABC transporter is responsible solely for sensing of the signal, whereas VraDE confers resistance against the inducing compounds, bacitracin and nisin (Hiron *et al.*, 2011). This complex interplay between the two transporters and the BraRS 2CS will be described in detail in a later section (see 'Regulatory shuffling: wiring and evolution of Bce-like systems').

S. epidermidis, similarly to *S. aureus*, also harbours two Bce-like detoxification systems and one additional ABC transporter. However, experimental data from this organism is restricted only to the GraRS system (here named Aps). It is induced by different antimicrobial peptides, including nisin, LL-37 and hBD3 (see Table 11.1) (Li *et al.*, 2007b). It also confers resistance against hBD3 and contributes to survival after uptake by human neutrophils (Cheung *et al.*, 2010).

Streptococcus mutans harbours only one BceRS-like system. MbrABRS was initially characterized as bacitracin resistance determinant in this organism (Tsuda *et al.*, 2002) and therefore termed *mbr* for mutans bacitracin resistance. However, it was recently shown that it also confers resistance to human defensins (Ouyang *et al.*, 2010). MbrRS controls the expression of a number of genes, mostly encoding membrane proteins of unknown functions (Ouyang *et al.*, 2010;

Kitagawa *et al.*, 2011). In the closely related species *S. pneumoniae*, the Pep7E-ABC transporter Spr0812–3 is involved in resistance to bacitracin, nisin, gramicidin and lincomycin (Becker *et al.*, 2009; Majchrzykiewicz *et al.*, 2010). However, in this organism no BceRS-like 2CS is located in the neighbourhood of the transporter.

In *Lactococcus lactis* the experimental data are also limited only to the ABC transporter YsaBC, while the role of BceRS-like 2CSs has not been elucidated. The YsaBC transporter was initially discovered in a nisin resistant mutant of *L. lactis*. Its expression is induced by nisin and, when overexpressed, it mediates resistance to this lantibiotic (Kramer *et al.*, 2006). Its genes are located upstream of genes encoding a BceRS-like 2CS (*llrG kinG*), separated by a gene encoding a VanZ-like protein. Interestingly, a VanZ protein from *Enterococcus faecium* confers resistance to teicoplanin, a peptide antibiotic interfering with lipid II cycle (Arthur *et al.*, 1995); its role in *L. lactis* is unknown.

In *Listeria monocytogenes* AnrAB transporter mediates resistance to different peptide antibiotics, including bacitracin, gallidermin, and nisin (Collins *et al.*, 2010). It is not located in the direct neighbourhood of any BceRS-like 2CS, but nisin-dependent induction of *anrAB* depends on the VirRS 2CS, located at another locus in the genome (Mandin *et al.*, 2005). VirRS 2CS regulates an array of processes, including adhesion and entry into eukaryotic cells (Mandin *et al.*, 2005). The set of genes under control of VirRS includes the *dltABCD* operon, *mprF* gene and the above mentioned ABC transporter *anrAB* (Camejo *et al.*, 2009) (see 'Extended Bce regulons and functions').

Common themes in Bce-like systems

Stimuli sensed by Bce-like systems

A common trait of all described BceRS-like 2CSs is the induction by a comparable set of compounds (Fig. 11.1). To our knowledge, inducers of BceRS-like systems are exclusively short antimicrobial peptides, both of bacterial origin and produced by insects or animals. Although they differ in structure and modifications, they all share the cationic and amphipathic properties and often have similar modes of action (including membrane damage and blockage of cell wall synthesis) (Yeaman and Yount, 2007). Inducers include peptides like bacitracin or gramicidin, often glycosylated (vancomycin, teicoplanin), with fatty acid modifications (enduracidin, polimyxin B, daptomycin) or lanthionine bridges (lantibiotics including nisin, gallidermin, actagardine, mersacidin, subtilin, Pep5). Other inducers include peptides that play a role in the immune response of higher organisms like cathelicidines (LL-37, indolicidin, ovispirin), and defensins (hBD3, brevinin, plectasin, dermaseptin). Although all inducing molecules belong to the group of small antimicrobial peptides, the exact nature of the signal is puzzling, as the sensors are able to bind compounds of very different structures, but also to distinguish between very similar ones. For example, the Psd system of *B. subtilis* reacts to both nisin, which is a lantibiotic, and enduracidin, a cyclic lipopeptide. However, another lipopeptide, ramoplanin, which is structurally very similar to enduracidin, does not trigger induction of the Psd system (Staroń *et al.*, 2011).

Role of the BceAB-like ABC transporters in sensing and detoxification

In a classical 2CS, a HK senses the stimulus, often via its extracytoplasmic domain. For example, PhoQ from *Salmonella enterica* harbours a PDC domain (Zhang and Hendrickson, 2010), which plays an important role in sensing antimicrobial peptides (Bader *et al.*, 2005). However, as mentioned before, BceS-like HKs have only a short extracytoplasmic loop that is enough to connect two transmembrane helices, but it is too short to harbour a substrate binding domain. And indeed, for some of the Bce-like systems, including BceRSAB itself (Bernard *et al.*, 2007; Rietkötter *et al.*, 2008), but also for PsdRSAB (Staroń and Mascher, unpublished), MbrABCD (Ouyang *et al.*, 2010), GraRS-VraFG (Meehl *et al.*, 2007), and BraRSAB (Hiron *et al.*, 2011), it has been shown that the kinases require the presence of the ABC transporter for signal transduction. Based on these findings, it can be assumed that this is a general feature of these detoxification systems and that all

BceS-like HKs are unable to sense the stimulus alone, but instead require the ABC transporter for induction.

While being responsible for signalling, the transporter proteins are also resistance determinants. The involvement of the ABC transporter in resistance against antimicrobial compounds has been shown for many systems including BceAB and PsdAB from *B. subtilis* (Ohki *et al.*, 2003; Staroń *et al.*, 2011), AnrAB from *L. monocytogenes* (Collins *et al.*, 2010), VraFG, BraAB, and VraDE from *S. aureus* (Li *et al.*, 2007a; Hiron *et al.*, 2011). Although much speculation on the mechanism of signalling and resistance can be found in the literature, there are no conclusive results so far. The available evidence shows that the actual transport of the substrate is important for both sensing of the compound and conferring resistance to it. Mutations in either Walker A or Walker B motifs, leading to defects in ATP binding and hydrolysis, render the BceRSAB (Bernard *et al.*, 2007; Rietkötter *et al.*, 2008), PsdRSAB (Staroń and Mascher, unpublished), BraRSAB and VraDE (Hiron *et al.*, 2011) systems dysfunctional. Moreover, a deletion of the large extracytoplasmic loop, that is a conserved feature of all BceB-like permeases, also abolishes both signalling and resistance (Bernard *et al.*, 2007; Rietkötter *et al.*, 2008). It was suggested that the large extracytoplasmic loop is the actual binding site for the antibiotic compounds (Rietkötter *et al.*, 2008; Hiron *et al.*, 2011). Indeed, in *S. aureus* an exchange of the loop of VraG (responsible for conferring resistance to colistin) for that of VraE (conferring resistance to bacitracin) resulted in a chimeric transporter with a changed specificity that was able to restore bacitracin resistance (Hiron *et al.*, 2011).

Signalling versus resistance in Bce-like detoxification modules

BceRS-like systems from *B. subtilis* and *S. aureus* have been extensively characterized with respect to the inducer spectrum and the compounds they confer resistance against. Interestingly, these two spectra of compounds do not fully overlap, and not every antibiotic that induces a Bce-like system can also be efficiently detoxified of by the same module.

The Psd system from *B. subtilis* is induced by actagardine, nisin, enduracidin and subtilin, but mediates resistance only against the last three compounds (Staroń *et al.*, 2011). Actagardine triggers a strong response, but the elevated expression of the ABC transporter does not lead to any change in the cell's resistance against it. Another very interesting example can be found in *S. aureus*, where the ABC transporter BraAB is responsible solely for signalling, and does not mediate resistance against inducing compounds (nisin and bacitracin) (Hiron *et al.*, 2011). Therefore, it seems that signalling and resistance determination are two distinct functions that can be separated. As mentioned before, not the sole presence of an ABC transporter, but also the ability to transport is important for signalling, as mutations abolishing ATP hydrolysis and therefore transport also impair signal transduction. One possible explanation of this phenomenon suggests that the rate of antimicrobial peptide (AMP) transport is not enough to confer resistance against it, but enough to trigger the signalling cascade, the transport rate being the primary stimulus (Gebhard and Mascher, 2011). However, this hypothesis remains to be proven experimentally.

Mechanism of stimulus perception in Bce-like systems: different possibilities

Despite the fast growing amount of experimental data on Bce-like systems, the exact mechanism of stimulus perception in BceRS-like 2CS is still unclear. Since both functional subunits of the ABC transporter, and hence transport, are necessary for signal transduction, three mechanisms of stimulus perception can be proposed (Fig. 11.3). The scenario favoured in the most recent studies on the subject proposes a direct contact between the HK and the ABC transporter as a necessary step in the signalling cascade (Fig. 11.3B). Upon the contact with the inducing peptide, possibly through the extracytoplasmic loop, the transporter changes its conformation, which in turn promotes the interaction with the HK and its subsequent activation. In support of this hypothesis, recent bioinformatic analysis demonstrated a coevolution of BceS-like histidine kinases and the cognate BceB-like permeases

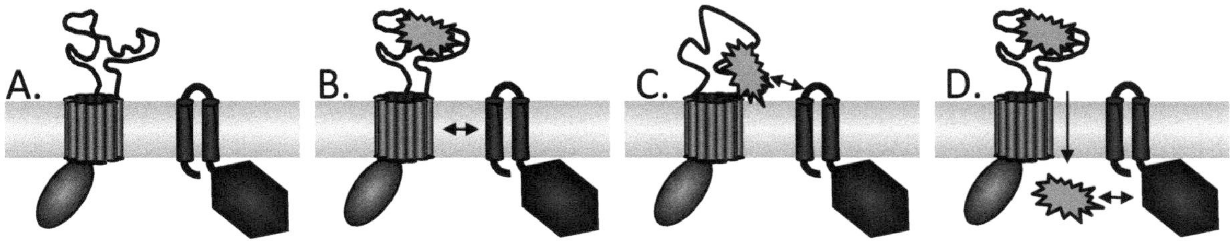

Figure 11.3 Proposed mechanisms of stimulus perception by Bce-like systems. (A) System in uninduced state. (B) Upon the contact with inducing peptide (here: through the extracytoplasmic loop) the transporter changes its conformation and activates the histidine kinase. (C) ABC transporter is presenting the inducing peptide to the histidine kinase, and the inducing compound is detected in a complex with ABC transporter. (D) BceAB transports the peptide to the cytoplasm where it is detected by BceS-like kinases.

(Dintner *et al.*, 2011). However, a coevolution of two proteins can be owing to other reasons besides a direct interaction, namely participation in the same process or interaction with the same ligand (Pazos and Valencia, 2008). Therefore, this hypothesis needs to be experimentally validated.

It cannot be ruled out that the ABC transporter is binding the inducing peptide and presenting it to the HK, so that the inducing compound can be detected only as a complex with ABC transporter (Fig. 11.3C). This detection could possibly happen through the short extracytoplasmic loop of the kinase. This hypothesis is supported by evidence obtained in studies on the GraS in *S. aureus* and ApsS in *S. epidermidis*. ApsS responds to hBD3, a compound that does not induce a GraS-dependent expression. An exchange of loops between these two kinases renders the *S. aureus* GraS responsive to hBD3 (Li *et al.*, 2007a). Moreover, antibodies against the short (nine residues) extracellular loop of the *S. epidermidis* HK GraS prevent up-regulation of target genes after induction by hDB3 (Li *et al.*, 2007b). The above mentioned results suggest that the extracellular loop of GraS/ApsS is responsible for signal recognition. However, the antibody, being a big molecule, may as well interfere with interaction of the ABC transporter with the HK. Moreover, it is not clear why this manner of stimulus perception should be energy dependent.

Lastly, it is also possible that the ABC transporter is an importer which transports the inducing compounds into the cytoplasm, where the stimulus is detected by the kinase in the cytoplasm (Fig. 11.3D). This hypothesis seems unlikely in the light of the above-mentioned data, but cannot be completely excluded at present.

Additional BceR-associated genes

A few operons for Bce-like detoxification modules are associated with open reading frames encoding additional proteins of unknown function. These genes are located either next to the ABC transporter (*yxeA* in *B. subtilis* and *B. licheniformis*, *spr0811a* in *S. pneumoniae*) or the 2CS (*apsX* and SA2419 in *S. aureus*, *bli04272* in *B. licheniformis*). Usually these are small proteins, ranging from 36 aa (Spr0811a) to 123 aa (Bli04272); only ApsX is significantly larger (307 aa). Although they do not show significant sequence conservation, most of them are transmembrane proteins, and some harbour a DUF1093 domain. Their function in signal perception or detoxification is not yet known. The only experimental data comes from the ApsRS system of *S. epidermidis* (homologous to GraRS in *S. aureus*), where the deletion of *apsX* leads to inactivation of the whole system (Li *et al.*, 2007b). However, it is not clear whether this observation can be extrapolated to other systems, as ApsX differs significantly from other detoxification module-associated proteins.

BceR-like binding sites

BceR-like RRs belong to the OmpR/PhoB subfamily, the largest subfamily of RRs. This subfamily accounts for half of RRs possessing a DNA binding domain (Galperin, 2006), which in this case consists of a winged helix–turn–helix motif. A study of RR binding sites in low-GC Gram positive bacteria predicted an operator for BceR-like proteins to include an inverted repeat (ACA-N_4-TGT) (de Been *et al.*, 2008). A following in-depth study identified motifs recognized by BceR-like RRs, and the overall consensus was found to be TNACA-N_4-TGTAA, with an AT-rich spacer (Dintner *et al.*, 2011). All binding sites that were experimentally identified support this consensus, even if the exact position in the identified region and nature of the repeat (inverted or direct) may not have been correctly recognized at the time. Binding sites were identified by promoter deletion analyses, EMSA and primer extension assays for Bce and Yxd systems in *B. subtilis* (Ohki *et al.*, 2003; Joseph *et al.*, 2004), Vir in *L. monocytogenes* (Mandin *et al.*, 2005), Bce in *S. mutans* (Ouyang *et al.*, 2010), and Bra and Vra systems in *S. aureus* (Hiron *et al.*, 2011).

The family of BceR-like RRs shows a high level of sequence conservation, and the same is true for the promoters that these RRs recognize. As there is often more than one BceR-like RR in one organism, an interesting question of promoter recognition specificity arises. Recent work from groups of Ann Stock and Michael Laub showed that there is a strong preference for very specific homodimerization of HKs (Ashenberg *et al.*, 2011), phosphotransfer to only cognate RRs (Skerker *et al.*, 2005), and also a preference for RRs homodimerization (Gao *et al.*, 2008). Most probably, there is also a preference for specific promoter recognition sites, especially as no cross-talk on the promoter level has been observed among Bce-like systems. However, it is not yet clear how the RR can recognize only the cognate binding site, given the high similarity of both RRs and the binding sites themselves.

Extended Bce regulons and functions

Some of the BceRS-like systems control not only the expression of one (or more) ABC transporters, but also the expression of additional genes, often including the *dltABCD* operon and the *mprF* gene. The *dlt* operon is reported to be regulated by YxdJK system of *B. subtilis* (Joseph *et al.*, 2004), VirRS of *L. monocytogenes* (Camejo *et al.*, 2009), ApsRS of *S. epidermidis* (Li *et al.*, 2007b), and GraRS of *S. aureus* (Li *et al.*, 2007a). The *mprF* gene is regulated by the same systems with the exception of YxdJK of *B. subtilis*. Products of both *dlt* and *mprF* genes are involved in changing the cell surface charge by modifying cell wall and cytoplasmic membrane components (Peschel, 2002).

Teichoic acids (TAs) are linear polyglycerol phosphate or polyribitol phosphate chains (Neuhaus and Baddiley, 2003), which are either covalently linked to peptidoglycan via a phosphodiester linkage (wall TAs) or linked to the cytoplasmic membrane via a terminal glycolipid moiety (lipoteichoic acids) (Neuhaus and Baddiley, 2003). TAs are the major source of the negative net charge of the Gram-positive envelope owing to intercalated PO_4 groups (Swoboda *et al.*, 2010). One of the modifications lowering the negative charge of TAs is D-alanylation (i.e. replacing the PO_4 groups by D-alanine), which is catalysed by products of genes encoded in the *dltABCD* operon (Peschel, 2002). Mutants in the *dlt* genes are pleiotropic, with phenotypes including increased autolysis (Nakao *et al.*, 2000; Cao and Helmann, 2004; Steen *et al.*, 2005), increased sensitivity to cationic antimicrobial peptides (CAMPs) (Peschel *et al.*, 1999), diminished adherence to phagocytic and non-phagocytic cells (Abachin *et al.*, 2002), and loss of acid tolerance (Boyd *et al.*, 2000). Consequently, many of these phenotypes have been observed in mutants of Bce-like systems, e.g. in *S. aureus* (Li *et al.*, 2007a; Meehl *et al.*, 2007).

Another source of the negative cell envelope net charge are phospholipids. One of the most common phospholipids found in bacterial membranes is the negatively charged phosphatidylglycerol (Ernst and Peschel, 2011). In order to lower the negative charge, the glycerol moiety of phosphatidylglycerol is modified by attaching lysine or alanine groups to it, thereby introducing positive charges into the membrane. These modifications are performed by the MprF protein. Mutants in *mprF* are more sensitive to various antimicrobial peptides, such as nisin, tachyplesin (Peschel *et al.*, 2001) or hBD3 (Nishi *et al.*, 2004) in *S. aureus* or LL-37 in *B. anthracis* (Samant et al., 2009).

Both D-alanylation of TAs and lysylination of phospholipids lower the negative cell surface charge (Peschel, 2002). CAMPs, which by definition carry an overall positive charge, have strong affinity towards negatively charged bacterial membranes. Therefore, introduction of positive charges into the envelope mediates resistance against CAMPs. While the elevated expression of ABC transporters confers a specific resistance against a small number of compounds, induction of *dlt* and *mprF* genes mediates more general resistance against CAMPs.

Regulatory shuffling: wiring and evolution of Bce-like systems

Many of the Bce-like modules constitute stand-alone autonomous systems that respond to and confer resistance against a certain subset of peptide antibiotics, as is the case in *B. subtilis* (Staroń *et al.*, 2011). However, the modular design of these systems allows shuffling of its subunits and creation of more complex regulatory circuits.

A very interesting and complex regulatory system can be found in *S. aureus* (Fig. 11.4B). This organism harbours two complete Bce-like systems GraRS-VraFG and BraRSAB, and an additional Pep7E-ABC transporter, VraDE. While GraRS-VraFG seems to be a stand-alone detoxification module, albeit with an extended regulon and an additional membrane protein, BraRSAB and VraDE show a surprising interplay. The ABC transporter BraAB is responsible only for stimulus detection, and the resulting induction of the BraRS system leads only to a slight increase in its expression (Hiron *et al.*, 2011). At the same time, it mediates a strong up-regulation of *vraDE* expression. VraDE then confers resistance to the inducing AMPs (nisin and bacitracin), but plays no role in signal transduction.

Another good example of such a complex system can be found in *L. monocytogenes* (Fig. 11.4A). As mentioned earlier, this organism harbours the AnrAB ABC transporter, which mediates resistance against a range of peptide antibiotics (Collins *et al.*, 2010). Its expression depends on the VirRS system (Mandin *et al.*, 2005). Consistent with this fact, there is a BceR-like binding site upstream of the *anrAB* operon. This situation is reminiscent of the typical Bce-like system, even though its components are located at different loci in the genome. However, the set-up is complicated by the fact that there is the second BceAB-like transporter (Lmo1747-6) located upstream of the VirRS 2CS. It is not preceded by a BceR-like binding site, and, consistent with that, its expression is not regulated by the VirRS 2CS (Mandin *et al.*, 2005¸Camejo *et al.*, 2009), at least not to the level that is detectable in a microarray experiment. It can be proposed that, similarly to *S. aureus*, Lmo1747-6 is responsible for sensing and

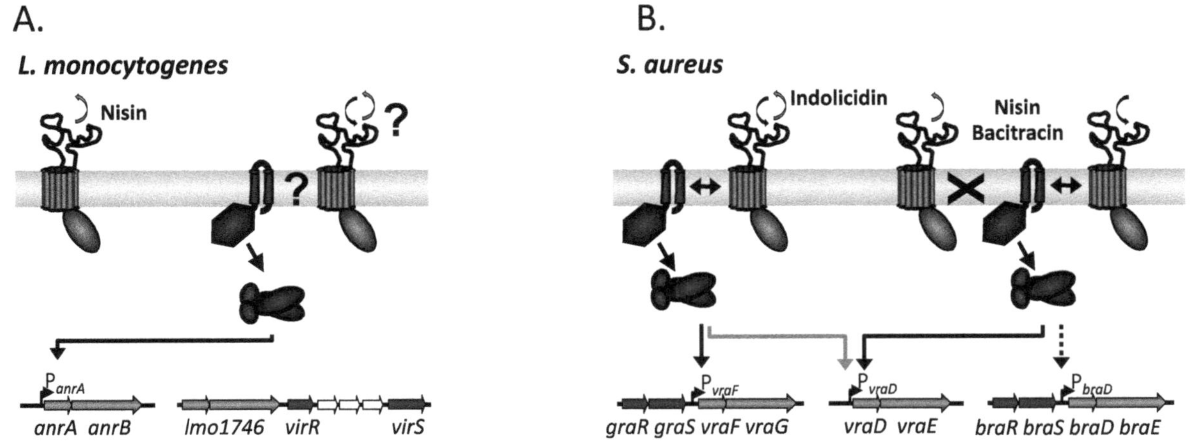

Figure 11.4 Schematic representation of Bce-like resistance modules of *L. monocytogenes* (A) and *S. aureus* (B). Genes and proteins are marked in dark grey (2CSs) and light grey (ABC transporters). Genes of unknown function are white. Inducible promoters are indicated by bent arrows. Substrate recognition by the ABC transporter is indicated by a curved downwards arrow, detoxification by a curved upwards arrow. Example substrates (if known) are shown above each ABC transporter. Phosphotransfer as well as strong transcriptional activation are shown by black arrows; moderate induction is indicated by a dotted arrow. Regulation of *vraDE* expression by GraR is shown by a light grey arrow, as contradictory results have been reported (see text for details).

AnrAB for resistance against AMPs, mediated by the VirRS 2CS.

Other types of complex arrangements can be found in many Firmicutes species, especially among bacilli and clostridia (Dintner *et al.*, 2011). Many genomes contain more BceAB-like ABC transporters than BceRS-like 2CS, sometimes located in one, sometimes in different loci, raising the question of cross-talk and division of labour between different ABC transporters. It seems that the modular design of these detoxification systems allows a certain degree of regulatory flexibility. Such subtle differences in transcriptional control in Bce-like systems can be observed in different strains of *S. aureus*. An orphan ABC transporter, VraDE, is responsible for conferring resistance against bacitracin and nisin. Its expression is controlled solely by the BraRS 2CS in strain HG001 (Hiron *et al.*, 2011), but also by GraRS in strain MW2 (Li *et al.*, 2007a). It is tempting to imagine that a duplication and subsequent specialization of an ABC transporter, coupled with the modulation of the transporter's expression level by changes in the respective RR-binding site gives bacteria an instrument for evolving adaption to altering environmental conditions.

LiaFSR-like systems

The paradigm: LiaFSR of *B. subtilis*

LiaFSR-like cell envelope stress-sensing signal transduction systems are widely conserved among the Firmicutes group of Gram-positive bacteria. The best studied system so far is LiaFSR of *B. subtilis*, which was originally identified in the course of investigations on the bacitracin stimulon in this organism (Mascher *et al.*, 2003) (Fig. 11.2B). LiaSR is one of four cell envelope stress sensing 2CSs in *B. subtilis*. Its name stands for 'lipid II cycle interfering antibiotic sensor and response regulator'. LiaSR does primarily respond to the presence of cell wall antibiotics that interfere with the lipid II cycle, such as bacitracin, vancomycin, ramoplanin, or cationic antimicrobial peptides (CAMP's) (Mascher *et al.*, 2004; Pietiäinen *et al.*, 2005) (Fig. 11.1), but also to detergents, organic solvents (e.g. ethanol, phenol), and to more unspecific stimuli like secretion stress, alkaline shock, and filamentous phage infection, although to a weaker degree (Petersohn *et al.*, 2001; Wiegert *et al.*, 2001; Mascher *et al.*, 2004; Hyyryläinen *et al.*, 2005; Pietiäinen *et al.*, 2005; Tam le *et al.*, 2006). A strong induction by membrane perturbation agents, such as daptomycin and rhamnolipids, was shown more recently (Hachmann *et al.*, 2009; Wecke *et al.*, 2009, 2011).

The HK LiaS also belongs to the subgroup of so-called IM-HKs possessing two membrane-spanning regions linked by a short extracytoplasmic loop of only 14 amino acids (Mascher, 2006; Mascher *et al.*, 2006). The second transmembrane helix is connected to a cytoplasmic HAMP domain, which is most likely involved in intramolecular signal transfer processes (Hulko *et al.*, 2006). The conserved HK core of LiaS contains the HisKA_3/dimerization (also DHp, standing for dimerization and histidine phosphotransfer) domain and the ATP-binding domain (HATPase_c = histidine kinase-like-ATPase). The invariant histidine residue of the autophosphorylation site is located within the HisKA_3 domain.

LiaR is a typical two-domain RR protein with a conserved N-terminal receiver domain that contains the invariant aspartate residue, and a variable C-terminal DNA-binding domain. It is a member of the NarL/FixJ family of RRs sharing the characteristic helix–turn–helix motif responsible for binding to DNA (Galperin, 2010).

The LiaSR 2CS is genetically and functionally linked to a third protein, LiaF. This membrane protein contains three to four membrane-spanning regions in its N-terminus. The functionally important C-terminus of LiaF features no obvious conserved domains. LiaF was identified as controlling the LiaSR-dependent signal transduction in a repressive manner (see section 'Stimulus perception by LiaS and LiaF'). Therefore, LiaF together with LiaSR constitutes a three-component system (Jordan *et al.*, 2006).

LiaFSR are encoded as the last three genes of the hexacistronic operon *liaIH–liaGFSR* (Fig. 11.2B), whose expression is controlled in an autoregulative manner. A basal expression level of the last four genes, *liaGFSR*, is ensured by a weak constitutive promoter upstream of *liaG* (P_{liaG}) (Jordan *et al.*, 2006). Inducing conditions lead to a strong LiaR-dependent activation of the promoter located directly upstream of *liaI* (P_{liaI}) resulting in an expression of two different transcripts: A major 1.1-kb transcript containing *liaIH* and a 4 kb transcript encompassing the entire locus. This pattern is due to a weak terminator structure located downstream of *liaH* (Mascher *et al.*, 2004). P_{liaI} seems to be the only relevant target promoter controlled by LiaR, but the physiological role of LiaIH remains unclear so far (Wolf *et al.*, 2010) (see section 'Regulon and physiological role of Lia-like systems'). LiaH is a member of the phage shock protein (Psp) family. LiaI and LiaG are membrane proteins of unknown function.

In the absence of any external stress, the *liaI* promoter is transiently induced at the onset of stationary phase (Jordan *et al.*, 2007) (Fig. 11.5). During this transition state from exponential to the stationary growth phase, *B. subtilis* cells undergo an intricate differentiation programme to adapt to non-optimal living conditions. A complex regulatory cascade is initiated at this time point that orchestrates this transition from vegetative cells to the formation of dormant endospores (Msadek, 1999; Phillips and Strauch, 2002; Errington, 2003). During vegetative growth, the transition state regulator AbrB represses P_{liaI} via direct binding to the promoter sequence, thereby preventing P_{liaI} activity. Moreover, the repressor protein LiaF keeps the LiaFSR system switched off in the absence of suitable extracellular stimuli. The transition to stationary phase leads to increasing amounts of phosphorylated Spo0A, the master regulator of sporulation, which inhibits *abrB* expression, thus releasing P_{liaI} from AbrB-dependent repression. Simultaneously, a so far unknown stimulus activates the HK LiaS (and/or its release from LiaF repression) that in turn activates LiaR, resulting in the expression of the *liaIH* operon. The observed induction is significantly weaker (approximately 10- to 15-fold) compared with the response to strong inducers (approximately 100-fold) (Jordan *et al.*, 2007).

In summary, LiaFSR-dependent gene expression is induced by cell envelope perturbating conditions and is also embedded in the transition

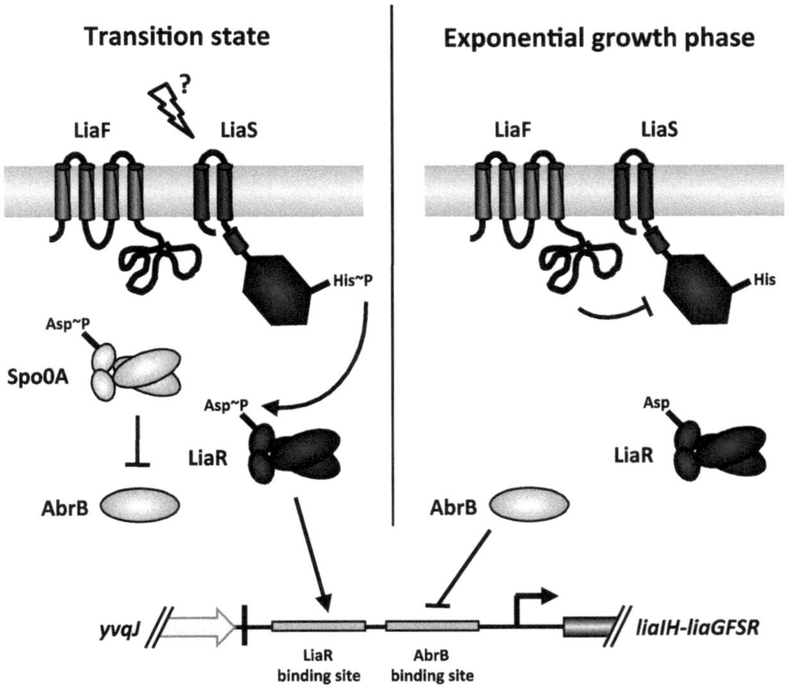

Figure 11.5 Transition state induction of P_{liaI} in *B. subtilis*. The involved regulatory proteins during exponential growth (right) and the transition state (left) are named and marked in light, medium, and dark grey. Arrows indicate activation, T-shaped lines repression. The genomic context of P_{liaI} is shown schematically below, including the LiaR and AbrB binding sites. See text for details.

state regulation of *B. subtilis* through the interplay of at least five regulatory proteins: LiaR, LiaS, LiaF, AbrB, and Spo0A (Jordan *et al.*, 2007) (Fig. 11.5). The biological significance of this induction is still unclear.

Distribution and conservation of LiaFSR-like regulatory systems

As already mentioned, LiaFSR-like systems are widely conserved within the group of Firmicutes bacteria, with the exception of the genera *Clostridium* and *Lactobacillus* (Jordan *et al.*, 2006, 2008). In all species possessing *liaSR* homologues, *liaF*-like genes are always located directly upstream of the 2CS genes. Two groups were described based on the genomic context of the corresponding loci (Jordan *et al.*, 2006; Mascher, 2006; Eldholm *et al.*, 2010). Group I includes *liaFSR* homologues found in *Bacillus* and *Listeria* species, which are regulatory linked to *liaIH* operons. The difference between these two genera is the location of the *liaIH*-like genes either as a part of the *liaIH-(G)FSR* locus (*Bacillus*) or genetically separated from the *liaFSR* operon, but still under the transcriptional control of LiaSR (*Listeria*). Additionally, *liaG* homologous genes are only found in *B. subtilis*, *B. licheniformis*, and *B. halodurans*.

Group II includes species that lack *liaIH* homologues. Here, only a *liaFSR* locus is conserved (Jordan *et al.*, 2006). In the following sections, the current state of knowledge about the LiaFSR-dependent cell envelope stress response in different Firmicutes species will be discussed. A detailed overview of these systems is given in Table 11.2.

LiaFSR-like systems of group I (*liaIH-liaFSR*) in other Firmicutes species

YvqEC of *B. licheniformis*

In-depth transcriptional profile analyses of *B. licheniformis* in response to cell wall antibiotics identified the LiaFSR homologous system YvqFEC (Wecke *et al.*, 2006). YvqFEC is strongly induced in the presence of bacitracin and to a weaker degree by vancomycin and D-cycloserine. Similar to *B. subtilis*, the YvqFEC system is encoded as part of the *yvqIH–yvqGFEC* locus, where *yvqIH* represents the *liaIH* and *yvqGFEC* the *liaGFSR* homologous genes. The expression of the *yvq* locus is controlled in an autoregulative manner by YvqC from a promoter upstream of *yvqIH*. In contrast to *B. subtilis*, three instead of two transcripts can be observed after induction, including the major transcript *yvqIH* (due to a stem loop structure downstream of *yvqH*), a transcript representing the whole locus, and an additional transcript of 2.5 kb in size (*yvqIH–yvqGF′*) caused by a second stem loop structure within the *yvqF* gene (Wecke *et al.*, 2006). Yvq mutants of *B. licheniformis* have not been analysed so far. Therefore, the physiological role of this system is unknown. Moreover, YvqC target genes other than the *yvq* locus itself have not been identified so far. However, the close homology between *B. subtilis* and *B. licheniformis* suggests that the two systems are very similar with regard to their behaviour, physiological role and hence regulons.

$LiaSR_{Lm}$ of *L. monocytogenes*

More information is available on the LiaFSR homologue of *L. monocytogenes*, which was named $LiaF_{Lm}$-$LiaSR_{Lm}$ (formerly Lmo1020–1021–1022). Detailed transcriptional studies revealed a massive up-regulation of this system by cell wall active antibiotics such as bacitracin and vancomycin. The regulon controlled by $LiaR_{Lm}$ is much larger than that of *B. subtilis*, including 29 genes organized in 16 transcriptional units (Fritsch *et al.*, 2011). In addition to the $liaIH_{Lm}$ and $liaFSR_{Lm}$ operons, transcription of genes encoding components of ABC transporters as well as predicted membrane or extracytoplasmic proteins was induced.

The most surprising difference between the Lia responses of *L. monocytogenes* and that of *B. subtilis* is a derepression of target genes in $liaS_{Lm}$ mutants. The lack of the HK protein in *B. subtilis* leads to a 'locked OFF' phenotype demonstrating a loss of response to adequate stimuli due to the lack of the sensor protein. In *L. monocytogenes*, it was postulated that the remaining $LiaR_{Lm}$ protein undergoes a non-specific phosphorylation via the small molecule phospho-donor acetyl phosphate, which can take place owing to the lack of $LiaS_{Lm}$ phosphatase activity (Fritsch *et al.*, 2011). In fact, we observed a similar behaviour for *B. subtilis*

LiaR, but only if LiaR is simultaneously overexpressed (Schrecke and Mascher, unpublished). The physiological role of the LiaFSR system in *L. monocytogenes* is so far unclear, since no resistance against inducing compounds could be detected (Fritsch *et al.*, 2011).

LiaFSR-like systems of group II (*liaFSR*) in other Firmicutes species

VraSR of *S. aureus*

The best studied LiaSR homologue, VraSR, was originally identified as one of the most strongly up-regulated systems after treatment with vancomycin in vancomycin-resistant *S. aureus* (VRSA) compared with a vancomycin-sensitive strain (VSSA) (Kuroda *et al.*, 2000). It is encoded as the last two genes of the tetracistronic autoregulated operon *orf1-yvqF-vraS-vraR* (Yin *et al.*, 2006). The VraSR system strongly responds to the presence of cell wall-interfering antibiotics like glycopeptides (e.g. vancomycin and teicoplanin), bacitracin, β-lactams, and D-cycloserine, but not to general stresses such as heat, high osmolarity, or pH shifts. A total of 46 genes were identified to be induced by VraR after exposure to vancomycin (Kuroda *et al.*, 2003). Some of them are associated with the cell wall biosynthesis, including *pbp2* and *sgtB* required for polymerization of peptidoglycan, and *murZ* crucial for murein monomer precursor synthesis.

The VraSR system plays a critical role in resistance against all of its inducers except D-cycloserine and levofloxiacin. The overproduction of penicillin-binding protein (PBP) 2 significantly increases the resistance to teicoplanin, showing that the reduction of observed teicoplanin resistance in *vraSR* null mutants is due to the loss of PBP2 induction (Kuroda *et al.*, 2003; Gardete *et al.*, 2006). Penicillin binds to PBPs, which are required for cell wall synthesis, and inactivates them. The resistance against the β-lactam antibiotic penicillin in almost all *S. aureus* strains is caused by the production of β-lactamases (Boyle-Vavra *et al.*, 2006). Accordingly, β-lactams that are insensitive to β-lactamase (e.g. methicillin, oxacillin) were used for treatment of *S. aureus* infections. However, methicillin-resistant *S. aureus* (MRSA) strains evolved quickly and had acquired the *mecA* gene, which encodes a new PBP variant, PBP2a, with a low affinity to β-lactams. PBP2 and PBP2a work together to mediate oxacillin resistance, but its expression is not sufficient for the resistance phenotype. The recovery of the oxacillin resistance of a *vraS* mutant is not warranted by the overexpression of *mecA* or *pbp2*, but rather by the complementation with the entire *vra* operon. This shows that VraS is also required even when *mecA* and *pbp2* are transcribed. It is most likely that VraSR modulate the expression of one or several factors other than PBP2 and PBP2a that influence the oxacillin resistance (Boyle-Vavra *et al.*, 2006). More recently, the biochemistry of VraSR-dependent signal transduction has been extensively studied (Belcheva and Golemi-Kotra, 2008), as described in detail below ('Catalytic activity of LiaS-like HK and phosphotransfer to LiaR').

CesSR of *L. lactis*

The CesSR system (formerly LlkinD–LlrD; also TCS-D) was originally identified in the *L. lactis* strain MG1363 as induced during the onset of stationary phase and involved in the resistance to salt and osmotic stress (O'Connell-Motherway *et al.*, 2000). Later, it was demonstrated that CesSR is strongly induced in the presence of the bacteriocin lactococcin 972 (Lcn972) (Martinez *et al.*, 2007), which is a ribosomally synthesized antimicrobial peptide produced by lactic acid bacteria. A genome-wide transcriptional analysis of *L. lactis* revealed 26 up-regulated genes after Lcn972 treatment, of which 21 are controlled by CesR (Martinez *et al.*, 2007). Some of these genes encode proteins involved in peptidoglycan biosynthesis and modification. CesR also regulates the transcription of its own tricistronic operon, which includes the *liaF* homologue *yjbB* (*llmg1650*). The highest up-regulation in CesSR-dependent gene expression was detected for *llmg0169* and the operon *llmg2164–2163* (Martinez *et al.*, 2007). While all three genes encode proteins with unknown function, Llmg2163 harbours an N-terminal PspC domain. PspC was originally described for the phage shock response of *E. coli* and proteins harbouring this domain are known to protect the cell against extracytoplasmic stress and to maintain the integrity of the cytoplasmic membrane (see section 'Regulon and physiological role'). The overexpression of *llmg2164–2163* was found to specifically protect *L. lactis* cells against Lcn972, but not against other cell wall antibiotics (Roces *et al.*, 2009).

The CesSR system is also induced in the presence of lipid II-interfering antibiotics like bacitracin, nisin, or plantaricin C. Deletion of *cesR* leads to an increased susceptibility to these agents, suggesting that CesSR mediates resistance against them (Martinez *et al.*, 2007). Additionally, peptidoglycan hydrolysis owing to the exposure to lysozyme is also detected by CesSR (Veiga *et al.*, 2007).

LiaSR of *S. mutans*

LiaSR (formerly HK11–RR11) of *S. mutans* was identified as one of several systems that affect the expression of virulence factors in this organism (Li *et al.*, 2002). LiaFSR was shown to be involved in biofilm formation and acid tolerance, and LiaS was suggested to be a pH sensor (Li *et al.*, 2002; Chong *et al.*, 2008).

LiaFSR of *S. mutans* is encoded within the pentacistronic operon *liaFSR–ppiB–pnpB*. The genes *ppiB* and *pnpB* encode a peptidyl-prolyl *cis/trans* isomerase and a polynucleotide phosphorylase, respectively. As in *B. subtilis*, the LiaFSR system of *S. mutans* is induced by lipid II-interfering antibiotics like bacitracin, vancomycin, and nisin, and by other cell wall antibiotics, such as chlorhexidine. Moreover, it is induced by low pH, high osmolarity, and ethanol. β-Lactams and D-cycloserine do not function as inducers (Suntharalingam *et al.*, 2009).

The LiaFSR system of *S. mutans* clearly differs from that of *B. subtilis* in terms of the time point of the stimulus-independent induction. The expression of *S. mutans lia* genes is 10-fold repressed in stationary phase and 2-fold in mid-logarithmic phase compared with the early-logarithmic growth phase. This observation suggests an involvement of LiaFSR in early log-phase growth, which is characterized by high growth rates, increased cell division, cell separation, and peptidoglycan biosynthesis (Suntharalingam *et al.*, 2009). Indeed, *S. mutans* LiaR directly and/or indirectly controls a large regulon consisting of 174 genes during biofilm growth encoding proteins involved in membrane protein synthesis and peptidoglycan biosynthesis, envelope chaperone/proteases, and transcriptional regulators (Perry *et al.*, 2008; Suntharalingam *et al.*, 2009).

The LiaFSR system of *S. mutans* was also associated with playing a role in activation of the essential WalRK-like 2CS VicRKX (also named CovRSX) (Dubrac *et al.*, 2008; Tremblay *et al.*, 2009), which controls the expression of fructosyltransferase, glucosyltransferases, and glucan-binding proteins (Lee *et al.*, 2004; Senadheera *et al.*, 2005). It is active during exponential growth at neutral pH, when a rapid delivery of cell wall precursors owing to fast rates of cell wall biosynthesis is required. Induction of VicRKX in the presence of cell wall antibiotics was shown to be controlled by LiaFSR rather then autoregulated (Tremblay *et al.*, 2009). Despite the wide distribution of WalRK-like and LiaFSR-like 2CSs, this is so far the only documented direct regulatory connection between the two systems.

LiaFSR of *S. pneumoniae*

The LiaFSR-homologous system TCS03 of *S. pneumoniae* was found to be up-regulated after exposure to vancomycin in both a vancomycin-sensitive and vancomycin-tolerant strain (Haas *et al.*, 2005). In a later study, bacitracin, nisin, and tunicamycin were described to induce the TCS03, whereas D-cycloserine and ampicillin did not (Eldholm *et al.*, 2010). Interestingly, the same authors did not detect any up-regulation of TCS03 after vancomycin treatment in laboratory strain R6, indicating strain-specific differences. In addition, cell envelope stress elicited by murein hydrolases and autolysins is also perceived by TCS03 (Eldholm *et al.*, 2010). TCS03 does not seem to provide resistance against the lipid II-interfering antibiotics, which act as inducers of the system. Instead of a resistance phenotype, a physiological role as protectant against self-lysis in competent *S. pneumoniae* cells was proposed (Eldholm *et al.*, 2010) (see section 'Regulon and physiological role of Lia-like systems'). This observation again indicates that LiaFSR-like systems, despite comparable inducer ranges, have adapted their very diverse physiological roles to the specific needs of the respective organism.

The TCS03-encoding operon consists of 10 genes (*spr0342*-*spr0351*) and is the most strongly induced target of the *S. pneumoniae* LiaR-dependent gene expression. The first three genes encode LiaFSR (*spr0342–0345*) (Eldholm *et al.*, 2010). A genome-wide transcriptional profiling revealed 18 genes that are up-regulated in a LiaR-dependent manner, ten of which are already part of the LiaFSR operon and some other target genes encode stress-related proteins, which are involved in the response to heat shock. Moreover, genes encoding a Spx homologue as well as the phage shock protein C (PspC) are induced, which were also found as part of the CesSR regulon in *L. lactis* and the LiaSR regulon in *S. mutans* (Martinez *et al.*, 2007; Suntharalingam *et al.*, 2009; Eldholm *et al.*, 2010).

Common themes in LiaFSR-like systems

Stimuli of Lia-like systems

LiaFSR-homologous systems respond to a relatively broad range of stimuli, which are all associated with cell envelope stress. Therefore, these systems presumably represent cell wall damage-sensing systems in contrast to drug-sensing detoxification modules like BceRSAB (Rietkötter *et al.*, 2008) (see Bce section). Despite a wide range of physiological functions associated with LiaFSR-like systems, the range of inducers is nevertheless well defined.

Three different classes of stimuli can be distinguished. The first class contains lipid II-cycle inhibitors like bacitracin, nisin, or vancomycin, which are common inducers of all described LiaFSR-like systems (Fig. 11.1). They belong to the class of AMPs which share a cationic and amphipathic nature as well as a similar mode of action, such as inhibiting the cell wall biosynthesis by binding to different moieties of either lipid II or undecaprenol pyrophosphate (UPP) (Rietkötter *et al.*, 2008; Schneider and Sahl, 2010; Staroń *et al.*, 2011). Bacitracin specifically binds to UPP and prevents its dephosphorylation/recycling thus blocking cell wall biosynthesis (Storm and Strominger, 1973). Nisin disrupts the membrane integrity by pore formation via initial binding to pyrophosphate of lipid II (Schneider and Sahl, 2010). Vancomycin inhibits the transpeptidation of peptidoglycan by binding to the D-Ala-D-Ala terminus of the pentapeptide chain of lipid II (Schneider and Sahl, 2010). The bacteriocin lactococcin 972 (Lcn972), which was described as strongest inducer of CesSR in *L. lactis*, inhibits the peptidoglycan synthesis by binding to lipid II at the level of septum formation, thus blocking cell division (Martinez *et al.*, 2008) (Fig. 11.1).

The second class of inducers includes agents that do not interfere with the lipid II cycle, but also influence the integrity of the cell envelope. Good examples are daptomycin, which affects the cell membrane, causing membrane depolarization or perforation (Silverman *et al.*, 2003; Straus and Hancock, 2006), rhamnolipids, which are biosurfactans that alter the cell surface leading to increased hydrophobicity and membrane permeability (Vasileva-Tonkova *et al.*, 2011), as well as more general stresses such as alkaline shock, detergents, phenol, ethanol, secretion stress and infections by filamentous phages (Petersohn *et al.*, 2001; Wiegert *et al.*, 2001; Mascher *et al.*, 2004; Hyyryläinen *et al.*, 2005; Pietiäinen *et al.*, 2005; Tam le *et al.*, 2006).

Class three contains inducers that are characteristic of the specific habitat or lifestyle of the different species, but all of them affect the integrity of the bacterial cell envelope. For example, the LiaFSR system of *S. mutans* responds to acidity (Li *et al.*, 2002), a typical condition in the human oral cavity. Furthermore, *B. subtilis* LiaFSR is embedded in the transition state regulation leading to the induction of this system not by an external but rather by a so far unknown internal stimulus (Jordan *et al.*, 2007) (Fig. 11.5). Another example is the induction of *S. pneumoniae* LiaFSR by murein hydrolases and autolysins in competent cells to protect against self-lysis (Eldholm *et al.*, 2010).

Stimulus perception by LiaS and LiaF

In contrast to the well-characterized inducer spectrum of the LiaFSR-like systems, the details of the mechanism of stimulus perception are basically unknown, similar to the situation described for Bce-like systems above. But based on the available data, it is most likely that two proteins are required for sensing the stimulus: The HK LiaS and the accessory protein LiaF.

Without exception, LiaF homologous proteins are always encoded directly upstream of *liaSR* in all species harbouring *lia*-like genes, but just a few of them have been characterized so far. In *B. subtilis*, *S. mutans*, and *L. monocytogenes*, LiaF was identified to act as a specific inhibitor of the LiaR-dependent gene expression in the absence of inducing conditions (Jordan *et al.*, 2006; Suntharalingam *et al.*, 2009; Fritsch *et al.*, 2011). Accordingly, disruption of the respective genes leads to high expression levels of LiaR target genes even without external stimuli. Here, the LiaSR system shows a 'locked-ON' phenotype. The only example of a putative positive modulation of the Lia system by LiaF was recently described for YvqF of *S. aureus* (McCallum *et al.*, 2011). Its deletion leads to a loss of induction of the VraSR system even in vancomycin-induced cultures and is comparable to the phenotypes observed for *vraS* or *vraR* null mutants. Therefore, YvqF seems to be essential for responding to cell envelope stress and resulting antibiotic tolerance in *S. aureus* (McCallum *et al.*, 2011).

No further mechanistic insight beyond the regulatory behaviour of *liaF* mutants is currently available. Preliminary studies of *B. subtilis* indicate that the negative function of LiaF resides in the cytoplasmic C-terminal domain, but that membrane localization is necessary for the full inhibitory activity (Jordan, unpublished).

The mechanism of how LiaFSR-like systems sense their respective stimuli is so far unknown. Typical HKs harbour an extracellular sensing domain located between two transmembrane helices, which is postulated to bind and/or detect suitable stimuli (Wolanin *et al.*, 2002). In contrast, LiaS homologues are IM-HKs, possessing only a small extracellular loop between the two TMRs, which is thought to be too small for ligand binding. It was initially proposed that these HKs detect the stimuli directly at the surface or within the cytoplasmic membrane (Mascher *et al.*, 2003, 2006; Mascher, 2006).

But recent evidence indicates that LiaS alone does not seem to be sufficient to coordinate a controlled response to its stimuli. Instead, the accessory protein LiaF seems to be involved in the stimulus sensing process, based on the behaviour of *liaF* mutants described in the previous section. A direct interaction between LiaS and LiaF has not been proven so far, but seems likely, based on initial data from *S. aureus* (McCallum *et al.*, 2011). Three possible mechanisms of stimulus perception by LiaS and LiaF can be envisaged (Fig. 11.6). First, a LiaF–LiaS-sensor complex is conceivable, in which both proteins participate in stimulus sensing (Fig. 11.6B). Thereby, LiaF modulates the kinase/phosphatase activity of LiaS according to the presence or absence of a stimulus. If LiaF is disrupted, this interplay is lost and LiaS activates LiaR constitutively. Second, LiaF may act as a repressor and stabilize the phosphatase activity of LiaS in the absence of a suitable stimulus, thereby keeping the system silent (Fig. 11.6C). In the presence of a stimulus, LiaF may then act as a sensor and release LiaS from repression, allowing phosphorylation of LiaR and subsequent activation of gene expression. Third, LiaF may function

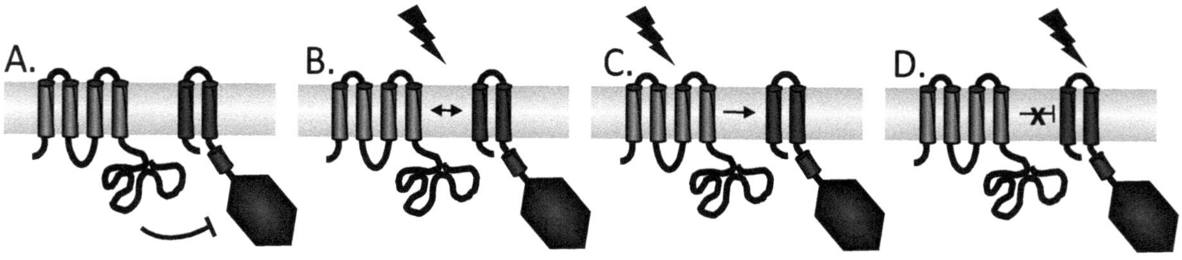

Figure 11.6 Proposed mechanisms of stimulus perception by LiaF- and LiaS-like proteins. (A) System in uninduced state. LiaF has a negative function on LiaS activity. (B) A LiaF-LiaS sensor complex is shown in which both proteins participate in stimulus perception. (C) LiaF acts as the sensor protein and releases LiaS from its repression in the presence of a stimulus. (D) LiaS is the sensor protein and undergoes a postulated conformational change after sensing the stimulus. This, in turn, leads to a disruption of the LiaF-dependent inhibition. See text for details.

just as a repressor in the absence of a stimulus and LiaS acts as a sensor (Fig. 11.6D). In the presence of a stimulus, LiaS undergoes a conformational change leading to a disruption of the direct interaction with LiaF and a subsequent activation of LiaR. This possibility is by far the most unlikely, because of the absence of extracytoplasmic sensor domains in LiaS-like HKs, but cannot be ruled out owing to the lack of experimental data. For all three alternatives, the default activity of LiaS would be 'kinase-ON', which is only inactivated by the presence of LiaF in the absence of a suitable trigger.

In *S. aureus*, the LiaF homologue YvqF seems to harbour a positive function on LiaS activity. Here, disruption of *yvqF* could result in a total loss of kinase function. Therefore, the default activity of VraS may be 'phosphatase-ON', which may need YvqF for its own activation and/or stimulus perception. In contrast to *B. subtilis*, a direct interaction between YyqF and VraS has been demonstrated, as mentioned above (McCallum *et al.*, 2011). Accordingly, the authors concluded that YyqF may be involved in sensing the stimulus, which is responsible for regulating signal transduction.

Catalytic activity of LiaS-like HKs and phosphotransfer to LiaR

In the presence of a suitable stimulus, LiaS autophosphorylates in an ATP-dependent manner, which may occur as a *cis* or *trans* reaction. Until recently, *trans*-autophosphorylation of a HK dimer where one HK monomer catalyses the phosphorylation of the second monomer was the only reaction described (Ninfa *et al.*, 1993; Stock *et al.*, 2000; Cai and Inouye, 2003; Trajtenberg *et al.*, 2010) and therefore thought to be a general characteristic of HKs. However, recent studies show that *cis*-autophosphorylation (HK monomer phosphorylates itself) can also commonly occur (Casino *et al.*, 2009, 2010; Pena-Sandoval and Georgellis, 2010). The HK DesK of *B. subtilis* was described to autophosphorylate in *trans* (Trajtenberg *et al.*, 2010). Since LiaS belongs to the same HK family like DesK (HPK7) (Grebe and Stock, 1999), a *trans*-autophosphorylation mechanism appears to be likely for LiaS as well.

For LiaS-like HKs, bifunctional activities composed of kinase and phosphatase activity were postulated and experimentally proven for VraS of *S. aureus* (Belcheva and Golemi-Kotra, 2008) and LiaS_{Lm} of *L. monocytogenes* (Fritsch *et al.*, 2011). A truncated version of VraS only consisting of the soluble C-terminal part was used to show an increasing VraS autophosphorylation over 20 minutes after incubation with ATP. A rapid phosphotransfer between VraS and VraR occurred within 30 seconds, with about 70% of the phosphate residues being transferred (Belcheva and Golemi-Kotra, 2008). VraR~P can only be dephosphorylated in the presence of VraS, clearly demonstrating the phosphatase activity of this protein. The kinetics of VraS autophosphorylation and phosphotransfer indicate a rapid *in vivo* response to cell wall damage (Belcheva and Golemi-Kotra, 2008). LiaS_{Lm} autophosphorylation was shown to occur within 15 minutes after ATP treatment. A LiaS_{Lm}-dependent phosphotransfer to the cognate RR LiaR_{Lm} was also observed in which phosphorylated LiaR_{Lm} protein could only be detected within the first 90 seconds of incubation with LiaS_{Lm}~P, indicating a rapid hydrolysis of LiaR_{Lm}~P due to the phosphatase activity of LiaS_{Lm} (Fritsch *et al.*, 2011).

The LiaS-dependent activation of LiaR-like RRs via phosphorylation is presumably followed by dimerization and subsequent activation of the output domain, leading to DNA binding and differential gene expression of the corresponding regulon. All LiaR-like RRs described so far seem to be transcriptional activators, leading to an up-regulation of target genes (Kuroda *et al.*, 2003; Mascher *et al.*, 2004; Martinez *et al.*, 2007; Suntharalingam *et al.*, 2009; Tremblay *et al.*, 2009; Eldholm *et al.*, 2010; Fritsch *et al.*, 2011). The only LiaR-dependent down-regulated genes were some metabolism-associated genes in *L. lactis*. But the relevance of this observation is disputable since this down-regulation could also be the result of a loss in cell viability after Lcn972 treatment (Martinez *et al.*, 2007).

LiaR binding sites

LiaR-like RRs belong to the NarL/FixJ subfamily, the second largest subfamily of RRs, whose output domains share a characteristic LuxR-like helix–turn–helix motif for DNA binding (Galperin, 2010). To identify LiaR binding sites, a sequence comparison of putative LiaR boxes located upstream of *liaI* homologues have been performed for members belonging to group I (*Bacillus* and *Listeria* species) (Jordan *et al.*, 2006). An imperfect inverted repeat consisting of seven nucleotides separated by two or four nucleotides were found to be a conformable motif. In a later study by Martinez *et al.* sequences of putative CesR binding sites upstream of CesR-depending target genes in *L. lactis* were compared, which all showed an inverted repeat with a 7-2-7 motif (Martinez *et al.*, 2007). But this motif significantly differs in its sequence pattern from the previous identified LiaR box (Jordan *et al.*, 2006), suggesting the existence of different LiaR-like RR subgroups with individual binding site motifs (Martinez *et al.*, 2007). Indeed, LiaR-like RRs of staphylococci, streptococci, and lactococci species phylogenetically cluster together and the CesR box was also found upstream of target genes of the orthologous systems in *S. aureus*, *S. mutans*, and *S. pneumoniae* (Martinez *et al.*, 2007; Belcheva *et al.*, 2009; Suntharalingam *et al.*, 2009; Eldholm *et al.*, 2010). In contrast, the LiaR motif is not present upstream of LiaR-/VraR-dependent target genes of *S. pneumoniae* and *S. aureus* (Jordan *et al.*, 2006).

A recent study summarized the current collection of LiaR binding sites (Eldholm *et al.*, 2010). Here again, *in silico* analyses revealed two distinct groups: (i) LiaR-like binding sites upstream of *liaIH* homologous operons in *Bacillus* and *Listeria* species showing a 7-2-7 inverted repeat; and (ii) CesR-like binding sites upstream of potential target genes in *Lactococcus* and *Streptococcus* species harbouring a 6-4-6 motif. The two groups differ in the sequence conservation pattern, whereas the core residues are identical (Eldholm *et al.*, 2010). Most of the RR binding motifs of other NarL-like RRs, characterized in the *Bacillus cereus* group and related Firmicutes, also show inverted repeats with a spacer of two to six nucleotides (de Been *et al.*, 2008). Therefore, this motif seems to be a common feature among this RR subgroup.

Regulon and physiological role of Lia-like systems

The number of LiaR-binding site and hence genes regulated by LiaFSR-like systems differs significantly between species. The LiaFSR-like systems of bacilli species (group I) encoded within the *liaIH-liaGFSR* operon mainly seem to regulate their own transcription with *liaIH* as the only target, whereas members of group II lacking *liaIH*-like genes (and also *Listeria* species) control a much larger number of target genes (Jordan *et al.*, 2008). This observation already indicates different physiological roles of LiaFSR-like systems.

B. subtilis is the only member of group I (except *Listeria*) that was intensively investigated with regard to its LiaR regulon. Of three potential targets, *liaIH*, *yhcYZ–yhdA*, and *ydhE*, only the *liaIH* operon is preceded by a well-conserved LiaR box and is bacitracin-inducible in the wild-type strain (Wolf *et al.*, 2010). Furthermore, the RR LiaR only binds to the *liaI* promoter shown in EMSA experiments (Wolf *et al.*, 2010). The other two loci seem to be artefacts only observed in a *liaF* mutant and are thought to be the result of unspecific binding under conditions of a significantly increased cellular amount of phosphorylated LiaR (Jordan *et al.*, 2006; Wolf *et al.*, 2010). In contrast, DNA microarray analyses identified 46, 21, 174 and 18 genes as regulated by group II LiaR-like RRs in *S. aureus*, *L. lactis*, *S. mutans*, and *S. pneumoniae* (Kuroda *et al.*, 2003; Martinez *et al.*, 2007; Perry *et al.*, 2008; Eldholm *et al.*, 2010).

Three general themes with regard to physiological roles can be extracted: (i) CESR-mediated resistance against envelope stress inducers, (ii) a Psp-like response to maintain the cell wall integrity, and (iii) habitat- and lifestyle-specific phenotypes.

Resistance phenotypes mediated by the LiaFSR system are only described for VraSR of *S. aureus*, CesSR of *L. lactis*, and LiaFSR of *S. mutans* (Kuroda *et al.*, 2003; Martinez *et al.*, 2007; Suntharalingam *et al.*, 2009). These species show an increased susceptibility to at least some of the cell wall antibiotics that act as inducers, if the LiaFSR system is disrupted. A more indirect effect on the resistance phenotype against peptidoglycan hydrolysis could be mediated by Spx, which is a transcriptional regulator found to be a member of the LiaR regulon in *L. lactis*, *S. mutans*, and *S. pneumoniae* (Martinez *et al.*, 2007; Suntharalingam *et al.*, 2009; Eldholm *et al.*, 2010). In *L. lactis*, *spxB* expression leads to an induction of *oatA*, encoding a peptidoglycan O-acetyltransferase that increases the O-acetylation of peptidoglycan and thus renders the cells more resistant to peptidoglycan hydrolysis (Veiga *et al.*, 2007). In *B. subtilis*, no altered antibiotic sensitivity was observed in a *liaIH* mutant, despite the strong induction of the *liaIH* operon in the wild-type strain by these agents. The only described exceptions are daptomycin, enduracidin and rhamnolipids, for which a slight increase in sensitivity was observed in a *liaH* mutant (Hachmann *et al.*, 2009; Wecke *et al.*, 2009, 2011; Wolf *et al.*, 2010).

Since *liaH* was described to be a homologue of *pspA*, a Psp-like response of LiaFSR to cell envelope stress was discussed (Wolf *et al.*, 2010). In proteobacteria, PspA is a central part of the Psp response which maintains the proton-motif-force under extracytoplasmic stress conditions that affect the membrane integrity. The expression of the *psp* genes from σ^{54}-dependent promoters is induced in the presence of envelope stress, such as filamentous phage infection, a mislocalization of envelope proteins, heat, osmotic shock, the presence of organic solvents and proton ionophores, and antibiotics that perturb the phospholipid biosynthesis (Brissette *et al.*, 1990; Bergler *et al.*, 1994; Hardie *et al.*, 1996; Model *et al.*, 1997; Darwin, 2005). Despite significant differences in mechanisms of signal transduction and gene regulation between the Lia and Psp response, there are also some similarities. The most obvious similarity is the ring structure of LiaH and PspA with a 9-fold rotational symmetry (Hankamer *et al.*, 2004; Wolf *et al.*, 2010). From a regulatory point, the most evident parallel is the overlap of the inducer spectra of the Lia and Psp responses. Both systems are induced by cell envelope stress and without any stimuli at the onset of stationary phase (Weiner and Model, 1994; Jordan *et al.*, 2007). Taken together, this supports the idea of a similar physiological role of *B. subtilis* LiaH and *E. coli* PspA (Wolf *et al.*, 2010).

Interestingly, a *liaH* mutant showed increased sensitivities to some agents that cause oxidative stress, e.g. menadione, plumbagin, sodium selenite, and hydrogen peroxide, even though these compounds do not function as inducers of the LiaFSR system (Wolf *et al.*, 2010). The significance of this finding is unclear.

Another link between the Lia- and Psp-response is the presence of PspC homologues as part of the LiaR regulon in *L. lactis* (Llmg2163), *S. mutans* (SMu753), and *S. pneumoniae* (Spr0810) (Martinez *et al.*, 2007; Suntharalingam *et al.*, 2009; Eldholm *et al.*, 2010). In *L. lactis*, Llmg2163 specifically mediates resistance against bacteriocin Lcn972, but not against other cell wall antibiotics (Roces *et al.*, 2009). The PspC-homologue Spr0810 of *S. pneumoniae* seems to play a role in maintaining the membrane integrity of competent cells (Eldholm *et al.*, 2010). The occurrence of *psp* homologous genes as targets for the LiaR-dependent gene expression in at least four different species (*B. subtilis, S. pneumoniae, L. lactis*, and *S. mutans*) further supports the hypothesis that a Psp-like response orchestrated by LiaFSR, is a common theme in Lia-dependent signalling.

Species-specific adaptations of LiaFSR-like systems in response to cell envelope stress have evolved in a number of Firmicutes bacteria, presumably to adapt to their specific habitat or lifestyle. In *S. pneumoniae*, the LiaFSR system seems to have a rather special function associated with natural competence development (Eldholm *et al.*, 2010). When reaching the competent state, pneumococcal cells produce fratricide-inducing proteins, e.g. the putative murein hydrolase CbpD that acts together with the autolysins LytA and LytC in killing non-competent sister cells of the same population via lysis (Guiral *et al.*, 2005; Kausmally *et al.*, 2005; Eldholm *et al.*, 2009). The competent cells avoid their own cell lysis by synthesizing the immunity protein ComM (Havarstein *et al.*, 2006). The LiaFSR system is induced in *comM* mutants most likely by sensing the cell envelope stress caused by CbpD, LytA, and LytC. Therefore, the LiaFSR response seems to be a second layer of self-protection against competence-induced cell lysis in addition to the major fratricide immunity protein ComM (Eldholm *et al.*, 2010). This protection is probably linked to the LiaR-regulated genes *pcpC* (*spr*0351) and *spr*0810. PcpC is a choline-binding protein which is encoded within the *lia* locus. It is a paralogue of CbpF that is described to abolish autolysis caused by LytC (Molina *et al.*, 2009). A similar function for PcpC is very likely due to the high homology.

The physiological role of Spr0810, which is a PspC homologue, was described in the previous section.

In the pathogenic bacteria *S. aureus* and *S. mutans*, virulence factors appeared to be the major targets controlled by VraR or LiaR, respectively (Li *et al.*, 2002; Kuroda *et al.*, 2003; Suntharalingam *et al.*, 2009). Both, the above described resistance against inducing cell wall antibiotics as well as the regulation of biofilm formation and acid tolerance are potent processes to colonize the human host.

In *B. subtilis*, it was shown that the LiaFSR-dependent gene expression is embedded in the transition state regulation (Jordan *et al.*, 2007) (see section 'The paradigm: LiaFSR of *B. subtilis*'). Therefore, a contribution of LiaIH to cellular differentiation seemed likely. However, none of the typical developmental processes, such as swarming, biofilm formation, competence development or endospore formation were affected by *liaIH* expression (Wolf *et al.*, 2010). The physiological significance of this species-specific behaviour of *B. subtilis* LiaFSR remains to be identified.

Taken together, LiaIH in *B. subtilis* and most likely other bacilli represents the only relevant LiaR target with a so far unclear physiological role. Recent studies raise the idea of an involvement in the protection of cells against envelope and oxidative stress in a Psp response-like manner. All other studied LiaFSR-like systems control a much larger regulon reflecting a more expanded response to cell envelope stress, as well as lifestyle- or habitat-specific traits.

Outlook: the central open questions

This chapter was aimed at presenting a detailed overview on our current state of knowledge on BceRS- and LiaRS-like 2CSs, which are instrumental in orchestrating the CESR in most Firmicutes bacteria. A lot of progress has been made in recent years on both systems from a range of Firmicutes bacteria, compared with the state of knowledge at the time of the last review on CESR in Gram-positive bacteria (Jordan *et al.*, 2008). But the more we generally learned, the less clear the details became. The two systems share the domain architecture of their respective HKs and an overlapping range of inducing peptide antibiotics that interfere with cell wall biosynthesis (Figs. 11.1 and 11.2). The currently available data seems to argue in favour of a 'damage-sensing' mechanism of stimulus perception in case of LiaRS-like systems, while BceRS-like systems presumably respond to peptide antibiotics directly as a compound via their associated BceAB-like ABC transporters ('drug-sensing'). But beyond this relatively well-supported hypothesis, the details remain largely unclear.

Both systems were initially identified and characterized in the Gram-positive model bacterium *B. subtilis*. While its CESR network remains the best characterized and most complex, there is growing evidence that at least the physiological role of both Bce- and Lia-like systems in this species is presumably only representative for *Bacillus* spp., but not for most of the other Firmicutes bacteria. In *B. subtilis*, both LiaRS and BceRS-like systems each specifically regulate a single target operon, encoding LiaIH and BceAB-like ABC transporters, respectively. In this species, the more general CESR is provided by ECF sigma factor-dependent signal transduction (Helmann, 2002; Butcher *et al.*, 2008). In contrast, both Bce- and Lia-like systems coordinate larger regulons (and therefore more general CESR) in most of the other species studied to date (Tables 11.1 and 11.2) that mostly lack a large repertoire of ECF sigma factors (Staroń *et al.*, 2009). It would be interesting to know, if Bce- and Lia-like systems became more specific in bacilli in the course of evolution, or whether these systems were rather adapted to broader functions in cocci and other Firmicutes bacteria with reduced genomes.

The most fascinating open question on both systems, and presumably a hard one to answer, regards the mechanism of stimulus perception and signal transduction facilitated by the two signal transfer HKs BceS and LiaS. Research in the last years has presented compelling evidence that the minimalistic architecture shared by these 'IM-HKs' is reflected by a common evolution to out-source the sensory component onto accessory membrane proteins. In case of BceS, the ABC transporter BceAB itself seems to be the true sensor, even though a participation of BceS cannot be safely ruled out, as discussed (Fig. 11.3). For LiaS-like HK, it remains to be seen if LiaF is the sole sensor (comparable to the postulated situation for Bce-like systems), or whether LiaF together with LiaS forms the sensor complex (Fig. 11.6). While all Bce-like systems seem to absolutely depend on BceAB-like transporters to be functional, the data on LiaF-like proteins is more complicated, since the corresponding mutants either render the systems constitutively 'ON' (*B. subtilis, L. monocytogenes, S. mutans*) or 'OFF' (*S. aureus*). Because of the clear sequence similarity between LiaF-like proteins, it seems rather unlikely that LiaF-like proteins from different species have adopted opposite functions (i.e. inhibitor vs. activator). Instead, initial biochemical and genetic evidence argue in favour of different default settings of LiaS-like bifunctional HKs (kinase vs. phosphatase), which are affected both by the presence of LiaF-like proteins and suitable stress conditions.

The unknown mechanism of stimulus perception is linked to the second major open question: What is the true nature/determinant of stimulus specificity? Studies on Lia-and Bce-like systems have benefited a lot from the ready availability of specific and strong inducers that helped to identify target genes and physiological role for many of these 2CSs. But despite a well-documented range of inducing compounds for both Bce- and Lia-like systems (Fig. 11.1), the exact nature of the signals sensed by these devices remains a mystery. How can it be possible, that Bce-like systems like BceRSAB or PsdRSAB from *B. subtilis* can distinguish between closely related peptide antibiotic pairs such as mersacidine/actagardine, or enduracidin/ramoplanin, while simultaneously responding to compounds as different as bacitracin and plectasin (Staroń *et al.*, 2011)? Especially, since Bce-like systems nevertheless seem to bind the inducing compounds directly.

This question is even more intriguing for Lia-like systems, especially in light of the apparent lack of a connection between inducer range and physiological role. LiaR-dependent signal transduction in *B. subtilis* is strongly induced by peptide antibiotics that interfere with the lipid II cycle of cell wall biosynthesis (Fig. 11.1) and no other cell wall antibiotics, yet the system does not confer any resistance against them (Wolf *et al.*, 2010). On the other hand, mutants in the Lia-system show increased sensitivities in the presence of some oxidative stress agents such as menadion, plumbagin, sodium selenite, hydrogen peroxide, which do not function as inducers. The only compounds so far which act as an inducer and also shows a sensitivity phenotype are enduracidin and daptomycin. Interestingly, the latter antibiotic does not interfere with cell wall biosynthesis, but rather with membrane integrity. While *B. subtilis* LiaSR remains the best-investigated Lia-like system, it still is the least well understood. In contrast, the data from other Lia-like systems support the idea that in all Firmicutes bacteria, with the exception of the bacilli, Lia-like systems orchestrate a relatively broad CESR (as discussed above).

The identification of the accessory membrane proteins LiaF and BceAB, respectively, has clearly changed our perception and understanding of Lia- and Bce-like systems and emphasized the role of intermolecular signal transfer as an important mechanism of IM-HK function. We hope that the overview and hypotheses described in this chapter will support and maybe even inspire future studies on these systems, which have displayed a surprising degree of complexity for 2CS-dependent signalling in a relatively short period of time. We are curious to see what future research will reveal.

Acknowledgements

The work in the authors' lab described in this chapter was supported by grants from the Deutsche Forschungsgemeinschaft (DFG-grant MA2837/1–3), the Fonds der Chemischen Industrie, and the Concept for the Future of the Karlsruhe Institute of Technology within the framework of the German Excellence Initiative (to TM). We would like to thank Susanne Gebhard for critical reading of the manuscript.

References

Abachin, E., Poyart, C., Pellegrini, E., Milohanic, E., Fiedler, F., Berche, P., and Trieu-Cuot, P. (2002). Formation of D-alanyl-lipoteichoic acid is required for adhesion and virulence of *Listeria monocytogenes*. Mol. Microbiol. *43*, 1–14.

Alba, B.M., and Gross, C.A. (2004). Regulation of the *Escherichia coli* sigma-dependent envelope stress response. Mol. Microbiol. *52*, 613–619.

Aravind, L., and Ponting, C.P. (1997). The GAF domain: an evolutionary link between diverse phototransducing proteins. Trends Biochem. Sci. *22*, 458–459.

Arthur, M., Depardieu, F., Molinas, C., Reynolds, P., and Courvalin, P. (1995). The *vanZ* gene of Tn1546 from *Enterococcus faecium* BM4147 confers resistance to teicoplanin. Gene *154*, 87–92.

Ashenberg, O., Rozen-Gagnon, K., Laub, M.T., and Keating, A.E. (2011). Determinants of homodimerization specificity in histidine kinases. J. Mol. Biol. *413*, 222–235.

Bader, M.W., Sanowar, S., Daley, M.E., Schneider, A.R., Cho, U., Xu, W., Klevit, R.E., Le Moual, H., and Miller, S.I. (2005). Recognition of antimicrobial peptides by a bacterial sensor kinase. Cell *122*, 461–472.

Becker, P., Hakenbeck, R., and Henrich, B. (2009). An ABC transporter of *Streptococcus pneumoniae* involved in susceptibility to vancoresmycin and bacitracin. Antimicrob. Agents Chemother. *53*, 2034–2041.

de Been, M., Bart, M.J., Abee, T., Siezen, R.J., and Francke, C. (2008). The identification of response regulator-specific binding sites reveals new roles of two-component systems in *Bacillus cereus* and closely related low-GC Gram-positives. Environ. Microbiol. *10*, 2796–2809.

Belcheva, A., and Golemi-Kotra, D. (2008). A close-up view of the VraSR two-component system. A mediator of *Staphylococcus aureus* response to cell wall damage. J. Biol. Chem. *283*, 12354–12364.

Belcheva, A., Verma, V., and Golemi-Kotra, D. (2009). DNA-binding activity of the vancomycin resistance associated regulator protein VraR and the role of phosphorylation in transcriptional regulation of the *vraSR* operon. Biochemistry *48*, 5592–5601.

Berdy, J. (2005). Bioactive microbial metabolites. J. Antibiot. *58*, 1–26.

Bergler, H., Abraham, D., Aschauer, H., and Turnowsky, F. (1994). Inhibition of lipid biosynthesis induces the expression of the *pspA* gene. Microbiology *140*, 1937–1944.

Bernard, R., Guiseppi, A., Chippaux, M., Foglino, M., and Denizot, F. (2007). Resistance to bacitracin in *Bacillus subtilis*: Unexpected requirement of the BceAB ABC transporter in the control of expression of its own structural genes. J. Bacteriol. *189*, 8636–8642.

Blake, K.L., Randall, C.P., and O'Neill, A.J. (2011). *In vitro* studies indicate a high resistance potential for the lantibiotic nisin in *Staphylococcus aureus* and define a genetic basis for nisin resistance. Antimicrob. Agents Chemother. *55*, 2362–2368.

Boyd, D.A., Cvitkovitch, D.G., Bleiweis, A.S., Kiriukhin, M.Y., Debabov, D.V., Neuhaus, F.C., and Hamilton, I.R. (2000). Defects in D-alanyl-lipoteichoic acid synthesis in *Streptococcus mutans* results in acid sensitivity. J. Bacteriol. *182*, 6055–6065.

Boyle-Vavra, S., Yin, S., and Daum, R.S. (2006). The VraS/VraR two-component regulatory system required for oxacillin resistance in community-acquired methicillin-resistant *Staphylococcus aureus*. FEMS Microbiol. Lett. *262*, 163–171.

Brissette, J.L., Russel, M., Weiner, L., and Model, P. (1990). Phage shock protein, a stress protein of *Escherichia coli*. Proc. Natl. Acad. Sci. U.S.A. *87*, 862–866.

Butcher, B.G., Mascher, T., and Helmann, J.D. (2008). Environmental sensing and the role of extracytoplasmic function (ECF) sigma factors. In Bacterial Physiology – a Molecular Approach, El-Sharoud, W.M., ed. (Springer-Verlag GmbH, Berlin and Heidelberg), pp. 233–261.

Cai, S.J., and Inouye, M. (2003). Spontaneous subunit exchange and biochemical evidence for trans-autophosphorylation in a dimer of *Escherichia coli* histidine kinase (EnvZ). J. Mol. Biol. *329*, 495–503.

Camejo, A., Buchrieser, C., Couve, E., Carvalho, F., Reis, O., Ferreira, P., Sousa, S., Cossart, P., and Cabanes, D. (2009). *In vivo* transcriptional profiling of *Listeria monocytogenes* and mutagenesis identify new virulence factors involved in infection. PLoS Pathog. *5*, e1000449.

Cao, M., and Helmann, J.D. (2004). The *Bacillus subtilis* extracytoplasmic-function σ^X factor regulates modification of the cell envelope and resistance to cationic antimicrobial peptides. J. Bacteriol. *186*, 1136–1146.

Casino, P., Rubio, V., and Marina, A. (2009). Structural insight into partner specificity and phosphoryl transfer in two-component signal transduction. Cell *139*, 325–336.

Casino, P., Rubio, V., and Marina, A. (2010). The mechanism of signal transduction by two-component systems. Curr. Opin. Struct. Biol. *20*, 763–771.

Cheung, G.Y., Rigby, K., Wang, R., Queck, S.Y., Braughton, K.R., Whitney, A.R., Teintze, M., DeLeo, F.R., and Otto, M. (2010). *Staphylococcus epidermidis* strategies to avoid killing by human neutrophils. PLoS Pathog. *6*, e1001133.

Chong, P., Drake, L., and Biswas, I. (2008). LiaS regulates virulence factor expression in *Streptococcus mutans*. Infect. Immun. *76*, 3093–3099.

Collins, B., Curtis, N., Cotter, P.D., Hill, C., and Ross, R.P. (2010). The ABC transporter AnrAB contributes to the innate resistance of *Listeria monocytogenes* to nisin, bacitracin, and various β-lactam antibiotics. Antimicrob. Agents Chemother. *54*, 4416–4423.

Cui, L., Lian, J.-Q., Neoh, H.-M., Reyes, E., and Hiramatsu, K. (2005). DNA microarray-based identification of genes associated with glycopeptide resistance in *Staphylococcus aureus*. Antimicrob. Agents Chemother. *49*, 3404–3413.

Darwin, A.J. (2005). The phage-shock-protein response. Mol. Microbiol. *57*, 621–628.

D'Costa, V.M., McGrann, K.M., Hughes, D.W., and Wright, G.D. (2006). Sampling the antibiotic resistome. Science *311*, 374–377.

Delcour, J., Ferain, T., Deghorain, M., Palumbo, E., and Hols, P. (1999). The biosynthesis and functionality of the cell-wall of lactic acid bacteria. Antonie van Leeuwenhoek *76*, 159–184.

Dintner, S., Staron, A., Berchtold, E., Petri, T., Mascher, T., and Gebhard, S. (2011). Co-evolution of

ABC-transporters and two-component regulatory systems as resistance modules against antimicrobial peptides in Firmicutes bacteria. J. Bacteriol. *193*, 3851–3862.

Dubrac, S., Bisicchia, P., Devine, K.M., and Msadek, T. (2008). A matter of life and death: cell wall homeostasis and the WalKR (YycGF) essential signal transduction pathway. Mol. Microbiol. *70*, 1307–1322.

Eldholm, V., Johnsborg, O., Haugen, K., Ohnstad, H.S., and Havarstein, L.S. (2009). Fratricide in *Streptococcus pneumoniae*: contributions and role of the cell wall hydrolases CbpD, LytA and LytC. Microbiology *155*, 2223–2234.

Eldholm, V., Gutt, B., Johnsborg, O., Bruckner, R., Maurer, P., Hakenbeck, R., Mascher, T., and Havarstein, L.S. (2010). The pneumococcal cell envelope stress-sensing system LiaFSR is activated by murein hydrolases and lipid II-interacting antibiotics. J. Bacteriol. *192*, 1761–1773.

Ernst, C.M., and Peschel, A. (2011). Broad-spectrum antimicrobial peptide resistance by MprF-mediated aminoacylation and flipping of phospholipids. Mol. Microbiol. *80*, 290–299.

Errington, J. (2003). Regulation of endospore formation in *Bacillus subtilis*. Nat. Rev. Microbiol. *1*, 117–126.

Foster, S.J., and Popham, D.L. (2002). Structure and synthesis of cell wall, spore cortex, teichoic acid, S-layers and capsules. In *Bacillus subtilis* and its Closest Relatives From Genes to Cells, Sonenshein, A.L., Hoch, J.A., and Losick, R., eds (ASM Press, Washigton, DC), pp. 21–41.

Fritsch, F., Mauder, N., Williams, T., Weiser, J., Oberle, M., and Beier, D. (2011). The cell envelope stress response mediated by the $LiaFSR_{Lm}$ three-component system of *Listeria monocytogenes* is controlled via the phosphatase activity of the bifunctional histidine kinase $LiaS_{Lm}$. Microbiology *157*, 373–386.

Galperin, M.Y. (2006). Structural classification of bacterial response regulators: diversity of output domains and domain combinations. J. Bacteriol. *188*, 4169–4182.

Galperin, M.Y. (2010). Diversity of structure and function of response regulator output domains. Curr. Opin. Microbiol. *13*, 150–159.

Gao, R., Tao, Y., and Stock, A.M. (2008). System-level mapping of *Escherichia coli* response regulator dimerization with FRET hybrids. Mol. Microbiol. *69*, 1358–1372.

Gardete, S., Wu, S.W., Gill, S., and Tomasz, A. (2006). Role of VraSR in antibiotic resistance and antibiotic-induced stress response in *Staphylococcus aureus*. Antimicrob. Agents Chemother. *50*, 3424–3434.

Gebhard, S., and Mascher, T. (2011). Antimicrobial peptide sensing and detoxification modules: unraveling the regulatory circuitry of *Staphylococcus aureus*. Mol. Microbiol. *81*, 581–587.

Grebe, T.W., and Stock, J.B. (1999). The histidine protein kinase superfamily. Adv. Microb. Physiol. *41*, 139–227.

Guiral, S., Mitchell, T.J., Martin, B., and Claverys, J.P. (2005). Competence-programmed predation of non-competent cells in the human pathogen *Streptococcus pneumoniae*: genetic requirements. Proc. Natl. Acad. Sci. U.S.A. *102*, 8710–8715.

Haas, W., Kaushal, D., Sublett, J., Obert, C., and Tuomanen, E.I. (2005). Vancomycin stress response in a sensitive and a tolerant strain of *Streptococcus pneumoniae*. J. Bacteriol. *187*, 8205–8210.

Hachmann, A.-B., Angert, E.R., and Helmann, J.D. (2009). Genetic analysis of factors affecting susceptibility of *Bacillus subtilis* to daptomycin. Antimicrob. Agents Chemother. *53*, 1598–1609.

Hankamer, B.D., Elderkin, S.L., Buck, M., and Nield, J. (2004). Organization of the AAA(+) adaptor protein PspA is an oligomeric ring. J. Biol. Chem. *279*, 8862–8866.

Hardie, K.R., Lory, S., and Pugsley, A.P. (1996). Insertion of an outer membrane protein in *Escherichia coli* requires a chaperone-like protein. EMBO J. *15*, 978–988.

Havarstein, L.S., Martin, B., Johnsborg, O., Granadel, C., and Claverys, J.P. (2006). New insights into the pneumococcal fratricide: relationship to clumping and identification of a novel immunity factor. Mol. Microbiol. *59*, 1297–1307.

Helmann, J.D. (2002). The extracytoplasmic function (ECF) sigma factors. Adv. Microb. Physiol. *46*, 47–110.

Herbert, S., Bera, A., Nerz, C., Kraus, D., Peschel, A., Goerke, C., Meehl, M., Cheung, A., and Götz, F. (2007). Molecular basis of resistance to muramidase and cationic antimicrobial peptide activity of lysozyme in staphylococci. PLoS Pathog. *3*, e102.

Hiron, A., Falord, M., Valle, J., Debarbouille, M., and Msadek, T. (2011). Bacitracin and nisin resistance in *Staphylococcus aureus*: a novel pathway involving the BraS/BraR two-component system (SA2417/SA2418) and both the BraD/BraE and VraD/VraE ABC transporters. Mol. Microbiol. *81*, 602–622.

Höltje, J.V. (1998). Growth of the stress-bearing and shape-maintaining murein sacculus of *Escherichia coli*. Microbiol. Mol. Biol. Rev. *62*, 181–203.

Hulko, M., Berndt, F., Gruber, M., Linder, J.U., Truffault, V., Schultz, A., Martin, J., Schultz, J.E., Lupas, A.N., and Coles, M. (2006). The HAMP domain structure implies helix rotation in transmembrane signaling. Cell *126*, 929–940.

Hyyryläinen, H.L., Sarvas, M., and Kontinen, V.P. (2005). Transcriptome analysis of the secretion stress response of *Bacillus subtilis*. Appl. Microbiol. Biotechnol. *67*, 389–396.

Hyyryläinen, H.L., Pietiäinen, M., Lunden, T., Ekman, A., Gardemeister, M., Murtomaki-Repo, S., Antelmann, H., Hecker, M., Valmu, L., Sarvas, M., *et al.*(2007). The density of negative charge in the cell wall influences two-component signal transduction in *Bacillus subtilis*. Microbiology *153*, 2126–2136.

Jordan, S., Junker, A., Helmann, J.D., and Mascher, T. (2006). Regulation of LiaRS-dependent gene expression in *Bacillus subtilis*: identification of inhibitor proteins, regulator binding sites and target genes of a conserved cell envelope stress-sensing two-component system. J. Bacteriol. *188*, 5153–5166.

Jordan, S., Rietkötter, E., Strauch, M.A., Kalamorz, F., Butcher, B.G., Helmann, J.D., and Mascher, T. (2007). LiaRS-dependent gene expression is embedded in transition state regulation in *Bacillus subtilis*. Microbiology *153*, 2530–2540.

Jordan, S., Hutchings, M.I., and Mascher, T. (2008). Cell envelope stress response in Gram-positive bacteria. FEMS Microbiol. Rev. *32*, 107–146.

Joseph, P., Guiseppi, A., Sorokin, A., and Denizot, F. (2004). Characterization of the *Bacillus subtilis* YxdJ response regulator as the inducer of expression for the cognate ABC transporter YxdLM. Microbiology *150*, 2609–2617.

Kausmally, L., Johnsborg, O., Lunde, M., Knutsen, E., and Havarstein, L.S. (2005). Choline-binding protein D (CbpD) in *Streptococcus pneumoniae* is essential for competence-induced cell lysis. J. Bacteriol. *187*, 4338–4345.

Kitagawa, N., Shiota, S., Shibata, Y., Takeshita, T., and Yamashita, Y. (2011). Characterization of MbrC involved in bacitracin resistance in *Streptococcus mutans*. FEMS Microbiol. Lett. *318*, 61–67.

Kramer, N.E., van Hijum, S.A.F.T., Knol, J., Kok, J., and Kuipers, O.P. (2006). Transcriptome analysis reveals mechanisms by which *Lactococcus lactis* acquires nisin resistance. Antimicrob. Agents. Chemother. *50*, 1753–1761.

Kraus, D., Herbert, S., Kristian, S.A., Khosravi, A., Nizet, V., Götz, F., and Peschel, A. (2008). The GraRS regulatory system controls *Staphylococcus aureus* susceptibility to antimicrobial host defenses. BMC Microbiol. *8*, 85.

Kuroda, M., Kuwahara-Arai, K., and Hiramatsu, K. (2000). Identification of the up- and down-regulated genes in vancomycin-resistant *Staphylococcus aureus* strains Mu3 and Mu50 by cDNA differential hybridization method. Biochem. Biophys. Res. Commun. *269*, 485–490.

Kuroda, M., Kuroda, H., Oshima, T., Takeuchi, F., Mori, H., and Hiramatsu, K. (2003). Two-component system VraSR positively modulates the regulation of cell-wall biosynthesis pathway in *Staphylococcus aureus*. Mol. Microbiol. *49*, 807–821.

Lee, S.F., Delaney, G.D., and Elkhateeb, M. (2004). A two-component *covRS* regulatory system regulates expression of fructosyltransferase and a novel extracellular carbohydrate in *Streptococcus mutans*. Infect. Immun. *72*, 3968–3973.

Li, M., Cha, D.J., Lai, Y., Villaruz, A.E., Sturdevant, D.E., and Otto, M. (2007a). The antimicrobial peptide-sensing system *aps* of *Staphylococcus aureus*. Mol. Microbiol. *66*, 1136–1147.

Li, M., Lai, Y., Villaruz, A.E., Cha, D.J., Sturdevant, D.E., and Otto, M. (2007b). Gram-positive three-component antimicrobial peptide-sensing system. Proc. Natl. Acad. Sci. U.S.A. *104*, 9469–9474.

Li, Y.H., Lau, P.C., Tang, N., Svensater, G., Ellen, R.P., and Cvitkovitch, D.G. (2002). Novel two-component regulatory system involved in biofilm formation and acid resistance in *Streptococcus mutans*. J. Bacteriol. *184*, 6333–6342.

McCallum, N., Meier, P.S., Heusser, R., and Berger-Bachi, B. (2011). Mutational analyses of open reading frames within the *vraSR* operon and their roles in the cell wall stress response of *Staphylococcus aureus*. Antimicrob. Agents Chemother. *55*, 1391–1402.

MacRitchie, D.M., Buelow, D.R., Price, N.L., and Raivio, T.L. (2008). Two-component signaling and Gram-negative envelope stress response systems. Adv. Exp. Med. Biol. *631*, 80–110.

Majchrzykiewicz, J.A., Kuipers, O.P., and Bijlsma, J.J.E. (2010). Generic and specific adaptive responses of *Streptococcus pneumoniae* to challenge with three distinct antimicrobial peptides, bacitracin, LL-37, and nisin. Antimicrob. Agents Chemother. *54*, 440–451.

Mandin, P., Fsihi, H., Dussurget, O., Vergassola, M., Milohanic, E., Toledo-Arana, A., Lasa, I., Johansson, J., and Cossart, P. (2005). VirR, a response regulator critical for *Listeria monocytogenes* virulence. Mol. Microbiol. *57*, 1367–1380.

Martinez, B., Zomer, A.L., Rodriguez, A., Kok, J., and Kuipers, O.P. (2007). Cell envelope stress induced by the bacteriocin Lcn972 is sensed by the lactococcal two-component system CesSR. Mol. Microbiol. *64*, 473–486.

Martinez, B., Bottiger, T., Schneider, T., Rodriguez, A., Sahl, H.G., and Wiedemann, I. (2008). Specific interaction of the unmodified bacteriocin Lactococcin 972 with the cell wall precursor lipid II. Appl. Environ. Microbiol. *74*, 4666–4670.

Mascher, T. (2006). Intramembrane-sensing histidine kinases: a new family of cell envelope stress sensors in Firmicutes bacteria. FEMS Microbiol. Lett. *264*, 133–144.

Mascher, T., Margulis, N.G., Wang, T., Ye, R.W., and Helmann, J.D. (2003). Cell wall stress responses in *Bacillus subtilis*: the regulatory network of the bacitracin stimulon. Mol. Microbiol. *50*, 1591–1604.

Mascher, T., Zimmer, S.L., Smith, T.A., and Helmann, J.D. (2004). Antibiotic-inducible promoter regulated by the cell envelope stress-sensing two-component system LiaRS of *Bacillus subtilis*. Antimicrob. Agents Chemother. *48*, 2888–2896.

Mascher, T., Helmann, J.D., and Unden, G. (2006). Stimulus perception in bacterial signal-transducing histidine kinases. Microbiol. Mol. Biol. Rev. *90*, 910–938.

Meehl, M., Herbert, S., Götz, F., and Cheung, A. (2007). Interaction of the GraRS two-component system with the VraFG ABC transporter to support vancomycin-intermediate resistance in *Staphylococcus aureus*. Antimicrob. Agents Chemother. *51*, 2679–2689.

Model, P., Jovanovic, G., and Dworkin, J. (1997). The *Escherichia coli* phage-shock-protein (*psp*) operon. Mol. Microbiol. *24*, 255–261.

Molina, R., Gonzalez, A., Stelter, M., Perez-Dorado, I., Kahn, R., Morales, M., Moscoso, M., Campuzano, S., Campillo, N.E., Mobashery, S., *et al.* (2009). Crystal structure of CbpF, a bifunctional choline-binding protein and autolysis regulator from *Streptococcus pneumoniae*. EMBO Reports *10*, 246–251.

Msadek, T. (1999). When the going gets tough: survival strategies and environmental signaling networks in *Bacillus subtilis*. Trends Microbiol. *7*, 201–207.

Nakao, A., Imai, S., and Takano, T. (2000). Transposon-mediated insertional mutagenesis of the D-alanyl-lipoteichoic acid (dlt) operon raises methicillin resistance in *Staphylococcus aureus*. Res. Microbiol. *151*, 823–829.

Neuhaus, F.C., and Baddiley, J. (2003). A continuum of anionic charge: structures and functions of D-alanyl-teichoic acids in Gram-positive bacteria. Microbiol. Mol. Biol. Rev. *67*, 686–723.

Ninfa, E.G. Atkinson, M.R., Kamberov, E.S., and Ninfa, A.J. (1993). Mechanism of autophosphorylation of *Escherichia coli* nitrogen regulator II (NRII or NtrB): trans-phosphorylation between subunits. J. Bacteriol. *175*, 7024–7032.

Nishi, H., Komatsuzawa, H., Fujiwara, T., McCallum, N., and Sugai, M. (2004). Reduced content of lysyl-phosphatidylglycerol in the cytoplasmic membrane affects susceptibility to moenomycin, as well as vancomycin, gentamicin, and antimicrobial peptides, in *Staphylococcus aureus*. Antimicrob. Agents Chemother. *48*, 4800–4807.

O'Connell-Motherway, M., van Sinderen, D., Morel-Deville, F., Fitzgerald, G.F., Ehrlich, S.D., and Morel, P. (2000). Six putative two-component regulatory systems isolated from *Lactococcus lactis* subsp. *cremoris* MG1363. Microbiology *146*, 935–947.

Ohki, R., Giyanto, Tateno, K., Masuyama, W., Moriya, S., Kobayashi, K., and Ogasawara, N. (2003). The BceRS two-component regulatory system induces expression of the bacitracin transporter, BceAB, in *Bacillus subtilis*. Mol. Microbiol. *49*, 1135–1144.

Ouyang, J., Tian, X.-L., Versey, J., Wishart, A., and Li, Y.-H. (2010). The BceABRS four-component system regulates the bacitracin-induced cell envelope stress response in *Streptococcus mutans*. Antimicrob. Agents Chemother. *54*, 3895–3906.

Pazos, F., and Valencia, A. (2008). Protein co-evolution, co-adaptation and interactions. EMBO J. *27*, 2648–2655.

Pena-Sandoval, G.R., and Georgellis, D. (2010). The ArcB sensor kinase of *Escherichia coli* autophosphorylates by an intramolecular reaction. J. Bacteriol. *192*, 1735–1739.

Perry, J.A., Levesque, C.M., Suntharaligam, P., Mair, R.W., Bu, M., Cline, R.T., Peterson, S.N., and Cvitkovitch, D.G. (2008). Involvement of *Streptococcus mutans* regulator RR11 in oxidative stress response during biofilm growth and in the development of genetic competence. Lett. Appl. Microbiol. *47*, 439–444.

Peschel, A. (2002). How do bacteria resist human antimicrobial peptides? Trends Microbiol. *10*, 179–186.

Peschel, A., Otto, M., Jack, R.W., Kalbacher, H., Jung, G., and Götz, F. (1999). Inactivation of the *dlt* operon in *Staphylococcus aureus* confers sensitivity to defensins, protegrins, and other antimicrobial peptides. J. Biol. Chem. *274*, 8405–8410.

Peschel, A., Jack, R.W., Otto, M., Collins, L.V., Staubitz, P., Nicholson, G., Kalbacher, H., Nieuwenhuizen, W.F., Jung, G., Tarkowski, A., *et al.* (2001). *Staphylococcus aureus* resistance to human defensins and evasion of neutrophil killing via the novel virulence factor MprF is based on modification of membrane lipids with L-lysine. J. Exp. Med. *193*, 1067–1076.

Petersohn, A., Brigulla, M., Haas, S., Hoheisel, J.D., Völker, U., and Hecker, M. (2001). Global analysis of the general stress response of *Bacillus subtilis*. J. Bacteriol. *183*, 5617–5631.

Phillips, Z.E., and Strauch, M.A. (2002). *Bacillus subtilis* sporulation and stationary phase gene expression. Cell. Mol. Life Sci. *59*, 392–402.

Pietiäinen, M., Gardemeister, M., Mecklin, M., Leskela, S., Sarvas, M., and Kontinen, V.P. (2005). Cationic antimicrobial peptides elicit a complex stress response in *Bacillus subtilis* that involves ECF-type sigma factors and two-component signal transduction systems. Microbiology *151*, 1577–1592.

Pietiäinen, M., Francois, P., Hyyryläinen, H.-L., Tangomo, M., Sass, V., Sahl, H.-G., Schrenzel, J., and Kontinen, V.P. (2009). Transcriptome analysis of the responses of *Staphylococcus aureus* to antimicrobial peptides and characterization of the roles of *vraDE* and *vraSR* in antimicrobial resistance. BMC Genomics *10*, 429.

Ponting, C.P., and Aravind, L. (1997). PAS: a multifunctional domain family comes to light. Curr. Biol. *7*, R674–R677.

Rietkötter, E., Hoyer, D., and Mascher, T. (2008). Bacitracin sensing in *Bacillus subtilis*. Mol. Microbiol. *68*, 768–785.

Roces, C., Campelo, A.B., Veiga, P., Pinto, J.P., Rodriguez, A., and Martinez, B. (2009). Contribution of the CesR-regulated genes llmg0169 and llmg2164–2163 to *Lactococcus lactis* fitness. Int. J. Food Microbiol. *133*, 279–285.

Rukmana, A., Morimoto, T., Takahashi, H., Giyanto, and Ogasawara, N. (2009). Assessment of transcriptional responses of *Bacillus subtilis* cells to the antibiotic enduracidin, which interferes with cell wall synthesis, using a high-density tiling chip. Genes Genet. Syst. *84*, 253–267.

Saier, M.H., Jr, Yen, M.R., Noto, K., Tamang, D.G., and Elkan, C. (2009). The transporter classification database: recent advances. Nucleic Acids Res. *37*, D274–D278.

Samant, S., Hsu, F.F., Neyfakh, A.A., and Lee, H. (2009). The *Bacillus anthracis* protein MprF is required for synthesis of lysylphosphatidylglycerols and for resistance to cationic antimicrobial peptides. J. Bacteriol. *191*, 1311–1319.

Sass, P., and Bierbaum, G. (2009). Native *graS* mutation supports the susceptibility of *Staphylococcus aureus* strain SG511 to antimicrobial peptides. Int. J. Med. Microbiol. *299*, 313–322.

Sass, V., Pag, U., Tossi, A., Bierbaum, G., and Sahl, H.G. (2008). Mode of action of human beta-defensin 3 against *Staphylococcus aureus* and transcriptional analysis of responses to defensin challenge. Int. J. Med. Microbiol. *298*, 619–633.

Schneider, T., and Sahl, H.-G. (2010). An oldie but a goodie – cell wall biosynthesis as antibiotic target pathway. Int. J. Med. Microbiol. *300*, 161–169.

Schneider, T., Kruse, T., Wimmer, R., Wiedemann, I., Sass, V., Pag, U., Jansen, A., Nielsen, A.K., Mygind, P.H., Raventos, D.S., *et al.*(2010). Plectasin, a fungal defensin, targets the bacterial cell wall precursor lipid II. Science *328*, 1168–1172.

Senadheera, M.D., Guggenheim, B., Spatafora, G.A., Huang, Y.C., Choi, J., Hung, D.C., Treglown, J.S., Goodman, S.D., Ellen, R.P., and Cvitkovitch, D.G. (2005). A VicRK signal transduction system in *Streptococcus mutans* affects *gtfBCD*, *gbpB*, and *ftf* expression, biofilm formation, and genetic competence development. J. Bacteriol. *187*, 4064–4076.

Shanks, R.M., Meehl, M.A., Brothers, K.M., Martinez, R.M., Donegan, N.P., Graber, M.L., Cheung, A.L., and O'Toole, G.A. (2008). Genetic evidence for an alternative citrate-dependent biofilm formation pathway in *Staphylococcus aureus* that is dependent on fibronectin binding proteins and the GraRS two-component regulatory system. Infect. Immun. *76*, 2469–2477.

Silverman, J.A., Perlmutter, N.G., and Shapiro, H.M. (2003). Correlation of daptomycin bactericidal activity and membrane depolarization in *Staphylococcus aureus*. Antimicrob. Agents Chemother. *47*, 2538–2544.

Skerker, J.M., Prasol, M.S., Perchuk, B.S., Biondi, E.G., and Laub, M.T. (2005). Two-component signal transduction pathways regulating growth and cell cycle progression in a bacterium: a system-level analysis. PLoS Biol. *3*, e334.

Staroń, A., and Mascher, T. (2010). Extracytoplasmic function σ factors come of age. Microbe *5*, 164–170.

Staroń, A., Sofia, H.J., Dietrich, S., Ulrich, L.E., Liesegang, H., and Mascher, T. (2009). The third pillar of bacterial signal transduction: classification of the extracytoplasmic function (ECF) σ factor protein family. Mol. Microbiol. *74*, 557–581.

Staroń, A., Finkeisen, D.E., and Mascher, T. (2011). Peptide antibiotic sensing and detoxification modules of *Bacillus subtilis*. Antimicrob. Agents Chemother. *55*, 515–525.

Steen, A., Palumbo, E., Deghorain, M., Cocconcelli, P.S., Delcour, J., Kuipers, O.P., Kok, J., Buist, G., and Hols, P. (2005). Autolysis of *Lactococcus lactis* is increased upon D-alanine depletion of peptidoglycan and lipoteichoic acids. J. Bacteriol. *187*, 114–124.

Stock, A.M., Robinson, V.L., and Goudreau, P.N. (2000). Two-component signal transduction. Ann. Rev. Biochem. *69*, 183–215.

Storm, D.R., and Strominger, J.L. (1973). Complex formation between bacitracin peptides and isoprenyl pyrophosphates. The specificity of lipid–peptide interactions. J. Biol. Chem. *248*, 3940–3945.

Storz, G., and Hengge-Aronis, R. (eds) (2000). Bacterial Stress Response (ASM press, Washington, DC).

Storz, G., and Hengge-Aronis, R. (eds) (2010). Bacterial Stress Response, 2nd edn (ASM press, Washington, DC).

Straus, S.K., and Hancock, R.E. (2006). Mode of action of the new antibiotic for Gram-positive pathogens daptomycin: comparison with cationic antimicrobial peptides and lipopeptides. Biochim. Biophys. Acta *1758*, 1215–1223.

Suntharalingam, P., Senadheera, M.D., Mair, R.W., Levesque, C.M., and Cvitkovitch, D.G. (2009). The LiaFSR system regulates the cell envelope stress response in *Streptococcus mutans*. J. Bacteriol. *191*, 2973–2984.

Swoboda, J.G., Campbell, J., Meredith, T.C., and Walker, S. (2010). Wall teichoic acid function, biosynthesis, and inhibition. Chembiochem. *11*, 35–45.

Tam le, T., Eymann, C., Albrecht, D., Sietmann, R., Schauer, F., Hecker, M., and Antelmann, H. (2006). Differential gene expression in response to phenol and catechol reveals different metabolic activities for the degradation of aromatic compounds in *Bacillus subtilis*. Environ. Microbiol. *8*, 1408–1427.

Trajtenberg, F., Grana, M., Ruetalo, N., Botti, H., and Buschiazzo, A. (2010). Structural and enzymatic insights into the ATP binding and autophosphorylation mechanism of a sensor histidine kinase. J. Biol. Chem. *285*, 24892–24903.

Tremblay, Y.D., Lo, H., Li, Y.H., Halperin, S.A., and Lee, S.F. (2009). Expression of the *Streptococcus mutans* essential two-component regulatory system VicRK is pH and growth-phase dependent and controlled by the LiaFSR three-component regulatory system. Microbiology *155*, 2856–2865.

Tsuda, H., Yamashita, Y., Shibata, Y., Nakano, Y., and Koga, T. (2002). Genes involved in bacitracin resistance in *Streptococcus mutans*. Antimicrob. Agents Chemother. *46*, 3756–3764.

Ulrich, L.E., Koonin, E.V., and Zhulin, I.B. (2005). One-component systems dominate signal transduction in prokaryotes. Trends Microbiol. *13*, 52–56.

Vasileva-Tonkova, E., Sotirova, A., and Galabova, D. (2011). The effect of rhamnolipid biosurfactant produced by *Pseudomonas fluorescens* on model bacterial strains and isolates from industrial wastewater. Curr. Microbiol. *62*, 427–433.

Veiga, P., Bulbarela-Sampieri, C., Furlan, S., Maisons, A., Chapot-Chartier, M.P., Erkelenz, M., Mervelet, P., Noirot, P., Frees, D., Kuipers, O.P., *et al.* (2007). SpxB regulates *O*-acetylation-dependent resistance of *Lactococcus lactis* peptidoglycan to hydrolysis. J. Biol. Chem. *282*, 19342–19354.

Wecke, T., and Mascher, T. (2011). Antibiotic research in the age of omics – from expression profiles to interspecies communication. J. Antimicrob. Chemother. *66*, 2689–2704.

Wecke, T., Veith, B., Ehrenreich, A., and Mascher, T. (2006). Cell envelope stress response in *Bacillus licheniformis*: Integrating comparative genomics, transcriptional profiling, and regulon mining to decipher a complex regulatory network. J. Bacteriol. *188*, 7500–7511.

Wecke, T., Bauer, T., Harth, H., Mäder, U., and Mascher, T. (2011). The rhamnolipid stress response of *Bacillus subtilis*. FEMS Microbiol. Lett. *323*, 113–123.

Wecke, T., Zühlke, D., Mäder, U., Jordan, S., Voigt, B., Pelzer, S., Labischinski, H., Homuth, G., Hecker, M., and Mascher, T. (2009). Daptomycin versus friulimicin B: In-depth profiling of *Bacillus subtilis* cell envelope stress responses. Antimicrob. Agents Chemother. *53*, 1619–1623.

Weiner, L., and Model, P. (1994). Role of an *Escherichia coli* stress-response operon in stationary-phase survival. Proc. Natl. Acad. Sci. U.S.A. *91*, 2191–2195.

Wiegert, T., Homuth, G., Versteeg, S., and Schumann, W. (2001). Alkaline shock induces the *Bacillus subtilis* σ^W regulon. Mol. Microbiol. *41*, 59–71.

Wolanin, P.M., Thomason, P.A., and Stock, J.B. (2002). Histidine protein kinases: key signal transducers outside the animal kingdom. Genome Biol. 3, REVIEWS3013.

Wolf, D., Kalamorz, F., Wecke, T., Juszczak, A., Mäder, U., Homuth, G., Jordan, S., Kirstein, J., Hoppert, M., Voigt, B., *et al.* (2010). In-depth profiling of the LiaR response of *Bacillus subtilis*. J. Bacteriol. *192*, 4680–4693.

Yeaman, M.R., and Yount, N.Y. (2007). Unifying themes in host defence effector polypeptides. Nat. Rev. Microbiol. *5*, 727–740.

Yin, S., Daum, R.S., and Boyle-Vavra, S. (2006). VraSR two-component regulatory system and its role in induction of *pbp2* and *vraSR* expression by cell wall antimicrobials in *Staphylococcus aureus*. Antimicrob. Agents Chemother. *50*, 336–343.

Yoshida, Y., Matsuo, M., Oogai, Y., Kato, F., Nakamura, N., Sugai, M., and Komatsuzawa, H. (2011). Bacitracin sensing and resistance in *Staphylococcus aureus*. FEMS Microbiol. Lett. *320*, 33–39.

Zhang, Z., and Hendrickson, W.A. (2010). Structural characterization of the predominant family of histidine kinase sensor domains. J. Mol. Biol. *400*, 335–353.

The CpxAR Two-component System Regulates a Complex Envelope Stress Response in Gram-negative Bacteria 12

Stefanie Vogt, Nicole Acosta, Julia Wong, Junshu Wang and Tracy Raivio

Abstract

The CpxA membrane-bound sensor kinase utilizes a periplasmic sensing domain to detect a wide variety of stresses to the bacterial envelope. This information is communicated via typical two-component phosphotransfer-mediated reactions to the response regulator CpxR. Phosphorylated CpxR binds upstream of numerous promoters to mediate adaptation. Initial studies of CpxA inducing signals and CpxR regulated genes demonstrated a role for this two-component system in responding to protein misfolding in the envelope. In this chapter, we discuss recent progress regarding the mechanisms of signal detection, transduction, and gene regulation employed by CpxA and CpxR. The data indicate that the majority of inducing cues are sensed through the periplasmic domain of CpxA and lead to the relief of one or more inhibitory controls that function to maintain Cpx pathway activity at low basal levels in the absence of envelope stress. Some activating signals also enter the pathway downstream of CpxA, in the cytoplasm. Analysis of the Cpx regulon in multiple organisms indicates that, in addition to regulating the production of well studied envelope protein folding and degrading factors, CpxA and CpxR also control the expression of cellular functions linked to cell wall modification, transport, and regulation, indicating that adaptation to envelope stress involves broad changes in cellular physiology. The Cpx two-component system has been shown to impact virulence in a number of pathogens, and our current knowledge of this field is discussed.

Introduction

The CpxA sensor kinase and the CpxR response regulator have been recognized since the 1980s as a typical bacterial two-component regulatory system that impacts the envelope. In this decade, Phillip Silverman's group first demonstrated that mutations affecting the *cpxA* gene impacted many diverse physiological functions, often associated with the envelope, and the genetically linked *cpxR* gene, encoding the cognate response regulator of CpxA, was identified. In the 1990s, Tom Silhavy's group discovered that activation of the CpxA-CpxR signalling pathway could rescue *Escherichia coli* mutants from the toxic effects of grossly misfolded and mislocalized envelope proteins, partly by activating expression of the periplasmic protease-chaperone DegP. These studies led to the proposal that CpxA and CpxR worked together to sense and mediate adaptation to envelope stresses, particularly those affecting protein folding in this compartment. The subsequent identification of the disulfide oxidase DsbA as a Cpx-regulated gene, and the demonstration that CpxA and CpxR participated in the standard phosphotransfer reactions characteristic of all two-component systems, solidified this concept.

In the last 15 years, our major knowledge gains have been an increased understanding about the mechanisms employed by CpxA to sense envelope stress and an appreciation that the Cpx response facilitates adaptation to envelope stress in multiple manners that affect many different cellular processes. Additionally, our view of the Cpx response has broadened to include other microbes, beyond *E. coli,* where CpxA and/or

CpxR have been shown to, in most cases, impact envelope-associated factors that are often involved in microbe–host interactions. It is these more recent advances that will be the focus of this chapter.

CpxAR signal transduction

The CpxA sensor kinase senses a number of diverse environmental signals and interacts with auxiliary regulators that communicate specific signals to CpxA through unknown mechanisms. These sensing events and interactions ultimately alter phosphotransfer communication with CpxR and subsequent expression of the Cpx regulon to enhance bacterial survival under adverse conditions that affect the bacterial envelope. Emerging evidence of CpxR-mediated gene expression that is independent of CpxA, and CpxR interaction with other transcription factors, ties the Cpx response to other networks of cellular regulation.

Inducing signals

The signals and conditions that have been demonstrated to induce the Cpx response are many and varied, but most are predicted to alter some aspect of the bacterial envelope, and many have an association with the inner membrane (IM) and/or protein misfolding at this cellular location (summarized in Fig. 12.1). The majority of Cpx inducers are detected and relayed by the sensor kinase CpxA, although some intracellular signals may act independently of CpxA to impact CpxR phosphorylation. Although we have identified many CpxA activating signals, the actual molecular nature of the inducer(s) remains mysterious.

Mutational induction

The first observed induction of the Cpx response involved constitutive gain-of-function alleles of the sensor kinase gene *cpxA*, the *cpxA** alleles (Cosma *et al.*, 1995; Snyder *et al.*, 1995). The *cpxA** alleles were isolated based on their ability to ameliorate the toxicity of the toxic fusion proteins LamB–LacZ–PhoA and LacZX90 in *E. coli* (Cosma *et al.*, 1995; Snyder *et al.*, 1995). Both of these lead to toxicity upon induction of their expression, and it is thought that this occurs upon their secretion across the IM into the periplasm, where they form noxious disulfide bonded aggregates (Snyder *et al.*, 1995). The *cpxA** mutations suppress this toxicity, in part through the up-regulation of the periplasmic chaperone protease DegP and other proteases and chaperones (Snyder *et al.*, 1995; van Stelten *et al.*, 2009). The *cpxA** mutations were found to affect a number of domains in CpxA, including the periplasmic and transmembrane domains, as well as the conserved transmitter domain (Raivio and Silhavy, 1997). Enzymatic assays revealed that the reason for the activated nature of these mutations stemmed from a large decrease in the ability of CpxA to dephosphorylate CpxR, leading to a substantial elevation in levels of phosphorylated CpxR (Raivio and Silhavy, 1997). Together, these data indicated that activation of the CpxAR two-component system is capable of yielding adaptation to a toxic envelope stress arising from misfolded proteins. Thus, these early studies provided the first clue that one component of the CpxA inducing signal may be related to the misfolding of secreted proteins.

Pilus protein overexpression

Further support for this hypothesis derived from the observations that overexpression of other misfolded proteins also resulted in Cpx activation, including that of the pilus proteins PapE and PapG in the absence of the cognate chaperone PapD (Jones *et al.*, 1997), and components of the bundle-forming pilus (BFP) from enteropathogenic *E. coli* (EPEC) (Nevesinjac and Raivio, 2005). PapE forms the tip fibrillar subunit of the P-pilus from uropathogenic *E. coli* and PapG is the adhesin (Kuehn *et al.*, 1992). Induction of the Cpx response by these cues was shown not to be a general effect of overexpressing an envelope protein, since only specific overexpression of either PapE or PapG, but not other P-pilus subunits, in the absence of chaperone PapD in *E. coli* K12, resulted in Cpx activation (Lee *et al.*, 2004). Further, while the Pap subunits are known to aggregate at the periplasmic face of the IM in the absence of the PapD chaperone (Jones *et al.*, 1997), Cpx response induction was not correlated with aggregation, and the N-terminal extension of PapE was required, but not sufficient, for the activation of the Cpx pathway (Lee *et al.*, 2004). These data suggest that the CpxA inducing signal

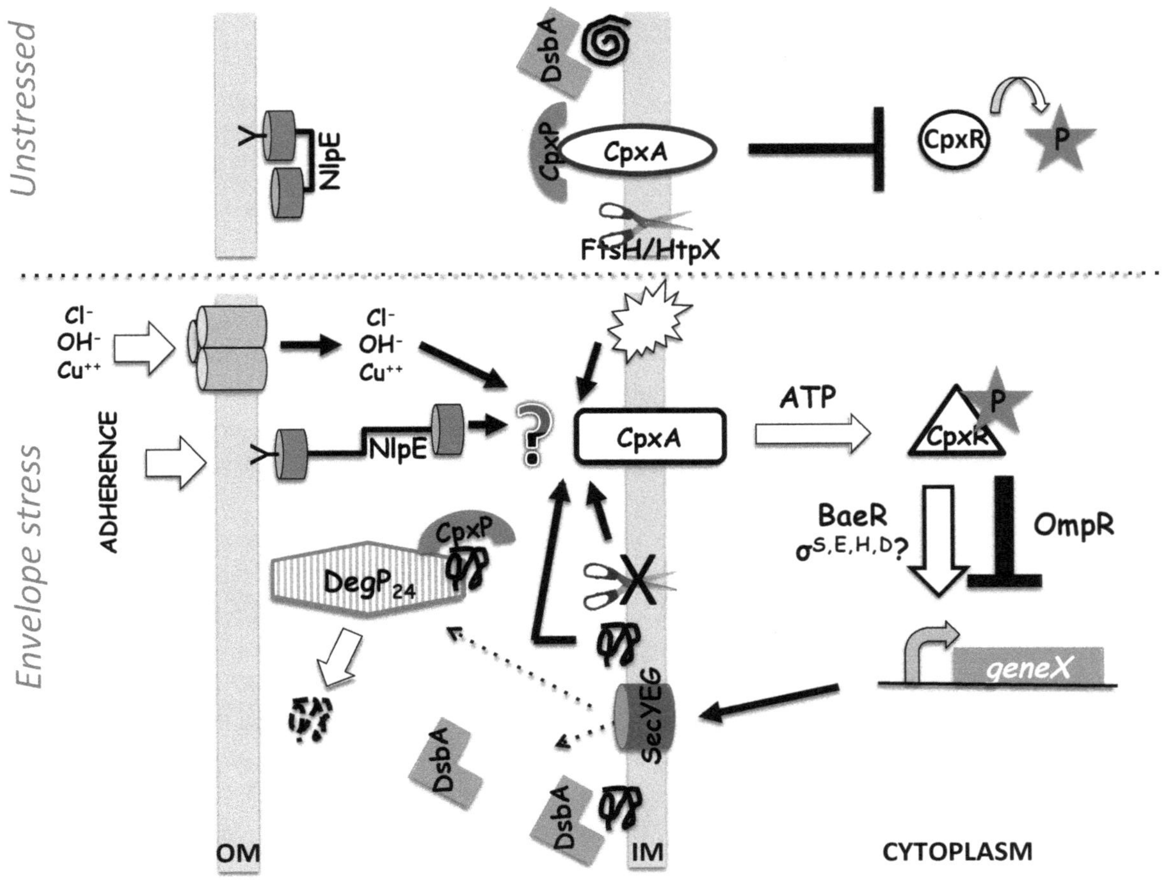

Figure 12.1 Signal transduction by the Cpx two-component system. In the absence of stress, CpxP likely interacts directly with the periplasmic domain of CpxA and CpxA acts as a phosphatase towards CpxR. Under conditions that are thought to cause envelope stress, such as: alkaline pH; accumulation of chloride ions; overexpression and misfolding/mislocalization of NlpE; misfolding of pilins such as PapE. PapG, or BfpA (folded pilin shown as spiral, misfolded pilin shown as squiggly line); loss of the inner membrane proteases FtsH or HtpX (scissors); changes to phospholipid composition of membranes (starburst); and exposure to heavy metal ions, CpxP is thought to be titrated away from CpxA by misfolded proteins, which it targets, together with itself, for degradation by the protease DegP. CpxA autophosphorylates and phosphorylates CpxR, which then represses or transcribes a number of genes encoding proteins that ameliorate envelope stress such as the protease DegP and the chaperones DsbA and CpxP. CpxR likely cooperates or competes with a number of transcription factors in the cytoplasm, including multiple sigma factors and other response regulators to modulate the transcription of a diverse number of genes, a subset of which are secreted and function to refold or degrade proteins (DsbA, DegP). CpxR autoregulates the *cpxRAP* operon, allowing for accumulation of CpxP in the periplasm once the stress has been cleared, a return of CpxP-mediated inhibition of CpxA kinase activity, and attenuation of the Cpx response. See section 1 for details.

conferred by PapG or PapE overexpression is highly specific.

Similarly, overexpression of the major structural subunit of the type IV BFP from EPEC, bundlin/BfpA, in *E. coli* K12, resulted in the activation of Cpx-regulated *lacZ* reporters (Nevesinjac and Raivio, 2005). Interestingly, BfpA overexpression is also predicted to lead to an accumulation of aggregated protein at the IM (Ramboarina *et al.*, 2005), suggesting that this may be a component of a CpxA inducing signal. Curiously, in the case of both PapE/G and bundlin overexpression, activation of the Cpx pathway was reported only during the heterologous overexpression of pilin proteins and not in their respective native strain backgrounds. It is possible that, in the presence of the cognate machinery for pilus assembly, these pilus proteins are not misfolded to an extent that induces the Cpx response. Alternatively, as has been proposed, perhaps during pilus assembly, 'off-pathway' misfolded subunits are generated which induce the Cpx response (Jones *et al.*, 1997). This model makes sense in light of the observation that a number of Cpx-regulated gene products are involved in both P pilus and BFP assembly (Jacob-Dubuisson *et al.*, 1994; Zhang and Donnenberg, 1996; Vogt *et al.*, 2010).

NlpE overexpression and surface adhesion

In addition to the overexpression of pilus subunits, the overexpression of the outer membrane (OM) lipoprotein NlpE also leads to induction of the Cpx response in a CpxA-dependent manner (Snyder *et al.*, 1995). Localization studies revealed that NlpE overexpression led to the production of some aberrantly localized NlpE, normally an OM associated lipoprotein, at the IM (Snyder *et al.*, 1995). Accordingly, it may be that a critical component of this inducing cue, as with the exogenously produced pilus proteins, consists of misfolding and/or aggregation at the periplasmic face of the IM. Alternatively, it may be that NlpE overexpression mimics another CpxA inducing cue: adhesion to abiotic surfaces. The Silhavy group showed that, in *E. coli*, bacterial adhesion to hydrophobic surfaces induced the Cpx pathway in an NlpE-dependent fashion (Otto and Silhavy, 2002). Thus, it is possible that over-production of NlpE leads to the generation of a signal normally generated upon bacterial surface adhesion. Although the mechanism by which NlpE may impact activity of the CpxA sensor kinase upon adhesion remains mysterious, some clues derive from recent structural studies (Hirano *et al.*, 2007) (see below in section 'Accessory signalling molecules').

pH and Cl^- ions

Experiments in *Shigella sonnei* showing that *cpxA* was required for the pH-dependent activation of the virulence factor regulator *virF* suggested CpxA could also be involved in sensing pH changes (Nakayama and Watanabe, 1995). Indeed, alkaline pH was shown to be a potent activator of Cpx-regulated gene expression (Danese and Silhavy, 1998) but the effects of alkaline pH on envelope homeostasis are not well understood. Changes to pH can affect the transmembrane electrical potential ($\Delta\psi$) and disrupt proton motive force but treatment with carbonyl cyanide *m*-chlorophenylhydrazone (CCCP) did not activate a Cpx-regulated reporter gene (Danese and Silhavy, 1998). Thus, the primary inducing cue for the Cpx pathway under alkaline conditions does not appear to be related to $\Delta\psi$. At alkaline pH (> 8.0), thiol groups in cysteine residues are deprotonated and disulfide bonds are oxidized (Huggins *et al.*, 1951; Collet and Bardwell, 2002). Thus, one possibility is that CpxA could sense free oxidized thiols in the periplasm. In support of this idea, DsbA is a direct target of the Cpx regulon (Pogliano *et al.*, 1997) and the *Vibrio cholerae* Cpx system appears to sense some aberrant disulfide-bonds in envelope proteins (Slamti and Waldor, 2009) (see 'Roles of the Cpx envelope stress response in other bacteria'). However, neither the auxiliary signalling molecule CpxP nor CpxA protein in *E. coli* contain any cysteine residues, therefore, a mechanism for sensing free thiols is not easily envisaged. Alkaline pH is also known to affect the net charge of lipopolysaccharide molecules (Niwa *et al.*, 1969; Nummila *et al.*, 1995), increasing OM permeability (Niwa *et al.*, 1969; Raetz and Whitfield, 2002). Thus, another possibility is that indirect changes in envelope physiology as a result of increased OM permeability result in a CpxA activating signal. How changing pH affects the physiology of the envelope and how CpxA monitors these changes remain to be determined experimentally.

High osmolarity in NaCl-containing media induced CpxR phosphorylation and caused repression of *csgD* expression in a CpxR-dependent manner, suggesting that CpxR represses curli expression in high osmotic conditions (Jubelin *et al.*, 2005). Deletion of *cpxR*, however, had no effect on gene expression upon sucrose-mediated osmolarity changes (Jubelin *et al.*, 2005). Further, a number of chloride salts (but not other solutes) also impacted CpxA signalling activities in an *in vitro* proteoliposome system (Fleischer *et al.*, 2007) and NaCl has been shown to induce the Cpx response in *Vibrio cholerae* and *Salmonella enterica* serovar Typhi (Leclerc *et al.*, 1998; Slamti and Waldor, 2009). Together, these studies suggest that chloride salts, but not osmolarity per se, may have a stimulatory effect on the signalling activities of CpxA.

Changes in membrane composition

The Cpx response is also induced by mutations that affect phospholipid composition or IM protein content. Specifically, the *pss-93* allele, which results in a lack of the major membrane constituent phosphatidylethanolamine (PE) from the envelope, caused increased transcription from a Cpx-regulated reporter gene (Mileykovskaya and Dowhan, 1997). Further, mutation of *pldA*, encoding the OM phospholipid degradation protein phospholipase A, also led to induction of the Cpx response (Langen *et al.*, 2001). Lastly, mutations that lead to the accumulation of the enterobacterial common antigen lipid II precursor in the IM activated the Cpx response (Danese *et al.*, 1998). The mechanism by which CpxA senses these changes is not known. Perhaps CpxA is sensitive to some specific aspect of the altered phospholipid composition of the membranes. Alternatively, it is possible that the altered phospholipid composition of the IM of these mutants acts directly on CpxA to affect its signalling properties (Mileykovskaya and Dowhan, 1997). Another possibility is that altered phospholipid concentration exerts an effect on envelope protein folding, which in turn is sensed by CpxA (Bogdanov *et al.*, 1996, 1999; Bogdanov and Dowhan, 1998).

Alterations to the protein content of the IM also impact CpxA signalling. Deletion of the IM protease FtsH resulted in Cpx activation (Shimohata *et al.*, 2002), though the mechanism by which CpxA senses FtsH loss is not clear. FtsH degrades a number of membrane and cytoplasmic proteins (Akiyama and Ito, 2000), including improperly assembled SecY subunits (Kihara *et al.*, 1995), as well as the $F0_a$ component of the major ATP synthase (Akiyama *et al.*, 1996). Thus, conceivably, in the *ftsH* mutant, abnormal or excessive proteins accumulate in the IM, ostensibly leading to CpxA activation (Shimohata *et al.*, 2002). A further link between the Cpx response and IM protein composition derives from the observation that the expression of *htpX* is Cpx-controlled (Shimohata *et al.*, 2002). HtpX is an integral IM protease with a cytoplasmic active site that degrades aberrantly folded IM-associated proteins (Kornitzer *et al.*, 1991). Similar to FtsH, HtpX can degrade SecY and casein *in vitro* and likely cooperates with FtsH to eliminate aberrant IM proteins (Sakoh *et al.*, 2005). Together, these data implicate the Cpx response in sensing and mediating adjustment to perturbations to the IM. These findings fit nicely with the previously described studies that demonstrated a connection between induction of the Cpx response and the overexpression of proteins that aggregate at the periplasmic face of the IM and also with new studies that have identified a strong connection between Cpx regulon members and the inner membrane (see 'Physiological role of the Cpx response').

CpxA-independent inducing signals

Recently, it has been proposed that CpxR may be the target of some intracellular signals that result in its CpxA-independent activation. DiGiuseppe and Silhavy (2003) showed that the Cpx response was induced during growth, and that this induction depended on a signal that enters the pathway downstream of CpxA. Similarly, excess glucose or pyruvate was shown to induce expression of the Cpx-regulated gene *cpxP* in a CpxA-independent manner (De Wulf *et al.*, 1999). Interestingly, although some response regulators, including CpxR, have been demonstrated to become phosphorylated by the small molecular weight phosphodonor acetyl-phosphate in the absence of their cognate kinases (Pogliano *et al.*, 1997; Raivio and Silhavy, 1997; Wolfe *et al.*, 2008), the effects of glucose and pyruvate were shown to occur even in the absence of acetyl phosphate (Wolfe *et al.*, 2008). Although the mechanism by which growth or carbon sources impact CpxR activity is not known, these results suggest that CpxR can integrate cellular signals independently of CpxA.

CpxA signal sensation and transduction

CpxA and CpxR constitute a typical two-component regulatory system capable of transducing changes detected via a periplasmic sensing domain in CpxA into changes in gene expression through well-described two-component phosphotransfer-mediated signalling events. CpxA is an IM sensor kinase with two transmembrane domains, a periplasmic domain required to sense signals, and a cytoplasmic kinase/phosphatase domain (Albin *et al.*, 1986; Weber and Silverman, 1988; Raivio and Silhavy, 1997). A cytoplasmic HAMP signalling domain is also recognizable between the second transmembrane domain and the conserved transmitter domain of CpxA (Appleman *et al.*, 2003). As with other two-component systems, evidence suggests that the major point of signal integration is the periplasmic domain of CpxA. Mutational analyses and studies of other sensor kinases support a model in which signal detection in all probability leads to movement of transmembrane domains relative to one another within a sensor kinase dimer, ultimately impacting HAMP domain signalling and the conformation of the transmitter to effect either phosphorylation of the conserved His in CpxA or hydrolysis of CpxR~P (Chen and Amster-Choder, 1998; Appleman *et al.*, 2003; Ayers and Moffat, 2008; Lee *et al.*, 2008; Zhang and Hendrickson, 2010). All three of these activities have been biochemically demonstrated (Raivio and Silhavy, 1997; Fleischer *et al.*, 2007; Keller and Hunke, 2009). In this section we focus on the most recent developments surrounding CpxA signal transduction – the identification and characterization of signalling molecules and events involved in sensing envelope stress.

Signal detection by CpxA

The first clue that the periplasmic domain of CpxA may be required for signal sensing came from analyses of constitutively activated *cpxA** mutants (Raivio and Silhavy, 1997). Several of these mutations affected the periplasmic domain of CpxA, and one, in particular, *cpxA24*, led to a deletion of 32 amino acids near the middle of this domain (Raivio and Silhavy, 1997). In addition to constitutively activating the Cpx response, some of these mutations also resulted in a signal blind phenotype, in which Cpx-inducing signals no longer lead to induction of the pathway (Raivio and Silhavy, 1997). Accordingly, it was proposed that the periplasmic domain of CpxA was necessary for signal detection.

Subsequent mutational analyses support this contention. Alanine substitutions of conserved single amino acids within the *cpxA24* deletion resulted in high levels of Cpx pathway activity, suggesting that the default state of the CpxA sensor kinase is 'ON' and that inhibition of autokinase activity and/or stimulation of phosphatase activity is crucial to regulation (Malpica and Raivio, unpublished). Intriguingly, although most mutants were attenuated with regard to the overall level of induction observed in response to alkaline pH or NlpE overexpression, no single substitution abolished induction of a Cpx-regulated reporter gene (Malpica and Raivio, unpublished). Together, these experiments suggest that the periplasmic region of CpxA deleted in the *cpxA24* mutant is critical for signal sensing but no single amino acid is responsible, perhaps suggesting that it is the overall fold of this domain that is critical.

In this regard, the solving of the CpxA sensing domain crystal structure could provide clues to the specific substrate or signal that regulates CpxA enzymatic activity and, ultimately, the molecular mechanism behind signal sensing by CpxA. Secondary structure predictions suggested that the CpxA sensing domain contains a mixed α/β structure (Malpica and Raivio, unpublished). Similar sensor kinases such as PhoQ and CitA possess α/β folds in their periplasmic sensing domains coined PAS (per-arnt-sim) or PDC (PhoQ-DcuS-CitA) domains that are responsible for mediating signal sensing and conformational changes in the kinase through ligand binding (Reinelt *et al.*, 2003; Cheung *et al.*, 2008). Thus, an attractive model that explains the signalling phenotypes of *cpxA** and Ala substitution mutants could be that the periplasmic domain of CpxA comprises a similar PAS/PDC structural fold where ligands are coordinated by a number of amino acid residues and no one residue is absolutely required for function.

Accessory signalling molecules

In agreement with the above signalling model supported by genetic data and structural predictions, two auxiliary regulatory molecules have been implicated in CpxA signalling, which may serve as ligands that act to regulate Cpx pathway activity. NlpE, mentioned above, is an OM lipoprotein that induces the Cpx response upon overexpression or bacterial adhesion in a fashion dependent on the periplasmic sensing domain of CpxA (Snyder *et al.*, 1995; Otto and Silhavy, 2002). CpxP, on the other hand, is a periplasmic protein that inhibits the Cpx response, again, in a manner requiring the signal integrating sensing domain of CpxA (Raivio *et al.*, 1999). These auxiliary factors represent another possible site of signal input.

Recent structural studies suggest a model for the mechanism by which NlpE may communicate adhesion to CpxA. The crystal structure of *E. coli* NlpE [NlpE(ec)] shows that it consists of an N-terminal Blc-resembling (bacterial lipocalin) domain and a C-terminal oligonucleotide/oligosaccharide-binding domain linked by a flexible region (Hirano *et al.*, 2007). One model of NlpE signalling to CpxA during adhesion involves the relative position of these two major domains. When cells adhere to hydrophobic surfaces, it has been proposed that NlpE anchored to the OM may sense changes at the bacterial cell surface and the linker between the two major domains could undergo structural changes resulting in the extension of the C-terminal domain. This extended protein is predicted to be long enough to span the distance between the OM and IM and would therefore enable NlpE to interact directly with the sensing domain of CpxA (Hirano *et al.*, 2007). Alternatively, the predicted Blc or oligonucleotide/oligosaccharide-binding domains of NlpE may be involved in interacting with a ligand differentially during bacterial adhesion (Bos *et al.*, 2007). Further, NlpE also possesses a CXXC motif and a protease inhibitor signature sequence in the N-terminal domain that are conserved across species. These highly conserved motifs may represent active sites involved in sensing adhesion and/or additional, unidentified environmental changes (Hirano *et al.*, 2007). The mechanism by which NlpE induces CpxA and the involvement of the various NlpE domains and motifs in sensing adhesion and/or other signals awaits further study.

In addition to NlpE, the Tokuda group identified a number of other lipoproteins that may act as auxiliary regulators of the Cpx pathway (Miyadai *et al.*, 2004). The overexpression of these lipoproteins resulted in the accumulation of the periplasmic protease DegP, which is regulated by CpxA and CpxR. The IM lipoprotein YafY resulted in the strongest induction of DegP expression in a CpxA-dependent manner (Miyadai *et al.*, 2004). By figuring out whether YafY plays a physiological stress responsive signalling role, it may be possible to identify new conditions to which the Cpx pathway is responsive. Interestingly, another IM lipoprotein, YfjS, shares more than 90% homology with YafY but did not induce DegP expression (Miyadai *et al.*, 2004). Mutations to the non-conserved residues in YafY and YfjS identified three amino acids that are important for DegP induction. These will be important clues to the study of the molecular signal generated by the overexpression of YafY that is recognized by CpxA.

A second identified auxiliary regulator of CpxA is the novel periplasmic protein CpxP (Raivio *et al.*, 1999). The *cpxP* gene is transcribed divergently from the *cpxRA* operon and inhibits the Cpx pathway in a manner that requires the periplasmic sensing domain of CpxA (Raivio *et al.*, 1999). CpxP-mediated inhibition is relieved in the presence of inducing cues (DiGiuseppe and Silhavy, 2003). The simplest signalling model suggested by these data posits a direct interaction of CpxP with the periplasmic sensing domain of CpxA that is relieved in the presence of inducers, though no direct interaction has been demonstrated between full-length or soluble proteins to date. The available data, however, support a direct interaction between CpxP and CpxA (Raivio *et al.*, 2000; Fleischer *et al.*, 2007; Thede *et al.*, 2011; Zhou *et al.*, 2011). When a CpxP fusion protein was tethered to the IM through a mutation in its signal sequence, it was retained in spheroplasts and also functioned to inhibit CpxA, while the corresponding wild-type construct was lost upon spheroplasting and also failed to inhibit CpxA (Raivio *et al.*, 2000). Further, reconstitution of CpxP and CpxA in proteoliposomes resulted in decreased CpxA autokinase activity and detection of both CpxA and CpxP in the lipid fractions of cells (Fleischer *et al.*, 2007). Lastly, experiments examining binding between 20mer CpxP or CpxA peptide oligomers and the purified CpxA sensing domain or CpxP, respectively, suggested that a negatively charged region in the C-terminus of the CpxA sensing domain may interact with a positively charged domain in CpxP (Zhou *et al.*, 2011). These three experiments suggest that CpxP interacts with the periplasmic domain of CpxA to inhibit CpxA autokinase activity and therefore Cpx pathway activity. Recent structural analyses of CpxP showed that it forms a bowl-shaped dimer with a prominent positively charged concave face, perhaps (together with the above studies) implicating this region in interactions with CpxA (Thede *et al.*, 2011; Zhou *et al.*, 2011). Contrary to this conclusion, however, loss-of-function mutations that negate the ability of CpxP to inhibit CpxA localize to one end of the edge of the bowl-shaped dimer (Buelow and Raivio, 2005). Co-immunoprecipitation or pull-down experiments with tagged full-length or soluble proteins coupled with allele-specific suppressors of the loss-of-function mutations within each gene will provide more direct evidence for the interaction between the two proteins and lend credence to the attractive model of inhibition by direct interaction.

How is CpxP-mediated inhibition of CpxA relieved upon induction? In some cases, including the overexpression of Pap pilus subunits and alkaline pH, CpxP has been shown to be degraded by the periplasmic protease DegP (Buelow and Raivio, 2005; Isaac *et al.*, 2005). These data suggest that Cpx inducing cues predispose CpxP to degradation by DegP. Interestingly, Isaac and colleagues demonstrated that in some strain backgrounds, an intact *cpxP* gene was required for degradation of misfolded, aggregated Pap subunits by DegP (Isaac *et al.*, 2005). Thus, an exciting possibility is that CpxP may function as a chaperone/adapter protein to target misfolded proteins, and itself, for degradation, thus alleviating CpxA inhibition (Isaac *et al.*, 2005). The bowl-shaped dimer formed by CpxP (Thede *et al.*, 2011; Zhou *et al.*, 2011) is virtually identical to its most closely related homologue, Spy (Quan *et al.*, 2011; Thede *et al.*, 2011; Zhou *et al.*, 2011). Since Spy acts as a novel chaperone protein (Quan *et al.*, 2011), these similarities lend support to a role for CpxP in binding misfolded proteins, although the manner in which this occurs remains unknown.

Deletion of *cpxP* led to induction of the expression of Cpx-regulated genes, but the Cpx response could still be further induced by all tested activating signals (Raivio *et al.*, 1999; DiGiuseppe and Silhavy, 2003). This finding demonstrated that CpxP is not required for induction of the Cpx response and suggested that other inhibitors of the Cpx pathway may exist that are responsive to envelope stress inputs (DiGiuseppe and Silhavy, 2003). Such an inhibitor could consist of the CpxA sensing domain itself, which may adopt a stable, inhibitory structure that is relieved by some undefined envelope stress condition. Alternatively, or in addition, other inhibitory signalling proteins may exist. Intriguingly, this situation is very similar to that of the unfolded protein response in eukaryotes. In yeast, a protein analogous to CpxA, Ire1, interacts with a small luminal protein, BiP, in the endoplasmic reticulum (ER) during normal homeostatic conditions. Under nutritional stress or during differentiation, the ER protein-folding machinery is overwhelmed and BiP binds unfolded proteins, dissociating from Ire1 and allowing Ire1 to oligomerize and affect differential splicing of the mRNA transcript for HAC1, a transcriptional activator of the unfolded protein response genes (Cox and Walter, 1996). In this system, BiP sequesters inactive Ire1 and prevents oligomerization, but BiP removal is not sufficient for maximal HAC1 splicing. When the BiP binding site is deleted from Ire1, basal levels of HAC1 splicing are higher but HAC1 mRNA splicing can be further induced by starvation (Pincus *et al.*, 2010). Thus, just as CpxP is likely not the sole regulator of CpxA, BiP is not the only negative regulator of Ire1 and Ire1 can self-oligomerize in the absence of the BiP binding site (Pincus *et al.*, 2010). Interestingly, a number of auxiliary regulators of sensor kinases have recently been identified suggesting that bacterial kinases may function as part of signalling complexes capable of integrating numerous parameters (Buelow and Raivio, 2010). Searches for other negative regulators of the Cpx pathway will provide greater insight into how CpxA acts as a node for the integration of a diverse set of envelope stresses.

Whether or not additional auxiliary regulators of CpxA exist, the question remains: what is the function of CpxP? As previously mentioned, it seems likely that CpxP plays a role in the folding and/or degradation of certain classes of misfolded envelope proteins. It is also possible that CpxP functions to relay a specific envelope stress signal to CpxA that has not yet been defined. Alternatively, CpxP might serve to 'fine-tune' the Cpx response. In this regard, the Cpx response has been shown to be autoregulated – phosphorylated

CpxR binds upstream of the *cpxRA* operon to induce further expression of the signalling proteins during envelope stress (Raivio *et al.*, 1999). At the same time as this autoinduction occurs, CpxP is also up-regulated, and presumably degraded in the presence of misfolded proteins. Accordingly, CpxP may serve as a barometer to measure when envelope stress has been relieved – once misfolded proteins stop accumulating, CpxP will no longer be degraded and can then function to inhibit CpxA and return Cpx response induction to its basal level.

Mechanism of transcription control by CpxR and cooperation with other transcription factors

Phosphorylation, DNA binding, and transcription activation

CpxR consists of a canonical two-component receiver domain, containing the conserved site of phosphorylation at D51, and an output domain bearing a winged helix–turn–helix DNA binding domain that is most similar to that of OmpR (Martínez-Hackert and Stock, 1997). Phosphorylation or dephosphorylation of CpxR impacts the ability of CpxR to bind DNA, with CpxR~P exhibiting enhanced DNA binding activity through unknown mechanisms (Raivio and Silhavy, 1997; DiGiuseppe and Silhavy, 2003). It is probable, based on studies of other response regulators, that phosphorylation leads to changes in oligomeric state and/or the orientation of the receiver domain with respect to that of the DNA binding domain to effect DNA binding (Robinson *et al.*, 2003; Friedland *et al.*, 2007). Mutational analyses confirmed that both phosphorylation and the predicted DNA recognition helix are required for normal CpxR function (DiGiuseppe and Silhavy, 2003).

A CpxR consensus binding site was arrived at through comparison of the upstream regions and DNA footprints of known CpxR-regulated genes and determined to be 5′-GTAAA-N5-GTAAA-3′ (De Wulf *et al.*, 2002), though the absence of a consensus sequence does not necessarily indicate a lack of CpxR-binding (Price and Raivio, 2009; Weatherspoon-Griffin *et al.*, 2011), suggesting that other CpxR binding sites, perhaps weaker, remain to be identified. Upon binding DNA CpxR may act as either a repressor or an activator (De Wulf *et al.*, 1999; Price and Raivio, 2009). The mechanism by which transcription is activated is not known, but based on the high similarity of CpxR to OmpR, it seems probable that the alpha loop of the winged helix DNA binding domain makes contact with RNA polymerase to facilitate transcription initiation (Slauch *et al.*, 1991; Pratt and Silhavy, 1994). This supposition is supported by the observation that CpxR-regulated genes containing a CpxR consensus binding site proximal to the promoter are the most sensitive to Cpx regulation (Price and Raivio, 2009). In contrast, neither the orientation of the binding site relative to the promoter nor its position on the coding versus non-coding strand affect whether a specific promoter is more highly up-regulated by CpxR (Price and Raivio, 2009). This observation might suggest that CpxR is capable of contacting RNA polymerase in more than one manner.

Interactions with other regulators

Binding of CpxR to specific targets may involve interaction or competition with other transcriptional regulators. The expression of Spy, a periplasmic chaperone, is regulated by both CpxR and BaeR and full induction of *spy* expression requires both Cpx and Bae systems (Raffa and Raivio, 2002). The BaeSR system regulates the expression of a number of genes involved in the resistance to compounds that could cause oxidative stress resulting in membrane damage (Makris and Rossiter, 2000; Leblanc *et al.*, 2011). Yamamoto and colleagues (2008) found that induction of *spy* expression was regulated by CpxR during exposure to external copper while *spy* induction during exposure to external zinc was dependent on BaeR. However, it has also been observed that the *baeSR* two-component system must be present in order for CpxR to exert an effect on the transcription of Bae-regulated genes (Hirakawa *et al.*, 2005). Thus, it appears that CpxR may interact with BaeR to increase Bae-regulated gene expression under conditions where both two-component systems are activated. For example, CpxR could co-regulate the expression of *spy* under conditions where toxic compounds that induce the BaeSR system, such as heavy metals, also cause envelope stress, thus inducing the Cpx system (Yamamoto *et al.*, 2008). Similarly, CpxR binding sites have been identified upstream of the promoter of *ompC*, a gene encoding a major OM porin (Nikaido and Rosenberg, 1983; Batchelor *et al.*, 2005). Interestingly, phosphorylated CpxR cannot activate *ompC* transcription in the absence of OmpR, suggesting that CpxR and OmpR bind *ompC* cooperatively to activate transcription (Batchelor *et al.*, 2005).

CpxR also binds to sites overlapping those of OmpR at the *ompF* promoter (Batchelor *et al.*, 2005). Here, it appears that CpxR may compete for binding with OmpR or RNA polymerase to repress *ompF* transcription (Batchelor *et al.*, 2005). In addition, CpxR appears to compete with OmpR for binding sites within the *csgD* promoter region to regulate the expression of curli during osmotic stress (Jubelin *et al.*, 2005). In fact, five transcription factors bind a total of 229 bp in the *csgD* promoter region in response to specific conditions to regulate curli expression and, therefore, biofilm formation (Ogasawara *et al.*, 2010). While CpxR and H-NS repress *csgD* expression, OmpR, RstA, and IHF activate expression (Ogasawara *et al.*, 2010). Thus, as with the *ompC* promoter, CpxR may compete with other activators for binding sites within the *csgD* promoter to effect repression of curli expression during envelope stress.

In addition to its cooperation or competition with the above regulators, CpxR has also been shown to regulate the expression of genes controlled by a number of different sigma factors. CpxR may cooperate with RpoS to activate *cpxRA* expression during entry into stationary phase (De Wulf *et al.*, 1999). In addition, CpxR has been shown to regulate and/or CpxR consensus binding sites have been identified upstream of, genes that are controlled by the general heat shock sigma factor, RpoH, as well as the envelope stress responsive sigma factor RpoE (De Wulf *et al.*, 1999; Price and Raivio, 2009). The fact that CpxR can regulate the expression of genes that are controlled by different sigma factor containing versions of RNA polymerase supports a model for CpxR-mediated transcription activation that involves contact of CpxR with some component of RNA polymerase other than sigma, such as the alpha subunit (see above). In any case, together, the above studies suggest that CpxR likely interacts with multiple transcription factors to integrate numerous cellular signals and modulate gene expression accordingly (Fig. 12.1).

Summary

The CpxA sensor kinase and CpxR response regulator function like a typical two-component system to transduce envelope stress signals into an adaptive change in gene expression (see Fig. 12.1 for a summary). The known envelope stress signals often impact protein folding at the IM and are sensed through a periplasmic sensing domain in CpxA. The auxiliary regulator NlpE relays an unknown signal upon adhesion through this sensing domain and the novel periplasmic adapter protein CpxP appears to be involved in modulating CpxA enzymatic activity in response to the presence of misfolded proteins, which it targets for proteolysis. Additional signals related to growth and carbon source are proposed to alter CpxR activity independently of CpxA. Inducing signals have been shown to increase CpxA autokinase and kinase activities, leading to phosphorylation of CpxR, while inhibitory cues (i.e. CpxP) may diminish CpxA autokinase activity. Although the phosphatase activity of CpxA has been demonstrated, no signals that alter this function have been identified. Phosphorylated CpxR binds upstream of Cpx-regulated genes to either activate or repress transcription and can work cooperatively, or in competition with, other transcription factors.

Physiological role of the Cpx response

In an effort to understand the physiological role of the Cpx envelope stress response in *E. coli*, numerous studies have attempted to identify the genes that comprise the Cpx regulon, using techniques such as random mutagenesis, computational predictions, and microarray analysis. These studies indicate that the Cpx regulon is large, containing dozens to hundreds of genes (Price and Raivio, unpublished). In addition, it appears that the number of genes up-regulated by CpxR is approximately equal to the number down-regulated

(Bury-Moné *et al.*, 2009; Price and Raivio, unpublished). The Cpx regulon members described thus far can be divided into six functional categories: envelope protein folding, envelope maintenance, transport, envelope-localized macromolecular complexes, regulators, and other cell maintenance and metabolic functions (summarized in Fig. 12.2).

Envelope protein folding

The first genes to be identified as Cpx-regulated encode a collection of envelope-localized protein folding and degrading factors. The identification of these Cpx regulon members supported the idea that the Cpx two-component system functions primarily as an envelope stress response, with these proteins playing a key role in clearing the envelope of misfolded proteins (Cosma *et al.*, 1995). Although many other categories of Cpx-regulated genes are now known, the genes encoding envelope protein folding factors have proven to be among the most strongly induced by the Cpx pathway (Price and Raivio, 2009), highlighting their important role in cell survival under Cpx-inducing conditions.

The bifunctional periplasmic protease and chaperone DegP was the first known member of the Cpx regulon (Danese *et al.*, 1995). CpxR enhances expression of *degP* in response to Cpx inducing cues such as NlpE overexpression and gain-of-function mutations in *cpxA* (Danese *et al.*, 1995); this regulation occurs through a direct binding of phosphorylated CpxR to the *degP* promoter (Pogliano *et al.*, 1997). DegP was originally characterized for its roles in degradation of misfolded envelope proteins and survival at high temperatures (Lipinska *et al.*, 1989; Strauch *et al.*, 1989). It was later shown that, in addition to its role as the primary protease in the *E. coli* periplasm, DegP also possesses chaperone activity (Spiess *et al.*, 1999). The protease activity of DegP is believed to be activated at higher temperatures, while its chaperone activity predominates at lower temperatures (Spiess *et al.*, 1999). A limited number of natural proteolytic substrates of DegP have been identified, including pilin subunits of several different types of pili (Jones *et al.*, 2002; Humphries *et al.*, 2010). Interestingly, the chaperone activity of DegP also plays a crucial role in the biogenesis of some pilins, such as the bundlin subunits of the EPEC BFP (Humphries *et al.*, 2010). DegP's chaperone function also plays a supporting role in the biogenesis of OM proteins (OMPs) (Sklar *et al.*, 2007). DegP therefore makes a major contribution to envelope protein quality control, by promoting the folding of and/or degrading misfolded versions of diverse categories of envelope proteins.

Another protein crucial for proper envelope protein folding is the disulfide bond

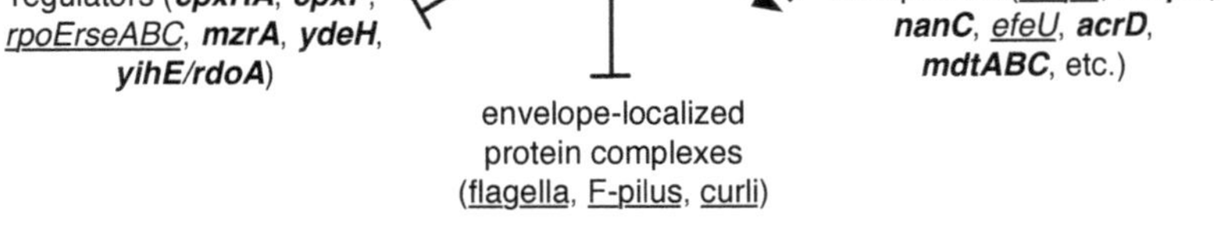

Figure 12.2 Overview of the Cpx regulon in *E. coli*. Phosphorylated CpxR acts as a transcription factor to activate (arrows) or repress (lines with bars) expression of genes involved in several processes: envelope protein folding, envelope maintenance, transport, biosynthesis of envelope-localized protein complexes, regulation, and other metabolic/maintenance functions. Target genes activated by CpxR-P are indicated in boldface; target genes repressed by CpxR-P are underlined. Gene names that are both bolded and underlined are regulated by CpxR-P in a strain-dependent manner.

oxidoreductase DsbA. CpxR directly activates the expression of *dsbA* by binding to the major promoter located upstream of the *yihE–dsbA* operon (Danese and Silhavy, 1997; Pogliano *et al.*, 1997). DsbA is the primary enzyme responsible for oxidizing cysteine residues of proteins that are translocated across the IM to form disulfide bonds, with substrates including a wide variety of OMPs, periplasmic enzymes, and various other envelope constituents (Kadokura *et al.*, 2003). In many cases, the DsbA-catalysed introduction of a disulfide bond is essential for stability of the substrate. For this reason, a functional DsbA is required for biogenesis of flagella (Dailey and Berg, 1993) as well as some components of both chaperone-usher and type IV pili (Jacob-Dubuisson *et al.*, 1994; Zhang and Donnenberg, 1996).

CpxR also directly activates the transcription of the gene *ppiA* (Pogliano *et al.*, 1997), encoding a peptidyl-prolyl isomerase (PPIase) that is responsible for *cis–trans* isomerization of Xaa-Pro peptide bonds. At this time, the specific substrates of PpiA are unknown. In spite of the fact that periplasmic extracts from a *ppiA* mutant possessed little remaining PPIase activity, no defects in the folding of periplasmic or OM proteins could be detected in this strain (Kleerebezem *et al.*, 1995). It was further shown that PpiA was not required for folding of the OMPs OmpA and LamB, nor did it play a role in biogenesis of chaperone-usher pili (Justice *et al.*, 2005). Interestingly, PPIase activity enhances the formation of disulfide bonds by disulfide bond oxidases (Schönbrunner and Schmid, 1992), suggesting that perhaps the Cpx response coordinately regulates both DsbA and PpiA as a means of increasing the efficiency of disulfide bond formation.

Expression of the periplasmic protein Spy, initially identified as a secreted protein synthesized in response to spheroplasting (Hagenmaier *et al.*, 1997), is also positively regulated by the Cpx response (Raivio *et al.*, 2000). The function of this small protein proved to be elusive for years after its discovery. In spite of the sequence homology between CpxP and Spy, Spy does not share CpxP's ability to inhibit Cpx pathway activation (Raivio *et al.*, 2000; Buelow and Raivio, 2005). Mutating *spy* resulted in an increase in activity of the σ^E envelope stress response, which is believed to primarily sense misfolded OMPs; it was therefore suggested that Spy could play a role in OMP biogenesis (Raivio *et al.*, 2000). A recent study has finally determined that Spy acts as a general periplasmic chaperone (Quan *et al.*, 2011). The authors performed a selection for *E. coli* mutations capable of stabilizing an unstable periplasmic fusion protein, and found several isolates that massively overexpressed Spy. *In vitro* experiments revealed that Spy reduced aggregation of substrate proteins, thereby confirming Spy's chaperone activity. Interestingly, Spy was even capable of accelerating *in vitro* protein refolding, a characteristic that is rare among ATP-independent chaperones (Quan *et al.*, 2011). Quan and colleagues also solved the crystal structure of Spy, revealing that Spy is a cradle-shaped dimer composed primarily of α-helices. Together, these results support a model where Spy's flexible structure and high abundance during envelope stress allow it to bind a variety of periplasmic substrates and prevent their aggregation and/or proteolysis (Quan *et al.*, 2011). These functions help the cell to survive and recover from envelope protein misfolding; indeed, Spy improved *E. coli*'s ability to grow in the presence of tannins, which are known to induce protein aggregation (Quan *et al.*, 2011).

Another Cpx regulon member whose chaperone activity has been recently discovered is the small periplasmic protein CpxP, which has an additional function as a negative regulator of Cpx signalling (see previous section). Transcription of *cpxP* is strongly up-regulated by the Cpx response (Danese and Silhavy, 1998); in fact, *cpxP* is one of the most strongly induced members of the Cpx regulon (DiGiuseppe and Silhavy, 2003; Price and Raivio, 2009). The stress-protective role of CpxP was first hinted at by the finding that *cpxP* mutants were more sensitive to alkaline pH and the expression of toxic envelope–localized fusion proteins than wild-type strains (Danese and Silhavy, 1998). CpxP was later proposed to act as a protease adaptor when it was shown that *degP* and *cpxP* mutants both accumulated toxic misfolded pilin subunits when the *papE* or *papG* genes were overexpressed in *E. coli* K-12 (Isaac *et al.*, 2005). The authors proposed a model wherein CpxP acts as an adaptor that improves DegP's ability to degrade some misfolded periplasmic proteins, with CpxP itself being degraded in this process (Isaac *et al.*, 2005). In contrast, several other studies have presented evidence that CpxP could act as a chaperone to enhance the folding of periplasmic proteins. Elaboration of the BFP was reduced in EPEC *cpxP* mutants, supporting a chaperone-like role for CpxP in BFP biogenesis (Vogt *et al.*, 2010). An *in vitro* assay showed that CpxP reduced the aggregation of citrate synthase; however, in contrast to Spy, CpxP could not refold citrate synthase to restore its activity (Zhou *et al.*, 2011). Quan *et al.* (2011) confirmed the *in vitro* chaperone activity of CpxP, and also demonstrated that CpxP could stabilize misfolded periplasmic fusion proteins *in vivo*, suggestive of a natural chaperone function. The two potential roles of CpxP as either a protease adaptor or a chaperone likely are not mutually exclusive, and further studies will hopefully reveal how the molecular function of CpxP is determined by periplasmic conditions.

Finally, it is now understood that DegP is not the only protease whose expression is enhanced by the Cpx response. The transcription of *htpX*, encoding an IM-localized zinc metalloprotease, is also up-regulated by CpxR (Shimohata *et al.*, 2002). HtpX is an endoprotease whose substrates include SecY, a component of the secretory apparatus, and likely other integral IM proteins (Sakoh *et al.*, 2005). Interestingly, topological studies revealed that the proteolytic active site of HtpX is located in the cytoplasm (Shimohata *et al.*, 2002). These findings demonstrate that the Cpx response has a broad role in envelope protein quality control, with HtpX assisting in the removal of misfolded IM proteins during envelope stress.

Envelope maintenance

In addition to up-regulating the expression of genes directly involved in envelope protein folding or degrading, the Cpx response also controls the expression of some genes involved in other aspects of envelope function. The Cpx regulon is now known to contain proteins with roles in the general secretory pathway, phospholipid biosynthesis, and peptidoglycan metabolism.

The Cpx-regulated proteins SecA and YccA both play important roles in ensuring that proteins are efficiently translocated across the IM. *secA*, whose transcription is positively regulated by Cpx (De Wulf *et al.*, 2002), encodes an ATPase that provides the energy driving protein translocation through the Sec complex (Papanikou *et al.*, 2007). Additionally, SecA acts as the IM docking site for the cytosolic chaperone SecB, which binds to and guides substrate preproteins to the secretory apparatus. SecA forms an essential part of the translocase, but owing to its interaction with SecB, SecA is especially important for translocation of soluble (periplasmic or OM) preproteins as compared with integral IM substrates of the Sec complex, which are targeted to the Sec complex by the signal recognition particle (SRP) rather than by SecB.

CpxR binds directly to the promoter of the *yccA* gene, causing a strong increase in its expression under Cpx-inducing conditions (Yamamoto and Ishihama, 2006; Price and Raivio, 2009). Interest in this gene originated with a study that showed that certain mutations in *yccA* acted to stabilize the translocon component SecY, owing to the inhibition of its proteolysis by FtsH (Kihara *et al.*, 1998). Wild-type YccA was found to be a proteolytic substrate of FtsH, while mutant YccA protein could bind to FtsH but could not be degraded, leading to a model wherein YccA competes with other substrates such as SecY for FtsH binding (Kihara *et al.*, 1998). A physiological role for YccA was recently described, when van Stelten and colleagues (2009) demonstrated that the overexpression of YccA ameliorated the toxicity of LamB–LacZ fusions, which are believed to jam the Sec complex owing to rapid folding of the LacZ domain inside the cytoplasm. Part of the toxicity of this fusion can be explained by the observation that LamB–LacZ caused FtsH-mediated proteolysis of SecY and, to a lesser extent, SecE (van Stelten *et al.*, 2009). YccA overexpression, however, prevented proteolysis of SecY, thereby improving the secretion of LamB–LacZ to the periplasm and increasing cell survival. It was therefore suggested that YccA provides the cell with a mechanism to prevent lethality due to protein translocation stress (van Stelten *et al.*, 2009). Considered together, the up-regulation of SecA and YccA by CpxR would be expected to improve the efficiency of protein translocation, particularly for periplasmic and OM proteins. This boosted secretion capacity could be important for ensuring that the other protein folding and degrading factors described above are efficiently delivered to the envelope to relieve stress.

Another Cpx regulon member with a role in envelope maintenance is *psd*, which encodes the enzyme phosphatidylserine decarboxylase. Psd is responsible for the conversion of the phospholipid phosphatidylserine into phosphatidylethanolamine (Voelker, 1997), which accounts for 70% of the membrane phospholipid content in *E. coli* (Mileykovskaya and Dowhan, 2005). The expression of *psd* is strongly up-regulated by CpxR (De Wulf *et al.*, 2002; Price and Raivio, 2009), which correlates well with the finding (described above) that the loss of phosphatidylethanolamine in an *E. coli pss-93* mutant induced the Cpx response (Mileykovskaya and Dowhan, 1997). Insufficient phosphatidylethanolamine production would be expected to damage the integrity of both the IM and the OM, given its role in the ethanolamine modification of LPS, as well as possibly to cause misfolding of integral IM proteins whose structure is affected by protein–lipid interactions (Mileykovskaya and Dowhan, 1997). Up-regulation of *psd* by the Cpx response is therefore likely to promote several aspects of envelope function.

In addition to its effects on the protein and lipid constituents of the envelope, emerging research indicates that the Cpx response also influences the peptidoglycan layer of the cell wall. CpxR was recently shown to positively regulate expression of *amiA* and *amiC*, encoding two *N*-acetylmuramyl-L-alanine amidases involved in cleaving peptide crosslinks in peptidoglycan (Weatherspoon-Griffin *et al.*, 2011). AmiA and AmiC play a key role in cell division by cleaving septal peptidoglycan to allow daughter cell separation (Heidrich *et al.*, 2001). In addition, these amidases promote OM integrity through a poorly understood mechanism, as *amiA* and *amiC* mutants have OM permeability defects (Ize *et al.*, 2003; Weatherspoon-Griffin *et al.*, 2011). Possibly related to this increase in OM permeability, *amiA* and *amiC* mutants were found to be sensitive to several cationic antimicrobial peptides (Weatherspoon-Griffin *et al.*, 2011). Partially due to this up-regulation of *amiA* and *amiC* expression, the Cpx response enhanced survival during treatment with the antimicrobial peptide protamine (Weatherspoon-Griffin *et al.*, 2011).

The list of Cpx regulon members involved in peptidoglycan metabolism is rapidly growing, and includes *ycfS*, which is strongly and directly up-regulated by CpxR (Yamamoto and Ishihama, 2006; Price and Raivio, 2009). YcfS is an L,D-transpeptidase that attaches peptidoglycan to the OM lipoprotein Lpp (Magnet *et al.*, 2007); YcfS may therefore complement the ability of AmiA and AmiC to promote OM integrity. A microarray analysing the effects of NlpE overproduction in *E. coli* (Price and Raivio, unpublished) found that several other peptidoglycan-related proteins were also positively regulated by the Cpx response, including DacC (PBP6), a D,D-carboxypeptidase (Baquero *et al.*, 1996); Slt, a membrane-bound lytic murein transglycosylase (Ehlert *et al.*, 1995); YcbB, an L,D-transpeptidase (Magnet *et al.*, 2008); and YgaU, which contains a LysM domain typically involved in peptidoglycan binding (Weber *et al.*, 2006). The up-regulation of this collection of genes may be indicative of peptidoglycan remodelling during the Cpx response, although the nature and purpose of this remodelling remains to be studied.

Transport

Another category of Cpx-regulated envelope proteins is those with a role in transport. A diverse array of OM- and IM-localized transport proteins are regulated by the Cpx response, some positively and others negatively.

OmpC and OmpF are two of *E. coli*'s major porins, forming largely non-specific channels in the OM through which molecules with a molecular mass of less than 600 Da can freely diffuse (Nikaido, 2003). Although the two proteins exhibit a high degree of structural similarity, the OmpF porin forms a slightly larger channel than the OmpC porin, making OmpF beneficial for the acquisition of nutrients in dilute environments, but detrimental in the presence of noxious substances like antibiotics and bile salts (Nikaido, 2003). Expression of *ompC* and *ompF* is reciprocally regulated by the Cpx pathway. Phosphorylated CpxR activates the expression of *ompC* while repressing the expression of *ompF* (De Wulf *et al.*, 2002; Batchelor *et al.*, 2005). DNase footprinting assays showed that CpxR bound directly to the promoters of both genes (Batchelor *et al.*, 2005). It should be noted, however, that CpxR regulates *ompF* much more strongly than *ompC*, as the slight increase in *ompC* expression under Cpx activating conditions observed during the previously mentioned studies could not be replicated using an *ompC-lux* reporter (Price and Raivio, 2009). Regardless, the increase in the ratio of OmpC to OmpF molecules mediated by CpxR likely protects the envelope by making the OM less permeable to harmful chemicals.

In addition to these non-specific porins, the Cpx response also regulates the expression of several transporters with a narrower substrate specificity. CpxR enhances the expression of *nanC*, which encodes an OM channel involved in the uptake of *N*-acetylneuraminic acid (Condemine *et al.*, 2005). The benefit of increased expression of *nanC* under conditions of envelope stress is currently unclear. On the other hand, the Cpx pathway represses the expression of *efeU*, which encodes an IM protein involved in the uptake of ferrous iron during growth in acidic medium (Cao *et al.*, 2007; Price and Raivio, 2009). This repression appears to be the result of CpxR binding directly to the *efeU* promoter (Yamamoto and Ishihama, 2006). Cpx repression of *efeU* may provide the cell with several benefits: first, lower levels of EfeU may reduce the cellular uptake of iron, which can damage proteins through the production of reactive oxygen species; second, Cpx repression prevents wasteful expression of the *efeU* operon at alkaline pH, when its substrate ferrous iron is poorly soluble and less likely to be available for uptake.

Several microarray studies have provided evidence that additional transporters may be regulated by the Cpx response. The Na^+/H^+ antiporter-encoding genes *chaA* (Oshima *et al.*, 2002) and *nhaB* (Price and Raivio, unpublished) were found to be positively regulated by CpxR; it is possible that these antiporters could help the cell to maintain an appropriate internal pH and/or osmolarity during envelope stress. In terms of peptide transporters, *sbmA* was up-regulated by CpxR (Price and Raivio, unpublished), while the di- and tripeptide importers *tppB*, *dppC*, and *dppD* were down-regulated by the Cpx response (Bury-Moné *et al.*, 2009; Price and Raivio, unpublished). The tryptophan importer *tnaB* was found to be Cpx-repressed in two different microarray experiments (Oshima *et al.*, 2002; Bury-Moné *et al.*, 2009). Interestingly, different microarrays showed Cpx regulation of different transport proteins, likely as a result of the different methods used to activate the Cpx response (overexpression of CpxR as compared with overexpression of NlpE). However, the categories of transporters listed above were usually represented, and transporters always made up a substantial portion of the Cpx-regulated genes. These preliminary results demonstrate a broad role for the Cpx response in modulating cell physiology, suggesting that altering the uptake of specific nutrients plays an important role in the survival of envelope stress. Alternatively, or in addition, the altered regulation of these various transporters upon Cpx response induction may act to improve the integrity of the IM or OM in conjunction with the other envelope changes that are effected.

In addition to proteins involved in the uptake of nutrients, the Cpx response also regulates the expression of transporters involved in the extrusion of toxic substances. CpxR activates the expression of two resistance-nodulation-division (RND)-type efflux pumps: *acrD* and *mdtABC* (Hirakawa *et al.*, 2003, 2005). AcrD plays a role in aminoglycoside resistance in *E. coli* (Rosenberg *et al.*, 2000), while MdtA, MdtB, and MdtC are components of an efflux pump that provides resistance to bile salt derivatives and the antibiotic novobiocin (Baranova and Nikaido, 2002; Nagakubo *et al.*, 2002). Expression of the *acrD* and *mdtABC* operons is also activated by the BaeSR two-component system (Hirakawa *et al.*, 2005). Indeed, it appears that BaeR is the primary regulator of these two operons, as the presence of *baeSR* was required for CpxR to affect transcription of *acrD* and *mdtABC*, while the reverse was not true (Hirakawa *et al.*, 2005). Additionally, there was no increase in expression of *mdtABC* and only a mild increase in *acrD* expression when *E. coli* was grown under conditions that activated the Cpx response but not BaeSR (Price and Raivio, 2009). CpxR therefore appears to be an accessory regulator of these efflux pumps, modulating their expression only under particular growth conditions that also activate BaeSR.

Envelope-localized macromolecular complexes

In order to interact effectively with their environment, *E. coli* K-12 strains produce several types of proteinaceous cellular appendages, including flagella and various kinds of pili or fimbriae. Expression of several of these envelope-localized protein complexes, including flagella, the F conjugal pilus, and curli, is repressed by the Cpx envelope stress response.

De Wulf and colleagues (1999) first demonstrated that the Cpx pathway decreased swarming motility in *E. coli* K-12. This decrease in motility was attributed to altered expression of numerous flagellar and chemotaxis genes, as it was shown that CpxR directly repressed transcription of the *motABcheAW*, *tsr*, and *aer* loci, encoding components of the flagellar motor and chemotaxis regulatory system, the serine chemoreceptor, and the aerotaxis receptor, respectively (De Wulf *et al.*, 1999; De Wulf *et al.*, 2002). A more recent microarray analysis indicated that expression of the flagellar master regulator FlhC was also down-regulated in response to Cpx pathway activation by overexpression of NlpE (Price and Raivio, unpublished). It therefore appears that the Cpx response reduces *E. coli's* motility through coordinated repression of many genes involved in both flagellar biogenesis and chemo- and aerotaxis.

Expression of several types of pili is also repressed by the Cpx response. Although down-regulation of F conjugal pilus expression was in fact the first known role of the Cpx response (McEwen and Silverman, 1980), more recent studies have shed light on the mechanism by which this repression occurs. Activation of the Cpx pathway via *cpxA** mutations or overexpression of NlpE reduces accumulation of the regulatory protein TraJ, the F-encoded activator of the *tra* genes involved in conjugal pilus synthesis (Silverman *et al.*, 1993; Gubbins *et al.*, 2002). Gubbins *et al.* (2002) concluded that the Cpx response reduced TraJ levels post-transcriptionally, since activation of the Cpx response had little effect on transcription of *traJ*, and because TraJ protein was less stable in *cpxA** strains compared with wild-type. It was later demonstrated that TraJ is a substrate for degradation by the Cpx-regulated protease complex HslVU both *in vivo* and *in vitro* (Lau-Wong *et al.*, 2008). This example, in which gene expression is modified through proteolysis of a regulatory protein, clearly illustrates the diverse mechanisms by which the Cpx response can alter cellular behaviour.

The Cpx pathway also reduces expression of curli, which are extracellular fibrils composed primarily of the protein CsgA that have been shown to play an important role in adherence to abiotic surfaces, other bacterial cells, and host organisms (Barnhart and Chapman, 2006). In addition to CsgA, curli biogenesis requires at least five other genes, arranged into two operons (*csgBA* and *csgDEFG*). CsgB is the nucleator of CsgA polymerization, while CsgE, CsgF, and CsgG all play roles in secretion and assembly of CsgA monomers. CsgD is a transcriptional activator that enhances expression of both *csg* operons. Phosphorylated CpxR, resulting from either mutation of *cpxA* or overexpression of *nlpE*, reduced the expression of *csgA* (Dorel *et al.*, 1999). Later studies revealed that CpxR bound directly to multiple sites in the promoters of both the *csgBA* and the *csgDEFG* operons (Prigent-Combaret *et al.*, 2001; Jubelin *et al.*, 2005; Ogasawara *et al.*, 2010). CpxR therefore affects curli expression in two ways: by directly reducing transcription of the *csgBA* operon, and by reducing levels of the transcriptional activator CsgD. It should be noted that the *csg* genes are not the only targets of CsgD regulation; for example, CsgD activates expression of *adrA*, which encodes a GGDEF protein that increases production of the biofilm matrix polysaccharide cellulose (Römling, 2005). Owing to its negative regulation of *csgD*, CpxR therefore inhibits production of cellulose as well as curli (Ma and Wood, 2009).

Several explanations have been proposed for the down-regulation of numerous envelope-localized protein complexes by the Cpx response. It is possible that reducing synthesis of these large structures could be a mechanism for conserving cellular energy during periods of stress (De Wulf *et al.*, 1999). Alternatively or in addition, this regulation could be a means for reducing non-essential protein traffic in the envelope when protein misfolding is already rampant. The inherent aggregative properties of some of these proteins, such as CsgA (Wang *et al.*, 2008), could make them particularly harmful during envelope stress. The down-regulation of flagella and pili could also play a role in biofilm development. There is evidence for a role of the Cpx response in biofilm development: for example, *cpxA* mutants of various *E. coli* strains had reduced adherence to polystyrene (Dorel *et al.*, 1999), and mutations in genes encoding several Cpx pathway components (*cpxP*, *cpxR*, *cpxA*, and *nlpE*) caused a reduction in biofilm biomass (Beloin *et al.*, 2004). The reduction in cellulose synthesis caused by Cpx pathway activation was shown to promote biofilm formation on hydrophobic surfaces, but inhibit biofilm formation on hydrophilic surfaces, as a consequence of the hydrophilic nature of cellulose (Ma and Wood, 2009). On the basis of results such as these, it was proposed that the cellular purpose of the Cpx response is to promote the sessile biofilm lifestyle (Dorel *et al.*, 2006). According to this view, the down-regulated expression of flagella and pili during the Cpx response would play an important role in biofilm maturation.

Regulatory connections

Bacteria such as *E. coli* possess multiple regulatory systems that promote stress survival and adaptation to different environmental conditions or lifestyles. It is therefore beneficial for the bacterium to have a means to activate several regulatory systems with complementary purposes simultaneously, or to be able to shut off regulatory pathways whose functions do not enhance survival under the prevailing conditions. In order to accomplish these goals, the Cpx regulon includes several genes encoding regulatory proteins.

The first regulatory genes whose expression was found to be controlled by the Cpx response were *cpxRA*. Several lines of evidence led to the conclusion that expression of *cpxRA* is autoregulated. De Wulf *et al.* (1999) noticed that a perfect CpxR consensus sequence was located upstream of the *cpxRA* operon, while Raivio *et al.* (1999) found that levels of CpxR and CpxA protein were elevated in a *cpxA** constitutively active background. Ultimately, both groups demonstrated that expression of a *cpxRA-lacZ* reporter was positively regulated by the Cpx pathway (De Wulf *et al.*, 1999; Raivio *et al.*, 1999). Interestingly, expression of the gene encoding the negative regulator of the Cpx pathway, *cpxP*, was also enhanced by CpxR (Danese and Silhavy, 1998). Although it may seem contradictory for the Cpx response to promote expression of both *cpxRA* and its negative regulator, it has been proposed that this regulatory scheme endows the system with both positive and negative feedback capabilities (Raivio *et al.*, 1999). The autoactivation of *cpxRA* allows the response to rapidly amplify in the presence of envelope stress. Expression of *cpxP* is also increased under these conditions, but CpxP is inactivated during envelope stress (DiGiuseppe and Silhavy, 2003). When the stress is relieved, the high level of *cpxP* expression will allow the inhibitor to accumulate and rapidly shut off the pathway.

In addition to regulating *cpxRA*, the Cpx response also affects the expression of the genes required for another envelope stress response: the σ^E pathway. The σ^E response detects misfolding of OMPs in the periplasmic space via proteolysis of its anti-sigma factor, RseA (Ades, 2008). In response, the alternative sigma factor σ^E activates the transcription of genes involved in folding and inserting OMPs into the OM (MacRitchie *et al.*, 2008a). Both σ^E and its anti-sigma factor are encoded in the operon *rpoErseABC*; expression of this operon is directly repressed by the binding of CpxR to its promoter (De Wulf *et al.*, 2002). The benefit of the Cpx response decreasing expression of genes encoding another, seemingly complementary envelope stress response is not currently understood. It was initially hypothesized that, under environmental conditions in which both the Cpx and σ^E pathways are activated, this mechanism would allow the Cpx response to take precedence (De Wulf *et al.*, 2002). Other possibilities are that it may be beneficial to the cell to limit the activation of similar stress responses, as a means of conserving limited energy, or that perhaps some σ^E regulon members are detrimental under Cpx-inducing conditions (Price and Raivio, 2009). In support of the latter hypothesis, a number of genes involved in the response to excess copper are reciprocally regulated by the Cpx and σ^E pathways (Price and Raivio, 2009), perhaps suggesting that the Cpx and σ^E responses perform some incompatible functions.

The Cpx response can also influence activity of the EnvZ/OmpR two-component system, which regulates expression of porins and other OMPs in response to changes in osmolarity (Pratt *et al.*, 1996; Guillier and Gottesman, 2006). This interconnection is achieved indirectly via a small, IM-localized protein called MzrA. *mzrA* was identified in a screen for multicopy suppressors of the conditionally lethal *bamB degP* double mutant, which has a severely reduced ability to fold and insert OMPs into the OM (Gerken *et al.*, 2009). The ability of a multicopy plasmid containing *mzrA* to suppress *bamB degP* lethality was attributed to decreased expression of several OMPs via the EnvZ-OmpR system. Further experiments revealed that transcription of *mzrA* was positively regulated by the Cpx response, and that MzrA physically interacted with EnvZ to increase expression of EnvZ/OmpR-dependent genes (Gerken *et al.*, 2009). Activation of the Cpx pathway therefore increases the expression of OmpR-regulated genes in a MzrA-dependent fashion. The mechanism by which MzrA activates EnvZ is not well understood. MzrA has been shown to interact with EnvZ via its periplasmic C-terminal region; several point mutations in this region of MzrA disrupt both dimerization and interaction with EnvZ, suggesting that MzrA acts as a dimer (Gerken and Misra, 2010). However, it is unclear which enzymatic activity of EnvZ (autokinase, OmpR kinase, or phospho-OmpR phosphatase) is altered by interaction with MzrA (Gerken and Misra, 2010). MzrA is a novel example of a 'connector' protein which allows the CpxAR two-component system to communicate with EnvZ/OmpR, which also regulates envelope contents in response to environmental parameters. For more information about connector proteins, see also Chapter 8.

In addition to regulating transcriptional regulators such as σ^E and EnvZ/OmpR, the Cpx response also influences the levels of the second messenger molecule cyclic-di-GMP (c-di-GMP) via YdeH. YdeH contains a GGDEF domain, typical of the diguanylate cyclase enzymes that produce c-di-GMP. The diguanylate cyclase activity of YdeH has been experimentally confirmed (Jonas *et al.*, 2008). c-di-GMP plays a role in regulating diverse cellular processes, including motility, surface adhesion, extracellular polysaccharide synthesis, and progression through the cell cycle (Mills *et al.*, 2011). YdeH seems to be involved in the repression of motility and adhesion, as no flagella or pili can be seen on the surface of *E. coli* cells overexpressing *ydeH* (Jonas *et al.*, 2008). In addition, YdeH promotes biofilm formation by increasing production of the extracellular polysaccharide poly-*N*-acetylglucosamine (Jonas *et al.*, 2008; Boehm *et al.*, 2009). *ydeH* is positively regulated at the transcriptional level by CpxR (Yamamoto and Ishihama, 2006; Price and Raivio, 2009). Activation of the Cpx response is therefore expected to enhance synthesis of c-di-GMP, which likely plays a role in the ability of the Cpx response to reduce expression of envelope-localized protein complexes and promote biofilm formation, as described above.

The final type of regulatory protein whose expression is known to be under the control of CpxR is a serine/threonine kinase, called YihE in *E. coli* and RdoA in *Salmonella*. The *yihE/rdoA* gene is encoded upstream of *dsbA* (Belin and Boquet, 1994). In both *E. coli* and *S. enterica* serovar Typhimurium, transcription of *yihE/rdoA* is positively regulated by the Cpx response (Pogliano *et al.*, 1997; Suntharalingam *et al.*, 2003). Structural and biochemical evidence indicates that YihE/RdoA is a novel protein serine/threonine kinase, although the only phosphorylation substrate that is currently known is YihE/RdoA itself (Zheng *et al.*, 2007). However, it has been suggested that YihE/RdoA plays a role in relaying signals from the Cpx response to several of its target genes, since an *S. enterica* serovar Typhimurium *rdoA* mutant had higher curli expression than the wild-type strain and was incapable of up-regulating *dsbA* and undergoing flagellar phase variation in response to Cpx activation (Suntharalingam *et al.*, 2003; Zheng *et al.*, 2007). Furthermore, mutation of *yihE* in *Shigella flexneri* affected the expression of more than 100 genes, including the *galETK* operon, which plays a key role in LPS biosynthesis (Li *et al.*, 2001; Edwards-Jones *et al.*, 2004). The large number of genes whose expression is altered by YihE/RdoA indicates that one of the phosphorylation targets of this kinase may be an important regulatory protein (Zheng *et al.*, 2007).

By controlling the expression of these other regulators, the Cpx response can expand its effect on cellular physiology. Some of these regulators control processes that are not known to be regulated by CpxR directly, but which may serve functions complementary to the Cpx response, such as YdeH control of exopolysaccharide production. Interestingly, some of the Cpx-controlled regulators also regulate the expression of the same or related genes as those directly controlled by CpxR itself; for example, MzrA increases EnvZ/OmpR activity and thereby increases *ompC* but decreases *ompF* expression, and YdeH represses motility. This could be a mechanism for amplifying the effect of Cpx activation upon expression of these targets.

Other Cpx-regulated genes

In addition to the above categories, the Cpx response also affects the expression of genes involved in diverse cellular processes. These include iron storage, amino acid metabolism, oxidative phosphorylation, and DNA repair.

The gene *ftnB* (*yecI*) encodes a putative ferritin-like protein. Expression of *ftnB* was found to be up-regulated in a Cpx-dependent fashion during exposure to copper (Yamamoto and Ishihama, 2005). Follow-up studies found that CpxR bound directly and with high affinity to the *ftnB* promoter (Yamamoto and Ishihama, 2006), and that *ftnB* was one of the genes most strongly up-regulated by Cpx pathway activation (Price and Raivio, 2009). Bacterial ferritins are important for storage of excess intracellular iron, allowing growth to continue when iron becomes limiting; some ferritins also detoxify iron, diminishing its ability to generate toxic oxygen radicals (Andrews *et al.*, 2003). Since *ftnB* is strongly up-regulated by the Cpx stress response, it seems likely that this ferritin may have a detoxifying function, but this hypothesis remains to be tested experimentally.

Two enzymes involved in aromatic amino acid biosynthesis are also regulated by the Cpx pathway, but in a strain-dependent fashion. *aroK* (encoding shikimate kinase I) and *aroG* (encoding 3-deoxy-D-arabino-heptulosonate 7-phosphate synthase) are Cpx-activated in *E. coli* MG1655 (De Wulf *et al.*, 2002; Yamamoto and Ishihama, 2006), but Cpx-repressed in strain MC4100 (Price and Raivio, 2009). CpxR has been shown to bind directly to the promoter of *aroG* (Yamamoto and Ishihama, 2006). Both enzymes are part of the shikimate pathway responsible for the synthesis of the aromatic amino acids tryptophan, tyrosine, and phenylalanine. The rationale for Cpx regulation of these genes is unclear, but altered expression of *aroK* and *aroG* could affect the rate of protein synthesis and therefore the volume of protein traffic entering the envelope (Price and Raivio, 2009). Alternatively, repression of these genes might help the cell to reduce energy expenditure, as aromatic amino acids are energetically expensive to synthesize.

There is preliminary evidence that the Cpx pathway also regulates energy generation by oxidative phosphorylation. Microarray analysis revealed that oxidative phosphorylation genes were enriched among Cpx-repressed genes (Price and Raivio, unpublished). Specifically, the *nuoA-N* operon, encoding components of the NADH dehydrogenase I complex, and the *sdhA–D* operon, encoding the succinate dehydrogenase complex, were down-regulated when NlpE was overexpressed under most strain and media combinations examined. Additionally, putative CpxR binding sequences are located upstream of both operons. *E. coli* may accrue several benefits from reducing oxidative phosphorylation during envelope stress. For example, a reduction in oxidative phosphorylation would be expected to slow growth, which may prevent the exacerbation of existing envelope stress (Price and Raivio, unpublished). In addition, a reduced rate of oxidative phosphorylation could potentially decrease oxidative damage to envelope proteins.

Finally, the Cpx pathway has been implicated in the regulation of a DNA repair enzyme, uracil *N*-glycosylase, which is encoded by the gene *ung*. De Wulf *et al.* (2002) found that *ung* expression was positively regulated by Cpx in the MG1655 background. However, a later study in the *E. coli* strain BW25113 showed that expression of *ung* was directly repressed by CpxR binding to its promoter (Ogasawara *et al.*, 2004). Furthermore, cells overexpressing *cpxR* were shown to have a lower uracil *N*-glycosylase activity and higher mutation rate than the vector control strain (Ogasawara *et al.*, 2004). In the strain MC4100, Cpx pathway activation had a weak positive or no effect on expression of an *ung-lux* reporter (Price and Raivio, 2009). As with the *aroK* and *aroG* genes above, these conflicting results may be the product of inherent strain differences. Interestingly, Cpx regulation of *ung* is not the only example of an extracytoplasmic stress response controlling DNA repair. The σ^E envelope stress response has been recently shown to promote stress-induced mutagenesis in *E. coli* (Gibson *et al.*, 2010). Such mechanisms may contribute to environmental adaptation by allowing microbes to increase their mutation rate when faced with stress.

Aside from *ftnB*, these genes involved in cellular maintenance and metabolism tend to be regulated weakly and/or in a strain-specific manner by the Cpx response. This could suggest that these targets are co-regulated with other regulatory proteins under specific strain and growth conditions, and may not represent a core function of the Cpx response.

Summary

Recent work has expanded the view of the Cpx response as a means for controlling the expression of envelope-localized protein folding and degrading factors. Some of the more recently identified functions include other aspects of envelope maintenance, such as peptidoglycan modification; regulation of various transporters; and interconnections with other signalling pathways (Fig. 12.2). Most of these roles still pertain to proper functioning of the envelope compartment and adaptation to stress, supporting the idea of Cpx as an envelope stress response, but revealing a multifaceted approach to stress survival. In addition, there are a large number of genes of unknown function whose expression is regulated by the Cpx pathway (Price and Raivio, unpublished). Further study of these genes will undoubtedly expand and clarify our understanding of the cellular role of the Cpx pathway.

A major outstanding question in this area relates to the significance of the Cpx response during the *E. coli* life cycle. Since CpxR regulates the expression of many genes associated with biofilm formation (Dorel *et al.*, 2006), it is possible that the Cpx response is activated when *E. coli* inhabits environments outside of a host organism. Alternatively, perhaps the Cpx response is activated within a host when conditions become unfavourable, helping to prepare the bacterium for transmission to a new host by decreasing the expression of surface structures anchoring it to the current host and increasing expression of genes required for stress survival. Further research on the role of the Cpx response in bacterial–host interaction will be required to answer this important question.

Roles of the Cpx envelope stress response in other bacteria

Until recently, the Cpx envelope stress response has been largely studied in *E. coli* K-12 laboratory strains. Early studies in *Shigella* spp. indicating that the Cpx response may impact bacterial traits involved in infection (Nakayama and Watanabe, 1995, 1998; Mitobe *et al.*, 2005) together with motility in *E. coli* (De Wulf *et al.*, 1999) prompted investigations of whether or not the CpxAR two-component system might impact other virulence phenotypes in other microbes. So far, these experiments indicate that, in the closely related enteric bacteria *E. coli*, *S. enterica* serovar Typhimurium, *Shigella* spp., and *Yersinia* spp., the Cpx response appears to play a predominantly negative role in the regulation of envelope-localized virulence determinants. In the less related *V. cholerae*, *Xenorhabdus nematophila* and *Legionella pneumophila*, the role(s) of the Cpx response appear to have diverged. Cpx-mediated changes in gene expression are involved in the induction of important virulence and/or symbiotic determinants in *X. nematophila*, and *L. pneumophila* (Gal-Mor and Segal, 2003; Murata *et al.*, 2006; Vincent *et al.*, 2006; Herbert *et al.*, 2007; Altman and Segal, 2008). Meanwhile, in *V. cholerae*, the cues that induce the Cpx response, although envelope-related, appear to be different in nature from those identified in *E. coli* (Slamti and Waldor, 2009).

A few studies have also linked the Cpx response to antibiotic resistance, another important trait during bacterial infection. In this section, we describe studies detailing these findings (see Table 12.1 for a summary).

The regulation of virulence factor expression in pathogenic *E. coli* strains

The Cpx envelope stress response system plays an important role in the regulation of envelope-localized virulence determinants (refer to 'Envelope-localized macromolecular complexes'). In pathogenic *E. coli* strains, these virulence determinants include some of the EPEC type III secretion system (T3SS) translocator and effector proteins (i.e. EspA, EspB, EspD, and the translocated bacterial receptor Tir) (MacRitchie *et al.*, 2008b) and the type IV BFP (Vogt *et al.*, 2010) as well as the uropathogenic *E. coli* (UPEC) Pap or P pilus (Jones *et al.*, 1997; Hung *et al.*, 2001; Hernday *et al.*, 2004).

The BFP is a type IV pilus necessary for the initial steps during the infection of epithelial cells (i.e. bacterial auto-aggregation and localized adherence) by EPEC (Girón *et al.*, 1991; Donnenberg *et al.*, 1992). Nevesinjac and Raivio (2005) reported that the Cpx pathway is involved in BFP expression and EPEC attachment to human epithelial cells. The elaboration of BFP in an *E.*

Table 12.1 Virulence factors in different pathogens regulated by the Cpx envelope stress response

Pathogen	CpxRA regulated proteins/virulence factors [a]	Role
EPEC	BFP	Auto-aggregation/adherence
	T3SS	Secretion of virulence proteins
UPEC	Pap or P pilus	Adhesion factor
Y. enterocolitica	T3SS	Secretion of Yops
	HtrA	Periplasmic protease
Y. pseudotuberculosis	T3SS	Secretion of Yops
	Invasin, pH 6 antigen and Ail	Surface-located adhesins
V. cholerae	DsbD	Isomerization of disulfide bonds
S. typhimurium	HilA	Virulence transcriptional regulator
	SipC	Invasion effector
	Curli	Adhesion factor
	AmiA and AmiC	Resistance to antimicrobial peptides
S. sonnei	VirF	Virulence master regulator
	InvE	Transcriptional activator
	Invasion plasmid antigen-B, C and D	Invasion effectors
L. pneumophila	T4SS	Translocation of virulence factors
H. ducreyi	LspB, LspA1/LspA2	Required for phagocytosis inhibition
	DsrA	Outer membrane serum resistance protein
	Flp1	Putative pilus
X. nematophila	NilA, -B and -C	Colonization factors in nematode host
	XaxA	Haemolysin activity
	PrtA	Protease activity
	LrhA	Positive regulator of lipase activity
	MrxA	Pilin subunit

[a]See text for references.
EPEC, enteropathogenic *E. coli*; UPEC, uropathogenic *E. coli*; BFP, bundle-forming pili; T3SS, type III secretion system; Yop, *Yersinia* outer proteins; T4SS, type IV secretion system.

coli lab strain, MC4100, carrying the entire *bfp* gene cluster under the control of an exogenous promoter, was only observed when the Cpx response was constitutively activated (i.e. *cpxA** background), suggesting that the Cpx pathway plays a post-transcriptional role in BFP expression (Nevesinjac and Raivio, 2005). Additional experiments in an EPEC *cpxR* mutant showed that, as in MC4100, the Cpx pathway was necessary for the efficient production of BfpA, the major subunit of the BFP. These data indicated that, at least in some strains, the Cpx pathway provides the folding and degrading factors necessary for BFP assembly (Nevesinjac and Raivio, 2005). Recent work by Vogt and collaborators (2010) confirmed that the Cpx pathway facilitates the biogenesis of the entire BFP apparatus in EPEC, by regulating the expression of protein folding factors (i.e. DsbA, DegP and CpxP). Accordingly, it seems likely that the Cpx response is necessary for the efficient assembly of BFP at the beginning of an infection.

Paradoxically, it was determined that constitutive, high-level activation of the Cpx pathway down-regulated the expression of the *bfp* operon at the transcriptional level in EPEC (Vogt *et al.*, 2010). This inhibition has been proposed to be a mechanism to limit protein traffic in the envelope during times of envelope stress. It seems likely that the role of the Cpx pathway in inhibiting the elaboration of the BFP occurs during the transmission phase of the EPEC life cycle and not during the initial steps in EPEC colonization of the human intestine (Vogt *et al.*, 2010).

Upon EPEC adherence to an epithelial cell, the bacterium initiates the elaboration of the T3SS. This secretion system is one of the major virulence factors in EPEC, and it is encoded in a 35-kb pathogenicity island called the locus of enterocyte effacement (LEE) (Elliott *et al.*, 1998). Translocators and EPEC secreted proteins (Esp) that are part of the T3SS are also encoded in the LEE locus. MacRitchie and collaborators (2008b) showed that the overexpression of the response regulator CpxR in EPEC diminished the expression of some Esps (e.g. EspA, EspB and EspD) and the translocated intimin receptor (Tir). Further analysis demonstrated that the transcription of the LEE operons (i.e. LEE4 and LEE5), which encode the *espADB* and *tir* genes, was inhibited when CpxR was overexpressed. Additional experiments in a *cpxA** mutant and a *cpxR* null strain confirmed these results and demonstrated that the induction of the Cpx response led to a partial down-regulation of LEE gene transcription. Most likely, the Cpx regulation of those translocators and secreted protein is independent of the LEE EPEC regulator, Ler, since overexpression of known activators of *ler* gene expression could not counter the inhibitory effect of the Cpx response (MacRitchie *et al.*, 2008b). Whether or not the Cpx response exerts its negative effect on *bfp* and LEE gene expression through direct binding of CpxR to the promoters of these genes remains to be determined, although it has been shown in a uropathogenic strain of *E. coli*, UPEC, that CpxR~P inhibited expression of the P or Pap pilus in that strain through direct binding of the DNA that is involved in phase variable expression of the *pap* gene cluster (Hernday *et al.*, 2004). Intriguingly, as for the BFP, the Cpx response also seems to facilitate P pilus expression under some conditions. In a laboratory strain of *E. coli*, an intact Cpx response was required for proper pilus assembly when the *pap* genes were expressed from an exogenous promoter (Hung *et al.*, 2001).

Taken together, these results indicate that the Cpx pathway plays a dual role in the regulation of envelope virulence determinant expression in *E. coli*. While an intact Cpx response seems to be necessary for maximally efficient assembly of pili, high level induction of this pathway eliminates pili and other virulence determinants from the envelope. Studies of the BFP and P pili suggest that the stimulatory role of the Cpx response on virulence determinant assembly derives from its regulation of envelope protein folding and degrading factors (Jones *et al.*, 1997; Nevesinjac and Raivio, 2005; Vogt *et al.*, 2010). Meanwhile, the inhibitory effect of the Cpx response appears to be mediated at both transcriptional and post-transcriptional levels. Evidence clearly shows that the BFP and P pili are inhibited at the transcriptional level (Hernday *et al.*, 2004; Vogt *et al.*, 2010); however, post-transcriptional events are also known to be involved in Cpx-mediated inhibitory effects. MacRitchie and Raivio (unpublished) showed that DegP is partly required for the inhibitory effect of the Cpx response on T3S in EPEC, and Lau-Wong *et al.* (2008) demonstrated that Cpx induction of the HslVU protease is necessary for the proteolytic degradation of the TraJ regulator of F pilus expression in laboratory strains of *E. coli*. Similar to these findings, we describe below experiments in which both stimulatory and inhibitory effects of the Cpx response on virulence determinant expression have been described in other bacterial pathogens (see below).

Y. enterocolitica and *Y. pseudotuberculosis*: bacterial growth, host cell contact and T3S-dependent cellular cytotoxicity

Y. enterocolitica and *Y. pseudotuberculosis* are human enteropathogenic bacteria involved in the development of infections in the intestinal tract and intestinal lymphoid system (Carniel, 2002). Studies in the Cpx pathway in these two *Yersinia* species showed that the Cpx pathway plays an important role in the regulation of virulence factors such as the T3SS as in *E. coli*. The first study in *Y. enterocolitica* regarding the Cpx pathway was focused on the HtrA protein (homologous to DegP in *E. coli*) (Heusipp *et al.*, 2004). HtrA is a protease that degrades misfolded proteins in the periplasm, and it is an important factor for intracellular survival and virulence in animal models (Li *et al.*, 1996). *Y. enterocolitica htrA* expression is regulated by the Cpx pathway, and Heusipp and colleagues (2004) provided evidence that its expression was stimulated in response to envelope stress induced by the overproduction of the OMP Ail. Overproduction of CpxR in *Y. enterocolitica* showed that, as in *E. coli*, the transcription of *cpxR* was autoregulated and *rpoE* was down-regulated (Rönnebäumer *et al.*, 2009). These data demonstrated that, as in *E. coli*, CpxR in *Y. enterocolitica* regulates DegP/HtrA expression and can act as both a positive and negative transcriptional regulator (De Wulf *et al.*, 2002). Moreover, it was shown that overproduction of CpxA or CpxR had a detrimental effect on growth (Rönnebäumer *et al.*, 2009). Finally, as in other organisms, it appears that elimination of the Cpx response has a minor or no effect on virulence, since a *cpxR* mutant was shown to be unaffected in the invasion of eukaryotic cells (Rönnebäumer *et al.*, 2009).

In contrast, studies in *Y. pseudotuberculosis* presented evidence that activation of the Cpx pathway can negatively affect bacterium-host cell contact (Carlsson *et al.*, 2007b). *cpxA* null mutants in *Y. pseudotuberculosis* were not able to interact with mammalian cells, and as a consequence these cells were not susceptible to the T3S-dependent injection of effector toxins (Carlsson *et al.*, 2007b). qRT-PCR analysis showed that Cpx-regulated genes were up-regulated in the *cpxA* mutant, indicating that in *Yersinia* spp., as has been noted in *E. coli*, CpxR can become phosphorylated in an uncontrolled fashion by small molecular weight phosphodonors when CpxA is absent (Danese and Silhavy, 1998). Further, the *cpxA* mutant exhibited a decrease in the expression in some of the major surface-located adhesins in *Y. pseudotuberculosis* [i.e. invasin (*inv*), pH 6 antigen (*psa*) and Ail] compared with the parental strain. This suggests that activation of the Cpx response leads to down-regulated expression of invasins in *Y. pseudotuberculosis* (Carlsson *et al.*, 2007b). In this study, the authors investigated if the lack of adhesion of a *Y. pseudotuberculosis cpxA* null mutant was due to an effect on the regulation of RovA, a transcriptional activator of invasion (*inv*) genes. Indeed, when RovA was expressed in a *cpxA* null background, it restored the ability of *Y. pseudotuberculosis* to adhere and activated the antiphagocytic properties of the T3SS (Carlsson *et al.*, 2007b). Indeed, CpxR was shown to directly inhibit the expression of RovA by binding to its promoter upon phosphorylation (Carlsson *et al.*, 2007b; Liu *et al.*, 2011). Consistent with the observation that induction of the Cpx response in a *cpxA* null mutant led to down-regulated invasin expression, in a *cpxR* null mutant strain of *Y. pseudotuberculosis*, there was more production of invasins and association with mammalian cells compared with wild-type cells (Carlsson *et al.*, 2007b).

Y. pseudotuberculosis and *Y. enterocolitica* harbour the plasmid-encoded *Yersinia* outer proteins (Yops), which make up a T3SS (Cornelis *et al.*, 1998). This secretion system plays an important role in the translocation of several toxic effectors into target cells, allowing bacterial resistance to phagocytic host defence mechanisms and replication within lymphoid tissues of their host

(Viboud and Bliska, 2005). As in *E. coli*, there is evidence that the Cpx pathway in *Y. pseudotuberculosis* regulates the T3SS. A *cpxA* null mutant strain (possessing an activated Cpx pathway, see above) showed a defect in Yop production and secretion (Carlsson *et al.*, 2007a). Consistent with this observation, the absence of CpxA in *Y. pseudotuberculosis* reduced the levels of late-secreted T3S substrates (i.e. YopE, YopH, YopK, and YopD) at both transcriptional and translational levels. Moreover, a HeLa cell cytotoxicity assay indicated that there was a defect in cytotoxicity of a *Y. pseudotuberculosis cpxA* null mutant towards target cells (Carlsson *et al.*, 2007a). These studies demonstrated that induction of the Cpx response in *Y. pseudotuberculosis* negatively impacts the translocation of Yop effectors into host cells (Carlsson *et al.*, 2007a). The detailed molecular mechanism by which CpxR-P negatively regulates pathogenicity in *Y. pseudotuberculosis* remains unknown, although it seems likely that the repression of transcription of the regulatory protein RovA is centrally involved (Carlsson *et al.*, 2007b; Liu *et al.*, 2011).

V. cholerae: salinity monitoring and sensing aberrant disulfide bond status

V. cholerae is a Gram-negative human pathogen that causes cholera, a diarrheal disease, in developing countries worldwide (Kosek *et al.*, 2003). Comparison between *V. cholerae* and *E. coli* Cpx system components (i.e. *cpxR*, *cpxA* and *cpxP*) showed that the *cpx* loci are organized similarly, but there are differences between the predicted amino acid sequences of the Cpx proteins (Slamti and Waldor, 2009). A genetic analysis in *V. cholerae* El Tor clinical isolate N16961 showed that unlike *E. coli*, there was no basal activation of the Cpx pathway. As well, the inducing cues that trigger this pathway in *V. cholerae* differ from *E. coli*. For example, there did not appear to be any basal activity of the Cpx response in *V. cholerae*, as measured using a *cpxP-lacZ* reporter gene, and deletion of *cpxP* did not induce the Cpx response (as it does in *E. coli*), nor did overexpression of NlpE (Slamti and Waldor, 2009). In contrast, as has been observed in *E. coli* and *S. enterica* serovar Typhi (see below), chloride ions did stimulate the *V. cholerae* Cpx pathway. Based on these observations, it was suggested that the Cpx response may function mainly to monitor salinity and perhaps be important for survival in the saline and brackish environments normally inhabited by *Vibrio* spp. in the environment (Slamti and Waldor, 2009). In agreement with this idea, intestinal colonization studies were performed and showed that the Cpx pathway in *V. cholerae* was not required for growth in the intestine (Slamti and Waldor, 2009). Finally, a genetic screen was performed to identify mutants in *V. cholerae* that activated the Cpx pathway. The results showed that more than 68% of the mutations that induced the Cpx pathway affected proteins that are localized in the cell envelope. This finding suggests that, despite differences between the *E. coli* and *V. cholerae* Cpx response, it serves an envelope stress role in both organisms. Interestingly, one of the mutations that induced the *V. cholerae* Cpx response occurred in the gene encoding DsbD, an IM protein that promotes isomerization of disulfide bonds in envelope proteins, suggesting the possibility that this pathway may sense misfolded envelope proteins that contain aberrant disulfide bonds (Slamti and Waldor, 2009). In support of this hypothesis, the oxidant $CuSO_4$ was shown to induce the Cpx response in *V. cholerae*; this induction could be inhibited by the addition of reductant to the media (Slamti and Waldor, 2009). Unlike *E. coli*, the CpxP signalling protein was required in *V. cholerae* for the induction of the Cpx pathway, at least in a *dsbC* mutant background (Slamti and Waldor, 2009). Cumulatively, these observations suggest that there are differences in the inducing cues, signalling mechanisms, and perhaps downstream targets of the Cpx response between *E. coli* and *V. cholerae*; however they also support a similar envelope stress responsive capacity for this regulatory response in both organisms. Future work examining the differences between these organisms will undoubtedly shed more light on the physiology of Cpx signalling and adaptation in both microbes.

S. enterica serovar Typhi and Typhimurium: adherence and invasion of host cells

S. enterica serovar Typhi is a facultative bacterium that invades the intestinal epithelium and causes typhoid fever. Leclerc and collaborators (1998) performed a transposon mutagenesis screen for mutants in *S. typhi* that were not able to adhere and invade small intestinal epithelial cells and found that a *cpxA* mutation compromised adhesion. This result suggested that the Cpx pathway could be involved in the early steps of *S. typhi* infection. Further, the authors showed that the expression of the *cpxA* gene was not impacted by low pH but was up-regulated by 0.3 M NaCl in *S. enterica* serovar Typhi (Leclerc *et al.*, 1998). In *S. enterica* serovar Typhimurium, the causative agent of human gastroenteritis, the HilA protein is a known activator involved in the regulation of invasion gene expression (Bajaj *et al.*, 1995). Nakayama *et al.* (2003) characterized the CpxRA regulon in order to detemine if, like in *S. enterica* serovar Typhi, the Cpx pathway had any effect on the regulation of invasion in *S. enterica* serovar Typhimurium. The results showed that the organization and direction of transcription of the *cpxRA* operon and *cpxP* gene were similar to their counterparts in *E. coli* (Nakayama *et al.*, 2003). When the authors mutated the *cpxR* and *cpxA* genes and then monitored *hilA* gene expression, it was demonstrated that *hilA* expression was dramatically decreased in the absence of *cpxA* at pH 6 but not in a *cpxR* background (Nakayama *et al.*, 2003). Moreover, the *S. enterica* serovar Typhimurium *cpxA* mutant showed a significant decrease in the expression of *sipC*, an effector necessary for invasion and regulated by HilA, at pH 6.0. Given that *cpxA* mutants in *Yersinia* and *Escherichia* spp. confer induction of the Cpx response, it seems highly probable that the *cpxA* mutant in these studies similarly leads to Cpx pathway activation. In this case, the data suggest that the Cpx response may act to inhibit expression of *sipC* at pH 6.0 through inhibitory effects on *hilA* expression (Nakayama *et al.*, 2003). This finding is consistent with the inhibitory roles the Cpx response is known to play with regard to virulence determinant expression in other enteropathogens (see above and below).

In further support of an inhibitory role for the Cpx response in virulence factor regulation in *S. enterica*, Humphreys *et al.* (2004) reported that constitutive activation of the Cpx pathway affects the pathogenesis of *S. enterica* serovar Typhimurium *in vivo*, by reducing its ability to adhere to host cells, rather than through an effect on invasion of eukaryotic cells as was previously described (see above). In this study, there was a reduction in the ability of *cpxA* or *cpxA** mutants to infect mice compared with the wild-type. Thus, as in *E. coli*, the Cpx pathway in *S. enterica* serovar Typhimurium appears to negatively regulate virulence factors involved in both adhesion and invasion. Finally, like *E. coli*, the Cpx pathway also negatively regulates the σ^E regulon in *S. enterica* serovar Typhimurium, which could partially explain the reduced virulence of the *cpxA** mutant (Humphreys *et al.*, 2004).

Shigella sonnei: invasion and T3SS

The *ipaBCD* and *mxi* genes are virulence determinants for the invasive phenotype of *Shigella* spp. towards epithelial cells of the colonic mucosa, and they are encoded on a large plasmid (Ménard *et al.*, 1993). VirF is the master regulator of this large plasmid, controlling gene expression in a pH-dependent manner in *S. sonnei* (Watanabe *et al.*, 1993). Similar to HilA in *S. enterica*, in *S. sonnei*, the Cpx pathway regulates *virF* expression at a transcriptional level (Nakayama and Watanabe, 1995, 1998). When *cpxA* was mutated in an *E. coli* laboratory strain carrying an *S. sonnei virF*-lacZ reporter gene, there was a disruption in the normal, pH-regulated expression of *virF*, suggesting that CpxA may play a role in the regulation of the *virF* gene (Nakayama and Watanabe, 1995). Similarly, *an E. coli cpxR* disruption mutant showed a loss of the expression of *virF*, and evidence has been published suggesting that the response regulator CpxR binds directly to the *virF* promoter region (Nakayama and Watanabe, 1998). These data suggest that the Cpx response may play a positive role in virulence factor expression in *Shigella* spp. In a subsequent study, Mitobe and colleagues (2005) provided evidence which seems to suggest that while both the response regulator CpxR and the sensor kinase CpxA are required for the expression of the invasion genes *ipaBCD* and their transcription factor *invE*, it appears that CpxR acts at a transcriptional level while CpxA impacts post-transcriptional expression of InvE. The manner in which these events occur is unknown, but despite the apparent complicated nature of this regulation, these data nevertheless suggest that the Cpx response in *Shigella* spp. plays some role in the expression of virulence determinants.

Legionella pneumophila: intracellular growth and T4SS

L. pneumophila is an intracellular pathogen that causes Legionnaires' disease in humans (Rowbotham, 1980). *L. pneumophila* has a repertory of virulence genes (i.e. *icm/dot* genes) to invade and multiply inside of human macrophages, which are involved in the elaboration of a type IV secretion system (T4SS) (Segal and Shuman, 1998). A genetic analysis in *L. pneumophila* showed that the CpxR response regulator positively controlled the expression of the *icmR* gene and two *icm-dot* operons (i.e. *icmV-dotA* and *icmW-icmX*). All of these genes are involved in human macrophage killing and intracellular multiplication. When an *E. coli* strain was transformed with a plasmid carrying a *icmR::lacZ* translational fusion and an overexpression vector for the *cpxR* gene of *L. pneumophila*, there was an increase in the level of expression of the reporter (Gal-Mor and Segal, 2003). These data suggested that *icmR* expression is controlled by the CpxR homologue of *L. pneumophila*. Further analysis of the regulation of *icmR* by CpxR demonstrated that CpxR bound to the *icmR* regulatory region and this binding was important for *icmR* expression. Curiously, even though the predicted CpxR binding site upstream of *icmR* was virtually identical to the CpxR consensus binding site identified in *E. coli*, the *E. coli* CpxR homologue did not confer increased expression of an *icmR* reporter gene when overexpressed (Gal-Mor and Segal, 2003). This observation remains unexplained. In addition, it was found that CpxR plays a role in the regulation of *icmF*, *icmV*, and *icmW* in *L. pneumophila* (Gal-Mor and Segal, 2003). Despite these observations, the *L. pneumophila* CpxR homologue was shown not to be required for intracellular growth, perhaps indicating that basal levels of *icm* gene expression are sufficient for this process. Alternatively, other regulators may compensate for the effect of CpxR on *icm* gene expression in *cpxR* mutants.

Recently, a study by Altman and Segal (2008) indicated that the *L. pneumophila* CpxR regulator is also an important regulator of several additional components of the *icm/dot* T4SS as well as effectors and genes of unknown function in this pathogen. CpxR activated or repressed the expression of nine genes (i.e. *legA10*, *legA11*, *ceg7*, *ceg18*, *ceg33*, *cegC1*, *cegC2*, *cegC3*, and *cegC4*); five of these were previously described as encoding putative *icm/dot* translocated substrates (de Felipe *et al.*, 2005; Brüggemann *et al.*, 2006; Zusman *et al.*, 2007). This ability of CpxR to both positively and negatively impact gene expression has been previously described in other organisms (see above) (De Wulf *et al.*, 2002; Rönnebäumer *et al.*, 2009). Altman and Segal (2008) confirmed that the effectors SidM and CegC4 were indeed translocated to HL-60-derived human macrophages. The depletion of *cpxR* diminished the translocation of those substrates regardless of whether they were expressed from their own or an exogenous promoter, indicating a requirement of CpxR for maximal expression of both the translocation apparatus itself as well as the effector proteins. These results demonstrated that the Cpx pathway is involved in the regulation of pathogenicity in *L. pneumophila*, however it is still unclear what Cpx pathway inducing signals may be important during infection in order to regulate different translocated substrates.

Two more studies showed evidence that the Cpx pathway, specifically CpxR, is involved in the regulation of virulence factors in *L. pneumophila*. The first of these used a transposon mutagenesis approach and found that a *cpxR* mutant was impaired in recruitment of the host GTPase Rab1 to the *Legionella*-containing vacuole (LCV) during macrophage infection (Murata *et al.*, 2006), which is required for the intracellular replication of *L. pneumophila* (Merriam *et al.*, 1997; Rubin *et al.*, 1999). Finally, Vincent *et al.* (2006) found that mutations in *cpxR* suppressed the lethality of the Δ*dotL* phenotype in *L. pneumophila*. The *dotL* gene encodes a component of the T4SS and it is essential for the viability of *L. pneumophila* (Buscher *et al.*, 2005). Since mutations in other components of the Dot/Icm T4SS suppress Δ*dotL* lethality, it is thought that an aberrant, toxic T4S complex is formed in the absence of DotL (Buscher *et al.*, 2005). It was shown that mutation of *cpxR* alleviates Δ*dotL* toxicity in the same fashion, by diminishing expression of Icm/Dot T4S complex components (Vincent *et al.*, 2006). Cumulatively, studies of the CpxR homologue in *L. pneumophila* support an important role for the CpxAR homologues in regulating the expression of components and effectors of the Dot/Icm T4S machinery and the ability of this pathogen to infect and survive within macrophages.

Haemophilus ducreyi: phagocytosis inhibition and virulence factors

H. ducreyi is one of the causative agents of genital ulcer disease (GUD) which involves the evasion of the host immune system by this pathogen. The *H. ducreyi* genome encodes the LspB, LspA1 and LspA2 proteins which comprise a two partner secretion system required for inhibition of phagocytosis by immune cells (Vakevainen *et al.*, 2003). A genetic analysis revealed that the Cpx pathway negatively regulated the expression of the *lspB* and *lspA2* genes but not *lspA1*. In this study, the operon encoding *cpxR* and *cpxA* was found to be down-regulated under conditions where the Lsp two partner secretion system was expressed and CpxR was found to down-regulate the expression of the *lspB/lspA2* operon by directly binding to the promoter region. Interestingly, *cpxR* and *cpxA* are found in an operon with two additional genes (i.e. *HD1468* and *HD1471*) in *H. ducreyi* 35000HP (Labandeira-Rey *et al.*, 2009). Further characterization of the CpxRA regulon in *H. ducreyi* showed that the Cpx pathway in this bacterium may be primarily involved in controlling the expression of genes not involved in the cell envelope stress response (Labandeira-Rey *et al.*, 2010). Deletion of *cpxA* in *H. ducreyi* 35000HP, which led to constitutive induction of the Cpx response, resulted in a different cellular protein profile compared with the wild-type strain, in which a number of known virulence factors (i.e. LspB, LspA2, LspA1, DsrA and Flp1) were down-regulated (Labandeira-Rey *et al.*, 2010). Interestingly, it was also demonstrated that CpxR may have a possible effect on the regulation of the levels of LspA1 at a posttranscriptional or translational level, because there was a reduction in its expression but no changes in *lspA1* transcript levels were detected (Labandeira-Rey *et al.*, 2010). Finally, a DNA microarray analysis performed in the *cpxA* mutant where the Cpx response is activated showed that the most highly Cpx up-regulated genes in *H. ducreyi* 35000HP were found in a putative fimbrial operon (*fimABCD*) and the most down-regulated were the members of the *flp* operon and *dsrA*, which encode a putative pilus and an OM serum resistance protein, respectively. Further analysis demonstrated that, indeed, CpxR binds directly to the promoter regions of *lspB*, *ompA2*, *ompP2A*, *flp*, and *dsrA*, suggesting that these genes are part of the Cpx regulon in *H. ducreyi* (Labandeira-Rey *et al.*, 2010). These results suggest that the Cpx response differentially impacts the expression of a number of virulence determinants in *H. ducreyi*, and support the previously identified role of the Cpx envelope stress response in regulating adhesion. Intriguingly, the Cpx response was shown not to control the expression of envelope protein folding and degrading factors in *H. ducreyi* that are positively regulated in *E. coli*, perhaps indicating a divergence in the role of this response in this organism.

Consistent with the data of Labandeira-Rey *et al.* (2010) indicating that the Cpx response inhibits expression of a number of key virulence determinants in *H. ducryei*, Spinola and collaborators (2010) reported that when human volunteers were inoculated with either *H. ducreyi* *cpxA* mutant or wild-type strains, the *cpxA* mutant exhibited a reduction in the formation of papules and pustules at the site of inoculation, indicative of a virulence defect. This is most likely due to diminished expression of the Lsp two-partner secretion system involved in resistance to phagocytosis, together with DsrA, the major serum resistance OMP. Another study with human volunteers revealed that elimination of CpxR had no effect on initial infection, formation of pustules, the expression of DsrA, or resistance to serum killing (Labandeira-Rey *et al.*, 2011).

These data indicate that in *H. ducreyi*, as in other organisms like *V. cholerae*, and *S. enterica* serovar Typhimurium, Cpx pathway basal activity is very low or unimportant for pathogenesis (Slamti and Waldor, 2009). Conversely, induction of the Cpx response in *H. ducreyi*, as in other enteropathogens, has a negative impact on infection through the repression of envelope-localized virulence determinants. It remains to be determined at what stages of a pathogen's life cycle these events may be important and how or if this response is naturally induced.

Xenorhabdus nematophila: mutualistic host–microbe interactions

X. nematophila is an entomopathogenic Gram-negative bacterium that exhibits both mutualistic and pathogenic behaviour during its life cycle. Herbert Tran and collaborators (2007) reported that the Cpx pathway is necessary for both of these behaviours of *X. nematophila*. A *cpxR* deletion in *X. nematophila* caused a decrease in the killing of an insect host model as well as the ability to colonize the nematode symbiotic host, *Steinernema carpocapsae* (Herbert Tran *et al.*, 2007). Moreover, when *cpxR* mutants were compared with the parental strain they showed an altered colony and cell morphology. Most likely, CpxR contributes to the regulation of *X. nematophila* symbiosis by facilitating the derepression of the *nil* genes (i.e. *nilA*, *nilB*, and *nilC*), which are necessary for the mutualistic colonization of the nematode host (Heungens *et al.*, 2002; Cowles and Goodrich-Blair, 2004). Indeed, Herbert Tran and collaborators (2009) reported that the *cpxR* mutant had a defect in *S. carpocapsae* colonization due to insufficient levels of Nil colonization factors.

CpxR also influences virulence gene regulation in *X. nematophila*. A profile of secreted activities in *X. nematophila* indicated that CpxR negatively regulates haemolysin activity together with antibiotic and protease activities. Further, qRT-PCR analysis confirmed that, indeed, CpxR regulates the expression of genes (e.g. *xaxA* and *prtA*) that are involved in these activities (Herbert Tran *et al.*, 2007). In addition, CpxR had a positive effect on lipase activity in *X. nematophila* by activating expression of the positive regulator *lrhA* (Richards *et al.*, 2008). Moreover, as in *E. coli* (Vogt *et al.*, 2010), CpxR negatively regulated pilus biogenesis in *X. nematophila* because the transcript levels of *mrxA*, which encodes a pilin subunit with pore-forming toxin activity (Banerjee *et al.*, 2006), were significantly higher in a *cpxR* mutant background compared with the wild-type. These data indicate that the main role of the Cpx pathway in *X. nematophila* is regulating the traffic of cell surface structures and secreted products (e.g. pili, protease and haemolysin) (Herbert Tran *et al.*, 2007). Recently, it was shown that the Cpx pathway in *X. nematophila* plays a role in host–pathogen interaction by affecting the immune response upon infection. *Manduca sexta* insects infected with *cpxR* mutants exhibited 21-fold higher expression of the antimicrobial peptide cecropin, indicating that the Cpx response is somehow involved in either immune suppression or induction, perhaps owing to changes in the cell envelope (Herbert Tran and Goodrich-Blair, 2009).

Cpx-regulated antibiotic resistance phenotypes

Several recent studies have implicated the Cpx response in antibiotic resistance. van Stelten *et al.* (2009) demonstrated that, through its up-regulation of the proteolytic regulatory factor YccA, the Cpx response could potentially affect resistance to antibiotics that inhibit translation elongation. YccA inhibits the degradation of jammed Sec translocator complexes in the IM by the protease FtsH (van Stelten *et al.*, 2009). Antibiotics that prevent translation elongation are expected to produce aberrant secreted proteins that jam Sec translocator complexes, leading to their degradation. This phenomenon may contribute to the bactericidal effect of these antibiotics. Induction of the Cpx response would thus be expected to result in a degree of antibiotic resistance, by preventing the degradation of the jammed Sec translocators (van Stelten *et al.*, 2009).

Another recent study aimed at identifying chromosomal loci in *S. enterica* serovar Typhimurium contributing to cationic antimicrobial peptide (CAMP) resistance identified a putative role for the Cpx envelope stress response in this process. Weatherspoon-Griffin *et al.* (2011) showed that *nlpE* overexpression, an inducing cue of the Cpx pathway, activated the transcription of the *amiA* and *amiC* genes, which are involved in resistance to some antimicrobial peptides, including protamine and the α-helical peptides magainin 2 and melittin. AmiA and AmiC are *N*-acetylmuramyl-L-alanine amidases involved in the cleavage of peptidoglycan during cell division (Heidrich *et al.*, 2001). Weatherspoon-Griffin *et al.* (2011) confirmed that the response regulator CpxR bound to the *amiA* and *amiC* promoter regions and that antimicrobial resistance stimulated by *nlpE* overexpression was CpxR-dependent. Therefore, the Cpx pathway in *S. enterica* serovar Typhimurium may contribute to the resistance to protamine and α-helical CAMPs, partly through an unknown mechanism that involves modulation of the integrity of the peptidoglycan layer by AmiA and AmiC (Weatherspoon-Griffin *et al.*, 2011). Intriguingly, a recent microarray analysis of the Cpx regulon in *E. coli* showed that when the Cpx pathway was induced by NlpE overexpression, it up-regulated some genes (i.e. *ycbB*, *ygaU*, *slt and dacC*) involved in cell wall metabolism (see 'Physiological role of the Cpx response'; Price and Raivio, unpublished). It will be interesting to see if these *E. coli* Cpx regulon members (or others) are also involved in resistance to antimicrobial peptides.

Additionally, the Cpx response appears to be involved in antimicrobial peptide resistance in *H. ducreyi*. Rinker and colleagues (2011) demonstrated that an *H. ducreyi cpxA* mutant (in which the Cpx response is induced, see above) exhibited diminished resistance to the cationic antimicrobial peptide LL-37, suggesting that, in *H. ducreyi*, contrary to other microbes, the Cpx response is responsible for inducing some mechanism that leads to antimicrobial peptide sensitivity. The mechanism by which this happens is known to be independent of the proton motive force but has otherwise not been defined (Rinker *et al.*, 2011).

Finally, the Cpx response has also been demonstrated to regulate the expression of multidrug efflux pumps involved in antibiotic resistance. NlpE overexpression was demonstrated to affect multidrug (i.e. oxacillin, cloxacillin, nafcillin, cefamandole, aztreonam, carbenicillin, sulbenicillin, carumonam, kanamycin, novobiocin, and deoxycholate) resistance in *E. coli* by inducing the expression of drug efflux pump systems (i.e. *acrD* and *mdtABC*) in a Cpx-dependent manner (Nishino *et al.*, 2010). Cumulatively, these studies indicate that Cpx-induced changes in the bacterial envelope impact antibiotic resistance. Some of these changes seem to involve undefined mechanisms dependent on cell wall modification enzymes, while others depend on changes in the expression of multidrug efflux pumps or unknown factors. It remains to be determined whether the Cpx response responds directly to the presence of antibiotics or whether these changes in resistance are secondary effects of changes in envelope physiology in response to protein folding, or other envelope stresses.

Summary

In summary, the Cpx pathway is involved in the regulation of a number of downstream targets that directly or indirectly impact pathogenesis and antibiotic resistance (Table 12.1). This regulation can be either positive or negative depending on the virulence factor or gene target and the organisms evaluated. The Cpx pathway regulates the expression of a number of virulence attributes including secretion systems (e.g. T3SS and T4SS in EPEC, *Y. pseudotuberculosis* and *L. pneumophila*), adherence (BFP and invasin in EPEC and *Y. pseudotuberculosis* respectively), invasion factors (HilA and InvE in *S. enterica* serotype Typhimurium), transcriptional virulence activators (VirF in *S. sonnei*), serum resistance (DsrA in *H. ducreyi*), pore-forming toxin (MrxA in *X. nematophila*) and immune response modulators (LspB/LspA2 in *H. ducreyi* and *X. nematophila*). The effects on pathogenesis and antibiotic resistance are mostly seen when the Cpx pathway is strongly activated. Most of the Cpx-regulated virulence determinants are localized in the cell envelope, and in the majority of cases induction of the Cpx response has a detrimental effect on virulence. These observations suggest that the aim of the down-regulation of virulence determinants in the cell envelope by the Cpx pathway could be to diminish protein traffic in this cellular compartment in order to limit damage in the presence of various potentially toxic envelope stress events (MacRitchie *et al.*, 2008a). This model makes the direct prediction that the Cpx response will play a more important role in environmental locales or at sites in the body where virulence factor expression is not needed or would be detrimental. Future studies will illuminate where in the normal life cycle of a given bacterium the Cpx response is induced and what the inducing signals may be. Finally, it will be of interest to determine if the impact of the Cpx response on virulence determinant expression in various pathogens can be exploited for the development of new therapeutics or methods to control the pathogenesis of different bacterial pathogens.

Conclusions and outlook

Initial reports on the CpxAR two-component system, some 25 years ago, led to the proposal that these regulators were involved in adaptation to envelope stress. A large body of work since then has confirmed this hypothesis and increased our understanding of how envelope stress signals are sensed, the mechanisms by which adaptation is conferred, and the impact that this adaptation has on microbial physiology, including the ability to cause disease. Important contributions have highlighted the mechanisms by which CpxA is induced, through the relief of inhibition, and the multiple aspects of cellular physiology that are altered to effect adaptation. The expansion of this field to study the Cpx envelope stress in a variety of microbes indicates that a number of horizontally acquired virulence determinants are differentially affected by CpxA and CpxR. This suggests that the activities of this two-component system have been co-opted differently in different bacteria to integrate horizontally acquired genetic elements into the cellular regulatory network. Despite our gains in understanding of the CpxAR two-component system, there are still considerable gaps in our knowledge. We have yet to elucidate the molecular nature of the inducing signal, and it is uncertain how all of the Cpx-regulated genes are involved in instilling adaptation to envelope stress. Although it has been demonstrated that the Cpx response impacts virulence determinant expression in a variety of pathogens, we still do not know where, during a microbe's life, the CpxAR two-component system is active and most important for survival. The answers to these questions and others will undoubtedly be facilitated by the application of new technologies to the study of the Cpx response in an increasing number of microbes.

References

Ades, S.E. (2008). Regulation by destruction: design of the σ^E envelope stress response. Curr. Opin. Microbiol. *11*, 535–540.

Akiyama, Y., and Ito, K. (2000). Roles of multimerization and membrane association in the proteolytic functions of FtsH (HflB). EMBO J. *19*, 3888–3895.

Akiyama, Y., Kihara, A., and Ito, K. (1996). Subunit *a* of proton ATPase F_0 sector is a substrate of the FtsH protease in *Escherichia coli*. FEBS Lett. *399*, 26–28.

Albin, R., Weber, R., and Silverman, P.M. (1986). The Cpx proteins of *Escherichia coli* K12. Immunologic detection of the chromosomal *cpxA* gene product. J. Biol. Chem. *261*, 4698–4705.

Altman, E., and Segal, G. (2008). The response regulator CpxR directly regulates expression of several *Legionella pneumophila icm/dot* components as well as new translocated substrates. J. Bacteriol. *190*, 1985–1996.

Andrews, S.C., Robinson, A.K., and Rodríguez-Quiñones, F. (2003) Bacterial iron homeostasis. FEMS Microbiol. Rev. *27*, 215–237.

Appleman, J.A., Chen, L.L., and Stewart, V. (2003). Probing conservation of HAMP linker structure and signal transduction mechanism through analysis of hybrid sensor kinases. J. Bacteriol. *185*, 4872–4882.

Ayers, R.A., and Moffat, K. (2008). Changes in quaternary structure in the signaling mechanisms of PAS domains. Biochemistry *47*, 12078–12086.

Bajaj, V., Hwang, C., and Lee, C.A. (1995), *hilA* is a novel *ompR/toxR* family member that activates the expression of *Salmonella typhimurium* invasion genes. Mol. Microbiol. *18*, 715–727.

Banerjee, J., Singh, J., Joshi, M.C., Ghosh, S., and Banerjee, N. (2006). The cytotoxic fimbrial structural subunit of *Xenorhabdus nematophila* is a pore-forming toxin. J. Bacteriol. *188*, 7957–7962.

Baquero, M.R., Bouzon, M., Quintela, J.C., Ayala, J.A., and Moreno, F. (1996). *dacD*, an *Escherichia coli* gene encoding a novel penicillin-binding protein (PBP6b) with DD-carboxypeptidase activity. J. Bacteriol. *178*, 7106–7111.,

Baranova, N., and Nikaido, H. (2002). The *baeSR* two-component regulatory system activates transcription of the *yegMNOB* (*mdtABCD*) transporter gene cluster in *Escherichia coli* and increases its resistance to novobiocin and deoxycholate. J. Bacteriol. *184*, 4168–4176.

Barnhart, M.M., and Chapman, M.R. (2006). Curli biogenesis and function. Annu. Rev. Microbiol. *60*, 131–147.

Batchelor, E., Walthers, D., Kenney, L.J., and Goulian, M. (2005). The *Escherichia coli* CpxA-CpxR envelope stress response system regulates expression of the porins OmpF and OmpC. J. Bacteriol. *187*, 5723–5731.

Belin, P., and Boquet, P.L. (1994). The *Escherichia coli dsbA* gene is partly transcribed from the promoter of a weakly expressed upstream gene. Microbiology *140*, 3337–3348.

Beloin, C., Valle, J., Latour-Lambert, P., Faure, P., Kzreminski, M., Balestrino, D., Haagensen, J.A., Moli, S., Prensier, G., Arbeille, B., *et al.* (2004). Global impact of mature biofilm lifestyle on *Escherichia coli* K-12 gene expression. Mol. Microbiol. *51*, 659–674.

Boehm, A., Steiner, S., Zaehringer, F., Casanova, A., Hamburger, F., Ritz, D., Keck, W., Ackermann, M., Schirmer, T., and Jenal, U. (2009). Second messenger signalling governs *Escherichia coli* biofilm induction upon ribosomal stress. Mol. Microbiol. *72*, 1500–1516.

Bogdanov, M., and Dowhan, W. (1998). Phospholipid-assisted protein folding: phosphatidylethanolamine is required at a late step of the conformational maturation of the polytopic membrane protein lactose permease. EMBO J. *17*, 5255–5264.

Bogdanov, M., Sun, J., Kaback, H.R., and Dowhan, W. (1996). A phospholipid acts as a chaperone in assembly of a membrane transport protein. J. Biol. Chem. *271*, 11615–11618.

Bogdanov, M., Umeda, M., and Dowhan, W. (1999). Phospholipid-assisted refolding of an integral membrane protein. Minimum structural features for phosphatidylethanolamine to act as a molecular chaperone. J. Biol. Chem. *274*, 12339–12345.

Bos, M.P., Robert, V., and Tommassen, J. (2007). Biogenesis of the Gram-negative bacterial outer membrane. Annu. Rev. Microbiol. *61*, 191–214.

Brüggemann, H., Cazalet, C., and Buchrieser, C. (2006). Adaptation of *Legionella pneumophila* to the host environment: role of protein secretion, effectors and eukaryotic-like proteins. Curr. Opin. Microbiol. *9*, 86–94.

Buelow, D.R., and Raivio, T.L. (2005). Cpx signal transduction is influenced by a conserved N-terminal domain in the novel inhibitor CpxP and the periplasmic protease DegP. J. Bacteriol. *187*, 6622–6630.

Buelow, D.R., and Raivio, T.L. (2010). Three (and more) component regulatory systems – auxiliary regulators of bacterial histidine kinases. Mol. Microbiol. *75*, 547–566.

Bury-Moné, S., Nomane, Y., Reymond, N., Barbet, R., Jacquet, E., Imbeaud, S., Jacq, A., and Bouloc, P. (2009). Global analysis of extracytoplasmic stress signaling in *Escherichia coli*. PLoS Genet. *5*, e1000651.

Buscher, B.A., Conover, G.M., Miller, J.L., Vogel, S.A., Meyers, S.N., Isberg, R.R., and Vogel, J.P. (2005). The DotL protein, a member of the TraG-coupling protein family, is essential for viability of *Legionella pneumophila* strain Lp02. J. Bacteriol. *187*, 2927–2938.

Cao, J., Woodhall, M.R., Alvarez, J., Cartron, M.L., and Andrews, S.C. (2007). EfeUOB (YcdNOB) is a tripartite, acid-induced and CpxAR-regulated, low-pH Fe^{2+} transporter that is cryptic in *Escherichia coli* K-12 but functional in *E. coli* O157:H7. Mol. Microbiol. *65*, 857–875.

Carlsson, K.E., Liu, J., Edqvist, P.J., and Francis, M.S. (2007a). Extracytoplasmic-stress-responsive pathways modulate type III secretion in *Yersinia pseudotuberculosis*. Infect. Immun. *75*, 3913–3924.

Carlsson, K.E., Liu, J., Edqvist, P.J., and Francis, M.S. (2007b). Influence of the Cpx extracytoplasmic-stress-responsive pathway on *Yersinia* sp.-eukaryotic cell contact. Infect. Immun. *75*, 4386–4399.

Carniel, E. (2002). Plasmids and pathogenicity islands of *Yersinia*. Curr. Top. Microbiol. Immunol. *264*, 89–108.

Chen, Q., and Amster-Choder, O. (1998). The different functions of BglF, the *E. coli* β-glucoside permease and sensor of the *bgl* system, have different structural requirements. Biochemistry *37*, 17040–17047.

Cheung, J., Bingman, C.A., Reyngold, M., Hendrickson, W.A., and Waldburger, C.D. (2008). Crystal structure of a functional dimer of the PhoQ sensor domain. J. Biol. Chem. *283*, 13762–13770.

Collet, J.F., and Bardwell, J.C. (2002). Oxidative protein folding in bacteria. Mol. Microbiol. *44*, 1–8.

Condemine, G., Berrier, C., Plumbridge, J., and Ghazi, A. (2005). Function and expression of an *N*-acetylneuraminic acid-inducible outer membrane channel in *Escherichia coli*. J. Bacteriol. *187*, 1959–1965.

Cornelis, G.R., Boland, A., Boyd, A.P., Geuijen, C., Iriarte, M., Neyt, C., Sory, M.P., and Stainier, I. (1998). The virulence plasmid of *Yersinia*, an antihost genome. Microbiol. Mol. Biol. Rev. *62*, 1315–1352.

Cosma, C.L., Danese, P.N., Carlson, J.H., Silhavy, T.J., and Snyder, W.B. (1995). Mutational activation of the Cpx signal transduction pathway of *Escherichia coli* suppresses the toxicity conferred by certain envelope-associated stresses. Mol. Microbiol. *18*, 491–505.

Cowles, C.E., and Goodrich-Blair, H. (2004). Characterization of a lipoprotein, NilC, required by *Xenorhabdus nematophila* for mutualism with its nematode host. Mol. Microbiol. *54*, 464–477.

Cox, J.S., and Walter, P. (1996). A novel mechanism for regulating activity of a transcription factor that controls the unfolded protein response. Cell *87*, 391–404.

Dailey, F.E., and Berg, H.C. (1993). Mutants in disulfide bond formation that disrupt flagellar assembly in *Escherichia coli*. Proc. Natl. Acad. Sci. U.S.A. *90*, 1043–1047.

Danese, P.N., and Silhavy, T.J. (1997). The σ^E and the Cpx signal transduction systems control the synthesis of periplasmic protein-folding enzymes in *Escherichia coli*. Genes Dev. *11*, 1183–1193.

Danese, P.N., and Silhavy, T.J. (1998). CpxP, a stress-combative member of the Cpx regulon. J. Bacteriol. *180*, 831–839.

Danese, P.N., Snyder, W.B., Cosma, C.L., Davis, L.J., and Silhavy, T.J. (1995). The Cpx two-component signal transduction pathway of *Escherichia coli* regulates transcription of the gene specifying the stress-inducible periplasmic protease, DegP. Genes Dev. *9*, 387–398.

Danese, P.N., Oliver, G.R., Barr, K., Bowman, G.D., Rick, P.D., and Silhavy, T.J. (1998). Accumulation of the enterobacterial common antigen lipid II biosynthetic intermediate stimulates *degP* transcription in *Escherichia coli*. J. Bacteriol. *180*, 5875–5884.

De Wulf, P., Kwon, O., and Lin, E.C. (1999). The CpxRA signal transduction system of *Escherichia coli*: growth-related autoactivation and control of unanticipated target operons. J. Bacteriol. *181*, 6772–6778.

De Wulf, P., McGuire, A.M., Liu, X., and Lin, E.C. (2002). Genome-wide profiling of promoter recognition by the two-component response regulator CpxR-P in *Escherichia coli*. J. Biol. Chem, *277*, 26652–26661.

DiGiuseppe, P.A., and Silhavy, T.J. (2003). Signal detection and target gene induction by the CpxRA two-component system. J. Bacteriol. *185*, 2432–2440.

Donnenberg, M.S., Girón, J.A., Nataro, J.P., and Kaper, J.B. (1992). A plasmid-encoded type IV fimbrial gene of enteropathogenic *Escherichia coli* associated with localized adherence. Mol. Microbiol. *6*, 3427–3437.

Dorel, C., Vidal, O., Prigent-Combaret, C., Vallet, I., and Lejeune, P. (1999). Involvement of the Cpx signal transduction pathway of *E. coli* in biofilm formation. FEMS Microbiol. Lett. *178*, 169–175.

Dorel, C., Lejeune, P., and Rodrigue, A. (2006). The Cpx system of *Escherichia coli*, a strategic signaling pathway for confronting adverse conditions and for settling biofilm communities? Res. Microbiol. *157*, 306–314.

Edwards-Jones, B., Langford, P.R., Kroll, J.S., and Yu, J. (2004). The role of the *Shigella flexneri yihE* gene in LPS synthesis and virulence. Microbiology *150*, 1079–1084.

Ehlert, K., Höltje, J.V., and Templin, M.F. (1995). Cloning and expression of a murein hydrolase lipoprotein from *Escherichia coli*. Mol. Microbiol. *16*, 761–768.

Elliott, S.J., Wainwright, L.A., McDaniel, T.K., Jarvis, K.G., Deng, Y.K., Lai, L.C., McNamara, B.P., Donnenberg, M.S., and Kaper, J.B. (1998). The complete sequence of the locus of enterocyte effacement (LEE) from enteropathogenic *Escherichia coli* E2348/69. Mol. Microbiol. *28*, 1–4.

de Felipe, K.S., Pampou, S., Jovanovic, O.S., Pericone, C.D., Ye, S.F., Kalachikov, S., and Shuman, H.A. (2005). Evidence for acquisition of *Legionella* type IV secretion substrates via interdomain horizontal gene transfer. J. Bacteriol. *187*, 7716–7726.

Fleischer, R., Heermann, R., Jung, K., and Hunke, S. (2007). Purification, reconstitution, and characterization of the CpxRAP envelope stress system of *Escherichia coli*. J. Biol. Chem. *282*, 8583–8593.

Friedland, N., Mack, T.R., Yu, M., Hung, L.W., Terwilliger, T.C., Waldo, G.S., and Stock, A.M. (2007). Domain orientation in the inactive response regulator *Mycobacterium tuberculosis* MtrA provides a barrier to activation. Biochemistry *46*, 6733–6743.

Gal-Mor, O., and Segal, G. (2003). Identification of CpxR as a positive regulator of *icm* and *dot* virulence genes of *Legionella pneumophila*. J. Bacteriol. *185*, 4908–4919.

Gerken, H., and Misra, R. (2010). MzrA–EnvZ interactions in the periplasm influence the EnvZ/OmpR two-component regulon. J. Bacteriol. *192*, 6271–6278.

Gerken, H., Charlson, E.S., Cicirelli, E.M., Kenney, L.J., and Misra, R. (2009). MzrA: a novel modulator of the EnvZ/OmpR two-component regulon. Mol. Microbiol. *72*, 1408–1422.

Gibson, J.L., Lombardo, M.J., Thornton, P.C., Hu, K.H., Galhardo, R.S., Beadle, B., Habib, A., Magner, D.B., Frost, L.S., Herman, C., *et al.* (2010). The σ^E stress response is required for stress-induced mutation and amplification in *Escherichia coli*. Mol. Microbiol. *77*, 415–430.

Girón, J.A., Ho, A.S., and Schoolnik, G.K. (1991). An inducible bundle-forming pilus of enteropathogenic *Escherichia coli*. Science *254*, 710–713.

Gubbins, M.J., Lau, I., Will, W.R., Manchak, J.M., Raivio, T.L., and Frost, L.S. (2002). The positive regulator, TraJ, of the *Escherichia coli* F plasmid is unstable in a *cpxA** background. J. Bacteriol. *184*, 5781–5788.

Guillier, M., and Gottesman, S. (2006). Remodelling of the *Escherichia coli* outer membrane by two small regulatory RNAs. Mol. Microbiol. *59*, 231–247.

Hagenmaier, S., Stierhof, Y.D., and Henning, U. (1997). A new periplasmic protein of *Escherichia coli* which is synthesized in spheroplasts but not in intact cells. J. Bacteriol. *179*, 2073–2076.

Heidrich, C., Templin, M.F., Ursinus, A., Merdanovic, M., Berger, J., Schwarz, H., de Pedro, M.A., and Höltje, J.V. (2001). Involvement of *N*-acetylmuramyl-L-alanine amidases in cell separation and antibiotic-induced autolysis of *Escherichia coli*. Mol. Microbiol. *41*, 167–178.

Herbert, E.E., Cowles, K.N., and Goodrich-Blair, H. (2007). CpxRA regulates mutualism and pathogenesis in *Xenorhabdus nematophila*. Appl. Environ. Microbiol. *73*, 7826–7836.

Herbert Tran, E.E., and Goodrich-Blair, H. (2009). CpxRA contributes to *Xenorhabdus nematophila* virulence through regulation of *lrhA* and modulation of insect immunity. Appl. Environ. Microbiol. *75*, 3998–4006.

Herbert Tran, E.E., Andersen, A.W., and Goodrich-Blair, H. (2009). CpxRA influences *Xenorhabdus nematophila* colonization initiation and outgrowth in *Steinernema carpocapsae* nematodes through regulation of the *nil* locus. Appl. Environ. Microbiol. *75*, 4007–4014.

Hernday, A.D., Braaten, B.A., Broitman-Maduro, G., Engelberts, P., and Low, D.A. (2004). Regulation of the Pap epigenetic switch by CpxAR: phosphorylated CpxR inhibits transition to the phase ON state by competition with Lrp. Mol. Cell *16*, 537–547.

Heungens, K., Cowles, C.E., and Goodrich-Blair, H. (2002). Identification of *Xenorhabdus nematophila* genes required for mutualistic colonization of *Steinernema carpocapsae* nematodes. Mol. Microbiol. *45*, 1337–1353.

Heusipp, G., Nelson, K.M., Schmidt, M.A., and Miller, V.L. (2004). Regulation of *htrA* expression in *Yersinia enterocolitica*. FEMS Microbiol. Lett. *231*, 227–235.

Hirakawa, H., Nishino, K., Hirata, T., and Yamaguchi, A. (2003). Comprehensive studies of drug resistance mediated by overexpression of response regulators of two-component signal transduction systems in *Escherichia coli*. J. Bacteriol. *185*, 1851–1856.

Hirakawa, H., Inazumi, Y., Masaki, T., Hirata, T., and Yamaguchi, A. (2005). Indole induces the expression of multidrug exporter genes in *Escherichia coli*. Mol. Microbiol. *55*, 1113–1126.

Hirano, Y., Hossain, M.M., Takeda, K., Tokuda, H., and Miki, K. (2007). Structural studies of the Cpx pathway activator NlpE on the outer membrane of *Escherichia coli*. Structure *15*, 963–976.

Huggins, C., Tapley, D.F., and Jensen, E.V. (1951). Sulphydryl-disulphide relationships in the induction of gels by urea. Nature *167*, 592–593.

Humphreys, S., Rowley, G., Stevenson, A., Anjum, M.F., Woodward, M.J., Gilbert, S., Kormanec, J., and Roberts, M. (2004). Role of the two-component regulator CpxAR in the virulence of *Salmonella enterica* serotype Typhimurium. Infect. Immun. *72*, 4654–4661.

Humphries, R.M., Griener, T.P., Vogt, S.L., Mulvey, G.L., Raivio, T., Donnenberg, M.S., Kitov, P.I., Surette, M., and Armstrong, G.D. (2010). *N*-acetyllactosamine-induced retraction of bundle-forming pili regulates virulence-associated gene expression in enteropathogenic *Escherichia coli*. Mol. Microbiol. *76*, 1111–1126.

Hung, D.L., Raivio, T.L., Jones, C.H., Silhavy, T.J., and Hultgren, S.J. (2001). Cpx signaling pathway monitors biogenesis and affects assembly and expression of P pili. EMBO J. *20*, 1508–1518.

Isaac, D.D., Pinkner, J.S., Hultgren, S.J., and Silhavy, T.J. (2005). The extracytoplasmic adaptor protein CpxP is degraded with substrate by DegP. Proc. Natl. Acad. Sci. U.S.A. *102*, 17775–17779.

Ize, B., Stanley, N.R., Buchanan, G., and Palmer, T. (2003). Role of the *Escherichia coli* Tat pathway in outer membrane integrity. Mol. Microbiol. *48*, 1183–1193.

Jacob-Dubuisson, F., Pinkner, J., Xu, Z., Striker, R., Padmanhaban, A., and Hultgren, S.J. (1994). PapD chaperone function in pilus biogenesis depends on oxidant and chaperone-like activities of DsbA. Proc. Natl. Acad. Sci. U.S.A. *91*, 11552–11556.

Jonas, K., Edwards, A.N., Simm, R., Romeo, T., Römling, U., and Melefors, O. (2008). The RNA binding protein CsrA controls cyclic di-GMP metabolism by directly regulating the expression of GGDEF proteins. Mol. Microbiol. 70, 236–257.

Jones, C.H., Danese, P.N., Pinkner, J.S., Silhavy, T.J., and Hultgren, S.J. (1997). The chaperone-assisted membrane release and folding pathway is sensed by two signal transduction systems. EMBO J. *16*, 6394–6406.

Jones, C.H., Dexter, P., Evans, A.K., Liu, C., Hultgren, S.J., and Hruby, D.E. (2002). *Escherichia coli* DegP protease cleaves between paired hydrophobic residues in a natural substrate: the PapA pilin. J. Bacteriol. *184*, 5762–5771.

Jubelin, G., Vianney, A., Beloin, C., Ghigo, J.M., Lazzaroni, J.C., Lejeune, P., and Dorel, C. (2005). CpxR/OmpR interplay regulates curli gene expression in response to osmolarity in *Escherichia coli*. J. Bacteriol. *187*, 2038–2049.

Justice, S.S., Hunstad, D.A., Harper, J.R., Duguay, A.R., Pinkner, J.S., Bann, J., Frieden, C., Silhavy, T.J., and Hultgren, S.J. (2005). Periplasmic peptidyl prolyl *cis-trans* isomerases are not essential for viability, but SurA is required for pilus biogenesis in *Escherichia coli*. J. Bacteriol. *187*, 7680–7686.

Kadokura, H., Katzen, F., and Beckwith, J. (2003). Protein disulfide bond formation in prokaryotes. Annu. Rev. Biochem. *72*, 111–135.

Keller, R.F., and Hunke, S. (2009). Misfolded maltose binding protein MalE219 induces the CpxRA envelope stress response by stimulating phosphoryl transfer from CpxA to CpxR. Res. Microbiol. *160*, 396–400.

Kihara, A., Akiyama, Y., and Ito, K. (1995). FtsH is required for proteolytic elimination of uncomplexed forms of SecY, an essential protein translocase subunit. Proc. Natl. Acad. Sci. U.S.A. *92*, 4532–4536.

Kihara, A., Akiyama, Y., and Ito, K. (1998). Different pathways for protein degradation by the FtsH/HflKC membrane-embedded protease complex: an implication from the interference by a mutant form of a new substrate protein, YccA. J. Mol. Biol. *279*, 175–188.

Kleerebezem, M., Heutink, M., and Tommassen, J. (1995). Characterization of an *Escherichia coli rotA* mutant, affected in periplasmic peptidyl-prolyl *cis/trans* isomerase. Mol. Microbiol. *18*, 313–320.

Kornitzer, D., Teff, D., Altuvia, S., and Oppenheim, A.B. (1991). Isolation, characterization, and sequence of an *Escherichia coli* heat shock gene, *htpX*. J. Bacteriol. *173*, 2944–2953.

Kosek, M., Bern, C., and Guerrant, R.L. (2003). The global burden of diarrhoeal disease, as estimated from studies published between 1992 and 2000. Bull. World Health Organ. *81*, 197–204.

Kuehn, M.J., Heuser, J., Normark, S., and Hultgren, S.J. (1992). P pili in uropathogenic *E. coli* are composite fibres with distinct fibrillar adhesive tips. Nature *356*, 252–255.

Labandeira-Rey, M., Mock, J.R., and Hansen, E.J. (2009). Regulation of expression of the *Haemophilus ducreyi* LspB and LspA2 proteins by CpxR. Infect. Immun. *77*, 3402–3411.

Labandeira-Rey, M., Brautigam, C.A., and Hansen, E.J. (2010). Characterization of the CpxRA regulon in *Haemophilus ducreyi*. Infect. Immun. *78*, 4779–4791.

Labandeira-Rey, M., Dodd, D., Fortney, K.R., Zwickl, B., Katz, B.P., Janowicz, D.M., Spinola, S.M., and Hansen, E.J. (2011). A *Haemophilus ducreyi* CpxR deletion mutant is virulent in human volunteers. J. Infect. Dis. *203*, 1859–1865.

Langen, G.R., Harper, J.R., Silhavy, T.J., and Howard, S.P. (2001). Absence of the outer membrane phospholipase A suppresses the temperature-sensitive phenotype of *Escherichia coli degP* mutants and induces the Cpx and σ^E extracytoplasmic stress responses. J. Bacteriol. *183*, 5230–5238.

Lau-Wong, I.C., Locke, T., Ellison, M.J., Raivio, T.L., and Frost, L.S. (2008). Activation of the Cpx regulon destabilizes the F plasmid transfer activator, TraJ, via the HslVU protease in *Escherichia coli*. Mol. Microbiol. *67*, 516–527.

Leblanc, S.K., Oates, C.W., and Raivio, T.L. (2011). Characterization of the induction and cellular role of the BaeSR two-component envelope stress response of *Escherichia coli*. J. Bacteriol. *193*, 3367–3375.

Leclerc, G.J., Tartera, C., and Metcalf, E.S. (1998). Environmental regulation of *Salmonella typhi* invasion-defective mutants. Infect. Immun. *66*, 682–691.

Lee, J., Tomchick, D.R., Brautigam, C.A., Machius, M., Kort, R., Hellingwerf, K.J., and Gardner, K.H. (2008), Changes at the KinA PAS-A dimerization interface influence histidine kinase function. Biochemistry *47*, 4051–4064.

Lee, Y.M., DiGiuseppe, P.A., Silhavy, T.J., and Hultgren, S.J. (2004) P pilus assembly motif necessary for activation of the CpxRA pathway by PapE in *Escherichia coli*. J. Bacteriol. *186*, 4326–4337.

Li, M.S., Kroll, J.S., and Yu, J. (2001). Influence of the *yihE* gene of *Shigella flexneri* on global gene expression: on analysis using DNA arrays. Biochem. Biophys. Res. Commun. *288*, 91–100.

Li, S.R., Dorrell, N., Everest, P.H., Dougan, G., and Wren, B.W. (1996). Construction and characterization of a *Yersinia enterocolitica* O:8 high-temperature requirement (*htrA*) isogenic mutant. Infect. Immun. *64*, 2088–2094.

Lipinska, B., Fayet, O., Baird, L., and Georgopoulos, C. (1989). Identification, characterization, and mapping of the *Escherichia coli htrA* gene, whose product is essential for bacterial growth only at elevated temperatures. J. Bacteriol. *171*, 1574–1584.

Liu, J., Obi, I.R., Thanikkal, E.J., Kieselbach, T., and Francis, M.S. (2011). Phosphorylated CpxR restricts production of the RovA global regulator in *Yersinia pseudotuberculosis*. PLoS One *6*, e23314.

Ma, Q., and Wood, T.K. (2009). OmpA influences *Escherichia coli* biofilm formation by repressing cellulose production through the CpxRA two-component system. Environ. Microbiol. *11*, 2735–2746.

McEwen, J., and Silverman, P. (1980). Chromosomal mutations of *Escherichia coli* that alter expression of conjugative plasmid functions. Proc. Natl. Acad. Sci. U.S.A. *77*, 513–517.

MacRitchie, D.M., Buelow, D.R., Price, N.L., and Raivio, T.L. (2008a), Two-component signaling and Gram-negative envelope stress response systems. Adv. Exp. Med. Biol. *631*, 80–110.

MacRitchie, D.M., Ward, J.D., Nevesinjac, A.Z., and Raivio, T.L. (2008b), Activation of the Cpx envelope stress response down-regulates expression of several locus of enterocyte effacement-encoded genes in enteropathogenic *Escherichia coli*. Infect. Immun. *76*, 1465–1475.

Magnet, S., Bellais, S., Dubost, L., Fourgeaud, M., Mainardi, J.L., Petit-Frère, S., Marie, A., Mengin-Lecreulx, D., Arthur, M., and Gutmann, L. (2007). Identification of the L,D-transpeptidases responsible for attachment of the Braun lipoprotein to *Escherichia coli* peptidoglycan. J. Bacteriol. *189*, 3927–3931.

Magnet, S., Dubost, L., Marie, A., Arthur, M., and Gutmann, L. (2008). Identification of the L,D-transpeptidases for peptidoglycan cross-linking in *Escherichia coli*. J. Bacteriol. *190*, 4782–4785.

Makris, D.P., and Rossiter, J.T. (2000). Heat-induced, metal-catalyzed oxidative degradation of quercetin and rutin (Quercetin 3-O-rhamnosylglucoside) in aqueous model systems. J. Agric. Food Chem. *48*, 3830–3838.

Martínez-Hackert, E., and Stock, A.M. (1997). Structural relationships in the OmpR family of winged-helix transcription factors. J. Mol. Biol. *269*, 301–312.

Ménard, R., Sansonetti, P.J., and Parsot, C. (1993). Nonpolar mutagenesis of the *ipa* genes defines IpaB, IpaC, and IpaD as effectors of *Shigella flexneri* entry into epithelial cells. J. Bacteriol. *175*, 5899–5906.

Merriam, J.J., Mathur, R., Maxfield-Boumil, R., and Isberg, R.R. (1997). Analysis of the *Legionella pneumophila fliI* gene: intracellular growth of a defined mutant defective for flagellum biosynthesis. Infect. Immun. *65*, 2497–2501.

Mileykovskaya, E., and Dowhan, W. (1997). The Cpx two-component signal transduction pathway is activated in *Escherichia coli* mutant strains lacking phosphatidylethanolamine. J. Bacteriol. *179*, 1029–1034.

Mileykovskaya, E., and Dowhan, W. (2005). Role of membrane lipids in bacterial division-site selection. Curr. Opin. Microbiol. *8*, 135–142.

Mills, E., Pultz, I.S., Kulasekara, H.D., and Miller, S.I. (2011). The bacterial second messenger c-di-GMP: mechanisms of signalling. Cell. Microbiol. *13*, 1122–1129.

Mitobe, J., Arakawa, E., and Watanabe, H. (2005). A sensor of the two-component system CpxA affects expression of the type III secretion system through posttranscriptional processing of InvE. J. Bacteriol. 187, 107–113.

Miyadai, H., Tanaka-Masuda, K., Matsuyama, S., and Tokuda, H. (2004). Effects of lipoprotein overproduction on the induction of DegP (HtrA) involved in quality control in the *Escherichia coli* periplasm. J. Biol. Chem. *279*, 39807–39813.

Murata, T., Delprato, A., Ingmundson, A., Toomre, D.K., Lambright, D.G., and Roy, C.R. (2006). The *Legionella pneumophila* effector protein DrrA is a Rab1 guanine nucleotide-exchange factor. Nat. Cell Biol. *8*, 971–977.

Nagakubo, S., Nishino, K., Hirata, T., and Yamaguchi, A. (2002). The putative response regulator BaeR stimulates multidrug resistance of *Escherichia coli* via a novel multidrug exporter system, MdtABC. J. Bacteriol. *184*, 4161–4167.

Nakayama, S., and Watanabe, H. (1995). Involvement of *cpxA*, a sensor of a two-component regulatory system, in the pH-dependent regulation of expression of *Shigella sonnei virF* gene. J. Bacteriol. *177*, 5062–5069.

Nakayama, S., and Watanabe, H. (1998). Identification of *cpxR* as a positive regulator essential for expression of the *Shigella sonnei virF* gene. J. Bacteriol. *180*, 3522–3528.

Nakayama, S., Kushiro, A., Asahara, T., Tanaka, R., Hu, L., Kopecko, D.J., and Watanabe, H. (2003). Activation of *hilA* expression at low pH requires the signal sensor CpxA, but not the cognate response regulator CpxR, in *Salmonella enterica* serovar Typhimurium. Microbiology *149*, 2809–2817.

Nevesinjac, A.Z., and Raivio, T.L. (2005). The Cpx envelope stress response affects expression of the type IV bundle-forming pili of enteropathogenic *Escherichia coli*. J. Bacteriol. *187*, 672–686.

Nikaido, H. (2003). Molecular basis of bacterial outer membrane permeability revisited. Microbiol. Mol. Biol. Rev. *67*, 593–656.

Nikaido, H., and Rosenberg, E.Y. (1983). Porin channels in *Escherichia coli*: studies with liposomes reconstituted from purified proteins. J. Bacteriol. *153*, 241–252.

Nishino, K., Yamasaki, S., Hayashi-Nishino, M., and Yamaguchi, A. (2010). Effect of NlpE overproduction on multidrug resistance in *Escherichia coli*. Antimicrob. Agents Chemother. *54*, 2239–2243.

Niwa, M., Milner, K.C., Ribi, E., and Rudbach, J.A. (1969). Alteration of physical, chemical, and biological properties of endotoxin by treatment with mild alkali. J. Bacteriol. *97*, 1069.

Nummila, K., Kilpeläinen, I., Zähringer, U., Vaara, M., and Helander, I.M. (1995). Lipopolysaccharides of polymyxin B-resistant mutants of *Escherichia coli* are extensively substituted by 2-aminoethyl pyrophosphate and contain aminoarabinose in lipid A. Mol. Microbiol. *16*, 271–278.

Ogasawara, H., Teramoto, J., Hirao, K., Yamamoto, K., Ishihama, A., and Utsumi, R. (2004). Negative regulation of DNA repair gene (*ung*) expression by the CpxR/CpxA two-component system in *Escherichia coli* K-12 and induction of mutations by increased expression of CpxR. J. Bacteriol. *186*, 8317–8325.

Ogasawara, H., Yamada, K., Kori, A., Yamamoto, K., and Ishihama, A. (2010). Regulation of the *Escherichia coli csgD* Promoter: Interplay between Five Transcription Factors. Microbiology *156*, 2470–2483.

Oshima, T., Aiba, H., Masuda, Y., Kanaya, S., Sugiura, M., Wanner, B.L., Mori, H., and Mizuno, T. (2002). Transcriptome analysis of all two-component regulatory system mutants of *Escherichia coli* K-12. Mol. Microbiol. *46*, 281–291.

Otto, K., and Silhavy, T.J. (2002). Surface sensing and adhesion of *Escherichia coli* controlled by the Cpx-signaling pathway. Proc. Natl. Acad. Sci. U.S.A. *99*, 2287–2292.

Papanikou, E., Karamanou, S., and Economou, A. (2007). Bacterial protein secretion through the translocase nanomachine. Nat. Rev. Microbiol. *5*, 839–851.

Pincus, D., Chevalier, M.W., Aragón, T., van Anken, E., Vidal, S.E., El-Samad, H., and Walter, P. (2010). BiP binding to the ER-stress sensor Ire1 tunes the homeostatic behavior of the unfolded protein response. PLoS Biol. *8*, e1000415.

Pogliano, J., Lynch, A.S., Belin, D., Lin, E.C., and Beckwith, J. (1997). Regulation of *Escherichia coli* cell envelope proteins involved in protein folding and degradation by the Cpx two-component system. Genes Dev. *11*, 1169–1182.

Pratt, L.A., and Silhavy, T.J. (1994). OmpR mutants specifically defective for transcriptional activation. J. Mol. Biol. *243*, 579–594.

Pratt, L.A., Hsing, W., Gibson, K.E., and Silhavy, T.J. (1996). From acids to *osmZ*: multiple factors influence synthesis of the OmpF and OmpC porins in *Escherichia coli*. Mol. Microbiol. *20*, 911–917.

Price, N.L., and Raivio, T.L. (2009). Characterization of the Cpx regulon in *Escherichia coli* strain MC4100. J. Bacteriol. *191*, 1798–1815.

Prigent-Combaret, C., Brombacher, E., Vidal, O., Ambert, A., Lejeune, P., Landini, P., and Dorel, C. (2001). Complex regulatory network controls initial adhesion and biofilm formation in *Escherichia coli* via regulation of the *csgD* gene. J. Bacteriol. *183*, 7213–7223.

Quan, S., Koldewey, P., Tapley, T., Kirsch, N., Ruane, K.M., Pfizenmaier, J., Shi, R., Hofmann, S., Foit, L., Ren, G. *et al.* (2011). Genetic selection designed to stabilize proteins uncovers a chaperone called Spy. Nat. Struct. Mol. Biol. *18*, 262–269.

Raetz, C.R., and Whitfield, C. (2002). Lipopolysaccharide endotoxins. Annu. Rev. Biochem. *71*, 635–700.

Raffa, R.G., and Raivio, T.L. (2002). A third envelope stress signal transduction pathway in *Escherichia coli*. Mol. Microbiol. *45*, 1599–1611.

Raivio, T.L., and Silhavy, T.J. (1997). Transduction of envelope stress in *Escherichia coli* by the Cpx two-component system. J. Bacteriol. *179*, 7724–7733.

Raivio, T.L., Popkin, D.L., and Silhavy, T.J. (1999). The Cpx envelope stress response is controlled by amplification and feedback inhibition. J. Bacteriol. *181*, 5263–5272.

Raivio, T.L., Laird, M.W., Joly, J.C., and Silhavy, T.J. (2000). Tethering of CpxP to the inner membrane prevents spheroplast induction of the Cpx envelope stress response. Mol. Microbiol. *37*, 1186–1197.

Ramboarina, S., Fernandes, P J., Daniell, S., Islam, S., Simpson, P., Frankel, G., Booy, F., Donnenberg, M S., and Matthews, S. (2005). Structure of the bundle-forming pilus from enteropathogenic *Escherichia coli*. J. Biol. Chem. *280*, 40252–40260.

Reinelt, S., Hofmann, E., Gerharz, T., Bott, M., and Madden, D.R. (2003). The structure of the periplasmic ligand-binding domain of the sensor kinase CitA reveals the first extracellular PAS domain. J. Biol. Chem. *278*, 39189–39196.

Richards, G.R., Herbert, E.E., Park, Y., and Goodrich-Blair, H. (2008). *Xenorhabdus nematophila lrhA* is necessary for motility, lipase activity, toxin expression, and virulence in *Manduca sexta* insects. J. Bacteriol. *190*, 4870–4879.

Rinker, S.D., Trombley, M.P., Gu, X., Fortney, K.R., and Bauer, M.E. (2011). Deletion of *mtrC* in *Haemophilus ducreyi* increases sensitivity to human antimicrobial peptides and activates the CpxRA regulon. Infect. Immun. *79*, 2324–2334.

Robinson, V.L., Wu, T., and Stock, A.M. (2003). Structural analysis of the domain interface in DrrB, a response regulator of the OmpR/PhoB subfamily. J. Bacteriol. *185*, 4186–4194.

Römling, U. (2005). Characterization of the rdar morphotype, a multicellular behaviour in Enterobacteriaceae. Cell. Mol. Life Sci. *62*, 1234–1246.

Rönnebäumer, K., Sander, G., Shutinoski, B., Schmidt, M.A., and Heusipp, G. (2009). Controlled activation of the Cpx system is essential for growth of *Yersinia enterocolitica*. FEMS Microbiol. Lett. *296*, 274–281.

Rosenberg, E.Y., Ma, D., and Nikaido, H. (2000). AcrD of *Escherichia coli* is an aminoglycoside efflux pump. J. Bacteriol. *182*, 1754–1756.

Rowbotham, T.J. (1980). Preliminary report on the pathogenicity of *Legionella pneumophila* for freshwater and soil amoebae. J. Clin. Pathol. *33*, 1179–1183.

Rubin, E.J., Akerley, B.J., Novik, V.N., Lampe, D.J., Husson, R.N., and Mekalanos, J.J. (1999). *In vivo* transposition of *mariner*-based elements in enteric bacteria and mycobacteria. Proc. Natl. Acad. Sci. U.S.A. *96*, 1645–1650.

Sakoh, M., Ito, K., and Akiyama, Y. (2005). Proteolytic activity of HtpX, a membrane-bound and stress-controlled protease from *Escherichia coli*. J. Biol. Chem. *280*, 33305–33310.

Schönbrunner, E.R., and Schmid, F.X. (1992). Peptidyl-prolyl *cis-trans* isomerase improves the efficiency of protein disulfide isomerase as a catalyst of protein folding. Proc. Natl. Acad. Sci. U.S.A. *89*, 4510–4513.

Segal, G., and Shuman, H.A. (1998). How is the intracellular fate of the *Legionella pneumophila* phagosome determined? Trends Microbiol. *6*, 253–255.

Shimohata, N., Chiba, S., Saikawa, N., Ito, K., and Akiyama, Y. (2002). The Cpx stress response system of *Escherichia coli* senses plasma membrane proteins and controls HtpX, a membrane protease with a cytosolic active site. Genes Cells *7*, 653–662.

Silverman, P.M., Tran, L., Harris, R., and Gaudin, H.M. (1993). Accumulation of the F plasmid TraJ protein in *cpx* mutants of *Escherichia coli*. J. Bacteriol. *175*, 921–925.

Sklar, J.G., Wu, T., Kahne, D., and Silhavy, T.J. (2007). Defining the roles of the periplasmic chaperones SurA, Skp, and DegP in *Escherichia coli*. Genes Dev. *21*, 2473–2484.

Slamti, L., and Waldor, M.K. (2009). Genetic analysis of activation of the *Vibrio cholerae* Cpx pathway. J. Bacteriol. *191*, 5044–5056.

Slauch, J.M., Russo, F.D., and Silhavy, T.J. (1991). Suppressor mutations in *rpoA* suggest that OmpR controls transcription by direct interaction with the α subunit of RNA polymerase. J. Bacteriol. *173*, 7501–7510.

Snyder, W.B., Davis, L.J., Danese, P.N., Cosma, C.L., and Silhavy, T.J. (1995). Overproduction of NlpE, a new outer membrane lipoprotein, suppresses the toxicity of periplasmic LacZ by activation of the Cpx signal transduction pathway. J. Bacteriol. *177*, 4216–4223.

Spiess, C., Beil, A., and Ehrmann, M. (1999). A temperature-dependent switch from chaperone to protease in a widely conserved heat shock protein. Cell 97, 339–347.

Spinola, S.M., Fortney, K.R., Baker, B., Janowicz, D.M., Zwickl, B., Katz, B.P., Blick, R.J., and Munson, R.S., Jr. (2010). Activation of the CpxRA system by deletion of *cpxA* impairs the ability of *Haemophilus ducreyi* to infect humans. Infect. Immun. *78*, 3898–3904.

van Stelten, J., Silva, F., Belin, D., and Silhavy, T.J. (2009). Effects of antibiotics and a proto-oncogene homolog on destruction of protein translocator SecY. Science *325*, 753–756.

Strauch, K.L., Johnson, K., and Beckwith, J. (1989). Characterization of *degP*, a gene required for proteolysis in the cell envelope and essential for growth of *Escherichia coli* at high temperature. J. Bacteriol. *171*, 2689–2696.

Suntharalingam, P., Spencer, H., Gallant, C.V., and Martin, N.L. (2003). *Salmonella enterica* serovar typhimurium *rdoA* is growth phase regulated and involved in relaying Cpx-induced signals. J. Bacteriol. *185*, 432–443.

Thede, G.L., Arthur, D.C., Edwards, R.A., Buelow, D.R., Wong, J.L., Raivio, T.L., and Glover, J.N. (2011). Structure of the periplasmic stress response protein CpxP. J. Bacteriol. *193*, 2149–2157.

Vakevainen, M., Greenberg, S., and Hansen, E.J. (2003). Inhibition of phagocytosis by *Haemophilus ducreyi* requires expression of the LspA1 and LspA2 proteins. Infect. Immun. *71*, 5994–6003.

Viboud, G.I., and Bliska, J.B. (2005). *Yersinia* outer proteins: role in modulation of host cell signaling responses and pathogenesis. Annu. Rev. Microbiol. *59*, 69–89.

Vincent, C.D., Buscher, B.A., Friedman, J.R., Williams, L.A., Bardill, P., and Vogel, J.P. (2006), Identification of non-*dot/icm* suppressors of the *Legionella pneumophila* Δ*dotL* lethality phenotype. J. Bacteriol. *188*, 8231–8243.

Voelker, D.R. (1997). Phosphatidylserine decarboxylase. Biochim. Biophys. Acta *1348*, 236–244.

Vogt, S.L., Nevesinjac, A.Z., Humphries, R., Donnenberg, M.S., Armstrong, G.D., and Raivio, T.L. (2010). The Cpx envelope stress response both facilitates and inhibits elaboration of the enteropathogenic *Escherichia coli* bundle-forming pilus. Mol. Microbiol. *76*, 1095–1110.

Wang, X., Hammer, N.D., and Chapman, M.R. (2008). The molecular basis of functional bacterial amyloid polymerization and nucleation. J, Biol, Chem, *283*, 21530–21539.

Watanabe, H., Nakayama, S., Ito, K., and Murai, T. (1993). Mutants affecting the pH dependent regulation of *Shigella sonnei virF* gene expression, In Proceedings of the 29th Joint Conference on Cholera and Related Diarrheal Disease. Bethesda, MD: U.S.-J.C.M.S. Program, ed., pp. 164–166.

Weatherspoon-Griffin, N., Zhao, G., Kong, W., Kong, Y., Morigen, M., Andrews-Polymenis, H., McClelland, M., and Shi, Y. (2011). The CpxR/CpxA two-component system up-regulates two Tat-dependent peptidoglycan amidases to confer bacterial resistance to antimicrobial peptide. J. Biol. Chem. *286*, 5529–5539.

Weber, A., Kögl, S.A., and Jung, K. (2006). Time-dependent proteome alterations under osmotic stress during aerobic and anaerobic growth in *Escherichia coli*. J. Bacteriol. *188*, 7165–7175.

Weber, R.F., and Silverman, P.M. (1988). The Cpx proteins of *Escherichia coli* K12. Structure of the CpxA polypeptide as an inner membrane component. J. Mol. Biol. *203*, 467–478.

Wolfe, A.J., Parikh, N., Lima, B.P., and Zemaitaitis, B. (2008). Signal integration by the two-component signal transduction response regulator CpxR. J. Bacteriol. *190*, 2314–2322.

Yamamoto, K., and Ishihama, A. (2005). Transcriptional response of *Escherichia coli* to external copper. Mol. Microbiol. *56*, 215–227.

Yamamoto, K., and Ishihama, A. (2006). Characterization of copper-inducible promoters regulated by CpxA/CpxR in *Escherichia coli*. Biosci. Biotechnol. Biochem. *70*, 1688–1695.

Yamamoto, K., Ogasawara, H., and Ishihama, A. (2008). Involvement of multiple transcription factors for metal-induced *spy* gene expression in *Escherichia coli*. J. Biotechnol. *133*, 196–200.

Zhang, H.Z., and Donnenberg, M.S. (1996). DsbA is required for stability of the type IV pilin of enteropathogenic *Escherichia coli*. Mol. Microbiol. *21*, 787–797.

Zhang, Z., and Hendrickson, W.A. (2010). Structural characterization of the predominant family of histidine kinase sensor domains. J. Mol. Biol. *400*, 335–353.

Zheng, J., He, C., Singh, V.K., Martin, N.L., and Jia, Z. (2007). Crystal structure of a novel prokaryotic Ser/Thr kinase and its implication in the Cpx stress response pathway. Mol. Microbiol. *63*, 1360–1371.

Zhou, X., Keller, R., Volkmer, R., Krauss, N., Scheerer, P., and Hunke, S. (2011). Structural basis for two-component system inhibition and pilus sensing by the auxiliary CpxP protein. J. Biol. Chem. *286*, 9805–9814.

Zusman, T., Aloni, G., Halperin, E., Kotzer, H., Degtyar, E., Feldman, M., and Segal, G. (2007). The response regulator PmrA is a major regulator of the *icm/dot* type IV secretion system in *Legionella pneumophila* and *Coxiella burnetii*. Mol. Microbiol. *63*, 1508–1523.

Cell Cycle and Developmental Regulation by Two-component Signalling Proteins in *Caulobacter crescentus*

13

Stephen C. Smith[1], Juan-Jesus Vicente and Kathleen R. Ryan

Abstract

The intricate cell division and developmental cycle of the α -Proteobacterium *Caulobacter crescentus* has been studied for four decades. During that time, elegant genetic screens and comprehensive post-genomic methods have uncovered a branched network of two-component signalling proteins that orchestrates *Caulobacter* cell cycle progression and morphological development. In addition to yielding the first and most detailed picture of bacterial cell cycle control, *Caulobacter* studies have revealed novel ways in which two-component proteins generate cellular outputs, interact with each other, and are themselves regulated.

Introduction

Both two-component signalling and other regulatory mechanisms play important roles in the *Caulobacter* life cycle, as reviewed in Curtis *et al.* (2010). Here we present a two-component centric view of the cell cycle, focused on the activities, locations, and regulation of two-component proteins. We will first describe key properties of the *Caulobacter* cell cycle and morphological development and explain how changes in the activity of a central response regulator, CtrA, drive cell cycle progression. We will then work our way upstream to the branched network of two-component proteins that regulates CtrA activity in time and space. We will also discuss the intimate linkage between the two-component network and the second messenger cyclic diguanylate (c-di-GMP), which is a key mediator of developmental processes. We will then describe two-component systems that function directly in the synthesis of polar organelles. Finally, we will briefly discuss two-component signalling proteins that are involved in selected *Caulobacter* environmental responses, because they exemplify unusual modes of action for histidine kinases or response regulators. In all cases, we will point out unanswered questions and areas of active research.

The *Caulobacter* genome encodes 62 histidine kinases and 44 response regulators. Of these 106 two-component proteins, 60 are encoded as 'orphans' without a cognate histidine kinase or response regulator in the same operon (Nierman *et al.*, 2001; Skerker *et al.*, 2005). In an attempt to delete each two-component gene individually, nine were found to be essential for viability under standard laboratory conditions (Table 13.1), and 39 in all contribute to some aspect of growth, cell cycle progression, or morphology (Skerker *et al.*, 2005). A previously unknown essential two-component system composed of the histidine kinase CenK and the DNA-binding response regulator CenR was uncovered in this comprehensive genomic approach. CenK and CenR are necessary for cell envelope integrity, as mutants depleted of these proteins are extensively blebbed, and cells overexpressing a phosphomimetic form of CenR become misshapen and lyse (Skerker *et al.*, 2005). The transcriptional targets of CenKR are as yet unknown, and connections between this two-component system and the cell cycle are still to be determined.

1 Stephen C. Smith and Juan-Jesus Vicente contributed equally to the writing of this chapter.

Table 13.1 Essential two-component signalling proteins in *Caulobacter crescentus*

Gene	Name	Type	Function	Reference
CC_0530	*cenK*	HK	Cell envelope biogenesis and integrity	Skerker *et al.* (2005)
CC_1078	*cckA*	HK	DNA replication, cell division	Jacobs *et al.* (1999)
CC_1743	*ntrX*	RR	(Not essential on minimal M2G medium)	Skerker *et al.* (2005)
CC_2931	*petR*	RR	–	Skerker *et al.* (2005)
CC_2932	–	HK	Putative kinase for PetR	Skerker *et al.* (2005)
CC_3035	*ctrA*	RR	DNA replication, cell division	Quon *et al.* (1996)
CC_2463	*divK*	RR	DNA replication, cell division	Hecht *et al.* (1995)
CC_3484	*divL*	HK	DNA replication, cell division	Wu *et al.* (1999)
CC_3743	*cenR*	RR	Cell envelope biogenesis and integrity	Skerker *et al.* (2005)

Gene numbers refer to the sequence of *Caulobacter crescentus* CB15 (Nierman *et al.*, 2001). Protein type is either histidine kinase (HK) or response regulator (RR). Each gene is essential for growth in rich (PYE) medium (Skerker *et al.*, 2005).

The *Caulobacter* cycle of cell division and morphogenesis

A detailed description of the *Caulobacter* cell cycle is available in Curtis *et al.* (2010). *Caulobacter* cells growing exponentially in rich medium have a generation time of ~90 minutes, and the populations are dimorphic. The motile swarmer cell has a single flagellum and multiple pili at one pole. The non-motile stalked cell lacks pili, and the polar flagellum is replaced by the stalk, a thin extension of the cell envelope. These distinct cell types are generated by asymmetric division of a polarized predivisional cell, in which one pole possesses the stalk and the other pole the flagellum and pili (Fig. 13.1A).

Swarmer and stalked cells not only have different shapes, they also occupy different phases of the cell cycle (Degnen and Newton, 1972). Unlike bacteria that can initiate more than one round of DNA replication per cell division, *Caulobacter* practices once-and-only-once chromosome replication (Marczynski, 1999). The stalked progeny is born into S-phase; after division it can immediately begin a new cycle of chromosome replication and cell division. As the stalked cell grows, it builds a flagellum at the pole opposite the stalk, creating an asymmetric predivisional cell. In contrast, swarmer cells are born into a pre-replication G1 phase, unable to initiate chromosome replication. When the swarmer cell enters S-phase, it also differentiates into a stalked cell by shedding the flagellum and building a stalk at the same pole (Fig. 13.1A). In wild-type, well-fed cultures of *Caulobacter*, morphological changes at the cell pole are coupled to cell cycle progression, but this linkage can be broken by mutations in regulatory genes (Sommer and Newton, 1989; Hottes *et al.*, 2005).

Oscillations in the activity of a key response regulator are necessary for orderly cell cycle progression

The essential response regulator CtrA is a critical hub connecting *Caulobacter* cell cycle progression with the development of distinct cell types (Quon *et al.*, 1996). CtrA is a member of the OmpR family of transcriptional regulators that activates or represses the transcription of 26% of the cell cycle-regulated genes in *Caulobacter*, including direct regulation of 95 genes (Laub *et al.*, 2000, 2002). Through its transcriptional targets, CtrA promotes critical events that occur late in the cell cycle, such as DNA methylation, flagellum and pilus biogenesis, and cell division, and several CtrA targets are themselves essential for viability. However, CtrA also blocks the initiation of DNA replication by binding to five sites within the *Caulobacter* origin of replication (*Cori*) (Quon *et al.*, 1998). If CtrA is constitutively active, *Caulobacter* cells arrest in the G1-phase of the cell cycle with one chromosome (Domian *et al.*, 1997). Therefore, CtrA activity must be

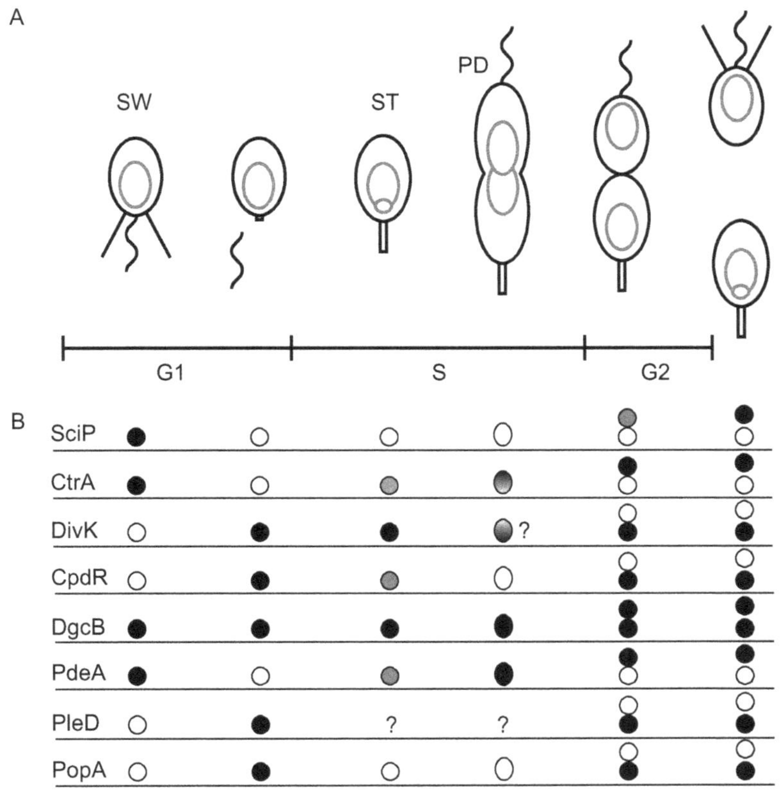

Figure 13.1 (A) Schematic diagram of the *Caulobacter crescentus* cell cycle. Wavy lines indicate the flagellum and straight lines indicate pili. The rectangular extension at the bottom of the cell denotes the stalk. Grey interior circles indicate the chromosome in successive stages of replication and segregation. Swarmer (SW), stalked (ST) and predivisional (PD) cells are indicated. (B) Activities of indicated proteins at successive stages during the cell cycle. Black circles/ovals indicate high activity, grey circles/ovals indicate intermediate activity, and white circles/ovals indicate low activity. Ovals containing gradient shading indicate that a gradient of CtrA activity has been observed, and a gradient of DivK activity is postulated to exist (denoted by a question mark). The direction of the gradient corresponds to the poles of the cell in panel A. SciP is thought to be active whenever it is present in the cell. CtrA and DivK are active when present and phosphorylated. CpdR is active when present and unphosphorylated. DgcB is always present and thought to be constitutively active. PdeA is thought to be active when present in the cell. PleD's DGC activity is low in swarmer cells, increases during the SW-ST transition, and is high again in newly born stalked cells. However, PleD's activity level in late stalked and early predivisional cells is unknown, indicated by question marks. PopA is active at promoting CtrA degradation at the stalked pole when c-di-GMP levels are high. PopA may have other activities, but these are not represented in the diagram. See text for references.

suppressed in any cell that is preparing to initiate chromosome replication. Conversely, cells harbouring conditional loss-of-function mutations in *ctrA* accumulate multiple chromosomes without dividing (Domian *et al.*, 1997). Because of CtrA's dual function, redundant mechanisms ensure that CtrA activity is eliminated at the G1-S transition and restored later in the cell cycle to drive polar morphogenesis and division (Fig. 13.1B).

CtrA activity oscillates as a result of cell cycle-regulated phosphorylation and proteolysis. CtrA is rapidly degraded during swarmer-to-stalked cell (SW-ST) differentiation and in the stalked compartment of the predivisional cell (Domian *et al.*, 1997). Regulated proteolysis by the ATP-dependent ClpXP protease at these times ensures that newly born stalked cells and cells proceeding through the SW-ST transition will be able to initiate chromosome replication (Jenal and Fuchs, 1998). Swarmer cells, which are born containing active CtrA, must wait until CtrA is eliminated at the SW-ST transition before beginning chromosome replication. CtrA regulates its own transcription, and later in the cell cycle, CtrA is replenished by a positive transcriptional feedback loop (Domian *et al.*, 1999). *ctrA* transcription also depends on the methylation state of the *ctrA* promoter, thus linking CtrA synthesis to the progress of chromosome replication (Reisenauer and Shapiro, 2002). If CtrA is artificially stabilized by mutation, so that the protein is present throughout the cell cycle, CtrA~P levels still oscillate (Domian *et al.*, 1997). CtrA phosphorylation is high in swarmer and predivisional cells and nearly undetectable in stalked cells. This variation in the level of CtrA~P is sufficient to allow chromosome replication to occur, and cells with stable CtrA protein have a near-normal life cycle.

CtrA function at cell cycle-regulated promoters and at the origin of replication

CtrA's mechanism of action at the origin of replication (*Cori*) and at promoters has been studied extensively. CtrA binds to the consensus sequence TTAA-N7-TTAA (Quon *et al.*, 1998), in which each TTAA half-site binds a CtrA monomer (Siam and Marczynski, 2000). Phosphorylation stimulates CtrA binding to full sites, but not if one of the corresponding half-sites is mutated to prevent recognition by CtrA, indicating that phosphorylation affects protein-protein contacts between CtrA molecules rather than the interaction between a single molecule and DNA (Siam and Marczynski, 2000).

Cori contains five full CtrA binding sites [a]–[e] (Quon *et al.*, 1998). The cloned replication origin supports the replication of *Cori* plasmids, and plasmid replication occurs with the same cell cycle timing as chromosome replication (Marczynski and Shapiro, 1992). When individual CtrA binding sites in *Cori* plasmids are mutated, plasmid copy number increases, supporting the model that CtrA binding at the origin represses DNA replication (Quon *et al.*, 1998; Bastedo and Marczynski, 2009). Chromatin immunoprecipitation experiments performed on synchronized *Caulobacter* populations indicate that CtrA occupancy of native *Cori* sites is high in swarmer cells and drops sharply during the first 20 minutes of the cell cycle, coincident with CtrA proteolysis and the onset of DNA replication (Taylor *et al.*, 2011).

CtrA~P binding to *Cori* sites likely blocks several steps necessary for the initiation of chromosome replication. First, the [a] and [b] sites overlap a strong promoter (*hemE* Ps), which directs transcription away from the *Cori* region. Transcription from this promoter is required for *Cori* plasmid replication (Marczynski *et al.*, 1995), and CtrA binding represses its transcription (Quon *et al.*, 1998). Second, the [c] site overlaps a binding site for the essential replication initiation protein DnaA, and the [e] site lies 4 bp from another DnaA binding site (Marczynski and Shapiro, 1992; Taylor *et al.*, 2011). DnaA has relatively weak affinity for its binding sites in *Cori*, and in footprinting assays, DnaA can be displaced from the DNA by CtrA (Taylor *et al.*, 2011). Finally, binding site [c] also overlaps a binding site for integration host factor (IHF) (Siam *et al.*, 2003), which in *E. coli* facilitates the melting of origin sequences by DnaA (Hwang and Kornberg, 1992). CtrA dephosphorylation and proteolysis at the G1-S transition, combined with increased expression of DnaA and IHF, simultaneously permit binding of DnaA and IHF to *Cori* sites and derepress transcription from *hemE* Ps, allowing chromosome replication to commence (Gober and Shapiro, 1992; Zweiger and Shapiro, 1994; Domian *et al.*, 1997; Jacobs *et al.*, 2003).

Despite the strong effect that CtrA~P has on blocking the initiation of DNA replication (Domian *et al.*, 1997), *Caulobacter* can survive under lab conditions without over-replicating its chromosome even when all five CtrA binding sites in *Cori* are mutated (Bastedo and Marczynski, 2009). This result contrasts with findings using *Cori* plasmids, where loss of CtrA binding sites increases copy number (Bastedo and Marczynski, 2009). A partial explanation for this phenomenon is that DnaA activity oscillates in *Caulobacter* cells independently of CtrA (Jonas *et al.*, 2011). DnaA activity is apparently low during the predivisional phase, because mutants lacking CtrA activity or CtrA binding sites do not reinitiate replication late within the same cell cycle (Jonas *et al.*, 2011). Lack of CtrA activity has little effect on the timing of replication initiation in stalked cells, but in swarmer cells, CtrA delays the onset of DNA replication, so that the swarmer cell cycle is longer than the stalked cycle (Jonas *et al.*, 2011). CtrA binding sites in *Cori* become essential for viability when *Caulobacter* is exposed to a combination of stresses, nutrient upshift and exposure to low levels of antibiotics (Bastedo and Marczynski, 2009). Thus, CtrA may have evolved to help generate non-identical progeny and to integrate environmental information into the control of chromosome replication.

Transcriptional regulation by CtrA is similarly complex, in that CtrA activates some promoters and represses others, and not all activated promoters fire at the same time in the cell cycle. This difference in cell cycle timing has been partially attributed to the strength of CtrA~P binding to individual promoters. For example, CtrA has 10- to-20-fold greater affinity for the promoter of *fliQ*, a class II flagellar gene which is activated early in the predivisional cell, than it does for the *ccrM* promoter, which is activated very late in the cell cycle, when more CtrA~P has accumulated (Reisenauer *et al.*, 1999). However, the strength of promoter binding does not entirely account for the transcription patterns of CtrA-regulated genes. While *fliQ* is transcribed in early predivisional cells, it is not transcribed in swarmer cells, which contain comparable levels of activated CtrA (Ouimet and Marczynski, 2000; Jacobs *et al.*, 2003). It was proposed that a swarmer cell-specific repressor of transcription could function at CtrA-activated promoters that are *not* expressed in swarmer cells (Ouimet and Marczynski, 2000), and such a regulator has recently been identified.

SciP, a small protein with homology to helix–turn–helix proteins, is present in the swarmer cell where it binds to CtrA, repressing the transcription of a large group of genes usually activated by CtrA late in the cell cycle. Genes normally repressed by CtrA are not affected by SciP (Gora *et al.*, 2010; Tan *et al.*, 2010). Electrophoretic mobility shift experiments demonstrated that SciP forms a complex with CtrA and DNA which prevents RNA polymerase from associating with the promoter of a CtrA-activated gene, *fliF*. To prevent SciP from inappropriately blocking the transcription of late cell cycle genes, SciP levels are strictly controlled. CtrA activates *sciP* transcription late in the cell cycle, but SciP protein is only abundant in swarmer cells, suggesting that its synthesis is post-transcriptionally regulated. In addition, SciP is very rapidly degraded at the swarmer-to-stalked cell transition, though the protease responsible has not yet been identified. The ability of SciP to inhibit CtrA-dependent transcription through direct contact with CtrA and without affecting CtrA's phosphorylation or DNA-binding ability represents a novel way of regulating the function of a DNA-binding response regulator (Gora *et al.*, 2010).

Spatiotemporal regulation of CtrA activity by an extended, branched two-component signalling network

CtrA phosphorylation on aspartate 51 (D51) is required for viability, and phosphorylation is accomplished by a phosphorelay comprising the hybrid histidine kinase CckA and the histidine phosphotransferase ChpT. CckA contains two transmembrane spans which anchor the protein in the cytoplasmic membrane. The cytoplasmic portion of CckA consists of a sensory region containing a predicted PAS (Per-Arnt-Sim) domain, a dimerization and histidine phosphotransfer domain (DHp), a catalytic domain (CA), and a receiver domain. A temperature-sensitive (ts) *cckA* mutant contains drastically less CtrA~P *in vivo*, and the transcriptional profiles of *ctrA* and *cckA* ts mutants are very highly correlated (Jacobs *et al.*, 1999, 2003). The phosphorelay was reconstituted *in vitro* after the discovery of ChpT, which shuttles the phosphoryl group between the receiver domains of CckA and CtrA (Biondi *et al.*, 2006a).

In ground-breaking experiments, CckA was shown to be the first two-component signalling protein located at a particular site within a bacterial cell (Jacobs *et al.*, 1999). CckA undergoes cell cycle-regulated changes in localization: it is delocalized in swarmer cells, but accumulates at the pole opposite the stalk (the swarmer or flagellar pole) during S-phase. A smaller focus of CckA is sometimes seen at the stalked pole of the predivisional cell. Shortly before cell division, CckA is released from the flagellar pole and delocalized around the swarmer compartment. Immediately after cell division, CckA is delocalized in the stalked progeny (Jacobs *et al.*, 1999; Angelastro *et al.*, 2010). These findings were intriguing, because

CckA's localization at the swarmer pole coincides loosely with periods of high CckA~P and CtrA~P in the cell (Jacobs *et al.*, 2003). Studies described below have further addressed the mechanism and consequences of CckA localization at the swarmer pole (Iniesta *et al.*, 2010a,b; Tsokos *et al.*, 2011).

The CckA-ChpT phosphorelay was identified by a pioneering technique called phosphotransfer profiling (Biondi *et al.*, 2006a). In a profiling experiment, one purified histidine kinase or phosphotransferase is used to phosphorylate multiple receiver domains in parallel. Histidine kinases are sometimes promiscuous *in vitro* and can phosphorylate response regulators other than their *in vivo* cognates. However, examining phosphotransfer after very short incubation times reveals the preferred partner(s) of a histidine kinase or phosphotransferase from among all the possible receiver domains encoded in an organism's genome (Laub *et al.*, 2007). Such an experiment showed that ChpT could act as a phosphodonor for either CtrA or a single-domain response regulator, CpdR.

This bifurcated phosphorelay not only regulates CtrA phosphorylation, but CtrA stability as well (Biondi *et al.*, 2006a; Iniesta *et al.*, 2006). The unphosphorylated form of CpdR is required to promote CtrA degradation during swarmer cell differentiation and in the stalked compartment of the predivisional cell (see below). This signalling logic dictates that when CckA is active as a kinase, phosphate flows onto both CtrA and CpdR, and CtrA is both activated and stabilized (Fig. 13.2A). However, phosphate can also flow backward from CtrA and CpdR via ChpT to the CckA receiver domain, where CckA phosphatase activity stimulates hydrolysis (Chen *et al.*, 2009). Thus, when CckA is in phosphatase mode, CtrA is both deactivated and destabilized (Fig. 13.2B). *In vivo*, a version of CckA harbouring kinase activity but lacking phosphatase activity (K^+P^-) causes only a mild cell cycle phenotype (Chen *et al.*, 2009). This result suggests that the autophosphatase activities of CtrA (Chen *et al.*, 2011) and CpdR are sufficient for their dephosphorylation, or that unknown phosphatases act on CtrA and CpdR to sharpen the G1-S cell cycle transition.

Since CckA activity determines the cellular level of CtrA~P, the next question is how CckA toggles between kinase and phosphatase modes. Although the mechanisms are not completely clear, recent studies have revealed that two other essential two-component proteins, DivL and DivK, modulate CckA activity.

DivL is a histidine kinase homologue composed of an N-terminal transmembrane anchor, a long domain of unknown function, and conserved DHp and CA domains. DivL is unusual because it contains a tyrosine residue (Y550) at the site normally occupied by the conserved, phosphorylatable histidine. Cells depleted of DivL or harbouring a ts mutation in *divL* are filamentous and contain more than one chromosome, resembling *ctrA* ts mutants (Wu *et al.*, 1999; Sciochetti *et al.*, 2005; Reisinger *et al.*, 2007). These phenotypes suggested that DivL could play a role in activating CtrA. DivL was at first thought to phosphorylate CtrA directly, but the DivL catalytic domain and Y550 residue are dispensable for viability, and mutants lacking these features of DivL have relatively mild phenotypes (Reisinger *et al.*, 2007). Recent studies show that DivL functions instead by a novel mechanism, stimulating CckA kinase activity. *divL* ts mutants contain lower levels of CtrA~P, CpdR~P, and CckA~P, and DivL is also required for CckA localization at the swarmer pole (Iniesta *et al.*, 2010a; Tsokos *et al.*, 2011). DivL and CckA can be co-immunoprecipitated from *Caulobacter* lysates, indicating that they exist in a complex *in vivo* (Iniesta *et al.*, 2010a). The current model encompassing these results is that DivL stimulates CckA kinase activity by contacting CckA directly (Fig. 13.2A).

DivK is a single-domain response regulator whose phosphorylation on D53 is required for *Caulobacter* viability (Hecht *et al.*, 1995; Lam *et al.*, 2003). At the non-permissive temperature, a cold-sensitive (cs) *divK* mutant forms long filaments arrested in the G1 phase of the cell cycle, with each cell containing one chromosome (Hecht *et al.*, 1995; Hung and Shapiro, 2002). This phenotype is strikingly similar to cells overexpressing a constitutively active form of CtrA. In fact, *divKcs* cells at the non-permissive temperature contain increased levels of CtrA~P and CckA~P, indicating that CckA kinase activity is up-regulated when DivK is compromised (Biondi *et al.*, 2006a; Tsokos *et al.*, 2011).

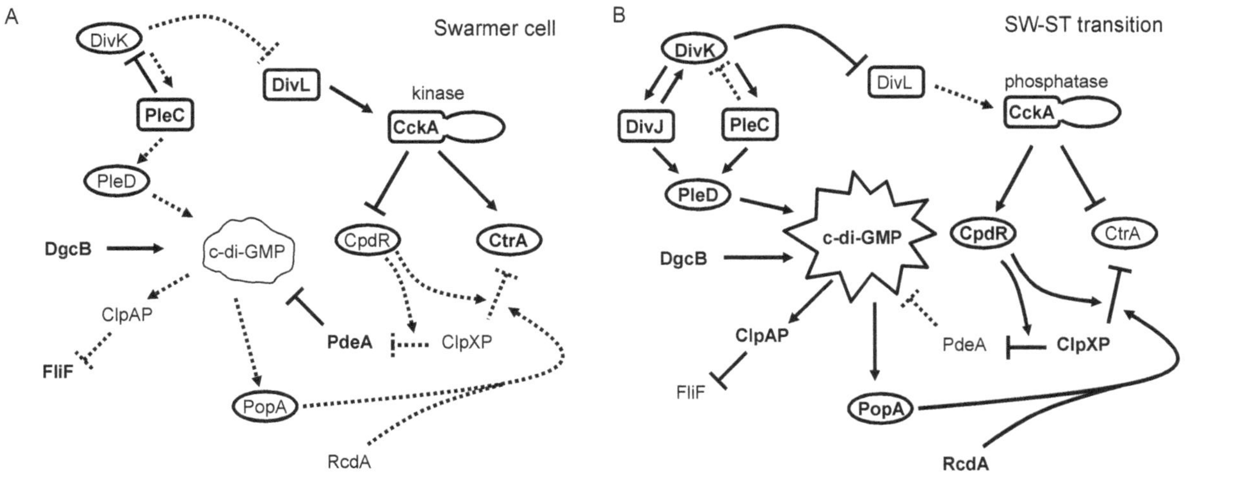

Figure 13.2 (A) Regulatory protein activities in the swarmer cell. (B) Regulatory protein activities during the swarmer-to-stalked cell (SW-ST) transition. Solid arrows and bars indicate processes that are occurring in that cell type, while dotted arrows and bars indicate processes that are not occurring. Bold type indicates components that are currently active, while regular type indicates components that are being proteolysed or are present but inactive for the processes specified in the diagram. DivJ levels are low in swarmer cells (A), but DivJ is present and active at the SW-ST transition (B). In kinase mode (A), CckA phosphorylates both CpdR and CtrA via ChpT, which activates CtrA and deactivates CpdR. In phosphatase mode (B), both CpdR and CtrA are dephosphorylated, which activates CpdR and deactivates CtrA. Arrows from CpdR denote stimulation of proteolysis by ClpXP. Bars from ClpXP and ClpAP indicate proteolysis of downstream components. Arrows from PopA and RcdA indicate stimulation of CtrA proteolysis. Arrows from DivJ and PleC to DivK or PleD denote phosphorylation, while the bar from PleC to DivK indicates dephosphorylation. Arrows from DivK to DivJ and PleC indicate allosteric stimulation of their kinase activities. PleC is thought to be active in both newly born swarmer cells and at the swarmer-to-stalked cell transition. In newly born swarmer cells, PleC functions as a DivK phosphatase (panel A). Later, during swarmer cell differentiation, PleC functions as a PleD kinase (panel B). The arrow from DivL to CckA denotes stimulation of CckA kinase activity via direct binding. The bar from DivK to DivL denotes inactivation of DivL by direct binding to DivK~P. Arrows from PleD and DgcB indicate diguanylate cyclase activity, while the bar from PdeA indicates c-di-GMP phosphodiesterase activity. In A, the cellular pool of c-di-GMP is smaller to indicate a lower concentration of this messenger in the swarmer cell. In B, the pool of c-di-GMP increases and is able to activate downstream proteins.

When DivK was used as the bait in a yeast two-hybrid screen, one of the interacting proteins recovered was DivL (Ohta and Newton, 2003). The meaning of this interaction is now becoming clear. DivK~P, but not unphosphorylated DivK, was shown by *in vitro* Förster resonance energy transfer (FRET) experiments to interact with the DHp and CA domains of DivL (Tsokos *et al.*, 2011). Amino acid substitutions in DivK or DivL that decrease the strength of their interaction have similar phenotypic effects, namely increasing the levels of CckA~P and blocking the initiation of chromosome replication. In contrast, an amino acid substitution in DivK that increases its interaction with DivL yields decreased levels of CckA~P and an increase in chromosome content. Thus, DivK~P appears to down-regulate CckA kinase activity *in vivo* by binding to DivL (Fig. 13.2B).

The DivK-DivL and DivL-CckA interactions represent exciting new mechanisms by which one two-component signalling protein regulates the activity of another. In the case of DivL and CckA, their interaction is thought to be direct, but this remains to be verified using purified proteins. The co-immunoprecipitation of DivL and CckA could be mediated by unknown proteins in the *Caulobacter* lysate. If DivL does contact CckA directly, what is the nature of the interaction? We do not yet know if DivL and CckA form heterodimers or a higher-order assembly. Based on the locations of mutations that alter DivK-DivL binding, DivK~P is predicted to interact with DivL in much the same way as a receiver domain interacts with a histidine kinase during phosphotransfer (Tsokos *et al.*, 2011). Because unphosphorylated DivK does not interact strongly with DivL, phosphorylation is predicted to cause a conformational change in DivK that permits better binding to DivL. With regard to DivK's effects on CckA, how does DivK~P binding alter DivL's ability to stimulate CckA kinase activity? DivK~P could compete with CckA for binding to DivL, disrupting the complex. Alternatively, it could cause a conformational change in DivL that prevents it from activating CckA without disrupting the DivL-CckA interaction.

The essential role of DivL is to (reversibly) stimulate CckA kinase activity, and this function can be performed by the N-terminal portion of DivL up to and including the DHp domain (Reisinger *et al.*, 2007; Iniesta *et al.*, 2010a). Since the CA domain is not essential, it is unclear if DivL autophosphorylates *in vivo*. However, mutants lacking the DivL CA domain are not phenotypically wild-type (Reisinger *et al.*, 2007). DivL may autophosphorylate at an undetectable level *in vivo*, or the CA domain may function to stabilize the interaction between DivL and DivK~P (Tsokos *et al.*, 2011). Point mutations designed to disrupt ATP binding or hydrolysis may help to clarify the function of the DivL CA domain.

Polarly localized histidine kinases with opposite effects on DivK generate asymmetry

How does DivK~P down-regulate CckA kinase activity via DivL specifically in the stalked progeny and not in the swarmer progeny? Two histidine kinases are known to regulate the phosphorylation state of DivK. DivJ contains five predicted transmembrane spans followed by DHp and CA domains. The PleC histidine kinase has two transmembrane spans flanking a periplasmic domain, followed by cytoplasmic DHp and CA domains. *In vitro*, both kinases can either phosphorylate or dephosphorylate DivK (Hecht *et al.*, 1995; Wu *et al.*, 1998), but *in vivo*, DivJ functions as a DivK kinase, while PleC functions as a DivK~P phosphatase (Wheeler and Shapiro, 1999; Matroule *et al.*, 2004) (Fig. 13.2). Cellular asymmetry of DivK activation is accomplished by the differential localization of DivJ and PleC: DivJ is located at the stalked pole, while PleC resides at the swarmer pole (Fig. 13.3).

PleC is located at the flagellated pole of the swarmer cell. During development into a stalked cell, the flagellum is shed, stalk biogenesis begins, and PleC is replaced at the stalked pole by DivJ. During S-phase, a new focus of PleC appears at the pole opposite the stalk, and it remains there until after cell division. DivJ resides at the stalked pole for the remainder of the cell cycle, and the progeny stalked cell is born with this focus of DivJ (Wheeler and Shapiro, 1999). DivK diffuses throughout the predivisional cell, being phosphorylated at the stalked pole by DivJ and dephosphorylated at the swarmer pole by PleC (Jacobs *et al.*, 2001; Matroule *et al.*, 2004) (Fig. 13.3A). *Caulobacter* predivisional cells form a diffusion barrier between the stalked and swarmer compartments several minutes before daughter cell separation (Judd *et al.*, 2003). Any DivK that is trapped in the stalked compartment becomes phosphorylated, while DivK trapped in the swarmer compartment becomes dephosphorylated (Lam *et al.*, 2003; Matroule *et al.*, 2004). As a result, the kinase activity of CckA residing in the stalked compartment is down-regulated by DivK~P, leading to CtrA dephosphorylation and proteolysis, while CckA and CtrA phosphorylation are preserved in the swarmer compartment (Domian *et al.*, 1997; Jacobs *et al.*, 2003). In this way, DivK acts as a cell division sensor that triggers different developmental programmes in the two progeny cells (Fig. 13.3B).

A recent study provided evidence for a gradient of CtrA activity in the *Caulobacter* predivisional cell, even before cytoplasmic compartmentalization (Chen *et al.*, 2011) (Fig. 13.1B). When division is inhibited by cephalexin or by FtsZ depletion, cells elongate for several hours without dividing. After a first round of DNA replication, each cell contains two chromosomes with an origin located at each pole. The non-dividing *Caulobacter* cells then begin a new round of DNA replication, but in this case, only the origin located at the stalked pole fires, creating a cell with three chromosomes

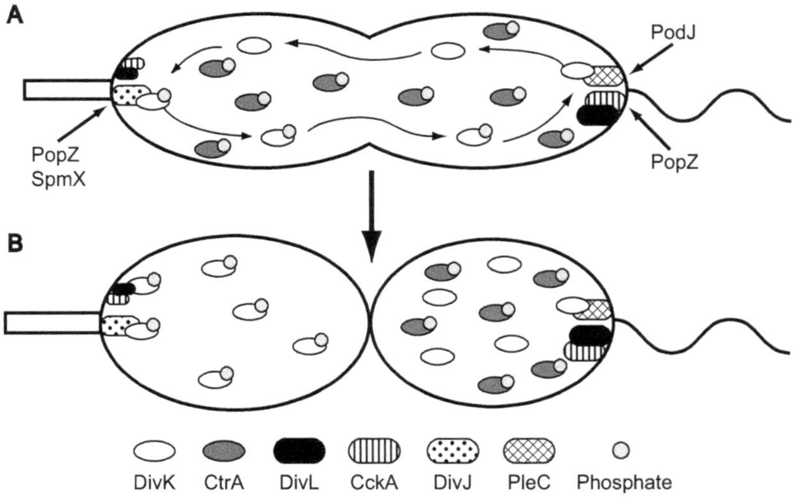

Figure 13.3 Localized histidine kinases with opposing activities create an asymmetric predivisional cell and distinct cell identities after compartmentalization. (A) Arrows leading from PopZ, SpmX, and PodJ denote that these factors are required for the polar localization of the indicated histidine kinases. In the late predivisional cell (A), DivK diffuses throughout the cell (interior arrows), being phosphorylated by DivJ at the stalked pole and dephosphorylated by PleC at the flagellar pole. At the swarmer pole, colocalization of PleC with DivL and CckA keeps DivK~P levels low and maintains CckA kinase activity. At the stalked pole, the smaller foci of DivL and CckA are not shielded from DivK~P, so CckA at that location acts as a phosphatase. This arrangement results in a gradient of CtrA~P, with higher CtrA activity at the flagellar pole. After a barrier to cytoplasmic diffusion has been established between the two compartments of the predivisional cell (B), DivK becomes phosphorylated in the stalked compartment and dephosphorylated in the swarmer compartment. DivK~P in the stalked compartment triggers c-di-GMP accumulation and CtrA proteolysis, while CtrA remains present and active in the swarmer compartment.

(Chen *et al.*, 2011). Why are the two origins treated differently within the same contiguous cytoplasm? The authors hypothesized that while CckA located at the swarmer pole acts as a kinase to phosphorylate CtrA, CckA located at the stalked pole may predominantly act as a CtrA phosphatase. If the kinetics of phosphorylation and dephosphorylation are sufficiently fast compared with diffusion of CtrA through the cytoplasm, this arrangement can generate a roughly linear gradient of CtrA~P along the length of the cell (Chen *et al.*, 2011). One test of the model was to repeat the experiment using cells whose CckA protein is specifically impaired in phosphatase activity (K+P–) (Chen *et al.*, 2009), so that the stalked pole is no longer a sink for CtrA~P. In these cells, the second round of DNA replication was no longer biased towards the stalked pole, but occurred with equal frequency at the stalked and swarmer poles of division-inhibited cells (Chen *et al.*, 2011). This indirect demonstration of a gradient of CtrA~P underscores the need for generalized methods to visualize the activated forms of response regulators *in vivo*.

Because DivK~P inhibits the kinase activity of CckA, the levels of DivK~P and CtrA~P are inversely related in swarmer and stalked cells. Surprisingly, both CtrA~P and DivK~P levels are high in predivisional cells (Jacobs *et al.*, 2001, 2003) (Fig. 13.1B). How is it that CckA is not down-regulated at this time in the cell cycle? Localization of CckA at the swarmer pole adjacent to PleC, the DivK~P phosphatase, is suggested to protect CckA from down-regulation by DivK~P (Tsokos *et al.*, 2011) (Fig. 13.3A). This model would account for the high level of CckA kinase activity in the face of high overall levels of DivK~P in predivisional cells. Phenotypes of *divK* and *pleC* mutants agree with the model, but it would be most convincing to examine CckA activity in cells where PleC was both present and able to dephosphorylate DivK~P, but mislocalized away from CckA. Below, we discuss the factors that mediate the polar localization of key histidine kinases. As this area of research advances, it will become easier to test the effects of protein mislocalization in *Caulobacter*.

Factors required for polar localization of signalling proteins

For this elegant mechanism to work, DivJ, PleC, CckA, and DivL must be located at the correct cell poles. Several studies have focused on the *cis* and *trans* factors required for localization of these histidine kinases, uncovering novel cellular organization proteins and links to other cell cycle processes. PleC localization to the pole opposite the stalk depends on PodJ, which occupies the pole before the arrival of PleC (Viollier *et al.*, 2002b; Hinz *et al.*, 2003) (Fig. 13.3A). In the absence of PodJ, PleC is delocalized or mislocalized to the stalked pole (Lawler *et al.*, 2006). *podJ* mutants have defects in chemotaxis, flagellar release, and pilus biosynthesis, though these phenotypes are less severe than in cells lacking PleC, suggesting that localization is necessary for some, but not all functions of PleC (Wang *et al.*, 1993). It is currently unknown how PodJ itself becomes localized at the flagellar pole, or if PodJ binds directly to PleC. During the SW-ST transition, PleC is released from the flagellar pole as it is remodelled into a stalked pole. PleC autophosphorylation is apparently required for its release from the pole, as the protein PleC-H610A lacking the conserved histidine residue is retained at the remodelled pole during differentiation (Viollier *et al.*, 2002a).

DivJ localization requires SpmX (Fig. 13.3A), which contains a periplasmic muramidase domain followed by two transmembrane spans (Radhakrishnan *et al.*, 2008). SpmX colocalizes with DivJ, and the two proteins can be immunoprecipitated together from *Caulobacter* lysates (Radhakrishnan *et al.*, 2008). It is not clear if SpmX is active as a muramidase enzyme, but polar localization of SpmX depends on its muramidase domain (Radhakrishnan *et al.*, 2008), suggesting that this domain may bind to a specific feature of stalked pole peptidoglycan. The entire SpmX protein is required for DivJ localization, so the interaction between them is most likely mediated by their respective transmembrane spanning domains (Sciochetti *et al.*, 2002; Radhakrishnan *et al.*, 2008).

DivL both stimulates CckA kinase activity and promotes CckA localization to the swarmer pole (Iniesta *et al.*, 2010a). Deletion studies have identified regions of DivL and CckA that are needed for their localization to the swarmer pole. DivL requires its single transmembrane spanning domain and sequences at the extreme C-terminus (Sciochetti *et al.*, 2005), while CckA requires its transmembrane spans, a PAS domain upstream of the DHp domain, and a loop within the CA domain for localization to the same pole (Jacobs *et al.*, 1999; Angelastro *et al.*, 2010). DivL and CckA only accumulate at the pole opposite the stalk after the initiation of DNA replication (Iniesta *et al.*, 2010b).

Upstream of both CckA and DivJ localization is the polar organizing protein PopZ (Fig. 13.3A), a 19-kDa protein that self-associates to form a reticular network *in vitro* (Bowman *et al.*, 2008). In *Caulobacter*, PopZ assembles in DNA-free zones at the cell poles and excludes ribosomes from these regions (Bowman *et al.*, 2008; Ebersbach *et al.*, 2008). PopZ is located at both poles of the predivisional cell; thus, both the swarmer and stalked progeny inherit a focus of PopZ at the 'old' pole. After the initiation of DNA replication in each daughter cell, a new focus of PopZ forms at the opposite cell pole, yielding bipolar PopZ (Bowman *et al.*, 2008; Ebersbach *et al.*, 2008).

CckA and DivJ co-immunoprecipitate with PopZ, and both proteins are mislocalized in a *popZ* mutant (Ebersbach *et al.*, 2008). In the case of CckA, these results suggest that DivL and CckA are recruited to the pole opposite the stalk by PopZ after the initiation of DNA replication. However, PopZ must not be sufficient to direct polar localization of DivL and CckA, because these proteins are only sometimes located at the stalked pole along with PopZ (Sciochetti *et al.*, 2005; Bowman *et al.*, 2008; Ebersbach *et al.*, 2008 Angelastro *et al.*, 2010). Both DivJ and SpmX depend on PopZ for stalked-pole localization (Ebersbach *et al.*, 2008; Bowman *et al.*, 2010). However, additional players are likely involved in this process as well, since neither protein follows PopZ to the pole opposite the stalk in predivisional cells. Furthermore, since its periplasmic muramidase domain is sufficient to localize SpmX to the stalked pole (Radhakrishnan *et al.*, 2008), it seems likely that another factor transmits the organizing effects of PopZ across the cytoplasmic membrane to SpmX.

Coordinated changes in two-component signalling and cyclic diguanylate link cell cycle progression and morphogenesis

Compartmentalization of the predivisional cell creates a sharp distinction between the biochemical environments in the two progeny, enabling immediate diversification of cell contents and functions. In contrast, swarmer cell differentiation occurs without the benefit of a sharp structural transition or separation of compartments. Nonetheless, during differentiation, many processes occur which fundamentally alter the identity of the cell or its position in the division cycle, such as deactivation of CtrA, the initiation of chromosome replication, ejection of the polar flagellum, and the beginning of stalk biogenesis. What molecular signals trigger the onset of DNA replication and polar morphogenesis in the developing swarmer cell? Studies in *Caulobacter* have been among the first to uncover tight linkages between two-component signalling and the second messenger cyclic c-di-GMP. In the current model, changes in the activity of key histidine kinases CckA, DivJ, and PleC are coordinated with an increase in the cellular level of c-di-GMP to trigger the G1-S transition and the remodelling of polar structures.

In the *Caulobacter* cell cycle, CtrA proteolysis is spatially and temporally restricted such that it occurs only during the SW-ST transition (Fig. 13.2B) and in the stalked compartment of the late predivisional cell (Domian *et al.*, 1997) (Fig. 13.3B). Yet ClpXP, the protease that degrades CtrA, is present throughout the cell cycle (Jenal and Fuchs, 1998). Determining how the stability of this key regulator is controlled is central to understanding cell cycle progression and morphogenesis in *Caulobacter*. At one level, CtrA proteolysis is understood: it is triggered by the inhibition of CckA kinase activity. In the unphosphorylated state, the single-domain response regulator CpdR promotes the degradation of a number of proteins, including CtrA. The activities of both CpdR and CtrA are regulated by the CckA-ChpT phosphorelay (Biondi *et al.*, 2006a). When CtrA is activated by phosphorylation, proteolysis of CtrA is simultaneously blocked by phosphorylation of CpdR (Iniesta *et al.*, 2006). CpdR itself is degraded by ClpXP shortly after proteolysis of CtrA, allowing for re-accumulation of CtrA after the initiation of chromosome replication (Iniesta and Shapiro, 2008). However, if we look closer, CtrA proteolysis becomes more complex, involving an additional response regulator and c-di-GMP.

In addition to CpdR and ClpXP, regulated CtrA proteolysis requires RcdA, a protein of unknown function, and PopA, a c-di-GMP-binding response regulator (McGrath *et al.*, 2006; Duerig *et al.*, 2009) (Fig. 13.2). Prior to CtrA degradation, all of these proteins are localized, either to the nascent stalked pole at the SW-ST transition, or to the stalked pole of the predivisional cell (Ryan *et al.*, 2002; Iniesta *et al.*, 2006; McGrath *et al.*, 2006; Duerig *et al.*, 2009). A branched dependency pathway mediates their localization, such that the polar organizing protein PopZ is needed for CpdR localization (Bowman *et al.*, 2010), and localization of ClpXP requires localized CpdR (Iniesta *et al.*, 2006; Bowman *et al.*, 2010). ClpX is one factor needed to localize RcdA, and CtrA localization is dependent on RcdA (McGrath *et al.*, 2006; Taylor *et al.*, 2009). The second factor required for RcdA localization is PopA, a response regulator that binds c-di-GMP (Duerig *et al.*, 2009). Despite the existence of this elaborate scheme, it is not clear how polar localization of the protease and accessory factors affects the recognition and degradation of CtrA by ClpXP. For example, some mutations in *rcdA* that prevent polar accumulation of CtrA still yield wild-type rates of CtrA proteolysis *in vivo* (Taylor *et al.*, 2009). Thus, the accessory factors seem not to stimulate CtrA proteolysis simply by increasing the local substrate concentration at the cell pole.

Co-immunoprecipitation experiments indicate that CtrA, ClpXP, and RcdA form a complex *in vivo* (McGrath *et al.*, 2006; Duerig *et al.*, 2009), and CpdR also forms a complex with ClpXP (Iniesta *et al.*, 2006). CpdR, RcdA or PopA could, singly or in concert, act as an adaptor that binds CtrA and physically delivers it to the protease at the appropriate time in the cell cycle (Schmidt *et al.*, 2009). We would expect such an adaptor to be required for CtrA proteolysis *in vitro*, or to stimulate the rate of degradation, and to interact directly with the substrate and protease. Currently none of the accessory proteins has been shown to perform these functions. First, although PopA-RcdA, CpdR-ClpX, and CpdR–RcdA interactions have been observed in bacterial adenylate cyclase two-hybrid (BACTH) assays, none of these proteins gives a positive result for interaction with CtrA (Duerig *et al.*, 2009). Second, CtrA can be degraded *in vitro* by purified ClpXP without any additional proteins, and adding RcdA does not accelerate the reaction (Chien *et al.*, 2007). The effects of PopA, c-di-GMP, and CpdR on *in vitro* CtrA proteolysis have not yet been reported. Proteolysis of CtrA by ClpXP alone is extremely surprising, because CtrA is stabilized in mutants lacking any one of the accessory proteins. More work is required to elucidate the mechanism of regulated CtrA proteolysis, the roles of individual accessory proteins, and the relationship between ClpXP localization and its activity towards CtrA.

As noted above, one protein that stimulates CtrA proteolysis is PopA, which contains two tandem receiver domains followed by a GGDEF domain. GGDEF domains often have diguanylate cyclase activity, synthesizing c-di-GMP (Schirmer and Jenal, 2009). However, PopA acts as a sensor of c-di-GMP levels, binding this second messenger at an allosteric site (I-site) in the GGDEF domain (Duerig *et al.*, 2009). The second receiver domain in PopA lacks an intact phosphoryl acceptor site, and mutational analysis showed that phosphorylation of the conserved aspartate residue in the first receiver domain is not required for PopA function. Instead, PopA is localized to the stalked pole and promotes CtrA degradation when c-di-GMP is bound to its I-site (Duerig *et al.*, 2009), underscoring the fact that two-component proteins can have critical inputs other than phosphorylation (Fig. 13.2B).

The regulatory scheme for PopA suggests that cellular levels of c-di-GMP rise at the G1-S transition, when CtrA proteolysis is triggered (Fig. 13.1B). In fact, the overall c-di-GMP concentration does oscillate during the *Caulobacter* cell cycle, with highest levels occurring coincident with CtrA degradation and remodelling of the flagellar pole into a stalked pole (Paul *et al.*, 2008). PopA binds c-di-GMP with a K_d of ~2 μM (Duerig *et al.*, 2009), and a mixed population of rapidly growing *Caulobacter* was found to contain 1.1 μM c-di-GMP (Christen *et al.*, 2005). In a different set of experiments using a FRET-based c-di-GMP biosensor, newly born stalked cells contained > 500 nM c-di-GMP, while newly born swarmer cells transiently contained low levels of c-di-GMP, below 100 nM (Christen *et al.*, 2010). Thus, although *in vivo* measurements of [c-di-GMP] vary slightly, it appears that PopA is poised to respond to physiological changes in c-di-GMP levels during the cell cycle and in different cell types.

Two-component proteins in *Caulobacter* are not only sensors of c-di-GMP, they also contribute to cellular levels of this second messenger, reflecting a tight interweaving of these signalling mechanisms in cell cycle regulation. One of these proteins is PleD, a paralogue of PopA containing two tandem receiver domains followed by a GGDEF domain. Structural and biochemical studies using PleD were the first to establish the diguanylate cyclase (DGC) activity of GGDEF domains and provided critical insights into the catalytic mechanism of c-di-GMP production (Chan *et al.*, 2004; Paul *et al.*, 2004). When phosphorylated on aspartate 53 in the first receiver domain, PleD dimerizes, creating a 2-fold symmetrical active site where two GTP molecules react to form one molecule of c-di-GMP (Paul *et al.*, 2007; Wassmann *et al.*, 2007). This activation mechanism provides a direct linkage between two-component signalling and c-di-GMP production (Fig. 13.2B).

pleD mutants are hypermotile, defective in stalk biogenesis, and delayed in the production of holdfast, an adhesive material located at the tip of the stalk, indicating that PleD is necessary for morphological changes during swarmer-to-stalked cell development (Hecht and Newton, 1995; Aldridge *et al.*, 2003; Levi and Jenal, 2006). PleD activity is excluded from swarmer cells and is stimulated during development by histidine kinases that regulate its phosphorylation state (see below and Fig. 13.2) (Paul *et al.*, 2008). During differentiation, c-di-GMP produced by PleD is necessary for the degradation of the flagellar MS-ring protein FliF by ClpAP, though the mechanism by which c-di-GMP affects this process is not yet understood (Aldridge and Jenal, 1999; Paul *et al.*, 2004). On the other hand, CtrA degradation by ClpXP proceeds normally in strains lacking PleD (Aldridge and Jenal, 1999). This result suggested that other DGCs contribute to the cellular pool of c-di-GMP that activates PopA and promotes CtrA degradation.

Recently two proteins with opposing enzymatic activities were found to work with PleD in modulating c-di-GMP levels during the SW-ST transition (Abel *et al.*, 2011). DgcB is a GGDEF protein with diguanylate cyclase activity that is present throughout the cell cycle, while PdeA is a swarmer cell-specific phosphodiesterase that hydrolyses c-di-GMP to 5′ phosphoguanylyl-(3′–5′)-guanosine (pGpG) (Christen *et al.*, 2005; Abel *et al.*, 2011). Mutants lacking PdeA are defective in swarming and have an increased propensity to attach to surfaces, owing to premature holdfast synthesis. These phenotypes can be rescued by *ΔdgcB*, but not by *ΔpleD*, indicating that unlike PleD, DgcB is active in the swarmer cell (Fig. 13.2A). PdeA maintains low c-di-GMP levels in the swarmer cell despite the presence of active DgcB, thus prolonging the motile G1-phase of the cell cycle. During swarmer cell differentiation, two events increase the levels of c-di-GMP: PleD activation and PdeA degradation by ClpXP (Fig. 13.2B). If both *pleD* and *dgcB* are deleted, or if PleD is absent and PdeA is stabilized, then c-di-GMP levels are not sufficient to activate PopA, and CtrA proteolysis is impaired (Abel *et al.*, 2011).

Through PleD activation and PdeA proteolysis, the increase in [c-di-GMP] during the swarmer-to-stalked cell transition is linked to the extended phosphorelay that also controls CtrA activation and stability. CpdR, the same response regulator that is necessary for CtrA degradation, interacts with PdeA and ClpX in BACTH assays, and its unphosphorylated form is required for PdeA proteolysis *in vivo* and *in vitro* (Abel *et al.*, 2011). This recent discovery is the first example of a single-domain response regulator acting as an adaptor protein to trigger proteolysis in a cell cycle-regulated manner. In this way, CpdR indirectly stimulates CtrA proteolysis via degradation of PdeA and the subsequent rise in c-di-GMP levels (Fig. 13.2B). However, it appears that CpdR also performs a second role in CtrA proteolysis, because deletion of *pdeA* cannot overcome the CtrA degradation defect of a *ΔcpdR* mutant (Abel *et al.*, 2011).

The essential single-domain response regulator DivK is a central node in coordinating cell cycle progression via CtrA elimination and polar morphogenesis triggered by increased c-di-GMP levels. Dephosphorylation of CpdR is stimulated when DivK~P down-regulates the CckA/ChpT phosphorelay (Biondi *et al.*, 2006a). DivK also stimulates PleD phosphorylation by allosterically modulating the activities of DivJ and PleC (Paul *et al.*, 2008). *divJ* and *pleC* mutants both contain less PleD~P than a wild-type strain, suggesting that both kinases contribute to PleD phosphorylation (Aldridge *et al.*, 2003). DivJ readily phosphorylates PleD *in vitro*, but phosphotransfer from PleC to PleD is inefficient unless DivK is added to the reaction (Paul *et al.*, 2004, 2008). DivK also stimulates DivJ autophosphorylation and subsequent phosphotransfer to PleD or to DivK itself (Paul *et al.*, 2008). These experiments were the first demonstration of allosteric regulation of a histidine kinase by its cognate response regulator, and the mechanism of stimulation is still under investigation.

In the newly born stalked cell, activation of DivJ by DivK is thought to result in a positive feedback loop that increases cellular levels of DivK~P and PleD~P, triggering immediate entry into S-phase. In the swarmer progeny, PleC dephosphorylates DivK, which maintains activity of the CckA/ChpT phosphorelay and blocks the proteolysis of CtrA and PdeA (Fig. 13.2A). After a delay, DivJ accumulates in the swarmer cell, and PleC and DivJ transiently occupy the developing pole together. Perhaps DivK phosphorylation by DivJ helps to switch PleC into kinase mode, which further activates both DivK and PleD, triggering the G1-S transition and differentiation into a stalked cell (Paul *et al.*, 2008) (Fig. 13.2B).

In this chapter, we have assumed that c-di-GMP functions as a pool of molecules that diffuses freely within a cell or cell compartment. However, a study in *Pseudomonas aeruginosa* showed that two DGCs have similar effects on overall cellular levels of c-di-GMP, but regulate two distinct processes, flagellar motility and production of extracellular polysaccharides (Merritt *et al.*, 2010). These results suggest that localized production or sensing of c-di-GMP may be involved in cellular outputs. In *Caulobacter*, the DGC PleD becomes localized at the nascent stalked pole during the SW-ST transition and remains at this location for the rest of the cell cycle (Paul *et al.*, 2004). PleD localization is dependent on phosphorylation of its N-terminal receiver domain by DivJ and PleC, so when PleD is activated to produce c-di-GMP, it is also localized to the stalked pole. PleD may thereby generate a high local concentration of c-di-GMP, which could stimulate downstream processes during the SW-ST transition and contribute to asymmetry in the predivisional cell.

It is unclear what external signals influence c-di-GMP levels in *Caulobacter*, but in other bacteria, c-di-GMP levels have been linked to blue light, antibiotics, and nutrient and oxygen availability (Jenal and Malone, 2006). Therefore, c-di-GMP signalling provides a potential mechanism to integrate the internal drivers of the *Caulobacter* cell cycle with information about the external environment. The phosphodiesterase PdeA is allosterically activated by GTP, providing one potential mechanism linking c-di-GMP to the cell's energy status (Christen *et al.*, 2005). Future work will afford a deeper understanding of how external stimuli might modulate the *Caulobacter* cell cycle through c-di-GMP signalling.

Regulation of flagellum and stalk biogenesis

In addition to being a key cell cycle regulator, CtrA is also at the top of a transcriptional cascade of flagellar gene expression (Smith and Hoover, 2009). In fact, the first *ctrA* mutant was identified in a screen to detect both misregulation of the Class II flagellar gene *fliQ* and temperature-sensitive lethality (Quon *et al.*, 1996). Transcription of Class II flagellar genes is limited to the predivisional phase of the cell cycle by requiring both high levels of CtrA~P (Reisenauer *et al.*, 1999) and the absence of SciP, a swarmer cell-specific protein that represses CtrA-dependent transcription (Gora *et al.*, 2010).

In the canonical view of flagellar biogenesis, a master Class I regulator (in this case CtrA) drives the transcription of Class II genes, which encode proximal components of the flagellum and regulators for the downstream Class III and Class IV genes. Class III and IV genes, encoding distal structural components and flagellins, require the prior transcription of Class II genes. CtrA was first found to recognize an element present in the promoters of Class II flagellar genes, as well as in the *ccrM* promoter and the origin of replication (Stephens *et al.*, 1995; Quon *et al.*, 1996, 1998). However, chromatin immunoprecipitation microarray (ChIP-chip) experiments revealed that CtrA also binds to the promoters of some Class III and Class IV flagellar genes (Laub *et al.*, 2002), suggesting that the current view of the flagellar transcriptional hierarchy may be oversimplified.

Among the Class II flagellar genes activated by CtrA is *flbD*, which encodes a DNA-binding response regulator with homology to the nitrogen regulatory protein NtrC (Ramakrishnan and Newton, 1990). Response regulators in the NtrC family have three domains, an N-terminal receiver domain, a central AAA+ ATPase domain, and a C-terminal DNA-binding domain. These proteins activate transcription in concert with RNA polymerase containing a specific sigma factor, σ^{54} (Rombel *et al.*, 1998). FlbD works with RNA polymerase containing the *Caulobacter* σ^{54} protein RpoN to transcribe Class III and IV flagellar genes, but FlbD can also behave as a transcriptional repressor (Ramakrishnan and Newton, 1990; Brun and Shapiro, 1992; Wingrove *et al.*, 1993; Benson *et al.*, 1994a). FlbD is phosphorylated *in vivo* and can be phosphorylated *in vitro* using *Caulobacter* extracts (Wingrove *et al.*, 1993), but FlbD is active in *in vitro* transcription assays without phosphorylation (Benson *et al.*, 1994b; Wu *et al.*, 1995). Moreover, a FlbD-specific histidine kinase has not been identified (Muir and Gober, 2001). Instead, FlbD activity is controlled by direct interaction with a trans-acting factor FliX that is thought to sense the assembly of the flagellar basal body (Muir and Gober, 2004).

Caulobacter fliX and *flbD* loss-of-function mutants are non-motile and fail to transcribe Class III and IV flagellar genes (Wingrove *et al.*, 1993; Mohr *et al.*, 1998; Muir *et al.*, 2001). In addition, mutations that block the assembly of early flagellar structures also prevent transcription of Class III/IV genes (Mangan *et al.*, 1995). Bypass mutations that restore Class III/IV gene expression to cells lacking Class II-encoded structures can be found in both *flbD* and *fliX*, and a bypass mutation in *flbD* can restore motility to a *fliX* null mutant (Mangan *et al.*, 1995; Muir *et al.*, 2001). These results suggest that FliX is not a structural component of the flagellum, but part of a checkpoint ensuring that early flagellar structures are complete before late flagellar proteins are synthesized. *In vitro*, FliX enhances FlbD-activated transcription by inducing FlbD to form higher-order oligomers (Dutton *et al.*, 2005), similar to activated NtrC (Porter *et al.*, 1993). Interestingly, overexpression of FliX in wild-type cells suppresses FlbD-activated transcription (Muir *et al., 2001).* Together, these findings suggest that FliX switches from a FlbD-repressing to a FlbD-activating mode upon completion of the Class II flagellar basal body structure. Regulation of a σ^{54}-dependent transcriptional activator by direct binding to another protein is novel (Muir and Gober, 2004), and the precise mechanism of FliX action is still being unravelled.

Caulobacter RpoN functions in concert with a different σ^{54}-dependent transcriptional activator, TacA, to promote stalk biogenesis (Brun and Shapiro, 1992; Skerker *et al.*, 2005; Biondi *et al.*, 2006b). Systematic deletion of each two-component gene in the *Caulobacter* genome revealed that two mutants, *ΔCC0138* and *ΔtacA* were stalkless in rich medium, suggesting that they function in a pathway to regulate stalk biogenesis (Skerker *et al.*, 2005). *CC0138* encodes a hybrid histidine kinase with N-terminal DHp and CA domains, two C-terminal receiver domains, and no obvious sensory domains. A bioinformatics approach identified a histidine phosphotransferase protein encoded by *CC1114*, and phosphotransfer profiling experiments demonstrated that CC1114 (ShpA) preferentially transfers a phosphoryl group between the second receiver domain of CC0138 (ShkA) and the receiver domain of TacA (Biondi *et al.*, 2006b). The constitutively active phosphomimetic protein TacA-D54E rescues the stalk biogenesis phenotype of *shkA* and *shpA* mutants (Biondi *et al.*, 2006b). Together these results indicate that ShkA and ShpA activate TacA through phosphorylation, leading to stalk biogenesis. Further studies are needed to determine the signals sensed by ShkA and when and where in the cell it is active.

Mutations in other two-component genes, notably *ctrA* and *pleC*, result in cells that are non-motile and stalkless when grown in rich medium, similar to cells lacking RpoN (Sommer and Newton, 1989; Brun and Shapiro, 1992; Quon *et al.*, 1996). These data can be rationalized by the facts that CtrA directly activates the transcription of *flbD*, *tacA*, and *rpoN* (Mohr *et al.*, 1998; Laub *et al.*, 2002), and PleC promotes CtrA activity by dephosphorylating DivK~P (Biondi *et al.*, 2006a; Tsokos *et al.*, 2011). The stalk biogenesis defects of *ctrA*, *pleC*, and *tacA* pathway mutants can be partially rescued by growth in minimal medium containing low amounts of phosphate, and low-phosphate conditions increase the stalk length of wild-type cells (Gonin *et al.*, 2000). Mutations affecting the Pst phosphate uptake system cause stalk elongation even in the presence of high levels of phosphate, while mutations in the gene encoding the response regulator PhoB prevent stalk elongation in low-phosphate medium (Gonin *et al.*, 2000). Since the *E. coli* Pst proteins deactivate PhoR, the cognate kinase of PhoB, when phosphate is in excess (Wanner, 1993), these results in *Caulobacter* suggest that the lack of phosphate is sensed by the Pst system, leading to phosphorylation of PhoB and a PhoB~P-dependent transcriptional response that promotes stalk elongation.

Despite knowledge of two regulatory pathways, developmental and starvation-induced, that affect the stalk, the relationship between these pathways has yet to be defined, and few proteins directly involved in stalk construction have been identified (Kuhn *et al.*, 2010). Recently, the first mutant to remain stalkless even in low-phosphate medium was discovered (Abel *et al.*, 2011).

Deletion of two diguanylate cyclases, *pleD* and *dgcB*, blocks both developmental and phosphate starvation-induced stalk biogenesis, suggesting that these developmental pathways converge on the second messenger c-di-GMP. Additional work is needed to uncover the mechanism linking the c-di-GMP signal to stalk elongation.

Two-component signalling systems responding to environmental conditions and stress

FixLJKT and oxygen

The FixLJKT signalling pathway is essential in species such as *Sinorhizobium meliloti* and *Bradyrhizobium japonicum* for the expression of nitrogen fixation (*nif*) genes in the low-oxygen environment of symbiotic root nodules. In *S. meliloti*, this system comprises the sensor histidine kinase FixL, the DNA-binding response regulator FixJ, a transcriptional regulator FixK, and a protein FixT, which inhibits the formation and/or accumulation of FixL~P (Dixon and Kahn, 2004). The PAS domain of FixL contains a haem prosthetic group that binds oxygen, inhibiting FixL kinase activity (Gilles-Gonzalez *et al.*, 1995). Under low oxygen concentrations, oxygen dissociates from the haem group, freeing FixL to autophosphorylate and transfer the phosphoryl group to the response regulator FixJ (Tuckerman *et al.*, 2002). FixJ~P activates the expression of both NifA, which drives the transcription of additional *nif* genes, and FixK, a transcriptional regulator that activates genes necessary for growth in a microaerobic environment (Dixon and Kahn, 2004). This pathway ensures that the oxygen-sensitive and energetically costly process of nitrogen fixation is limited to the low-oxygen environment of root nodules.

The Fix pathway exists in *Caulobacter*, where it is used not for nitrogen fixation, but to respond to low oxygen levels (Crosson *et al.*, 2005). The Fix signalling pathway is activated under low-oxygen conditions and controls the expression of genes for respiratory terminal oxidases, cytochrome *c* biogenesis, haem biosynthesis, and alternative sugar and amino acid metabolism (Crosson *et al.*, 2005). In *Caulobacter*, FixJ~P directly activates genes involved in haem and cytochrome *c* biogenesis. FixJ~P also promotes the transcription of *fixK*, which controls the expression of alternative terminal oxidase genes. FixK positively regulates the expression of high-affinity terminal oxidases and represses the expression of those with low oxygen affinity. The result of Fix pathway activation is to allow *Caulobacter* growth under low oxygen conditions (Crosson *et al.*, 2005).

Caulobacter FixK drives the transcription of *fixL*, generating a positive feedback loop that increases the number of FixL molecules in the cell (Crosson *et al.*, 2005). This regulatory scheme is thought to allow a rapid response to a decrease in oxygen concentration. However, FixK also promotes the expression of a distant homologue of FixT, encoded by *CC0753*, which is predicted to down-regulate the FixLJ pathway. In a Δ*CC0753* mutant, *fixK* transcription is derepressed under high oxygen concentrations, supporting the role of *Caulobacter* FixT in inhibiting the FixLJ pathway. Coordinate activation of *fixL* and its inhibitor *fixT* is proposed to prevent runaway expression of FixL (Crosson *et al.*, 2005). Interestingly, although FixT was first reported to be unrelated to known proteins, it has weak homology to response regulator receiver domains and contains all of the conserved residues that catalyse phosphate acquisition from histidine kinases (Foussard *et al.*, 1997; Garnerone *et al.*, 1999; Crosson *et al.*, 2005). Future work will determine if FixT down-regulates the FixLJ pathway by competing with FixJ for phosphorylation by or binding to FixL.

PhyKR and extracytoplasmic stress

Bacterial σ-factors are subunits of RNA polymerase that mediate the recognition of specific sets of promoters. In addition to a 'housekeeping' σ-factor that transcribes many genes in an organism's genome, bacteria also contain alternative σ-factors that recognize smaller sets of promoters. Alternative σ-factors direct the transcription of subsets of genes that allow adaptation to specific external conditions (Grainger and Busby, 2008; Balleza *et al.*, 2009). In particular, the subfamily of extracytoplasmic function (ECF) σ-factors directs transcription of genes relating to cell envelope functions such as secretion, transport, and responding to extracytoplasmic stress (Helmann, 2002). To restrict ECF σ-factor activity to the appropriate conditions, anti-σ-factors can bind to them and prevent their association with RNA polymerase. In the presence of the correct stimulus, the anti-σ-factor releases the σ-factor so that it can associate with RNA polymerase and mediate a transcriptional response. In some cases, an anti-anti-σ-factor binds to the anti-σ-factor to release the σ-factor (Hughes and Mathee, 1998).

The protein PhyR, first described in *Methylobacterium extorquens*, directly links two global regulatory systems, two-component signalling and the use of alternative σ-factors. PhyR has an N-terminal domain similar to ECF σ-factors and a C-terminal receiver domain (Gourion *et al.*, 2006). The σ-like N-terminal domain has significant similarity to σ^{EcfG}, the general stress regulator in *M. extorquens*, but it does not function as a classical σ-factor. Instead PhyR controls gene expression by interacting with the anti-σ-factor NepR (Francez-Charlot *et al.*, 2009). NepR binds to σ^{EcfG}, restricting its activity in unstressed cells. From these studies, stress-induced phosphorylation of the PhyR receiver domain was hypothesized to stimulate its binding to NepR, releasing σ^{EcfG}. This model has been supported and extended by recent studies of PhyR in *Caulobacter*.

M. extorquens σ^{EcfG} is orthologous to σ^{T}, which mediates the *Caulobacter* response to oxidative and osmotic stress (Alvarez-Martinez *et al.*, 2007). Genes adjacent to *sigT* encode homologues of NepR and PhyR (Herrou *et al.*, 2010). Mutants overexpressing *nepR*, lacking *sigT* or *phyR*, or expressing a mutant form of PhyR lacking the conserved site of phosphorylation (PhyR-D129A) have reduced survival after osmotic stress (Herrou *et al.*, 2010). The crystal structure of unphosphorylated PhyR revealed an intramolecular interface between the σ-like and receiver domains. This interface includes secondary elements in the PhyR receiver domain equivalent to the α4–β5–α5 surface of isolated receiver domains (Herrou *et al.*, 2010). Importantly, this interface is distinct from the surface required for binding and phosphotransfer between a receiver domain and its cognate histidine kinase (Skerker *et al.*, 2008; Casino *et al.*, 2009), so the intramolecular interaction would not preclude phosphorylation of PhyR. Furthermore, in the consensus model of receiver domain activation, phosphorylation creates the largest conformational changes at the α4-β5-α5 surface (Bourret, 2010; Gao and Stock, 2010), suggesting that phosphorylation-driven rearrangements could stimulate binding between the PhyR σ-like domain and NepR, thereby releasing σ^{T} to direct transcription.

In fact, *Caulobacter* σ^{T} and PhyR each bind to NepR *in vitro*, and PhyR binding requires phosphorylation by acetyl phosphate (Lourenço *et al.*, 2011). *In vivo*, phosphorylation of PhyR depends on the histidine kinase PhyK, encoded downstream of *sigT*. Like cells lacking σ^{T} or PhyR, *phyK* mutants have ~1000-fold lower survival after osmotic or oxidative stress, and they fail to activate σ^{T}-dependent genes. PhyK contains a periplasmic sensory domain, and mutation of a conserved cysteine residue in this domain produces a *phyK* null phenotype (Lourenço *et al.*, 2011). It will be of great interest to learn how osmotic and oxidative stresses activate PhyK.

LovKR, light and redox

Bacteria can sense and respond to light in the visible and near-infrared region of the electromagnetic spectrum through various proteins that respond to different wavelengths of light (Losi and Gartner, 2008; Purcell and Crosson, 2008). LOV (light, oxygen or voltage) domains are a subclass of the PAS domain family that bind a flavin cofactor and can detect light in the blue spectrum (440 to 490 nm). When the flavin cofactor is illuminated with blue light, it forms a covalent bond with a cysteine residue in the LOV domain that activates an attached output domain. Around 50% of bacterial LOV domains are associated with histidine kinase domains, while ~20% are associated with GGDEF or EAL domains that synthesize or hydrolyse c-di-GMP (Herrou and Crosson, 2011). Bacterial LOV proteins regulate such various cellular processes as stress responses (Avila-Pérez *et al.*, 2006), virulence (Swartz *et al.*, 2007), and surface attachment (Purcell *et al.*, 2007).

Caulobacter senses and responds to light through a two-component system composed of the histidine kinase LovK and the single-domain response regulator LovR (Purcell *et al.*, 2007). *lovK* and *lovR* are adjacent in the chromosome and are maximally expressed during the swarmer-to-stalked cell transition. Deletion of *lovK* or *lovR* reduces *Caulobacter* surface attachment, while simultaneous overexpression of both genes provokes a light-dependent increase in cell–cell attachment via holdfast material at the tips of stalks (Purcell *et al.*, 2007). The conserved sites of phosphorylation in LovK and LovR are required for the overexpression- and light-dependent increase in attachment, as is a conserved cysteine residue in the LOV domain of LovK that forms a covalent adduct with the flavin cofactor upon light absorption (Purcell *et al.*, 2007). These data indicate that the LovKR pathway regulates cell adhesion via light sensing and phosphotransfer, but it is unclear how the single-domain response regulator LovR generates this output.

How does *Caulobacter* benefit by sensing light through LovKR? This system might provide a growth advantage by anchoring the cell to nutrient-containing detritus found near the surface of dilute freshwater environments (Purcell *et al.*, 2007). Additionally, LovK could act as a sensor of cellular redox potential, similar to the NifL-NifA system in *Azotobacter vinelandii* (Martinez-Argudo *et al.*, 2004). The reduction potential for the LovK flavin is around −258 mV, close to the average cytoplasmic reduction potential during exponential growth, −260 to −280 mV (Purcell *et al.*, 2010). Thus, some LovK molecules in the cytoplasm are likely to be reduced, and the ratio of oxidized to reduced LovK may be sensitive to fluctuations in the redox state of the cell. Blue light absorption and the formation of the cysteine–flavin adduct require the oxidized form of the flavin cofactor, so reduction of LovK flavins via changes in the cytoplasmic reduction potential are predicted to affect the cell's ability to respond to blue light via LovK. In this way, LovK can integrate external and internal signals to optimize surface attachment (Purcell *et al.*, 2010).

Conclusions and outlook

The last decade of research on two-component proteins in *Caulobacter* has revealed interesting new features that deepen our understanding of how these signalling systems operate *in vivo*. The localization of histidine kinases to particular positions in the cell was first observed in *Caulobacter* (Jacobs *et al.*, 1999), and it has since been shown to be functionally important for *Caulobacter* cell polarity and asymmetric division. The field would greatly benefit from methods to specifically visualize the phosphorylated species of response regulators, which would enable the mapping of two-component signalling activity onto living cells. The *Caulobacter* cell cycle and developmental signalling network is highly branched and involves non-canonical interactions among two-component proteins, in which histidine kinase activity can be modulated by response regulators or by other histidine kinases (Paul *et al.*, 2008; Iniesta *et al.*, 2010a; Tsokos *et al.*, 2011). Additional work will continue to refine our understanding of these regulatory mechanisms and will likely demonstrate new types of interactions among two-component proteins. Finally, work in *Caulobacter* has underscored the intimate connection between two-component signal transduction and the second messenger molecule c-di-GMP. Specific response regulator proteins in the cell cycle network have been shown to generate c-di-GMP, bind directly to c-di-GMP, or mediate the timed degradation of enzymes that hydrolyse c-di-GMP (Paul *et al.*, 2007; Duerig *et al.*, 2009; Abel *et al.*, 2011). Coordination between these two signalling modes is required to orchestrate asymmetric cell division and the morphological development of a swarmer cell into a stalked cell. We anticipate that future studies will uncover new connections between two-component signalling and c-di-GMP and will focus on the localized production of c-di-GMP as a means of regulating biochemical pathways in specific regions within the cell.

Acknowledgements

This work was supported by NSF award MCB 0920619.

References

Abel, S., Chien, P., Wassmann, P., Schirmer, T., Kaever, V., Laub, M.T., Baker, T.A., and Jenal, U. (2011). Regulatory cohesion of cell cycle and cell differentiation through interlinked phosphorylation and second messenger networks. Mol. Cell *43*, 550–560.

Aldridge, P., and Jenal, U. (1999). Cell cycle-dependent degradation of a flagellar motor component requires a novel-type response regulator. Mol. Microbiol. *32*, 379–391.

Aldridge, P., Paul, R., Goymer, P., Rainey, P., and Jenal, U. (2003). Role of the GGDEF regulator PleD in polar development of *Caulobacter crescentus*. Mol. Microbiol. *47*, 1695–1708.

Alvarez-Martínez, C.E., Lourenço, R.F., Baldini, R.L., Laub, M.T., and Gomes, S.L. (2007). The ECF sigma factor σ^T is involved in osmotic and oxidative stress responses in *Caulobacter crescentus*. Mol. Microbiol. *66*, 1240–1255.

Angelastro, P.S., Sliusarenko, O., and Jacobs-Wagner, C. (2010). Polar localization of the CckA histidine kinase and cell cycle periodicity of the essential master regulator CtrA in *Caulobacter*. J. Bacteriol. *192*, 539–552.

Avila-Pérez, M., Hellingwerf, K.J., and Kort, R. (2006). Blue light activates the sigmaB-dependent stress response of *Bacillus subtilis* via YtvA. J. Bacteriol. *188*, 6411–6414.

Balleza, E., Lopez-Bojorquez, L.N., Martinez-Antonio, A., Resendis-Antonio, O., Lozada-Chavez, I., Balderas-Martinez, Y.I., Encamacion, S., and Collado-Vides, J. (2009). Regulation by transcription factors in bacteria: beyond description. FEMS Microbiol. Rev. *33*, 133–151.

Bastedo, D.P., and Marczynski, G.T. (2009). CtrA response regulator binding to the *Caulobacter* chromosome replication origin is required during nutrient and antibiotic stress as well as during cell cycle progression. Mol. Microbiol. *72*, 139–154.

Benson, A.K., Ramakrishnan, G., Ohta, N., Feng, J., Ninfa, A.J., and Newton, A. (1994a). The *Caulobacter crescentus* FlbD protein acts at *ftr* sequence elements both to activate and to repress transcription of cell cycle-regulated flagellar genes. Proc. Natl. Acad. Sci. U.S.A. *91*, 4989–4993.

Benson, A.K., Wu, J., and Newton, A. (1994b). The role of FlbD in regulation of flagellar gene transcription in *Caulobacter crescentus*. Res. Microbiol. *145*, 420–430.

Biondi, E.G. Reisinger, S.J., Skerker, J.M., Arif, M., Perchuk, B.S., Ryan, K.R., and Laub, M.T. (2006a). Regulation of the bacterial cell cycle by an integrated genetic circuit. Nature *444*, 899–904.

Biondi, E.G. Skerker, J.M., Arif, M., Prasol, M.S., Perchuk, B.S., and Laub, M.T. (2006b). A phosphorelay system controls stalk biogenesis during cell cycle progression in *Caulobacter crescentus*. Mol. Microbiol. *59*, 386–401.

Bourret, R.B. (2010). Receiver domain structure and function in response regulator proteins. Curr. Opin. Microbiol. *13*, 142–149.

Bowman, G.R., Comolli, L.R., Zhu, J., Eckart, M., Koenig, M., Downing, K.H., Moerner, W.E., Earnest, T., and Shapiro, L. (2008). A polymeric protein anchors the chromosomal origin/ParB complex at a bacterial cell pole. Cell *134*, 945–955.

Bowman, G.R., Comolli, L.R., Gaietta, G.M., Fero, M., Hong, S.H., Jones, Y., Lee, J.H., Downing, K.H., Ellisman, M.H., Mcadams, H.H., *et al.* (2010). *Caulobacter* PopZ forms a polar subdomain dictating sequential changes in pole composition and function. Mol. Microbiol. *76*, 173–189.

Brun, Y.V., and Shapiro, L. (1992). A temporally controlled sigma-factor is required for polar morphogenesis and normal cell division in *Caulobacter*. Genes Dev. *6*, 2395–2408.

Casino, P., Rubio, V., and Marina, A. (2009). Structural insight into parter specificity and phosphoryl transfer in two-component signal transduction. Cell *139*, 325–336.

Chan, C., Paul, R., Samoray, D., Amiot, N.C., Giese, B., Jenal, U., and Schirmer, T. (2004). Structural basis of activity and allosteric control of diguanylate cyclase. Proc. Natl. Acad. Sci. U.S.A. *101*, 17084–17089.

Chen, Y.E., Tsokos, C.G., Biondi, E.G. Perchuk, B.S., and Laub, M.T. (2009). Dynamics of two phosphorelays controlling cell cycle progression in *Caulobacter crescentus*. J. Bacteriol. *191*, 7417–7429.

Chen, Y.E., Tropini, C., Jonas, K., Tsokos, C.G., Huang, K.C., and Laub, M.T. (2011). Spatial gradient of protein phosphorylation underlies replicative asymmetry in a bacterium. Proc. Natl. Acad. Sci. U.S.A. *108*, 1052–1057.

Chien, P., Perchuk, B.S., Laub, M.T., Sauer, R.T., and Baker, T.A. (2007). Direct and adaptor-mediated substrate recognition by an essential AAA+ protease. Proc. Natl. Acad. Sci. U.S.A. *104*, 6590–6595.

Christen, M., Christen, B., Folcher, M., Schauerte, A., and Jenal, U. (2005). Identification and characterization of a cyclic di-GMP-specific phosphodiesterase and its allosteric control by GTP. J. Biol. Chem. *280*, 30829–30837.

Christen, M., Kulasekara, H.D., Christen, B., Kulasekara, B.R., Hoffman, L.R., and Miller, S.I. (2010). Asymmetrical distribution of the second messenger c-di-GMP upon bacterial cell division. Science *328*, 1295–1297.

Crosson, S., McGrath, P.T., Stephens, C., McAdams, H.H., and Shapiro, L. (2005). Conserved modular design of an oxygen sensory/signaling network with species-specific output. Proc. Natl. Acad. Sci. U.S.A. *102*, 8018–8023.

Degnen, S.T., and Newton, A. (1972). Chromosome replication during development in *Caulobacter crescentus*. J. Mol. Biol. *64*, 671–680.

Dixon, R., and Kahn, D. (2004). Genetic regulation of biological nitrogen fixation. Nat. Rev. Microbiol. *2*, 621–631.

Domian, I.J., Quon, K.C., and Shapiro, L. (1997). Cell type-specific phosphorylation and proteolysis of a transcriptional regulator controls the G1-to-S transition in a bacterial cell cycle. Cell *90*, 415–424.

Domian, I.J., Reisenauer, A., and Shapiro, L. (1999). Feedback control of a master bacterial cell-cycle regulator. Proc. Natl. Acad. Sci. U.S.A. *96*, 6648–6653.

Duerig, A., Abel, S., Folcher, M., Nicollier, M., Schwede, T., Amiot, N., Giese, B., and Jenal, U. (2009). Second messenger-mediated spatiotemporal control of protein degradation regulates bacterial cell cycle progression. Genes Dev. *23*, 93–104.

Dutton, R.J., Xu, Z., and Gober, J.W. (2005). Linking structural assembly to gene expression: a novel mechanism for regulating the activity of a sigma54 transcription factor. Mol. Microbiol. *58*, 743–757.

Ebersbach, G., Briegel, A., Jensen, G.J., and Jacobs-Wagner, C. (2008). A self-associating protein critical for chromosome attachment, division, and polar organization in *Caulobacter*. Cell *134*, 956–968.

Foussard, M., Garnerone, A.M., Ni, F., Soupene, E., Boistard, P., and Batut, J. (1997). Negative autoregulation of the *Rhizobium meliloti fixK* gene is indirect and requires a newly identified regulator, FixT. Mol. Microbiol. *25*, 27–37.

Francez-Charlot, A., Frunzke, J., Reichen, C., Ebneter, J.Z., Gourion, B., and Vorholt, J.A. (2009). Sigma factor mimicry involved in regulation of general stress response. Proc. Natl. Acad. Sci. U.S.A. *106*, 3467–3472.

Gao, R., and Stock, A.M. (2010). Molecular strategies for phosphorylation-mediated regulation of response regulator activity. Curr. Opin. Microbiol. *13*, 160–167.

Garnerone, A.M., Cabanes, D., Foussard, M., Boistard, P., and Batut, J. (1999). Inhibition of the FixL sensor kinase by the FixT protein in *Sinorhizobium meliloti*. J. Biol. Chem. *274*, 32500–32506.

Gilles-Gonzalez, M.A., Gonzalez, G., and Perutz, M.F. (1995). Kinase activity of oxygen sensor FixL depends on the spin state of its heme iron. Biochemistry *34*, 232–236.

Gober, J.W., and Shapiro, L. (1992). A developmentally regulated *Caulobacter* flagellar promoter is activated by 3′ enhancer and IHF binding elements. Mol. Biol. Cell *3*, 913–926.

Gonin, M., Quardokus, E.M., O'Donnol, D., Maddock, J., and Brun, Y.V. (2000). Regulation of stalk elongation by phosphate in *Caulobacter crescentus*. J. Bacteriol. *182*, 337–347.

Gora, K.G., Tsokos, C.G., Chen, Y.E., Srinivasan, B.S., Perchuk, B.S., and Laub, M.T. (2010). A cell-type-specific protein–protein interaction modulates transcriptional activity of a master regulator in *Caulobacter crescentus*. Mol. Cell *39*, 45–467.

Gourion, B., Rossignol, M., and Vorholt, J.A. (2006). A proteomic study of *Methylobacterium extorquens* reveals a response regulator essential for epiphytic growth. Proc. Natl. Acad. Sci. U.S.A. *103*, 13186–13191.

Grainger, D.C., and Busby, S.J.W. (2008). Global regulators of transcription in *Escherichia coli*: mechanisms of action and methods for study. Adv. Appl. Microbiol. *65*, 93–113.

Hecht, G.B., and Newton, A. (1995). Identification of a novel response regulator required for the swarmer-to-stalked-cell transition in *Caulobacter crescentus*. J Bacteriol *177*, 6223–6229.

Hecht, G.B., Lane, T., Ohta, N., Sommer, J.M., and Newton, A. (1995). An essential single domain response regulator required for normal cell division and differentiation in *Caulobacter crescentus*. EMBO J. *14*, 3915–3924.

Helmann, J.D. (2002). The extracytoplasmic function (ECF) sigma factors. Adv. Microb. Physiol. *46*, 47–110.

Herrou, J., and Crosson, S. (2011). Function, structure and mechanism of bacterial photosensory LOV proteins. Nat. Rev. Microbiol. *9*, 713–723.

Herrou, J., Foreman, R., Fiebig, A., and Crosson, S. (2010). A structural model of anti-anti-σ inhibition by a two-component receiver domain: the PhyR stress response regulator. Mol. Microbiol. *78*, 290–304.

Hinz, A.J., Larson, D.E., Smith, C.S., and Brun, Y.V. (2003). The *Caulobacter crescentus* polar organelle development protein PodJ is differentially localized and is required for polar targeting of the PleC development regulator. Mol. Microbiol. *47*, 929–941.

Hottes, A.K., Shapiro, L., and McAdams, H.H. (2005). DnaA coordinates replication initiation and cell cycle transcription in *Caulobacter crescentus*. Mol. Microbiol. *58*, 1340–1353.

Hughes, K.T., and Mathee, K. (1998). The anti-sigma factors. Annu. Rev. Microbiol. *78*, 290–304.

Hung, D., and Shapiro, L. (2002). A signal transduction protein cues proteolytic events critical to *Caulobacter* cell cycle progression. Proc. Natl. Acad. Sci. U.S.A. *99*, 13160–13165.

Hwang, D.S., and Kornberg, A. (1992). Opening of the replication origin of *Escherichia coli* by DnaA protein with protein HU or IHF. J. Biol. Chem. *267*, 23083–23086.

Iniesta, A.A., and Shapiro, L. (2008). A bacterial control circuit integrates polar localization and proteolysis of key regulatory proteins with a phospho-signaling cascade. Proc. Natl. Acad. Sci. U.S.A. *105*, 16602–16607.

Iniesta, A.A., McGrath, P.T., Reisenauer, A., McAdams, H.H., and Shapiro, L. (2006). A phospho-signaling pathway controls the localization and activity of a protease complex critical for bacterial cell cycle progression. Proc. Natl. Acad. Sci. U.S.A. *103*, 10935–10940.

Iniesta, A.A., Hillson, N.J., and Shapiro, L. (2010a). Cell pole-specific activation of a critical bacterial cell cycle kinase. Proc. Natl. Acad. Sci. U.S.A. *107*, 7012–7017.

Iniesta, A.A., Hillson, N.J., and Shapiro, L. (2010b). Polar remodeling and histidine kinase activation, which is essential for *Caulobacter* cell cycle progression, are dependent on DNA replication initiation. J. Bacteriol. *192*, 3893–3902.

Jacobs, C., Domian, I.J., Maddock, J.R., and Shapiro, L. (1999). Cell cycle-dependent polar localization of an essential bacterial histidine kinase that controls DNA replication and cell division. Cell *97*, 111–120.

Jacobs, C., Hung, D., and Shapiro, L. (2001). Dynamic localization of a cytoplasmic signal transduction response regulator controls morphogenesis during the *Caulobacter* cell cycle. Proc. Natl. Acad. Sci. U.S.A. *98*, 4095–4100.

Jacobs, C., Ausmees, N., Cordwell, S.J., Shapiro, L., and Laub, M.T. (2003). Functions of the CckA histidine kinase in *Caulobacter* cell cycle control. Mol. Microbiol. *47*, 1279–1290.

Jenal, U., and Fuchs, T. (1998). An essential protease involved in bacterial cell-cycle control. EMBO J. *17*, 5658–5669.

Jenal, U., and Malone, J. (2006). Mechanisms of cyclic-di-GMP signaling in bacteria. Annu. Rev. Genet. *40*, 385–407.

Jonas, K., Chen, Y.E., and Laub, M.T. (2011). Modularity of the bacterial cell cycle enables independent spatial and temporal control of DNA replication. Curr. Biol. *21*, 1092–1101.

Judd, E.M., Ryan, K.R., Moerner, W.E., Shapiro, L., and McAdams, H.H. (2003). Fluorescence bleaching reveals asymmetric compartment formation prior to cell division in *Caulobacter*. Proc. Natl. Acad. Sci. U.S.A. *100*, 8235–8240.

Kuhn, J., Briegel, A., Morschel, E., Kahnt, J., Leser, K., Wick, S., Jensen, G.J., and Thanbichler, M. (2010). Bactofilins, a ubiquitous class of cytoskeletal proteins mediating polar localization of a cell wall synthase in *Caulobacter crescentus*. EMBO J. *29*, 327–339.

Lam, H., Matroule, J.-Y., and Jacobs-Wagner, C. (2003). The asymmetric spatial distribution of bacterial signal transduction proteins coordinates cell cycle events. Dev. Cell *5*, 149–159.

Laub, M.T., McAdams, H.H., Feldblyum, T., Fraser, C.M., and Shapiro, L. (2000). Global analysis of the genetic network controlling a bacterial cell cycle. Science *290*, 2144–2148.

Laub, M.T., Chen, S.L., Shapiro, L., and McAdams, H.H. (2002). Genes direcly controlled by CtrA, a master regulator of the *Caulobacter* cell cycle. Proc. Natl. Acad. Sci. U.S.A. *99*, 4632–4637.

Laub, M.T., Biondi, E.G., and Skerker, J.M. (2007). Phosphotransfer profiling: systematic mapping of two-component signal transduction pathways and phosphorelays. Methods Enzymol. *423*, 531–548.

Lawler, M.L., Larson, D.E., Hinz, A.J., Klein, D., and Brun, Y.V. (2006). Dissection of functional domains of the polar localization factor PodJ in *Caulobacter crescentus*. Mol. Microbiol. *59*, 301–316.

Levi, A., and Jenal, U. (2006). Holdfast formation in motile swarmer cells optimizes surface attachment during *Caulobacter crescentus* development. J. Bacteriol. *188*, 5315–5318.

Losi, A., and Gärtner, W. (2008). Bacterial bilin- and flavin-binding photoreceptors. Photochem. Photobiol. Sci. *7*, 1168–1178.

Lourenço, R.F., Kohler, C., and Gomes, S.L. (2011). A two-component system, an anti-sigma factor and two paralogous ECF sigma factors are involved in the control of general stress response in *Caulobacter crescentus*. Mol. Microbiol. *80*, 1598–1612.

McGrath, P.T., Iniesta, A.A., Ryan, K.R., Shapiro, L., and McAdams, H.H. (2006). A dynamically localized protease complex and a polar specificity factor control a cell cycle master regulator. Cell *124*, 535–547.

Mangan, E.K., Bartamian, M., and Gober, J.W. (1995). A mutation that uncouples flagellum assembly from transcription alters the temporal pattern of flagellar gene expression in *Caulobacter crescentus*. J. Bacteriol. *177*, 3176–3184.

Marczynski, G.T. (1999). Chromosome methylation and measurement of faithful, once and only once per cell cycle chromosome replication in *Caulobacter crescentus*. J. Bacteriol. *181*, 1984–1993.

Marczynski, G.T., and Shapiro, L. (1992). Cell-cycle control of a cloned chromosomal origin of replication from *Caulobacter crescentus*. J. Mol. Biol. *226*, 959–977.

Marczynski, G.T., Lentine, K., and Shapiro, L. (1995). A developmentally regulated chromosomal origin of replication uses essential transcription elements. Genes Dev. *9*, 1543–1557.

Martinez-Argudo, I., Little, R., Shearer, N., Johnson, P., and Dixon, R. (2004). The NifL-NifA system: a multidomain transcriptional regulatory complex that integrates environmental signals. J. Bacteriol. *186*, 601–610.

Matroule, J.-Y., Lam, H., Burnette, D.T., and Jacobs-Wagner, C. (2004). Cytokinesis monitoring during development: Rapid pole-to-pole shuttling of a signaling protein by localized kinase and phosphatase in *Caulobacter*. Cell *118*, 579–590.

Merritt, J.H., Ha, D.-G., Cowles, K.N., Lu, W., Morales, D.K., Rabinowitz, J., Gitai, Z., and O'Toole, G.A. (2010). Specific control of *Pseudomonas aeruginosa* surface-associated behaviors by two c-di-GMP diguanylate cyclases. mBio *1*, e00183–10.

Mohr, C.D., MacKichan, J.K., and Shapiro, L. (1998). A membrane-associated protein, FliX, is required for an early step in *Caulobacter* flagellar assembly. J. Bacteriol. *180*, 2175–2185.

Muir, R.E., and Gober, J.W. (2001). Regulation of late flagellar gene transcription and cell division by flagellum assembly in *Caulobacter crescentus*. Mol. Microbiol. *41*, 117–130.

Muir, R.E., and Gober, J.W. (2004). Regulation of FlbD activity by flagellum assembly is accomplished through direct interaction with the trans-acting factor, FliX. Mol. Microbiol. *54*, 715–730.

Muir, R.E., O'Brien, T.M., and Gober, J.W. (2001). The *Caulobacter crescentus* flagellar gene, *fliX*, encodes a novel trans-acting factor that couples flagellar assembly to transcription. Mol. Microbiol. *39*, 1623–1637.

Nierman, W.C., Feldblyum, T.V., Laub, M.T., Paulsen, I.T., Nelson, K.E., Eisen, J., Heidelberg, J.F., Alley, M.R., Ohta, N., Maddock, J.R., *et al.* (2001). Complete genome sequence of *Caulobacter crescentus*. Proc. Natl. Acad. Sci. U.S.A. *98*, 4136–4141.

Ohta, N., and Newton, A. (2003). The core dimerization domains of histidine kinases contain recognition specificity for the cognate response regulator. J. Bacteriol. *185*, 4424–4431.

Ouimet, M.-C., and Marczynski, G.T. (2000). Analysis of a cell-cycle promoter bound by a response regulator. J. Mol. Biol. *302*, 761–775.

Paul, R., Weiser, S., Amiot, N.C., Chan, C., Schirmer, T., Giese, B., and Jenal, U. (2004). Cell cycle-dependent dynamic localization of a bacterial response regulator with a novel di-guanylate cyclase output domain. Genes Dev. *18*, 715–727.

Paul, R., Abel, S., Wassmann, P., Beck, A., Heerklotz, H., and Jenal, U. (2007). Activation of the diguanylate cyclase PleD by phosphorylation-mediated dimerization. J. Biol. Chem. *282*, 29170–29177.

Paul, R., Jaeger, T., Abel, S., Wiederkehr, I., Folcher, M., Biondi, E.G. Laub, M.T., and Jenal, U. (2008). Allosteric regulation of histidine kinases by their cognate response regulator determines cell fate. Cell *133*, 452–461.

Porter, S.C., North, A.K., Wedel, A.B., and Kustu, S. (1993). Oligomerization of NTRC at the *glnA* enhancer is required for transcriptional activation. Genes Dev. *7*, 2258–2273.

Purcell, E.B., and Crosson, S. (2008). Photoregulation in prokaryotes. Curr. Opin. Microbiol. *11*, 168–178.

Purcell, E.B., Siegal-Gaskins, D., Rawling, D.C., Fiebig, A., and Crosson, S. (2007). A photosensory two-component system regulates bacterial cell attachment. Proc. Natl. Acad. Sci. U.S.A. *104*, 18241–18246.

Purcell, E.B., McDonald, C.A., Palfey, B.A., and Crosson, S. (2010). An analysis of the solution structure and signaling mechanism of LovK, a sensor histidine kinase integrating light and redox signals. Biochemistry *49*, 6761–6770.

Quon, K.C., Marczynski, G.T., and Shapiro, L. (1996). Cell cycle control by an essential bacterial two-component signal transduction protein. Cell *84*, 83–93.

Quon, K.C., Yang, B., Domian, I.J., Shapiro, L., and Marczynski, G.T. (1998). Negative control of bacterial DNA replication by a cell cycle regulatory protein that binds at the chromosome origin. Proc. Natl. Acad. Sci. U.S.A. *95*, 120–125.

Radhakrishnan, S.K., Thanbichler, M., and Viollier, P.H. (2008). The dynamic interplay between a cell fate determinant and a lysozyme homolog drives the asymmetric division cycle of *Caulobacter crescentus*. Genes Dev. *22*, 212–225.

Ramakrishnan, G., and Newton, A. (1990). FlbD of *Caulobacter crescentus* is a homologue of the NtrC (NRI) protein and activates sigma 54-dependent flagellar gene promoters. Proc. Natl. Acad. Sci. U.S.A. *87*, 2369–2373.

Reisenauer, A., and Shapiro, L. (2002). DNA methylation affects the cell cycle transcription of the CtrA global regulator in *Caulobacter*. EMBO J. *21*, 4969–4977.

Reisenauer, A., Quon, K.C., and Shapiro, L. (1999). The CtrA response regulator mediates temporal control of gene expression during the *Caulobacter* cell-cycle. J. Bacteriol. *181*, 2430–2439.

Reisinger, S.J., Huntwork, S., Viollier, P.H., and Ryan, K.R. (2007). DivL performs critical cell cycle functions in *Caulobacter crescentus* independent of kinase activity. J. Bacteriol. *189*, 8308–8320.

Rombel, I., North, A., Hwang, I., Wyman, C., and Kustu, S. (1998). The bacterial ehnancer-binding protein NtrC as a molecular machine. Cold Spring Harb. Symp. Quant. Biol. *63*, 157–166.

Ryan, K.R., Judd, E.M., and Shapiro, L. (2002). The CtrA response regulator essential for *Caulobacter crescentus* cell-cycle progression requires a bipartite degradation signal for temporally controlled proteolysis. J. Mol. Biol. *324*, 443–455.

Schirmer, T., and Jenal, U. (2009). Structural and mechanistic determinants of c-di-GMP signalling. Nat. Rev. Microbiol. *7*, 724–735.

Schmidt, R., Bukau, B., and Mogk, A. (2009). Principles of general and regulatory proteolysis by AAA+ proteases in *Escherichia coli*. Res. Microbiol. *160*, 629–636.

Sciochetti, S.A., Lane, T., Ohta, N., and Newton, A. (2002). Protein sequences and cellular factors required for polar localization of a histidine kinase in *Caulobacter crescentus*. J. Bacteriol. *184*, 6037–6049.

Sciochetti, S.A., Ohta, N., and Newton, A. (2005). The role of polar localization in the function of an essential *Caulobacter crescentus* tyrosine kinase. Mol. Microbiol. *56*, 1467–1480.

Siam, R., and Marczynski, G.T. (2000). Cell cycle regulator phosphorylation stimulates two distinct modes of binding at a chromosome replication origin. EMBO J. *19*, 1138–1147.

Siam, R., Brassinga, A.K.C., and Marczynski, G.T. (2003). A dual binding site for integration host factor and the response regulator CtrA inside the *Caulobacter crescentus* replication origin. J. Bacteriol. *185*, 5563–5572.

Skerker, J.M., Prasol, M.S., Perchuk, B.S., Biondi, E.G., and Laub, M.T. (2005). Two-component signal transduction pathways regulating growth and cell cycle progression in a bacterium: A systems-level analysis. PLoS Biol. *3*, e334.

Skerker, J.M., Perchuk, B.S., Siryaporn, A., Lubin, E.A., Ashenberg, O., Goulian, M., and Laub, M.T. (2008). Rewiring the specificity of two-component signal transduction systems. Cell *133*, 1043–1054.

Smith, T.G., and Hoover, T.R. (2009). Deciphering bacterial flagellar gene regulatory networks in the genomic era. Adv. Appl. Microbiol. *67*, 257–295.

Sommer, J.M., and Newton, A. (1989). Turning off flagellum rotation requires the pleiotropic gene *pleD*: *pleA*, *pleC*, and *pleD* define two morphogenic pathways in *Caulobacter crescentus*. J. Bacteriol. *171*, 392–401.

Stephens, C.M., Zweiger, G., and Shapiro, L. (1995). Coordinate cell cycle control of a *Caulobacter* DNA methyltransferase and the flagellar genetic hierarchy. J. Bacteriol. *177*, 1662–1669.

Swartz, T.E., Tseng, T.-S., Fredrickson, M.A., Paris, G., Comerci, D.J., Rajashekara, G., Kim, J.-G., Mudgett, M.B., Splitter, G.A., Ugalde, R.A., *et al.* (2007). Blue-light-activated histidine kinases: two-component sensors in bacteria. Science *317*, 1090–1093.

Tan, M.H., Kozdon, J.B., Shen, X., Shapiro, L., and McAdams, H.H. (2010). An essential transcription factor, SciP, enhances robustness of *Caulobacter* cell cycle regulation. Proc. Natl. Acad. Sci. U.S.A. *107*, 18985–18990.

Taylor, J.A., Wilbur, J.D., Smith, S.C., and Ryan, K.R. (2009). Mutations that alter RcdA surface residues decouple protein localization and CtrA proteolysis in *Caulobacter crescentus*. J. Mol. Biol. *394*, 46–60.

Taylor, J.A., Ouimet, M.-C., Wargachuk, R., and Marczynski, G.T. (2011). The *Caulobacter crescentus* chromosome replication origin evolved two classes of weak DnaA binding sites. Mol. Microbiol. *82*, 312–326.

Tsokos, C.G., Perchuk, B.S., and Laub, M.T. (2011). A dynamic complex of signaling proteins uses polar localization to regulate cell-fate asymmetry in *Caulobacter crescentus*. Dev. Cell *20*, 329–341.

Tuckerman, J.R., Gonzalez, G., Dioum, E.M., and Gilles-Gonzalez, M.A. (2002). Ligand and oxidation-state specific regulation of the heme-based oxygen sensor FixL from *Sinorhizobium meliloti*. Biochemistry *41*, 6170–6177.

Viollier, P.H., Sternheim, N., and Shapiro, L. (2002a). A dynamically localized histidine kinase controls the asymmetric distribution of polar pili proteins. EMBO J. *21*, 4420–4428.

Viollier, P.H., Sternheim, N., and Shapiro, L. (2002b). Identification of a localization factor for the polar positioning of bacterial structural and regulatory proteins. Proc. Natl. Acad. Sci. U.S.A. *99*, 13831–13836.

Wang, S.P., Sharma, P.L., Schoenlein, P.V., and Ely, B. (1993). A histidine protein kinase is involved in polar organelle development in *Caulobacter crescentus*. Proc. Natl. Acad. Sci. U.S.A. *90*, 630–634.

Wanner, B.L. (1993). Gene regulation by phosphate in enteric bacteria. J. Cell. Biochem. *51*, 47–54.

Wassmann, P., Chan, C., Paul, R., Beck, A., Heerklotz, H., Jenal, U., and Schirmer, T. (2007). Structure of BeF^{3-}-modified response regulator PleD: implications for diguanylate cyclase activation, catalysis, and feedback inhibtion. Structure *15*, 915–927.

Wheeler, R.T., and Shapiro, L. (1999). Differential localization of two histidine kinases controlling bacterial cell differentiation. Mol. Cell *4*, 683–694.

Wingrove, J.A., Mangan, E.K., and Gober, J.W. (1993). Spatial and temporal phosphorylation of a transcriptional activator regulates pole-specific gene expression in *Caulobacter*. Genes Dev. *7*, 1979–1992.

Wu, J., Benson, A.K., and Newton, A. (1995). Global regulation of a sigma 54-dependent flagellar gene family in *Caulobacter crescentus* by the transcriptional activator FlbD. J. Bacteriol. *177*, 3241–3250.

Wu, J., Ohta, N., and Newton, A. (1998). An essential, multicomponent signal transduction pathway required for cell cycle regulation in *Caulobacter*. Proc. Natl. Acad. Sci. U.S.A. *95*, 1443–1448.

Wu, J., Ohta, N., Zhao, J.-L., and Newton, A. (1999). A novel bacterial tyrosine kinase essential for cell division and differentiation. Proc. Natl. Acad. Sci. U.S.A. *96*, 13068–13073.

Zweiger, G., and Shapiro, L. (1994). Expression of *Caulobacter dnaA* as a function of the cell cycle. J. Bacteriol. *176*, 401–408.

Two-component Systems Involved in Regulation of Motility and Development in *Myxococcus xanthus* 14

Daniela Keilberg, Stuart Huntley and Lotte Søgaard-Andersen

Abstract

The *Myxococcus xanthus* life cycle is characterized by many social interactions. In particular, *M. xanthus* forms cooperatively spreading colonies in the presence of nutrients and multicellular, spore-filled fruiting bodies in the absence of nutrients. Formation of both cellular patterns depends on two intact motility systems. Moreover, fruiting body formation depends on intercellular communication and temporally regulated gene expression. The *M. xanthus* genome encodes a staggering 272 putative proteins of two-component system and most aspects of the *M. xanthus* life cycle are regulated by one or more of these proteins. Interestingly, many of the corresponding genes encoding two-component system proteins possess an unusual organization in complex genes clusters and as orphan genes. However, major strides have been made in our understanding of a large number of these proteins. Here, we focus on the function of well-studied proteins of two-component systems in motility and development in *M. xanthus*.

Introduction

Most, if not all, bacteria exist under non-static conditions. Therefore, bacteria must be able to sense and respond to environmental changes to optimize their chances of survival. Different bacterial species have adopted different survival strategies to respond to changes in their environments. The various strategies are played out in their entire splendour in response to starvation with strategies ranging from adaptive changes in gene expression to moving away from nutrient poor conditions to differentiation with the formation of specialized cell types with novel properties. One of the most complex survival strategies in response to starvation have been adopted by the myxobacteria with the formation of multicellular spore-filled fruiting bodies. Fruiting body formation depends on regulation of motility, intercellular communication, and temporally regulated changes in gene expression. *Myxococcus xanthus* has emerged as the model myxobacterium to understand how these processes are regulated and coordinated in time and space. Over the past two decades, evidence has emerged that proteins of two-component systems have essential functions in these three processes. In this chapter, we specifically focus on the importance of two-component systems in the regulation of motility and development in *M. xanthus*.

The life cycle of *M. xanthus*

Myxobacteria are often referred to as 'social bacteria'. The social behaviours include the formation of three different cellular patterns, (1) spreading colonies, (2) ripples, and (3) fruiting bodies, depending on the nutritional status of the cells. The anti-social behaviour involves predation on other microorganisms by the secretion of hydrolytic enzymes and antimicrobials (Konovalova *et al.*, 2010). The life cycle of myxobacteria consists of two separate cycles. In the presence of nutrients the motile, rod-shaped vegetative cells grow and divide. Myxobacteria move by gliding and if the cells are present on a solid surface they form cooperatively feeding spreading colonies with large flares at the edge (Fig. 14.1). In the presence

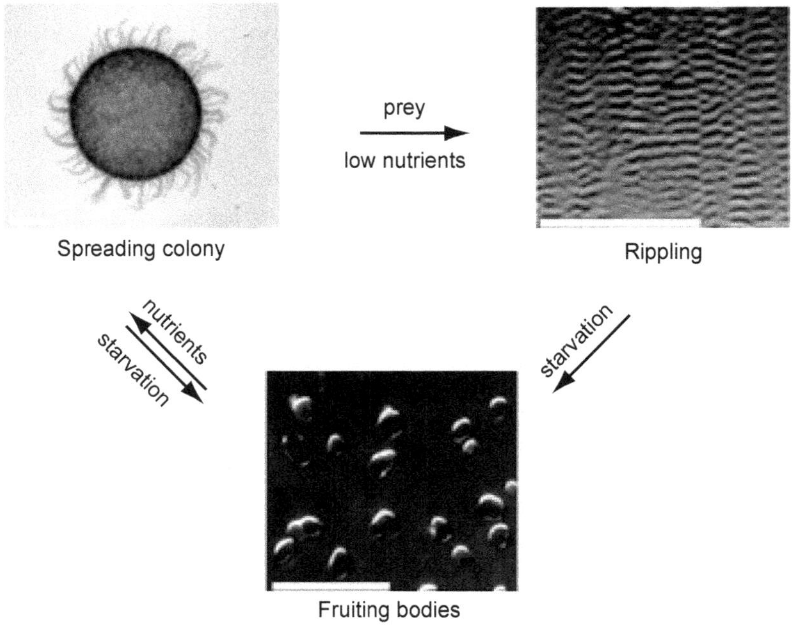

Figure 14.1 The three cellular patterns formed during the *M. xanthus* life cycle. *M. xanthus* cells organize into three types of patterns in response to nutrient levels. Scale bars: 1 mm. A colour version of this figure is located in the plate section at the back of the book.

of prey microorganisms or if nutrient levels are low, cells organize to form ripples (Shimkets and Kaiser, 1982; Berleman *et al.*, 2006). During rippling cells align to maximize side-to-side contacts and organize into elongated structures, i.e. ripples, that move rhythmically back-and-forth (Fig. 14.1). Fruiting body formation is initiated in response to the absence of nutrients, at a high cell density and on a solid surface. If these three conditions are met, growth ceases and the developmental programme that culminates in the formation of the multicellular, spore-filled fruiting bodies is initiated (Fig. 14.1). Under the appropriate conditions, the spores in the fruiting bodies germinate to give rise to the growing, motile and metabolically active vegetative cells. Each fruiting body consists of 10^5 cells and, therefore, the simultaneous germination of 10^5 spores on the same location immediately gives rise to a colony. Subsequently, such colonies initiate the formation of a new cooperatively feeding spreading colony.

Two-component systems in *M. xanthus*

Two-component systems are wide spread signal transduction systems in bacteria (Stock *et al.*, 2000). Two-component systems are involved in regulating most aspects of bacterial cell functions including regulation of gene expression, motility, cell division, virulence, metabolism, biofilm formation and cell differentiation (Stock *et al.*, 2000). The simple two-component system consists of a histidine protein kinase that regulates the activity of a cognate, partner response regulator (Stock *et al.*, 2000) (Fig. 14.2). The histidine kinase has a modular structure with an N-terminal variable sensor or input domain and a C-terminal kinase domain. The variable sensor domain is the part of the kinase that senses a particular intercellular or intracellular signal. Moreover, this part of the kinase may contain one or more transmembrane helices that anchor the kinase in the cytoplasmic membrane. In response to the presence of the relevant signal, the sensor domain signals to the kinase domain leading to autophosphorylation of a conserved histidine residue in the kinase domain using ATP as a phosphoryl donor. Subsequently, this phosphoryl is transferred to a conserved aspartate residue in the receiver domain of the cognate response regulator. Response regulators also have a modular structure with an N-terminal receiver domain and a C-terminal output domain. Phosphorylation of the response regulator leads

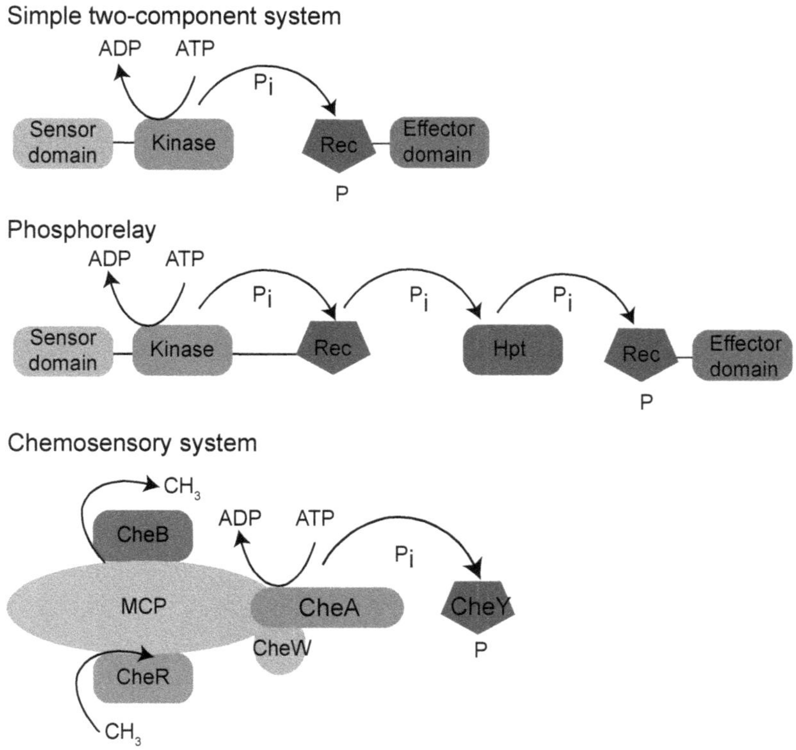

Figure 14.2 Structure of generic two-component signal transduction pathways. Schematics show structure and phosphotransfer reactions in a simple two-component system, in a phosphorelay, and in a chemosensory type of two-component system. A colour version of this figure is located in the plate section at the back of the book.

to activation of the output domain and the generation of an output response. The output response depends on the nature of the output domain and often involves changes in gene expression via a DNA-binding output domain, changes in enzymatic activity via an output domain with an enzymatic function, or in protein–protein interactions via a response regulator without an output domain (Jenal and Galperin, 2009; Galperin, 2010). Eventually, the output response is terminated by the spontaneous dephosphorylation or a phosphatase-dependent dephosphorylation of the response regulator.

The phosphorelay constitute a structurally more complex type of two-component systems with phosphotransfer occurring in three sequential steps (Appleby *et al.*, 1996) (Fig. 14.2). Overall, the domain organization of these systems is similar to that of the simple two-component systems. After autophosphorylation of the kinase, the phosphoryl group is transferred from the histidine kinase to the conserved aspartate in a receiver domain, then to a conserved histidine in a phosphotransferase (HPt) domain and, finally, to the conserved aspartate in the second response regulator. The domain organization of proteins in phosphorelays is also highly modular, i.e. all four domains involved in phosphotransfer may reside on separate proteins (Burbulys *et al.*, 1991), the kinase and first receiver may be present in the same protein (Mok *et al.*, 2003) or the kinase, the first receiver and the HPt domain may be present in the same protein (Uhl and Miller, 1996). Histidine protein kinases with a covalently linked receiver domain are referred to as hybrid kinases (Stock *et al.*, 2000).

Chemosensory systems constitute a third type of two-component systems (Kirby, 2009; Porter *et al.*, 2011) (Fig. 14.2). In chemosensory systems, a signal is detected by a methyl-accepting chemotaxis protein (MCP), which signals to the histidine protein kinase CheA in a complex also containing the CheW adaptor protein. The phosphoryl group on CheA is subsequently transferred to the single domain response regulator CheY, which generates an output response that often involves changes in the rotational direction of one or more flagella. A special feature of chemosensory system is their ability to display adaptation to the presence or absence of a signal. Adaptation depends on methylation of MCPs by the methyltransferase CheR and demethylation by the methylesterase CheB.

Comparative genomics approaches have documented that most bacterial genomes encode proteins of two-component systems and with a recent survey showing that 864 out of 899 completely sequenced bacterial genomes encode such proteins (Wuichet *et al.*, 2010). Generally, the number of two-component proteins encoded by a genome correlates positively with genome size and the total number of encoded proteins (Galperin, 2005). *M. xanthus* has a genome size of 9.1 Mb (Goldman *et al.*, 2006) and is predicted to encode an estimated 7400 proteins including 272 proteins of two-component systems (Shi *et al.*, 2008; Whitworth and Cock, 2008). Among these, 21 are part of eight chemosensory systems (Zusman *et al.*, 2007).

Genes for cognate histidine kinases and response regulators are often found as pairs next to each other, encoded in an operon, and are thought to have co-evolved (Alm *et al.*, 2006) (Table 14.1). Interestingly, this pattern is not observed in *M. xanthus* (Table 14.1). In the case of *M. xanthus*, only 29% of these genes have a paired organization, while 16% are part of complex gene clusters encoding more than one response regulator or histidine kinase, and 55% are found as orphan genes (Shi *et al.*, 2008; Whitworth and Cock, 2008) (Fig. 14.3A).

Domain analyses of two-component proteins encoded by paired, complex and orphan genes in *M. xanthus* revealed that these three groups of proteins are significantly different and, therefore, likely to have different functions (Fig. 14.3B and C). First, 75% of histidine kinases encoded by paired genes have one or more trans-membrane helices and are likely involved in monitoring an external signal. On the contrary, the majority of histidine kinases encoded by genes in complex clusters or by orphan genes do not contain trans-membrane helices and are likely cytoplasmic and involved in sensing cytoplasmic signals. Moreover, the majority of these kinases are hybrid kinases containing one or more C-terminal receiver domains (Fig. 14.3B). Similarly, response regulators encoded by paired genes are enriched for proteins containing a DNA binding output domain (Fig. 14.3C). In contrast, response regulators without output domains are overrepresented among proteins encoded by genes in complex gene clusters and orphan genes. The most frequently occurring output domains in these response regulators are those involved in DNA binding, cyclic-di-GMP metabolism (GGDEF and PilZ domains) and

Table 14.1 Quantitative comparison of the genetic organization of genes encoding two-component proteins in *M. xanthus* to that in other model bacteria

Species[a]	Genome size (Mb)	Total number of TCS genes	Paired TCS (%)	Complex TCS (%)	Orphan TCS (%)
Myxococcus xanthus	9.1	251	29	16	55
Streptomyces coelicolor	8.7	190	57	16	27
Pseudomonas aeruginosa	6.3	149	51	17	32
Escherichia coli	4.6	63	65	14	21
Bacillus subtilis	4.2	68	73	9	18
Caulobacter crescentus	4.0	111	22	18	60

[a]References for genomes: *Myxococcus xanthus* DK1622 (Goldman *et al.*, 2006), *Streptomyces coelicolor* A3(2) (Bentley *et al.*, 2002), *Pseudomonas aeruginosa* PAO1 (Stover *et al.*, 2000), *Escherichia coli* K-12 MG1655 (Blattner *et al.*, 1997), *Bacillus subtilis* 168 (Kunst *et al.*, 1997), and *Caulobacter crescentus* CB15 (Nierman *et al.*, 2001).
TCS, genes encoding two-component proteins.

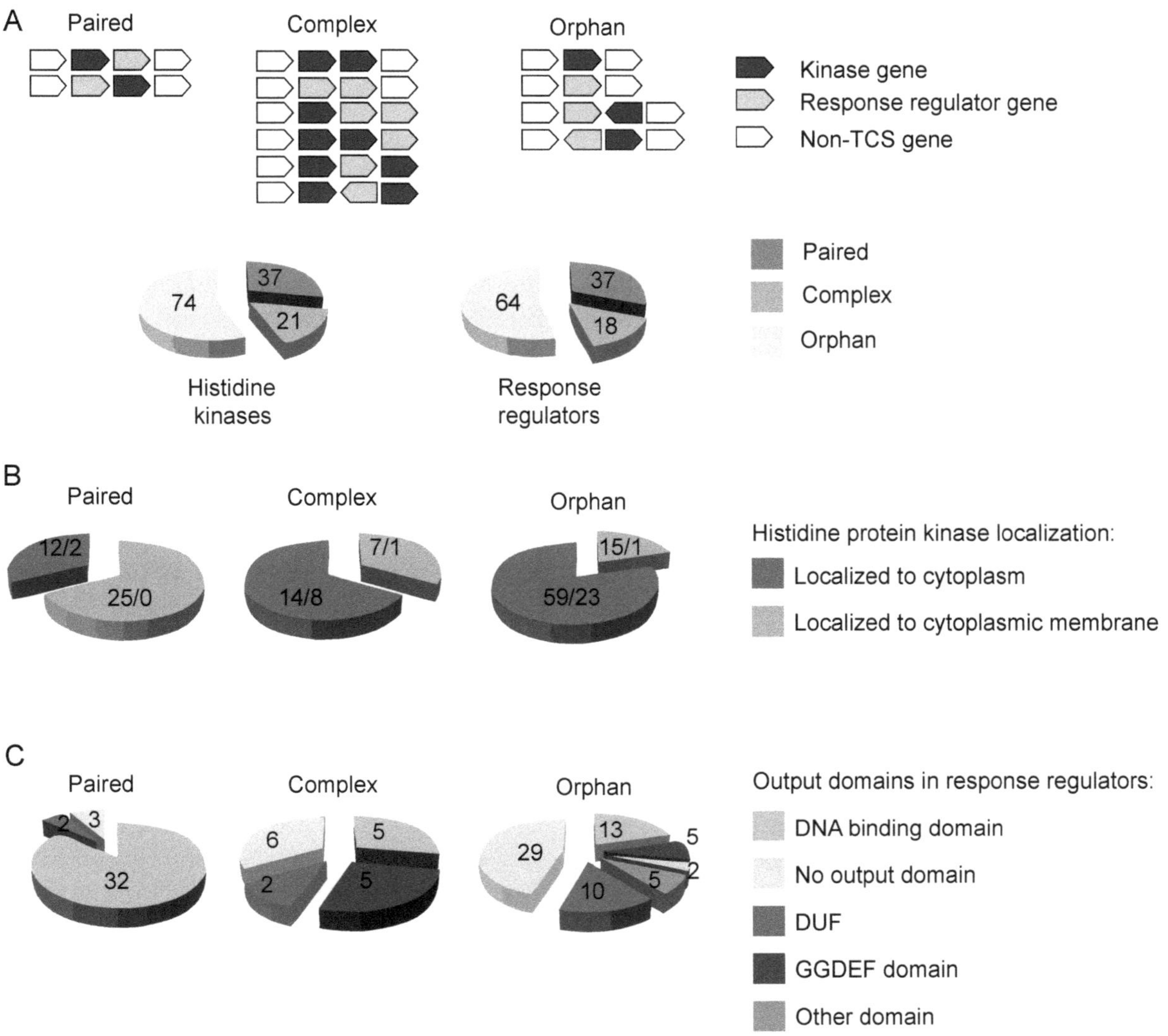

Figure 14.3 Two-component systems of *M. xanthus*. (A) Definition of genetic organization of genes of two-component systems. The pie-diagrams show the distribution of histidine kinase and response regulators encoded by genes in the different configurations. (B) Localization of histidine kinases according to genetic organization. The second number indicates the number of hybrid kinases. (C) Output domains in response regulators according to genetic organization. DUF, domains of unknown function; other domains indicate domains of known function. A colour version of this figure is located in the plate section at the back of the book.

unknown functions. Thus, the picture that emerges from these comparisons is one in which paired two-component genes encode proteins that are part of simple two-component systems involved in sensing extracellular signals and generating an output that involves changes in gene expression. Similarly, orphan two-component genes or genes in complex gene clusters would encode systems involved in sensing intracellular signals and generating a more diverse set of outputs. Finally, the high abundance of hybrid kinases and single domain response regulators encoded by orphan genes and complex gene clusters suggest that these proteins are part of phosphorelays or branched pathways. Phosphorelays would depend on the presence of HPt domain-containing proteins. It has been argued that HPt domains are difficult to identify owing to their low sequence conservation (Biondi *et al.*, 2006). In agreement with this study, only two proteins containing HPt domains have been identified in the *M. xanthus* genome (Shi *et al.*, 2008).

Two-component systems involved in motility of *M. xanthus*

M. xanthus cells move by gliding, which is the movement of a rod-shaped cell in the direction of its long axis on a surface in the absence of a flagellum (Henrichsen, 1972). *M. xanthus* are slow-movers and only move with a velocity of 2 to 4 µm/min (Jelsbak and Søgaard-Andersen, 1999, 2002). *M. xanthus* cells move by means of two different motility systems, the A- and S-motility systems (Hodgkin and Kaiser, 1979b) and they control their direction of movement by occasionally undergoing reversals (Blackhart and Zusman, 1985; Jelsbak and Søgaard-Andersen, 2002). The two motility systems function synergistically to generate force in the same direction (Kaiser and Crosby, 1983) and show different selective advantages on different surfaces (Shi and Zusman, 1993). Generally, S-motility is cell-to-cell contact-dependent and, therefore, mostly only functional when cells are within contact distance of each other, while A-motility functions in single cells as well as in groups of cells (Hodgkin and Kaiser, 1979a,b). During a cellular reversal, the polarity of the cell is inverted with the old leading cell pole becoming the new lagging cell pole. Reversals are highly regulated events and induced by the Frz chemosensory system (Blackhart and Zusman, 1985), and mutants, which are not able to regulate the reversal frequency, have defects in the formation of spreading colonies (Blackhart and Zusman, 1985), in rippling (Sager and Kaiser, 1994), and in fruiting body formation (Blackhart and Zusman, 1985). Three chemosensory systems have functions in motility, Dif, Frz and Che4 (Vlamakis *et al.*, 2004). The mechanism and exact function of Che4 is largely unknown and here we focus on the Dif and Frz systems.

A-motility – gliding by multiple motors along the cell body

The mechanism of the A-motility system has remained enigmatic since its initial discovery in 1979 (Hodgkin and Kaiser, 1979a,b). Recently, the molecular motor that propels cells by means of A-motility is beginning to be uncovered (Fig. 14.4A). The break-through observation was provided in a cell biological approach in which it was observed that the AglZ protein, which is required for A-motility, localizes to a cluster at the leading cell pole as well as to focal adhesion complexes (FACs) that are distributed along the cell body (Mignot *et al.*, 2007). In a moving cell, FACs remain stationary with respect to the substratum on which the cell is moving and, therefore, move from the leading to the lagging cell pole (Mignot *et al.*, 2007). Recently, it was shown that traction force is generated at the sites of the focal adhesion complexes and that the A-motility motor in the FACs consists of AglR, which is a MotA/TolQ/ExbB homologue, and AglQ and AglS, which are MotB/TolR/ExbD homologues (Sun *et al.*, 2011). According to the current model, FACs are assembled at the leading cell pole, distributed along the cell body, and disassemble at the lagging cell pole (Nan *et al.*, 2011; Sun *et al.*, 2011). Each FAC is thought to consist of a multiprotein complex that spans the cell envelope (Mauriello *et al.*, 2010; Nan *et al.*, 2010, 2011; Sun *et al.*, 2011). Three A-motility proteins have been studied in details, AglZ (Mignot *et al.*, 2007), RomR (Leonardy *et al.*, 2007) and AgmU (Nan *et al.*, 2010, 2011).

AglZ is an orphan pseudo-response regulator with an N-terminal receiver domain that lacks the conserved phosphorylatable aspartate (Yang *et al.*, 2004). Between reversals AglZ localizes in a cluster at the leading cell pole and in FACs (Mignot *et al.*, 2007) (Fig. 14.4A). During reversals, AglZ localization is dynamic and the cluster at the old leading cell pole relocates to the new leading cell pole. Simultaneously, FACs disassemble and then reassemble after a reversal. RomR is an orphan response regulator with a conserved receiver domain and an unusual proline-rich output domain (Leonardy *et al.*, 2007). Between reversals, RomR localizes in an asymmetric bipolar pattern with a large cluster at the lagging cell pole and a small cluster at the leading cell pole. During reversals, RomR localization is dynamic and the large cluster relocates from the old lagging pole to the new lagging pole (Fig. 14.4A). Interestingly, RomR phosphorylation is likely required for the correct reversal frequency, because the mutant RomR^{D53E} protein, in which the phosphorylatable aspartate in the receiver domain has been substituted with a glutamate and likely mimics the

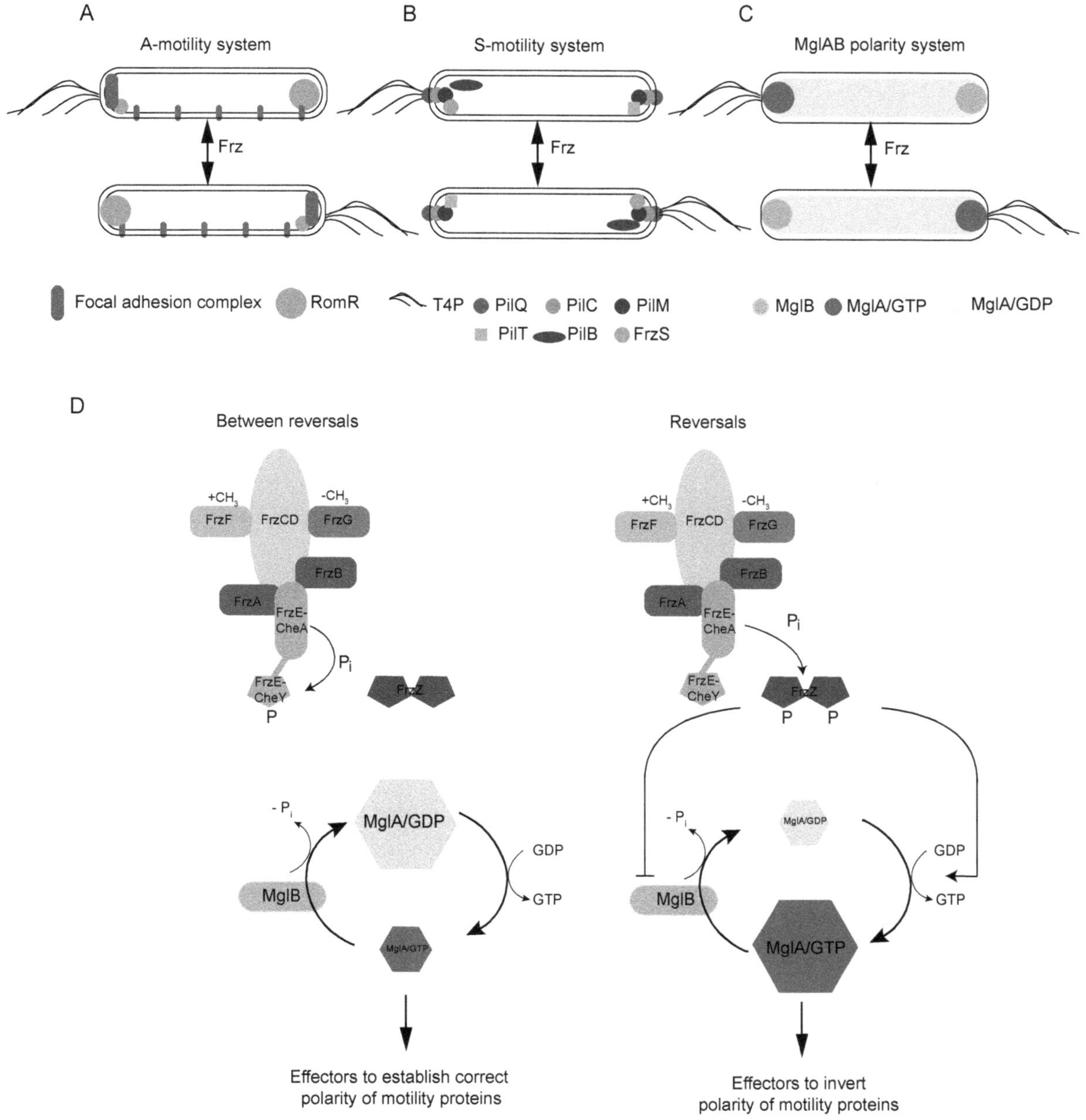

Figure 14.4 The chemosensory Frz system in combination with the MglA GTPase regulate the leading/lagging cell pole polarity axis. (A). Key components of the A-motility system. T4P indicate the leading cell pole. Focal adhesion complexes contain the proteins AglZ, AgmU, AglQ, AglR and AglS. These complexes are assembled at the leading pole and disassemble at the lagging pole. The polar clusters of RomR are indicated. The two cells demonstrate the localization of A-motility proteins before and after a reversal. (B) Key components of the S-motility system. T4P indicate the leading cell pole. The six T4P proteins that have been localized are shown in the indicated colour code. The two cells demonstrate the localization of T4P-proteins before and after a reversal. (C) Spatial regulation of MglA and MglB. Between reversals MglA/GTP binds at the leading pole and MglB at the lagging cell pole. The two proteins bind to opposite pole by a mechanism that involves mutual exclusion. MglA/GDP is diffusely localized to the cytoplasm. During a reversal and in a Frz-dependent manner, MglA/GTP is released from the leading cell pole and relocates to the lagging cell pole. Subsequently, MglB relocates to the new lagging cell pole. (D) Model of Frz chemosensory system. The two proposed alternative phosphate flows from the kinase domain of FrzE (FrzE-CheA) to either the CheY domain in FrzE (FrzE-CheY) or to the receiver-domains in FrzZ are shown. The signal that governs the direction of the phosphate flow is unknown. MglA is a nucleotide-dependent molecular switch and MglB its cognate GAP. Between reversals, MglB functions as a GAP. During reversals, FrzZ-PP is thought to either directly or indirectly act as a GEF or to inhibit GAP activity of MglB to stimulate accumulation of MglA/GTP. A colour version of this figure is located in the plate section at the back of the book.

phosphorylated form of the protein, causes hyper-reversals, whereas cells containing the mutant $RomR^{D53N}$ protein, which cannot be phosphorylated, only rarely reverse. It should be noted that the cognate RomR kinase has not been identified. AgmU does not contain conserved domains and is thought to form a helix in the cytoplasm and periplasm (Nan *et al.*, 2010, 2011). It has been suggested that the direction of rotation of this helix changes during a reversal (Nan *et al.*, 2011). For all three proteins, the change in localization during a reversal depends on the Frz chemosensory system (see below).

S-motility – motility by means of type IV pili

In the S-motility system, force generation depends on type IV pili (T4P) (Kaiser, 1979; Wu and Kaiser, 1995). T4P are long, thin and flexible filaments consisting of the PilA subunit (Wu and Kaiser, 1995). T4P are highly dynamic structures and undergo cycles of extensions, surface attachment, and retractions; and, during a retraction, the cell is pulled forward (Skerker and Berg, 2001; Maier *et al.*, 2002). *M. xanthus* cells contain 5–10 T4P at the leading cell pole (Sun *et al.*, 2000; Mignot *et al.*, 2005; Clausen *et al.*, 2009). Assembly and retraction of T4P depends on 10–15 conserved proteins that form a multi–protein complex that spans the cytoplasm, inner and outer membrane and periplasm (Nudleman *et al.*, 2006; Jakovljevic *et al.*, 2008; Pelicic, 2008; Bulyha *et al.*, 2009). In addition to T4P, S-motility depends on lipopolysaccharide O-antigen (Bowden and Kaplan, 1998) and the extracellular matrix (Arnold and Shimkets, 1988), which consists of expolysaccharides and proteins (Behmlander and Dworkin, 1994a,b). It is not known how lipopolysaccharide O-antigen is involved in T4P motility. The connection between the extracellular matrix and T4P-dependent motility was recently clarified by the observation that the polysaccharide portion of ECM triggers retraction of T4P (Li *et al.*, 2003).

The cell pole at which T4P are assembled switches during a reversal (Sun *et al.*, 2000; Mignot *et al.*, 2005; Bulyha *et al.*, 2009) (Fig. 14.4B). The mechanism underlying the switch in polarity of T4P has been analysed by determining the localization of proteins required for T4P function. These analyses demonstrated that several T4P proteins including the PilQ secretin in the outer membrane, PilC in the inner membrane and PilM in the cytoplasm localize to symmetrical clusters at both cell poles (Nudleman *et al.*, 2006; Bulyha *et al.*, 2009). In contrast, FrzS, which is an orphan pseudo-response regulator with an unknown function in T4P function (Mignot *et al.*, 2005), and PilB, which is the ATPase that stimulates T4P extension, localize mainly to the leading cell pole (Bulyha *et al.*, 2009); and, PilT which is the ATPase that stimulates T4P retraction, localizes mainly to the lagging cell pole (Bulyha *et al.*, 2009). During reversals, FrzS, PilB and PilT localization is dynamic and the three proteins relocate between the cell poles. For all three proteins, the change in localization during a reversal depends on the Frz chemosensory system.

Cellular reversals – directional control and cell polarity

Two-component systems have essential functions in motility mainly by regulating synthesis of the extracellular matrix and by regulating cellular reversals. On average, cells undergo cellular reversals every 10–15 minutes and reversals are induced by the Frz chemosensory system (Blackhart and Zusman, 1985). The *frz* genes form a single gene cluster of seven genes and encode homologues of the chemosensory proteins found in many bacteria (McBride *et al.*, 1989; Trudeau *et al.*, 1996) (Fig. 14.4D): The Frz system consists of a cytoplasmic MCP, FrzCD; two CheW homologues, FrzA and FrzB; FrzE, possessing both a CheA histidine kinase domain and a CheY-like receiver domain; a methyltransferase FrzF, which methylates FrzCD; a methylesterase FrzG, which demethylates FrzCD; and, FrzZ, composed of two CheY receiver domains.

To date, the input signals that regulate the activity of the Frz system are not known. According to current models, signals are sensed by either FrzCD or FrzF (Bustamante *et al.*, 2004; Scott *et al.*, 2008). The output of the Frz system is generated via autophosphorylation of FrzE on a conserved histidine residue in the FrzE-CheA domain (Inclan *et al.*, 2007, 2008). *In vitro* phosphorylation assays have demonstrated direct transfer of this phosphoryl group to both of the two receiver domains of FrzZ (Inclan *et al.*, 2007). Moreover, genetic evidence suggests that this phosphoryl group can also be transferred to the receiver domain of FrzE (FrzE-CheY) (Li *et al.*, 2005). However, biochemical evidence suggests that the His-phosphoryl on FrzE is not transferred to FrzE-CheY and that FrzE-CheY inhibits FrzE autophosphorylation (Inclan *et al.*, 2008). A model that incorporates these observations suggests that in the absence of FrzE stimulation, FrzE-CheY either inhibits FrzE autophosphorylation or that FrzE autophosphorylates and then transfers the phosphate to FrzE-CheY (Fig. 14.4D). In the presence of a stimulating signal, FrzE autophosphorylates, and the phosphoryl group is transferred to FrzZ to generate FrzZ~PP, which would then stimulate reversals.

Emerging evidence is pointing in the direction that a small Ras-like GTPase could be the downstream target of the Frz system. Both *M. xanthus* motility systems depend on correct polar localization of motility proteins between reversals (Fig. 14.4A and B) and the dynamic relocation of several of these proteins during reversals. MglA is required for the functioning of both motility systems (Hodgkin and Kaiser, 1979b; Hartzell and Kaiser, 1991). Recently, it was shown that the small Ras-like GTPase MglA in combination with its cognate GTPase activating protein (GAP) MglB act to regulate both motility systems (Leonardy *et al.*, 2010; Mauriello *et al.*, 2010; Patryn *et al.*, 2010; Zhang *et al.*, 2010). Specifically, MglA acts as a nucleotide-dependent molecular switch to set up the correct polarity of motility proteins between reversals and to induce their relocation during reversals (Leonardy *et al.*, 2010; Zhang *et al.*, 2010) (Fig. 14.4D). MglA cycles between an inactive GDP-bound form and an active GTP-bound form. While a cell is moving, MglA/GTP is restricted to the leading pole, MglA/GDP, the inactive form of MglA, is diffusely localized, and MglB is at the lagging pole. The binding of MglA/GTP and MglB at opposite poles seems to involve a mutual exclusion mechanism and defines the leading/lagging cell pole polarity axis (Fig. 14.4C). For reversals, the Frz chemosensory system induces the relocation of MglA/GTP from the old leading pole to the new leading pole and, as a consequence, MglB relocates from the old lagging pole to the new lagging pole (Fig. 14.4C). In total, the relocation of MglA and MglB causes an inversion of the leading/lagging polarity axis. In this model, FrzZ~PP is thought to either function as a guanine-nucleotide-exchange factor (GEF) to stimulate accumulation of MglA/GTP or to inhibit GAP activity of MglB (Fig. 14.4D). Thus, in this model there is an interesting intersection between proteins of two-component systems and regulation by a small Ras-like GTPase. It is not known how MglA/GTP establishes the correct polarity of motility proteins between reversals and causes their relocation during reversal. However, by homology to eukaryotic systems, MglA/GTP is thought to interact with effector proteins, which then execute these functions.

Extracellular matrix accumulation – required for S-motility

The Dif chemosensory system has an essential function in regulation of the extracellular matrix and is encoded in a single operon (Yang *et al.*, 1998, 2000). The *dif* gene cluster contains five genes (Fig. 14.5): DifA is an MCP homologue in the cytoplasmic membrane (Xu *et al.*, 2011), DifE a homologue of CheA kinases, DifD of CheY response regulators, DifC of CheW coupling proteins, and DifG of CheC phosphatases (Yang *et al.*, 1998; Black and Yang, 2004). *difA*, *difC* and *difE* mutants are unable to accumulate exopolysaccharides and the FibA metalloprotease, which is generally used as readout for the protein content of the extracellular matrix (Yang *et al.*, 2000; Bellenger *et al.*, 2002), whereas *difD* and *difG* mutants show an overproduction of exopolysaccharide (Black and Yang, 2004); the effect of DifD and DifG on FibA accumulation is unknown (Black and Yang, 2004). According to the current model (Black and Yang, 2004; Yang *et al.*, 2008; Black *et al.*, 2010), DifE stimulates extracellular matrix accumulation by phosphorylation of an unknown response regulator, DifX. Moreover, DifD is thought to act as a phosphate sink, draining the phosphoryl group on DifE away

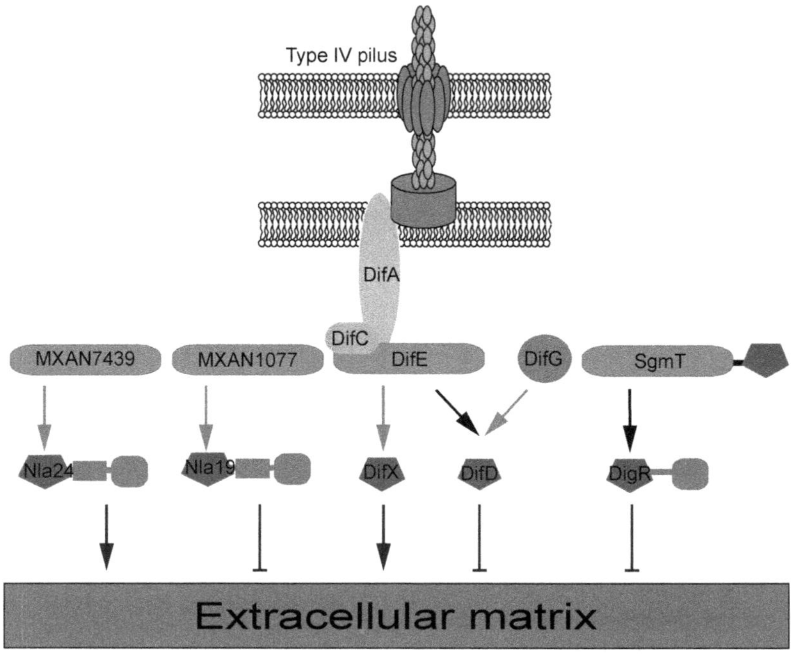

Figure 14.5 Regulation of extracellular matrix accumulation by proteins of two-component systems. A T4P is shown spanning the cell envelope and potentially interacting with DifACE proteins. From the DifE histidine protein kinase, the phosphoryl group is thought to be transferred to an unknown response regulator (DifX) to stimulate ECM accumulation or to DifD to inhibit ECM accumulation. Histidine protein kinase domains are indicated in light brown boxes; receiver domains are indicated by red pentagons. Nla24 and Nla19 are NtrC-like enhancer binding proteins. DigR is a DNA binding response regulator. Arrows or inhibitions indicate positive or negative effects on extracellular matrix accumulation. Grey arrows indicate that phosphotransfer has not been shown. A colour version of this figure is located in the plate section at the back of the book.

from DifX and in this way inhibiting extracellular matrix accumulation. Moreover, DifG is thought to be involved in regulating extracellular matrix accumulation by dephosphorylating DifD~P. Interestingly, mutants unable to assemble T4P accumulate reduced amounts of extracellular matrix and a *pilT* mutant, which assembles non-functional T4P, accumulate increased amounts of extracellular matrix (Black *et al.*, 2006). Moreover, *dif* mutations are epistatic to mutations affecting T4P assembly and function (Black *et al.*, 2006). Therefore, it has been proposed that T4P may serve as a sensor for the Dif system (Black *et al.*, 2006).

In addition to the Dif chemosensory system several other regulators are involved in regulation of extracellular matrix accumulation as illustrated in (Fig. 14.5). Nla24 is an NtrC-like enhancer binding protein (Caberoy *et al.*, 2003) [also referred to as EpsI (Lu *et al.*, 2005)] encoded with a paired histidine kinase MXAN7439. An *nla24* mutant accumulates reduced amounts of EPS and normal levels of FibA (Lancero *et al.*, 2004). Lack of the NtrC-like enhancer binding protein Nla19 (also referred to as SpdR; Hager *et al.*, 2001) causes increased accumulation of EPS (Lancero *et al.*, 2005). Nla19 is encoded with the paired histidine kinase MXAN1077 (also referred to as SpdS; Hager *et al.*, 2001). DigR is an orphan DNA binding response regulator (Overgaard *et al.*, 2006). Lack of DigR causes increased accumulation of EPS while FibA accumulation is reduced. SgmT is an orphan histidine protein kinase. Lack of SgmT causes the same phenotype as lack of DigR, and SgmT is the cognate kinase of DigR (Petters *et al.*, 2012). The NtrC-like enhancer binding proteins Nla1 (encoded in a paired configuration with the histidine kinase gene MXAN5852) and Nla23 (also known as PilR2), which is encoded in a paired configuration with the kinase gene MXAN5778, are both required for normal S-motility (Caberoy *et al.*, 2003); however, the underlying mechanisms are not yet known.

Fruiting body formation

M. xanthus feeds on amino acids, which are used as a source of carbon, nitrogen, and energy (Dworkin, 1962). In response to starvation for amino acids, *M. xanthus* initiates the stringent response resulting in accumulation of the second messenger (p) ppGpp (Manoil and Kaiser, 1980). Initiation of the stringent response is essential for fruiting body formation (Harris *et al.*, 1998) and sufficient for the initiation of this developmental programme (Singer and Kaiser, 1995). The process of fruiting body formation takes approximately 72 hours and depends on intercellular signalling, temporally regulated changes in gene expression, and coordinated changes in organized cell movements from spreading to aggregation (Søgaard-Andersen *et al.*, 2003, 2004). Proteins of two-component systems have essential functions in the decision to initiate fruiting body formation as well as in these three processes. 4–6 hours after the onset of starvation cells begin to aggregate to form small aggregation centres. As more cells accumulate the aggregation centres gradually increase in size. After 24 hours the fruiting body formation is complete, and each fruiting body contains approximately 10^5 cells, which differentiate into spherical, non-motile spores. While the aggregation process takes about 24 hours to complete, spore-maturation is concluded 72 hours after onset of starvation. Of the initial population, approximately 10% differentiate into spores, 30% of the cells remain outside the fruiting bodies as peripheral rods (O'Connor and Zusman, 1991a,b), and the remaining 60% undergo cell lysis (Rosenbluh *et al.*, 1989; Nariya and Inouye, 2008).

Two-component systems in development

Intercellular signalling

Five intercellular signals have been defined genetically that are important for fruiting body formation (Hagen *et al.*, 1978; Downard *et al.*, 1993). These signals are referred to as the A-, B-, C-, D- and E-signals. Among these signals, only two, the A-signal and the C-signal, have been characterized biochemically and functionally. These two signals fulfil different functions during fruiting body formation. The A-signal system is important early during fruiting body formation and functions as a type of quorum sensing systems to ensure that fruiting body formation does not initiate until starving cells are present at a sufficiently high density (Kaplan and Plamann, 1996). Synthesis of the A-signal depends on the stringent response (Harris *et al.*, 1998). The A-signal consists of amino acids and peptides that are thought to be generated by extracellular proteases digesting surface proteins in response to starvation. If a threshold concentration of the A-signal is reached, fruiting body formation proceeds. In total, eight genes (*nla4* and *nla18*, *asgABCDE* and *sigD*) have been identified that are required for the synthesis of the A-signal (Fig. 14.6) (Kaplan and Plamann, 1996; Cho and Zusman, 1999a; Garza *et al.*, 2000; Diodati *et al.*, 2006; Viswanathan *et al.*, 2006; Ossa *et al.*, 2007). Nla4 and Nla18 are both enhancer binding proteins – Nla4 of the NtrC-like type with an N-terminal receiver domain (Ossa *et al.*, 2007), whereas Nla18 has an N-terminal forkhead-associated (FHA) domain, which is typically involved in protein–protein interactions and preferentially binds phosphothreonine residues in their targets (Li *et al.*, 2000; Diodati *et al.*, 2006) – and indirectly affect A-signal accumulation by stimulating stringent response. Nla4 is encoded by an orphan response regulator gene (Shi *et al.*, 2008). Among the remaining six gene products important for A-signal accumulation, two are two-component proteins: AsgA is a hybrid histidine kinase with a C-terminal receiver domain (Plamann *et al.*, 1995; Li and Plamann, 1996) encoded upstream of a response regulator (MXAN2671) with an N-terminal receiver domain and a C-terminal domain of unknown function (Shi *et al.*, 2008). The function of MXAN2671 – if any – in A-signal synthesis has not been analysed. AsgD is an orphan hybrid histidine kinase with an N-terminal receiver domain (Cho and Zusman, 1999a; Shi *et al.*, 2008). It is currently not known whether AsgA and AsgD act in the same pathway to generate A-signal. A-signal induces the expression of genes that are normally induced after two hours of starvation; the regulators involved have not been identified. However, it has been speculated that the orphan kinase SasS (Yang and Kaplan, 1997), which likely localizes

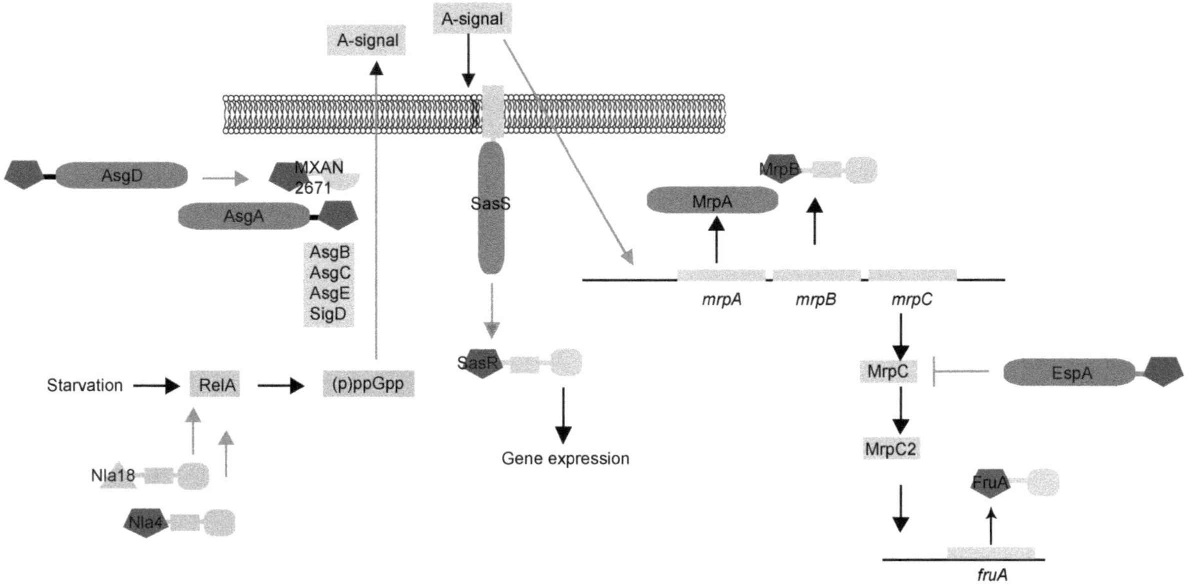

Figure 14.6 A-signal transduction pathway. A-signal synthesis is induced by the RelA-dependent stringent response with the accumulation of (p)ppGpp. By an unknown mechanism, this results in A-signal synthesis. The six proteins required for A-signal synthesis are indicated. Histidine protein kinase domains are indicated in light brown boxes; receiver domains are indicated by red pentagons; other domains are indicated in grey. Nla4, Nla18, SasR and MrpB are enhancer binding proteins; FruA is a DNA-binding response regulator. Light red indicates components of the stringent response; light green indicates components involved in synthesis and response to the A-signal; and, light blue indicates components in the C-signal transduction pathway. It is not known if AsgD acts in a pathway with the paired response regulator MXAN2671. Potential interactions are indicated by grey arrows. A colour version of this figure is located in the plate section at the back of the book.

to the cytoplasmic membrane, and the orphan NtrC-like enhancer binding protein SasR (Guo *et al.*, 2000) make up a cognate pair that regulates the expression of some A-signal-dependent genes and that the paired MrpAB two-component system regulates the expression of other A-signal-dependent genes (Sun and Shi, 2001a,b).

The C-signalling system becomes important for fruiting body formation after 6 hours of starvation (Kroos and Kaiser, 1987) and functions in a threshold manner to induce and ensure the correct order of aggregation and sporulation (Kim and Kaiser, 1991; Li *et al.*, 1992; Kruse *et al.*, 2001) with low levels of C-signalling inducing aggregation and higher levels inducing sporulation. C-signal transmission is contact-dependent, i.e. signal transmission depends on direct cell-to-cell contacts (Kim, 1990a,b,c). This characteristic provides the C-signal with the ability to provide cells with positional information, ensuring the spatial coupling of aggregation of cells into fruiting bodies and sporulation of cells that have accumulated inside fruiting bodies. The C-signal is a 17-kDa protein (p17) and its synthesis depends on the *csgA* gene (Kim and Kaiser, 1990b,d; Lobedanz and Søgaard-Andersen, 2003). The *csgA* gene codes for a 25-kDa protein (p25) (Kruse *et al.*, 2001), which is proteolytically cleaved to generate p17 (Lobedanz and Søgaard-Andersen, 2003) (Fig. 14.7A). p25 accumulates in the outer membrane of vegetative cells (Lobedanz and Søgaard-Andersen, 2003); however, it is only cleaved to generate p17, which is also anchored in the outer membrane, during starvation (Kruse *et al.*, 2001). PopC, a subtilisin-like protease, is the protease that cleaves p25 (Rolbetzki *et al.*, 2008): PopC accumulates in the cytoplasm of vegetative cells and is only secreted in starving cells in a RelA-dependent manner (A. Konovalova and L. Sogaard-Andersen, unpublished). Therefore, PopC and p25 are only present in the same cell compartment in starving cells, in that way, limiting p25 cleavage to starving cells. p17 is anchored

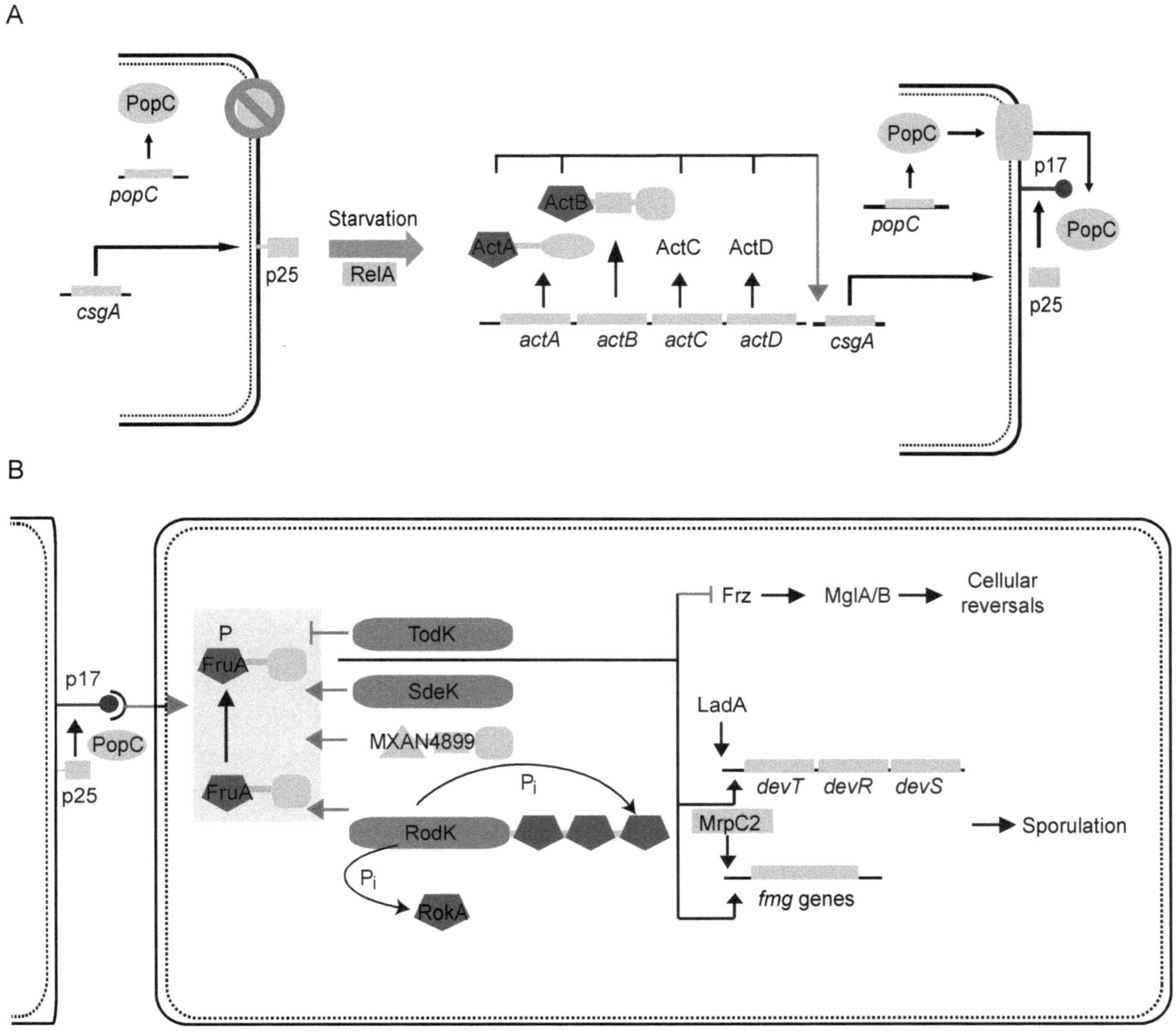

Figure 14.7 C-signal transduction pathway. (A) Synthesis of p17, the C-signal. The left-hand part of the schematic illustrates a vegetative cell and the right-hand part a starving cell. p25 and p17 are shown anchored in the outer membrane. The secretion system important for PopC secretion is not known. (B). The C-signal transduction pathway. The signalling event in the pathway is thought to be the interaction between p17 and a receptor on a neighbouring cell. This receptor remains to be identified. FruA is shown to be phosphorylated in response to C-signalling; genetic evidence supports this model, however, phosphorylation of FruA has not been demonstrated and a cognate kinase not identified. The grey box behind FruA and FruA-P indicates the level of the pathway where the signal transduction pathways defined by SdeK, TodK, RodK/RokA and MXAN4899 converge with the C-signal transduction pathway. Histidine protein kinase domains are indicated in light brown boxes; receiver domains are indicated by red pentagons. Potential interactions and indirect interactions are indicated by grey arrows. MrpC2 is marked on light green to indicate that it is induced by the A-signal pathway. MXAN4899 is an enhancer binding protein with an N-terminal FHA domain. A colour version of this figure is located in the plate section at the back of the book.

in the outer membrane and is non-diffusible (Shimkets and Rafiee, 1990; Lobedanz and Søgaard-Andersen, 2003). In addition to regulation by proteolysis, p17 accumulation by an unknown mechanism depends on the four gene products of the *actABCD* operon (see below; Fig. 14.7B) (Gronewold and Kaiser, 2001).

In total, the C-signal induces three responses during development, i.e. developmental gene expression, and aggregation, which are separated temporally and spatially. To understand how one signal may induce three events separated in time and space, many efforts have been dedicated to understanding how the C-signal transduction pathway is structured (Fig. 14.7B). C-signal transmission is thought to depend on the interaction between p17 located on one cell and a p17-receptor, which remains to be identified, located on an adjacent cell. A key protein in the C-signal transduction pathway is the orphan DNA binding response regulator FruA, which consists of an N-terminal receiver domain and a C-terminal DNA binding domain (Ogawa *et al.*, 1996; Ellehauge *et al.*, 1998). FruA activity is regulated at the transcriptional and, most likely, also at post-translational levels (Figs. 14.6 and 14.7B).

The expression of *fruA* is induced during development (Ogawa *et al.*, 1996; Ellehauge *et al.*, 1998). The key regulator in *fruA* transcription is a truncated version (MrpC2) of the MrpC protein, which is a homologue of the cyclic AMP receptor protein (CAP) in *Escherichia coli* (Sun and Shi, 2001a) (Fig. 14.6). MrpC2 binds directly to the *fruA* promoter and induces *fruA* transcription (Ueki and Inouye, 2003). Expression of *mrpC*, in turn, is activated by the MrpAB proteins, which constitute a two-component system with MrpA being a cytoplasmic histidine protein kinase and MrpB an enhancer binding protein of the NtrC type (Sun and Shi, 2001a,b) (Fig. 14.6). Expression of *mrpAB* by an unknown mechanism is induced by the A-signal (Sun and Shi, 2001b). Thus, a five-tiered cascade, i.e. stringent response, gives A-signal, gives *mrpAB* expression, gives *mrpC* expression, gives *fruA* expression, ensures *fruA* transcription and connects *fruA* transcription to earlier events during development. In addition, translation of the *mrpC* mRNA or MrpC degradation has been suggested to be inhibited or activated, respectively by the orphan hybrid kinase EspA (Cho and Zusman, 1999b; Higgs *et al.*, 2008) (Fig. 14.7). The regulatory mechanism that acts at the post-translational level to activate FruA likely involves the phosphorylation of the conserved aspartate residue in the N-terminal receiver domain (Ellehauge *et al.*, 1998). Moreover, genetic evidence suggests that FruA phosphorylation is induced by the C-signal (Søgaard-Andersen and Kaiser, 1996; Ellehauge *et al.*, 1998). However, the cognate FruA histidine protein kinase(s) has yet to be identified.

Downstream from activated FruA the C-signal transduction pathway contains a branch point, one branch results in aggregation and a second branch results in gene expression and sporulation (Fig. 14.7B). The Frz chemosensory system is a part of the aggregation branch (Søgaard-Andersen and Kaiser, 1996; Søgaard-Andersen *et al.*, 1996). At the molecular level, C-signalling results in methylation of the MCP homologue FrzCD (Søgaard-Andersen and Kaiser, 1996) and at the cellular level, reversals are inhibited (Jelsbak and Søgaard-Andersen, 2002). The C-signal-dependent inhibition of cellular reversals results in a change in the motility behaviour of starving cells. Therefore, after the initiation of C-signalling cells switch behaviour from occasionally reversing to a behaviour in which cells move in a unidirectional manner.

The second branch downstream from phosphorylated FruA leads to sporulation, and in this branch FruA acts as a transcriptional regulator (Fig. 14.7B). Recent analyses have shown that FruA and the LysR-type transcriptional regulator Lad in combination regulate expression of the *dev* operon, which is essential for sporulation (Viswanathan *et al.*, 2007). In the case of several other developmentally activated promoters, FruA acts in combination with MrpC2 to stimulate transcription (Yoder-Himes and Kroos, 2006; Mittal and Kroos, 2009a,b; Lee *et al.*, 2011; Son *et al.*, 2011). Interestingly, in the promoters jointly regulated by MrpC2 and FruA the arrangement of binding sites for the two regulators is highly variable. To date, it is not known whether transcriptional regulation by FruA involves unphosphorylated or phosphorylated FruA.

Genetic evidence suggests that several signal transduction pathways converge with the C-signal transduction pathway to regulate aggregation, sporulation and gene expression (Fig. 14.7B). For all pathways, genetic evidence suggests that they by unknown mechanisms converge with the C-signalling pathway downstream from FruA accumulation and upstream from the Frz system and *dev* operon. Three of these pathways are defined by histidine protein kinases. The orphan, cytoplasmic SdeK histidine protein kinase, is synthesized in a RelA-dependent manner immediately after the initiation of starvation, and the SdeK pathway converges with the C-signal transduction pathway to stimulate aggregation, sporulation and gene expression (Garza *et al.*, 1998; Pollack and Singer, 2001). The pathway defined by the orphan cytoplasmic TodK kinase inhibits aggregation, sporulation and gene expression (Rasmussen and Søgaard-Andersen, 2003). Synthesis of TodK is inhibited by starvation in a RelA-independent manner (Rasmussen and Søgaard-Andersen, 2003). The cytoplasmic protein RodK is the third kinase that defines a pathway that converges with the C-signal transduction pathway (Rasmussen *et al.*, 2005, 2006; Wegener-Feldbrügge and Søgaard-Andersen, 2009). RodK is encoded in an operon with the single domain response regulator RokA. RodK is structurally highly unusual and contains three C-terminal receiver domains that are all required for function. Based on *in vivo* and *in vitro* analyses it has been suggested that early during development the RodK kinase domain engages in phosphotransfer to its third receiver domain. Later in development, the specificity of the RodK kinase domain changes resulting in phosphotransfer to RokA. How an output response is generated by RodK or RokA remains to be elucidated. The fourth pathway that converges with the C-signal transduction pathway to regulate aggregation, sporulation and C-signal-dependent gene expression is defined by the enhancer binding protein MXAN4899 that contains an N-terminal FHA-domain (Jelsbak *et al.*, 2005).

Developmental gene expression

Initiation of the stringent response is essential for fruiting body formation and it depends on the RelA protein (Harris *et al.*, 1998). One branch of the stringent response is synthesis of the intercellular A-signal (Fig. 14.6) and secretion of the PopC protease that results in synthesis of the C-signalling molecule p17 (Fig. 14.7A). These two pathways converge at the level of *fruA* transcription and post-translational activation of FruA. A third pathway also functions to connect initiation of stringent response and C-signalling. This pathway consists of six enhancer binding proteins that together form a cascade important for developmental gene expression (Giglio *et al.*, 2011) (Fig. 14.8). The first set in the cascade consists of Nla4 and Nla18, which both stimulate induction of the stringent response during the transition from vegetative growth to development (see above) (Diodati *et al.*, 2006; Ossa *et al.*, 2007). Nla4 and Nla18 directly activate transcription of *nla6* but not that of *nla4* and *nla18*. Nla6, in turn, activate transcription of *nla28*. Nla6 and Nla28 constitute the second set of enhancer binding proteins in the cascade. Both are of the NtrC-type and are likely encoded in operons with a histidine protein kinase (MXAN4043 for Nla6 and MXAN1166 for Nla28) (Shi *et al.*, 2008). Nla6 and Nla28 are both required for development (Caberoy *et al.*, 2003) and directly activate transcription of the enhancer binding protein encoding gene *actB* but not that of *nla4* or *nla18*. ActB is of the NtrC-type and is essential for development, gene expression after 6 hours of starvation, and correct timing of C-signal accumulation (Gronewold and Kaiser, 2001, 2002) (Fig. 14.7A). ActB directly activates transcription of MXAN4899, which also encodes an enhancer binding protein and contains a FHA domain instead of a receiver domain (see above) and is also essential for development and gene expression after 6 hours (Jelsbak *et al.*, 2005). ActB is encoded in a complex gene cluster with ActA, which is a response regulator with a GGDEF domain, but without a histidine kinase (Gronewold and Kaiser, 2001). In addition, Nla6 and Nla28 and ActB directly activate transcription of their own genes. In total, these six enhancer binding proteins establish a gene regulatory cascade that interface with other signalling modules: ActB is essential for correct timing of C-signal accumulation and MXAN4899 converges on the C-signal transduction pathway downstream of FruA accumulation (Fig. 14.7).

Generally, chemosensory systems regulate motility behaviour of bacterial cells (Porter and Armitage, 2002; Kirby, 2009). The Che3 system of *M. xanthus* is a notable exception to this general rule. The Che3 system is encoded in an operon and consists of two MCP homologues, one CheW homologue, a CheA homologue, as well as a CheB and a CheR homologue. Upstream from the *che3* operon and transcribed in the opposite direction is the gene for CrdA, which is an enhancer binding protein of the NtrC-type (Kirby and Zusman, 2003). Mutations in the genes encoding any of the two MCPs or the CheA homologue caused early aggregation of starving cells and increased expression of developmental genes, a mutation

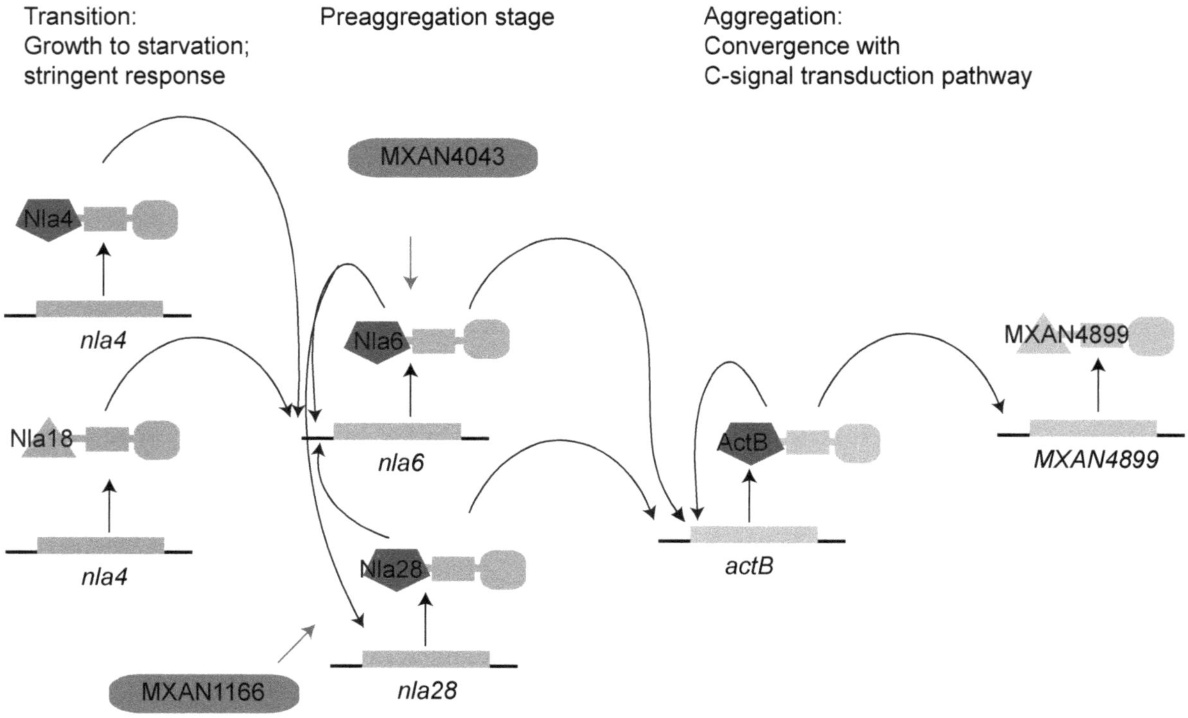

Figure 14.8 Enhancer binding protein cascade. Nla 4 and Nla18 are shown in light red to indicate their connection to the stringent response; Nla6 and Nla28 are shown in light orange and are not known to converge with other pathways than the one shown; ActB and MXAN4899 are shown in light blue to indicate that they connect to the C-signal transduction pathway. It is not known if MXAN1166 and MXAN4043 are the cognate kinases of Nla28 and Nla6, respectively. Histidine protein kinase domains are indicated in light brown boxes; receiver domains are indicated by red pentagons. A colour version of this figure is located in the plate section at the back of the book.

of *crdA* caused the opposite phenotype, and a double mutant had the *crdA* phenotype. On the basis of these observations and because CrdA and the CheA homologue interact in the yeast-two-hybrid system, it has been suggested that the primary function of the Che3 system is to regulate developmental gene expression. An advantage of regulating gene expression with a chemosensory system instead of a standard two-component system could be the ability of a chemosensory system to display adaptation to a persistent/persisting signal.

Formation of spore-filled fruiting bodies is a costly survival strategy because only 10–20% of cells differentiate to spores. Moreover, spores have not matured until 72 hours of starvation. Therefore, at least theoretically, it would 'make sense' if *M. xanthus* cells harboured systems to inhibit fruiting body formation in case nutrients become available before fruiting body formation including sporulation. Such systems have not been shown to exist; however, several systems have been identified which, when inactivated, cause early aggregation, sporulation and also sporulation of cells outside fruiting bodies suggesting that they act as inhibitors of development. Interestingly, most of these inhibitory systems are built on proteins of two-component systems including the orphan histidine kinase EspA (Cho and Zusman, 1999b; Higgs *et al.*, 2008); the orphan histidine kinases EspC (Lee *et al.*, 2005) and TodK (Rasmussen and Søgaard-Andersen, 2003); the Che3 system (Kirby and Zusman, 2003); the highly complex 'four-component' two-component system RedCDEF (Higgs *et al.*, 2005; Jagadeesan *et al.*, 2009); and, the NtrC-type enhancer binding protein SpdR (Hager *et al.*, 2001). The precise signals or parameters monitored by these systems are not known. Similarly, in most cases, the mechanism by which inhibition of development is achieved at the molecular level is not known. The many of negative regulators of development indicates that cells are continuously monitoring their nutritional status and possibly also other conditions to decide whether or not fruiting body formation is the best survival strategy.

Conclusions and outlook

The *M. xanthus* genome encodes a staggering 272 proteins of two-component system. Although far from all of these proteins having been analysed, major strides have been made and genetic analyses show that most aspects of the life cycle of *M. xanthus* are regulated by one or more of these proteins. Adding to the complexity of understanding signalling by two-component proteins in *M. xanthus* is the unusual genetic organization of the corresponding genes with many of these genes being orphans or localized in complex gene clusters and the domain structure of some of the proteins such as the RodK kinase that has one kinase domain and three receiver domains that are all required for function. Interestingly, the number of proteins of ligand-regulated one-component systems, which consist of single protein molecules containing both a sensing domain and an output domain (Ulrich *et al.*, 2005), is lower than expected and the number of Ser/Thr kinases much higher than expected based on the *M. xanthus* genome size (Goldman *et al.*, 2006) suggesting that signalling modalities based on phosphorylation/dephosphorylation by means of Ser/Thr protein kinases and histidine protein kinases are the dominating signalling systems in *M. xanthus*. The overriding question regarding two-component systems in *M. xanthus* is undoubtedly their connectivity, i.e. how are the orphan proteins connected? And, how are the proteins encoded by complex gene clusters connected? Because the connectivity of two-component proteins cannot yet be predicted based on sequence conservation alone (Skerker *et al.*, 2005, 2008), exciting experimental work lies ahead of us.

Acknowledgement

Work in the authors' laboratories is supported by the Max Planck Society, the German Research Council within the framework of the Graduate School 'Intra- and Intercellular Transport and Communication' and the LOEWE Research Centre for Synthetic Microbiology.

References

Alm, E., Huang, K., and Arkin, A. (2006). The evolution of two-component systems in bacteria reveals different strategies for niche adaptation. PLoS Comp. Biol. *2*, e143.

Appleby, J.L., Parkinson, J.S., and Bourret, R.B. (1996). Signal transduction via the multi-step phosphorelay: not necessarily a road less traveled. Cell *86*, 845–848.

Arnold, J.W., and Shimkets, L.J. (1988). Cell surface properties correlated with cohesion in *Myxococcus xanthus*. J. Bacteriol. *170*, 5771–5777.

Behmlander, R.M., and Dworkin, M. (1994a). Biochemical and structural analyses of the extracellular matrix fibrils of *Myxococcus xanthus*. J. Bacteriol. *176*, 6295–6303.

Behmlander, R.M., and Dworkin, M. (1994b). Integral proteins of the extracellular matrix fibrils of *Myxococcus xanthus*. J. Bacteriol. *176*, 6304–6311.

Bellenger, K., Ma, X., Shi, W., and Yang, Z. (2002). A CheW homologue is required for *Myxococcus xanthus* fruiting body development, social gliding motility, and fibril biogenesis. J. Bacteriol. *184*, 5654–5660.

Bentley, S.D., Chater, K.F., Cerdeno-Tarraga, A.M., Challis, G.L., Thomson, N.R., James, K.D., Harris, D.E., Quail, M.A., Kieser, H., Harper, D., *et al.* (2002). Complete genome sequence of the model actinomycete *Streptomyces coelicolor* A3(2). Nature *417*, 141–147.

Berleman, J.E., Chumley, T., Cheung, P., and Kirby, J.R. (2006). Rippling is a predatory behavior in *Myxococcus xanthus*. J. Bacteriol. *188*, 5888–5895.

Biondi, E.G. Reisinger, S.J., Skerker, J.M., Arif, M., Perchuk, B.S., Ryan, K.R., and Laub, M.T. (2006). Regulation of the bacterial cell cycle by an integrated genetic circuit. Nature *444*, 899–904.

Black, W.P., and Yang, Z. (2004). *Myxococcus xanthus* chemotaxis homologs DifD and DifG negatively regulate fibril polysaccharide production. J. Bacteriol. *186*, 1001–1008.

Black, W.P., Xu, Q., and Yang, Z. (2006). Type IV pili function upstream of the Dif chemotaxis pathway in *Myxococcus xanthus* EPS regulation. Mol. Microbiol. *61*, 447–456.

Black, W.P., Schubot, F.D., Li, Z., and Yang, Z. (2010). Phosphorylation and dephosphorylation among Dif chemosensory proteins essential for exopolysaccharide regulation in *Myxococcus xanthus*. J. Bacteriol. *192*, 4267–4274.

Blackhart, B.D., and Zusman, D.R. (1985). "Frizzy" genes of *Myxococcus xanthus* are involved in control of frequency of reversal of gliding motility. Proc. Natl. Acad. Sci. U.S.A. *82*, 8771–8774.

Blattner, F.R., Plunkett, G., Bloch, C.A., Perna, N.T., Burland, V., Riley, M., Collado-Vides, J., Glasner, J.D., Rode, C.K., Mayhew, G.F., *et al.* (1997). The complete genome sequence of *Escherichia coli* K-12. Science *277*, 1453–1462.

Bowden, M.G., and Kaplan, H.B. (1998). The *Myxococcus xanthus* lipopolysaccharide O-antigen is required for social motility and multicellular development. Mol. Microbiol. *30*, 275–284.

Bulyha, I., Schmidt, C., Lenz, P., Jakovljevic, V., Höne, A., Maier, B., Hoppert, M., and Søgaard-Andersen, L. (2009). Regulation of the type IV pili molecular machine by dynamic localization of two motor proteins. Mol. Microbiol. *74*, 691–706.

Burbulys, D., Trach, K.A., and Hoch, J.A. (1991). Initiation of sporulation in *B. subtilis* is controlled by a multicomponent phosphorelay. Cell *64*, 545–552.

Bustamante, V.H., Martinez-Flores, I., Vlamakis, H.C., and Zusman, D.R. (2004). Analysis of the Frz signal transduction system of *Myxococcus xanthus* shows the importance of the conserved C-terminal region of the cytoplasmic chemoreceptor FrzCD in sensing signals. Mol. Microbiol. *53*, 1501–1513.

Caberoy, N.B., Welch, R.D., Jakobsen, J.S., Slater, S.C., and Garza, A.G. (2003). Global mutational analysis of NtrC-like activators in *Myxococcus xanthus*: Identifying activator mutants defective for motility and fruiting body development. J. Bacteriol. *185*, 6083–6094.

Cho, K., and Zusman, D.R. (1999a). AsgD, a new two-component regulator required for A-signalling and nutrient sensing during early development of *Myxococcus xanthus*. Mol. Microbiol. *34*, 268–281.

Cho, K., and Zusman, D.R. (1999b). Sporulation timing in *Myxococcus xanthus* is controlled by the *espAB* locus. Mol. Microbiol. *34*, 714–725.

Clausen, M., Jakovljevic, V., Søgaard-Andersen, L., and Maier, B. (2009). High force generation is a conserved property of type IV pilus systems. J. Bacteriol. *191*, 4633–4638.

Diodati, M.E., Ossa, F., Caberoy, N.B., Jose, I.R., Hiraiwa, W., Igo, M.M., Singer, M., and Garza, A.G. (2006). Nla18, a key regulatory protein required for normal growth and development of *Myxococcus xanthus*. J. Bacteriol. *188*, 1733–1743.

Downard, J., Ramaswamy, S.V., and Kil, K.S. (1993). Identification of *esg*, a genetic locus involved in cell–cell signaling during *Myxococcus xanthus* development. J. Bacteriol. *175*, 7762–7770.

Dworkin, M. (1962). Nutritional requirements for vegetative growth of *Myxococcus xanthus*. J. Bacteriol. *288*, 250–257.

Ellehauge, E., Nørregaard-Madsen, M., and Søgaard-Andersen, L. (1998). The FruA signal transduction protein provides a checkpoint for the temporal co-ordination of intercellular signals in *M. xanthus* development. Mol. Microbiol. *30*, 807–817.

Galperin, M. (2005). A census of membrane-bound and intracellular signal transduction proteins in bacteria: Bacterial IQ, extroverts and introverts. BMC Microbiol. *5*, 35.

Galperin, M.Y. (2010). Diversity of structure and function of response regulator output domains. Curr. Opin. Microbiol. *13*, 150–159.

Garza, A.G., Pollack, J.S., Harris, B.Z., Lee, A., Keseler, I.M., Licking, E.F., and Singer, M. (1998). SdeK is required for early fruiting body development in *Myxococcus xanthus*. J. Bacteriol. *180*, 4628–4637.

Garza, A.G., Harris, B.Z., Pollack, J.S., and Singer, M. (2000). The *asgE* locus is required for cell–cell signalling during *Myxococcus xanthus* development. J. Bacteriol. *35*, 812–824.

Giglio, K.M., Caberoy, N., Suen, G., Kaiser, D., and Garza, A.G. (2011). A cascade of coregulating enhancer binding proteins initiates and propagates a multicellular developmental program. Proc. Natl. Acad. Sci. U.S.A. *108*, E431–E439.

Goldman, B.S., Nierman, W.C., Kaiser, D., Slater, S.C., Durkin, A.S., Eisen, J.A., Ronning, C.M., Barbazuk, W.B., Blanchard, M., Field, C., *et al.* (2006). Evolution of sensory complexity recorded in a myxobacterial genome. Proc. Natl. Acad. Sci. U.S.A. *103*, 15200–15205.

Gronewold, T.M.A., and Kaiser, D. (2001). The *act* operon controls the level and time of C-signal production for *Myxococcus xanthus* development. Mol. Microbiol. *40*, 744–756.

Gronewold, T.M.A., and Kaiser, D. (2002). *act* operon control of developmental gene expression in *Myxococcus xanthus*. J. Bacteriol. *184*, 1172–1179.

Guo, D., Wu, Y., and Kaplan, H.B. (2000). Identification and characterization of genes required for early *Myxococcus xanthus* developmental gene expression. J. Bacteriol. *182*, 4564–4571.

Hagen, D.C., Bretscher, A.P., and Kaiser, D. (1978). Synergism between morphogenetic mutants of *Myxococcus xanthus*. Dev. Biol. *64*, 284–296.

Hager, E., Tse, H., and Gill, R.E. (2001). Identification and characterization of *spdR* mutations that bypass the BsgA protease-dependent regulation of developmental gene expression in *Myxococcus xanthus*. Mol. Microbiol. *39*, 765–780.

Harris, B.Z., Kaiser, D., and Singer, M. (1998). The guanosine nucleotide (p)ppGpp initiates development and A-factor production in *Myxococcus xanthus*. Genes Dev. *12*, 1022–1035.

Hartzell, P., and Kaiser, D. (1991). Function of MglA, a 22-kilodalton protein essential for gliding in *Myxococcus xanthus*. J. Bacteriol. *173*, 7615–7624.

Henrichsen, J. (1972). Bacterial surface translocation: a survey and a classification. Bacteriol. Rev. *36*, 478–503.

Higgs, P.I., Cho, K., Whitworth, D.E., Evans, L.S., and Zusman, D.R. (2005). Four unusual two-component signal transduction homologs, RedC to RedF, are necessary for timely development in *Myxococcus xanthus*. J. Bacteriol. *187*, 8191–8195.

Higgs, P.I., Jagadeesan, S., Mann, P., and Zusman, D.R. (2008). EspA, an orphan hybrid histidine protein kinase, regulates the timing of expression of key developmental proteins of *Myxococcus xanthus*. J. Bacteriol. *190*, 4416–4426.

Hodgkin, J., and Kaiser, D. (1979a). Genetics of gliding motility in *Myxococcus xanthus* (Myxobacterales): genes controlling movement of single cells. Mol. Gen. Genet. *171*, 167–176.

Hodgkin, J., and Kaiser, D. (1979b). Genetics of gliding motility in *Myxococcus xanthus* (Myxobacterales): Two gene systems control movement. Mol. Gen. Genet. *171*, 177–191.

Inclan, Y.F., Vlamakis, H.C., and Zusman, D.R. (2007). FrzZ, a dual CheY-like response regulator, functions as an output for the Frz chemosensory pathway of *Myxococcus xanthus*. Mol. Microbiol. *65*, 90–102.

Inclan, Y.F., Laurent, S., and Zusman, D.R. (2008). The receiver domain of FrzE, a CheA–CheY fusion protein, regulates the CheA histidine kinase activity and downstream signalling to the A- and S-motility systems of *Myxococcus xanthus*. Mol. Microbiol. *68*, 1328–1339.

Jagadeesan, S., Mann, P., Schink, C.W., and Higgs, P.I. (2009). A novel "Four-component" two-component signal transduction mechanism regulates developmental progression in *Myxococcus xanthus*. J. Biol. Chem. *284*, 21435–21445.

Jakovljevic, V., Leonardy, S., Hoppert, M., and Søgaard-Andersen, L. (2008). PilB and PilT are ATPases acting antagonistically in type IV pili function in *Myxococcus xanthus*. J. Bacteriol. *190*, 2411–2421.

Jelsbak, L., and Søgaard-Andersen, L. (1999). The cell surface-associated intercellular C-signal induces behavioral changes in individual *Myxococcus xanthus* cells during fruiting body morphogenesis. Proc. Natl. Acad. Sci. U.S.A. *96*, 5031–5036.

Jelsbak, L., and Søgaard-Andersen, L. (2002). Pattern formation by a cell surface-associated morphogen in *Myxococcus xanthus*. Proc. Natl. Acad. Sci. U.S.A. *99*, 2032–2037.

Jelsbak, L., Givskov, M., and Kaiser, D. (2005). Enhancer-binding proteins with a forkhead-associated domain and the sigma54 regulon in *Myxococcus xanthus* fruiting body development. Proc. Natl. Acad. Sci. U.S.A. *102*, 3010–3015.

Jenal, U., and Galperin, M.Y. (2009). Single domain response regulators: molecular switches with emerging roles in cell organization and dynamics. Curr. Opin. Microbiol. *12*, 152–160.

Kaiser, D. (1979). Social gliding is correlated with the presence of pili in *Myxococcus xanthus*. Proc. Natl. Acad. Sci. U.S.A. *76*, 5952–5956.

Kaiser, D., and Crosby, C. (1983). Cell movements and its coordination in swarms of *Myxococcus xanthus*. Cell Motility *3*, 227–245.

Kaplan, H.B., and Plamann, L. (1996). A *Myxococcus xanthus* cell density-sensing system required for multicellular development. FEMS Microbiol. Lett. *139*, 89–95.

Kim, S.K., and Kaiser, D. (1990a). Cell motility is required for the transmission of C factor, an intercellular signal that coordinates fruiting body morphology of *Myxococcus xanthus*. Genes Dev. *4*, 896–904.

Kim, S.K., and Kaiser, D. (1990b). C-factor: a cell–cell signaling protein required for fruiting body morphogenesis of *M. xanthus*. Cell *61*, 19–26.

Kim, S.K., and Kaiser, D. (1990c). Cell alignment required in differentiation of *Myxococcus xanthus*. Science *249*, 926–928.

Kim, S.K., and Kaiser, D. (1990d). Purification and properties of *Myxococcus xanthus* C-factor, an intercellular signaling protein. Proc. Natl. Acad. Sci. U.S.A. *87*, 3635–3639.

Kim, S.K., and Kaiser, D. (1991). C-factor has distinct aggregation and sporulation thresholds during *Myxococcus* development. J. Bacteriol. *173*, 1722–1728.

Kirby, J.R. (2009). Chemotaxis-like regulatory systems: Unique roles in diverse bacteria. Annu. Rev. Microbiol. *63*, 45–59.

Kirby, J.R., and Zusman, D.R. (2003). Chemosensory regulation of developmental gene expression in *Myxococcus xanthus*. Proc. Natl. Acad. Sci. U.S.A. *100*, 2008–2013.

Konovalova, A., Petters, T., and Søgaard-Andersen, L. (2010). Extracellular biology of *Myxococcus xanthus*. FEMS Microbiol. Rev. *34*, 89–106.

Kroos, L., and Kaiser, D. (1987). Expression of many developmentally regulated genes in *Myxococcus* depends on a sequence of cell interactions. Genes Dev. *1*, 840–854.

Kruse, T., Lobedanz, S., Berthelsen, N.M.S., and Søgaard-Andersen, L. (2001). C-signal: A cell surface-associated morphogen that induces and coordinates multicellular fruiting body morphogenesis and sporulation in *M. xanthus*. Mol. Microbiol. *40*, 156–168.

Kunst, F., Ogasawara, N., Moszer, I., Albertini, A.M., Alloni, G., Azevedo, V., Bertero, M.G., Bessieres, P., Bolotin, A., Borchert, S., *et al.* (1997). The complete genome sequence of the Gram-positive bacterium *Bacillus subtilis*. Nature *390*, 249–256.

Lancero, H., Caberoy, N.B., Castaneda, S., Li, Y., Lu, A., Dutton, D., Duan, X.-y., Kaplan, H.B., Shi, W., and Garza, A.G. (2004). Characterization of a *Myxococcus xanthus* mutant that is defective for adventurous motility and social motility. Microbiology *150*, 4085–4093.

Lancero, H.L., Castaneda, S., Caberoy, N.B., Ma, X., Garza, A.G., and Shi, W. (2005). Analysing protein–protein interactions of the *Myxococcus xanthus* Dif signalling pathway using the yeast two-hybrid system. Microbiology *151*, 1535–1541.

Lee, B., Higgs, P.I., Zusman, D.R., and Cho, K. (2005). EspC is involved in controlling the timing of development in *Myxococcus xanthus*. J. Bacteriol. *187*, 5029–5031.

Lee, J.S., Son, B., Viswanathan, P., Luethy, P.M., and Kroos, L. (2011). Combinatorial regulation of *fmgD* by MrpC2 and FruA during *Myxococcus xanthus* development. J. Bacteriol. *193*, 1681–1689.

Leonardy, S., Freymark, G., Hebener, S., Ellehauge, E., and Søgaard-Andersen, L. (2007). Coupling of protein localization and cell movements by a dynamically localized response regulator in *Myxococcus xanthus*. EMBO J. *26*, 4433–4444.

Leonardy, S., Miertzschke, M., Bulyha, I., Sperling, E., Wittinghofer, A., and Sogaard-Andersen, L. (2010). Regulation of dynamic polarity switching in bacteria by a Ras-like G-protein and its cognate GAP. EMBO J. *29*, 2276–2289.

Li, J., Lee, G.I., Van Doren, S.R., and Walker, J.C. (2000). The FHA domain mediates phosphoprotein interactions. J. Cell. Sci. *113*, 4143–4149.

Li, S., Lee, B.-U., and Shimkets, L.J. (1992). *csgA* expression entrains *Myxococcus xanthus* development. Genes Dev. *6*, 401–410.

Li, Y., and Plamann, L. (1996). Purification and *in vitro* phosphorylation of *Myxococcus xanthus* AsgA protein. J. Bacteriol. *178*, 289–292.

Li, Y., Bustamante, V.H., Lux, R., Zusman, D., and Shi, W. (2005). Divergent regulatory pathways control A and S motility in *Myxococcus xanthus* through FrzE, a CheA–CheY fusion protein. J. Bacteriol. *187*, 1716–1723.

Lobedanz, S., and Søgaard-Andersen, L. (2003). Identification of the C-signal, a contact dependent morphogen coordinating multiple developmental responses in *Myxococcus xanthus*. Genes Dev. *17*, 2151–2161.

Lu, A., Cho, K., Black, W.P., Duan, X.-Y., Lux, R., Yang, Z., Kaplan, H.B., Zusman, D.R., and Shi, W. (2005). Exopolysaccharide biosynthesis genes required for social motility in *Myxococcus xanthus*. Mol. Microbiol. *55*, 206–220.

McBride, M.J., Weinberg, R.A., and Zusman, D.R. (1989). "Frizzy" aggregation genes of the gliding bacterium *Myxococcus xanthus* show sequence similarities to the chemotaxis genes of enteric bacteria. Proc. Natl. Acad. Sci. U.S.A. *86*, 424–428.

Maier, B., Potter, L., So, M., Long, C.D., Seifert, H.S., and Sheetz, M.P. (2002). Single pilus motor forces exceed 100 pN. Proc. Natl. Acad. Sci. U.S.A. *99*, 16012–16017.

Manoil, C., and Kaiser, D. (1980). Guanosine pentaphosphate and guanosine tetraphosphate accumulation and induction of *Myxococcus xanthus* fruiting body development. J. Bacteriol. *141*, 305–315.

Mauriello, E.M.F., Mouhamar, F., Nan, B., Ducret, A., Dai, D., Zusman, D.R., and Mignot, T. (2010). Bacterial motility complexes require the actin-like protein, MreB and the Ras homologue, MglA. EMBO J. *29*, 315–326.

Mignot, T., Merlie, J.P., and Zusman, D.R. (2005). Regulated pole-to-pole oscillations of a bacterial gliding motility protein. Science *310*, 855–857.

Mignot, T., Shaevitz, J.W., Hartzell, P.L., and Zusman, D.R. (2007). Evidence that focal adhesion complexes power bacterial gliding motility. Science *315*, 853–856.

Mittal, S., and Kroos, L. (2009a). A combination of unusual transcription factors binds cooperatively to control *Myxococcus xanthus* developmental gene expression. Proc. Natl. Acad. Sci. U.S.A. *106*, 1965–1970.

Mittal, S., and Kroos, L. (2009b). Combinatorial regulation by a novel arrangement of FruA and MrpC2 transcriptionfactors during *Myxococcus xanthus* development. J. Bacteriol. *191*, 2753–2763.

Mok, K.C., Wingreen, N.S., and Bassler, B.L. (2003). *Vibrio harveyi* quorum sensing: a coincidence detector for two autoinducers controls gene expression. EMBO J. *22*, 870–881.

Nan, B., Mauriello, E.M.F., Sun, I.-H., Wong, A., and Zusman, D.R. (2010). A multi-protein complex from *Myxococcus xanthus* required for bacterial gliding motility. Mol. Microbiol. *76*, 1539–1554.

Nan, B., Chen, J., Neu, J.C., Berry, R.M., Oster, G., and Zusman, D.R. (2011). Myxobacteria gliding motility requires cytoskeleton rotation powered by proton motive force. Proc. Natl. Acad. Sci. U.S.A. *108*, 2498–2503.

Nariya, H., and Inouye, M. (2008). MazF, an mRNA interferase, mediates programmed cell death during multicellular *Myxococcus* development. Cell *132*, 55–66.

Nierman, W.C., Feldblyum, T.V., Laub, M.T., Paulsen, I.T., Nelson, K.E., Eisen, J., Heidelberg, J.F., Alley, M.R.K., Ohta, N., Maddock, J.R., *et al.* (2001). Complete genome sequence of *Caulobacter crescentus*. Proc. Natl. Acad. Sci. U.S.A. *98*, 4136–4141.

Nudleman, E., Wall, D., and Kaiser, D. (2006). Polar assembly of the type IV pilus secretin in *Myxococcus xanthus*. Mol. Microbiol. *60*, 16–29.

O'Connor, K.A., and Zusman, D.R. (1991a). Behaviour of peripheral rods and their role in the life cycle of *Myxococcus xanthus*. J. Bacteriol. *173*, 3342–3355.

O'Connor, K.A., and Zusman, D.R. (1991b). Development in *Myxococcus xanthus* involves differentiation into two cell types, peripheral rods and spores. J. Bacteriol. *173*, 3318–3333.

Ogawa, M., Fujitani, S., Mao, X., Inouye, S., and Komano, T. (1996). FruA, a putative transcription factor essential for the development of *Myxococus xanthus*. Mol. Microbiol. *22*, 757–767.

Ossa, F., Diodati, M.E., Caberoy, N.B., Giglio, K.M., Edmonds, M., Singer, M., and Garza, A.G. (2007). The *Myxococcus xanthus* Nla4 protein is important for expression of stringent response-asociated genes, ppGpp accumulation, and fruiting body development. J. Bacteriol. *189*, 8474–8483.

Overgaard, M., Wegener-Feldbrügge, S., and Søgaard-Andersen, L. (2006). The orphan response regulator DigR is required for synthesis of extracellular matrix fibrils in *Myxococcus xanthus*. J. Bacteriol. *188*, 4384–4394.

Patryn, J., Allen, K., Dziewanowska, K., Otto, R., and Hartzell, P.L. (2010). Localization of MglA, an essential gliding motility protein in *Myxococcus xanthus*. Cytoskeleton *67*, 322–337.

Pelicic, V. (2008). Type IV pili: e pluribus unum? Mol. Microbiol. *68*, 827–837.

Petters, T., Zhang, X., Nesper, J., Treuner-Lange, A., Gomez-Santos, N., Hoppert, M., Jenal, U., and Søgaard-Andersen, L. (2012). The orphan histidine protein kinase SgmT is a c-di-GMP receptor and regulates composition of the extracellular matrix together with the orphan DNA binding response regulator DigR in *Myxococcus xanthus*. Mol. Microbiol. *84*, 147–165.

Plamann, L., Li, Y., Cantwell, B., and Mayor, J. (1995). The *Myxococcus xanthus asgA* gene encodes a novel signal transduction protein required for multicellular development. J. Bacteriol. *177*, 2014–2020.

Pollack, J.S., and Singer, M. (2001). SdeK, a histidine kinase required for *Myxococcus xanthus* development. J. Bacteriol. *183*, 3589–3596.

Porter, S.L., and Armitage, J.P. (2002). Phosphotransfer in *Rhodobacter sphaeroides* chemotaxis. J. Mol. Biol. *324*, 35–45.

Porter, S.L., Wadhams, G.H., and Armitage, J.P. (2011). Signal processing in complex chemotaxis pathways. Nat. Rev. Microbiol. *9*, 153–165.

Rasmussen, A.A., and Søgaard-Andersen, L. (2003). TodK, a putative histidine protein kinase, regulates timing of fruiting body morphogenesis in *Myxococcus xanthus*. J. Bacteriol. *185*, 5452–5464.

Rasmussen, A.A., Porter, S.L., Armitage, J.P., and Søgaard-Andersen, L. (2005). Coupling of multicellular morphogenesis and cellular differentiation by an unusual hybrid histidine protein kinase during fruiting body morphogenesis in *Myxococcus xanthus*. Mol. Microbiol. *56*, 1358–1372.

Rasmussen, A.A., Wegener-Feldbrügge, S., Porter, S.L., Armitage, J.P., and Søgaard-Andersen, L. (2006). Four signalling domains in the hybrid histidine protein kinase RodK of *Myxococcus xanthus* are required for activity. Mol. Microbiol. *60*, 525–534.

Rolbetzki, A., Ammon, M., Jakovljevic, V., Konovalova, A., and Søgaard-Andersen, L. (2008). Regulated secretion of a protease activates intercellular signaling during fruiting body formation in *M. xanthus*. Dev. Cell *15*, 627–634.

Rosenbluh, A., Nir, R., Sahar, E., and Rosenberg, E. (1989). Cell-density-dependent lysis and sporulation of *Myxococcus xanthus* in agarose beads. J. Bacteriol. *171*, 4923–4929.

Sager, B., and Kaiser, D. (1994). Intercellular C-signaling and the traveling waves of *Myxococcus*. Genes Dev. *8*, 2793–2804.

Scott, A.E., Simon, E., Park, S.K., Andrews, P., and Zusman, D.R. (2008). Site-specific receptor methylation of FrzCD in *Myxococcus xanthus* is controlled by a tetra-trico peptide repeat (TPR) containing regulatory domain of the FrzF methyltransferase. Mol. Microbiol. *69*, 724–735.

Shi, W., and Zusman, D.R. (1993). The two motility systems of *Myxococcus xanthus* show different selective advantages on various surfaces. Proc. Natl. Acad. Sci. U.S.A. *90*, 3378–3382.

Shi, X., Wegener-Feldbrügge, S., Huntley, S., Hamann, N., Hedderich, R., and Søgaard-Andersen, L. (2008). Bioinformatics and experimental analysis of proteins of two-component systems in *Myxococcus xanthus*. J. Bacteriol. *190*, 613–624.

Shimkets, L.J., and Kaiser, D. (1982). Induction of coordinated movement of *Myxococcus xanthus* cells. J. Bacteriol. *152*, 451–461.

Shimkets, L.J., and Rafiee, H. (1990). CsgA, an extracellular protein essential for *Myxococcus xanthus* development. J. Bacteriol. *172*, 5299–5306.

Singer, M., and Kaiser, D. (1995). Ectopic production of guanosine penta- and teraphosphate can initiate early developmental gene expression in *Myxococcus xanthus*. Genes Dev. *9*, 1633–1644.

Skerker, J.M., and Berg, H.C. (2001). Direct observation of extension and retraction of type IV pili. Proc. Natl. Acad. Sci. U.S.A. *98*, 6901–6904.

Skerker, J.M., Prasol, M.S., Barrett, S.P., Biondi, E.G., and Michael, T. Laub, M.T. (2005). Two-component signal transduction pathways regulating growth and cell cycle progression in a bacterium: A system-level analysis. PLoS Biol. *3*, e334.

Skerker, J.M., Perchuk, B.S., Siryaporn, A., Lubin, E.A., Ashenberg, O., Goulian, M., and Laub, M.T. (2008). Rewiring the specificity of two-component signal transduction systems. Cell *133*, 1043–1054.

Søgaard-Andersen, L. (2004). Cell polarity, intercellular signalling and morphogenetic cell movements in *Myxococcus xanthus*. Curr. Opin. Microbiol. *7*, 587–593.

Søgaard-Andersen, L., and Kaiser, D. (1996). C factor, a cell-surface-associated intercellular signaling protein, stimulates the cytoplasmic Frz signal transduction system in *Myxococcus xanthus*. Proc. Natl. Acad. Sci. U.S.A. *93*, 2675–2679.

Søgaard-Andersen, L., Slack, F.J., Kimsey, H., and Kaiser, D. (1996). Intercellular C-signaling in *Myxococcus xanthus* involves a branched signal transduction pathway. Genes Dev. *10*, 740–754.

Søgaard-Andersen, L., Overgaard, M., Lobedanz, S., Ellehauge, E., Jelsbak, L., and Rasmussen, A.A. (2003). Coupling gene expression and multicellular morphogenesis during fruiting body formation in *Myxococcus xanthus*. Mol. Microbiol. *48*, 1–8.

Son, B., Liu, Y., and Kroos, L. (2011). Combinatorial regulation by MrpC2 and FruA involves three sites in the *fmgE* promoter region during *Myxococcus xanthus* development. J. Bacteriol. *193*, 2756–2766.

Stock, A.M., Robinson, V.L., and Goudreau, P.N. (2000). Two-component signal transduction. Annu. Rev. Biochem. *69*, 183–215.

Stover, C.K., Pham, X.Q., Erwin, A.L., Mizoguchi, S.D., Warrener, P., Hickey, M.J., Brinkman, F.S.L., Hufnagle, W.O., Kowalik, D.J., Lagrou, M., *et al.* (2000). Complete genome sequence of *Pseudomonas aeruginosa* PAO1, an opportunistic pathogen. Nature *406*, 959–964.

Sun, H., and Shi, W. (2001a). Analyses of *mrp* genes during *Myxococcus xanthus* development. J. Bacteriol. *183*, 6733–6739.

Sun, H., and Shi, W. (2001b). Genetic studies of *mrp*, a locus essential for cellular aggregation and sporulation of *Myxococcus xanthus*. J. Bacteriol. *183*, 4786–4795.

Sun, H., Zusman, D.R., and Shi, W. (2000). Type IV pilus of *Myxococcus xanthus* is a motility apparatus controlled by the *frz* chemosensory system. Curr. Biol. *10*, 1143–1146.

Sun, M., Wartel, M., Cascales, E., Shaevitz, J.W., and Mignot, T. (2011). Motor-driven intracellular transport powers bacterial gliding motility. Proc. Natl. Acad. Sci. U.S.A. *108*, 7559–7564.

Trudeau, K.G., Ward, M.J., and Zusman, D.R. (1996). Identification and characterization of FrzZ, a novel response regulator necessary for swarming and fruiting-body formation in *Myxococcus xanthus*. Mol. Microbiol. *20*, 645–655.

Ueki, T., and Inouye, S. (2003). Identification of an activator protein required for the induction of *fruA*, a gene essential for fruiting body development in *Myxococcus xanthus*. Proc. Natl. Acad. Sci. U.S.A. *100*, 8782–8787.

Uhl, M.A., and Miller, J.F. (1996). Central role of the BvgS receiver as a phosphorylated intermediate in a complex two-component regulatory phosphorelay. J. Biol. Chem. *271*, 33176–33180.

Ulrich, L.E., Koonin, E.V., and Zhulin, I.B. (2005). One-component systems dominate signal transduction in prokaryotes. Trends Microbiol. *13*, 52–56.

Viswanathan, P., Singer, M., and Kroos, L. (2006). Role of sigmaD in regulating genes and signals during *Myxococcus xanthus* development. J. Bacteriol. *188*, 3246–3256.

Viswanathan, P., Ueki, T., Inouye, S., and Kroos, L. (2007). Combinatorial regulation of genes essential for *Myxococcus xanthus* development involves a response regulator and a LysR-type regulator. Proc. Natl. Acad. Sci. U.S.A. *104*, 7969–7974.

Vlamakis, H.C., Kirby, J.R., and Zusman, D.R. (2004). The Che4 pathway of *Myxococcus xanthus* regulates type IV pilus-mediated motility. Mol. Microbiol. *52*, 1799–1811.

Wegener-Feldbrügge, S., and Søgaard-Andersen, L. (2009). The atypical hybrid histidine protein kinase RodK in *Myxococcus xanthus*: Spatial proximity supersedes kinetic preference in phosphotransfer reactions. J. Bacteriol. *191*, 1765–1776.

Whitworth, D.E., and Cock, P.J.A. (2008). Two-component systems of the myxobacteria: structure, diversity and evolutionary relationships. Microbiology *154*, 360–372.

Wu, S.S., and Kaiser, D. (1995). Genetic and functional evidence that type IV pili are required for social gliding motility in *Myxococcus xanthus*. Mol. Microbiol. *18*, 547–558.

Wuichet, K., Cantwell, B.J., and Zhulin, I.B. (2010). Evolution and phyletic distribution of two-component signal transduction systems. Curr. Opin. Microbiol. *13*, 219–225.

Xu, Q., Black, W.P., Nascimi, H.M., and Yang, Z. (2011). DifA, a methyl-accepting chemoreceptor protein-like sensory protein, uses a novel signaling mechanism to regulate exopolysaccharide production in *Myxococcus xanthus*. J. Bacteriol. *193*, 759–767.

Yang, C., and Kaplan, H.B. (1997). *Myxococcus xanthus sasS* encodes a sensor histidine kinase required for early developmental gene expression. J. Bacteriol. *179*, 7759–7767.

Yang, R., Bartle, S., Otto, R., Stassinopoulos, A., Rogers, M., Plamann, L., and Hartzell, P. (2004). AglZ is a filament-forming coiled-coil protein required for adventurous motility of *Myxococcus xanthus*. J. Bacteriol. *186*, 6168–6178.

Yang, Z., Geng, Y., Xu, D., Kaplan, H.B., and Shi, W. (1998). A new set of chemotaxis homologues is essential for *Myxococcus xanthus* social motility. Mol. Microbiol. *30*, 1123–1130.

Yang, Z., Ma, X., Tong, L., Kaplan, H.B., Shimkets, L.J., and Shi, W. (2000). *Myxococcus xanthus dif* genes are required for biogenesis of cell surface fibrils essential for social gliding motility. J. Bacteriol. *182*, 5793–5798.

Yang, Z., Duan, X., Esmaeiliyan, M., and Kaplan, H.B. (2008). Composition, structure, and function of the *Myxococcus xanthus* cell envelope. In Myxobacteria: Multicellularity and Differentiation, Whitworth, D.E., ed. (ASM Press, Washington, DC), pp. 229–240.

Yoder-Himes, D.R., and Kroos, L. (2006). Regulation of the *Myxococcus xanthus* C-signal-dependent omega4400 promoter by the essential developmental protein FruA. J. Bacteriol. *188*, 5167–5176.

Zhang, Y., Franco, M., Ducret, A., and Mignot, T. (2010). A bacterial Ras-like small GTP-binding protein and its cognate GAP establish a dynamic spatial polarity axis to control directed motility. PLoS Biol. *8*, e1000430.

Zusman, D.R., Scott, A.E., Yang, Z., and Kirby, J.R. (2007). Chemosensory pathways, motility and development in *Myxococcus xanthus*. Nat. Rev. Microbiol. *5*, 862–872.

Two-component Systems in *Streptomyces*

15

Juan-Francisco Martín, Alberto Sola-Landa and Antonio Rodríguez-García

Abstract

Two-component systems (TCS) play a very important role in the regulation of metabolism in *Streptomyces* species in response to different nutritional or environmental signals. *Streptomyces* are Gram-positive soil-dwelling filamentous bacteria with large genomes that have the ability to produce thousands of different secondary metabolites. *Streptomyces* genomes contain a large number of paired two-component systems (usually more than 70) and some additional orphan sensor kinases and response regulators. Several of these systems have been studied in detail in the model species *Streptomyces coelicolor*. Particular attention has been paid to the PhoR/PhoP and the orphan GlnR systems owing to their relevance in the control of primary metabolism and secondary metabolite biosynthesis. The PhoP binding sequence in many phosphate regulated promoters is formed by 11 nucleotide direct-repeats. A cross-talk between PhoP and other global regulators such as AfsR or GlnR has been found. Other two-component systems, particularly AbsA1/AbsA2, also control antibiotic biosynthesis in *S. coelicolor*, while others control chitinase synthesis, stress responses or cellular differentiation. Finally, some orphan response regulators named 'atypical response regulators' (e.g. RedZ in *S. coelicolor* and JadR1 in *Streptomyces venezuelae*) bind as ligands the final product of the antibiotic biosynthetic pathway and act as feedback regulators of the biosynthesis of these secondary metabolites.

Two-component systems: basic features and DNA binding ability

Bacterial two-component systems typically consist of a membrane-bound histidine kinase which acts as a sensor (SK) and is able to autophosphorylate at a conserved histidine residue in response to an environmental or nutritional stimulus and subsequently transfers this phosphoryl group onto an aspartic acid residue on the receiver region of the cytoplasmic response regulator (RR). As a result of this transphosphorylation, a conformational change is produced in the response regulator which activates its output or effector domain, frequently a C-terminal DNA binding domain, thus regulating the expression of the target genes (see below) (Hakenbeck and Stock, 1996). Finally, the RR~P can be dephosphorylated leading to a down-modulation of the response and allowing the recycling of the RR for a new transphosphorylation round. In this last step some aspartyl-phosphate phosphatases may be involved in addition to the intrinsic phosphatase activity of SKs (Silversmith, 2010).

Histidine kinases are sensor proteins that have been identified in bacteria, archaea, filamentous fungi, yeasts, and some plants. The prototypical histidine kinase is a homodimeric integral membrane protein (Goulian, 2010). Usually, the sensor domain is formed by an extracytoplasmic loop flanked by two membrane-spanning segments. A so-called transmitter domain is located in the C-terminal region following the last transmembrane segment which protrudes into the cytoplasm (Cheung and Hendrickson, 2010). Some histidine kinases deviate from the prototype and may have the sensor domains embedded within the membrane or in the cytoplasmic portion of the protein (Cheung and Hendrickson, 2010). An extracytoplasmic C-terminal sensor domain has been proposed for one of the TCS of *Streptomyces coelicolor* (McKenzie and Nodwell, 2009) (see below). The subcellular localization of the domains determines whether the sensing system works from the outside or the inside of the cytoplasmic membrane. The SKs have been classified into five different groups by aligning the 16 amino acids surrounding the conserved histidine residue, which is the site of autophosphorylation (Fabret *et al.*, 1999).

The SKs perform three kinds of reactions: autophosphorylation in response to an appropriate signal, phosphoryl group transfer to the response regulator and dephosphorylation (phosphatase activity) of the phosphorylated response regulator (Stewart, 2010).

A key role in the regulation by TCSs is played by the partner response regulators that most frequently are cytoplasmic DNA binding proteins that act as transcriptional regulators. RRs contain a conserved phosphoacceptor (also named receiver, REC) domain, but they differ in their DNA-binding domains. The high number and diversity of RRs is correlated with the diversity of functions that they perform (Galperin, 2010).

Two-component signal transduction systems affect expression of many genes (Hsieh and Wanner, 2010; Silversmith, 2010), primarily through transcriptional, post-transcriptional and post-translational regulation of gene expression, but also through a variety of protein–protein interactions (Goulian, 2010).

Genome size and number of TCSs

The relationship between the number of genes coding for TCSs and the genome size is approximately linear for most bacteria (Whitworth and Cock, 2008), although the percentage of TCS in relation to the total number of genes is higher for free-living bacteria than for obligate pathogens (Kim and Forst, 2001).

The *S. coelicolor* (a model *Streptomyces* species) genome encodes 84 SKs and 80 RRs including 67 sensor–regulator pairs (Hutchings *et al.*, 2004), which highlights the importance of signal sensing in this microorganism. According to the MiST database (Ulrich and Zhulin, 2010) this number increases up to 138 SK, 2 hybrid-SK, and 87 RRs.

Streptomyces metabolism: adaptation to diverse and changing environmental conditions

Streptomyces are Gram-positive bacteria that inhabit a variety of soils and decay plant materials. They have to adapt to substantial changes in nutritional and environmental conditions, particularly in the organic matter-content, salinity, temperature, water activity, pH, calcium and iron content (Hsiao and Kirby, 2008). Two-component systems play a key role in the control of gene expression in *Streptomyces* that allows adaptation to the ever changing environmental and nutritional conditions.

Streptomyces are able to produce thousands of different secondary metabolites, many of them showing important biological or pharmacological activities (Challis and Hopwood, 2003; Demain, 2009). More than 40% of microbial products with pharmacological activity are obtained from *Streptomyces* and related actinomycetes (Sánchez *et al.*, 2012). More than 2500 *Streptomyces* strains have been described based on their ribosomal RNA sequences (Ribosomal Database Project; www.rdpcinemsu.edu). However, their classification into species still remains confusing (Martínez-Castro *et al.*, 2010). One of the reasons for this problem is that pigments and other secondary metabolites have been used for species identification, but many of them are not produced in a species- but in a strain-specific manner, and, therefore, they are not useful for taxonomic purposes.

The ability of *Streptomyces* species to produce such a large number of secondary metabolites is correlated with their large genome size, which is in the range of 8–10 Mb (see below). In those *Streptomyces* genomes sequenced so far, there are 20 to 25 gene clusters per genome involved in the biosynthesis of secondary metabolites and many of them result in the formation of structurally related compounds (a so-called family) (Bentley *et al.*, 2002; Medema *et al.*, 2010). Knowledge about the control of gene expression in *Streptomyces* is therefore of utmost interest for the generation of new secondary metabolites (Martín *et al.*, 2011a,b). Moreover, global regulators of secondary metabolism play an important role in the adaptation of *Streptomyces* to different habitats (see below).

Two-component systems in *Streptomyces*

As indicated above the *Streptomyces* genome size is large compared with other bacteria: 8,667,507 bp for *S. coelicolor* (Bentley *et al.*, 2002), 9,025,608 bp for *S. avermitilis* (Ikeda *et al.*, 2003), 8,545,929 bp for *S. griseus* (Ohnishi *et al.*, 2008) and 10,148,695 bp for *S. scabies* (accession number FN554889). That means that *Streptomyces* possesses one of the highest numbers of TCS, probably only overpassed by some cyanobacteria and myxobacteria (Whitworth and Cock, 2008). In fact, in *S. coelicolor* 67 adjacent TCS, 13 orphan response regulators and 17 orphan sensor kinases have been found (Bentley *et al.*, 2002; Hutchings *et al.*, 2004), and in *S. avermitilis* 53 adjacent TCS, 15 orphan RRs and 14 orphan SKs have been identified (Wei *et al.*, 2007).

An exhaustive analysis of TCSs present in *S. coelicolor, S. avermitilis, S. scabies* and *S. griseus* shows that a significant number of cognate TCSs conserved across the four species have orthologues in other actinomycetes. Interestingly, these TCSs are located mainly in the central region of the linear chromosomes. On the other hand, the presence of orphan SKs and RRs orthologues in other actinomycetes is less common and they are located preferentially in the terminal regions of the chromosomes (Hsiao and Kirby, 2009). This distribution is in agreement with the overall structure described for a *Streptomyces* chromosome, with a central core containing conserved syntenous genes across many actinomycetes, and two arms with specific genes, including those required for the secondary metabolism (Hsiao and Kirby, 2008; Kirby, 2011).

Most streptomycete TCSs that so far have been experimentally studied belong to the model species *S. coelicolor*. The major focus of most of these studies regards the regulation of the secondary metabolism. Here, we confine our description to those TCSs for which experimental evidence of their biological role exists.

Global positive regulators that may act as negative regulators in specific cases

The PhoR–PhoP system

One of the most studied TCS of *Streptomyces* is the PhoR/PhoP system, which is involved in the control of primary and secondary metabolism in response to phosphate availability. Although the negative effect of phosphate on the biosynthesis of an impressive number of secondary metabolites is known since a long time (Martín and Demain, 1980; Masuma *et al.*, 1986; Doull and Vining, 1990; Asturias *et al.*, 1994; Martín, 2004), only in the last decade the molecular mechanism has been largely elucidated. The PhoR-PhoP system was first described in *S. lividans*, where mutants deleted in either the regulator PhoP or the entire PhoR-PhoP system were defective in production of alkaline phosphatase (PhoA), while they overproduced the secondary metabolites actinorhodin and undecylprodigiosin (Sola-Landa *et al.*, 2003; Ghorbel *et al.*, 2006). Later, the relevance of PhoR/PhoP for the secondary metabolism was also shown in other *Streptomyces* species, such as *S. coelicolor* or *S. natalensis* (Mendes *et al.*, 2007; Santos-Beneit *et al.*, 2009a). When the phosphate concentration in the medium decreases, PhoP is phosphorylated by PhoR and activates the expression of the genes of the *pho* regulon by direct binding to the so-called PHO boxes present in the respective promoters (Sola-Landa *et al.*, 2005, 2008; Apel *et al.*, 2007; Santos-Beneit *et al.*, 2008, 2009a,b; Rodríguez-García *et al.*, 2009).

In most cases PhoP~P acts as a positive transcriptional factor, but there are a few promoters in which PhoP binds to the −10 region thereby acting as a repressor (see below). The PHO boxes are formed by direct repeat units (DRus) consisting of eleven nucleotides, where the first seven nucleotides are well conserved (consensus sequence GTTCACC), whereas the remaining four nucleotides are less conserved. When the number of experimentally well defined PhoP~P binding sites increased, the frequency of the nucleotides in the consensus sequence (see model 2 in Fig. 15.1) changed slightly with respect to the originally published frequency (Sola-Landa *et al.*, 2005), but the change did not affect the consensus box (Fig. 15.1). Shortening of a DRu from 11 to 10 nucleotides prevents binding of PhoP~P (Sola-Landa *et al.*, 2005, 2008). The number, organization and degree of conservation of the DRus vary between the different promoters. We have proposed a model in which a PhoP monomer binds to a DRu, thus at least two DRus are required for PhoP~P binding (Sola-Landa *et al.*, 2008; Martín *et al.*, 2011a). Two (or more) well conserved units form the core of a binding site, and they are recognized by PhoP~P in first instance. The other less conserved DRus present in different promoters are bound by PhoP~P only when the PhoP~P concentration increases, favoured by protein–protein interaction (Martín *et al.*, 2011a).

The genes directly regulated by PhoP are mainly related to the primary metabolism, but the total number of genes whose expression is controlled through PhoR/PhoP regulation after

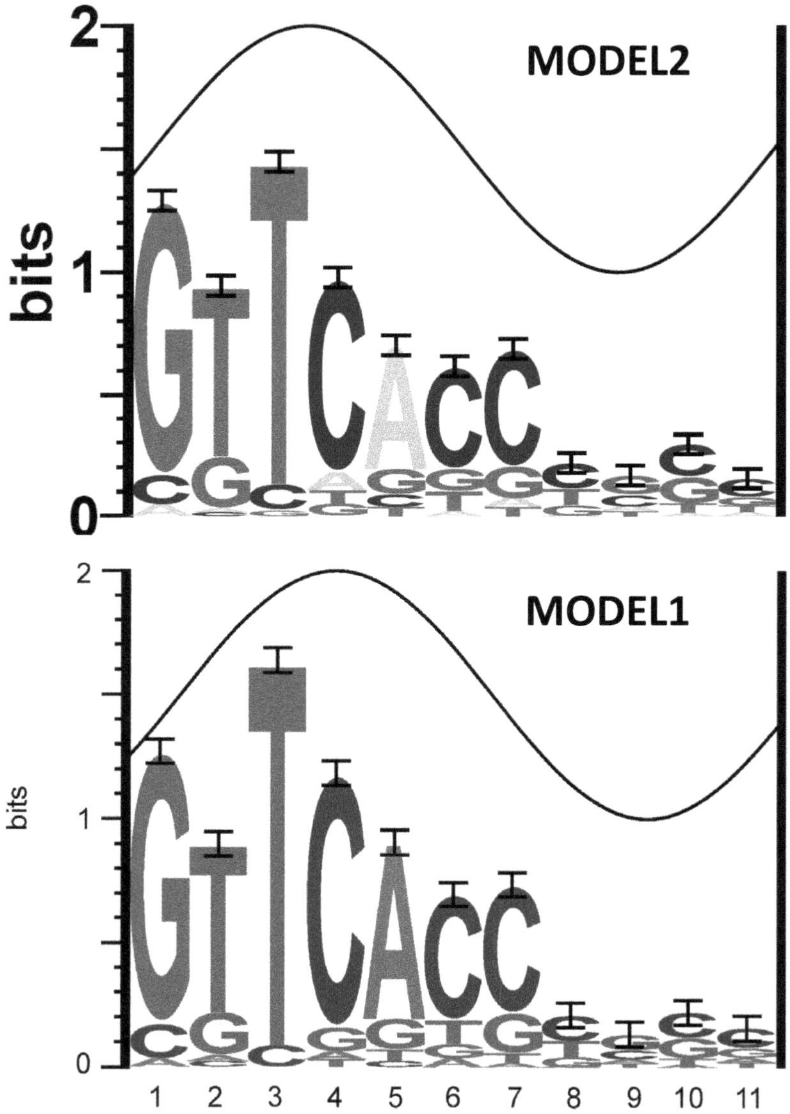

Figure 15.1 Sequence logo of the 11 nucleotide direct repeats that form the PhoP binding sequence (PHO box) in *Streptomyces* species. Model 1, taken from Sola-Landa *et al.* (2008); model 2 is an adjusted sequence taking into account all additional *S. coelicolor* genes to which PhoP binding has been confirmed (Sola-Landa, Rodríguez-García, Santos-Beneit and Martín, unpublished).

phosphate depletion is much larger, as shown by transcriptomic and proteomic analysis in *S. coelicolor* (Rodríguez-García *et al.*, 2007). These genes include those related to secondary metabolism, although no direct binding of PhoP to the promoters neither of the biosynthetic genes nor of the pathway-specific regulatory genes such as *actII*-ORF4, *redD* or *atrA* has been found so far (Santos-Beneit *et al.*, 2009a). The PhoP-mediated regulation of the secondary metabolism takes place through the control of the expression of *afsS*, a small σ-factor-like protein, whose amplification in a high-copy number plasmid leads to an overproduction of actinorhodin and undecylprodigiosin in both *S. coelicolor* and *S. lividans* (Vögtli *et al.*, 1994; Matsumoto *et al.*, 1995; Floriano and Bibb, 1996). The direct regulator of *afsS* is AfsR, a member of a family of high molecular weight regulators (Tanaka *et al.*, 2007). PhoP down-regulates the expression of *afsS* by direct competition with AfsR, since both regulators recognize overlapping sequence in the *afsS* promoter (Santos-Beneit *et al.*, 2009a, 2011b). In addition, PhoP affects expression of the promoter of the *S. coelicolor* RNA polymerase omega factor gene *rpoZ*, required for normal biosynthesis of actinorhodin and undecylprodigiosin (Santos-Beneit *et al.*, 2011a).

PhoP can be considered as a global regulator in *Streptomyces* since in addition to its role on phosphate uptake and secondary metabolism it also affects other nutritional regulators (Martín *et al.*, 2011b). In fact, an important interaction is that between PhoP and the genes involved in nitrogen metabolism. In this case the main regulator is GlnR that controls the expression of numerous genes directly or indirectly (Wray and Fisher, 1993; Reuther and Wohlleben, 2007; Tiffert *et al.*, 2008, 2011; Pullan *et al.*, 2011). Previously it was shown that some genes related to nitrogen metabolism are overexpressed in a δ*phoP* mutant after a phosphate downshift (Rodríguez-García *et al.*, 2007). More recently, it was shown that PhoP~P binds directly to the promoters of many genes involved in nitrogen assimilation, including that of *glnR* (Rodríguez-García *et al.*, 2009). This cross-regulation does not happen in both directions, since GlnR does not bind to promoters of genes belonging to the *pho* regulon (Sola-Landa and Martín, unpublished).

Although PhoP is usually a transcriptional activator in a few cases it may work as a repressor. This is the case of PhoP repression of the *hrdA* and *rpoZ* genes, encoding a σ-factor and the omega factor of the RNA polymerase (Sola-Landa *et al.*, 2008; Santos-Beneit *et al.*, 2011a). This repression is due to a 'road-block' effect when PhoP binds to the –10 to +1 region of the promoter thus interfering with RNA polymerase binding.

PhoP is extremely well conserved in *Streptomyces* species

The important role of the PhoR-PhoP system in *Streptomyces* gene regulation can be inferred from the very high sequence conservation of the PhoP proteins of all the species of this genus. Searching in the databases, we can find at least 25 complete and four partial sequences of PhoP in 29 different *Streptomyces* species, with only small differences between them. The length of the protein varies between 223 or 226 amino acids, depending on the translational start codon predicted. Taking the same start codon for all of them (corresponding to the Met1 of *S. coelicolor* which has a length of 223 amino acids), the sequence identity ranges from 96 to 100%. This is a very high value in a genus in which the proteins rarely show an identity above 70%. For example, the sequence identity of the major vegetative σ-factor HrdB between different *Streptomyces* species is lower than 90% (Marcos *et al.*, 1995; Bentley *et al.*, 2002). However, the sensor protein PhoR is not so well conserved, suggesting that although the network under the control of PhoP might be important, the relevance of phosphate for regulation in the different species may vary (Hsiao and Kirby, 2009).

Fidelity between partners in the PhoR/PhoP system: monogamy or polygamy?

The fidelity between the two cognate partners in TCSs is a subject of great interest (Szumant and Hoch, 2010). The structural organization of the *phoR-phoP* operon in *Streptomyces* indicates that both genes are very likely co-transcribed, since they share a common promoter. In fact, this transcriptional organization of the *phoR-phoP* operon has been confirmed experimentally in *S. lividans* (Ghorbel *et al.*, 2006). However, because of this operon structure, it had been difficult so far to study the role of the sensor kinase PhoR via a deletion mutant. In other bacteria, where disruption of the sensor kinase was less complicated owing to the genetic arrangement in an operon in which the SK gene is located downstream of the RR (e.g. PhoP–PhoR in *Bacillus*), it has been shown that other sensor proteins and/or molecules are able to compensate for the loss of PhoR. For example, in *E. coli* the sensor kinase CreC (also named PhoM), which is involved in carbon metabolism, is able to phosphorylate the RR PhoB in the absence of PhoR (Amemura *et al.*, 1990). At the same time, it has also been observed that acetyl phosphate, an intermediate of the carbon metabolism, was able to activate PhoB, thus indicating a certain cross-talk between the phosphate and carbon metabolic pathways (McCleary and Stock, 1994). Furthermore, other studies have revealed that at least other six different kinases are endowed with the ability to phosphorylate PhoB *in vivo* in *E. coli* (Fisher *et al.*, 1995; Zhou *et al.*, 2005). The promiscuity shown by PhoB in *E. coli* is probably due to the need to generate complex interconnected networks that potentially lead to cross-talk and cross-regulation, particularly between carbon and phosphate regulatory cascades. This promiscuity can be beneficial or even necessary when simultaneous integration of multiple signals is required for the appropriate response. On the other hand, *B. subtilis* PhoP seems to require specific phosphorylation by its cognate kinase PhoR, indicating a more stringent control of the activity of this RR (Sun *et al.*, 1996). The *B. subtilis* PhoP–PhoR system, which belongs to a more intricate multicomponent network involved in phosphate signal transduction (Sun *et al.*, 1996), seems to be very specific and although PhoR is able to phosphorylate a non-cognate partner, YycF, the reciprocal cross-talk does not occur. Thus, PhoR appears to be the only kinase able to phosphorylate PhoP in *B. subtilis*. Recently, we found a similar situation also in *S. coelicolor*: PhoR appears to be the only SK able to phosphorylate PhoP, thus providing solid evidence of TCS fidelity amongst these cognate partners in *Streptomyces* (Fernández-Martínez and Martín, unpublished). Some environmental or nutritional signals require a unique response in *Streptomyces*, thus promoting exclusivity within particular signalling cascades. The recent attainment of crystallized structures for a number of SK–RR complexes has given us an insight into how this specificity is achieved, i.e. mainly through conserved contact residues within the primary sequence of both proteins (Szurmant and Hoch, 2010).

Our results reveal that PhoR is necessary for the complete induction of the *pho* regulon. However, a low level of expression of genes of the *pho* regulon was detected in a *S. coelicolor* δ*phoR* background. This could be due to weak unspecific phosphorylation of PhoP by other SKs or acetyl phosphate. Alternatively, the low expression level of some of the *pho* regulon genes in the *S. coelicolor* Δ*phoR* mutant may indicate the existence of a basal constitutive level of expression of these genes from alternative transcription start points that are independent of the activation by the TCS. Indeed, this seems to be the case regarding the autoregulation of expression of the *phoR-phoP* operon, both in *S. coelicolor* and *B. subtilis*.

Expression of the *phoR-phoP* operon from the common *phoRP* promoter is activated by PhoP~P. Interestingly, a certain level of expression of the *phoR* transcript was detectable in an *S. coelicolor* δ*phoP* background. In *B. subtilis*, the well characterized *phoPR* promoter region contains five different starts which respond to different environmental stimuli, some of them giving a basal expression level which can be enhanced by PhoP~P binding (Paul *et al.*, 2004). Constitutive basal expression from the *phoRP* promoter region of *S. coelicolor* has been observed in a similar manner (Sola-Landa *et al.*, 2008). This guarantees that a basal level of PhoR-PhoP ready to be activated when phosphate becomes scarce is already available within the cell.

Other positive regulators

The AfsQ1/AfsQ2 system

One of the first reported TCSs in *Streptomyces*, AfsQ1/AfsQ2, was identified by cloning a DNA fragment that conferred the ability to produce A-factor and the pigmented antibiotics actinorhodin and undecylprodigiosin when introduced in *Streptomyces lividans*, a species which does not produce these antibiotics under normal culture conditions (Ishizuka *et al.*, 1992). This approach has been extensively used to identify secondary metabolism regulatory genes. The cloned DNA fragment contained a gene, named *afsQ1* (acronym for A-factor synthesis), which showed strong homology with response regulators. Cloning of the downstream sequence revealed the gene of the putative sensor kinase *afsQ2*, the start codon of which overlapped the TGA stop codon of *afsQ1*. Phosphorylation of the RR AfsQ1 is a prerequisite for its activity because a *S. lividans* strain expressing an AfsQ1 mutant with a D52E exchange was defective in antibiotics production. Despite of these results, in *S. coelicolor* (a species with some regulatory differences with respect to *S. lividans*) the disruption of the *afsQ1* or *afsQ2* genes did not alter the secondary metabolism or sporulation. However, in a *S. coelicolor absA* mutant altered in a different TCS that failed to produce the pigmented antibiotics (see below), the introduction of the homologous *afsQ1* gene in a plasmid restored the production of both, actinorhodin and undecylprodigiosin (Ishizuka *et al.*, 1992). Recently, the function of AfsQ1/AfsQ2 on secondary metabolism and morphological development was shown to be culture medium-dependent (Shu *et al.*, 2009). These results indicate that the biosynthesis of actinorhodin and undecylprodigiosin responds to nutrients (i.e. they are medium-dependent) through the mediation of different response regulators. The *afsQ1/Q2* deletion mutant lacked any observed phenotype in complex media, but when grown on defined medium using glutamate as the sole nitrogen source, the mutant was impaired in antibiotics production (actinorhodin, undecylprodigiosin, and calcium-dependent antibiotic CDA). By contrast, formation of aerial mycelium was faster in the mutant than in the parental strain.

Upstream of *afsQ1*, but transcribed in the opposite orientation, the σ-factor gene *sigQ* is located. As occurs with the Δ*afsQ* mutation, deletion of *sigQ* only caused a distinct phenotype in defined medium. This phenotype was, however, opposite to that observed for the *afsQ* mutant, i.e. it caused precocious overproduction of antibiotics and poor sporulation. The function of the σ^Q-factor appears to be antagonistic to, and also dependent on, the contiguous TCS (Shu *et al.*, 2009). Thus, expression of *sigQ* was reduced in the *afsQ* mutant. The triple deletion mutant *sigQ-afsQ1/Q2* had the same phenotype as the *afsQ1/Q2* mutant; introduction of both *sigQ* and *afsQ1/Q2* genes in the triple mutant restored the parental phenotype (normal production), whilst introduction of only *afsQ1/Q2* led to antibiotics overproduction (*sigQ* phenotype). A putative lipoprotein, AfsQ3, is located downstream of this TCS (Hutchings *et al.*, 2004). There are not experimental data on the function of this protein, but a connection with the TCS function is indicated by the fact that there is translational coupling with the upstream *afsQ2* gene. In the genome of *S. coelicolor* other five TCS are associated to lipoprotein-encoding genes (Hutchings, 2007), such as *cseA* and *lpqB*.

Global negative regulators

The AbsA1/AbsA2 system

Among the negative regulators of secondary metabolism, the most studied is the AbsA1/AbsA2 system (Brian *et al.*, 1996), a TCS that is required for the timely onset of antibiotic production dependent on the growth phase. The locus *absA* (for antibiotic synthesis deficient) was identified after selection of UV-induced mutants which lost pigmentation but not the normal morphological differentiation (Abs⁻ phenotype) (Adamidis *et al.*, 1990). It should be noted that morphological differentiation in *S. coelicolor* coincides with production of secondary metabolites in solid medium, and that there are regulatory links between both processes (Bibb, 2005). In fact, these mutants were defective in production of the four *S. coelicolor* antibiotics (the coloured actinorhodin and undecylprodigiosin, and the unpigmented methylenomycin and CDA), although the TCS was located within the CDA biosynthesis cluster. Two mutant alleles were sequenced and point mutations were identified in conserved domains of the SK *absA1* gene (Anderson *et al.*, 1999). Mutations in the *absA1* gene that cause the Abs⁻ phenotype are expected to impair the phosphatase activity of this SK, as shown by biochemical studies (Sheeler *et al.*, 2005). Either disruption or deletion mutants lacking the transmitter domain of AbsA1, or disruption of the *absA2* gene, showed the same precocious hyperproduction of antibiotics phenotype (Pha) for both, actinorhodin and undecylprodigiosin pigments. The Pha phenotype was also shown for CDA using bioassays. That the phosphorylated form of AbsA2 is the negative regulator of antibiotics production was confirmed by site-directed mutations in the Asp codon of the RR that were expected to prevent phosphorylation (D54E, D54A, and D54N). The three mutants showed the Pha phenotype. Indeed, overexpression of *absA1* alone also resulted in the Pha phenotype; this has been interpreted to be caused by an increased phosphatase over kinase activity of the AbsA1 protein (Anderson *et al.*, 2001).

Using chromatin immunoprecipitation (ChIP), the RR AbsA2 was shown to bind the promoters of the pathway-specific activators *actII-ORF4* of actinorhodin, *cdaR* of CDA, and *redZ* of undecylprodigiosin, but not the promoters of other 23 global or pathway-specific genes that regulate the secondary metabolism. Moreover, the controlled expression of the *actII-ORF4* and *redZ* by means of an inducible promoter caused pigment production in a hyperrepressive *absA* mutant (McKenzie and Nodwell, 2007). The operon promoter, located upstream of *absA1*, increases its activity throughout the growth phase, previous to the onset of antibiotic production (Anderson *et al.*, 2001). The system appears to be self-regulated since the transcription of the *absA* promoter is increased in an Abs$^-$ mutant and decreased in a Pha mutant (Anderson *et al.*, 2001; Ryding *et al.*, 2002), yet the ChIP analysis failed to detect AbsA2 binding to the promoter sequence (McKenzie and Nodwell, 2007).

The transmembrane topology of the sensor kinase has been investigated using enhanced GFP (EGFP) fusions (McKenzie and Nodwell, 2009). The AbsA1 protein lacks the extracellular N terminal domain but has an unusual extracellular C terminal domain that is proposed to participate in the stimulus detection. This topology of the SK is different from that of other bacterial SKs in which the C-terminal end is proposed to be internal to the membrane. Several residues of the C terminal sequence are conserved in other sensor kinases from actinomycetes. At least three of these SKs are located near biosynthetic secondary metabolism genes. As yet, the nature of the input signal for the AbsA1 sensor domain remains unknown.

Yepes *et al.* (2011) obtained replacement mutants of five TCSs that were selected by sequence homology with the AbsA1/AbsA2 system. Only the mutants of two TCSs (named as *abr* for antibiotic regulator) showed significant differences in both, antibiotic production and morphological differentiation under the culture conditions assayed. The AbrA1Abr/A2 system acted as a negative regulator of both, antibiotic production and morphological differentiation. By contrast, the system AbrC1/AbrC2/AbrC3, composed of two histidine kinases and one response regulator, had positive effects on both processes. Microarray analyses of the *abrC1/C2/C3* mutant detected down-regulation of *actII*-ORF4 and *cdaR* genes, with respect to the wild-type expression level. These results indicate a clearly higher specificity of the Abs TCS as compared with the Abr TCS in the regulation of antibiotics biosynthesis.

The CutR/CutS system

Another TCS that negatively controls secondary metabolism is CutR/CutS. This TCS was first identified in *S. lividans* (Chang *et al.*, 1996). Transcription at the single operon promoter initiates at the transition phase and extended to the stationary phase of the culture. The deletion of *cutR* or of *cutS* in *S. lividans* caused overproduction of actinorhodin. This effect was suppressed in both cases by the introduction of the RR gene. The heterologous expression of the TCS in *S. coelicolor* had a negative regulatory influence on actinorhodin production. A Blast search demonstrated that this TCS, and in particular the CutR protein, is highly conserved across *Streptomyces* and other actinobacterial strains.

Positive regulators of specific secondary metabolites

So far we have described global regulators in *Streptomyces*. However, some TCSs appear to serve for the control of specific biosynthetic pathways. The time series microarray analysis performed by Huang *et al.* (2001) revealed two TCSs, EcrA1/EcrA2 and EcrE1/EcrE2, that followed the transcription profile of the undecylprodigiosin biosynthesis genes (expression coordinated with *red*), but mapped at distant sites from the *red* cluster. Deletion of either *ecrA1* or *ecrA2*, or of *ecrE1* or *ecrE2* caused only a decrease in undecylprodigiosin production, but did not affect actinorhodin production (Li *et al.*, 2004; Wang *et al.*, 2007).

The ChiS/ChiR system

The ChiS/ChiR system of *S. thermoviolaceus* was found to increase the chitinase activity when introduced on a multicopy plasmid (Tsujibo *et al.*, 1999). Kormanec *et al.* (2000) located the *S. coelicolor* orthologous *chiS/chiR* genes in a divergent arrangement to the chitinase *chiC* gene. Disruption of the *chiR* gene, however, failed to affect antibiotics production, morphology or chitinase activity.

The CseA/CseB/CseC system: linkage to SigE

The cell envelope is the first and major defence against physical, chemical and biological threats from the environment. Indeed, cell wall biosynthesis is a common target of antibiotics. The response to stress factors is commonly regulated by a combination of TCSs and extracytoplasmic function (ECF) σ-factors (Jordan *et al.*, 2008). In *S. coelicolor*, analysis of null mutants of the ECF σ-factor gene *sigE* demonstrated that this factor is required for normal cell wall function (Paget *et al.*, 1999a). The *sigE* gene is mainly transcribed monocistronically, but also as part of the operon *sigE-cseABC* (Paget *et al.*, 1999b). CseB and CseC represent a RR and a SK, respectively. Transcription from the *sigE* promoter is totally dependent on CseB. The indistinguishable phenotypes of *cseB* and *sigE* null mutants indicate that the *sigE* promoter is the only target of CseB (Paget *et al.*, 1999b). The *sigE-cseABC* operon also encodes a lipoprotein, CseA, which is localized to the external face of the membrane. An in-frame deletion of *cseA* caused up-regulation of the *sigE* promoter (Hutchings *et al.*, 2006b). In the *S. coelicolor* genome, other five SK genes, including *afsQ2*, are next to lipoprotein genes with which are putatively co-transcribed (Hutchings *et al.*, 2006b). One of these cases is the *mtrAB-lpqB* operon that is conserved across actinobacteria. A modulating role for these lipoproteins is suggested, such as the lipoprotein LpqB on the activity of the SK MtrB (Hoskisson and Hutchings, 2006).

The VanR/VanS system

The glycopeptide antibiotic vancomycin inhibits cell wall formation by binding with high affinity the D-Ala-D-Ala terminus of the lipid-linked disaccharide pentapeptide. The alternative synthesis of peptidoglycan precursors that terminate in D-Ala-D-Lactate is the predominant mechanism of resistance (Walsh *et al.*, 1996). *S. coelicolor* contains a resistance gene cluster homologous to that of the vancomycin producer *S. toyocaensis* (Pootoolal *et al.*, 2002; Hong *et al.*, 2004). Expression of the cluster depends on the TCS VanR/VanS. VanR is a response regulator and VanR~P activates the transcription of the cluster operons *vanRS–vanJ–vanK–vanHAX* (Hong *et al.*, 2004, Hutchings *et al.*, 2006a). It has been found that VanR can be phosphorylated unspecifically by acetyl phosphate or specifically by its cognate sensor kinase VanS. In absence of vancomycin, the phosphatase activity of VanS ensures that the *van* genes are turned off (Hutchings *et al.*, 2006a,b). Exposition to vancomycin induces the kinase activity of VanS (Hong *et al.*, 2005) leading to phosphorylation of VanR and transcriptional activation of the resistance regulon.

The OsaA/OsaB/OsaC system

The atypical response regulator OsaB is required for the osmoadaptation response. When cultured under continuous hyperosmotic conditions, a *S. coelicolor osaB* mutant did not develop aerial hyphae and overproduced undecylprodigiosin and actinorhodin. The protein lacks a DNA binding domain, although possible interactions with other proteins were suggested because of the presence of a coiled-coil motif. Mutation of the upstream gene, *osaA*, which encodes a hybrid histidine kinase, resulted only in a delayed formation of aerial hyphae (Bishop *et al.*, 2004). It has been suggested that the presence of a similar orphan hybrid histidine kinase, SCO7327, located at a distant position in the chromosome might account for the lack of a clear phenotype of the *osaA* mutant. A third gene, *osaC*, is encoded in the neighbourhood of the *osaA* gene, but transcribed divergently. OsaC represents a new family of streptomycete regulatory proteins which is characterized by its domain architecture. These proteins are composed of an N-terminal kinase domain related to anti-σ–factors, sensory PAS and GAF domains, and a C-terminal phosphatase domain. The proposed role of OsaC is the modulation of the activity of the σ-factor σ^B, the master regulator of the response to both osmotic and oxidative stress (Fernández-Martínez *et al.*, 2009). According to the model proposed by these authors, a sudden increase in osmolarity triggers the σ^B regulon, which includes *osaB*. It must be noted, however, that σ^B is not required for the distinct osmoadaptation response; in that case *osaB* is transcribed from a second promoter. Physiological recovery from the osmotic stress induces the kinase activity of OsaC leading to phosphorylation of a predicted OsaC antagonist protein. Once OsaC is released it can associate with σ^B and stop the continuous expression of the regulon.

The HbpS/SenS/SenR system from *S. reticuli*

The TCS SenS/SenR has been studied in the cellulose degrader *S. reticuli* (Ortiz de Orué-Lucana *et al.*, 2005). The *senS* gene encodes a predicted membrane protein endowed with five transmembrane regions. SenS has been shown to autophosphorylate at a conserved histidine residue and to be able to transfer the phosphate group to its cognate response regulator SenR. Upon phosphorylation SenR binds with high affinity to the upstream regions of the *senS–senR* operon and the gene *hbpS*, which encodes a secreted haem-binding protein (HbpS). The unphosphorylated SenR binds upstream of the *furS–cpeB* operon coding for the redox regulator FurS and a catalase/peroxidase, CpeB, respectively (Bogel *et al.*, 2007). The SenS/SenR system participates in the sensing of redox signals mediated by iron (Ortiz de Orué-Lucana *et al.*, 2005; Bogel *et al.*, 2008).

HbpS is an accessory protein of the SenS/SenR system, which has been shown to be secreted in a Tat (twin-arginine translocation)-dependent manner (Berks *et al.*, 2000; Schaerlaekens *et al.*, 2001; Ortiz de Orué-Lucana *et al.*, 2004). The presence of HbpS increases the synthesis of the highly active catalase peroxidase CpeB *in vivo*. Therefore, it was proposed that HbpS interacts with an extracellular, membrane-associated or membrane-integrated protein(s) involved in a signal transduction cascade regulating *cpeB* transcription. Protein–protein interaction studies revealed that both, native *S. reticuli* HbpS and recombinant HbpS, interact specifically with the iron-sensing histidine kinase SenS *in vitro* (Ortiz de Orué-Lucana *et al.*, 2005). The SenS N-terminal domain (containing a predicted extracytoplasmic and transmembrane regions) is essential for the interaction with HbpS (Bogel *et al.*, 2008).

The RapA1/RapA2 system

Lu *et al.* (2007) screened a dozen of TCS replacement mutants and identified a mutant, M145-Δ*rapA1/A2*, which showed a visible decrease of actinorhodin production. The introduction of the TCS *rapA1/rapA2* genes reverted the M145-Δ*rapA1/rapA2* mutant to the wild-type phenotype. Transcription analysis by semiquantitative RT-PCR indicated that the *rapA1/rapA2* genes are co-transcribed and that their expression is growth-dependent, reaching maximum expression at the transition phase. By means of proteomic analysis and RT-PCR, the authors detected in the M145-Δ*rapA1/rapA2* mutant reduced expression levels of several biosynthetic genes of actinorhodin and also of the cluster encoding the cryptic type I polyketide (*cpk*) genes. Its positive effect is apparently exerted through the corresponding pathway-specific regulatory genes *actII*-ORF4 and *kasO*, since they were down-regulated in the mutant.

GlnR and its role on nitrogen regulation

An important orphan RR is GlnR that has been identified in several *Streptomyces* species as well as in the related actinobacteria *Amycolatopsis* and *Mycobacterium*. GlnR is a member of the OmpR family of RRs with a conserved winged HTH DNA-binding domain. GlnR is a global regulator known to activate several genes involved in nitrogen metabolism including *glnA* and *glnII* that encode two different glutamine synthetases in *S. coelicolor*, and *amtB* encoding an ammonium transporter that is co-transcribed with two other genes, *glnK* and *glnD*, involved in modification and regulation of nitrogen metabolism enzymes

(Wray *et al.*, 1991; Fink *et al.*, 1992; Hesketh *et al.*, 2002). Interestingly, there is a second response regulator, GlnRII, with strong similarities to GlnR that is linked to the *glnII* gene. GlnRII binds to the same nucleotide sequence as GlnR, but its role in nitrogen regulation is obscure. The bioinformatically proposed GlnR-binding sequence was confirmed experimentally by footprinting assays of the above mentioned nitrogen metabolism genes (Tiffert *et al.*, 2008).

A total of 20 *S. coelicolor* genes were proposed to be under control of GlnR (Tiffert *et al.*, 2008). However, surprisingly GlnR was found to bind to the *nas* gene promoter encoding a nitrate reductase which lacks the previously defined GlnR binding box (Wang and Zhao, 2009). These results suggest that the GlnR binding sequence may not be well defined so far. Recently, a global analysis of the GlnR-regulated genes in *S. venezuelae* has been performed using a ChIP-chip approach (Pullan *et al.*, 2011). These authors identified 36 binding sequences in the promoter regions adjacent to genes playing different biological roles. Some of these genes coincide with the genes of *S. coelicolor* reported to be regulated by GlnR, but others were novel, suggesting that there might be differences between *Streptomyces* species.

RamR: role in morphological differentiation

The orphan response regulator RamR is considered to be a key activator of the development of the aerial hyphae (Nguyen *et al.*, 2002), although several other TCSs affect *Streptomyces* differentiation indirectly. Mutants blocked in the initiation of aerial hyphae formation are known as *bld* mutants (the bald phenotype) because they produce colonies with a distinctive shiny, non-sporulating surface. The products of the *bld* genes are involved in an extracellular signalling cascade leading to the production of hydrophobic secreted proteins (Chater and Chandra, 2006). These proteins (so-called chaplins) and SapB, serve as surfactants to allow the hyphae to brake the air–water interface and grow upright (Elliot and Talbot, 2004). SapB is a lantibiotic-like peptide derived from the protein product of *ramS*, part of the *ramCSABR* cluster. The regulator RamR binds the promoter of the *ramCSAB* operon and activates its transcription (O'Connor *et al.*, 2002; 2005). Microarray comparison among the *ramR* mutant, the wild type and the strain harbouring multicopies of the *ramR* gene identified four <u>r</u>*amR*-<u>a</u>ctivated genes (*rag*). These genes form an operon that includes two components of an ABC transporter, *ragA/ragB*, and a TCS *ragK/ragR* (SanPaolo *et al.*, 2006); this operon is activated by direct binding of RamR. The operon null mutant produced abnormal aerial hyphae and showed a severe delay in sporulation. Constitutive overexpression of both *ragK/ragR* genes epistatically restored the morphological differentiation in *ramR* and *ramS* development mutants. The role of this TCS in differentiation is independent of the SapB pathway, since this protein is also detected by immunoblots in the null *ragABKR* mutant (SanPaolo *et al.*, 2006).

Atypical response regulators with ligands: role in the regulation of secondary metabolite biosynthesis

Five conserved amino acids are important for phosphorylation of the RRs at the aspartic acid site. However, a subset of the orphan response regulators in different bacteria lack two or three of these conserved amino acids (Hutchings *et al.*, 2004; Hutchings, 2007). However, their overall structure is similar to that of the classical response regulators. Such proteins have been named 'atypical' response regulators (ARR) (Wang *et al.*, 2009). It is unclear if these ARRs are really phosphorylated *in vivo*, and the hypothesis that they may recognize and interact with other ligands was proposed (Guthrie *et al.*, 1998; Ainsa *et al.*, 1999; Pflock *et al.*, 2007; Tian *et al.*, 2007; Wang *et al.*, 2009). These ARRs control a variety of processes including virulence, cell differentiation, iron transport and secondary metabolism (Guthrie *et al.*, 1998; Ulijasz *et al.*, 2004; Yang *et al.*, 2004; Pflock *et al.*, 2007; Tian *et al.*, 2007).

The mechanism of regulation by these ARRs is different from the standard one. Whereas in the classical TCSs the SK autophosphorylates at a conserved histidine in response to an extracellular stimulus, and then the phosphorylated SK transfers its phosphate group to an aspartic acid residue in the RR, the mechanism of regulation by the ARRs may involve the participation of a small ligand (Galperin, 2006).

One of this ARRs in *S. coelicolor* is RedZ belonging to the NarL family. It is an orphan response regulator encoded by a gene located in the *red* gene cluster encoding undecylprodigiosin biosynthesis genes. RedZ has been shown to regulate expression of the *red* gene cluster by binding to the promoter region of the *redD* gene, a SARP-type pathway-specific regulator. Binding of purified RedZ to the *redD* promoter was inhibited by undecylprodigiosin, whereas this compound alone was unable to bind directly to the DNA even at very high concentrations (Wang *et al.*, 2009). This system is an example of an ARR modulator that controls expression of antibiotic biosynthetic genes in *Streptomyces* using as a regulatory ligand the final product of the pathway.

A similar mode of action was described for the ARR JadR1, an OmpR-type regulator of *S. venezuelae*, that activates the transcription of jadomycin biosynthetic genes in this microorganism. Jadomycin B was found to bind to the N-terminal region of JadR1 causing its release from the target promoter. Thus, antibiotics biosynthesis is controlled by a feedback autoregulatory mechanism when high levels of jadomycin have accumulated (Wang *et al.*, 2009). For additional information about atypical response regulators, see also Chapter 6.

Evolutionary relationships between *Streptomyces* two-component systems

Both, in *S. coelicolor* and *S. avermitilis* most of the SKs belong to groups II and IIIa of the Fabret classification (Fabret *et al.*, 1999), with only a few cases of SKs belonging to groups I, IIIb and IV. It is noteworthy that in both species all paired members of group II are adjacent to an RR of the NarL family, whereas most paired members of group IIIa are linked to regulators of the OmpR family (Hutchings *et al.*, 2004; Wei *et al.*, 2007). The only exceptions are two SKs in *S. coelicolor* that are adjacent to regulators lacking a DNA binding domain (Hutchings *et al.*, 2004). The members belonging to group I (only one in each species) and IIIb (two SKs in *S. coelicolor* and one in *S. avermitilis*) are unpaired, whereas the members of group IV are unpaired or linked to regulators of the winged helix–turn–helix family (Hutchings *et al.*, 2004; Wei *et al.*, 2007). It has been proposed that the SKs and RRs belonging to the same group have evolved together as a unit from a common ancestor (Fabret *et al.*, 1999). This hypothesis is supported by the fact that in both *S. coelicolor* and *S. avermitilis* all the paired TCSs within a group keep the gene order in the transcription unit with only few exceptions (Hutchings *et al.*, 2004; Wei *et al.*, 2007). However, the SK structures show a wide domain variety. Thus, there are (i) some SKs without any transmembrane (TM) domain, probably being activated by other proteins, as shown for CheA of *E. coli* (Levit *et al.*, 1996); (ii) SKs with a very short sensor domain (flanked by two TM domains) of less than 20 amino acids, belonging to a SKs subfamily sensing changes in membrane structure or topology (Mascher *et al.*, 2003); (iii) SKs containing GAF or PAS domains, the first one probably acting as a binding site for small ligands (Aravind and Ponting, 1997) and the second one responding to oxygen, redox potential and light (Taylor and Zhulin, 1999). The presence of hybrid kinases has been detected in *S. coelicolor*, *S. avermitilis* and *S. griseus*, some of them are well conserved across the three species suggesting that these kinases might be functionally important (Wei *et al.*, 2007; Hsiao and Kirby, 2009).

Evolution of orphan proteins

The presence of orphan SKs and RRs is quite common in most bacteria, although their relative percentage can range between 6% in *E. coli* to 57% in *Caulobacter crescentus* (Raghavan and Groisman, 2010). *Streptomyces* species are no exception and there are around 20% of TCS orphan proteins in their genomes. These proteins probably have evolved by duplication of an ancestral TCS and they may have more than one interaction partner (Alm *et al.*, 2006; Whitworth and Cock, 2009). In addition, in *Streptomyces* orphan proteins sometimes present an atypical structure, lacking conserved residues involved in phosphorylation but nevertheless keeping their regulatory activity. As it has been explained before, they are usually involved in regulation of morphological differentiation processes and secondary metabolism (Hutchings *et al.*, 2004; Hutchings, 2007; Wang *et al.*, 2009; Lu *et al.*, 2011). The ability of these

ARRs to bind ligands and to act as repressors of antibiotic biosynthetic pathways is an interesting evolutionary innovation with respect to other response regulators.

Future outlook

A few TCSs have been characterized in the model actinomycete *S. coelicolor*, but many others remain to be studied. Some of these TCSs like PhoR/PhoP and AbsA1/AbsA2 exert a global effect on primary and/or secondary metabolism in *S. coelicolor*, but others like ChiS/ChiR appear to be more specific controlling particular genes. It will be interesting to investigate the role of these uncharacterized systems and their possible interactions with other global or specific regulatory systems. Another aspect of future interest is the study of the role of lipoproteins or ABC transport systems linked to TCS genes and their possible role as connecting proteins between extracytoplasmic signal-sensing proteins and the TCSs.

A promising research field is emerging based on the fact that the activity of atypical response regulators may be controlled by small ligands binding to them. Taking into account the number of ARRs and the wealth of secondary metabolites produced by *Streptomyces* species, it is extremely interesting to correlate the ligands bound by each of these ARR and the feedback control mechanisms of the different biosynthetic pathways for secondary metabolites.

Finally, there are more than two thousand *Streptomyces* species, most of them poorly explored at the molecular genetics level. Some of the global TCSs such as PhoR/PhoP are well conserved in most *Streptomyces* species, but many others appear to be species-specific. The study of the diversity of TCSs in different *Streptomyces* in relation to their habitats will provide important information to understand the role of these TCSs in the molecular ecology of these bacteria.

Acknowledgements

This article was supported by grants of the CICYT (BIO2006-14853-C02-01) and BIO2010-16094), Ministry of Science and Innovation of Spain and the European Union (ERA-IB EUI2008-03656). We acknowledge the excellent assistance of A. Mulero in preparing the manuscript.

References

Adamidis, T., Riggle, P., and Champness, W. (1990). Mutations in a new *Streptomyces coelicolor* locus which globally block antibiotic biosynthesis but not sporulation. J. Bacteriol. *172*, 2962–2969.

Ainsa, J.A., Parry, H.D., and Chater, K.F. (1999). A response regulator-like protein that functions at an intermediate stage of sporulation in *Streptomyces coelicolor* A(3). Mol. Microbiol. *34*, 607–619.

Alm, E., Huang, K., and Arkin, A. (2006). The evolution of two-component systems in bacteria reveals different strategies for niche adaptation. PLoS Comput. Biol. *2*, e143.

Amemura, M., Makino, K., Shinagawa, H., and Nakata, A. (1990). Cross talk to the phosphate regulon of *Escherichia coli* by PhoM protein: PhoM is a histidine protein kinase and catalyzes phosphorylation of PhoB and PhoM-open reading frame 2. J. Bacteriol. *172*, 6300–6007.

Anderson, T.B., Brian, P., Riggle, P., Kong, R., and Champness, W. (1999). Genetic suppression analysis of non-antibiotic-producing mutants of the *Streptomyces coelicolor absA* locus. Microbiology *145*, 2343–2353.

Anderson, T.B., Brian, P., and Champness, W.C. (2001). Genetic and transcriptional analysis of *absA*, an antibiotic gene cluster-linked two-component system that regulates multiple antibiotics in *Streptomyces coelicolor*. Mol. Microbiol. 39, 553–566.

Apel, A.K., Sola-Landa, A., Rodríguez-García, A., and Martín, J.F. (2007). Phosphate control of *phoA*, *phoC* and *phoD* gene expression in *Streptomyces coelicolor* reveals significant differences in binding of PhoP to their promoter regions. Microbiology *153*, 3527–3537.

Aravind, L., and Ponting, C.P. (1997). The GAF domain: An evolutionary link between diverse phototransducing proteins. Trends Biochem. Sci. *22*, 458–459.

Asturias, J.A., Martín, J.F., and Liras, P. (1994). Biosynthesis and phosphate control of candicidin by *Streptomyces acrimycini* JI2236: effect of amplification of the *pabAB* gene. J. Ind. Microbiol. *13*, 183–189.

Bentley, S.D., Chater, K.F., Cerdeno-Tarraga, A.M., Challis, G.L., Thomson, N.R., James, K.D., Harris, D.E., Quail, M.A., Kieser, H., Harper, D., *et al.* (2002). Complete genome sequence of the model actinomycete *Streptomyces coelicolor* A3(2). Nature *417*, 141–147.

Berks, B.C., Sargent, F., and Palmer, T. (2000). The Tat protein export pathway. Mol. Microbiol. 35, 260–274.

Bibb, M.J. (2005). Regulation of secondary metabolism in streptomycetes. Curr. Opin. Microbiol. *8*, 208–215.

Bishop, A., Fielding, S., Dyson, P., and Herron, P. (2004). Systematic insertional mutagenesis of a streptomycete genome: a link between osmoadaptation and antibiotic production. Genome Res. *14*, 893–900.

Bogel, G., Schrempf, H., and Ortiz de Orué Lucana, D. (2007). DNA-binding characteristics of the regulator SenR in response to phosphorylation by the sensor histidine autokinase SenS from *Streptomyces reticuli*. FEBS J. *274*, 3900–3913.

Bogel, G., Schrempf, H., and Ortiz de Orué Lucana, D. (2008). The hemebinding protein HbpS regulates the activity of the *Streptomyces reticuli* iron-sensing histidine kinase SenS in a redox-dependent manner. Amino Acids. *37*, 681–691.

Brian, P., Riggle, F.J., Santos, R.A., and Champness, W.C. (1996). Global negative regulation of *Streptomyces coelicolor* antibiotic synthesis mediated by an *absA*-encoded putative signal transduction system J. Bacteriol. *178*, 3221–3231.

Challis, G.L., and Hopwood, D.A. (2003). Synergy and contingency as driving forces for the evolution of multiple secondary metabolite production by *Streptomyces* species. Proc. Natl. Acad. Sci. U.S.A. *100*, 14555–14561.

Chang, H.M., Chen, M.Y., Shieh, Y.T., Bibb, M.J., and Chen, C.W. (1996). The *cutRS* signal transduction system of *Streptomyces lividans* represses the biosynthesis of the polyketide antibiotic actinorhodin. Mol. Microbiol. *21*, 1075–1085.

Chater, K.F., and Chandra, G. (2006). The evolution of development in *Streptomyces* analysed by genome comparisons FEMS Microbiol. Rev. *30*, 651–672.

Cheung, J., and Hendrickson, W.A. (2010). Sensor domains of two-component regulatory systems. Curr. Opin. Microbiol. *13*, 116–123.

Demain, A.L. (2009). Antibiotics: natural products essential to human health. Med. Res. Rev. *29*,821–842.

Doull, J.L., and Vining, L.C. (1990). Nutritional control of actinorhodin production by *Streptomyces coelicolor* A3(2): suppressive effects of nitrogen and phosphate. Appl. Microbiol. Biotechnol. *32*, 449–454.

Elliot, M.A., and Talbot, N.J. (2004). Building filaments in the air: aerial morphogenesis in bacteria and fungi. Curr. Opin. Microbiol. *7*, 594–601.

Fabret, C., Feher, V.A., and Hoch, J.A. (1999). Two-component signal transduction in *Bacillus subtilis*: how one organism sees its world. J. Bacteriol. *181*, 1975–1983.

Fernández-Martínez, L., Bishop, A., Parkes, L., Del Sol, R., Salerno, P., Sevcikova, B., Mazurakova, V., Kormanec, J., and Dyson, P. (2009). Osmoregulation in *Streptomyces coelicolor*: modulation of SigB activity by OsaC. Mol. Microbiol. *71*, 1250–1262.

Fink, D., Weissschuh, N., Reuther, J., Wohlleben, W., and Engels, A. (2002). Two transcriptional regulators GlnR and GlnRII are involved in regulation of nitrogen metabolism in *Streptomyces coelicolor* A3(2). Mol. Microbiol. *46*, 319–330.

Fisher, S.L., Jiang, W., Wanner, B.L., and Walsh, C.T. (1995). Cross-talk between the histidine protein kinase VanS and the response regulator PhoB. Characterization and identification of a VanS domain that inhibits activation of PhoB. J. Biol. Chem. *270*, 23143–23149.

Floriano, B., and Bibb, M. (1996). *afsR* is a pleiotropic but conditionally required regulatory gene for antibiotic production in *Streptomyces coelicolor* A3(2). Mol. Microbiol. *21*, 385–396.

Galperin, M.Y. (2006). Structural classification of bacterial response regulators: diversity of output domains and domain combinations. J. Bacteriol. *188*, 4169–4182.

Galperin, M.Y. (2010). Diversity of structure and function of response regulator output domains. Curr. Opin. Microbiol. *13*, 150–159.

Ghorbel, S., Smirnov, A., Chouayekh, H., Sperandio, B., Esnault, C., Kormanec, J., and Virolle, M.-J. (2006). Regulation of *ppk* expression and *in vivo* function of Ppk in *Streptomyces lividans* TK24. J. Bacteriol. *188*, 6269–6276.

Goulian, M. (2010). Two-component signaling circuit structure and properties. Curr. Opin. Microbiol. *13*, 184–189.

Guthrie, E.P., Flaxman, C.S., White, J., Hodgson, D.A., Bibb, M.J., and Chater, K.F. (1998). A response-regulator-like activator of antibiotic synthesis from *Streptomyces coelicolor* A3(2) with an amino-terminal domain that lacks a phosphorylation pocket. Microbiology *144*, 727–738.

Hakenbeck, R., and Stock, J.B. (1996). Analysis of two-component signal transduction systems involved in transcriptional regulation. Methods Enzymol. *273*, 281–300.

Hesketh, A., Fink, D., Gust, B., Rexer, H., Scheel, B., Chater, K., and Wohlleben, W. (2002). The GlnD and GlnK homologues of *Streptomyces coelicolor* A(3) are functionally dissimilar to their nitrogen regulatory system counterparts from enteric bacteria. Mol. Microbiol. *46*, 319–330.

Hong, H.-J., Hutchings, M.I., Neu, J.M., Wright, G.D., Paget, M.S.B., and Buttner, M.J. (2004). Characterization of an inducible vancomycin resistance system in *Streptomyces coelicolor* reveals a novel gene (*vanK*) required for drug resistance. Mol. Microbiol. *52*, 1107–1121.

Hong, H.-J., Hutchings, M.I., Hill, L.M., and Buttner, M.J. (2005). The role of the novel Fem protein VanK in vancomycin resistance in *Streptomyces coelicolor*. J. Biol. Chem. *280*, 13055–13061.

Hoskisson, P.A., and Hutchings, M.I. (2006). MtrAB-LpqB: a conserved three-component system in actinobacteria? Trends Microbiol. *14*, 444–449.

Hsiao, N.H., and Kirby, R. (2008). Comparative genomics of *Streptomyces avermitilis*, *Streptomyces cattleya*, *Streptomyces maritimus* and *Kitasatospora aureofaciens* using a *Streptomyces coelicolor* microarray system. Antonie Van Leeuwenhoek *93*, 1–25.

Hsiao, N.H., and Kirby, R. (2009). Two-component signal transduction systems in *Streptomyces* and related organisms studied using DNA comparative microarray analysis. Antonie Van Leeuwenhoek *95*, 189–206.

Hsieh, Y.-J., and Wanner, B.L. (2010). Global regulation by the seven-component P_i signaling system. Curr. Opin. Microbiol. *13*, 198–203.

Huang, J., Lih, C.J., Pan, K.H., and Cohen, S.N. (2001). Global analysis of growth phase responsive gene expression and regulation of antibiotic biosynthetic pathways in *Streptomyces coelicolor* using DNA microarrays. Genes Dev. *15*, 3183–3192.

Hutchings, M.I. (2007). Unusual two-component signal transduction pathways in the actinobacteria. Adv. Appl. Microbiol. *61*, 1–26.

Hutchings, M.I., Hoskisson, P.A., Chandra, G., and Buttner, M.J. (2004). Sensing and responding to diverse extracellular signals? Analysis of the sensor kinases and response regulators of *Streptomyces coelicolor* A3(2). Microbiology *150*, 2795–2806.

Hutchings, M.I., Hong, H.-J., and Buttner, M.J. (2006a). The vancomycin resistance VanRS two-component signal transduction system of *Streptomyces coelicolor*. Mol. Microbiol. *59*, 923–935.

Hutchings, M.I., Hong, H.-J., Leibovitz, E., Sutcliffe, I.C., and Buttner, M.J. (2006b). The sigma(E) cell envelope stress response of *Streptomyces coelicolor* is influenced by a novel lipoprotein, CseA. J. Bacteriol. *188*, 7222–7229.

Ikeda, H., Ishikawa, J., Hanamoto, A., Shinose, M., Kikuchi, H., Shiba, T., Sakaki, Y., Hattori, M., and Omura, S. (2003). Complete genome sequence and comparative analysis of the industrial microorganism *Streptomyces avermitilis*. Nat. Biotechnol. *21*, 526–531.

Ishizuka, H., Horinouchi, S., Kieser, H.M., Hopwood, D.A., and Beppu, T. (1992). A putative two-component regulatory system involved in secondary metabolism in *Streptomyces* spp. J. Bacteriol. *174*, 7585–7594.

Jordan, S., Hutchings, M.I., and Mascher, T. (2008). Cell envelope stress response in Gram-positive bacteria. FEMS Microbiol. Rev. *32*, 107–146.

Kim, D., and Forst, S. (2001). Genomic analysis of the histidine kinase family in bacteria and archaea. Microbiology *147*, 1197–1212.

Kirby, R. (2011). Chromosome diversity and similarity within the Actinomycetales. FEMS Microbiol. Lett. *319*, 1–10.

Kormanec, J., Sevcíková, B., and Homérová, D. (2000). Cloning of a two-component regulatory system probably involved in the regulation of chitinase in *Streptomyces coelicolor* A3(2). Folia Microbiol. (Praha) *45*, 397–406.

Levit, M., Liu, Y., Surette, M., and Stock, J. (1996). Active site interference and asymmetric activation in the chemotaxis protein histidine kinase CheA. J. Biol. Chem. *271*, 32057–32063.

Li, Y., Cen, P., fei Chen, S., Wu, D., and Zheng, J. (2004). A pair of two-component regulatory genes *ecrA1/A2* in *S. coelicolor*. J. Zhejiang Univ. Sci. *5*, 173–179.

Lu, Y., Wang, W., Shu, D., Zhang, W., Chen, L., Qin, Z., Yang, S., and Jiang, W. (2007). Characterization of a novel two-component regulatory system involved in the regulation of both actinorhodin and a type I polyketide in *Streptomyces coelicolor*. Appl. Microbiol. Biotechnol. *77*, 625–635.

Lu, Y., He, J., Zhu, H., Yu, Z., Wang, R., Chen, Y., Dang, F., Zhang, W., Yang, S., and Jiang, W. (2011). An orphan histidine kinase, OhkA, regulates both secondary metabolism and morphological differentiation in *Streptomyces coelicolor*. J. Bacteriol. *193*, 3020–3032.

McCleary, W.R., and Stock, J.B. (1994). Acetyl phosphate and the activation of two-component response regulators. J. Biol. Chem. *269*, 31567–31572.

McKenzie, N.L., and Nodwell, J.R. (2007). Phosphorylated AbsA2 negatively regulates antibiotic production in *Streptomyces coelicolor* through interactions with pathway-specific regulatory gene promoters. J. Bacteriol. *189*, 5284–5292.

McKenzie, N.L., and Nodwell, J.R. (2009). Transmembrane topology of the AbsA1 sensor kinase of *Streptomyces coelicolor*. Microbiology *155*, 1812–1818.

Marcos, A.T., Gutiérrez, S., Díez, B., Fernández, F.J., Oguiza, J.A., and Martín, J.F. (1995). Three genes *hrd*B, *hrd*D and *hrd*T of *Streptomyces griseus* IMRU 3570, encoding sigma factor-like proteins, are differentially expressed under specific nutritional conditions. Gene *153*, 41–48.

Martín, J.F. (2004). Phosphate control of the biosynthesis of antibiotics and other secondary metabolites is mediated by the PhoR-PhoP system: an unfinished story. J. Bacteriol. *186*, 5197–5201.

Martín, J.F., and Demain, A. (1980). Control of antibiotic biosynthesis. Microbiol. Rev. *44*, 230–251.

Martín, J.F., Sola-Landa, A., Santos-Beneit, F., and Rodríguez-García, A. (2011a). Network mechanisms of phosphate control of primary and secondary metabolism. In *Streptomyces*: Molecular Biology and Biotechnology, Dyson, P., ed. (Caister Academic Press, Norwich, UK), pp. 137–149.

Martín, J.F., Sola-Landa, A., Santos-Beneit, F., Fernández-Martínez, L.T., Prieto, C., and Rodríguez-García, A. (2011b). Cross-talk of global nutritional regulators in the control of primary and secondary metabolism in *Streptomyces*. Microb. Biotechnol. *4*, 165–174.

Martínez-Castro, M., Barreiro, C., Romero, F., Fernández-Chimeno, R.I., and Martín, J.F. (2010). *Streptomyces tacrolimicus* sp. nov., a low producer of the immunosuppressant tacrolimus (FK506). Int. J. Syst. Evol. Microbiol. *61*, 1084–1088.

Mascher, T., Margulis, N.G., Wang, T., Ye, R.W., and Helmann, J.D. (2003). Cell wall stress responses in *Bacillus subtilis*: The regulatory network of the bacitracin stimulon. Mol. Microbiol. *50*, 1591–1604.

Masuma, R., Tanaka, Y., Tanaka, H., and Omura, S. (1986). Production of nanaomycin and other antibiotics by phosphate-depressed fermentation using phosphate-trapping agents. J. Antibiot. (Tokyo) *39*, 1557–1564.

Matsumoto, A., Ishizuka, H., Beppu, T., and Horinouchi, S. (1995). Involvement of a small ORF downstream of the *afsR* gene in the regulation of secondary metabolism in *Streptomyces coelicolor* A3(2). Actinomycetologica *9*, 37–43.

Medema, M.H., Trefzer, A., Kovalchuk, A., van den Berg, M., Müller, U., Heijne, W., Wu, L., Alam, M.T., Ronning, C.M., Nierman, W.C., *et al.* (2010). The sequence of a 1.8-mb bacterial linear plasmid reveals a rich evolutionary reservoir of secondary metabolic pathways. Genome Biol. Evol. *12*, 212–224.

Mendes, M.V., Tunca, S., Antón, N., Recio, E., Sola-Landa, A., Aparicio, J.F., and Martín, J.F. (2007). The two-component *phoR-phoP* system of *Streptomyces natalensis*: Inactivation or deletion of *phoP* reduces the negative phosphate regulation of pimaricin biosynthesis. Metabol. Engineer. *9*, 217–227.

Nguyen, K.T., Willey, J.M., Nguyen, L.D., Nguyen, L.T., Viollier, P.H., and Thompson, C.J. (2002). A central regulator of morphological differentiation in the multicellular bacterium *Streptomyces coelicolor*. Mol. Microbiol. *46*, 1223–1238.

O'Connor, T.J., and Nodwell, J.R. (2005). Pivotal roles for the receiver domain in the mechanism of action of the response regulator RamR of *Streptomyces coelicolor* J. Mol. Biol. *351*, 1030–1047.

O'Connor, T.J., Kanellis, P., and Nodwell, J.R. (2002). The *ramC* gene is required for morphogenesis in *Streptomyces coelicolor* and expressed in a cell type-specific manner under the direct control of RamR. Mol. Microbiol. *45*, 45–57.

Ohnishi, Y., Ishikawa, J., Hara, H., Suzuki, H., Ikenoya, M., Ikeda, H., Yamashita, A., Hattori, M., and Horinouchi, S. (2008). Genome sequence of the streptomycin-producing microorganism *Streptomyces griseus* IFO 13350. J. Bacteriol. *190*, 4050–4060.

Ortiz de Orué Lucana, D., Schaa, T., and Schrempf, H. (2004). The novel extracellular *Streptomyces reticuli* haem-binding protein HbpS influences the production of the catalase-peroxidase CpeB. Microbiology *150*, 2575–2585.

Ortiz de Orué Lucana, D., Zou, P., Nierhaus, M., and Schrempf, H. (2005). Identification of a novel two-component system SenS/SenR modulating the production of the catalase-peroxidase CpeB and the haem-binding protein HbpS in *Streptomyces reticuli*. Microbiology *151*, 3603–3614.

Paget, M.S., Chamberlin, L., Atrih, A., Foster, S.J., and Buttner, M.J. (1999a). Evidence that the extracytoplasmic function sigma factor sigmaE is required for normal cell wall structure in *Streptomyces coelicolor* A3(2). J. Bacteriol. *181*, 204–211.

Paget, M.S., Leibovitz, E., and Buttner, M.J. (1999b). A putative two-component signal transduction system regulates sigma(E), a sigma factor required for normal cell wall integrity in *Streptomyces coelicolor* A3(2). Mol. Microbiol. *33*, 97–107.

Paul, S., Birkey, S., Liu, W., and Hulett, F.M. (2004). Autoinduction of *Bacillus subtilis phoPR* operon transcription results from enhanced transcription from EsigmaA- and EsigmaE-responsive promoters by phosphorylated PhoP. J. Bacteriol. *186*, 4262–4275.

Pflock, M., Bathon, M., Schär, J., Müller, S., Mollenkopf, H., Meyer, T.F., and Beier, D. (2007). The orphan response regulator HP1021 of *Helicobacter pylori* regulates transcription of a gene cluster presumably involved in acetone metabolism. J. Bacteriol. *189*, 2339–2349.

Pootoolal, J., Thomas, M.G., Marshall, C.G., Neu, J.M., Hubbard, B.K., Walsh, C.T., and Wright, G.D. (2002). Assembling the glycopeptide antibiotic scaffold: The biosynthesis of A47934 from *Streptomyces toyocaensis* NRRL15009. Proc. Natl. Acad. Sci. U.S.A. *99*, 8962–8967.

Pullan, S.T., Chandra, G., Bibb, M.J., and Merrick, M. (2011). Genome-wide analysis of the role of GlnR in *Streptomyces venezuelae* provides new insights into global nitrogen regulation in actinomycetes. BMC Genomics *12*, 175–189.

Raghavan, V., and Groisman, E.A. (2010). Orphan and hybrid two-component system proteins in health and disease. Curr. Opin. Microbiol. *13*, 226–231.

Reuther, J., and Wohlleben, W. (2007). Nitrogen metabolism in *Streptomyces coelicolor*: transcriptional and post-translational regulation. J. Mol. Microbiol. Biotechnol. *12*, 139–146.

Rodríguez-García, A., Barreiro, C., Santos-Beneit, F., Sola-Landa, A., and Martín, J.F. (2007). Genome-wide transcriptomic and proteomic analysis of the primary response to phosphate limitation in *Streptomyces coelicolor* M145 and in a Δ*phoP* mutant. Proteomics 7, 2410–2429.

Rodríguez-García, A., Sola-Landa, A., Apel, K., Santos-Beneit, F., and Martín, J.F. (2009). Phosphate control over nitrogen metabolism in *Streptomyces coelicolor*: direct and indirect negative control of *glnR*, *glnA*, *glnII* and *amtB* expression by the response regulator PhoP. Nucleic Acids Res. *37*, 3230–3242.

Ryding, N.J., Anderson, T.B., and Champness, W.C. (2002). Regulation of the *Streptomyces coelicolor* calcium-dependent antibiotic by *absA*, encoding a cluster-linked two-component system. J. Bacteriol. *184*, 794–805.

Sánchez, S., Gúzman-Trampe, S., Avalos, M., Ruiz, B., Rodríguez-Sanoja, R., and Jimenez-Estrada, M. (2012). Microbial natural products: chemical diversity. In Wiley Encyclopedia of Chemical Biology, Begley, T., ed. (John Wiley and Sons, Hoboken, NJ). doi:10.1002/978/0470048672.wecb721.

SanPaolo, S.S., Huang, J., Cohen, S.N., and Thompson, C.J. (2006). *rag* genes: novel components of the RamR regulon that trigger morphological differentiation in *Streptomyces coelicolor*. Mol. Microbiol. *61*, 1167–1186.

Santos-Beneit, F., Rodríguez-García, A., Franco-Domínguez, E., and Martín, J.F. (2008). Phosphate-dependent regulation of the low- and high-affinity transport systems in the model actinomycete *Streptomyces coelicolor*. Microbiology *154*, 2356–2370.

Santos-Beneit, F., Rodríguez-García, A., Sola-Landa, A., and Martín, J.F. (2009a). Cross-talk between two global regulators in *Streptomyces*: PhoP and AfsR interact in the control of *afsS*, *pstS* and *phoRP* transcription. Mol. Microbiol. *72*, 53–68.

Santos-Beneit, F., Rodríguez-García, A., Apel, A.K., and Martín, J.F. (2009b). Phosphate and carbon source regulation of two PhoP-dependent glycerophosphodiester phosphodiesterase genes of *Streptomyces coelicolor*. Microbiology *155*, 1800–1811.

Santos-Beneit, F., Barriuso-Iglesias, M., Fernández-Martínez, L., Martínez-Castro, M., Sola-Landa, A., Rodríguez-García, A., and Martín, J.F. (2011a). The RNA polymerase omega factor RpoZ is not essential for growth but is required for normal actinorhodin and undecylprodigiosin biosynthesis in *Streptomyces coelicolor*: Binding of PhoP to its promoter region. Appl. Environ. Microbiol. doi:10.1128/AEM.00465-11.

Santos-Beneit, F., Rodríguez-García, A., and Martín, J.F. (2011b). Complex transcriptional control of the antibiotic regulator *afsS* in *Streptomyces*: PhoP and AfsR are overlapping, competitive activators. J. Bacteriol. *193*, 2242–2251.

Schaerlaekens, K., Schierova, M., Lammertyn, E., Geukens, N., Anne, J., and van Mellaert, L. (2001). Twin-arginine translocation pathway in *Streptomyces lividans*. J. Bacteriol. *183*, 6727–6732.

Sheeler, N.L., MacMillan, S.V., and Nodwell, J.R. (2005). Biochemical activities of the *absA* two-component system of *Streptomyces coelicolor*. J. Bacteriol. *187*, 687–696.

Shu, D., Chen, L., Wang, W., Yu, Z., Ren, C., Zhang, W., Yang, S., Lu, Y., and Jiang, W. (2009). *afsQ1-Q2-sigQ* is a pleiotropic but conditionally required signal transduction system for both secondary metabolism and morphological development in *Streptomyces coelicolor*. Appl. Microbiol. Biotechnol. *81*, 1149–1160.

Silversmith, R.E. (2010). Auxiliary phosphatases in two-component signal transduction. Curr. Opin. Microbiol. *13*, 177–183.

Sola-Landa, A., Moura, R.S., and Martín, J.F. (2003). The two-component PhoR-PhoP system controls both primary metabolism and secondary metabolite biosynthesis in *Streptomyces lividans*. Proc. Natl. Acad. Sci. U.S.A. *100*, 6133–6138.

Sola-Landa, A., Rodríguez-García, A., Franco-Domínguez, E., and Martín, J.F. (2005). Binding of PhoP to promoters of phosphate-regulated genes in *Streptomyces coelicolor*: identification of PHO boxes. Mol. Microbiol. *56*, 1373–1385.

Sola-Landa, A., Rodríguez-García, A., Apel, A.K., and Martín, J.F. (2008). Target genes and structure of the direct repeats in the DNA binding sequences of the response regulator PhoP in *Streptomyces coelicolor*. Nucleic Acids Res. *36*, 1358–1368.

Stewart, R.C. (2010). Protein histidine kinases: assembly of active sites and their regulation in signaling pathways. Curr. Opin. Microbiol. *13*, 133–141.

Sun, G., Birkey, S.M., and Hulett, F.M. (1996). Three two-component signal-transduction systems interact for Pho regulation in *Bacillus subtilis*. Mol. Microbiol. *19*, 941–948.

Szurmant, H., and Hoch, J.A. (2010). Interaction fidelity in two-component signaling. Curr. Opin. Microbiol. *13*, 190–197.

Tanaka, A., Takano, Y., Ohnishi, Y., and Horinouchi, S. (2007). AfsR recruits RNA polymerase to the *afsS* promoter: a model for transcriptional activation by SARPs. J. Mol. Biol. *369*, 322–333.

Taylor, B.L., and Zhulin, I.B. (1999). PAS domains: internal sensors of oxygen, redox potential, and light. Microbiol. Mol. Biol. Rev. *63*, 479–506.

Tian, Y., Fowler, K., Findlay, K., Tan, H., and Chater, K.F. (2007) An unusual response regulator influences sporulation at early and late stages in *Streptomyces coelicolor*. J. Bacteriol. *189*, 2873–2885.

Tiffert, Y., Supra, P., Wurm, R., Wohlleben, W., Wagner, R., and Reuther, J. (2008). The *Streptomyces coelicolor* GlnR regulon: identification of new GlnR targets and evidence for a central role of GlnR in nitrogen metabolism in actinomycetes. Mol. Microbiol. *67*, 861–880.

Tiffert, Y., Franz-Wachtel. M., Fladerer, C., Nordheim, A., Reuther, J., Wohlleben, W., and Mast, Y. (2011). Proteomic analysis of the GlnR-mediated response to nitrogen limitation in *Streptomyces coelicolor* M145. Appl. Microbiol. Biotechnol. *89*, 1149–1159.

Tsujibo, H., Hatano, N., Okamoto, T., Endo, H., Miyamoto, K., and Inamori, Y. (1999). Synthesis of chitinase in *Streptomyces thermoviolaceus* is regulated by a two-component sensor-regulator system. FEMS Microbiol. Lett. *181*, 83–90.

Ulijasz, A.T., Andes, D.R., Glasner, J.D., and Weisblum, B. (2004). Regulation of iron transport in *Streptococcus pneumoniae* by RitR, an orphan response regulator. J. Bacteriol. *186*, 8123–8136.

Ulrich, L.E., and Zhulin, I.B. (2010). The MiST2 database: a comprehensive genomics resource on microbial signal transduction. Nucleic Acids Res. *38*, 401–407.

Vögtli, M., Chang, P.C., and Cohen, S.N. (1994). *afsR2*: a previously undetected gene encoding a 63-amino acid protein that stimulates antibiotic production in *Streptomyces lividans*. Mol. Microbiol. *14*, 643–653.

Walsh, C.T., Fisher, S.L., Park, I.S., Prahalad, M., and Wu, Z. (1996). Bacterial resistance to vancomycin: five genes and one missing hydrogen bond tell the story. Chem. Biol. *3*, 21–28.

Wang, C., Ge, H., Dong, H., Zhu, C., Li, Y., Zheng, J., and Cen, P. (2007). A novel pair of two-component signal transduction system. Biologia (Bratisl.) *62*, 511–516.

Wang, J., and Zhao, G.-P., (2009). GlnR positively regulates *nasA* transcription in *Streptomyces coelicolor*. Biochem. Biophys. Res. Commun. *386*, 77–81.

Wang, L., Tian, X., Wang, J., Yang, H., Fan, K., Xu, G., Yang, K., and Tan, H. (2009). Autoregulation of antibiotic biosynthesis by binding of the end product to an atypical response regulator. Proc. Natl. Acad. Sci. U.S.A. *106*, 8617–8622.

Wei, W., Wang, W., Cao, Z., Yu, H., Wang, X., Zhao, J., Tan, H., Xu, H., Jiang, W., and Li, Y. (2007). Comparative analysis of two-component signal transduction system in two streptomycete genomes. Acta Biochim. Biophys. Sin. (Shanghai) *39*, 317–325.

Whitworth, D.E., and Cock, P.J. (2008). Two-component systems of the myxobacteria: structure, diversity and evolutionary relationships. Microbiology *154*, 360–372.

Whitworth, D.E., and Cock, P.J. (2009). Evolution of prokaryotic two-component systems: insights from comparative genomics. Amino Acids. *37*, 459–466.

Wray, L.V., and Fisher, S.H. (1993). The *Streptomyces coelicolor glnR* gene encodes a protein similar to other bacterial response regulators. Gene *130*, 145–150.

Wray, L.V., Atkinson, M., and Fisher, S. (1991). Identification and cloning of the *glnR* locus, which is required for transcription of the *glnA* gene in *Streptomyces coelicolor* A3(2). J. Bacteriol. *173*, 7351–7360.

Yang, R., Bartle, S., Otto, R., Stassinopoulos, A., Rogers, M., Plamann, L., and Hartzell, P. (2004) AglZ is a filament-forming coiled-coil protein required for adventurous gliding motility of *Myxococcus xanthus*. J. Bacteriol. *186*, 6168–6178.

Yepes, A., Rico, S., Rodríguez-García, A., Santamaría, R.I., and Díaz, M. (2011). Novel two-component systems implied in antibiotic production in *Streptomyces coelicolor*. PLoS One *6*, e19980.

Zhou, L., Gregori, G., Blackman, J.M., Robinson, J.P., and Wanner, B.L. (2005). Stochastic activation of the response regulator PhoB by noncognate histidine kinases. J. Int. Bioinf. *2*, 11–24.

The Rcs Phosphorelay: Biofilm Formation and Virulence in the *Enterobacteriaceae*

16

David J. Clarke

Abstract

The Rcs phosphorelay is a complex signalling network that is restricted in distribution to the *Enterobacteriaceae*. The core Rcs phosphorelay is composed of three proteins: the sensor kinase RcsC, the HPt domain protein RcsD and the response regulator RcsB. In addition to these core components the Rcs phosphorelay also contains some auxiliary proteins that are involved in mediating the inputs to and outputs from this signalling pathway. Therefore RcsF is a lipoprotein involved in signal perception, IgaA inhibits the activity of the phosphorelay and RcsA is required for the expression of some of the genes that are regulated by the phosphorelay. Whilst initially identified as a positive regulator of capsule production in *Escherichia coli* the Rcs phosphorelay is now recognized as a key regulator of motility, biofilm formation and virulence in many members of the *Enterobacteriaceae*.

Introduction

The Rcs phosphorelay is perhaps one of the best-characterized signalling pathways in *Escherichia coli*. The well-studied nature of this phosphorelay is in no small part owing to the fact that it plays an important role in the regulation of key cellular processes such as swimming motility, swarming motility, biofilm formation and virulence. The Rcs phosphorelay first appeared in the literature in 1985 in a report describing a nitrosoguanidine-induced mutant screen designed to look for positive and negative regulators of the expression of the *cps* genes (Gottesman *et al.*, 1985). For further details on the history of the Rcs phopshorelay the reader is directed to a recent review (Majdalani and Gottesman, 2005). The *cps* operon is composed of approximately 20 genes that are required for the production and export of the extracellular polysaccharide (EPS), colanic acid. The *cps* operon has been renamed to the *wca* operon as part of an effort to standardize the nomenclature of all genes involved in the production of different polysaccharides (including O antigen and other EPS) (Reeves *et al.*, 1996). The activation of the Rcs phosphorelay is associated with the appearance of mucoid colonies on agar plates, a phenotype resulting from the overproduction of colanic acid.

The Rcs phosphorelay is an unusually (for *E. coli*) complex signalling pathway that is composed of three proteins, RcsC, RcsD and RcsB (Majdalani and Gottesman, 2005; Clarke, 2010). These proteins are considered to be the core components of the phosphorelay that are predicted to be directly involved in carrying the phosphoryl group. RcsC is the senor kinase that, upon receiving a signal, autophosphorylates on a histidine residue. The phosphoryl group is transferred to RcsD, which acts as an intermediary protein to transfer the phosphoryl group to the DNA-binding response regulator RcsB. Phosphorylated RcsB (RcsB~P) then binds DNA and changes the expression of a large number of genes called the Rcs regulon, which will be discussed in more detail below. The involvement of three proteins in the signal transduction pathway is unique to the Rcs phosphorelay in *E. coli* and other enterics and this is the reason why the Rcs system is referred to as a phosphorelay rather than a two-component pathway (Majdalani and Gottesman,

2005; Clarke, 2010). The complexity of the Rcs phosphorelay is further enhanced by a number of accessory components that, whilst not directly involved in the phosphorelay, play an important role in modulating the signalling activity of the core components. These accessory components include RcsA, TviA, RcsF and IgaA/YrfF. A more detailed account of both the core and accessory components of the Rcs phosphorelay will be provided below.

Phylogenetic distribution of the Rcs phopshorelay

The genes encoding the core components of the Rcs phosphorelay are found at a single locus on the *E. coli* genome. The *rcsD* and *rcsB* genes appear to form a single transcriptional unit that converges with the *rcsC* gene that is transcribed in the opposite direction to the *rcsDB* transcript (Fig. 16.1A). Using this distinctive synteny (in addition to simple nucleotide homolegy) as a search parameter the Rcs phosphorelay can be readily found in all sequenced members of the *Enterobacteriaceae* (with the exception of obligate endosymbionts with extremely reduced genome sizes such as *Buchnera* and *Wigglesworthia*) (Erickson and Detweiler, 2006; Huang *et al.*, 2006). Moreover homologues of *rcsA*, *rcsF* and *igaA* (also annotated as *yrfF*) can be found in all of the genomes where the Rcs phosphorelay is predicted to be present. Comparative studies have shown that the function of the RcsC protein, at least, appears to be conserved in those homologues that have been tested (Huang *et al.*, 2009). There are some instances in the literature stating that the Rcs pathway controls *cup* expression in *Pseudomonas* not belonging to the *Enterobacteriaceae*. However, the *rcsC* and *rcsB* genes in *Pseudomonas* were originally annotated on the basis that their best hit was with the *rcsC* and *rcsB* genes in *Salmonella* and closer examination of this locus (e.g. synteny and the absence of a *rcsD* homologue) strongly suggests that the *rcs* locus in *Pseudomonas* is not homologous to the *rcs* locus in the enterics (He *et al.*, 2004).

The Rcs phosphorelay, therefore, appears to be restricted to the *Enterobacteriaceae* suggesting that it may be important in some aspect of the life-style of this family of bacteria. One characteristic feature of the *Enterobacteriaceae* is their ability to grow on MacConkey agar, a phenotype that is associated with the ability of these bacteria to flourish in the intestinal tracts of many animals. Indeed, many *Enterobacteriaceae* are important gut pathogens. It has been suggested that the Rcs phosphorelay may provide some important fitness advantage during growth in the gut (Erickson and Detweiler, 2006). A major part of this argument is based on the observation that the Rcs phosphorelay is not present in strains of *Buchnera* and *Wigglesworthia*, bacteria that are endosymbionts of aphids and the Tsetse fly, respectively (Zientz *et al.*, 2004; Oliver *et al.*, 2010). These endosymbionts have adapted to grow inside specialized cells in so-called bacteriomes and in the sexual organs of their host and are vertically transmitted from mother to offspring thus removing any direct exposure of these bacteria to the gut of their host. On the other hand, analysis of the genome of *Sodalis glossinidius*, a secondary endosymbiont in the tsetse fly, appears to suggest that the genes encoding a complete Rcs phosphorelay are present, whilst the *rcsC* and *rcsD* genes in *Serratia symbiotica*, a newly described secondary endosymbiont of the aphid, have been shown to be pseudogenes (intriguingly the *rcsB* gene remains intact in this bacterium) (Toh *et al.*, 2006; Burke and Moran, 2011). Therefore it may be, that at some point during the formation of an obligate endosymbiont, the role of the Rcs phosphorelay becomes redundant and, consequently, neutral selection results in the loss of the *rcs* genes from the genomes of these bacteria. In fact there is some recent evidence suggesting that some *rcs* genes might be negatively selected under certain circumstances. *Yersinia pestis*, the causative agent of plague, has recently diverged from the mammalian gut pathogen, *Y. pseudotuberculosis* (Achtman *et al.*, 1999). However, unlike *Y. pseudotuberculosis*, *Y. pestis* infects its mammalian host via the blood. Genomic studies have identified a relatively large number of pseudogenes on the genome of *Y. pestis* and these were thought to be indicative of genome degradation as an adaptation to the life-cycle of *Y. pestis* (Wren, 2003). One of the pseudogenes identified was in *rcsA* and inserting an active copy of *rcsA* into *Y.*

pestis was shown to have serious consequences on the transmission of this bacterium (Sun *et al.*, 2008). The degradation of the genes encoding the Rcs phosphorelay in bacteria that no longer infect the gut of their hosts supports the hypothesis that the Rcs phosphorelay may play an important role during life in the gut of the host. However, it is important to remember that not all members of the *Enterobactericeae* have an obvious association with the gut (e.g. *Erwinia* and other phytopathogens) and an alternative hypothesis whereby the Rcs phosphorelay may play an important role during survival in the environment is equally feasible.

Components of the Rcs phosphorelay

Core components

RcsC

RcsC is a hybrid sensor kinase with a unique domain organization. RcsC is predicted to be 949 amino acids long and localized in the inner membrane with a large N-terminal portion exposed in the periplasm (i.e. the input domain) and a large C-terminal portion, containing the signalling domains, in the cytoplasm. The RcsC protein is a hybrid kinase meaning that it contains

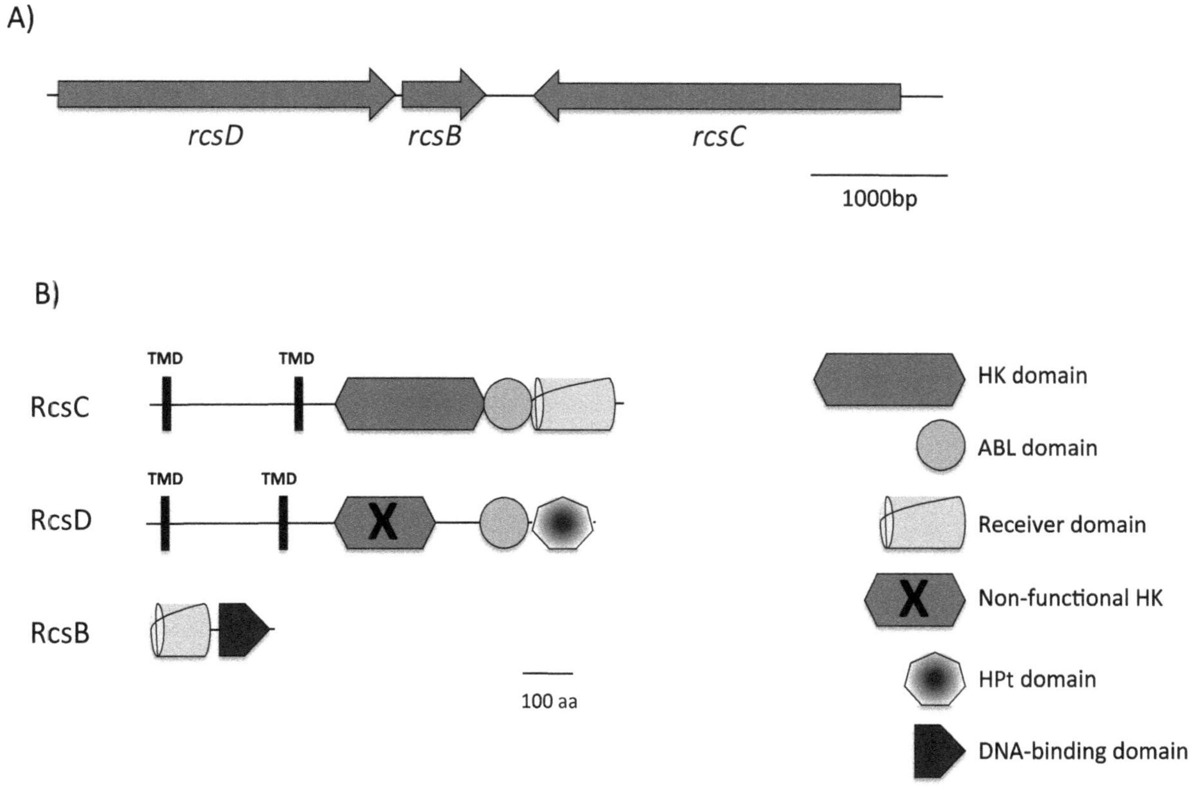

Figure 16.1 (A) Schematic representation of the *rcs* locus in *Enterobacteriaceae*. Both *rcsD* and *rcsB* are transcribed in the same direction, probably as an operon, whilst *rcsC* is transcribed in the opposite direction resulting in converging transcripts. (B) Schematic representation of domain organization in the RcsC, RcsD and RcsB proteins. The key for the domains is as indicated. TMD: transmembrane domain. (C) The Rcs phosphorelay is a complex signalling network that involves the core components of RcsC, RcsD and RcsB and the auxiliary proteins, IgaA/YrfF, RcsF and RcsA. Upon receiving an environmental signal, the RcsC protein is predicted to initiate a phosphorelay by autophosphorylating on His479 and transferring the phosphoryl group to an Asp875 residue in the C-terminal receiver domain. The phosphoryl group is subsequently transferred to the HPt domain of RcsD (His842) before ending up on the Asp56 residue found in the receiver domain of RcsB. Once phosphorylated the RcsB protein can affect the expression of a number of genes, called the Rcs regulon, in a manner that can be dependent or independent on the auxiliary protein RcsA. Whilst RcsF is involved in signal perception, the role of the IgaA/YrfF protein appears to be to inhibit signalling through the Rcs phosphorelay.

two signalling domains, a typical transmitter domain followed by a CheY-like receiver domain (Fig. 16.1B). The structure of the RcsC receiver domain has been shown to be composed of a central 5-stranded parallel beta sheet surrounded by five α-helices (Rogov *et al.*, 2008). Both the transmitter domain and the receiver domain have been shown to be required for signalling although, as is the case with many similar systems, it has not yet been confirmed biochemically that these domains are involved in carrying phosphoryl groups (Clarke *et al.*, 2002). Recent structural studies have identified a topologically unique α–β loop (ABL) domain that is located between the transmitter and the receiver domain (Rogov *et al.*, 2006). This ABL domain is conserved in all RcsC homologues and appears to be similar in structure (although not necessarily sequence) to receiver domains. However, this domain is not thought to be directly involved in phosphotransfer, but rather it is proposed to have a structural role with one possibility being that the RcsC-ABL domain interacts with an, as yet, unidentified protein partner (Rogov *et al.*, 2006).

Many sensor kinases have been shown to have both kinase and phosphatase activities and it seems that RcsC is no different. Mutations in the key conserved amino acids in the transmitter and receiver domains of RcsC (His479 and Asp875, respectively) completely prevented the expression of the *wca* genes supporting the idea that these residues are required for the kinase activity of RcsC (Clarke *et al.*, 2002). Nonetheless it appears that the phosphatase activity associated with RcsC is dominant under normal physiological conditions. Deletion mutants of *rcsC* in *E. coli* have been shown to have higher basal levels of expression of several Rcs-regulated genes (particularly the small non-coding RNA, RprA) suggesting that there is an increased level of RcsB~P in the cell under these conditions as would be expected in the absence of phosphatase activity (Majdalani *et al.*, 2002). In *E. coli* the *rcsC137* allele has also been used as evidence of phosphatase activity although, in this case, the evidence is less straightforward. The *rcsC137* allele encodes a sensor protein that drives high levels of expression of the *wca* operon and other Rcs-regulated genes in the absence of any signal, and this constitutive activity is owing to a point mutation in the receiver domain (A904V) (Majdalani *et al.*, 2005). Importantly the *rcsC137* allele is recessive to the *in trans* expression of wild-type *rcsC*, suggesting that the A904V alteration is a loss of function mutation (Clarke *et al.*, 2002; Majdalani and Gottesman, 2005). Similarly recessive *rcsC* constitutive mutants were also identified in the receiver domain of the RcsC protein in *Salmonella enterica* (Garcia-Calderon *et al.*, 2005). Further work has shown that, in *E. coli*, complementation of the *rcsC137* allele only requires the RcsC receiver domain but is independent of Asp875 suggesting that the kinase and phosphatase activities of RcsC might actually be independent (Clarke *et al.*, 2002). However, as has been pointed out previously, the *rcsC137* phenotype cannot be simply explained by the loss of phosphatase activity (Majdalani and Gottesman, 2005). A deletion mutant in *rcsC* would also be expected to have no phosphatase activity and yet this mutant does not exhibit the constitutive activity associated with *rcsC137*. Therefore, it seems more likely that the *rcsC137* mutation might alleviate a negative effect on the kinase activity of RcsC by virtue of a protein–protein interaction. Perhaps the A904V mutation affects inter- or intra-molecular interactions between RcsC and/or RcsD? Indeed, NMR titration experiments have shown that, upon mixing the receiver domain of RcsC with the HPt domain of RcsD the biggest conformational changes in the RcsC receiver domain were observed at the C-terminus of strand β4, exactly where the Ala904 residue is found. Therefore this residue is likely to play an important role in the interaction between RcsC and RcsD (Rogov *et al.*, 2008).

RcsD

RcsD is predicted to be a protein of 890 amino acids that topologically, at least, resembles RcsC, i.e. RcsD has a large N-terminal portion that is exposed to the periplasm and a large C-terminal portion, containing the signalling domains that are found in the cytoplasm. RcsD was originally annotated as YojN and was only confirmed to be part of the Rcs phosphorelay in 2001 (Takeda *et al.*, 2001). A deletion of *rcsD* has the same phenotype as a deletion of *rcsC* suggesting that these proteins work together during signalling

(Ferrieres *et al.*, 2009). However, in contrast to RcsC, RcsD does not have a transmitter or a receiver domain but rather this protein has a histidine-phosphotransfer (HPt) domain located at the extreme C-terminus. The HPt-domain consists of five α-helices, four of which form a four-helix bundle with an up-and-down topology and contains His842, which is required for phosphotransfer (Rogov *et al.*, 2004). The HPt domain interacts with the receiver domains of both RcsC and RcsB, and phosphotransfer between the HPt domain of RcsD and RcsB has also been confirmed (Takeda *et al.*, 2001; Rogov *et al.*, 2004). The HPt domain is immediately preceded by an ABL domain that functions to increase the affinity of RcsD for RcsB by interacting with the DNA-binding domain of the response regulator (Schmoe *et al.*, 2011). Therefore, although RcsD does not have any kinase activity, it is intimately involved in the transfer of the phosphoryl group between RcsC and RcsB.

RcsB

The response regulator, RcsB, is the final core component of the Rcs phosphorelay. RcsB is predicted to be a 216 amino acid protein with a typical response regulator domain organization, i.e. an N-terminal receiver domain containing the phosphoryl-accepting aspartate residue (Asp56) and a C-terminal DNA-binding output domain with homology to the LuxR-family of transcriptional regulators. The structure of the DNA-binding domain of RcsB from *Erwinia amylovora* has been resolved and shown to be composed of a 4-helix bundle with the two central helices forming the DNA-binding helix–turn–helix motif (Pristovsek *et al.*, 2003). RcsB is phosphorylated by the HPt domain of RcsD, although it is not clear whether the phosphoryl group is transferred to the Asp56 residue (Takeda *et al.*, 2001). Nonetheless an Asp56Glu mutation has been used to produce constitutively active RcsB providing genetic support for the involvement of this residue in the activation of RcsB (Boulanger *et al.*, 2005).

Genetic studies have shown that, in the different species of bacteria examined, RcsB regulates the expression of a relatively large regulon (the composition and role of the Rcs regulon will be discussed in more detail below). When phosphorylated, RcsB can repress or activate gene expression depending on the target gene. For example, in *E. coli*, RcsB~P represses the expression of *flhDC* and, therefore, motility, but increases the expression of *rprA* and other genes (Majdalani *et al.*, 2002; Ferrieres and Clarke, 2003; Francez-Charlot *et al.*, 2003). RcsB has been shown to directly bind to DNA, and attempts to identify a RcsB consensus sequence (or RcsB box) upstream from regulated genes have resulted in the identification of a 14bp piece of DNA with the sequence KMRGAWTMWYCTGS (where W=A or T; K=G or T; M=A or C; R=A or G; W=A or T; Y=C or T; S=C or G) (Davalos-Garcia *et al.*, 2001; Sturny *et al.*, 2003). However, whether RcsB binds to this target sequence by itself (either as a monomer or a dimer) or, in some cases, as a heteroduplex with accessory proteins such as RcsA (see below) remains to be fully clarified.

The phosphorylation of RcsB by RcsCD is not the only way that the activity of this DNA-binding protein is regulated. RcsB can also be phosphorylated by the small phosphodonor molecule acetyl~P thus linking signalling through the Rcs phosphorelay with central metabolism (Wolfe *et al.*, 2003). Cells secrete acetate (i.e. acetate dissimilation) when growing on a variety of carbon sources, and the same cells assimilate the secreted acetate when nutrients become more limiting. This switch from acetate dissimilation to assimilation is called the acetate switch (Wolfe, 2005). An important regulatory molecule in the acetate switch is acetyl~P, and the level of acetyl~P increases in the cell immediately preceding the acetate switch. Therefore this molecule can be used as an indication of the metabolic status of the cell and, as such, it has been implicated in the regulation of a number of important cellular processes (Wolfe, 2010). The level of acetyl~P is controlled by the activity of two enzymes, phosphotransacetylase (encoded by the *pta* gene) and acetate kinase (encoded by the *ack* gene). In *E. coli* a mutation in the *ack* gene results in the accumulation of acetyl~P in the cell and, in this background, colanic acid production is increased and motility is decreased in an RcsB-dependent manner (Wolfe *et al.*, 2003). Interestingly there is evidence that other small phosphodonors also interact with RcsB (and other response regulators) thus potentially linking the regulation of gene expression by two-component pathways with many cellular physiological processes (D.J. Clarke, unpublished).

Acetyl-coenzyme A synthetase, encoded by the *acs* gene, plays an important role in the assimilation of low concentrations of acetate and Acs activity is post-translationally controlled (in an ON–OFF manner) by acetylation of a lysine residue (Lys609). Therefore acetylation of Acs by the enzyme Pat (protein acetyltransferase) inactivates Acs, whilst deacetylation by the sirtuin, CobB, activates Acs and thus acetate assimilation (Starai *et al.*, 2002; Starai and Escalante-Semerena, 2004). Acs can be used by Pat as donor of acetyl groups to other proteins and molecules in the cell such as CheY (Barak *et al.*, 2004). Interestingly it has recently been shown that Pat can acetylate RcsB in *S. enterica* on a lysine residue, Lys180, which is found in the DNA-binding domain of this regulator (Thao *et al.*, 2011). Therefore, acetylation interferes with the ability of RcsB to bind DNA and prevents the regulation of gene expression. Although it has not been tested whether Acs can act as the acetyl donor for RcsB, it still remains possible that RcsB activity may be post-translationally controlled at two levels by products associated with acetate metabolism. However, is this regulatory interaction with acetate metabolism unique to RcsB? As we have previously mentioned acetyl~P can phosphorylate many response regulators and further studies are required to determine if the acetylation of response regulators is more widespread than originally appreciated.

Expression of the rcsDB and rcsC genes

As previously mentioned the genes encoding the core components of the Rcs phosphorelay are organized in a single genetic locus as two converging transcriptional units, *rcsDB* and *rcsC*. In *S. enterica rcsD* and *rcsB* appear to be co-transcribed, although there is some evidence that, under some circumstances, *rcsB* may also be transcribed independently (Mercedes Pescaretti *et al.*, 2010). Very little is known about how the expression of the *rcsDB* and *rcsC* genes is regulated. In *E. coli* there is some evidence that *rcsC* expression may be regulated by oxidative stress in an OxyR-dependent manner. In a screen for H_2O_2 sensitive mutants one λ*placMu53* insertion was found to be in the *rcsC* gene. The expression of the *rcsC*::λ*placMu53* fusion was nearly 3-fold induced by H_2O_2 and there was a 3-fold reduction in expression in an *oxyR* mutant (Mukhopadhyay and Schellhorn, 1997). It has also been reported that the expression of *rcsC-gfp* and *rcsD–gfp* fusions in *Salmonella* is induced on transit from LB broth to minimal medium with a pH = 5 (Detweiler *et al.*, 2003). This suggests that the Rcs phosphorelay may be involved in the cellular response to acid stress, and this has recently been shown to be the case (Johnson *et al.*, 2011). Interestingly Detweiler *et al.* (2003) also reported that the expression of *rcsD* requires OmpR in *Salmonella* suggesting that there is regulatory cross-talk between the Rcs phosphorelay and other two-component systems.

Many two-component systems auto-regulate (either positively or negatively) the expression of their genes as a mechanism to adjust their response to a particular environmental signal. For example, auto-regulation may allow the cell to increase or decrease the amount of response regulator in order to regulate different sets of genes on the basis of parameters such as the affinity of promoters for the response regulator (Bijlsma and Groisman, 2003). In *S. enterica rcsB* appears to be negatively regulated by the Rcs phosphorelay whilst *rcsA* is positively regulated (Mariscotti and Garcia-del Portillo, 2009). Unsurprisingly, given the co-transcribed nature of the genes, the expression of *rcsD* also appears to be negatively regulated by activation of RcsB (Mercedes Pescaretti *et al.*, 2010). In contrast, in *E. coli*, it has been reported that both *rcsB* and *rcsA* are positively regulated by the Rcs phosphorelay suggesting that there are significant differences in the operation of this phosphorelay even in closely related bacteria (Ferrieres and Clarke, 2003).

Accessory components of the Rcs phosphorelay

RcsA

Mutations in the *lon* gene in *E. coli*, encoding the Lon protease, result in the overproduction of colanic acid and, therefore, *lon* mutants grow as mucoid colonies on agar plates. This mucoid phenotype was shown to be due to an increase in the level of RcsA, a normally unstable protein that accumulates in the *lon* mutant owing to the absence of the Lon protease (Trisler and Gottesman, 1984). RcsA has also been shown to be sensitive to the ClpQV protease, and there is evidence that Lon and ClpQV have largely redundant functions in the cell (Wu *et al.*, 1999). RcsA is 207 amino acids long with a C-terminal region that has homology to the LuxR family of DNA-binding domains. However, RcsA does not appear to be able to bind DNA by itself but, rather, it functions by modulating the DNA-binding activity of RcsB by forming a heterodimer with this protein (Stout *et al.*, 1991). RcsA does not alter the DNA-binding specificity of RcsB, but the interaction appears to stabilize the RcsB–DNA interaction (Pristovsek *et al.*, 2003). Interestingly RcsA only affects the expression of a subset of all of the genes that are regulated by the Rcs phosphorelay, most significantly the *wca* operon and *rcsA* itself. Therefore the availability of RcsA can be, in itself, an important regulatory checkpoint for the expression of some genes and this is tightly linked to the activity and/or level of cellular proteases such as Lon (and, to a lesser extent, ClpQV). Lon is responsible for the turnover of approximately 50% of all abnormal cellular proteins (Van Melderen and Aertsen, 2009) and, therefore, conditions that might increase the competition for these proteases are likely to result in a net increase in the level of RcsA. One possible role for RcsA may be to simply link the expression of certain genes that are regulated by the Rcs phosphorelay to Lon-dependent processes in the cell. However, RcsA levels have also been linked to the cell-density-dependent induction of prophage excision in *E. coli* suggesting a more diverse role for this protein (Ghosh *et al.*, 2009). The *alpA* gene encodes the Alp protease and is located within a P4-like prophage found in *E. coli* K-12 genomes called CP4–57 (X. Wang *et al.*, 2009). Interestingly overproduction of the Alp protease has been shown to rescue the mucoidy phenotype associated with a *lon* mutation in *E. coli* (Trempy *et al.*, 1994). Presumably this complementation is due to the degradation of RcsA by the Alp protease. Production of the Alp protease also leads to excision of the CP4-57 prophage. The expression of *alpA* (and therefore prophage excision) has been shown to be controlled by the autoinducer AI-2 during biofilm formation suggesting that under these environmental conditions there may be an AI-2-dependent increase in AlpA levels that might also affect the level of RcsA in the cell (Herzberg *et al.*, 2006). Significantly, the engineered deletion of the CP4-57 prophage from the genome of *E. coli* MG1655 did affect biofilm formation suggesting that prophages can make a contribution to host physiology (X. Wang *et al.*, 2009, 2010).

An important virulence factor in *S. enterica* Serovar *typhi*, the causative agent of typhoid fever, is the capsular Vi antigen, encoded by the *viaB* gene cluster (Hashimoto *et al.*, 1993; Virlogeux *et al.*, 1995). The expression of the *viaB* gene cluster is dependent on RcsB and another protein called TviA, a homologue of RcsA (Virlogeux *et al.*, 1996; Arricau *et al.*, 1998). It has recently been suggested that the utility of TviA is to facilitate the introduction of an additional regulatory input (in this case osmolarity) into the RcsB-dependent control of via*B* expression (Winter *et al.*, 2009). Finally, *Klebsiella pneumoniae* CG43 is a highly mucoid clinical isolate that carries the *rmpA* gene on a large virulence plasmid. Although there is no significant sequence homology with RcsA, RmpA has been shown to have functional homology as RmpA interacts with RcsB and is involved in the positive regulation of the expression of genes involved in K2 capsular polysaccharide production and virulence in mice (Cheng *et al.*, 2010). Interestingly, the expression of *rmpA* appears to be repressed by the ferric uptake regulator Fur, suggesting that RmpA may function to link the RcsB-dependent regulation of the K2 capsule genes with the level of iron available (Lin *et al.*, 2011).

In addition to the post-translational control of RcsA, mediated primarily by the Lon protease, RcsA levels in the cell can also be controlled by simply regulating the expression of the *rcsA* gene. The expression of *rcsA* appears to be positively auto-regulated by the RcsB/RcsA heterodimer (Ebel and Trempy, 1999). Therefore, even if there are only very low levels of RcsA in the cell, activation of the Rcs phosphorelay will result in an increase in the expression of *rcsA* thus ensuring that RcsA is present in sufficient levels to affect the expression of its target genes. H-NS is a global regulator in bacteria that represses the expression of *rcsA*, and this repression can be alleviated by the overproduction of *dsrA*, a small non-coding RNA that binds to the H-NS protein, thus preventing this regulator from binding to DNA (Sledjeski and Gottesman, 1995). The expression of *rcsA* is also controlled by c-di-GMP and acylhomoserine lactone (AHL)-mediated quorum sensing. High levels of c-di-GMP in *E. coli* (achieved by overproducing the YddV protein) have been shown to increase the expression of *rcsA* over 30-fold (Mendez-Ortiz *et al.*, 2006). YddV is a haem-coupled oxygen sensor that is involved in controlling the biosynthesis of curli, fibre-like surface proteins that contribute to biofilm formation, and the polysaccharide, poly-*N*-acetylglucosamine (PNAG) (Tagliabua *et al.*, 2010a,b). Therefore RcsA levels can link oxygen levels with the production of important surface appendages in *E. coli*. Finally the quorum sensing-mediated induction of EPS production in the plant-pathogenic bacterium *Pantonea* has been shown to be dependent on RcsA (Minogue *et al.*, 2005). EsaR is a LuxR-like transcriptional regulator that binds to the AHL produced by these bacteria. At low cell densities EsaR represses the expression of *rcsA* and thus prevents the production of EPS, an important virulence factor. At high cell densities the AHL accumulates in the environment, binds to EsaR, alleviates repression of *rcsA* expression and allows the production of EPS.

Therefore RcsA and its functional homologues appear to play an important role in linking the regulation of a particular subset of Rcs-regulated genes with regulatory inputs that are independent of the core Rcs phosphorelay e.g. accumulation of abnormal proteins and quorum sensing.

RcsF and IgaA

Whilst the accessory protein RcsA plays an important role in modulating the output of the Rcs phosphorelay, two other accessory proteins, RcsF and IgaA, are involved in the perception of, and response to, environmental signals. RcsF is a 134 amino acid outer membrane lipoprotein that is required for the perception of many, but not all, of the signals that have been shown to activate the Rcs phosphorelay (Table 16.1) (Castanie-Cornet *et al.*, 2006). The *rcsF* gene was first identified as a multicopy suppressor of the temperature sensitive *ftsZ84* mutation (Gervais and Drapeau, 1992). Suppression was shown to be due to the RcsB-dependent increase in the expression of this allele (Gervais *et al.*, 1992). Overproduction of RcsF also results in mucoid colonies on plates, a phenotype that is dependent on *rcsC*, *rcsD* and *rcsB* (Majdalani *et al.*, 2005; Castanie-Cornet *et al.*, 2006). Epistatic studies have shown that RcsF acts upstream from RcsC suggesting that RcsF connects some environmental signals with the core Rcs phosphorelay (Majdalani *et al.*, 2005). Although RcsF has limited sequence homology with other proteins, recent structural studies have determined that RcsF contains a central four-stranded β-sheet, a structural motif that is conserved in other proteins, e.g. in the copper binding domain of the amyloid precursor protein (Leverrier *et al.*, 2011). Interestingly, RcsF also has two disulphide bonds that are, unusually, formed between non-consecutive cysteine residues, i.e. Cys74/Cys118 and Cys109/Cys124. Although both disulphide bonds appear to be critical for the structural integrity and normal activity of RcsF, there is some evidence that Cys109/Cys124 is the more important of the two bonds (Farris *et al.*, 2010; Leverrier *et al.*, 2011; Rogov *et al.*, 2011). As the disulphide bonds are between non-consecutive Cys residues, RcsF function requires DsbC, a disulphide isomerase, that can rearrange disulphide bonds in proteins (Leverrier *et al.*, 2011). The rerouting of RcsF to the inner membrane (where RcsC and RcsD are located) or the production of RcsF as a soluble periplasmic protein resulted in the constitutive activation of the Rcs phosphorelay whilst a deletion mutation in *rcsF* did not (Farris *et al.*, 2010). Therefore RcsF is not blocking or preventing signalling through the Rcs phosphorelay, but, rather, RcsF appears to positively affect the activity of RcsC in a manner that is probably dependent on changes in the availability and/or structure of RcsF (Farris *et al.*, 2010).

IgaA, on the other hand, is a 711 amino acid inner membrane protein that appears to block signalling through the Rcs phosphorelay. The *igaA*

Table 16.1 Conditions that activate the Rcs phosphorelay in the *Enterobacteriaceae*

Signal	Organism	Alleles	RcsF dependence	References
Mutants				
LPS biosynthesis	*E. coli*	*rfa-1*	Yes	Parker *et al.* (1992), Majdalani *et al.* (2005)
ECA structure	*Serratia marcescens*	*wec*	Yes	Castelli and Vescovi (2011)
Folding of OMPs	*E. coli*	*surA*	Yes	Castanie-Cornet *et al.* (2006)
Mdo	*E. coli*	*mdo*	Yes	Ebel *et al.* (1997)
Opg	*Dickeyia*	*opg*	?	Bouchart *et al.* (2010)
Acidic phospholipids	*E. coli*	*pgsA*	Yes	Shiba *et al.* (2004)
Metabolic stress	*E. coli*	*pgi*	?	El-Kazzaz *et al.* (2004)
Protein overproduction				
DjlA	*E. coli*	–	No	Clarke *et al.* (1996), Kelley and Georgopolous (1997), Castanie-Cornet *et al.* (2006)
HU	*E. coli*	*hupA*, *hupB*	?	Painbeni *et al.* (1993)
YedR	*E. coli*	–	?	Bury-Mone *et al.* (2009)
LolA/OmpG	*E. coli*	–	?	Chen *et al.* (2001)
WzxC/DnaK/TolB/Spot 42/YeiU	*E. coli*	–	Yes	Majdalani *et al.* (2005)
Environmental signal				
Surface contact	*E. coli*	–	No	Ferrieres and Clarke (2003)
Peptidoglycan synthesis				
β-Lactam antibiotics	*E. coli*	–	Yes	Laubacher and Ames (2008) Joseleau-Petis *et al.* (2007)
L-Forms	*E. coli*	–	Yes	Joseleau-Petis *et al.* (2007) Glover *et al.* (2009)
Lysozyme	*E. coli*	–	Yes	Callewaert *et al.* (2009)
Osmolarity up-shift	*E. coli*	–	?	Sledjeski and Gottesamn (1996)
Blue-light	*E. coli*	–	?	Tschowri *et al.* (2009)
Copper stress	*S. enterica*	–	Yes	Pontel *et al.* (2010)
Zn^{2+}/low T/glucose	*E. coli*	–	Yes	Hagiwara *et al.* (2003)
CAMP	*S. enterica*	–	Yes	Farris *et al.* (2010)
Fatty acids	*P. mirabilis*	–	?	Liaw *et al.* (2004)

?, unknown.

gene, i.e. intracellular growth attenuator, was originally described in *Salmonella* as an allele (*igaA1*) that permitted *Salmonella* to replicate within fibroblasts, a cell-type that is normally restrictive for intracellular *Salmonella* growth (Cano *et al.*, 2001). The presence of the *igaA1* allele resulted in mucoidy colonies on agar and mucoidy was suppressed by mutations in *rcsC*, *rcsD* or *rcsB*. Attempts to produce a deletion mutant in *igaA* in *Salmonella* were originally unsuccessful leading to the assumption that the gene was essential (Cano *et al.*, 2001). However, *igaA* is not essential, rather the deletion of this gene is very toxic and mutants have very strong growth defects. Indeed, *igaA* deletion mutants accumulate suppressor mutations that have been shown to be either deletions in the *rcsDBC* locus or mutations that alter the DNA around the promoters of *rcsD* or *rcsC*. Both possibilities result in a decrease in the expression of these genes and a lower responsiveness of this pathway to stimuli (Mariscotti and Garcia-Del Portillo, 2008). In *E. coli* the *igaA* homologue is *yrfF* and the construction of a deletion mutation in *yrfF* does not have the same deleterious effects on growth that were observed in *Salmonella*. However, Δ*yrfF* mutants still produce mucoid colonies suggesting that YrfF inhibits the activity of the Rcs phosphorelay in *E. coli* (Meberg *et al.*, 2001). Therefore, IgaA/YrfF appears to negatively interact with the Rcs phosphorelay to repress the activity of this signalling pathway. One possible function of this interaction may be to ensure that the Rcs phosphorelay has a very low basal level of activity in the absence of the appropriate signal. Interestingly, IgaA has also been reported to regulate the activity of the PhoPQ two-component pathway, and it has been suggested that IgaA represses the activity of the Rcs phosphorelay to allow full expression of the PhoPQ regulon (Tierrez and Garcia-del Portillo, 2004). It is also possible that cells may actively alter the level of IgaA to control the activity of the Rcs phosphorelay. In *Salmonella* and *E. coli igaA/yrfF* is the first gene in a four gene operon (*igaA/yrfF yrfGHI*), and expression of this operon is predicted to be driven by a σ^{70}-dependent promoter. In *Salmonella* an *igaA–lacZ* fusion was used to show that MviA is a positive regulator of *igaA* transcription. MviA is an orphan response regulator that participates in the degradation of RpoS (the *E. coli* homologue of MviA is RssB). The decrease in the level of *igaA* expression in the *mvi*A mutant background is thought to be owing to an increase in σ^S levels (Garcia-Calderon *et al.*, 2009). This might explain why IgaA levels decrease on entry into stationary phase suggesting that there may be a temporal regulation of Rcs phosphorelay activity (Cano *et al.*, 2001).

The Rcs phosphorelay: inputs and outputs

One of the most important goals in the study of any two-component pathway is the identification of the environmental signal(s) that activates signalling through the pathway i.e. the input(s). Identification of an appropriate signal allows for the conditional activation of a pathway that, in turn, facilitates genetic analyses to identify the genes that are regulated by a two-component pathway i.e. the outputs. Knowledge of the inputs and outputs of a two-component pathway plays an important part in enhancing our understanding of the role of a two-component pathway in a bacterium. The current consensus is that the Rcs phosphorelay is involved in monitoring some aspect of the bacterial cell envelope to regulate important cellular phenomena such as motility, biofilm formation and virulence. Genetic studies have described a role for the Rcs phosphorelay in ten members of the *Enterobacteriaceae*, although the vast majority of studies on the Rcs phosphorelay have been undertaken in either *E. coli* or *S. enterica*.

Inputs

A relatively large number of signals have been identified that activate the Rcs phosphorelay in the different *Enterobacteriaceae* (Table 16.1). Essentially, the signals or activating conditions can be divided into three different groups: (1) mutations that generate an activating signal, (2) protein overproduction as an activating signal and (3) environmental conditions that result in activation.

Mutations that generate an activating signal

Many mutations that have resulted in mucoid colonies on plates have been identified over the years and, in some cases, it has been established that the increase in mucoidy was dependent on the Rcs phosphorelay. One of the first such mutants described was the Δ*rfa1* deletion mutation in *E. coli*, a deletion that removed the *rfaGPBI* genes and resulted in the production of truncated LPS, i.e. a deep rough mutant (Parker *et al.*, 1992). Indeed many of the Rcs phosphorelay activating mutations are in genes that encode proteins involved in the maintenance of outer membrane function and/or integrity. For example, a deletion of the *surA* gene in *E. coli* has been shown to modestly increase the expression of the *wca* operon in a manner that is dependent on *rcsC* and *rcsF* (Missiakas *et al.*, 1996; Castanie-Cornet *et al.*, 2006). The *surA* gene encodes an important periplasmic chaperone that is required for normal outer membrane protein folding and assembly and mutations in *surA* result in outer membrane defects (Behrens-Kneip, 2010). Mutations in the *tol/pal* system (encoding the *tolQAR* and *pal* genes involved in outer membrane integrity) also activated the Rcs phosphorelay in *E. coli* and resulted in mucoid colonies at low temperature (Clavel *et al.*, 1996). In addition, a deletion of *pgsA*, encoding phosphatidylglycerophosphate synthase, resulted in the absence of the major acidic phospholipids (phosphatidylglycerol and cardiolipin) from the membranes of *E. coli*. Although this mutant was viable it exhibited poor growth at 37°C and this poor growth phenotype could be rescued by null mutations in *rcsC*, *rcsF* or *rcsA* (Shiba *et al.*, 2004; Nagahama *et al.*, 2006). Therefore it has been suggested that poor growth is a consequence of the constitutive activation of the Rcs phosphorelay in the *pgsA* mutant background. In *E. coli* acidic phospholipids are also required for the production of membrane-derived oligosaccharides (Mdo), osmoregulated glucans that are found in the periplasm. Mdo are a heterogeneous family of closely related oligosaccharides containing glucose as the sole sugar and substituted with phospho-*sn*-1-glycerol, phosphoethanolamine, and *O*-succinyl ester residues (Kennedy, 1996). A mutation in *mdoH*, encoding an inner membrane glycosyltransferase involved in Mdo synthesis in *E. coli*, resulted in an Rcs phosphorelay-dependent increase in the expression of the *wca* operon (Ebel *et al.*, 1997). Interestingly, mutations in *opgH*, the *mdoH* homologue in *Erwinia chrysanthemi* (*Dickeya dadantii*), also resulted in an increased production of EPS and a complete loss of virulence (Page *et al.*, 2001). The dependence of these phenotypes on the Rcs phosphorelay has recently been confirmed (Bouchart *et al.*, 2010). How do Mdo affect Rcs activity? The absence of Mdo can affect the osmolarity of the periplasm and this may have a direct effect on the activity of the Rcs phosphorelay (see below). On the other hand, mutants affected in Mdo biosynthesis are also affected in outer membrane protein production and this can also act on the activity of the Rcs phosphorelay (Kennedy, 1996). A mutation in *pgi*, encoding phosphoglucose isomerase, has been shown to transiently increase the activity of the Rcs phosphorelay when glucose is added to the medium, but only when present in the same genetic background as a null mutation in *mlc* (El-Kazzaz *et al.*, 2004). Mlc represses the expression of a number of genes involved in glucose uptake and metabolism whilst Pgi is involved in an early step of glycolysis (Plumbridge, 2002). Together the *pgi mlc* mutations result in an increased flux of glucose to dTDP-glucose, which acts as the metabolic signal for Rcs phosphorelay activation. Interestingly dTDP-glucose is an important precursor of O antigen and enterobacterial common antigen (ECA) biosynthesis, both important components of the outer membrane in *E. coli*. Accordingly, it has been speculated that an overflow in the biosynthetic pathway of these surface molecules may be a signal that activates the Rcs phosphorelay (El-Kazzaz *et al.*, 2004). Finally, mutations in the *wec* gene cluster of *Serratia marcescens* that result in alterations in the structure of the ECA activate signalling through the Rcs phosphorelay (Castelli and Vescovi, 2011). Therefore all of the activating mutants so far described appear to, in some way, affect the structure and/or integrity of the outer membrane implying a potential role for the Rcs phosphorelay in maintaining this important structural barrier.

Protein overproduction as an activating signal

Many studies have identified genes that, when present in multiple copies, activate the Rcs phosphorelay resulting in mucoid colonies on agar plates (Table 16.1). The best characterized of these genes is probably *djlA*, encoding a DnaJ-like chaperone. When moderately overproduced from a plasmid DjlA activates the Rcs phosphorelay in an *rcsF*-independent manner (Clarke *et al.*, 1996; Kelley and Georgopoulos, 1997). DjlA is localized to the inner membrane by a single N-terminal transmembrane domain (TMD) that also functions as a dimerization domain (Toutain *et al.*, 2003). Membrane localization and dimerization of DjlA are required for the activation of the Rcs phosphorelay (Clarke *et al.*, 1997). Moreover, DjlA has, at the C-terminal end of the protein, a J-domain. J-domains are found in all DnaJ-like (Hsp40) proteins where they are required to mediate the interaction with the chaperone

DnaK (Hsp70) (Kampinga and Craig, 2010). The DnaJ–DnaK (Hsp40–Hsp70) chaperone machine plays an important and conserved role in protein folding and stress response throughout the prokaryotes and eukaryotes. Genetic analyses have indicated that the J-domain of DjlA is functional and interacts with DnaK, and it has been shown that activation of the Rcs phosphorelay by DjlA requires a functional DnaK suggesting that chaperone activity is important for this activity (Kelley and Georgopoulos, 1997; Gur *et al.*, 2004). The role of DjlA in *E. coli* is not known but it is notable that the expression of *djlA* is positively regulated by σ^E suggesting that the level of DjlA may be controlled in response to extracytoplasmic stress (Bury-Mone *et al.*, 2009). Moreover, there is some literature suggesting a role for DjlA in the virulence of some bacteria (Ohnishi *et al.*, 2004; Lakhal *et al.*, 2008). However, it has not been determined whether the Rcs phosphorelay is involved in this aspect of DjlA activity.

Other proteins have also been shown to activate the Rcs phosphorelay when overproduced in *E. coli*, but these proteins will not be discussed in any detail here (for a list refer to Table 16.1). However, one protein, *yedR*, is notable as it encodes a predicted inner membrane protein with no known function that, when overproduced, generates highly mucoid colonies that were *rcsC* and *rcsB* dependent (Bury-Mone *et al.*, 2009). Interestingly, *yedR* overproduction also strongly stimulated the Psp pathway, an important cellular response to membrane stress, again highlighting the potential link between the Rcs phosphorelay and the cellular membrane (Bury-Mone *et al.*, 2009).

Environmental conditions that result in activation

The activating conditions detailed above are, in many ways, artificial and would not be expected to play a role in the normal control of Rcs phosphorelay activity in the cell. However, when considered together, these activating conditions do strongly suggest that membrane integrity may be an important parameter controlling the activity of this phosphorelay. Many studies have tried to identify activating signals that might be considered to be more physiological. For example, the Rcs phosphorelay is transiently activated following an up-shift in the osmolarity of the growth medium or by the transfer of planktonic *E. coli* to a surface (Sledjeski and Gottesman, 1996; Ferrieres and Clarke, 2003). Both of these conditions would be expected to change the structure and/or composition of the cellular membranes. Interestingly the signal that was generated by the transfer of the planktonic *E. coli* to a solid surface, although unidentified, might be specific for the Rcs phosphorelay, as neither the EnvZ/OmpR (responds to changes in osmolarity) nor the CpxAR (responds to membrane perturbations) two-component pathways were activated under the same conditions (Ferrieres and Clarke, 2003). Moreover, treatment of *E. coli* with the antipsychotic drug chlorpromazine was also shown to transiently increase the expression of Rcs-regulated genes in an *rcsC*- and *rcsB*-dependent manner. Indeed the Rcs phosphorelay was essential for *E. coli* to be able to overcome chlorpromazine-induced stress (Conter *et al.*, 2002). Chlorpromazine is a cationic amphipathic molecule that alters membrane topology by inserting into the inner leaflet of the membrane bilayer. In *S. enterica* the Rcs phosphorelay is activated by the addition of polymyxin B, a cationic antimicrobial peptide (CAMP), that functions by inserting into the membrane bilayer of the target cell (Farris *et al.*, 2010). It has also been reported that the Rcs phosphorelay is required for resistance to CAMPs in *S. enterica* and *E. amylovora* (Detweiler *et al.*, 2003; Erickson and Detweiler, 2006; D. Wang, *et al.*, 2009). Although it has not been tested whether polymyxin B functions as a signal in *E. coli*, another CAMP called colistin (polymyxin E) has been shown to induce the expression of the *wca* operon suggesting that this signal is conserved (Sailer *et al.*, 2003). Therefore alterations in membrane structure are associated with changes in Rcs phosphorelay activity. However, it is possible that gross changes in membrane structure will affect the activity of many sensor proteins and, therefore, might not be a specific signal associated with the Rcs phosphorelay.

A number of reports have recently implicated peptidoglycan (PG) as having a significant role in controlling the activity of the Rcs phosphorelay. It was originally shown that some β-lactam antibiotics induced the expression of a *cps–lacZ* fusion in *E. coli* suggesting that the inhibition of PG biosynthesis may activate the Rcs phosphorelay (Sailer *et al.*, 2003). The level of *cps-lacZ* induction appeared to be dependent on the antibiotic suggesting that it may not be a general response to PG damage (Sailer *et al.*, 2003). More recently, it was shown that the Rcs phosphorelay is activated in a strain of *E. coli* following exposure to the PG-degrading enzyme lysozyme (Callewaert *et al.*, 2009). In addition, the aberrant localization of AmiA and AmiC, amidases involved in splitting the PG during cell division, also activate the Rcs phosphorelay (Ize *et al.*, 2004). Therefore blocking PG biosynthesis and enhancing PG degradation both result in an activation of the Rcs phosphorelay. The treatment of *E. coli* with high levels of certain β-lactam antibiotics results in the formation of cells that grow as spherical cells, as they have little or no PG in their cell wall. These cells are called L-forms and they are capable of forming colonies under very stringent growth conditions. L-forms form mucoid colonies on the surface of an agar plate, and recent genetic analyses have shown that the Rcs phosphorelay is essential for the viability of L-forms, probably as it is required to drive the production of colanic acid (Joseleau-Petit *et al.*, 2007; Glover *et al.*, 2009). Therefore the colony-forming ability of L-forms is dependent on an active Rcs phosphorelay. Although L-forms are formed as a consequence of the absence (or near absence) of PG, these cells have many other significant morphological and compositional changes in their membranes. Therefore, whether the Rcs-activating signal is due to the accumulation of some PG precursor and/or degradation product or the changes in the membrane associated with perturbations in PG synthesis remains to be established.

All the available evidence suggests that the Rcs phosphorelay is one of several systems that respond to membrane or extracytoplasmic damage including the CpxAR and BaeSR two-component pathways, the Psp regulon and the σ^E regulon (MacRitchie *et al.*, 2008). The CpxAR and σ^E pathways are generally involved in the response to misfolded outer membrane proteins (Alba and Gross, 2004; Raivio, 2005; Hayden and Ades, 2008; Price and Raivio, 2009). The BaeSR pathway has recently been shown to increase the expression of an efflux pump in response to the presence of specific envelope-damaging compounds, e.g. flavonoids and sodium tungstate and the Psp pathway appears to respond to a dissipation of the proton motive force (Joly *et al.*, 2010; Leblanc *et al.*, 2011). Extracytoplasmic stress might occur whilst a bacterium is in the gut of the host, e.g. as a result of exposure to bile or high temperatures. However, it has also been suggested that the production of some Rcs-regulated capsules in *E. coli*, i.e. colanic acid and Group 1 capsules, may be important for survival in environments outside of the animal host, e.g. marine or freshwater lakes (Power *et al.*, 2005). In these aquatic niches the sensing of blue-light will be an important environmental cue and many bacteria have been shown to contain photoreceptors that contain the appropriate blue-light sensing domains, i.e. LOV, PYP and BLUF domains (Gomelsky and Hoff, 2011). YcgF is a BLUF-domain containing protein in *E. coli* that has been shown to affect the activity of the Rcs phosphorelay (Tschowri *et al.*, 2009). Upon irradiation with blue-light YcgF binds to the MerR-like repressor protein, YcgE, preventing it from binding DNA and resulting in the strong induction (or derepression) of eight proteins. The increased production of two of these proteins, YmgA and YmgB, results in the activation of the Rcs phosphorelay by an unknown mechanism. Interestingly, the expression of the operon containing *ymgA* and *ymgB* is dependent on σ^S, whilst the expression of *ycgE* and *ycgF* is increased at low temperature (Tschowri *et al.*, 2009). Therefore, this system links several cues that would be important in aquatic environments (i.e. blue-light, starvation and low temperature) to the activation of the Rcs phosphorelay.

Outputs

An understanding of the range of genes that are controlled by a particular two-component pathway (i.e. the regulon) can provide a powerful insight into the role of this regulatory pathway in the cell. Nowadays, the characterization of a regulon is a relatively straightforward exercise in transcriptomics using microarrays or RNA-seq. However, prior to the widespread application of transcriptomics, only four genetic loci had been shown to be regulated by the Rcs phosphorelay, i.e. the *wca* operon, *rprA*, *flhDC* and *osmC*. The first Rcs-regulated operon to be identified and studied was, of course, the *wca* operon, encoding all of the genes required for the production and export of the EPS, colanic acid (Gottesman *et al.*, 1985; Brill *et al.*, 1988; Stout and Gottesman, 1990). The expression of the *wca* operon appears to be completely dependent on the activity of the Rcs phosphorelay, and this feature has been extremely useful as a tool to study the activity of this phosphorelay. Another Rcs-regulated gene is the small non-coding RNA, *rprA* (Majdalani *et al.*, 2002). This small RNA increases the translatability of the *rpoS* mRNA resulting in an increase in the level of σ^S in the cell. The expression of *rprA* is positively controlled by the Rcs phosphorelay and regulation is independent of RcsA. Measuring RprA expression also appears to be a very sensitive reporter of Rcs phosphorelay activity. The *flhDC* operon encodes the master regulator of flagellar motility and this locus is another target of the Rcs phosphorelay. In one of the few studies to demonstrate direct binding of RcsB to a target gene it was shown that *flhDC* expression is repressed by the binding of RcsB to an RcsAB box located downstream from the *flhDC* promoter. Therefore binding of RcsB to this operator, which is potentiated by RcsA, would be expected to block transcription initiation (Francez-Charlot *et al.*, 2003). Finally the expression of *osmC*, encoding an osmotically induced peroxidase, is controlled by the activity of two promoters, *osmCp1* and *osmCp2*. The osmotic inducibility of this gene is likely to be owing to the Rcs phosphorelay, as RcsB has been shown to stimulate the recruitment of RNA polymerase to the *osmCp1* promoter (Davalos-Garcia *et al.*, 2001; Sturny *et al.*, 2003).

In 2003, two independent transcriptomic studies described a more extensive Rcs regulon. The first study used the overproduction of DjlA as the inducing signal and reported that 149 genes were regulated by the Rcs phosphorelay in *E. coli* with 83 genes up- and 66 genes down-regulated (Ferrieres and Clarke, 2003). During the same study the regulon was further extended by five genes using a complementary approach based on a plasmid-based *lacZ* fusion and λ*placMu53* mutagenesis (Ferrieres and Clarke, 2003). The reason why these five genes were not originally identified in the array might suggest the involvement of post-transcriptional regulation and highlights the importance of taking several complementary approaches in this type of work. In the second study, the authors used a different activating signal (culturing the cells in the presence of 1 mM $ZnCl_2$ and 0.4% (w/v) glucose at 20°C), and they reported a much smaller regulon consisting of just 35 genes (32 of which were also identified in the study using DjlA as the activating signal) (Hagiwara *et al.*, 2003). As expected, many of the genes identified in both of the array studies were involved in the production of colanic acid. Targets previously identified as being regulated by the Rcs phosphorelay were also identified, e.g. *osmC*, although *flhDC* was not identified in either study. In the DjlA-overproducing study this can be simply explained, as this work was done in *E. coli* MC4100, a strain that is non-motile owing to a +1 frameshift mutation in the *flhDC* operon.

Analysis of the full spectrum of genes that are controlled by the Rcs phosphorelay led to the conclusion that the activation of this phosphorelay will result in a significant remodelling of the cell surface (Ferrieres and Clarke, 2003). In addition to regulating the production of colanic acid and flagella, the Rcs phosphorelay also controls the expression of genes involved in PG biosynthesis and turnover, curli biosynthesis, adhesins such as antigen 43, lipoproteins and a large number of hypothetical genes predicted to encode proteins that are localized in either the cytoplasmic or outer membranes (Ferrieres and Clarke, 2003; Vianney *et al.*, 2005). Furthermore, the Rcs phosphorelay has also been shown to positively regulate the production of Type I fimbriae by inversely controlling the expression of the genes encoding the recombinases that determine the orientation of the *fim* switch, i.e. *fimB* and *fimE* (Schwan *et al.*, 2007). The Rcs regulons in *S. enterica* and *Y. pseudotuberculosis* have also been described, and remodelling of the cell surface appears to be a conserved consequence of the activation of the Rcs phosphorelay (Erickson and Detweiler, 2006; Hinchliffe *et al.*, 2008).

Function of the Rcs phosphorelay

Many studies have reported that the Rcs phosphorelay negatively regulates swimming and swarming motility in a number of species including *E. coli*, *S. enterica*, *P. mirabilis*, *D. dadantii*, *Y. pseudotuberculosis* and *S. marcescens* (Belas *et al.*, 1998; Takeda *et al.*, 2001; Wang *et al.*, 2007; Hinchliffe *et al.*, 2008; Bouchart *et al.*, 2010; Castelli and Vescovi, 2011). This observation can be mechanistically explained owing to the negative regulation of *flhDC* by the Rcs phosphorelay. In contrast the Rcs phosphorelay appears to positively regulate motility in *Y. enterocolitica* (Hinchliffe *et al.*, 2008). The Rcs phosphorelay is also required for inducible acid resistance in *E. coli* K-12 and *E. coli* O157::H7 (Johnson *et al.*, 2011). Indeed, the role of RcsB in acid resistance appears to be mediated by GadE, the key regulator of the glutamate-dependent (AR2) acid stress response, at two levels: (a) RcsB is required for the expression of *gadE* and (b) RcsB forms a heterodimer with GadE to regulate gene expression (Castanie-Cornet *et al.*, 2007; Krin *et al.*, 2010; Johnson *et al.*, 2011). The Rcs phosphorelay has also been implicated in biofilm formation, as null mutations in either *rcsC* or *rcsD* were shown to affect the ability of *E. coli* to attach to plastic (Ferrieres and Clarke, 2003). Indeed this phenotype, coupled with the observation that genes involved in attachment were repressed by the phosphorelay (i.e. gene encoding flagella, curli and adhesins) and genes involved in maturation were induced by the phosphorelay (i.e. colanic acid), has led to the hypothesis that the Rcs phosphorelay may play an important role as a switch that controls the transition from early to late biofilm (Huang *et al.*, 2006; Clarke, 2010). In support of this hypothesis it has been shown that (a) the *wca* operon is induced during biofilm formation and (b) colanic acid is important for the maturation of *E. coli* biofilms (Prigent-Combaret *et al.*, 1999; Danese *et al.*, 2000). However, it is still unclear whether signalling through the Rcs phosphorelay is required for normal biofilm formation in *E. coli*. A deletion of *rcsB* has been shown to have no biofilm phenotype, whilst the defect in biofilm formation observed in the *rcsC* null mutant appears to be a consequence of the increased expression of RprA (and therefore the increased level of σ^S) in this background (Ferrieres *et al.*, 2009). More detailed experiments using flow cells and fluorescently tagged cells, for example, are required in order to address this issue.

The Rcs phosphorelay is involved in the pathogenicity of a number of enterics including enterohaemorrhagic *E. coli* (EHEC), *S. enterica*, *Y. pseudotuberculosis*, *Y. enterocolitica*, *P. mirabilis*, *D. dadantii*, *E. amylovora* and *Pectobacterium carotovora* (Liaw *et al.*, 2004; Tobe *et al.*, 2005; Venecia and Young, 2005; Andresen *et al.*, 2007; Hinchliffe *et al.*, 2008; D. Wang *et al.*, 2009; Andresen *et al.*, 2010; Bouchart *et al.*, 2010). In many of these strains the relationship between the Rcs phosphorelay and virulence is directly correlated with an effect on the production of an important virulence factor (Table 16.2). However, the role of the Rcs phosphorelay in virulence is best characterized in *S. enterica* where this signalling system has been shown to have an important role during the systemic infection of mice (Detweiler *et al.*, 2003). Mutations that increase the level of RcsB~P in the cell, e.g. either a deletion in *rcsC* or mutations that constitutively activating the RcsC kinase activity, strongly attenuated the virulence of *S. enterica* in mice, and attenuation was shown to be partially due to the overproduction of colanic acid (Mouslim *et al.*, 2004; Garcia-Calderon *et al.*, 2005). This is similar to what was observed with the *opg* mutant in *D. dadantii* where constitutive activation of the Rcs phosphorelay resulted in the complete loss in virulence (Bouchart *et al.*, 2010). On the other hand, the Rcs phosphorelay is required for systemic infection and growth and persistence of *S. enterica* within the macrophage (Detweiler *et al.*, 2003). Therefore it appears that the Rcs phosphorelay must be activated at some point during infection, although, if activation occurs too early, the infection will not succeed. This implies an important role for this phosphorelay in the temporal development of an infection.

A subsequent study has identified a specific Rcs-regulated gene, called *ydeI*, that is required for the *in vivo* persistence of *S. enterica* in mice. The *ydeI* gene encodes a lipoprotein that interacts with OmpD and is required for resistance of *S. enterica* to CAMPs (Erickson and Detweiler, 2006). The expression of *ydeI* is regulated by both the PhoPQ and PmrAB two-component pathways in *S. enterica* (Erickson and Detweiler, 2006). Indeed,

Table 16.2 The Rcs phosphorelay and virulence

Species	Potential role in pathogenicity	Reference
EHEC	Regulation of genes in LEE pathogenicity island	Tobe *et al.* (2005)
S. enterica	Resistance to CAMPs, regulation of *ydeI*	Erickson and Detweiler (2006), Farris *et al.* (2010)
Y. enterocolitica	Regulation of Ysa type 3 secretion system, attachment to mammalian cells	Venecia and Young (2005), Hinchcliffe *et al.* (2008)
Y. pseudotuberculosis	Attachment to mammalian cells	Hinchliffe *et al.* (2008)
P. mirabilis	Swarming motility, haemolysin secretion	Liaw *et al.* (2005)
D. dadantii	n.d.[a]	Bouchart *et al.* (2010)
E. amylovora	Resistance to CAMPs	Wang *et al.* (2009a)
P. carotovorum	Regulation of plant cell wall degrading enzymes	Andresen *et al.* (2007, 2010)

[a]Mutations in either *rcsC*, *rcsD* or *rcsB* restores virulence in an *opg* mutant suggesting that the constitutive activation of the Rcs phosphorelay blocks virulence, a phenomenon that has also been reported in *S. enterica*.
CAMP, cationic antimicrobial peptide; LEE, locus for enterocyte effacement; Ysa, Yersinia secretion apparatus.

there appears to be significant overlap between the Rcs phosphorelay and the PhoPQ pathway with regard to the regulation of genes involved in virulence and, as previously mentioned, it has been suggested that the repression of Rcs phosphorelay activity by IgaA is required for the full expression of the PhoPQ regulon (Tierrez and Garcia-del Portillo, 2004; Garcia-Calderon *et al.*, 2007). Therefore, although the Rcs and Pho regulons do appear to overlap in *S. enterica*, there may be some kind of regulation, potentially mediated by IgaA, ensuring that the two signalling pathways are not activated at the same time. The Rcs–Pho network also appears to be present in *E. coli*, although it is not as well characterized in this species. The Rcs-dependent increase in the expression of the *wca* operon observed when cells were grown in the presence of Zn^{2+} and glucose at low temperatures was shown to require the PhoPQ two-component pathway (Hagiwara *et al.*, 2003). Indeed, the increase in *wca* expression was also shown to be dependent on the EnvZ/OmpR two-component pathway and the ArcA response regulator, hinting at the presence of very complex regulatory networks that involve the co- and cross-regulation of genes by different signalling pathways (Hagiwara *et al.*, 2003).

Conclusions and perspectives

From its humble beginnings nearly 30 years ago as a regulator of colanic acid production, the Rcs phosphorelay has become a complex signalling network with a key role in the regulation of many important processes in the bacterial cell including swimming and swarming motility, biofilm formation and pathogenicity. Although genetics and structural studies have provided us with some excellent insights into the Rcs phosphorelay, our understanding remains far from being complete. Indeed, it has still not been conclusively shown that there is a phosphorelay between RcsC, RcsD and RcsB and, although there is a growing consensus that PG is somehow involved in the activation of this phosphorelay, we still have not identified the activating signal(s). The Rcs phosphorelay has been shown to negatively regulate the production of flagella and, therefore, motility in all of the *Enterobacteriaceae* that have been tested. This suggests that the repression of motility is a key (and probably conserved) function of the Rcs phosphorelay. However, in both *E. coli* and *S. enterica*, the Rcs phosphorelay has also been implicated with an important role during the later stages of biofilm formation and virulence, respectively. Indeed the evidence suggests that the promiscuous activation of the Rcs phosphorelay is detrimental for both biofilm formation and virulence (Mouslim *et al.*, 2004; Garcia-Calderon *et al.*, 2005; Ferrieres *et al.*, 2009). Therefore, the Rcs phosphorelay may play an important role in the temporal control of lifestyle decisions in bacteria, e.g. controlling the timing of the transition from life as a free-living, planktonic, swimming cell to life as a sessile cell growing as part of a biofilm.

References

Achtman, M., Zurth, K., Morelli, G., Torrea, G., Guiyoule, A., and Carniel, E. (1999). *Yersinia pestis*, the cause of plague, is a recently emerged clone of *Yersinia pseudotuberculosis*. Proc. Natl. Acad. Sci. U.S.A. *96*, 14043–14048.

Alba, B.M., and Gross, C.A. (2004). Regulation of the *Escherichia coli* sigma-dependent envelope stress response. Mol. Microbiol. *52*, 613–619.

Andresen, L., Koiv, V., Alamae, T., and Mae, A. (2007). The Rcs phosphorelay modulates the expression of plant cell wall degrading enzymes and virulence in *Pectobacterium carotovorum* ssp. *carotovorum*. FEMS Microbiol. Lett. *273*, 229–238.

Andresen, L., Sala, E., Koiv, V., and Mae, A. (2010). A role for the Rcs phosphorelay in regulating expression of plant cell wall degrading enzymes in *Pectobacterium carotovorum* subsp. *carotovorum*. Microbiology *156*, 1323–1334.

Arricau, N., Hermant, D., Waxin, H., Ecobichon, C., Duffey, P.S., and Popoff, M.Y. (1998). The RcsB-RcsC regulatory system of *Salmonella typhi* differentially modulates the expression of invasion proteins, flagellin and Vi antigen in response to osmolarity. Mol. Microbiol. *29*, 835–850.

Barak, R., Prasad, K., Shainskaya, A., Wolfe, A.J., and Eisenbach, M. (2004). Acetylation of the chemotaxis response regulator CheY by acetyl-CoA synthetase purified from *Escherichia coli*. J. Mol. Biol. *342*, 383–401.

Behrens-Kneip, S. (2010). The role of SurA factor in outer membrane protein transport and virulence. Int. J. Med. Microbiol. *300*, 421–428.

Belas, R., Schneider, R., and Melch, M. (1998). Characterization of *Proteus mirabilis* precocious swarming mutants: identification of *rsbA*, encoding a regulator of swarming behavior. J. Bacteriol. *180*, 6126–6139.

Bijlsma, J.J., and Groisman, E.A. (2003). Making informed decisions: regulatory interactions between two-component systems. Trends Microbiol. *11*, 359–366.

Bouchart, F., Boussemart, G., Prouvost, A.F., Cogez, V., Madec, E., Vidal, O., Delrue, B., Bohin, J.P., and Lacroix, J.M. (2010). The virulence of a *Dickeya dadantii* 3937 mutant devoid of osmoregulated periplasmic glucans is restored by inactivation of the RcsCD-RcsB phosphorelay. J. Bacteriol. *192*, 3484–3490.

Boulanger, A., Francez-Charlot, A., Conter, A., Castanie-Cornet, M.P., Cam, K., and Gutierrez, C. (2005). Multistress regulation in *Escherichia coli*: expression of *osmB* involves two independent promoters responding either to sigmaS or to the RcsCDB His-Asp phosphorelay. J. Bacteriol. *187*, 3282–3286.

Brill, J.A., Quinlan-Walshe, C., and Gottesman, S. (1988). Fine-structure mapping and identification of two regulators of capsule synthesis in *Escherichia coli* K-12. J. Bacteriol. *170*, 2599–2611.

Burke, G.R., and Moran, N.A. (2011). Massive genomic decay in *Serratia symbiotica*, a recently evolved symbiont of aphids. Genome Biol. Evol. *3*, 195–208.

Bury-Mone, S., Nomane, Y., Reymond, N., Barbet, R., Jacquet, E., Imbeaud, S., Jacq, A., and Bouloc, P. (2009). Global analysis of extracytoplasmic stress signaling in *Escherichia coli*. PLoS Genet. *5*, e1000651.

Callewaert, L., Vanoirbeek, K.G., Lurquin, I., Michiels, C.W., and Aertsen, A. (2009). The Rcs two-component system regulates expression of lysozyme inhibitors and is induced by exposure to lysozyme. J. Bacteriol. *191*, 1979–1981.

Cano, D.A., Martinez-Moya, M., Pucciarelli, M.G., Groisman, E.A., Casadesus, J., and Garcia-Del Portillo, F. (2001). *Salmonella enterica* serovar Typhimurium response involved in attenuation of pathogen intracellular proliferation. Infect. Immun. *69*, 6463–6474.

Castanie-Cornet, M.P., Cam, K., and Jacq, A. (2006). RcsF is an outer membrane lipoprotein involved in the RcsCDB phosphorelay signaling pathway in *Escherichia coli*. J. Bacteriol. *188*, 4264–4270.

Castanie-Cornet, M.P., Treffandier, H., Francez-Charlot, A., Gutierrez, C., and Cam, K. (2007). The glutamate-dependent acid resistance system in *Escherichia coli*: essential and dual role of the His-Asp phosphorelay RcsCDB/AF. Microbiology *153*, 238–246.

Castelli, M.E., and Vescovi, E.G. (2011). The Rcs signal transduction pathway is triggered by enterobacterial common antigen structure alterations in *Serratia marcescens*. J. Bacteriol. *193*, 63–74.

Cheng, H.Y., Chen, Y.S., Wu, C.Y., Chang, H.Y., Lai, Y.C., and Peng, H.L. (2010). RmpA regulation of capsular polysaccharide biosynthesis in *Klebsiella pneumoniae* CG43. J. Bacteriol. *192*, 3144–3158.

Clarke, D.J. (2010). The Rcs phosphorelay: more than just a two-component pathway. Future Microbiol. *5*, 1173–1184.

Clarke, D.J., Jacq, A., and Holland, I.B. (1996). A novel DnaJ-like protein in *Escherichia coli* inserts into the cytoplasmic membrane with a type III topology. Mol. Microbiol. *20*, 1273–1286.

Clarke, D.J., Holland, I.B., and Jacq, A. (1997). Point mutations in the transmembrane domain of DjlA, a membrane-linked DnaJ-like protein, abolish its function in promoting colanic acid production via the Rcs signal transduction pathway. Mol. Microbiol. *25*, 933–944.

Clarke, D.J., Joyce, S.A., Toutain, C.M., Jacq, A., and Holland, I.B. (2002). Genetic analysis of the RcsC sensor kinase from *Escherichia coli* K-12. J. Bacteriol. *184*, 1204–1208.

Clavel, T., Lazzaroni, J.C., Vianney, A., and Portalier, R. (1996). Expression of the *tolQRA* genes of *Escherichia coli* K-12 is controlled by the RcsC sensor protein involved in capsule synthesis. Mol. Microbiol. *19*, 19–25.

Conter, A., Sturny, R., Gutierrez, C., and Cam, K. (2002). The RcsCB His-Asp phosphorelay system is essential to overcome chlorpromazine-induced stress in *Escherichia coli*. J. Bacteriol. *184*, 2850–2853.

Danese, P.N., Pratt, L.A., and Kolter, R. (2000). Exopolysaccharide production is required for development of *Escherichia coli* K-12 biofilm architecture. J. Bacteriol. *182*, 3593–3596.

Davalos-Garcia, M., Conter, A., Toesca, I., Gutierrez, C., and Cam, K. (2001). Regulation of *osmC* gene expression by the two-component system *rcsB-rcsC* in *Escherichia coli*. J. Bacteriol. *183*, 5870–5876.

Detweiler, C.S., Monack, D.M., Brodsky, I.E. Mathew, H., and Falkow, S. (2003). *virK*, *somA* and *rcsC* are important for systemic *Salmonella enterica* serovar Typhimurium infection and cationic peptide resistance. Mol. Microbiol. *48*, 385–400.

Ebel, W., and Trempy, J.E. (1999). *Escherichia coli* RcsA, a positive activator of colanic acid capsular polysaccharide synthesis, functions to activate its own expression. J. Bacteriol. *181*, 577–584.

Ebel, W., Vaughn, G.J., Peters, H.K., 3rd, and Trempy, J.E. (1997). Inactivation of *mdoH* leads to increased expression of colanic acid capsular polysaccharide in *Escherichia coli*. J. Bacteriol. *179*, 6858–6861.

El-Kazzaz, W., Morita, T., Tagami, H., Inada, T., and Aiba, H. (2004). Metabolic block at early stages of the glycolytic pathway activates the Rcs phosphorelay system via increased synthesis of dTDP-glucose in *Escherichia coli*. Mol. Microbiol. *51*, 1117–1128.

Erickson, K.D., and Detweiler, C.S. (2006). The Rcs phosphorelay system is specific to enteric pathogens/commensals and activates *ydeI*, a gene important for persistent *Salmonella* infection of mice. Mol. Microbiol. *62*, 883–894.

Farris, C., Sanowar, S., Bader, M.W., Pfuetzner, R., and Miller, S.I. (2010). Antimicrobial peptides activate the Rcs regulon through the outer membrane lipoprotein RcsF. J. Bacteriol. *192*, 4894–4903.

Ferrieres, L., and Clarke, D.J. (2003). The RcsC sensor kinase is required for normal biofilm formation in *Escherichia coli* K-12 and controls the expression of a regulon in response to growth on a solid surface. Mol. Microbiol. *50*, 1665–1682.

Ferrieres, L., Thompson, A., and Clarke, D.J. (2009). Elevated levels of sigmaS inhibit biofilm formation in *Escherichia coli*: a role for the Rcs phosphorelay. Microbiology *155*, 3544–3553.

Francez-Charlot, A., Laugel, B., Van Gemert, A., Dubarry, N., Wiorowski, F., Castanie-Cornet, M.P., Gutierrez, C., and Cam, K. (2003). RcsCDB His-Asp phosphorelay system negatively regulates the *flhDC* operon in *Escherichia coli*. Mol. Microbiol. *49*, 823–832.

Garcia-Calderon, C.B., Garcia-Quintanilla, M., Casadesus, J., and Ramos-Morales, F. (2005). Virulence attenuation in *Salmonella enterica rcsC* mutants with constitutive activation of the Rcs system. Microbiology *151*, 579–588.

Garcia-Calderon, C.B., Casadesus, J., and Ramos-Morales, F. (2007). Rcs and PhoPQ regulatory overlap in the control of *Salmonella enterica* virulence. J. Bacteriol. *189*, 6635–6644.

Garcia-Calderon, C.B., Casadesus, J., and Ramos-Morales, F. (2009). Regulation of *igaA* and the Rcs system by the MviA response regulator in *Salmonella enterica*. J. Bacteriol. *191*, 2743–2752.

Gervais, F.G., and Drapeau, G.R. (1992). Identification, cloning, and characterization of *rcsF*, a new regulator gene for exopolysaccharide synthesis that suppresses the division mutation *ftsZ84* in *Escherichia coli* K-12. J. Bacteriol. *174*, 8016–8022.

Gervais, F.G., Phoenix, P., and Drapeau, G.R. (1992). The *rcsB* gene, a positive regulator of colanic acid biosynthesis in *Escherichia coli*, is also an activator of *ftsZ* expression. J. Bacteriol. *174*, 3964–3971.

Ghosh, D., Roy, K., Williamson, K.E., Srinivasiah, S., Wommack, K.E., and Radosevich, M. (2009). Acyl-homoserine lactones can induce virus production in lysogenic bacteria: an alternative paradigm for prophage induction. Appl. Environ. Microbiol. *75*, 7142–7152.

Glover, W.A., Yang, Y., and Zhang, Y. (2009). Insights into the molecular basis of L-form formation and survival in *Escherichia coli*. PLoS One *4*, e7316.

Gomelsky, M., and Hoff, W.D. (2011). Light helps bacteria make important lifestyle decisions. Trends Microbiol. *19*, 441–448.

Gottesman, S., Trisler, P., and Torres-Cabassa, A. (1985). Regulation of capsular polysaccharide synthesis in *Escherichia coli* K-12: characterization of three regulatory genes. J. Bacteriol. *162*, 1111–1119.

Gur, E., Biran, D., Shechter, N., Genevaux, P., Georgopoulos, C., and Ron, E.Z. (2004). The *Escherichia coli* DjlA and CbpA proteins can substitute for DnaJ in DnaK-mediated protein disaggregation. J. Bacteriol. *186*, 7236–7242.

Hagiwara, D., Sugiura, M., Oshima, T., Mori, H., Aiba, H., Yamashino, T., and Mizuno, T. (2003). Genome-wide analyses revealing a signaling network of the RcsC-YojN-RcsB phosphorelay system in *Escherichia coli*. J. Bacteriol. *185*, 5735–5746.

Hashimoto, Y., Li, N., Yokoyama, H., and Ezaki, T. (1993). Complete nucleotide sequence and molecular characterization of ViaB region encoding Vi antigen in *Salmonella typhi*. J. Bacteriol. *175*, 4456–4465.

Hayden, J.D., and Ades, S.E. (2008). The extracytoplasmic stress factor, sigmaE, is required to maintain cell envelope integrity in *Escherichia coli*. PLoS One *3*, e1573.

Herzberg, M., Kaye, I.K., Peti, W., and Wood, T.K. (2006). YdgG (TqsA) controls biofilm formation in *Escherichia coli* K-12 through autoinducer 2 transport. J. Bacteriol. *188*, 587–598.

Hinchliffe, S.J., Howard, S.L., Huang, Y.H., Clarke, D.J., and Wren, B.W. (2008). The importance of the Rcs phosphorelay in the survival and pathogenesis of the enteropathogenic yersiniae. Microbiology *154*, 1117–1131.

Huang, Y.H., Ferrieres, L., and Clarke, D.J. (2006). The role of the Rcs phosphorelay in *Enterobacteriaceae*. Res. Microbiol. *157*, 206–212.

Huang, Y.H., Ferrieres, L., and Clarke, D.J. (2009). Comparative functional analysis of the RcsC sensor kinase from different *Enterobacteriaceae*. FEMS Microbiol. Lett. *267*, 30–37.

Ize, B., Porcelli, I., Lucchini, S., Hinton, J.C., Berks, B.C., and Palmer, T. (2004). Novel phenotypes of *Escherichia coli tat* mutants revealed by global gene expression and phenotypic analysis. J. Biol. Chem. *279*, 47543–47554.

Johnson, M.D., Burton, N.A., Gutierrez, B., Painter, K., and Lund, P.A. (2011). RcsB is required for inducible acid resistance in *Escherichia coli* and acts at *gadE*-dependent and -independent promoters. J. Bacteriol. *193*, 3653–3656.

Joly, N., Engl, C., Jovanovic, G., Huvet, M., Toni, T., Sheng, X., Stumpf, M.P., and Buck, M. (2010). Managing membrane stress: the phage shock protein (Psp) response, from molecular mechanisms to physiology. FEMS Microbiol. Rev. *34*, 797–827.

Joseleau-Petit, D., Liebart, J.C., Ayala, J.A., and D'Ari, R. (2007). Unstable *Escherichia coli* L forms revisited: growth requires peptidoglycan synthesis. J. Bacteriol. *189*, 6512–6520.

Kampinga, H.H., and Craig, E.A. (2010). The HSP70 chaperone machinery: J proteins as drivers of functional specificity. Nat. Rev. Mol. Cell. Biol. *11*, 579–592.

Kelley, W.L., and Georgopoulos, C. (1997). Positive control of the two-component RcsC/B signal transduction network by DjlA: a member of the DnaJ family of molecular chaperones in *Escherichia coli*. Mol. Microbiol. *25*, 913–931.

Kennedy, E.P. (1996). Membrane-derived oligosaccharides (periplasmic beta-D-glucans) of *Escherichia coli*. In *Escherichia coli* and *Salmonella typhimurium*: Cellular and Molecular Biology, Neidhardt, F.C., Ingraham, J.L., Low, K.B., Magasanik, B., Schaechter, M., and Umbarger, H.E., eds (ASM Press, Washington, DC), pp. 1064–1071.

Krin, E., Danchin, A., and Soutourina, O. (2010). RcsB plays a central role in H-NS-dependent regulation of motility and acid stress resistance in *Escherichia coli*. Res. Microbiol. *161*, 363–371.

Lakhal, F., Bury-Mone, S., Nomane, Y., Le Goic, N., Paillard, C., and Jacq, A. (2008). DjlA, a membrane-anchored DnaJ-like protein, is required for cytotoxicity of clam pathogen *Vibrio tapetis* to hemocytes. Appl. Environ. Microbiol. *74*, 5750–5758.

Leblanc, S.K., Oates, C.W., and Raivio, T.L. (2011). Characterization of the induction and cellular role of the BaeSR two-component envelope stress response of *Escherichia coli*. J. Bacteriol. *193*, 3367–3375.

Leverrier, P., Declercq, J.-P., Denoncin, K., Vertommen, D., Hiniker, A., Cho, S.-H., and Collet, J.F. (2011). Crystal structure of the outer membrane protein RcsF, a new substrate for the periplasmic protein-disulphide isomerase DsbC. J. Biol. Chem. *286*, 16734–16742.

Liaw, S.J., Lai, H.C., and Wang, W.B. (2004). Modulation of swarming and virulence by fatty acids through the RsbA protein in *Proteus mirabilis*. Infect. Immun. *72*, 6836–6845.

Lin, C.T., Wu, C.C., Chen, Y.S., Lai, Y.C., Chi, C., Lin, J.C., Chen, Y., and Peng, H.L. (2011). Fur regulation of the capsular polysaccharide biosynthesis and iron-acquisition systems in *Klebsiella pneumoniae* CG43. Microbiology *157*, 419–429.

MacRitchie, D.M., Buelow, D.R., Price, N.L., and Raivio, T.L. (2008). Two-component signaling and Gram-negative envelope stress response systems. Adv. Exp. Med. Biol. *631*, 80–110.

Majdalani, N., and Gottesman, S. (2005). The Rcs phosphorelay: a complex signal transduction system. Annu. Rev. Microbiol. *59*, 379–405.

Majdalani, N., Hernandez, D., and Gottesman, S. (2002). Regulation and mode of action of the second small RNA activator of RpoS translation, RprA. Mol. Microbiol. *46*, 813–826.

Majdalani, N., Heck, M., Stout, V., and Gottesman, S. (2005). Role of RcsF in signaling to the Rcs phosphorelay pathway in *Escherichia coli*. J. Bacteriol. *187*, 6770–6778.

Mariscotti, J.F., and Garcia-Del Portillo, F. (2008). Instability of the *Salmonella* RcsCDB signalling system in the absence of the attenuator IgaA. Microbiology *154*, 1372–1383.

Mariscotti, J.F., and Garcia-del Portillo, F. (2009). Genome expression analyses revealing the modulation of the *Salmonella* Rcs regulon by the attenuator IgaA. J. Bacteriol. *191*, 1855–1867.

Meberg, B.M., Sailer, F.C., Nelson, D.E., and Young, K.D. (2001). Reconstruction of *Escherichia coli mrcA* (PBP 1a) mutants lacking multiple combinations of penicillin binding proteins. J. Bacteriol. *183*, 6148–6149.

Mendez-Ortiz, M.M., Hyodo, M., Hayakawa, Y., and Membrillo-Hernandez, J. (2006). Genome-wide transcriptional profile of *Escherichia coli* in response to high levels of the second messenger 3',5'-cyclic diguanylic acid. J. Biol. Chem. *281*, 8090–8099.

Mercedes Pescaretti, M., Lopez, F., Morero, R., and Delgado, M.A. (2010). Transcriptional autoregulation of the RcsCDB phosphorelay system in *Salmonella enterica* serovar Typhimurium. Microbiology *156*, 3513–3521.

Minogue, T.D., Carlier, A.L., Koutsoudis, M.D., and von Bodman, S.B. (2005). The cell density-dependent expression of stewartan exopolysaccharide in *Pantoea stewartii* is a function of EsaR-mediated repression of the *rcsA* gene. Mol. Microbiol. *56*, 189–203.

Missiakas, D., Betton, J.M., and Raina, S. (1996). New components of protein folding in extracytoplasmic compartments of *Escherichia coli* SurA, FkpA and Skp/OmpH. Mol. Microbiol. *21*, 871–884.

Mouslim, C., Delgado, M., and Groisman, E.A. (2004). Activation of the RcsC/YojN/RcsB phosphorelay system attenuates *Salmonella* virulence. Mol. Microbiol. *54*, 386–395.

Mukhopadhyay, S., and Schellhorn, H.E. (1997). Identification and characterization of hydrogen peroxide-sensitive mutants of *Escherichia coli*: genes that require OxyR for expression. J. Bacteriol. *179*, 330–338.

Nagahama, H., Sakamoto, Y., Matsumoto, K., and Hara, H. (2006). RcsA-dependent and -independent growth defects caused by the activated Rcs phosphorelay system in the *Escherichia coli pgsA* null mutant. J. Gen. Appl. Microbiol. *52*, 91–98.

Ohnishi, H., Mizunoe, Y., Takade, A., Tanaka, Y., Miyamoto, H., Harada, M., and Yoshida, S. (2004). *Legionella dumoffii* DjlA, a member of the DnaJ family, is required for intracellular growth. Infect. Immun. *72*, 3592–3603.

Oliver, K.M., Degnan, P.H., Burke, G.R., and Moran, N.A. (2010). Facultative symbionts in aphids and the horizontal transfer of ecologically important traits. Annu. Rev. Entomol. *55*, 247–266.

Page, F., Altabe, S., Hugouvieux-Cotte-Pattat, N., Lacroix, J.M., Robert-Baudouy, J., and Bohin, J.P. (2001). Osmoregulated periplasmic glucan synthesis is required for *Erwinia chrysanthemi* pathogenicity. J. Bacteriol. *183*, 3134–3141.

Parker, C.T., Kloser, A.W., Schnaitman, C.A., Stein, M.A., Gottesman, S., and Gibson, B.W. (1992). Role of the *rfaG* and *rfaP* genes in determining the lipopolysaccharide core structure and cell surface properties of *Escherichia coli* K-12. J. Bacteriol. *174*, 2525–2538.

Plumbridge, J. (2002) Regulation of gene expression in the PTS in *Escherichia coli*: the role and interactions of Mlc. Curr. Opin. Microbiol. 5:187–193.

Power, M.L., Littlefield-Wyer, J., Gordon, D.M., Veal, D.A., and Slade, M.B. (2005). Phenotypic and genotypic characterization of encapsulated *Escherichia coli* isolated from blooms in two Australian lakes. Environ. Microbiol. *7*, 631–640.

Price, N.L., and Raivio, T.L. (2009). Characterization of the Cpx regulon in *Escherichia coli* strain MC4100. J. Bacteriol. *191*, 1798–1815.

Prigent-Combaret, C., Vidal, O., Dorel, C., and Lejeune, P. (1999). Abiotic surface sensing and biofilm-dependent regulation of gene expression in *Escherichia coli*. J. Bacteriol. *181*, 5993–6002.

Pristovsek, P., Sengupta, K., Lohr, F., Schafer, B., von Trebra, M.W., Ruterjans, H., and Bernhard, F. (2003). Structural analysis of the DNA-binding domain of the *Erwinia amylovora* RcsB protein and its interaction with the RcsAB box. J. Biol. Chem. *278*, 17752–17759.

Raivio, T.L. (2005). Envelope stress responses and Gram-negative bacterial pathogenesis. Mol. Microbiol. *56*, 1119–1128.

Reeves, P.R., Hobbs, M., Valvano, M.A., Skurnik, M., Whitfield, C., Coplin, D., Kido, N., Klena, J., Maskell, D., Raetz, C.R., *et al.* (1996). Bacterial polysaccharide synthesis and gene nomenclature. Trends Microbiol. *4*, 495–503.

Rogov, V.V., Bernhard, F., Lohr, F., and Dotsch, V. (2004). Solution structure of the *Escherichia coli* YojN histidine-phosphotransferase domain and its interaction with cognate phosphoryl recever domains. J. Mol. Biol. *343*, 1035–1048.

Rogov, V.V., Rogova, N.Y., Bernhard, F., Koglin, A., Lohr, F., and Dotsch, V. (2006). A new structural domain in the *Escherichia coli* RcsC hybrid sensor kinase connects histidine kinase and phosphoreceiver domains. J. Mol. Biol. *364*, 68–79.

Rogov, V.V., Schmoe, K., Lohr, F., Rogova, N.Y., Bernhard, F., and Dotsch, V. (2008). Modulation of the Rcs-mediated signal transfer by conformational flexibility. Biochem. Soc. Trans. *36*, 1427–1432.

Rogov, V.V., Rogova, N.Y., Bernhard, F., Lohr, F., and Dotsch, V. (2011). A disulphide bridge network within the soluble periplasmic domain determines structure and function of the outer membrane protein RcsF. J. Biol. Chem. *286*, 18775–18783.

Sailer, F.C., Meberg, B.M., and Young, K.D. (2003). beta-Lactam induction of colanic acid gene expression in *Escherichia coli*. FEMS Microbiol. Lett. *226*, 245–249.

Schmoe, K., Rogov, V.V., Rogova, N.Y., Lohr, F., Guntert, P., Bernhard, F., and Dotsch, V. (2011). Structural insights into the Rcs phosphotransfer: th enewly identified RcsD-ABL domain enhances interaction with the response regulator RcsB. Structure *19*, 577–587.

Schwan, W.R., Shibata, S., Aizawa, S., and Wolfe, A.J. (2007). The two-component response regulator RcsB regulates type 1 piliation in *Escherichia coli*. J. Bacteriol. *189*, 7159–7163.

Shiba, Y., Yokoyama, Y., Aono, Y., Kiuchi, T., Kusaka, J., Matsumoto, K., and Hara, H. (2004). Activation of the Rcs signal transduction system is responsible for the thermosensitive growth defect of an *Escherichia coli* mutant lacking phosphatidylglycerol and cardiolipin. J. Bacteriol. *186*, 6526–6535.

Sledjeski, D., and Gottesman, S. (1995). A small RNA acts as an antisilencer of the H-NS-silenced rcsA gene of *Escherichia coli*. Proc. Natl. Acad. Sci. U.S.A. *92*, 2003–2007.

Sledjeski, D.D., and Gottesman, S. (1996). Osmotic shock induction of capsule synthesis in *Escherichia coli* K-12. J. Bacteriol. *178*, 1204–1206.

Starai, V.J., and Escalante-Semerena, J.C. (2004). Identification of the protein acetyltransferase (Pat) enzyme that acetylates acetyl-CoA synthetase in *Salmonella enterica*. J. Mol. Biol. *340*, 1005–1012.

Starai, V.J., Celic, I., Cole, R.N., Boeke, J.D., and Escalante-Semerena, J.C. (2002). Sir2-dependent activation of acetyl-CoA synthetase by deacetylation of active lysine. Science *298*, 2390–2392.

Stout, V., and Gottesman, S. (1990). RcsB and RcsC: a two-component regulator of capsule synthesis in *Escherichia coli*. J. Bacteriol. *172*, 659–669.

Stout, V., Torres-Cabassa, A., Maurizi, M.R., Gutnick, D., and Gottesman, S. (1991). RcsA, an unstable positive regulator of capsular polysaccharide synthesis. J. Bacteriol. *173*, 1738–1747.

Sturny, R., Cam, K., Gutierrez, C., and Conter, A. (2003). NhaR and RcsB independently regulate the *osm*Cp1 promoter of *Escherichia coli* at overlapping regulatory sites. J. Bacteriol. *185*, 4298–4304.

Sun, Y.C., Hinnebusch, B.J., and Darby, C. (2008). Experimental evidence for negative selection in the evolution of a *Yersinia pestis* pseudogene. Proc. Natl. Acad. Sci. U.S.A. *105*, 8097–8101.

Tagliabua, L., Antoniani, D., Maciag, A., Bocci, P., Raffaelli, N., and Landini, P. (2010a). The diguanylate cyclase YddVcontrols production of he exopolysaccharide poly-N-acetylglucosamine (PNAG) through regulation of the PNAG biosynthetic *pgaABCD* operon. Microbiology *156*, 2901–2911.

Tagliabua, L., Maciag, A., Antoniani, D., and Landini, P. (2010b). The *yddV-dos* operon controls biofilm formation through the regulation of genes encoding curli fibers' subunits in aerobically growing *Escherichia coli*. FEMS Immunol. Med. Microbiol. *59*, 477–484.

Takeda, S., Fujisawa, Y., Matsubara, M., Aiba, H., and Mizuno, T. (2001). A novel feature of the multistep phosphorelay in *Escherichia coli*: a revised model of the RcsC --> YojN --> RcsB signalling pathway implicated in capsular synthesis and swarming behaviour. Mol. Microbiol. *40*, 440–450.

Thao, S., Chen, C.-S., Zhu, H., and Escalante-Semerena, J.C. (2011). N-lysine acetylation of a bacterial transcriptional factor inhibits its DNA-binding activity. PLoS One *5*, e15123.

Tierrez, A., and Garcia-del Portillo, F. (2004). The *Salmonella* membrane protein IgaA modulates the activity of the RcsC-YojN-RcsB and PhoP–PhoQ regulons. J. Bacteriol. *186*, 7481–7489.

Tobe, T., Ando, H., Ishikawa, H., Abe, H., Tashiro, K., Hayashi, T., Kuhara, S., and Sugimoto, N. (2005). Dual regulatory pathways integrating the RcsC-RcsD-RcsB signalling system control enterohaemorrhagic *Escherichia coli* pathogenicity. Mol. Microbiol. *58*, 320–333.

Toh, H., Weiss, B.L., Perkin, S.A., Yamashita, A., Oshima, K., Hattori, M., and Aksoy, S. (2006). Massive genome erosion and functional adaptations provide insights into the symbiotic lifestyle of *Sodalis glossinidius* in the tsetse host. Genome Res. *16*, 149–156.

Toutain, C.M., Clarke, D.J., Leeds, J.A., Kuhn, J., Beckwith, J., Holland, I.B., and Jacq, A. (2003). The transmembrane domain of the DnaJ-like protein DjlA is a dimerisation domain. Mol. Genet. Genomics *268*, 761–770.

Trempy, J.E., Kirby, J.E., and Gottesman, S. (1994). Alp suppression of Lon: dependence on the *slpA* gene. J. Bacteriol. *176*, 2061–2067.

Trisler, P., and Gottesman, S. (1984). *lon* transcriptional regulation of genes necessary for capsular polysaccharide synthesis in *Escherichia coli* K-12. J. Bacteriol. *160*, 184–191.

Tschowri, N., Busse, S., and Hengge, R. (2009). The BLUF-EAL protein YcgF acts as a direct anti-repressor in a blue-light response of *Escherichia coli*. Genes Dev. *23*, 522–534.

Van Melderen, L., and Aertsen, A. (2009). Regulation and quality control by Lon-dependent proteolysis. Res. Microbiol. *160*, 645–651.

Venecia, K., and Young, G.M. (2005). Environmental regulation and virulence attributes of the Ysa type III secretion system of *Yersinia enterocolitica* biovar 1B. Infect. Immun. *73*, 5961–5977.

Vianney, A., Jubelin, G., Renault, S., Dorel, C., Lejeune, P., and Lazzaroni, J.C. (2005). *Escherichia coli tol* and *rcs* genes participate in the complex network affecting curli synthesis. Microbiology *151*, 2487–2497.

Virlogeux, I., Waxin, H., Ecobichon, C., and Popoff, M.Y. (1995). Role of the via*B* locus in synthesis, transport and expression of *Salmonella typhi* Vi antigen. Microbiology *141*, 3039–3047.

Virlogeux, I., Waxin, H., Ecobichon, C., Lee, J.O., and Popoff, M.Y. (1996). Characterization of the *rcsA* and *rcsB* genes from *Salmonella* typhi: *rcsB* through tviA is involved in regulation of Vi antigen synthesis. J. Bacteriol. *178*, 1691–1698.

Wang, D., Korban, S.S., and Zhao, Y. (2009). The Rcs phosphorelay system is essential for pathogenicity in *Erwinia amylovora*. Mol. Plant Pathol. *10*, 277–290.

Wang, Q., Zhao, Y., McClelland, M., and Harshey, R.M. (2007). The RcsCDB signaling system and swarming motility in *Salmonella enterica* serovar Typhimurium: dual regulation of flagellar and SPI-2 virulence genes. J. Bacteriol. *189*, 8447–8457.

Wang, X., Kim, Y., and Wood, T.K. (2009). Control and benefits of CP4–57 prophage excision in *Escherichia coli* biofilms. ISME J. *3*, 1164–1179.

Wang, X., Kim, Y., Ma, Q., Hong, S.H., Pokusaeva, K., Sturino, J.M., and Wood, T.K. (2010). Cryptic prophages help bacteria cope with adverse environments. Nat. Comm. *1*, 147.

Winter, S.E., Winter, M.G., Thiennimitr, P., Gerriets, V.A., Nuccio, S.P., Russmann, H., and Bäumler, A.J. (2009). The TviA auxiliary protein renders the *Salmonella enterica* serotype Typhi RcsB regulon responsive to changes in osmolarity. Mol. Microbiol. *74*, 175–193.

Wolfe, A.J. (2005). The acetate switch. Microbiol. Mol. Biol. Rev. *69*, 12–50.

Wolfe, A.J. (2010). Physiologically relevant small phosphodonors link metabolism to signal transduction. Curr. Opin. Microbiol. *13*, 204–209.

Wolfe, A.J., Chang, D.E., Walker, J.D., Seitz-Partridge, J.E., Vidaurri, M.D., Lange, C.F., Pruss, B.M., Henk, M.C., Larkin, J.C., and Conway, T. (2003). Evidence that acetyl phosphate functions as a global signal during biofilm development. Mol. Microbiol. *48*, 977–988.

Wren, B.W. (2003). The yersiniae--a model genus to study the rapid evolution of bacterial pathogens. Nat. Rev. Microbiol. *1*, 55–64.

Wu, W.F., Zhou, Y., and Gottesman, S. (1999). Redundant *in vivo* proteolytic activities of *Escherichia coli* Lon and the ClpYQ (HslUV) protease. J. Bacteriol. *181*, 3681–3687.

Zientz, E., Dandekar, T., and Gross, R. (2004). Metabolic interdependence of obligate intracellular bacteria and their insect hosts. Microbiol. Mol. Biol. Rev. *68*, 745–770.

Chemotactic Signal Transduction in *Helicobacter pylori* 17

Paphavee Lertsethtakarn, Jenny Draper and Karen M. Ottemann

Abstract

The gastric pathogenic bacterium *Helicobacter pylori* has a relatively simple chemotaxis system, with four chemoreceptors plus a set of signal transduction proteins consisting of the CheA kinase, CheW and CheV coupling proteins, a CheY response regulator, the CheZ phosphatase and an unusual protein termed ChePep. *H. pylori* chemotaxis response has proved challenging to study, but studies have established that *H. pylori* uses its chemotaxis system to move away from low pH and autoinducer-2, and to respond to cellular energy status along with several chemicals. During stomach infection, *H. pylori* chemotaxis plays multiple roles including promoting early infection and critical epithelial interactions that drive a wild-type immune response.

Introduction

Bacterial chemotaxis is the ability to move in response to environmental cues (Wadhams and Armitage, 2004; Miller *et al.*, 2009). Bacteria use this ability to find nutrients, avoid toxins and occupy favourable niches. *Helicobacter pylori* is a bacterial pathogen that uses chemotaxis to colonize its only known niche, the stomach. *H. pylori* infection leads to gastric and duodenal ulcers in about 10% of infected people, and is a strong risk factor for gastric cancer (Wroblewski *et al.*, 2010). *H. pylori* has emerged as a model for studies on the role of chemotaxis in mammalian colonization, as well as the variations on chemotactic signal transduction.

The role of chemotaxis in stomach infection has been studied using non-chemotactic (Che^-) *H. pylori* in mouse and gerbil oral infection models. These studies have identified several functions for chemotaxis in infection. Chemotaxis is critical for initial tissue colonization, promoting efficient access to the stomach epithelial lining and rapid bacterial multiplication (Foynes *et al.*, 2000; Terry *et al.*, 2005). Chemotaxis is even more vital when Che^- *H. pylori* compete with wild-type for a place in the stomach. In this situation, Che^- mutants are ~2000-fold outcompeted by wild-type (Terry *et al.*, 2005). The observation that wild-type bacteria exacerbate the Che^- colonization defect suggests that chemotaxis may direct the microbe towards limiting nutrients or allow it to avoid immune cells. Additionally, Che^- mutants do not nestle intimately against host epithelial cells as do wild-type *H. pylori* (Williams *et al.*, 2007; Howitt *et al.*, 2011), and elicit much less gastric inflammation, which is the canonical *H. pylori* disease (Williams *et al.*, 2007). Thus chemotaxis seems to help *H. pylori* with both nutrient acquisition and proper host–cell interactions, allowing the microbe the ability to persist for the life of its mammalian host.

Chemotaxis is studied in *H. pylori* using spatial and temporal assays

Chemotactic responses can be assayed in a variety of ways (Miller *et al.*, 2009). Chemotaxis assays are either spatial in nature, in which bacteria migrate to particular areas in response to a gradient of chemoeffectors, or temporal in nature, in which flagellar switching is monitored. Several of these approaches have been used successfully for *H. pylori* and are described below. At least one assay, in contrast, gives false results for *H. pylori* and its relatives; this assay, the chemical-in-plug assay, should be used with caution (Kanungpean *et al.*, 2010; Li *et al.*, 2010).

The most widely used assay for *H. pylori* and generally, is the spatial soft-agar migration assay. This assay analyses movement of bacterial cells over time in low percentage agar plates (Wolfe and Berg, 1989; Suerbaum *et al.*, 1993; Beier *et al.*, 1997). Substrates are homogeneously distributed throughout the agar, and gradients are created by bacterial consumption of nutrients and excretion of waste products. *H. pylori* requires rich media and serum for robust growth, so this assay is typically done using brucella broth with additional fetal bovine serum. This requirement prevents analysis of single chemotactic agents, as opposed to *E. coli*, which can grow in minimal media with only one chemoattractant agent. The soft-agar assay as used thus allows one to assess general chemotactic ability, but not response to any individual chemical.

Two other spatial assays that have been used with *H. pylori* are the capillary and the Ibidi μ slide chemotaxis assays (Mizote *et al.*, 1997; Cerda *et al.*, 2003; Schweinitzer *et al.*, 2008). Both assays compare the number of bacteria that migrate into a compartment that contains an attractant versus those that enter a control chamber with plain buffer. Bacterial motility can be gleaned by examining the number of bacteria that enter the capillary in basal buffer that lacks attractants or repellents. Both published capillary assay reports use phosphate buffers that work well for *E. coli* but, in our experience, create non-motile *H. pylori*. When bacteria are non-motile, fewer than 100 bacteria will enter the capillary. If the bacteria are motile, in contrast, approximately 5000 bacteria will enter the capillary (Adler, 1973). The low numbers of *H. pylori* reported to enter the capillary even with the strongest attractant, urea, ~200, are consistent with idea that these bacteria are actually non-motile (Mizote *et al.*, 1997). A variant on the capillary assay that uses Ibidi μ-slides for chemotaxis has also been employed, using *H. pylori* in growth medium to ensure motility (Schweinitzer *et al.*, 2008). These spatial assays thus can work well but need to be used carefully to ensure that *H. pylori* retains motility in the media employed.

Microscopic filming of bacteria in liquid media has also been used to assess chemotactic response of many bacteria, including *H. pylori* (Berg and Brown, 1972; Terry *et al.*, 2006; Schweinitzer *et al.*, 2008; Lowenthal *et al.*, 2009b; Howitt *et al.*, 2011; Rader *et al.*, 2011). One strength of this method is that it does not require bacterial metabolism or growth, and can be used to monitor responses to individual components. A challenge, however, is that bacterial swimming behaviour can be complex and varied. *E. coli*, for example, combines straight swims and direction changes or tumbles. *H. pylori*, on the other hand, exhibits swims, reversals, stops and slow downs. Thus for *H. pylori* work, scientists have reported stops (Schweinitzer *et al.*, 2008; Rader *et al.*, 2011), reversals (Terry *et al.*, 2006), a combination of all behaviours that contribute to net migration (Lowenthal *et al.*, 2009b), and measurements of bacterial path linearity (Karim *et al.*, 1998; Foynes *et al.*, 2000).

H. pylori responds chemotactically to numerous chemicals and parameters

Using a variety of assays, *H. pylori* has been found to respond chemotactically to both mixtures of compounds as well as individual chemicals. Given some of the challenges of working with *H. pylori* motility mentioned above, further verification of the chemotactic response to many of these compounds is still needed. With these caveats in mind, we summarize what has been shown in Table 17.1. *H. pylori* appears to respond to several nutrients as attractants, including numerous amino acids and cholesterol (Table 17.1). *H. pylori* also responds to compounds that benefit its growth, including urea, bicarbonate, energy and pH. Urea was the first reported *H. pylori* attractant, although subsequent work showed that some *H. pylori* strains were actually repelled by urea (Table 17.1). One report showed that the *H. pylori* chemoreceptor TlpA could mediate the urea response, but whether that response was attractant or repellent depended on whether TlpA was expressed in its native *H. pylori* or heterologously in *E. coli* (Cerda

Table 17.1 Reported *H. pylori* chemoattractants and repellents

Compound	Attractant (A) or repellent (R)	Method	Receptor	Strain	Comments	Citation
Mixtures						
1% Plasma	A	Capillary	None	Clinical	–	Worku *et al.* (2004)
Hog gastric mucin	A	Capillary	None	N6	–	Foynes *et al.* (2000)
Bile	R	Capillary	None	Clinical	–	Worku *et al.* (2004)
Single compounds						
Organic chemicals						
Urea	A	Capillary	None	CPY3401	Motility not verified	Mizote *et al.* (1997)
Urea	A	Capillary	TlpA	26695	Motility not verified	Cerda *et al.* (2003)
Urea	A	Capillary	None	Clinical	–	Worku *et al.* (2004)
Urea	R	Capillary	TlpA	43504 26695	Motility not verified	Cerda *et al.* (2003)
N-(diaminophosphinyl)–4-fluorobenzamie (urease inhibitor)	A	Capillary	None	CPY3401	Motility not verified	Mizote *et al.* (1997)
Hydroxyurea (urea analogue)	A	Capillary	None	CPY3401	Motility not verified	Mizote *et al.* (1997)
Acetohydroxamic acid	A	Capillary	None	CPY3401	Motility not verified	Mizote *et al.* (1997)
Bicarbonate (Na or K)	A	Capillary	None	CPY3401	Motility not verified	Mizote *et al.* (1997)
Bicarbonate (Na or K)	A	Capillary	TlpA	43504	Motility not verified	Cerda *et al.* (2003)
Bicarbonate	R	Capillary	None	26695	Motility not verified	Cerda *et al.* (2003)
Cholesterol	A	Spatial (three tube)	None	P12	–	Wunder *et al.* (2006)
HQNQ electron transport chain block	A	Temporal (Filming)	TlpD	N6	–	Schweinitzer *et al.* (2008)
Myxothiazol electron transport chain block	A	Temporal (Filming)	TlpD	N6	–	Schweinitzer *et al.* (2008)
Specific bile acids[c]	R	Capillary	None	Clinical	–	Worku *et al.* (2004)
Amino acids						
L-Alanine[b]	A	Capillary	None	Clinical	–	Worku *et al.* (2004)
L-Arginine	A	Capillary	TlpA	43504	Motility not verified	Cerda *et al.* (2003)
L-Arginine	A	Capillary	None	Clinical	–	Worku *et al.* (2004)
L-Asparagine	A	Capillary	None	Clinical	–	Worku *et al.* (2004)

Table 17.1 Continued

Compound	Attractant (A) or repellent (R)	Method	Receptor	Strain	Comments	Citation
L-Aspartate	A	Capillary	None	43504 26695	Motility not verified	Cerda *et al.* (2003)
L-Aspartate	R	Capillary	None	Clinical	–	Worku *et al.* (2004)
L-Glutamine[a]	A	Capillary	None	Clinical	–	Worku *et al.* (2004)
Glycine[a]	A	Capillary	None	Clinical	–	Worku *et al.* (2004)
L-Histidine[b]	A	Capillary	None	Clinical	–	Worku *et al.* (2004)
L-Leucine[b]	A	Capillary	None	Clinical	–	Worku *et al.* (2004)
L-Proline	A	Capillary	None	Clinical	–	Worku *et al.* (2004)
L-Valine	A	Capillary	None	Clinical	–	Worku *et al.* (2004)
Inorganics/pH						
$NiCl_2$	R	Capillary	None	43504 26695	Motility not verified	Cerda *et al.* (2003)
Acid pH	R?	Filming, wall assay	TlpB	SS1	Not clear if behaviour is attractant or repellent	Croxen *et al.* (2006)
Acid pH	R	Filming	TlpB	G27	–	Rader *et al.* (2011)

[a]Indicates that Mizote *et al.* (1997) found these compounds to be neutral and not chemoattractants.
[b]Indicates that Cerda *et al.* (2003) found these were not chemoattractants for their two strains and additionally determined that there was no response to cysteine, glutamate, methionine, or tryptophan
[c]Specific bile acids include glycocholic acid, taurocholic acid, glycodeoxycholic acid, taurodeoxycholic acid, glychochenodeoxycholic acid, taurochenodeoxycholic acid.

et al., 2003). This finding suggests that chemotactic response might thus depend on which other chemoreceptors are present (Cerda *et al.*, 2003). *H. pylori* is repelled by the quorum sensing molecule AI-2, via the TlpB chemoreceptor (Rader *et al.*, 2011). This function may help *H. pylori* colonize new regions of the stomach, by driving it away from regions of high bacterial load. *H. pylori* also uses TlpB to sense low pH, a response that is thought to drive *H. pylori*'s localization above the gastric epithelial cells (Schreiber *et al.*, 2004; Croxen *et al.*, 2006). Interestingly, our lab recently discovered that low pH can act as either a repellent or an attractant, depending on whether strains express TlpC in addition to TlpB (L.M. Collison and K.M. Ottemann, unpublished). The implications of possible variations on acid response are not known. Similarly, the bicarbonate response appears to differ between strains (Table 17.1). *H. pylori* uses the TlpD chemoreceptor to respond to some aspect of cellular energy status that is influenced by flow through the electron transport chain (Schweinitzer *et al.*, 2008); this so-called energy taxis is a response found in many bacteria (Alexandre, 2010). Interestingly, $NiCl_2$ is a repellent for *H. pylori* like it is for *E. coli* (Cerda *et al.*, 2003; Englert *et al.*, 2010). *H. pylori* requires nickel, but apparently it is found in high enough concentrations that *H. pylori* does not need chemotaxis to track it down.

The *H. pylori* molecular chemotaxis components differ from those of *E. coli*

Chemotactic signal transduction has been extensively studied in the *E. coli* system (Wadhams and Armitage, 2004; Miller *et al.*, 2009). Chemotaxis starts when chemoreceptors detect specific environmental characteristics and transmit this information to a signal-transduction cascade that affects flagellar rotation (Fig. 17.1). The core set of signal transduction proteins are conserved across all microbes (Wuichet and Zhulin, 2010) and consists of (1) coupling proteins (CheW or CheV) that physically connect the receptors to (2) the CheA kinase, and (3) the CheY response regulator (Fig. 17.2). CheY can exist in either a phosphorylated or non-phosphorylated form. The phosphorylated form, called CheY-P, binds

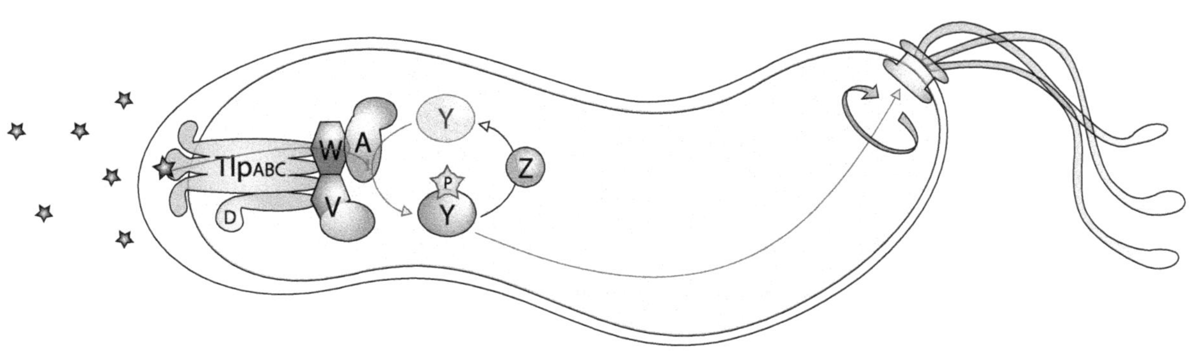

Figure 17.1 *Chemotactic signal transduction in H. pylori.* Chemotatic signalling proceeds from the chemoreceptors (TlpABCD) via coupling proteins CheW (W) or CheV (V) to the histidine kinase CheA (A). CheA phosphorylates CheY (Y) which interacts with the flagellar motor, and is dephosphorylated by the CheZ phosphatase (Z).

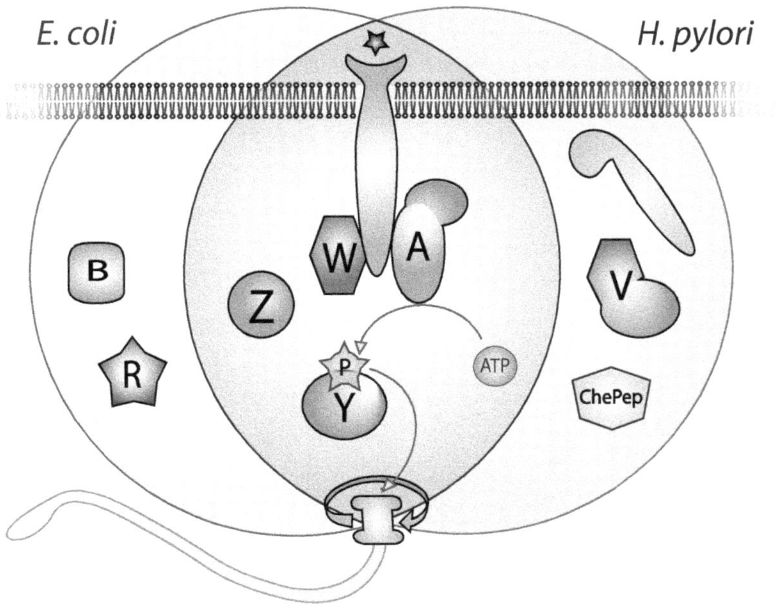

Figure 17.2 *H. pylori and* E. coli *share many but not all chemotaxis proteins.* Venn diagram showing proteins unique to *E. coli* (CheB (B) and CheR (R)) at the left, *H. pylori* at the right (CheV (V), ChePep and a soluble chemoreceptor) and common in the intersection area. Abbreviations as: B, CheB; R, CheR; Z, CheZ; W, CheW; A, CheA; Y, CheY; V, CheV. Receptors are shown without letters and as extended grey (gold in colour version) shapes.

to the flagellar motor and causes it to rotate in a clockwise orientation that promotes direction changes. The amount of CheY-P is regulated by ligand binding to the chemoreceptors. When an attractant ligand binds to the chemoreceptor, the kinase activity of CheA is squelched and non-phosphorylated CheY predominates. In this form, CheY fails to interact with the flagellar motor, the flagella rotate counter clockwise and the bacteria swim. When a chemoreceptor has no ligand, CheA is active, CheY is phosphorylated and thus interacts with the flagellar motor. With CheY-P bound, the bacteria randomly reorient. CheY has intrinsic auto-dephosphorylation activity, but most microbes have proteins, such as CheZ, that accelerate this reaction. The final chemotaxis components are the CheR and CheB enzymes that control adaptation to sustained ligand levels by reversibly methylating specific amino acids on the chemoreceptor.

H. pylori contains most of the chemotaxis proteins mentioned above, with some intriguing exceptions (Fig. 17.2). Most striking is the lack of CheR and CheB; most bacteria possess these proteins and are thought to utilize methylation-dependent adaptation (Alexander *et al.*, 2010; Wuichet and Zhulin, 2010). There are other forms of adaptation, however, including the CheC/CheD phosphatase and CheV coupling proteins (Rao *et al.*, 2008). All *Helicobacter* spp. possess at least one of these systems (Table 17.2). Another unusual aspect is that *H. pylori* has three CheV proteins while most systems have at most one (Szurmant and Ordal, 2004). Finally, *H. pylori* was recently found to have a new chemotaxis protein, ChePep, that does not exist in other chemotaxis systems (Howitt *et al.*, 2011). These proteins are described in the next sections.

H. pylori has four chemoreceptors

H. pylori's chemoreceptors are the key to defining the conditions that this microbe senses. The four *H. pylori* chemoreceptors, TlpA, TlpB, TlpC and TlpD each contribute to life in the stomach (Andermann *et al.*, 2002; Croxen *et al.*, 2006; Williams *et al.*, 2007). Each one consists of a conserved region called the 'MA domain' that interacts with the CheW or CheV coupling proteins, and a sensing domain that detects a particular ligand either directly or indirectly via interaction with a ligand-bound periplasmic binding protein (Zhulin, 2001; Wuichet *et al.*, 2007). Many chemoreceptors span the membrane and carry an extracellular signal-sensing region, but variations on this theme exist (Storch *et al.*, 1999; Hou *et al.*, 2000); for instance, *H. pylori's* TlpD lacks a transmembrane domain and is soluble (Fig. 17.3) (Schweinitzer *et al.*, 2008).

H. pylori has three integral membrane chemoreceptors, TlpA, TlpB, and TlpC (Tomb *et al.*, 1997; Alm *et al.*, 1999) that sense various conditions. TlpA has been proposed to sense arginine, urea and bicarbonate (Table 17.1) (Cerda *et al.*, 2003). TlpB was shown to mediate a response to AI-2 and to low pH, using both temporal and spatial assays (Croxen *et al.*, 2006; Rader *et al.*, 2011). Movement away from low pH is widespread amongst bacteria (Seymour and Doetsch, 1973; Kihara and Macnab, 1981), suggesting this response is not specific to the stomach niche. Nonetheless, the repellent response to low pH has been proposed to account for *H. pylori's* observed localization within the stomach (Schreiber *et al.*, 2004). No responses have been linked to TlpC.

The soluble chemoreceptor TlpD was shown to mediate a tactic response to bacterial energy levels (Schweinitzer *et al.*, 2008). Specifically, treating *H. pylori* with chemicals that disrupt electron transport cause the bacteria to swim without direction changes as if responding to an attractant. Adding back nutrients or electron donors boosted the frequency of direction switches for wild-type *H. pylori* but not for strains lacking *tlpD* suggesting that TlpD senses these conditions, although it is not yet known what this receptor specifically senses. TlpD was recently found to bear a conserved C-terminal zinc-binding domain that is shared amongst many chemoreceptors but does not yet have a defined function (Draper *et al.*, 2011).

H. pylori is thought to live only in the stomach, and so one would expect that chemoreceptors would play a role in this setting. Mutants lacking each chemoreceptor have been analysed for

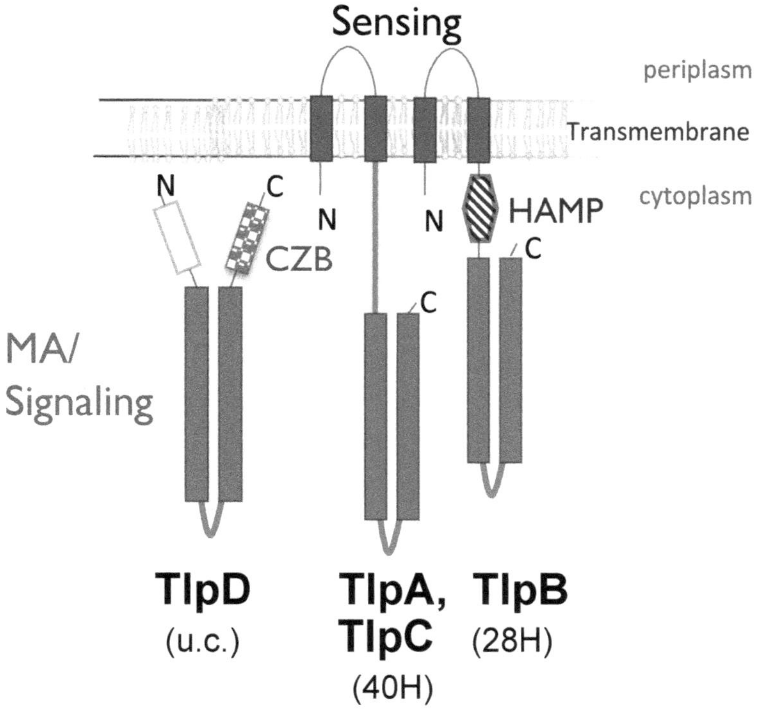

Figure 17.3 *H. pylori has four chemoreceptors.* Shown are representations of the four *H. pylori* chemoreceptors, their key domains and classification according to Alexander and Zhulin (2007). TlpD is a member of the unclassified group, is cytoplasmic but membrane associated (Schweinitzer *et al.*, 2008), and has a conserved C-terminal zinc binding domain termed CZB (Draper *et al.*, 2011). TlpA and TlpC are highly related to each other (Andermann *et al.*, 2002), and are both integral membrane proteins of the 40H class. TlpB is an integral membrane of the 28H class; it is the only *H. pylori* chemoreceptor that bears an identifiable HAMP domain.

ability to infect either mice or gerbils. *tlpA*, *tlpC* or *tlpD* mutants colonize mice normally when any is the sole infecting *H. pylori* strain. Each mutant is outcompeted, however, by wild-type *H. pylori* during co-infections confirming that each chemoreceptor is required for stomach colonization (Andermann *et al.*, 2002; Williams *et al.*, 2007; S.M. Williams and K.M. Ottemann, unpublished). The *in vivo* phenotype associated with TlpB is less clear, as two studies report that *tlpB* mutants have no colonization defect in either mice or gerbils (McGee *et al.*, 2005; Williams *et al.*, 2007), while one reports that *tlpB* mutants have a defect in single-strain colonization assays in a transgenic mouse strain that is particularly permissive for *H. pylori* infection (Croxen *et al.*, 2006). Future experiments that analyse the phenotype of Che$^-$ mutants in this model would help to understand why *tlpB* mutants might be defective under these conditions.

The proteins of the *H. pylori* chemotaxis signal transduction system

The CheW and CheV coupling proteins

Chemoreceptors are connected to the kinase via coupling or scaffold proteins. In *H. pylori*, there are two types, CheW and CheV (Figs. 17.1 and 17.2). CheW is a simple protein that consists only of a coupling domain, while CheV proteins fuse an N-terminal CheW coupling domain to a C-terminal phosphorylated REC domain (Alexander *et al.*, 2010). REC domains are found in two-component effector proteins such as CheY and are typically phosphorylated to allow phosphorylation-based control of themselves or proteins to which they are attached. CheV proteins thus differ from CheW in that some aspect of their function can be modulated by phosphorylation.

Table 17.2 Chemotaxis proteins encoded in genomes of sequenced *Helicobacter* genera

	CheAY			CheV		
		Y-like	CheW	CheV1	CheV2	CheV3
H. acinonychis str. Sheeba	Hac_0450	X	Hac_0449[a]	Hac_1689	Hac_0832	Hac_0451
H. bilis ATCC 43879	HRAG_01041	X	HRAG_01042	HRAG_00687	HRAG_01164	HRAG_01040
H. bizzozeronii CIII-1	HBZC1_03830	X	HBZC1_03840	HBZC1_03380	HBZC1_14450	HBZC1_03820
H. canadensis MIT 98–5491	HCAN_00975	X	HCAN_00974	HCAN_01097	–	HCAN_00976
H. cinaedi CCUG 18818	HCCG_00486	X	HCCG_00487	HCCG_01829	HCCG_01318	HCCG_00485
H. felis ATCC 49179	HFELIS_07460	X	HFELIS_07450	HFELIS_02360	HFELIS_04990	HFELIS_07470
H. mustelae 12198	HMU03770	X	HMU03760	HMU14170	HMU06240	–
H. hepaticus ATCC 51449	HH0672	X	HH0673	HH00819	HH1802	HH0671
H. pullorum MIT 98–5489	HPMG_00306	X	HPMG_00305	HPMG_00266	–	HPMG_00307
H. pylori 26695	HP0392	X	HP0391	HP0019	HP0616	HP0393
H. pylori 98–10	HP9810_491g14	X	HP9810_491g13	HP9810_880g44	HP9810_905g79	HP9810_491g15
H. pylori B128	HPB128_142g16	X	HPB128_142g15	HPB128_155g78	HPB128_187g42	HPB128_142g17
H. pylori B38	HELPY_1033	X	HELPY_1034	HELPY_0017	HELPY_0756	HELPY_1032
H. pylori B8	HPB8_442	X	HPB8_441	HPB8_1606	HPB8_814	HPB8_444
H. pylori G27	HPG27_1005	X	HPG27_1006	HPG27_0018	HPG27_0576	HPG27_1004
H. pylori HPAG1	HPAG1_1000	X	HPAG1_1001	HPAG1_0019	HPAG1_0597	HPAG1_0999
H. pylori HPKX_438_AG0C1	HpylHP_010100003740	X	HpylHP_010100003735	HpylHP_010100002278	HpylHP_010100006145	HpylHP_010100003745
H. pylori HPKX_438_CA4C1	HpylH_010100002807*	ND	HpylH_010100013675*	–	–	HpylH_010100004784
H. pylori J99	jhp0989	X	jhp0990	jhp0017	jhp0559	jhp0988
H. pylori P12	HPP12_1028	X	HPP12_1029	HPP12_0017	HPP12_0625	HPP12_1027
H. pylori PeCan4	HPPC_05145	X	HPPC_05150	HPPC_00095	HPPC_03105	HPPC_05140
H. pylori Shi470	HPSH_05445	X	HPSH_05445	HPSH_00100	HPSH_03790	HPSH_05440
H. pylori SJM180	HPSJM_0525	X	HPSJM_05230	HPSJM_00105	HPSJM_03120	HPSJM_0520
H. suis HS1	HSUHS1_1003	X	HSUHS1_1002	HSUHS1_0538	HSUHS1_0846	HSUHS1_1004
H. suis HS5	–	–	HSUHS5_0775	HSUHS5_0706	HSUHS5_0081	HSUHS5_0649
H. winghamensis ATCC BAA-430	HWAG_00712	X	HWAG_00713	HWAG_00724	–	HWAG_00711

Helicobacter chemotaxis proteins were identified by performing BLAST using *H. pylori* 26695 protein sequences as queries against sequenced *Helicobacter* genomes in NCBI, except in the case of CheR and CheB, in which case sequences from *H. hepaticus* were used. 'X' in 'Y-like column' indicates the presence of REC domain in

H. pylori has one CheW and three CheV proteins, but they do not appear to be functionally redundant as deletion of *cheW* renders *H. pylori* non-chemotactic (Pittman *et al.*, 2001; Terry *et al.*, 2005). CheW of *H. pylori* has been shown to have the predicted function of increasing CheA kinase activity (Schaer, 2003), and furthermore many of the known critical residues are conserved between well-studied CheW proteins and those in *H. pylori* and *C. jejuni* (Lowenthal *et al.*, 2009b; Alexander *et al.*, 2010).

H. pylori has three CheV proteins termed CheV1, CheV2, and CheV3. All three CheV proteins in *H. pylori* have been shown to have roles in chemotaxis but their exact functions have not been determined (Lowenthal *et al.*, 2009b). Deletion of *cheV1* in *H. pylori* has the most severe chemotaxis phenotype while deletion of *cheV2* or *cheV3* has only subtle chemotaxis effects (Pittman *et al.*, 2001; Lowenthal *et al.*, 2009b). Motility analysis of these mutants showed that *cheW*, *cheV1* and *cheV2* null mutants have smooth-swimming behaviour as do *cheY* or *cheA* deletions, while a *cheV3* mutant had the opposite hyper-switching behaviour (Lowenthal *et al.*, 2009b). This finding suggests that most of these proteins contribute

CheY	CheZ	FliY	ChePep	CheR	CheB
Hac_1178	Hac_0351	Hac_1134	Hac_0999	–	–
HRAG_01033	HRAG_00793	HRAG_00567	–	HRAG_00662	HRAG_00661
HBZC1_08260	HBZC1_13970	HBZC1_05640	–	–	–
HCAN_0093 and HCAN_01058	HCAN_01191	HCAN_01353	–	HCAN_00900	HCAN_00901
HCCG_01824	HCCG_01888	HCCG_00808	–	HCCG_02133	HCCG_02134
HFELIS_12550	HFELIS_03990	HFELIS_13900	HFELIS_15280[a]	–	–
HMU02170	–	HMU12769	–	–	–
HH0825	HH0659[a]	HH1148	–	HH0455	HH0456
HPMG_00762	HPMG_00114	HPMG_01449	HPMG_01476[a]	HPMG_00430	HPMG_00431
HP1067	HP0170	HP1030	HP0322	–	–
HP9810_9g80	HP9810_886g55	HP9810_9g41	HP9810_1g1[a]	–	–
HPB128_21g165	HPB128_3g49	HPB128_21g205	HPB128_21g105[a] and 106[a]	–	–
HELPY_0383	HELPY_0173	HELPY_0423	HELPY_0325	–	–
HPB8_1186	HPB8_1396	HPB8_1148	HPB8_1243	–	–
HPG27_0361	HPG27_155	HPG27_0398	HPG27_0303	–	–
HPAG1_0380	HPAG1_0166	HPAG1_0417	HPAG1_0325	–	–
HpylHP_010100007221	HpylHP_010100007346	HpylHP_010100005613	HpylHP_010100010469[a] and 06825[a]	–	–
–	HpylH_010100010449[a]	HpylH_010100010964[a]	HpylH_010100007316	–	–
jhp0358	jhp0156	jhp0394	jhp0305	–	–
HPP12_0377	HPP12_0167	HPP12_0414	HPP12_0319	–	–
HPPC_01915	HPPC_00860	HPPC_02105	HPPC_01630	–	–
HPSH_01985	HPSH_00855	HPSH_02165	HPSH_01665	–	–
HPSJM_02010	HPSJM_00915	HPSJM_2205	HPSJM_01725	–	–
HSUHS1_0299	HSUHS1_0562	HSUHS1_0609	–	–	–
HSUHS5_0250	HSUHS5_0571	HSUHS5_0182	HSUHS5_0485	–	–
HWAG_00421	HWAG_01277	HWAG_00539	–	HWAG_00743	HWAG_00742

CheA. ND indicates not determined based on sequences available. *H. pylori* 98–10 and B128 are not annotated, and *H. pylori* HPKX_438_AG0C1 and *H. pylori*_CA4C1 are not completely annotated.
[a]The identified protein is only a partial match to the *H. pylori* 26695 protein sequence.

to the formation of CheY-P, such that loss of any one results in less CheY-P. CheV3, on the other hand, appears to function to remove CheY-P, such that loss of *cheV3* creates cells with abnormally high amounts of CheY-P. This latter phenotype matches that of CheY-P phosphatase mutants, as described below.

The *H. pylori* CheA kinase possesses an added C-terminal REC domain

CheA is a histidine kinase that uses ATP to autophosphorylate and transfer a phosphoryl group to the response regulator CheY (Figs. 17.1 and 17.2). Mutants lacking *cheA* in *H. pylori* are non-chemotactic *in vitro*, confirming this protein's central role (Foynes *et al.*, 2000; Terry *et al.*, 2005). The CheA protein of *H. pylori* appears to have four of the five canonical CheA domains, but like many other CheA proteins, lacks the so-called P2 CheY binding domain (Wuichet *et al.*, 2007). The area corresponding to the missing P2 region, however, is able to bind CheY suggesting that there are as-yet-unknown sequences that determine CheY binding (Lam *et al.*, 2010). CheA in *H. pylori* has an added REC domain fused at its

C-terminus (Foynes *et al.*, 2000). CheA proteins with an added REC domain are sometimes called CheAY, and are found in many bacterial groups and all ε-*Proteobacteria* (Table 17.2). Wuichet *et al.* observed that CheA proteins that lack a well-conserved P2 domain typically are CheAY proteins, but the reason for this co-occurrence is not known (Wuichet *et al.*, 2007). The function of the CheA REC domain is not yet known for any CheAY protein.

The response regulator CheY of *H. pylori*

In the chemotaxis system, phosphorylated CheY plays the key role by binding to the flagella switch proteins, FliM and FliN (Wadhams and Armitage, 2004; Sarkar *et al.*, 2010) and in turn causing the flagellar motor to rotate in the non-default direction that drives tumbling/direction changes. As in model systems, *cheY* null mutants in *H. pylori* are non-chemotactic (Beier *et al.*, 1997) and as expected for a non-chemotactic strain, colonize animals poorly (Foynes *et al.*, 2000; McGee *et al.*, 2005; Terry *et al.*, 2005).

CheY proteins in *H. pylori* contain a predicted conserved aspartate for phosphorylation, Asp53. *H. pylori* CheY functions as expected in that it can be phosphorylated (Jimenez-Pearson *et al.*, 2005) and interacts with FliM in phosphorylation-dependent manner (Lowenthal *et al.*, 2009a). The crystal structure of a phosphorylated mimic of *H. pylori* CheY, activated with BeF_3^-, shows that it has the general structure of a typical REC domain, a $(\beta/\alpha)_5$ fold, although there were some differences noted between *H. pylori* CheY and that of others (Lam *et al.*, 2010).

H. pylori encodes two predicted CheY-P phosphatases – CheZ and FliY

CheY-P must be dephosphorylated for efficient chemotaxis. Microorganisms use various mechanisms for CheY-P dephosphorylation, including several types of phosphatases (Szurmant and Ordal, 2004). *H. pylori* encodes the CheZ and FliY phosphatases; these type of proteins dephosphorylate CheY-P by enhancing the endogenous CheY autodephosphorylation activity (Zhao *et al.*, 2002; Szurmant *et al.*, 2003). A gene encoding CheZ was not originally annotated in the genome sequence of *H. pylori* or its relatives owing to its poor conservation, but instead was identified based on a chemotaxis phenotype associated with inactivation of *cheW* (Terry *et al.*, 2006). *H. pylori* CheZ ($CheZ_{HP}$) is very distant from known CheZ proteins, and only conserves residues corresponding to the CheZ active site and CheY-P binding regions, although apparently these similarities are enough to allow it to function as a CheY-P phosphatase (Terry *et al.*, 2006; Lertsethtakarn and Ottemann, 2010). $cheZ_{HP}$ null mutants create cells that switch direction frequently, consistent with elevated CheY-P (P. Lertsethtakarn and K.M. Ottemann, unpublished).

FliY is another protein encoded by *H. pylori* that can have CheY phosphatase activity (Szurmant *et al.*, 2003). FliY is a component of the flagellar switch, along with FliM, FliG and FliN. Microbes typically encode only three flagellar switch proteins (FliM, FliG and either FliY or FliN), but the ε-*Proteobacteria* encode all four for an unknown reason (Lowenthal *et al.*, 2009a). It is not yet known whether *H. pylori* FliY has phosphatase activity. The protein lacks one hallmark of the FliY family, a CheY-P binding domain (Lowenthal *et al.*, 2009a), but retains the other hallmark, amino acids that match the weakly conserved active site consensus sequence of D/S-X_3-E-X_2-N-X_{22}-P (Muff and Ordal, 2008; P. Lertsethtakarn and K.M. Ottemann, unpublished). Analysis of *fliY* mutants for a phosphatase phenotype is complicated by the fact that *H. pylori fliY* mutants are non-motile, so chemotaxis phenotypes cannot be studied (Lowenthal *et al.*, 2009a).

Unique *H. pylori* chemotaxis proteins: ChePep

Recently, Howitt *et al.* (2011) described a new protein that appears to play a role in *H. pylori* chemotaxis. This protein, termed ChePep, is not found in other chemotaxis systems outside of the ε-*Proteobacteria*, and thus represents the first unique chemotaxis protein to this group of microbes. The ChePep protein bears an N-terminal REC domain and a C-terminal domain that consists of a preponderance of glutamic acids (Howitt *et al.*, 2011). Null mutations of *chePep* result in bacteria that switch direction frequently, and even swim backwards a significant percentage of the time, suggesting the protein acts to normally decrease the amount of CheY-P similarly to the CheZ phosphatase (Howitt *et al.*, 2011).

Phosphate flow in the *H. pylori* chemotaxis system

The chemotaxis system controls the CheA kinase that in turn phosphorylates target REC domain proteins. *H. pylori* has multiple chemotaxis proteins that bear REC domains: CheY, CheV1, CheV2, CheV3, ChePep and CheA. Two studies have demonstrated that CheA phosphorylates CheY and furthermore that CheY is the preferred target over the CheA REC or any CheV (Jimenez-Pearson *et al.*, 2005; Lertsethtakarn and Ottemann, 2010). Several experiments have shown that the *H. pylori* CheA REC domain can be phosphorylated by CheA (Jimenez-Pearson *et al.*, 2005). Additionally, CheY can retrophosphorylate CheA and the phosphoryl group is transferred to the CheA REC domain (Jimenez-Pearson *et al.*, 2005). One possible function of the CheA REC domain is to remove phosphoryl groups from CheY, as a so-called phosphate sink, although such a model has not been proven and is inconsistent with the straight-swim biased swimming behaviour associated with loss of only the CheA REC domain (Foynes *et al.*, 2000).

Whether CheA phosphorylates any of the CheV proteins is not entirely clear. No stable phosphorylated CheV proteins have been detected (Pittman *et al.*, 2001; Jimenez-Pearson *et al.*, 2005; Lertsethtakarn and Ottemann, 2010), although all of the *H. pylori* CheV proteins catalyse the removal of phosphate from CheA (Jimenez-Pearson *et al.*, 2005). This observation suggests that either CheV-phosphate has a very short half life or that CheV is a poor CheA substrate. ChePep has not yet been tested for phosphorylation.

REC domain phosphorylation is terminated either by autodephosphorylation, or via one of the phosphatases. Likely all of the *H. pylori* REC proteins autodephosphorylate, as has been shown for *H. pylori* CheY, CheA, and each CheV (Jimenez-Pearson *et al.*, 2005; Lertsethtakarn and Ottemann, 2010). Three REC-domain proteins are additionally targeted by $CheZ_{HP}$: CheY, CheV2 and the CheA REC domain (Lertsethtakarn and Ottemann, 2010). $CheZ_{HP}$ is thus the first CheZ family member known to have substrates beyond CheY, which suggests it can affect phosphate flow at numerous levels.

Is there adaptation in *H. pylori?*

Adaptation during chemotaxis allows bacteria to respond to a wide range of chemical concentrations that would otherwise saturate the system. Chemotaxis adaptation is typically performed by the CheB methylesterase and the CheR methyltransferase; however, the genes encoding these proteins are completely missing from *H. pylori* (Table 17.2), suggesting that this bacterium relies on other proteins that function in adaptation (Rao *et al.*, 2008). The only known adaptation proteins present in *H. pylori* are the CheV proteins. It is not yet known whether *H. pylori* adapts to sustained chemotactic ligands, however analysis of some chemotaxis mutants suggests that there is adaptation in this microbe. Specifically, mutants lacking *cheV1* or *cheV2* swim with very few direction changes, similar to the behaviour of *cheY*, *cheW* or *cheA* mutants (Lowenthal *et al.*, 2009b). *cheV1* or *cheV2* mutants, however, are able to migrate well through soft agar (Lowenthal *et al.*, 2009b); if they retained their no-direction change behaviour, they would not migrate efficiently as is seen for *cheY*, *cheA* or *cheW* mutants (Beier *et al.*, 1997; Pittman *et al.*, 2001; McGee *et al.*, 2005). This observation suggests that these *cheV* mutants can change their behaviour under some conditions, consistent with adaptation (Lowenthal *et al.*, 2009b).

Transcriptional control of the *H. pylori* chemotaxis protein genes

One interesting aspect of the *H. pylori* chemotaxis system genes is that they are generally not spatially linked on the chromosome (Fig. 17.4). Each gene is part of an operon, but they are not all part of the same operon. This organization is typical for *H. pylori* but very different from other model microbes that often place their chemotaxis

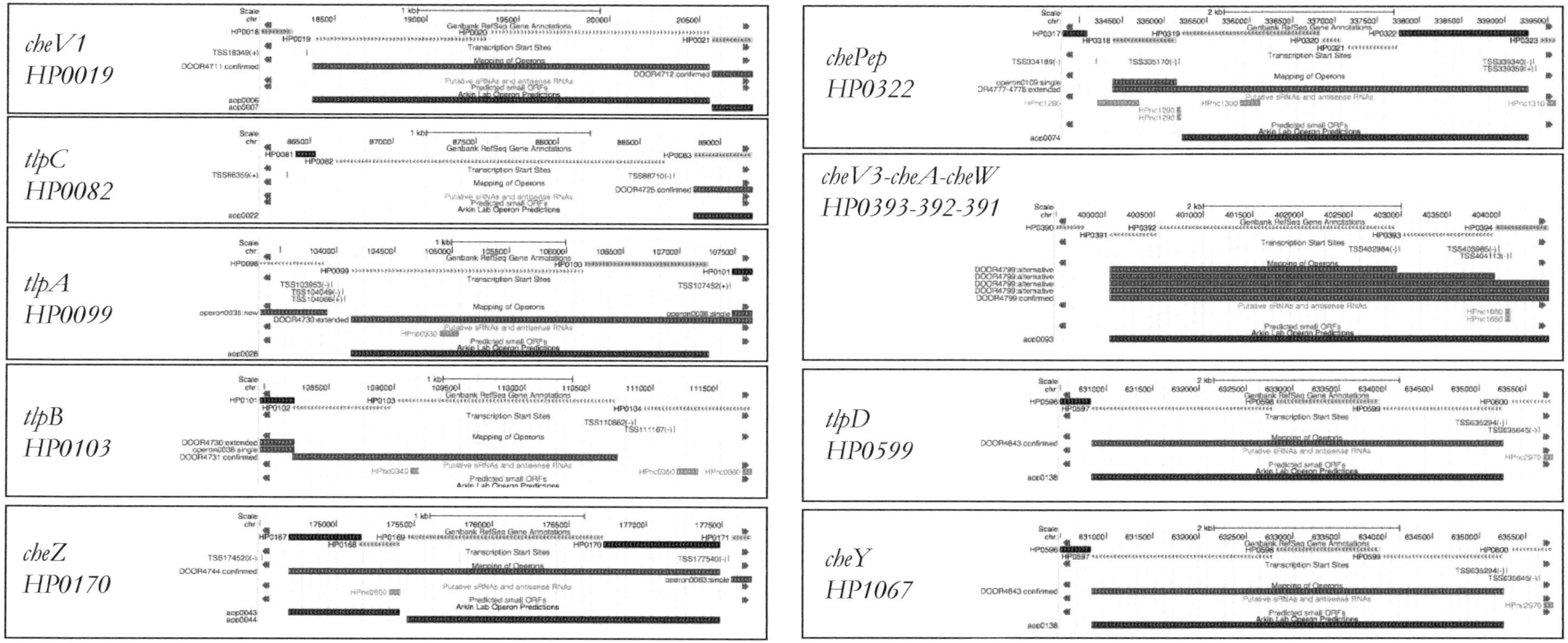

Figure 17.4 *Chemotaxis genes are organized into many operons in H. pylori*. Shown are images from the *H. pylori* genome browser for strain 26695 at http://hpylori.ucsc.edu/. Genes or sets of genes are indicated to the left of each image, with common name and 26695 gene number. In each image, the top line shows the genome sequence position, and then immediately below are the relevant open reading frames indicated as bars with adjacent numbers (e.g. HP0019) and arrows inside to indicate orientation. Below that is indicated (1) the mapped transcription start sites as a vertical bar, from Sharma *et al*. (2010); (2) DOOR (Database of Prokaryotic Operons; Mao *et al*., 2009) predicted operons as described in Sharma *et al*. (2010); (3) Any small RNAs or small ORFs detected in Sharma *et al*. (2010); (4) operon predictions from Microbes Online http://www.microbesonline.org/(Dehal *et al*., 2009).

genes in the same operon. Nearly all of the *H. pylori* chemotaxis genes appear to be under the control of the housekeeping sigma factor, sigma 70, with the exception of *cheV1* which is controlled by the flagellar sigma, sigma 28 (Niehus *et al.*, 2004; Sharma *et al.*, 2010). Additionally, some proteins are regulated by other transcriptional regulators or conditions. One chemotaxis gene, *cheV2* is under the control of Fur and was found to be repressed by iron-free Fur (Ernst *et al.*, 2005). *cheY* is regulated by copper, but the specific regulatory protein is not known (Beier *et al.*, 1997). No regulator or critical conditions have been reported for *cheZ*, or the *cheV3-A-W* operon.

Of the chemoreceptors, only *tlpB* has identified regulation. *tlpB* is regulated by the essential response regulator HP1043 based on the observation that the HP1043 protein binds to the *tlpB* promoter region upstream of the –10 box. There are only two genes in the HP1043 regulon, *tlpB* and *hp1043* (Delany *et al.*, 2002); the cellular role of HP1043 is unclear. *tlpB* is also regulated by a small RNA, such that the small RNA decreases the levels of TlpB (Sharma *et al.*, 2010). *tlpB* expression is down-regulated in stationary phase (Delany *et al.*, 2002) and by iron starvation (Merrell *et al.*, 2003).

Chemotaxis systems of other *Helicobacter* spp.

This review has focused on *H. pylori*, but all members of the *Helicobacter* genus are motile and possess genome sequences that suggest they are chemotactic. However, there have been no studies of chemotaxis in these other *Helicobacter* spp. All *Helicobacter* spp. possess the core chemotaxis proteins: CheA (called CheAY to indicate it has a REC domain), CheW and CheY (Table 17.2). Nearly all contain three CheV proteins, although *H. canadensis* MIT 98-5491 and *H. pullorum* seem to contain only two (Table 17.2). *Campylobacter jejuni* contains only one CheV protein (Marchant *et al.*, 2002), suggesting possibly that *Helicobacter* spp. have successively gained *cheV* genes throughout evolution. CheR and CheB are not found in any sequenced gastric *Helicobacter* spp., including *H. pylori, H. acinonychis, H. bizzozeronii, H. felis, H. mustelae* and *H. suis*, but are retained in enterohepatic species as well as in *C. jejuni* (Solnick and Vandamme, 2001; Marchant *et al.*, 2002), suggesting that this form of adaptation is not needed for gastric colonization. A few *Helicobacter* spp. appear to lack ChePep, which is somewhat surprising since it was found in other very diverse *ε-Proteobacteria* including *C. jejuni* and *Caminibacter mediatlanticus* (Howitt *et al.*, 2011); one possibility is that it is poorly conserved and thus difficult to identify in these species.

The number of chemoreceptors found in *Helicobacter* genus members varies greatly, from two in a few *H. pylori* strains, to a high of 29 in *H. felis* (Table 17.3). In most cases these numbers have not been verified experimentally. This variation in number may reflect the diversity of signal inputs into each chemotaxis system, but this issue remains to be addressed experimentally.

Conclusions

H. pylori uses its chemotaxis ability during stomach colonization, and also to promote gastric interactions that drive disease (Foynes *et al.*, 2000; McGee *et al.*, 2005; Terry *et al.*, 2005; Williams *et al.*, 2007; Lowenthal *et al.*, 2009b; Howitt *et al.*, 2011). Great progress has been made on studies of the *H. pylori* chemotaxis system proteins in the past ~15 years, since the first reported chemotaxis gene in 1997 (Beier *et al.*, 1997). We now know many players, with two unexpected chemotaxis proteins found after genome annotation owing to poor conservation (Terry *et al.*, 2006) or unappreciated roles (Howitt *et al.*, 2011). Additional studies are needed to discern how several atypical aspects of the *H. pylori* chemotaxis system operate, including adaptation in the context of only CheV proteins, as well as phosphate flow within the system. Studies on the chemoreceptors have recently born fruit, with specific roles in sensing pH, energy and AI-2 being mapped to particular chemoreceptors (Croxen *et al.*, 2006; Schweinitzer *et al.*, 2008; Rader *et al.*, 2011). Future work will undoubtedly identify additional chemotaxis ligands or conditions, and further our understanding of how chemotaxis is used in disease.

Table 17.3 Annotated chemoreceptors in *Helicobacter* spp. genomes. Chemoreceptors were identified by a BLAST search for homologues of the *H. pylori* 26695 chemoreceptors (TlpA, TlpB, TlpC), combined with a search of the 'Protein Table' of each sequenced genome in NCBI using the keywords 'chemotaxis', 'tlp', and 'MCP' to identify additional less-homologous chemoreceptors. Soluble chemoreceptors are included in the total number

	Number of annotated chemoreceptors
H. acinonychis str. Sheeba	4
H. bilis ATCC 43879	20
H. bizzozeronii CIII-1	23
H. canadensis MIT 98–5491	9
H. cinaedi CCUG 18818	14
H. felis ATCC 49179	29
H. mustelae 12198	3
H. hepaticus ATCC 51449	10
H. pullorum MIT 98-5489	8
H. pylori 26695	4
H. pylori 98-10	2
H. pylori B128	2
H. pylori B38	4
H. pylori B8	4
H. pylori G27	3
H. pylori HPAG1	4
H. pylori HPKX_438_AG0C1	10
H. pylori HPKX_438_CA4C1	13
H. pylori J99	4
H. pylori P12	4
H. pylori PeCan4	4
H. pylori Shi470	4
H. pylori SJM180	4
H. suis HS1	8
H. suis HS5	7
H. winghamensis ATCC BAA-430	12

References

Adler, J. (1973). A method for measuring chemotaxis and use of the method to determine optimum conditions for chemotaxis by *Escherichia coli*. J. Gen. Microbiol. *74*, 77–91.

Alexander, R.P., and Zhulin, I.B. (2007). Evolutionary genomics reveals conserved structural determinants of signaling and adaptation in microbial chemoreceptors. Proc. Natl. Acad. Sci. U.S.A. *104*, 2885–2890.

Alexander, R.P., Lowenthal, A.C., Harshey, R.M., and Ottemann, K.M. (2010). CheV: CheW-like coupling proteins at the core of the chemotaxis signaling network. Trends Microbiol. *18*, 494–503.

Alexandre, G. (2010). Coupling metabolism and chemotaxis-dependent behaviours by energy taxis receptors. Microbiology *156*, 2283–2293.

Alm, R.A., Ling, L.S., Moir, D.T., King, B.L., Brown, E.D., Doig, P.C., Smith, D.R., Noonan, B., Guild, B.C., Dejonge, B.L., *et al.* (1999). Genomic-sequence comparison of two unrelated isolates of the human gastric pathogen *Helicobacter pylori*. Nature *397*, 176–180.

Andermann, T.M., Chen, Y.-T., and Ottemann, K.M. (2002). Two predicted chemoreceptors promote *Helicobacter pylori* infection. Infect. Immun. *70*, 5877–5881.

Beier, D., Spohn, G., Rappuoli, R., and Scarlato, V. (1997). Identification and characterization of an operon of *Helicobacter pylori* that is involved in motility and stress adaptation. J. Bacteriol. *179*, 4676–4683.

Berg, H.C., and Brown, D.A. (1972). Chemotaxis in *Escherichia coli* analysed by three-dimensional tracking. Nature *239*, 500–504.

Cerda, O., Rivas, A., and Toldeo, H. (2003). *Helicobacter pylori* strain ATCC700392 encodes a methyl-accepting chemotaxis receptor protein (MCP) for arginine and sodium bicarbonate. FEMS Microbiol. Lett. *224*, 175–181.

Croxen, M.A., Sisson, G., Melano, R., and Hoffman, P.S. (2006). The *Helicobacter pylori* chemotaxis receptor TlpB (HP0103) is required for pH taxis and for colonization of the gastric mucosa. J. Bacteriol. *188*, 2656–2665.

Dehal, P.S., Joachimiak, M.P., Price, M.N., Bates, J.T., Baumohl, J.K., Chivian, D., Friedland, G.D., Huang, K.H., Keller, K., Novichkov, P.S., *et al.* (2009). MicrobesOnline: an integrated portal for comparative and functional genomics. Nucleic Acids Res. *38*, D396–D400.

Delany, I., Spohn, G., Rappuoli, R., and Scarlato, V. (2002). Growth phase-dependent regulation of target gene promoters for binding of the essential orphan response regulator HP1043 of *Helicobacter pylori*. J. Bacteriol. *184*, 4800–4810.

Draper, J., Karplus, K., and Ottemann, K.M. (2011). Identification of a chemoreceptor zinc-binding domain common to cytoplasmic bacterial chemoreceptors. J. Bacteriol. *193*, 4338–4345.

Englert, D.L., Adase, C.A., Jayaraman, A., and Manson, M.D. (2010). Repellent taxis in response to nickel ion requires neither Ni^{2+} transport nor the periplasmic NikA binding protein. J. Bacteriol. *192*, 2633–2637.

Ernst, F.D., Bereswill, S., Waidner, B., Stoof, J., Mäder, U., Kusters, J.G., Kuipers, E.J., Kist, M., van Vliet, A.H.M., and Homuth, G. (2005). Transcriptional profiling of *Helicobacter pylori* Fur- and iron-regulated gene expression. Microbiology *151*, 533–546.

Foynes, S., Dorrell, N., Ward, S.J., Stabler, R.A., McColm, A.A., Rycroft, A.N., and Wren, B.W. (2000). *Helicobacter pylori* possesses two CheY response regulators and a histidine kinase sensor, CheA, which are essential for chemotaxis and colonization of the gastric mucosa. Infect. Immun. *68*, 2016–2023.

Hou, S., Larsen, R.W., Boudko, D., Riley, C.W., Karatan, E., Zimmer, M., Ordal, G.W., and Alam, M. (2000). Myoglobin-like aerotaxis transducers in Archaea and Bacteria. Nature *403*, 540–544.

Howitt, M.R., Lee, J.Y., Lertsethtakarn, P., Vogelmann, R., Joubert, L.M., Ottemann, K.M., and Amieva, M.R. (2011). ChePep controls *Helicobacter pylori* infection of the gastric glands and chemotaxis in the Epsilonproteobacteria. mBio *2*, e00098–11.

Jimenez-Pearson, M.A., Delany, I., Scarlato, V., and Beier, D. (2005). Phosphate flow in the chemotactic response system of *Helicobacter pylori*. Microbiology *151*, 3299–3311.

Kanungpean, D., Kakuda, T., and Takai, S. (2010). False positive responses of *Campylobacter jejuni* when using the chemical-in-plug chemotaxis assay. V. Med. Sci. *73*, 389–391.

Karim, Q.N., Logan, R.P., Puels, J., Karnholz, A., and Worku, M.L. (1998). Measurement of motility of *Helicobacter pylori*, *Campylobacter jejuni*, and *Escherichia coli* by real time computer tracking using the Hobson BacTracker. J. Clin. Pathol. *51*, 623–628.

Kihara, M., and Macnab, R.M. (1981). Cytoplasmic pH mediates pH taxis and weak-acid repellent taxis of bacteria. J. Bacteriol. *145*, 1209–1221.

Lam, K.H., Ling, T.K., and Au, S.W. (2010). Crystal structure of activated CheY1 from *Helicobacter pylori*. J. Bacteriol. *192*, 2324–2334.

Lertsethtakarn, P., and Ottemann, K.M. (2010). A remote CheZ orthologue retains phosphatase function. Mol. Microbiol. 77, 225–235.

Li, J., Go, A.C., Ward, M.J., and Ottemann, K.M. (2010). The chemical-in-plug bacterial chemotaxis assay is prone to false positive responses. BMC Res. Notes 3, 77.

Lowenthal, A.C., Hill, M., Sycuro, L.K., Mehmood, K., Salama, N.R., and Ottemann, K.M. (2009a). Functional analysis of the *Helicobacter pylori* flagellar switch proteins. J. Bacteriol. *191*, 7147–7156.

Lowenthal, A.C., Simon, C., Fair, A.S., Mehmood, K., Terry, K., Anastasia, S., and Ottemann, K.M. (2009b). A fixed-time diffusion analysis method determines that the three *cheV* genes of *Helicobacter pylori* differentially affect motility. Microbiology *155*, 1181–1191.

Mao, F., Dam, P., Chou, J., Olman, V., and Xu, Y. (2009). DOOR: a database for prokaryotic operons. Nucleic Acids Res. *37*, D459–D463.

McGee, D.J., Langford, M.L., Watson, E.L., Carter, J.E., Chen, Y.-T., and Ottemann, K.M. (2005). Colonization and inflammation deficiencies in Mongolian gerbils infected by *Helicobacter pylori* chemotaxis mutants. Infect. Immun. *73*, 1820–1827.

Marchant, J., Wren, B., and Ketley, J. (2002). Exploiting genome sequence: predictions for mechanisms of *Campylobacter* chemotaxis. Trends Microbiol. *10*, 155–159.

Merrell, D.S., Thompson, L.J., Kim, C.C., Mitchell, H., Tompkins, L.S., Lee, A., and Falkow, S. (2003). Growth phase-dependent response of *Helicobacter pylori* to iron starvation. Infect. Immun. *71*, 6510–6525.

Miller, L.D., Russell, M.H., and Alexandre, G. (2009). Diversity in bacterial chemotactic responses and niche adaptation. Adv. Appl. Microbiol. *66*, 53–75.

Mizote, T., Yoshiyama, H., and Nakazawa, T. (1997). Urease-independent chemotactic responses of *Helicobacter pylori* to urea, urease inhibitors, and sodium bicarbonate. Infect. Immun. *65*, 1519–1521.

Muff, T.J., and Ordal, G.W. (2008). The diverse CheC-type phosphatases: chemotaxis and beyond. Mol. Microbiol. *70*, 1054–1061.

Niehus, E., Gressmann, H., Ye, F., Schlapbach, R., Dehio, M., Dehio, C., Stack, A., Meyer, T.F., Suerbaum, S., and Josenhans, C. (2004). Genome-wide analysis of transcriptional hierarchy and feedback regulation in the flagellar system of *Helicobacter pylori*. Mol. Microbiol. *52*, 947–961.

Pittman, M.S., Goodwin, M., and Kelly, D.J. (2001). Chemotaxis in the human gastric pathogen *Helicobacter pylori*: different roles for CheW and the three CheV paralogues, and evidence for CheV2 phosphorylation. Microbiology *147*, 2493–2504.

Rader, B.A., Wreden, C., Hicks, K.G., Sweeney, E.G. Ottemann, K.M., and Guillemin, K. (2011). *Helicobacter pylori* perceives the quorum-sensing molecule AI-2 as a chemorepellant via the chemoreceptor TlpB. Microbiology *157*, 2445–2455.

Rao, C.V., Glekas, G.D., and Ordal, G.W. (2008). The three adaptation systems of *Bacillus subtilis* chemotaxis. Trends Microbiol. *16*, 480–487.

Sarkar, M.K., Paul, K., and Blair, D. (2010). Chemotaxis signaling protein CheY binds to the rotor protein FliN to control the direction of flagellar rotation in *Escherichia coli*. Proc. Natl. Acad. Sci. U.S.A. *107*, 9370–9375.

Schaer, A. (2003). Studies on the Chemotaxis Network in *Halobacterium salinarum* and *Helicobacter pylori*. Dissertation. University of Munich, Munich, Germany.

Schreiber, S., Konradt, M., Groll, C., Scheid, P., Hanauer, G., Werling, H.O., Josenhans, C., and Suerbaum, S. (2004). The spatial orientation of *Helicobacter pylori* in the gastric mucus. Proc. Natl. Acad. Sci. U.S.A. *101*, 5024–5029.

Schweinitzer, T., Mizote, T., Ishikawa, N., Dudnik, A., Inatsu, S., Schreiber, S., Suerbaum, S., Aizawa, S.I., and Josenhans, C. (2008). Functional characterization and mutagenesis of the proposed behavioral sensor TlpD of *Helicobacter pylori*. J. Bacteriol. *190*, 3244–3255.

Seymour, F.W.K., and Doetsch, R.N. (1973). Chemotactic responses by motile bacteria. J. Gen. Microbiol. *78*, 287–296.

Sharma, C.M., Hoffmann, S., Darfeuille, F., Reignier, J., Findeiss, S., Sittka, A., Chabas, S., Reiche, K., Hackermuller, J., Reinhardt, R., *et al.* (2010). The primary transcriptome of the major human pathogen *Helicobacter pylori*. Nature *464*, 250–255.

Solnick, J.V., and Vandamme, P. (2001). Taxonomy of the *Helicobacter* genus. In *Helicobacter pylori*: Physiology and Genetics, Mobley, H.L.T., Mendz, G.L., and Hazell, S.L., eds (ASM Press, Washington, DC), pp. 39–51.

Storch, K.F., Rudolph, J., and Oesterhelt, D. (1999). Car: a cytoplasmic sensor responsible for arginine chemotaxis in the archaeon *Halobacterium salinarum*. Embo J. *18*, 1146–1158.

Suerbaum, S., Josenhans, C., and Labigne, A. (1993). Cloning and genetic characterization of the *Helicobacter pylori* and *Helicobacter mustelae flaB* flagellin genes and construction of *H. pylori flaA*- and *flaB*-negative mutants by electroporation-mediated allelic exchange. J. Bacteriol. *175*, 3278–3288.

Szurmant, H., and Ordal, G.W. (2004). Diversity in chemotaxis mechanisms among the bacteria and archaea. Microbiol. Mol. Biol. Rev. *68*, 301–319.

Szurmant, H., Bunn, M.W., Cannistraro, V.J., and Ordal, G.W. (2003). *Bacillus subtilis* hydrolyzes CheY-P at the location of its action, the flagellar switch. J. Biol. Chem. *278*, 48611–48616.

Terry, K., Williams, S.M., Connolly, L., and Ottemann, K.M. (2005). Chemotaxis plays multiple roles in *Helicobacter pylori* animal infection. Infect. Immun. *73*, 803–811.

Terry, K., Go, A.C., and Ottemann, K.M. (2006). Proteomic mapping of a suppressor of nonchemotactic *cheW* mutants reveals that *Helicobacter pylori* contains a new chemotaxis protein. Mol. Microbiol. *61*, 871–882.

Tomb, J.-F., White, O., Kerlavage, A.R., Clayton, R.A., Sutton, G.G., Fleischmann, R.D., Ketchum, K.A., Klenk, H.P., Gill, S., Dougherty, B.A., *et al.* (1997). The complete genome sequence of the gastric pathogen *Helicobacter pylori*. Nature *388*, 539–547.

Wadhams, G.H., and Armitage, J.P. (2004). Making sense of it all: bacterial chemotaxis. Nat. Rev. Mol. Cell. Biol. *5*, 1024–1037.

Williams, S.M., Chen, Y.T., Andermann, T.M., Carter, J.E., McGee, D.J., and Ottemann, K.M. (2007). *Helicobacter pylori* chemotaxis modulates inflammation and bacterium-gastric epithelium interactions in infected mice. Infect. Immun. *75*, 3747–3757.

Wolfe, A.J., and Berg, H.C. (1989). Migration of bacteria in semisolid agar. Proc. Natl. Acad. Sci. U.S.A. *86*, 6973–6977.

Worku, M.L., Karim, Q.N., Spencer, J., and Sidebotham, R.L. (2004). Chemotactic response of *Helicobacter pylori* to human plasma and bile. J. Med. Microbiol. *53*, 807–811.

Wroblewski, L.E., Peek, R.M., and Wilson, K.T. (2010). *Helicobacter pylori* and gastric cancer: factors that modulate disease risk. Clin. Microbiol. Rev. *23*, 713–739.

Wuichet, K., and Zhulin, I.B. (2010). Origins and diversification of a complex signal transduction system in prokaryotes. Sci. Signal. *3*, ra50.

Wuichet, K., Alexander, R.P., and Zhulin, I.B. (2007). Comparative genomic and protein sequence analysis of a complex system controlling bacterial chemotaxis. Methods Enzymol. *422*, 1–31.

Wunder, C., Churin, Y., Winau, F., Warnecke, D., Vieth, M., Lindner, B., Zähringer, U., Mollenkopf, H.-J., Heinz, E., and Meyer, T.F. (2006). Cholesterol glucosylation promotes immune evasion by *Helicobacter pylori*. Nat. Med. *12*, 1030–1038.

Zhao, R., Collins, E.J., Bourret, R.B., and Silversmith, R.E. (2002). Structure and catalytic mechanism of the *E. coli* chemotaxis phosphatase CheZ. Nat. Struct. Biol. *9*, 570–575.

Zhulin, I.B. (2001). The superfamily of chemotaxis transducers: from physiology to genomics and back. Adv. Microb. Physiol. *45*, 157–198.

Two-component Regulators in the *Vibrio fischeri–Euprymna scolopes* Symbiosis

18

Valerie A. Ray and Karen L. Visick

Abstract

The symbiotic relationship between the marine bioluminescent bacterium *Vibrio fischeri* and its host, the Hawaiian bobtail squid *Euprymna scolopes*, depends upon the ability of the two partners to sense and respond to each other. *V. fischeri* colonizes a specialized squid organ called the light organ in three general stages: initiation, accommodation, and persistence. To respond to the different environments encountered during these stages of colonization, *V. fischeri* utilizes specialized two-component signal transduction systems to regulate processes such as biofilm formation, motility and chemotaxis, and luminescence. In this chapter, we discuss in detail the two-component systems that regulate these processes and how they impact successful colonization of the squid host.

Introduction

Bacteria are ubiquitous in nature and can be found free-living or associated with a host. One subset of associations of bacteria with hosts leads to disease and sometimes death. Another important subset of associations involves beneficial, or symbiotic, bacteria–host interactions. Pathogenic and symbiotic relationships share certain host association characteristics, and thus studies of symbiosis can inform our thinking about pathogenic associations. One simple, yet elegant model system used to study bacteria–host interactions is the symbiotic relationship between the marine bioluminescent bacterium *Vibrio fischeri* and the Hawaiian bobtail squid *Euprymna scolopes* (reviewed in Ruby, 1996). Some advantages of this model system include the following: (1) the symbiosis is highly specific: *V. fischeri* is the only bacterium capable of colonizing the squid in a specialized squid organ called the light organ (Fig. 18.1), and thus this one-to-one relationship can be probed in detail; (2) the symbiosis can be manipulated: it initiates anew with each generation of squid and both partners can be maintained separately, and then placed together to evaluate bacteria–host interactions; thus, the process by which the two organisms recognize and respond to one another can be readily assessed; (3) colonization can be monitored non-invasively by measuring bioluminescence, a natural product of this symbiosis; only productive colonization results in high levels of bioluminescence; and (4) the stages of colonization (Fig. 18.2) parallel those that occur in pathogenic associations, permitting an understanding of the commonalities and differences between pathogenic and symbiotic associations (Ruby, 1996; Stabb, 2006). Such studies will promote our ability to generate more specific therapies that will target detrimental pathogenic bacteria over the beneficial symbiotic bacteria generally present in the host.

The *V. fischeri–E. scolopes* symbiosis

Studies on the symbiosis between *V. fischeri* and *E. scolopes* have provided a wealth of information about the process of colonization and the environments experienced by the bacteria during colonization (Ruby, 1996; Nyholm and McFall-Ngai, 2004; Stabb, 2006). These studies have identified three main stages of colonization: initiation, accommodation, and persistence

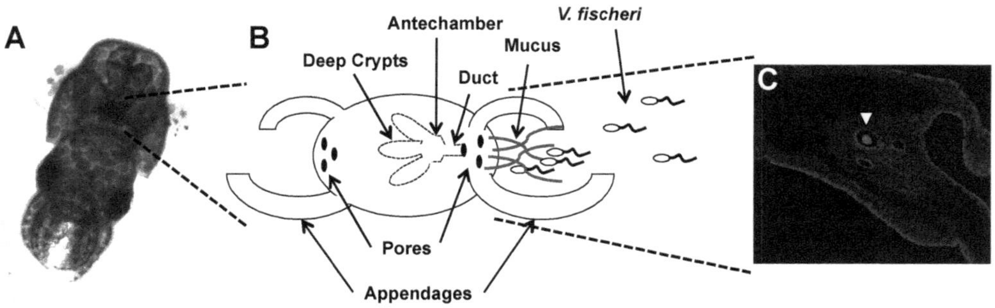

Figure 18.1 The light organ of juvenile *E. scolopes*. (A) Juvenile *E. scolopes*. The light organ can be seen as a dark mass in the centre of the body (mantle cavity) between the dotted lines. (B) A cartoon depiction of the light organ. This organ is bi-lobed, with a set of three pores located on each side at the base of two ciliated epithelial appendages. Each pore connects to a deep crypt space via a duct and antechamber (only one is represented for simplicity). *V. fischeri* cells are represented as white ovals with a curved line representing flagella. The bacteria aggregate outside of the light organ in the squid-secreted mucus (lines extending from the light organ surface). After aggregation, these cells migrate to the pores, through the ducts and the antechambers to the deep crypt spaces where they grow to high cell densities and bioluminesce. (C) A confocal microscopy image from (Yip *et al*., 2006), showing one side of the squid light organ with both appendages on the right side of the picture (top and bottom). The three pores can be seen in the middle of the image, with an aggregate of *V. fischeri* located above one of the pores (indicated by the white triangle).

(Fig. 18.2A) (Ruby, 1996). During the initiation stage, newly hatched juvenile squid (Fig. 18.1A) obtain their symbiotic partner from the surrounding environment. Seawater containing *V. fischeri* and other bacteria is continuously flushed through the body cavity and, as a result, across the light organ surface. The presence of bacteria and bacterial products induces the squid to secrete mucus to the light organ surface (Fig. 18.1B) (Nyholm *et al.*, 2000). Within about 3 hours after exposure of squid to *V. fischeri*, the bacteria form biofilm-like aggregates in the secreted mucus (Figs. 18.1C and 18.2B) (Nyholm *et al.*, 2000, 2002). The ability of *V. fischeri* to aggregate in the mucus is critical for colonization: mutants that cannot aggregate exhibit severe defects in colonizing the light organ (Yip *et al.*, 2006; Morris *et al.*, 2011). Within a few hours, *V. fischeri* cells transition from a biofilm-like state to a planktonic state and migrate towards the light organ pores (three on each side of the organ, six total) (Figs. 18.1B and 18.2B) (Nyholm *et al.*, 2000). This migration may be due to some chemotactic gradient associated with the pores. *V. fischeri* can sense and swim towards numerous attractants, including *N*-acetylneuraminic acid (a component of the mucus) and nucleosides (DeLoney-Marino *et al.*, 2003). Thus, these components, or some other attractant, may serve as a signal for entry into the light organ.

Once *V. fischeri* cells enter the light organ, they migrate through ducts and antechambers (Fig. 18.1B), both of which are non-permissive to colonization, to the deep crypt spaces where colonization and growth occur. The ducts and antechambers contain high concentrations of nitric oxide (NO) and other antimicrobial compounds, which *V. fischeri* is able to sense and respond to accordingly (Davidson *et al.*, 2004; Dunn *et al.*, 2010; Wang *et al.*, 2010a,b). Furthermore, the light organ is patrolled by macrophage-like immune cells, known as haemocytes (Nyholm and McFall-Ngai, 1998), yet *V. fischeri* is able to reside within the deep crypts. A recent study suggests that these haemocytes become 'educated' to the presence of *V. fischeri* within the light organ, and this apparent 'tolerance', mediated by both host and bacterial factors, prevents removal of the symbiont (Nyholm *et al.*, 2009).

Once in the deep crypt spaces, *V. fischeri* grows to high cell densities. This is the accommodation stage. The squid supplies the bacteria with nutrients, including amino acids in the form of small peptides, which permit rapid bacterial growth (Graf and Ruby, 1998). At high cell densities, the bacteria initiate production of bioluminescence (Fig. 18.2B), which the squid uses to avoid

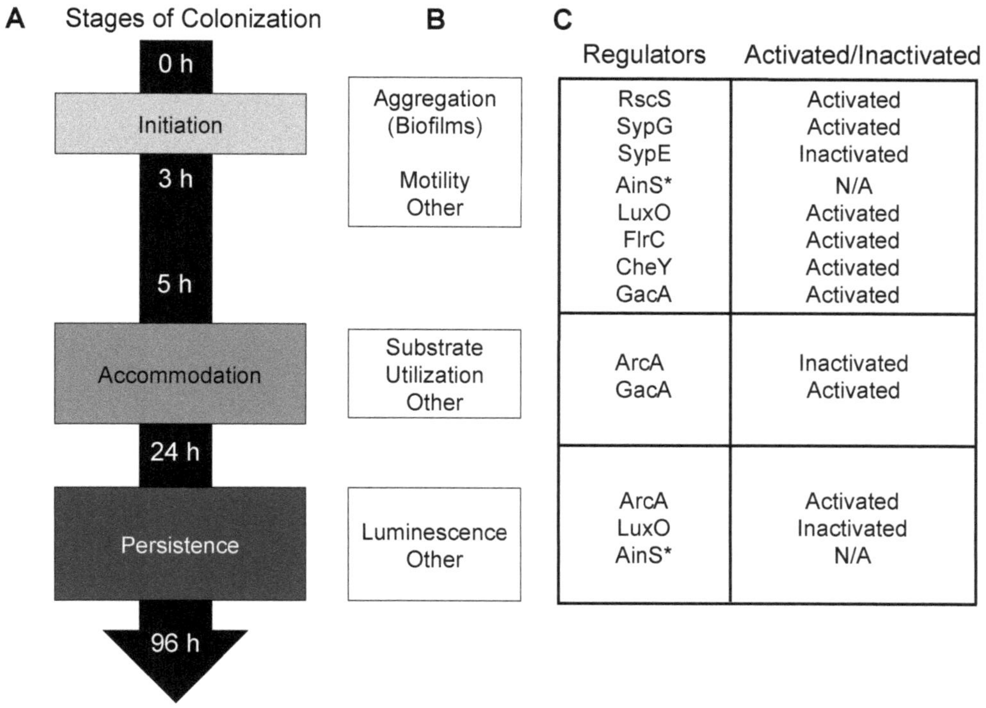

Figure 18.2 Model of the stages, processes, and predicted activities of a subset of regulators involved in colonization. (A) Time line and stages of colonization with the approximate times of and between each stage. At the initiation stage, the bacteria aggregate outside of the light organ and then migrate to the pores to enter and passage through the ducts and antechambers to reach the deep crypts (between hours 1–5 after hatching). At the accommodation stage, the bacteria reach the deep crypt spaces and grow to high cell densities and produce bioluminescence (between hours 5–24 after hatching). At the persistence stage, the bacteria must be maintained throughout the life of the squid, re-populating the light organ each day after the daily expulsion at dawn (between hours 24–96+ after hatching). (B) A brief list of the processes required for colonization that are discussed in this chapter. The approximate timing of these processes correspond to the time line and stages shown in panel A. (C) TCS regulators associated with the stages and processes required for colonization are listed along with their corresponding putative state of activation (activated or inactivated). For SypE, there is experimental evidence that strongly supports the hypothesis that this protein is inactivated at the indicated stage; for the other proteins, the putative state of activation is based on whether a mutant defective for the regulator exhibits a colonization defect at that stage. *AinS is not a TCS regulator, but it produces a signal proposed to be sensed by a TCS regulator.

detection by predators: the light, which is directed downwards, disrupts the shadow that would otherwise result from the down-welling moonlight shining on the animal as it forages for food in shallow bays at night (Jones and Nishiguchi, 2004). Motility, which is essential for entry, appears unnecessary at this stage of colonization, as most bacteria lack flagella in the deep crypts (Ruby and Asato, 1993).

The last stage of colonization is termed persistence. At dawn every morning, the squid expels approximately 95% of the bacteria into the seawater (Lee and Ruby, 1994; Boettcher *et al.*, 1996). The remaining *V. fischeri* cells then re-populate the light organ within a few hours. Thus, this partnership is dynamic, yet it is maintained for the life of the squid.

It is clear from this brief description of the initiation of the *V. fischeri*–squid symbiosis that the bacteria experience numerous different environments as they transit from seawater to symbiosis, adhering to the surface of the light organ, migrating through the ducts and antechambers and into the deep crypts, and growing rapidly once in the right location. Thus, it seems likely that *V. fischeri* recognizes and responds to specific signals in different environments to enter into a productive relationship with its squid host. A well-studied and important mechanism by which many bacteria, including *V. fischeri*, assess the environment is through two-component signal transduction systems (TCSs). Here, we will describe the known TCS encoded by *V. fischeri*, with a focus on those known to play a role in symbiosis.

Two-component signal transduction systems

Bacteria use TCSs to adapt to changes in their environment. A 'simple' TCS is typically composed of a membrane-bound histidine sensor kinase (SK), which senses some environmental or cellular signal (e.g. membrane stress), and a response regulator (RR), which elicits the appropriate cellular

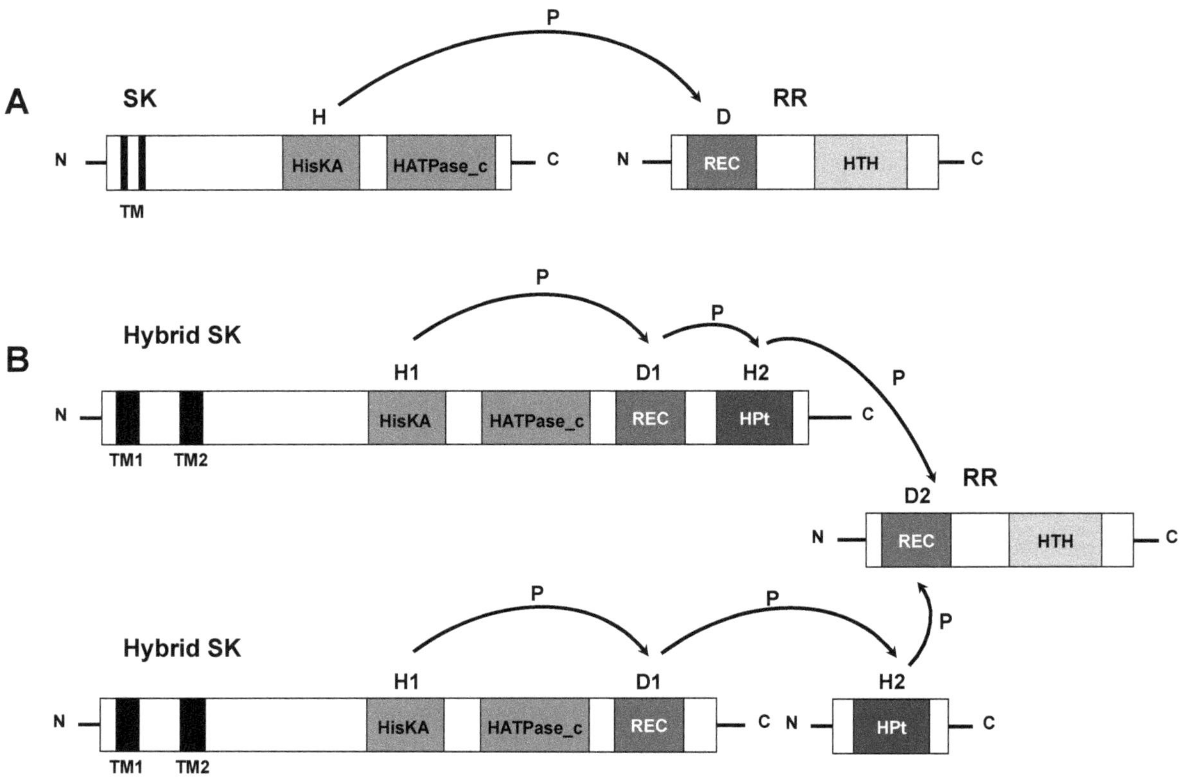

Figure 18.3 Two-component systems. (A) A simple TCS composed of a sensor kinase (SK) and a response regulator (RR). A SK typically contains: (1) a periplasmic signal sensing loop, with at least two transmembrane spanning regions (depicted by the black boxes), and (2) HisKA and HATPase-c domains (light grey boxes with black lettering); the latter domain binds ATP and promotes phosphorylation of a conserved histidine residue located in the HisKA domain (depicted as an H above the HisKA domain) upon signal receipt. SKs can also contain cytoplasmically located signalling domains, such as PAS or HAMP domains (described in the text) (not shown). A RR typically contains: (1) a receiver (REC) domain (depicted as a darker grey box with white lettering), which catalyses the transfer of the phosphoryl group from the SK to a conserved aspartate residue (indicated by a D above the REC domain) within this domain, and (2) an attached effector domain, depicted here as a helix–turn–helix (HTH) DNA binding domain (very light grey box with black lettering); effector domains are not always DNA binding domains. (B) A more complex TCS is known as a phosphorelay, and typically contains between two and four proteins and a total of four domains with residues that become phosphorylated. These domains, known as HisKA, REC, HPt (histidine phosphotransferase), and REC, contain H1, D1, H2 and D2, respectively. Two representative phosphorelays are depicted. At the top, a hybrid SK is depicted that contains three of the four domains involved in phosphotransfer to a RR, while at the bottom, a hybrid SK with two of the four domains is shown along with a separate phosphotransferase protein containing an HPt domain that serves as the phosphodonor to the RR. As with the simple TCS system, a signal stimulates the hybrid SK to autophosphorylate on H1. The phosphoryl group is transferred intramolecularly to D1 and, in either the same protein or a second protein, to H2. The phosphorelay is completed by donation of the phosphoryl group to D2 on the RR.

response (e.g. transcriptional regulation of a gene or set of genes) (Fig. 18.3A). More specifically, when the SK senses a signal, it autophosphorylates on a conserved histidine residue, located within its transmitter domain (HisKA), and serves as a phospho-donor to its partner RR (Stock *et al.*, 2000; West and Stock, 2001). The RR then catalyses the transfer of the phosphoryl group to a conserved aspartate residue, located within its receiver (REC) domain (Fig. 18.3A) (Bourret *et al.*, 1990). RR phosphorylation is thought to stabilize the protein in a particular conformation, typically activating an attached effector domain (e.g. DNA binding domain), which promotes the necessary response (Stock *et al.*, 2000).

A more complex type of TCS system is known as a phosphorelay (Fig. 18.3B). A typical phosphorelay relies on multiple phosphotransfer events, between two or up to four separate proteins. One common type of phosphorelay involves a hybrid SK, which contains multiple residues involved in phospho-transfer, and a partner RR (Fig. 18.3B) (West and Stock, 2001). The hybrid SK usually contains a conserved histidine residue (H1) located within its transmitter domain (HisKA), a conserved aspartate residue (D2) located within a REC domain, and a second conserved histidine (H2) located within a histidine phosphotransferase (HPt) domain; alternatively, the HPt domain may be contained within a separate protein (Fig. 18.3B). The RR contains the second conserved aspartate residue (D2), located within its REC domain. Upon sensing a signal, the hybrid SK autophosphorylates on H1 and subsequently transfers the phosphoryl group intra-molecularly to D1 and then H2 (Fig. 18.3B) (West and Stock, 2001). The RR then catalyses the transfer of the phosphoryl group from H2 to D2 (Fig. 18.3B), which promotes RR activity (West and Stock, 2001).

Two-component regulators in *V. fischeri*

V. fischeri contains 40 genes that encode proteins with a predicted, conserved REC domain (Hussa *et al.*, 2007). The larger of *V. fischeri's* two chromosomes contains about half of the RR genes, with the remaining genes located on the smaller chromosome. These 40 RR proteins can be grouped into four classes: class one contains 15 non-DNA binding RRs, including CheY, a protein similar to the well-studied chemotaxis regulator in *E. coli* (Bourret *et al.*, 1990; Silversmith and Bourret, 1999); class two contains six RRs that are NarL-like owing to the presence of a classical helix–turn–helix (HTH) DNA binding domain (Gunsalus *et al.*, 1989; Baikalov *et al.*, 1998); class three contains 13 RRs that contain a winged helix–turn–helix (wHTH) domain, making them OmpR-like (Martinez-Hackert and Stock, 1997); and class four contains six NtrC-like RRs, which contain three domains, including an N-terminal REC domain, a centrally located σ^{54}-interaction domain, and a C-terminal DNA binding domain (Popham *et al.*, 1989; Weiss *et al.*, 1991). Of the 40 RR genes, 30 have been linked, by function or location, to a SK gene. A similar survey, however, has not yet been performed for SK genes.

Of the 40 RRs, 35 were disrupted and assessed for various phenotypes, including bioluminescence, motility, and colonization capability in the presence of wild-type *V. fischeri* (i.e. competitive colonization) (Hussa *et al.*, 2007). In addition to this global survey, a number of other studies have investigated the roles of specific two-component regulators in the *V. fischeri*–squid symbiosis. Most of these studies have focused on RRs (i.e. SypG, SypE, VpsR, FlrC, CheY, LuxO, ArcA, and GacA) (Figs. 18.2C and 18.4A), although a few have characterized hybrid SKs (i.e. RscS and SypF) (Figs. 18.2C and 18.4B). These studies have uncovered important roles of specific regulators in controlling biofilm formation (RscS, SypG, SypE, SypF, and VpsR), motility and chemotaxis (FlrC and CheY), and bioluminescence (LuxO and ArcA); one of these proteins (GacA) contributes to numerous phenotypes and thus is an important global regulator. The remainder of this review will focus on discussing the known roles of these regulators, both with respect to their control of the indicated phenotypes and to their importance in symbiosis.

Regulation of biofilm formation

Biofilms are populations of bacteria adherent to a surface and embedded in a matrix, typically self-produced and composed of polysaccharides,

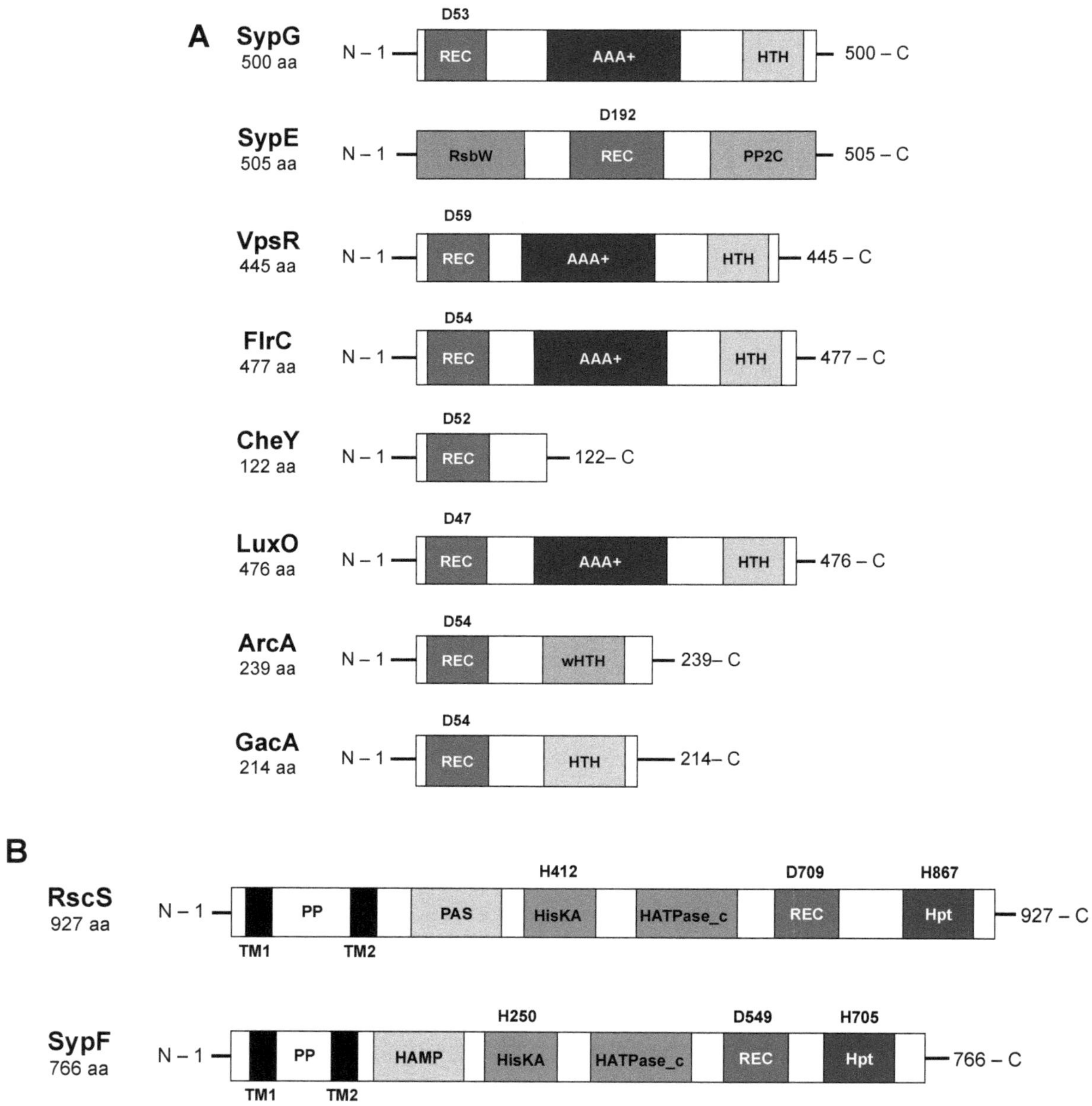

Figure 18.4 Domain structures of the RR and (hybrid) SK proteins described in detail in this chapter. (A) All of the RRs contain a REC domain (grey boxes with white lettering) with the predicted, conserved aspartate residue numbered and shown above this domain. All of these RRs, with the exception of SypE and CheY, contain a HTH or wHTH DNA binding domain (light grey boxes with black lettering). Four of these RRs (SypG, FlrC, LuxO, and VpsR) contain an AAA+ domain (dark grey box with white lettering), which is involved in interactions with RNA polymerase carrying the alternative sigma factor σ^{54}. SypE contains an N-terminal putative RsbW-like serine kinase domain and a C-terminal PP2C-like serine phosphatase domain, and CheY lacks these other domains. (B) Each SK contains a cytoplasmically located signal domain (either PAS or HAMP, described in the text; very light grey box with black lettering) and the HisKA and HATPase-c domains (light grey boxes with black lettering), with the predicted, conserved histidine numbered and shown above the HisKA domain. RscS and SypF are hybrid SKs and contain a REC (dark grey box with white lettering) and HPt (very dark grey box with white lettering) domain; the predicted, conserved aspartate and histidine residues are numbered and shown above these domains, respectively. The hybrid SKs each have two predicted transmembrane spanning regions (black boxes) flanking a periplasmic loop likely involved in the detection of an external signal.

proteins, and DNA, although the exact composition varies depending on the microbe and its environment (Flemming and Wingender, 2010). Biofilms promote adherence of the bacteria to themselves and/or a surface. Furthermore, bacteria in biofilms exhibit an altered physiological state that permits increased resistance to antimicrobials such as antibiotics. As a result, biofilm formation is an important survival and colonization strategy used by bacteria. To date, few models have been described that allow the study of biofilm formation during host colonization; the *V. fischeri*–squid symbiosis is one of these. During the initiation stage of colonization, *V. fischeri* cells form a biofilm-like aggregate in mucus on the surface of the light organ (Fig. 18.1C), and subsequently disperse from the aggregate to enter the organ (Nyholm *et al.*, 2000). For *V. fischeri*, it seems likely that this aggregate is a form of biofilm, as its formation depends upon genes with known roles in biofilm formation in culture (Fig. 18.2B) (Yip *et al.*, 2006; Morris *et al.*, 2011). Whether or not the bacteria exist in a biofilm at a subsequent stage of colonization remains to be determined. Surprisingly, under traditional growth conditions in laboratory culture, *V. fischeri* fails to form any substantial amount of biofilm, although some adherence to the test tube surface can be detected (Yip *et al.*, 2005; Hussa *et al.*, 2008). Thus, it is possible that *V. fischeri* recognizes some host-specific signal(s) in its environment to promote aggregation on the surface of the light organ. Here, we describe the TCS regulators known to control biofilm formation in *V. fischeri*.

RscS

The gene *rscS* (regulator of symbiotic colonization – sensor) encodes a hybrid SK that was first identified in a screen for mutants unable to colonize juvenile *E. scolopes* (Fig. 18.4B) (Visick and Skoufos, 2001). The mutant defective for *rscS* exhibited a severe defect in initiating colonization: when exposed to the *rscS* mutant, most squid remained uncolonized. It was subsequently determined that this colonization defect resulted from the inability of the *rscS* mutant to aggregate on the surface of the light organ (Yip *et al.*, 2006). These experiments revealed for the first time a bacterial determinant necessary for aggregate formation, thus supporting the identification of this process as a stage critical to colonization. They also demonstrated that, rather than being passively taken up by the squid, *V. fischeri* plays an active role in promoting colonization. More recently, the importance of RscS was further emphasized when it was discovered that the gene that encodes this hybrid SK was missing in the fish symbiont MJ11, a strain of *V. fischeri* incapable of colonizing squid (Mandel *et al.*, 2009). When *rscS* was introduced into MJ11, the resulting strain became competent to colonize squid (Mandel *et al.*, 2009). Thus, this work determined that *rscS* – a single regulatory gene – was able to specify the interaction between a bacterium and its host. Together, these data demonstrated that RscS is a critical factor necessary for colonization, owing to its ability to promote aggregation.

RscS is now known to regulate the symbiosis polysaccharide (*syp*) locus, which is composed of 18 genes predicted to be involved in the regulation, production, and export of a polysaccharide necessary for biofilm formation (Fig. 18.5) (Yip *et al.*, 2005). Overexpression of *rscS* from either of two increased activity alleles, termed *rscS1* and *rscS2*, caused a marked increase in *syp* transcription (Yip *et al.*, 2006). Overexpression of these alleles also led to the production of distinctive *syp*-dependent phenotypes associated with biofilm formation (Fig. 18.6). These phenotypes included wrinkled colony formation on solid media, pellicle formation at the air–liquid interface of static cultures, cell aggregation in liquid cultures grown with shaking, increased attachment to glass surfaces, and increased hydrophobicity. Furthermore, examination of the wrinkled colonies formed by *rscS1*-containing cells using scanning and transmission electron microscopy revealed the presence of an extracellular matrix between the cells and at the colony surface. This polysaccharide matrix contained, in part, glucose and/or α-linked mannose sugars (Yip *et al.*, 2006).

These dramatic biofilm phenotypes begged the question, are they relevant to symbiosis? The answer was yes, as an *rscS1* overexpression strain showed a dramatic increase in the size of the symbiotic aggregate relative to the control (est. 50–200 μm vs. 10–20 μm in diameter) (Yip *et al.*, 2006). Furthermore, aggregate formation

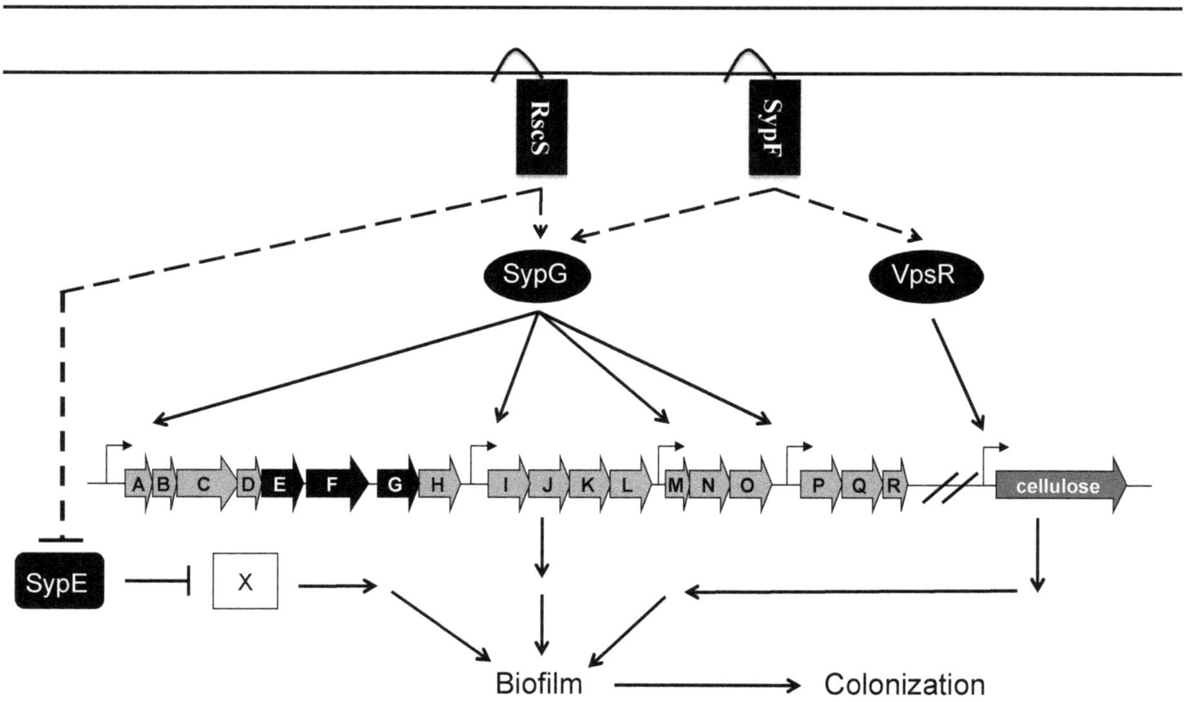

Figure 18.5 Model of *syp*-dependent biofilm formation. The hybrid SK RscS is predicted to activate SypG via phosphorylation, thus promoting transcription of each of four operons in the *syp* locus (from promoters indicated by the bent arrows). The *syp* genes encode proteins involved in the production and transport of a polysaccharide involved in biofilm formation. SypE is predicted to work downstream of *syp* transcription and inhibit biofilm formation in its unphosphorylated form. When RscS becomes active, the inhibitory activity of SypE is inactivated, promoting biofilm formation. SypF also appears to act upstream of *syp* transcription, as well as upstream of the RR VpsR, which is necessary for the production of cellulose, another component of biofilm formation. Biofilm formation in culture correlates with aggregation of *V. fischeri* within the squid secreted mucus, which promotes subsequent colonization events. The signals sensed by RscS and SypF are currently unknown. Furthermore, it is unknown whether cellulose promotes colonization.

depended on *syp*: loss of *sypN* abolished the increase in aggregate formation induced by RscS overexpression. Finally, cells overexpressing *rscS1* substantially out-competed the vector-containing control strain during mixed inoculation experiments, indicating that the increased ability to aggregate conferred an advantage to the cells during colonization (Yip *et al.*, 2006). Thus, these studies reveal a clear correlation between biofilm formation in culture and both symbiotic aggregation (i.e. biofilm formation) and colonization in an animal model.

What are *rscS1* and *rscS2*, and why do they have increased activity? These alleles were generated in a study that sought to determine the function of RscS. It was predicted that the signal sensed by RscS might exist only in the context of the symbiosis, and thus, a signal-insensitive mutant might be necessary to evaluate the role of RscS in culture. It was expected that the increased-activity alleles isolated, *rscS1* and *rscS2*, would contain changes to the coding sequence that made the RscS protein independent of the inducing signal. Instead, however, each increased activity allele contained a mutation in or near the putative ribosome binding site (RBS), while the *rscS1* allele also contained a silent mutation at codon Leu_{25} (Yip *et al.*, 2006; Geszvain and Visick, 2008a). Consistent with their locations, these mutations caused an increase in the level of the RscS protein, but did not impact the *rscS* transcript level (Geszvain and Visick, 2008a). Importantly, no RscS protein was detected when the wild-type allele was expressed from the chromosome or overexpressed from a plasmid lacking the mutations (Geszvain and Visick, 2008a), suggesting that this protein is normally expressed poorly or not at all under standard culturing conditions. Surprisingly,

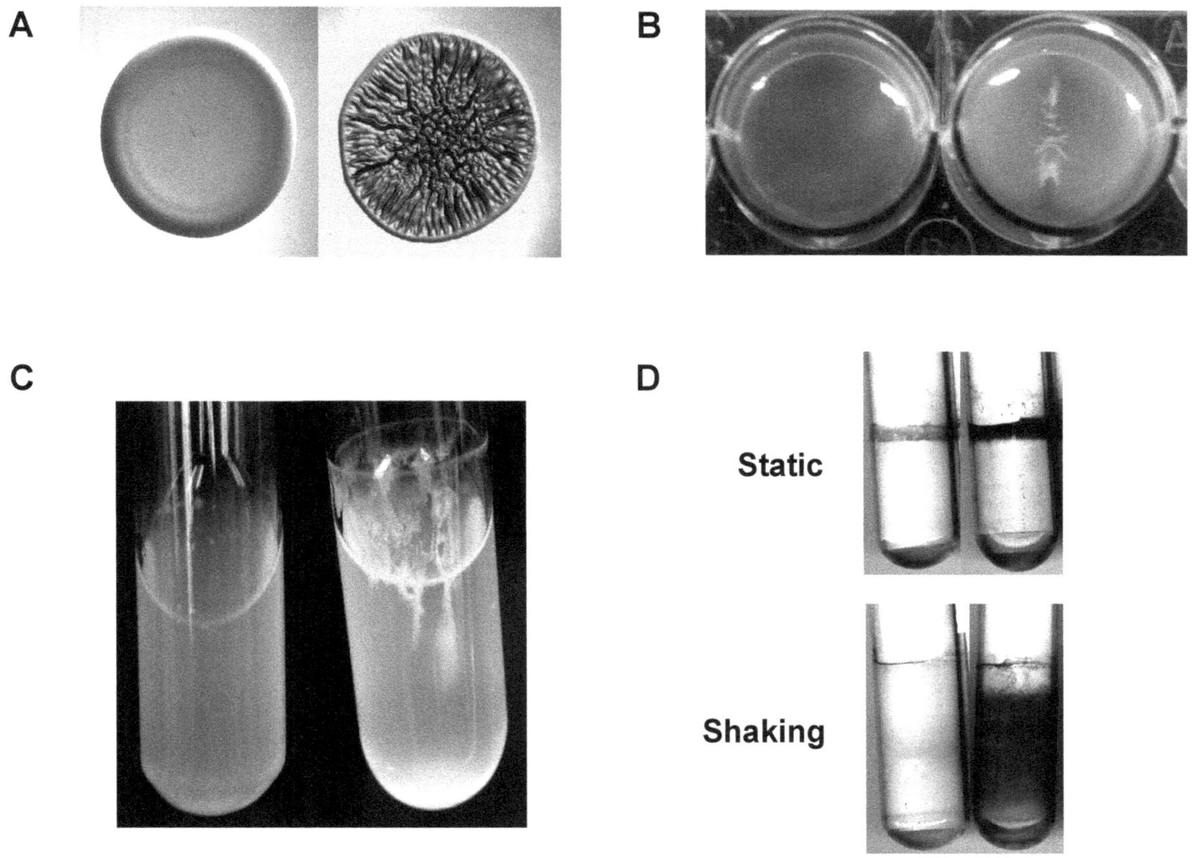

Figure 18.6 *V. fischeri* biofilm phenotypes. The left image in each section is of a non-biofilm forming *V. fischeri* strain, while the right image is of a biofilm-forming strain. (A) Wrinkled colony formation of spotted cultures on solid agar media. (B) Pellicle formation at the air–liquid interface of a static culture. The pellicle can be seen as a 'bunched' cell mass, generated to promote visualization by dragging a sterile pipette tip over the surface. (C) Bacterial aggregation of shaking cultures. 'Stringy' material, observed for the biofilm forming strain, consists of clumps of adherent cells. (D) Glass attachment of either statically grown cultures or shaking cultures, visualized following staining with crystal violet. Cells in static cultures attach at the air–liquid interface of the tube, while cells in shaking cultures can attach throughout the tube. These images are compiled from a variety of strains that produced or lacked biofilms.

when the effects of the mutations in *rscS1* were separately assessed, the silent mutation at codon Leu_{25} appeared to exert a greater impact on RscS activity than the RBS-linked mutation. The silent mutation substitutes a rare Leu codon (used in 4% of Leu codons in *V. fischeri*) for a more common Leu codon (used in ~20% of Leu codons). This Leu codon occurs within a region ($L_{23}ML_{25}TRN_{28}$) that contains, with one exception, all rare codons (Geszvain and Visick, 2008a). Thus, there may be multiple levels at which the production of RscS protein is controlled. These studies concluded that the increase in protein production likely accounts for the increase in RscS activity and subsequent biofilm phenotypes when *rscS1* and *rscS2* are overexpressed. It will be of interest to determine the extent of regulatory control over RscS translation in *V. fischeri*. Furthermore, future studies should address whether simple overexpression is sufficient to overcome a need for a specific, potentially host-associated signal, or if the signal naturally exists in laboratory culture but depends upon sufficient protein production.

The identification of phenotypes for both RscS and the *syp* locus paved the way for understanding the specific roles of RscS and other biofilm regulators. RscS is an orphan hybrid SK (*rscS* is physically unlinked to a RR gene), and thus its partner(s) was initially unknown. However, the *syp* locus encodes two RRs, SypE and SypG (Fig. 18.5), and since RscS regulates the *syp* locus, it was possible that one of these RRs could serve as a partner for RscS. SypE is a unique RR that contains a centrally located REC domain, flanked by putative domains of opposing function (an N-terminal serine kinase domain and a C-terminal serine phosphatase domain; Figure 18.4A) (Hussa *et al.*, 2008; Morris and Visick, 2010); however, this protein lacks any sequences that resemble a DNA-binding motif. In contrast, SypG is a member of the well-studied NtrC family of RRs: it contains three predicted domains, an N-terminal REC domain, a central σ^{54} interaction domain, and a C-terminal DNA binding domain (Fig. 18.4A). Like RscS, SypG activates the *syp* locus (Yip *et al.*, 2005), making this RR a likely partner for RscS. In support of this idea, deletion of *sypG* disrupts all of the RscS-induced phenotypes, including *syp* transcription (Hussa *et al.*, 2008). Thus, it appears that RscS works upstream of SypG to induce *syp* transcription and thus biofilm formation (Fig. 18.5). The current knowledge of the roles of SypE and SypG in biofilm formation and colonization will be described later in this review.

Structurally, RscS resembles the hybrid SKs ArcB and BvgS (Visick and Skoufos, 2001). These proteins contain three conserved domains (HisKA, REC, and HPt) with residues predicted to be involved in phosphotransfer (H412, D709, and H867 in RscS) (Fig. 18.4B). Investigation of the roles of the conserved residues in RscS supported the function of RscS as a hybrid SK: mutations of H1 (H412Q), D1 (D709A), and H2 (H867Q) abolished (H1 and D1) or diminished (H2) induction of RscS-dependent biofilm phenotypes (Geszvain and Visick, 2008b). Since H2 was not essential for RscS-induced phenotypes, the signal transduction pathway may be complicated. For example, it is possible that the phosphorelay is branched, with RscS donating its phosphoryl group to another HPt domain-containing protein, or that phosphotransfer to the RR could occur directly from H1 to D2. Such events are not unprecedented [e.g. ArcA/B (Georgellis *et al.*, 1997)], but further studies are needed to fully dissect the dispensability of the Hpt domain for RscS function.

RscS also contains a large periplasmic (PP) loop and a PAS domain, both of which are implicated in signal recognition (Fig. 18.4B). The PP loop is ~200 amino acids in length and is flanked on either side by two transmembrane (TM) regions. The PAS domain is located in the cytoplasm after the second TM region (Fig. 18.4B). In other SKs, the PP loop is responsible for sensing an environmental signal, and transmitting that information through TM regions to the cytoplasmic signalling portion of the protein (Mascher *et al.*, 2006). This appears to be the case for RscS, as disruption of the PP loop or the first transmembrane domain promoted RscS activity (Geszvain and Visick, 2008b), indicating that these regions may serve to negatively regulate RscS function. In contrast, disruption of the cytoplasmically localized PAS domain abolished RscS function (Geszvain and Visick, 2008b), indicating that this domain may positively regulate RscS function. PAS domains can sense signals such as oxygen, redox potential, light, and small molecules such as ATP, and frequently require binding of a cofactor for signal transduction (Taylor and Zhulin, 1999). RscS may sense an FAD cofactor, since its PAS domain shares homology with the PAS domain of NifL from *Azotobacter vinelandii*, which senses an FAD cofactor (Key *et al.*, 2007). In support of this idea, conserved residues required for FAD binding are essential for RscS activity (Geszvain and Visick, 2008b). Taken together, it seems likely that RscS senses multiple signals using the PP loop and PAS domain, which regulate its function accordingly. Thus, determining what these signals are and understanding how RscS responds to them will provide a better understanding of how biofilm formation is regulated in the context of the squid host.

SypG

Like *rscS* mutants, *sypG* mutants exhibited severe defects in colonization as well (Hussa *et al.*, 2007). As noted above, SypG is a member of the NtrC family of RRs and contains three predicted domains: an N-terminal REC domain, a C-terminal DNA binding domain, and between those two domains, a σ^{54} interaction domain (Fig. 18.4B). This latter domain is predicted to provide the ATPase activity necessary for transcription: RNA polymerase containing σ^{54} binds to DNA and forms a closed complex, but cannot form the transcriptionally active open complex without the help of an activator protein to provide the energy

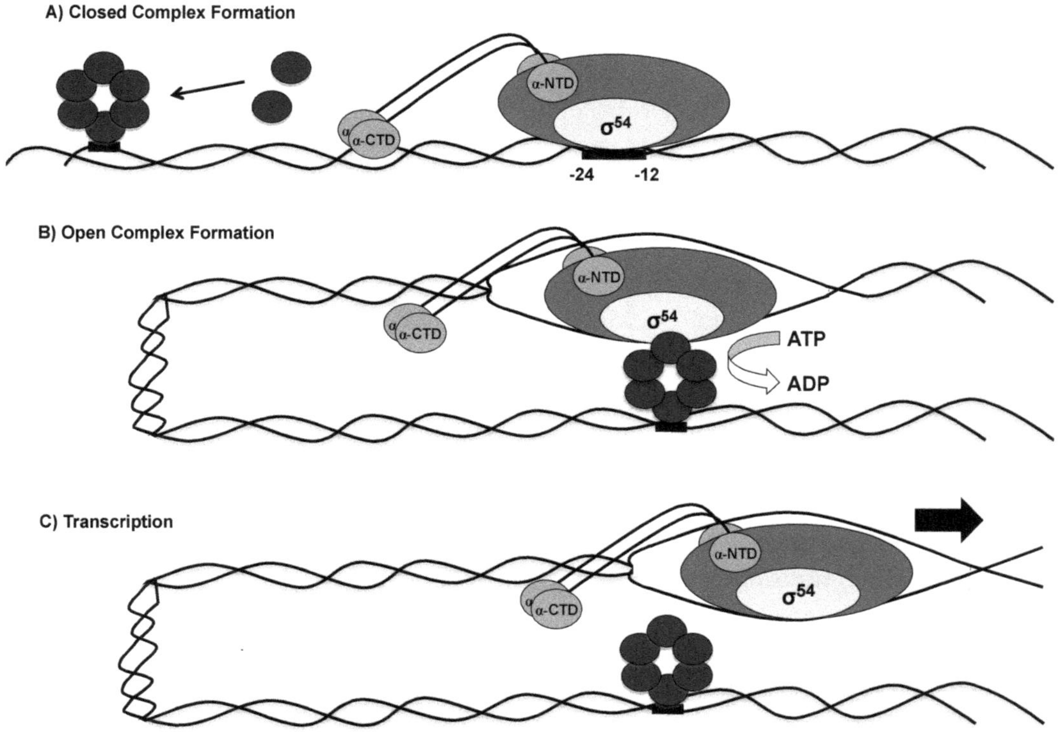

Figure 18.7 Model for σ^{54}-dependent transcriptional activation. (A) σ^{54}-containing RNA polymerase (shaded grey circles) binds to the promoter region (typically at conserved sequences centred at −12 and −24 relative to the start of transcription; large black box), promoting closed complex formation. Concurrently, activation of a σ^{54}-dependent activator protein (dark grey circles) typically promotes the formation of a hexamer, which binds to an enhancer sequence upstream of the promoter region. (B) Bending of the DNA enables interaction of the σ^{54}-dependent activator with σ^{54}. The σ^{54}-dependent activator then hydrolyses ATP to ADP, providing the energy for open complex formation. (C) With the energy provided by the σ^{54}-dependent activator protein and formation of the open complex, RNA polymerase is able to promote transcription (indicated by the black arrow). [The steps of transcription promoted by σ^{54}-containing RNA polymerase and an activator protein are adapted from Wigneshweraraj *et al.* (2008).]

(Fig. 18.7) (Buck *et al.*, 2000; Wigneshweraraj *et al.*, 2008). Intriguingly, the *syp* locus contains four promoters (Fig. 18.5) with σ^{54} recognition sequences (Barrios *et al.*, 1999; Yip *et al.*, 2005); three of the four promoters were confirmed by primer extension analyses (Yip *et al.*, 2005). Located upstream of each of the σ^{54} promoters is a conserved 22-bp sequence that could serve as a binding site for a σ^{54}-dependent activator (Fig. 18.7). Since *sypG* encodes such a protein, Yip *et al.* (2005) hypothesized that SypG could activate transcription of the *syp* locus. Although there was little *syp* transcription in wild-type cells, overexpression of *sypG* from a multicopy plasmid caused a 37- to 70-fold increase in *syp* transcription over that of the vector control strain. This SypG-induced transcription depended upon the presence of σ^{54}, as disruption of the *rpoN* gene, which encodes σ^{54}, abolished transcriptional activation of *syp* (Yip *et al.*, 2005). Together, these data suggest that SypG, in conjunction with σ^{54}, promotes transcription from the *syp* locus (Fig. 18.5). To date, however, it remains unknown whether SypG binds to the 22-bp sequences located upstream of the σ^{54}-dependent promoters within the *syp* locus.

Coincident with induction of *syp* transcription, overexpression of *sypG* induced biofilm formation (Yip *et al.*, 2005). Specifically, *sypG* overexpression promoted a 3.5-fold increase in glass attachment (e.g. Figure 18.6D) when cells were grown statically. Furthermore, when grown under shaking conditions, cells that overexpressed *sypG* exhibited a > 30-fold increase in glass attachment relative to the vector control strain. Biofilms formed under either condition (static or shaking) depended upon a functional copy of *rpoN* (Yip *et al.*, 2005). Together, these data demonstrated that SypG activates *syp* transcription, and thus biofilm formation, in a σ^{54}-dependent manner.

Although both RscS and SypG induce *syp* transcription and biofilm formation, and activity of the SK clearly depends on the RR, the specific biofilm phenotypes displayed by the two overexpression strains were dissimilar: overexpression of *rscS* resulted in the formation of wrinkled colonies and a strong pellicle (Yip *et al.*, 2006), while overexpression of *sypG* only led to weak pellicle formation (Hussa *et al.*, 2008). Two possibilities could account for these differences: (1) when overexpressed in the absence of its SK, SypG is not sufficiently activated to promote transcription of the genes required for strong biofilm phenotypes, or (2) RscS signals through a second RR to either inactivate a negative regulator or activate a positive regulator to promote biofilm formation. To assess the first possibility, a constitutively active version of SypG (SypG*) was generated, in which the conserved aspartate residue (D53) within the REC domain was changed to a glutamate (D53E); these types of mutations have been shown to promote RR activity in the absence of phosphorylation in other RRs [e.g. CheY, NtrC, and LuxO (Sanders *et al.*, 1989, 1992; Freeman and Bassler, 1999)]. Overexpression of *sypG** led to an increase in *syp* transcription, but could not induce wrinkled colony formation or enhance pellicle formation (Hussa *et al.*, 2008). Support for the second possibility came from an unexpected result: when *sypG* was overexpressed in a mutant lacking SypE, the second predicted RR encoded within the *syp* locus (Fig. 18.5), the cells formed wrinkled colonies and pellicles that were indistinguishable from those induced by RscS (Hussa *et al.*, 2008). Subsequent analyses, described in the SypE section below, supported the idea that RscS promotes inactivation of an inhibitory activity of SypE (Morris *et al.*, 2011). These results thus strengthen the hypothesis that RscS and SypG function together to promote *syp*-dependent biofilm formation (Fig. 18.5), but suggest that control over biofilm formation is complex and extends beyond transcriptional activation of the *syp* locus.

In summary, SypG plays a key role in controlling *syp* transcription and, thus, biofilm formation and colonization. At least one SK, RscS, appears to signal through SypG to control biofilm formation (Fig. 18.5). The function of another SK, SypF, which is encoded adjacent to the *sypG* gene, also appears to partially depend on SypG (discussed below in the SypF and VpsR section) (Fig. 18.5). Thus, from these studies it is clear that *V. fischeri* expresses numerous regulators that exert significant control over biofilm formation, indicating that it is important to *V. fischeri* to prevent biofilm formation when it is not needed—perhaps to restrict it to times when *V. fischeri* encounters its host.

SypE

The *syp*-encoded RR SypE is unusual. It contains a centrally located REC domain that is flanked by putative effector domains with apparently opposing enzymatic functions (Fig. 18.4A) (Morris and Visick, 2010). The N-terminal domain of SypE exhibits sequence similarity to SK-like serine kinases of the GHKL (Gyrase, Hsp90, HK, MutL) superfamily. These types of proteins contain a conserved asparagine residue, D52 in SypE (Morris and Visick, 2010), necessary for ATP binding (Dutta and Inouye, 2000). The C-terminal domain exhibits sequence similarity to the PP2C family of serine phosphatases, which contain invariant aspartate residues, D443 and D495 in SypE (Morris and Visick, 2010), involved in catalytic activity (Adler *et al.*, 1997; Jackson *et al.*, 2003). To date, no other RR protein with this unique domain structure has been characterized.

Consistent with the presence of two domains of opposing activity, SypE appears to exert both positive and negative control over biofilm formation. First, as described above, SypE appears to antagonize SypG-induced biofilm formation: *sypG* overexpression induced robust biofilms only when *sypE* was disrupted (Hussa *et al.*, 2008), indicating that SypE functions as a negative regulator of biofilm formation. Second, SypE enhances biofilms produced by RscS overexpression: overexpression of *rscS* in a strain lacking SypE resulted in wrinkled colony formation that was delayed relative to the *sypE*$^{+}$ control (Hussa *et al.*, 2008; Morris *et al.*, 2011). Taken together, these data suggest that SypE can function to both promote and inhibit biofilm formation.

How is this dual regulatory activity controlled? The answer to this question came from an elegant study that examined the function and interplay between the three domains of SypE and their impact on biofilm formation and host colonization. This study assessed the ability of a variety of SypE mutants to control biofilm formation induced by RscS (Morris *et al.*, 2011). It was found that the C-terminal putative serine phosphatase domain was responsible for the positive regulatory activity of SypE (i.e. promoting biofilm formation), while the N-terminal putative serine kinase domain was responsible for the negative regulatory activity of SypE (i.e. inhibiting biofilm formation). The negative and positive functions of the two domains depended upon conserved residues predicted to be necessary for serine kinase and serine phosphatase activities, respectively. Finally, mutation of the predicted site of phosphorylation in the REC domain (D192) to an alanine 'locked' SypE into an inhibitory state. This type of mutation can result in a protein that mimics the unphosphorylated state, since the residue can no longer become phosphorylated (Freeman and Bassler, 1999). Thus, it appears that D192 and, presumably, phosphorylation is necessary for switching the activity of SypE from negative to positive.

Which, if any, of these activities is important for colonization? A deletion of *sypE* exerted little effect on colonization, suggesting that the positive activity is not critical. However, expression of *sypE*^{D192A}, the REC domain mutant that is 'locked' into the inhibitory state, severely impaired colonization (Morris *et al.*, 2011). For example, in single strain colonization experiments, most squid remained uncolonized, while those that became colonized contained up to 1000-fold fewer bacteria than wild-type colonized squid. This colonization defect could be attributed to a defect at the aggregation stage: whereas RscS-overexpressing wild-type cells formed large aggregates above the light organ pores, RscS-overexpressing *sypE*^{D192A} cells were unable to aggregate (Morris *et al.*, 2011). These data, along with previous studies of RscS, demonstrate the importance of symbiotic aggregation in promoting colonization, as well as indicate that SypE naturally becomes inactivated during this stage of host colonization.

To date, no studies have confirmed phosphorylation of SypE, but it seems reasonable to expect that D192 becomes phosphorylated in response to some as-yet-unknown signal. The next question is, what SK works upstream of SypE? Morris *et al.* (2011) reasoned that RscS functions upstream of SypE (Fig. 18.5), because overexpression of RscS but not SypG seemed to turn off the inhibitory activity of SypE. Thus, RscS would have two activities: (1) activating SypG and (2) inactivating the negative regulatory activity of SypE. As a result of the latter activity, SypE would function as a negative regulator of colonization in an *rscS* mutant. If so, then deletion of *sypE* should suppress at least part of the colonization defect of an *rscS* mutant. Indeed, an *rscS sypE* double mutant colonized better than the single *rscS* mutant (Morris *et al.*, 2011), supporting the hypothesis that SypE functions as a negative regulator in the absence of RscS. Thus, during symbiotic colonization, SypE may serve to restrict colonization until RscS is activated (Fig. 18.5). Whether RscS directly or indirectly influences the phosphorylation state of SypE remains to be determined.

In summary, SypE is a positive and negative regulator of biofilm formation and, at least, a negative regulator of host colonization. Many questions remain unanswered, such as, at what level does SypE regulate biofilm formation and what is the target of SypE-mediated activity? For the first question, current data suggest that SypE works downstream of *syp* transcription (Hussa *et al.*, 2008), though the exact level remains to be determined. As for the target of SypE-mediated activity, one likely candidate is SypA, a putative anti-σ-factor antagonist with a conserved STAS (anti-σ-factor antagonist and sulfate transporter) domain; these types of proteins require a conserved serine residue (S56 on SypA) to regulate their activity (Morris and Visick, 2010). However, whether SypA functions to control biofilm formation, and whether it functions downstream of SypE remain unknown. As SypE is conserved in a number of distantly related bacteria, understanding its mechanism of action will be of benefit in a variety of fields.

SypF and VpsR

The *syp* locus contains a gene, located between the two RR genes (*sypE* and *sypG*), that encodes a putative hybrid SK, SypF (Fig. 18.5). SypF contains three conserved domains (HisKA, REC, and HPt) and residues predicted to be involved in phosphotransfer (H250, D549, and H705, respectively) (Fig. 18.4B). Similar to RscS, SypF contains two additional putative signalling regions, a PP loop and a cytoplasmic domain (Fig. 18.4B). Whereas RscS contains a PAS domain, SypF contains a HAMP domain. Similar to PAS domains, HAMP domains are often involved in signal transduction. HAMP domains contain a highly conserved helix–turn–helix fold, a motif also common in histidine kinases, adenylyl cyclases, methyl-accepting chemotaxis proteins, and phosphatases; the fold but not the amino acid sequence is conserved, making it difficult to identify these domains (Kishii *et al.*, 2007). While the HAMP domain and periplasmic loop of SypF are likely involved in signal transduction, the signal(s) sensed by SypF are unknown.

Like RscS, SypG, and SypE, SypF also appears to serve as a regulator of biofilm formation in *V. fischeri*. Overexpression of a signal-independent allele of *sypF*, *sypF1*, resulted in a 10-fold increase in *syp* transcription and four distinct biofilm phenotypes: (1) cell aggregation (the clumping of cells in liquid culture under shaking growth conditions, not to be confused with symbiotic aggregation), (2) wrinkled colony formation, (3) pellicle formation, and (4) increased glass attachment (3-fold increase vs. the wild-type strain) (Fig. 18.6) (Darnell *et al.*, 2008). Sequence analysis of the *sypF1* allele showed that this allele contains two amino acid substitutions, one at residue 247 (S247F) and the other at residue 439 (V439I). Of these two mutations, the S247F substitution, located in the HisKA domain three residues N-terminal to the conserved histidine (H250), appeared to be more important for the increased activity of *sypF1* (Darnell *et al.*, 2008). Owing to the proximity of this residue to the conserved histidine, it is likely that this mutation impacts the predicted kinase activity of SypF.

Because SypF is a hybrid SK, it is predicted to exert its effect by acting through a downstream RR. In the chromosome *sypF* is adjacent to *sypG* (see the SypG section above; Figure 18.5), thus, it seemed reasonable to expect that SypF worked by activating SypG. Indeed, disruption of *sypG* eliminated the 10-fold induction of *syp* transcription caused by overexpression of *sypF1* (Darnell *et al.*, 2008). Furthermore, disruption of *sypG* also diminished SypF1-induced cell aggregation, a known *syp*-dependent biofilm phenotype. Surprisingly, however, the SypF1-induced wrinkled colony and pellicle formation phenotypes were only diminished, not eliminated, by loss of *sypG*. In addition, loss of *sypG* did not impact SypF1-induced glass attachment. These results suggested that some of the SypF1-induced phenotypes are SypG-dependent (*syp* transcription and cell aggregation), while other phenotypes are only partially dependent upon SypG (wrinkled colony and pellicle formation), and yet others were SypG-independent (glass attachment). Overall, while SypF appears to work upstream of SypG to regulate the *syp* locus (Fig. 18.5), it also appears to work through another regulator(s) to control biofilm formation.

What is the identity of the other regulator(s)? Of the 40 RRs encoded by *V. fischeri* (Hussa *et al.*, 2007), a prime candidate was VpsR, which is an NtrC-like σ^{54}-dependent activator (Figs. 18.4A and 18.7). In a related vibrio, *V. cholerae*, VpsR controls biofilm formation through activating expression of the *vps* polysaccharide locus (Yildiz *et al.*, 2001). In *V. fischeri*, a *vpsR* mutant produces colonies that are mucoid and opaque, distinct from the smooth and yellowish colonies formed by the wild-type strain (Darnell *et al.*, 2008), suggesting that VpsR might play a role in biofilm formation. Indeed, overexpression of *sypF1* in a *vpsR* mutant led to diminished wrinkled colony and pellicle formation, similar to when *sypG* was disrupted, as well as a total loss of glass attachment (which was unaffected by loss of *sypG*). However, disruption of *vpsR* did not impact cell aggregation (which was diminished by loss of *sypG*). Thus, wrinkled colony and pellicle formation appear to depend on both *syp* and *vpsR*, while *syp* transcription and cell aggregation (i.e. cell–cell interactions) are *sypG*-dependent phenotypes, and glass attachment (i.e. cell–surface interactions) is a *vpsR*-dependent phenotype. When *sypF1* was overexpressed in a double *sypG vpsR* mutant, all SypF1-induced biofilm phenotypes were lost (Darnell *et al.*, 2008), indicating that *syp* and *vpsR* together account for all of these phenotypes. Thus, SypF appears to modulate biofilm formation through regulation of the *syp* locus and an unknown, *vpsR*-dependent pathway (Fig. 18.5).

To begin to elucidate the role of VpsR in biofilm formation, a *vpsR* overexpression construct was generated. Overexpression of *vpsR* increased glass attachment under static growth conditions and allowed for weak pellicle formation, but did not induce wrinkled colony formation (Darnell *et al.*, 2008), suggesting that VpsR plays a role in biofilm formation that is distinct from *syp*. However, despite the role of the *V. cholerae* homologue in controlling the *vps* polysaccharide locus, VpsR-induced biofilm formation did not require the *vps*-like locus present in *V. fischeri* (Darnell *et al.*, 2008; Grau *et al.*, 2008; Yildiz and Visick, 2009). Instead, disruption of another polysaccharide locus, responsible for the production of cellulose, eliminated glass attachment and diminished pellicle formation. It also eliminated another phenotype associated with VpsR and SypF1: overexpression of either regulator caused an increase in the binding of the dye Congo Red, a phenotype which is associated with cellulose production (Teather and Wood, 1982). Together, these data suggest that SypF works upstream of VpsR to regulate cellulose biosynthesis and implicate cellulose as a player in *V. fischeri* biofilm formation and, specifically, cell–surface interactions (Fig. 18.5).

Intriguingly, there's another connection between *syp* and cellulose. Just downstream of the *syp* locus is a gene, *binA*, that encodes a c-di-GMP phosphodiesterase (PDE) involved in the degradation of c-di-GMP. In many bacteria, increased levels of c-di-GMP correspond to increased biofilm formation, in some cases owing to increased cellulose production (Ross *et al.*, 1987; Römling *et al.*, 2005). Consistent with that idea, disruption of *binA* resulted in increased c-di-GMP levels and increased cellulose-dependent biofilm formation (Bassis and Visick, 2010), findings that implicate BinA as a negative regulator of biofilm formation. Interestingly, transcription of *binA* increased upon overexpression of *sypG*, likely owing to read-through from the *syp* locus. However, BinA did not appear to impact *syp*-dependent biofilm formation (Bassis and Visick, 2010). Thus, BinA is both physically and functionally connected to two polysaccharide loci, potentially serving to down-regulate one when the other is up-regulated.

Of these regulators (SypF and VpsR), only VpsR has been investigated for a role in host colonization. When juvenile squid were exposed to a mixture of the *vpsR* mutant and wild-type *V. fischeri*, the resulting symbiotic animals contained a higher percentage of the wild-type strain (Hussa *et al.*, 2007), suggesting that the *vpsR* mutant exhibited a slight colonization defect. However, since this mutant exhibited multiple phenotypes (e.g. colony morphology and cellulose biosynthesis) (Hussa *et al.*, 2007; Darnell *et al.*, 2008), it is unclear whether the slight colonization defect resulted from a lack of one of these processes, from the loss of a combination of them, or from an as-yet-unknown factor. Further studies are needed to better assess the roles of SypF and VpsR in host colonization.

In conclusion, these studies (1) further demonstrate the complexity of biofilm formation in *V. fischeri*; (2) indicate that *vpsR* is conserved, but serves a distinct role in biofilm formation in *V. fischeri* relative to *V. cholerae*; (3) suggest that cell–surface interactions (glass attachment) depend upon cellulose, while cell–cell interaction (pellicles and cell aggregation) depend upon *syp*; and (4) suggest that the cellulose biosynthesis is regulated by multiple proteins, including those encoded in/near the *syp* locus (*sypF* and *binA*). Many unanswered questions remain, including the identity of the environmental signal(s) that stimulate the formation of *syp*- and cellulose-dependent biofilm formation and the role, if any, of cellulose in promoting host colonization.

Motility and chemotaxis

How does *V. fischeri* find its way to and into the squid? The answer to that is unknown, but one part of the puzzle is clearly motility and, likely, chemotaxis, the random biased movement towards attractants and away from repellents. *V. fischeri* is lophotrichous (i.e. contains a tuft of polar flagella) and is predicted to encode about 40 different chemoreceptors that could direct its movement (DeLoney-Marino *et al.*, 2003, Ruby *et al.*, 2005). Early studies demonstrated that this flagella-based motility is essential for colonization, as non-motile strains of *V. fischeri* were unable to colonize *E. scolopes* (Graf *et al.*, 1994; Millikan and Ruby, 2002, 2003, 2004; Wolfe *et al.*, 2004). The defect, however, was not at the aggregation stage, as flagella mutants formed normal aggregates (Nyholm *et al.*, 2000). Thus, some subsequent event, potentially entering and/or navigation through the ducts and antechambers, depends on motility and perhaps, chemotaxis (Fig. 18.8A). Once *V. fischeri* cells have colonized the deep crypt spaces, the bacteria no longer contain flagella (Fig. 18.8A) (Ruby and Asato, 1993). Finally, upon release, the bacteria once again become flagellated (Fig. 18.8A). The signal(s) and regulators that control flagella expression in symbiosis remain unknown. Furthermore, whether the presence of flagella on colonized bacteria is detrimental or merely unnecessary remains to be determined. Therefore, an understanding of the regulatory mechanisms controlling the production of flagella is important to facilitate studies of mechanisms of control during symbiosis.

FlrC

In many bacteria, the production of flagella depends upon multiple regulators that control transcription at different levels within a transcriptional hierarchy (McCarter, 2006). For example, in *V. cholerae* there are four classes of flagellar genes: I, II, III, and IV (Fig. 18.8B) (reviewed by McCarter, 2001). Class I contains one gene, *flrA*, which encodes the master flagellar regulator (Fig. 18.8B) (Prouty *et al.*, 2001). FlrA is a σ^{54}-dependent transcriptional activator (Fig. 18.7) that promotes transcription of class II genes, which encode the SK/RR pair FlrB/FlrC, along with other genes involved in flagellar synthesis (Fig. 18.8B). FlrB, a cytoplasmic SK, serves as a phospho-donor to the RR FlrC (Correa *et al.*,

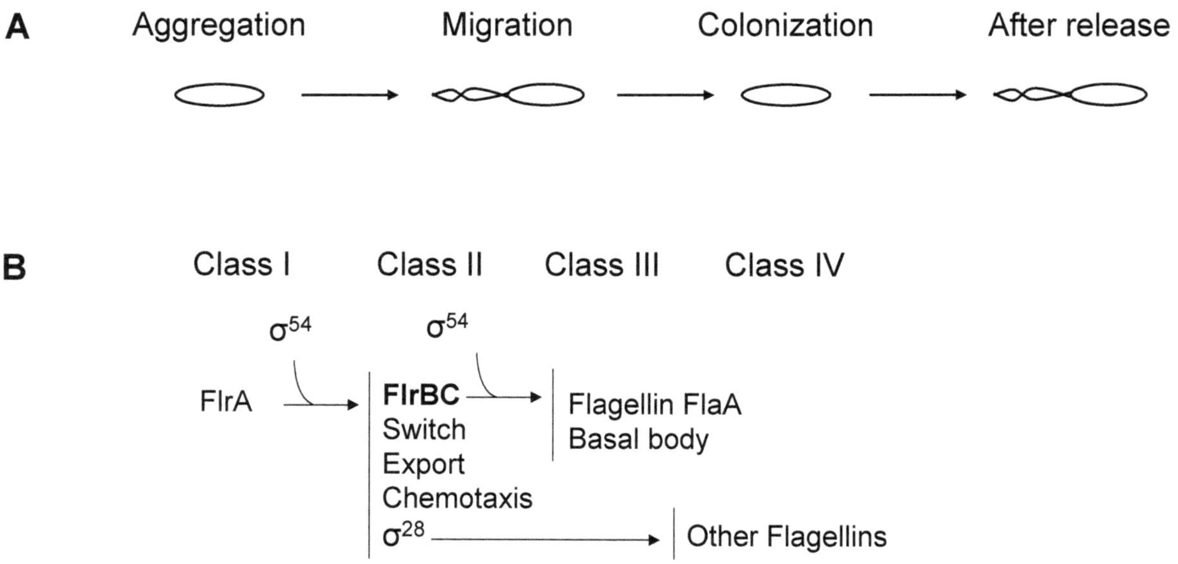

Figure 18.8 Control of *V. fischeri* motility. (A) *V. fischeri* adapts to its environment by the expression of flagella or lack thereof at various stages of colonization (flagella are represented by wavy lines coming off of the bacteria). Flagella are not necessary for aggregate formation, but are critical for some early stage in initiating symbiosis (migration). Once *V. fischeri* colonizes the squid, most cells lack flagella (colonization). Finally, upon release from the light organ, *V. fischeri* cells rapidly re-grow their flagella (after release). (B) Transcriptional hierarchy controlling flagella gene expression in *V. cholerae*, which serves as a model for *V. fischeri*. Class I includes the *flrA* gene (designated as FlrA to indicate its activity in controlling transcription of Class II genes), and the other three classes contain numerous genes involved in the production and rotation of flagella; these are listed simply as ring, switch, export, flagellin, and basal body (for a review, see McCarter, 2006). Of note is Class II, which includes the *flrB* and *flrC* TCS regulatory genes (designated as FlrBC to indicate their role in activating transcription of Class III genes). Both σ^{54} and σ^{28} contribute to regulation of flagellar genes. Panel B was adapted from Prouty *et al.* (2001) and Correa *et al.* (2005).

2000). Upon phosphorylation, FlrC promotes transcription of class III genes by RNA polymerase carrying the alternative sigma factor σ^{54}. The last class of genes, class IV, are transcribed by RNA polymerase carrying σ^{28}. Class III and IV genes encode the remaining components necessary to complete flagellar synthesis (Fig. 18.8B). Genes for many of the *V. cholerae* regulators are conserved in the squid symbiont, including *flrA* and *flrB*/*flrC* (Millikan and Ruby, 2003); thus, it seems likely that this type of hierarchical regulation also occurs in *V. fischeri*.

The FlrB and FlrC proteins of *V. fischeri* are 72% and 84% identical, respectively, to their homologues in *V. cholerae* (Millikan and Ruby, 2003). FlrB is predicted to contain an N-terminal PAS domain and a C-terminal HisKA domain. FlrC is predicted to contain an N-terminal REC domain, with a predicted, conserved site of phosphorylation (D54), a centrally located σ^{54}-interaction domain, and a C-terminal helix–turn–helix DNA binding domain (Fig. 18.4A) (Millikan and Ruby, 2003). Consistent with the predicted role of FlrC as a motility regulator, a *flrC* mutant of *V. fischeri* was non-motile (Hussa *et al.*, 2007). The mutant also exhibited a couple of other phenotypes: it consistently reached a higher optical density in culture than the wild-type strain and reached levels of luminescence that were slightly, but reproducibly, decreased (Hussa *et al.*, 2007). The reason(s) for these additional phenotypes remains unclear. FlrC also appears to play a role in biofilm formation: the *flrC* mutant exhibited a 2.3-fold decrease in glass attachment as compared with the wild-type strain, and exhibited diminished pellicle formation (Hussa *et al.*, 2008). Since biofilm formation was assessed under RscS-inducing conditions, these data suggest that motility and/or another *flrC*-controlled gene is necessary for *syp*-dependent biofilm formation.

Since motility is critical for *V. fischeri* to colonize *E. scolopes* (Graf *et al.*, 1994; Millikan and Ruby, 2002, 2003, 2004; Wolfe *et al.*, 2004), it is not surprising that the *flrC* mutant exhibited a severe colonization defect: juvenile squid inoculated with a 1:1 mixture of the *flrC* mutant and wild-type strains contained only the wild-type strain (Hussa *et al.*, 2007). Likely, this defect is due to the mutant's motility defect. However, because the *flrC* mutant exhibited additional phenotypes in the limited survey that was performed, it is possible that other FlrC-controlled traits could contribute to colonization proficiency. At this time, it is not possible to distinguish impacts on motility from the other effects of the *flrC* mutation.

Further studies of FlrC, as well as of its SK FlrB, will aid in our understanding of the role that motility, and its loss, play in symbiotic colonization. The signal sensed by FlrB to activate FlrC is unknown, although it is presumably present in normal laboratory conditions; potentially, this signal is absent in the deep crypts. It would be of interest to determine whether mutants that express a constitutively active version of FlrC (i.e. signal independent) would retain their flagella in the deep crypts. If not, then the control over flagellation likely occurs at a distinct level, such as at a higher level of transcriptional control (at FlrA or above) or, potentially, post-transcriptionally. Assessing the levels of flagellar transcripts from symbiotic cells would provide important insights into the level of control exerted over flagellation in symbiosis (Fig. 18.8). Analyses of how both bacteria and squid transcriptomes change during symbiosis are beginning to be performed, but the question of flagella control remains to be addressed (Jones and Nishiguchi, 2006; Wier *et al.*, 2010).

CheY

Chemotaxis is a random biased movement towards favourable substrates (e.g. nutrients) and away from unfavourable compounds (e.g. toxic molecules). In many bacteria, chemotaxis controls the direction of rotation of the flagella such that they alternate between counter-clockwise (CCW) and clockwise (CW) (Silversmith and Bourret, 1999). CCW rotation promotes swimming in a relatively linear path (i.e. 'smooth swimming'), while CW rotation generally disrupts the bundle of flagella such that they dissociate, causing the bacteria to tumble; tumbling permits re-orientation of the cell. In general, cells that are heading towards an attractant or away from a repellent suppress CW rotation and thus tumbling, and on average swim for extended amounts of time. As one might expect, the ability to switch the direction of flagellar rotation involves an intricate network of regulators. One such regulator, the RR CheY, appears to be critical in the ability to switch flagellar rotation from CCW to CW. Upon phosphorylation, CheY binds at the base of the flagellar motor to the 'switch' complex, increasing the likelihood that the flagella will switch from CCW to CW rotation (Bren and Eisenbach, 1998). In *E. coli*, *cheY* mutants exhibit a smooth swimming phenotype with relatively few 'tumbles', since the switch to CW rotation is absent (reviewed in Silversmith and Bourret, 1999). As a result, the cells are unable to re-orient to follow a gradient of attractant.

In *V. fischeri*, a *cheY* mutant also exhibits a 'smooth swimming' phenotype and is unable to migrate through soft agar (Hussa *et al.*, 2007). Furthermore, this mutant is defective during colonization: when juvenile squid were inoculated with a mixture of *cheY* mutant and wild-type strains, only the wild-type strain was recovered from the animals (Hussa *et al.*, 2007). Together, these data suggest that *V. fischeri* utilizes chemotaxis to promote migration into the light organ and/or to reach the deep crypt spaces (Fig. 18.2B and C). *V. fischeri* has been shown to swim towards *N*-acetylneuraminic acid (a component of the mucus) and nucleosides (DeLoney-Marino *et al.*, 2003), but whether either of these molecules promotes directed migration into the light organ crypts remains to be determined. The question of how *V. fischeri* finds its way into the light organ thus remains an exciting outstanding question to be explored.

Control of bioluminescence

The major product of the *V. fischeri*–squid symbiosis is bioluminescence, which the squid uses to avoid detection by predators (Jones and Nishiguchi, 2004). The light organ comprises tissues whose function it is to control light emission (McFall-Ngai and Montgomery, 1990; Montgomery and McFall-Ngai, 1992, 1993). For example, a reflector tissue directs the light downwards through a lens, which diffuses the light (Montgomery and McFall-Ngai, 1992; Crookes *et al.*, 2004). In addition, the wavelength of the light emission is thought to be modified by passage through an accessory tissue to match that of the down-welling moonlight (McFall-Ngai and Montgomery, 1990). Furthermore, the light can be concealed by a muscle-controlled ink sac, a mechanism that permits the squid to adapt rapidly to prevent light emission when there's no moonlight, such as when the moon is obscured by clouds (McFall-Ngai and Montgomery, 1990). In addition to these mechanisms, the squid is thought to control light production in two other ways by: (1) controlling the amount of oxygen delivered to the bacteria; luminescence production requires oxygen (Boettcher *et al.*, 1996); and (2) controlling the size and growth of the bacterial population through the daily expulsion of *V. fischeri* (reviewed in Stabb, 2006). Finally, in contrast to many bioluminescent bacteria, and even other strains of *V. fischeri*, the bacteria isolated from squid are non-visibly bioluminescent, although light production can be measured (Stabb *et al.*, 2008). In fact, the amount of light per cell (specific luminescence) is about 1000-fold higher in symbiosis than under standard laboratory conditions (Boettcher and Ruby, 1990), suggesting that symbiotic conditions favour induction of bioluminescence. Given the role of luminescence and the evidence of the various levels of control over its production and emission, it is perhaps not surprising that the ability to bioluminesce is important for *V. fischeri* to colonize its squid host: mutants defective for genes involved in the production and regulation of bioluminescence exhibit initiation and/or persistence defects during colonization (Stabb *et al.*, 2008). In this section, we describe the TCSs involved in bioluminescence production, and their roles in symbiosis.

Lux

Production of bioluminescence requires the *lux* operon, *luxCDABEG*. The *luxA* and *luxB* genes encode the two subunits of the enzyme luciferase, which utilizes long-chain aliphatic aldehyde, reduced flavomononucleotide ($FMNH_2$), and oxygen to produce light, aliphatic acid, oxidized flavomononucleotide (FMN) and water (Ziegler and Baldwin, 1981). The other *lux* genes encode an aliphatic acid reductase complex (LuxC, LuxD, and LuxE), involved in recycling aliphatic acid to aldehyde (Boylan *et al.*, 1989), and a protein that reduces FMN to $FMNH_2$ using NADH (LuxG) (Zenno and Saigo, 1994). These proteins together are sufficient to permit a non-bioluminescent organism, such as *E. coli*, to produce light (Engebrecht *et al.*, 1983; Engebrecht and Silverman, 1984).

V. fischeri utilizes a complex network of regulators, including two-component regulators, to control light production in a cell density-dependent manner (i.e. quorum sensing): bioluminescence is repressed at low cell densities and induced at high cell densities. Cell densities are sensed through the use of autoinducers (AIs), small molecules that are secreted from the cell and accumulate at high cell density. Specifically, two separate AI-based regulatory circuits control bioluminescence: the LuxR/LuxI (LuxR/I) circuit and the Lux phosphorelay (Fig. 18.9C).

LuxR is a transcription factor that becomes competent to activate transcription when bound to the AI *N*-3-oxo-hexanoyl homoserine lactone (3-O-C_6-HSL), synthesized by LuxI (Sitnikov *et al.*, 1995). During growth in culture, luminescence is initially low, as is the concentration of 3-O-C_6-HSL, but as the cell density increases, so does the concentration of this AI. This increased concentration of AI favours its interaction with LuxR, which becomes competent to induce *luxICDABEG* transcription, leading to the production of bioluminescence (Fig. 18.9C) (Engebrecht and Silverman, 1984; Meighen, 1991; Stevens *et al.*, 1994). As a result of this regulatory mechanism, *luxR* and *luxI* are required for luminescence: *luxR* and *luxI* mutants of visibly luminescent strains (e.g. the non-squid symbiont MJ1) are dark in culture (Dunlap and Kuo, 1992). In ES114, a non-visibly luminescent strain isolated from squid, *luxR* and *luxI* mutants exhibited only slight decreases in luminescence in culture (Visick *et al.*, 2000). However, during squid colonization, the *luxR* and *luxI* mutants of ES114 produced no detectable light at any time, indicating that these regulators are, in fact, critical for luminescence induction as predicted from the literature. Furthermore, the *luxR* and *luxI* mutants exhibited a colonization defect: they were unable to sustain the high colonization levels of the wild-type strain (Visick *et al.*, 2000). These data indicate that these regulators are necessary for *V. fischeri* to persist within the light organ. Similarly, mutants defective for the luciferase gene *luxA* (Visick *et al.*, 2000) or the entire *lux* operon (Bose *et al.*, 2008) exhibited a defect in persistence as well, suggesting that the colonization defects of *luxR* and *luxI* mutants can be attributed to a lack of bioluminescence. Taken together, these data indicate that both regulation of bioluminescence and bioluminescence itself are important for *V. fischeri* to colonize *E. scolopes*.

The second *lux* regulatory circuit in *V. fischeri* consists of the Lux phosphorelay, which also reports on cell density but does not rely on the AI produced by LuxI. The Lux phosphorelay is predicted to be similar to the well-studied pathway in *V. harveyi* (Ng and Bassler, 2009), which consists of three AI synthase proteins (LuxS, LuxM, and CqsA), which produce the signals detected by three hybrid SKs (LuxQ, in association with LuxP, LuxN, and CqsS, respectively), the HPt protein LuxU, and the RR LuxO (Fig. 18.9A,B). Briefly, at low cell densities when the AI concentrations are low, the hybrid SKs function as kinases, donating phosphoryl groups to LuxU, which serves as a phospho-donor to LuxO. Phosphorylated LuxO directs the transcription of five small RNAs (sRNAs) (*qrr1–5*), which bind to and destabilize the mRNA transcript of a transcription factor known as LuxR and which we will designate as $LuxR_{Vh}$; the *V. harveyi* protein is not homologous to LuxR from *V. fischeri*. $LuxR_{Vh}$ serves as the direct transcriptional activator of the *luxCDABEGH* operon. Thus, at low cell densities, little $LuxR_{Vh}$ is produced owing to transcript instability, and luminescence is not induced (Fig. 18.9A). As the cell density increases, AI signalling induces the hybrid SKs to switch from kinases to phosphatases, promoting the removal of the phosphoryl group from LuxO, through LuxU. Unphosphorylated LuxO is no longer active, leading to increased $LuxR_{Vh}$ protein and subsequent induction of luminescence (Fig. 18.9B).

The Lux phosphorelay in *V. fischeri* is similar to that of *V. harveyi*, and is predicted to function in largely the same way (Fig. 18.9C). *V. fischeri* contains homologues of LuxP, LuxQ, LuxU, LuxO, and LuxS, as well as LuxN and LuxM (called AinR and AinS, respectively). However, there are a few distinct differences between the Lux phosphorelays in *V. harveyi* and *V. fischeri*: (1) LuxM

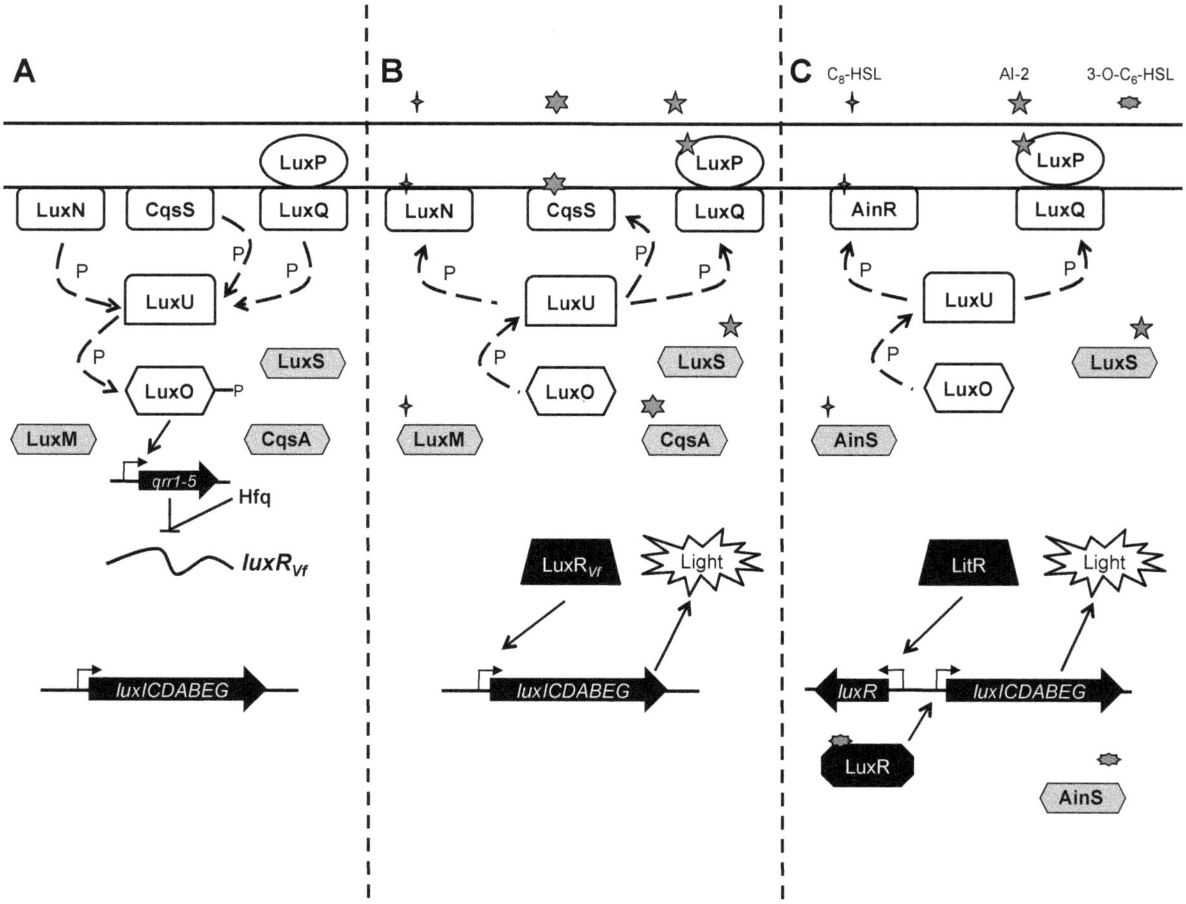

Figure 18.9 Models of bioluminescence regulation by Lux in *V. harveyi* and *V. fischeri*. (A) In *V. harveyi*, at low cell densities the AI concentration is low and the hybrid SKs (LuxN, CqsS, and LuxQ) act as kinases, serving as phospho-donors to the HPt protein LuxU, which serves as the phospho-donor to the RR LuxO. LuxO becomes active when phosphorylated and promotes the transcription of five sRNAs (*qrr1–5*), which, in conjunction with Hfq, bind to and destabilize the mRNA transcript of the transcriptional activator $LuxR_{Vh}$. Without $LuxR_{Vh}$, the *lux* operon is not transcribed and luminescence is not induced. (B) In *V. harveyi*, at high cell densities the AI concentration increases (represented by the diamond, sun, and star), and thus these molecules can interact with their respective hybrid SK and induce a conformational change that promotes phosphatase over kinase activity (in the case of LuxQ, AI-2 binds to LuxP, a periplasmic associated protein, which promotes the conformational change in LuxQ). As phosphatases, the hybrid SKs promote the removal of phosphoryl groups from LuxO, through LuxU. Unphosphorylated LuxO is no longer able to promote transcription of *qrr1–5*, in which case $LuxR_{Vh}$ is produced. $LuxR_{Vh}$ binds to the promoter region of the *lux* operon and promotes transcription, thus inducing luminescence. (C) In *V. fischeri*, Lux regulation at low (not shown) and high (shown) cell densities is predicted to be similar to that of *V. harveyi*, except: (1) there is only one *qrr* sRNA, which, in conjunction with Hfq, binds to and destabilizes the mRNA transcript for the transcriptional activator protein LitR, which is similar to $LuxR_{Vh}$; (2) at high cell densities, LitR is produced and activates transcription of *luxR* in *V. fischeri*, which, when bound by its AI 3-O-C6-HSL promotes transcription of the *lux* operon in *V. fischeri*. *V. harveyi* does not contain this LuxR/I circuit, while *V. fischeri* does not contain CqsS or CqsA. The models in this figure are based on the work reviewed in Ng and Bassler (2009) and Stabb *et al.* (2008).

and AinS do not produce the same AI (Cao and Meighen, 1989; Kuo *et al.*, 1994); (2) *V. fischeri* does not encode CqsS or CqsA; and (3) *V. harveyi* does not contain the LuxR/I circuit. In addition there is a nomenclature difference: $LuxR_{Vh}$ is homologous to LitR from *V. fischeri*, which is the transcriptional regulator for *luxR* from *V. fischeri*. Similar to $luxR_{Vh}$, *litR* is subject to LuxO-mediated regulation (Fig. 18.9C), which will be discussed further below.

At the current time, no studies have investigated the roles in symbiosis of the *V. fischeri* HPt protein LuxU or the hybrid SKs LuxQ and AinR, although one (Lyell *et al.*, 2010) examined the role of AinR in the control of bioluminescence in the squid symbiont. Furthermore, no genetic studies have verified that LuxQ and AinR work upstream of LuxU and LuxO in *V. fischeri*. Thus, for the remainder of this section, we will focus solely on the RR LuxO, the phenotypes it controls, and the input signals that control its activity. For a more comprehensive review of the Lux regulators in *V. fischeri* and their roles in bioluminescence and symbiosis, we recommend the review by Stabb *et al.* (2008).

LuxO-mediated regulation of bioluminescence

LuxO was initially identified as a regulator of bioluminescence in *V. harveyi* (Bassler *et al.*, 1994). LuxO homologues were subsequently identified in two *V. fischeri* strains: ES114, a squid symbiont, and MJ1, an organism that does not colonize squid, but in which luminescence has been extensively studied (Ruby and Nealson, 1976; Miyamoto *et al.*, 2000). In both strains, LuxO exerts negative control over bioluminescence: *V. fischeri luxO* mutants exhibited an increase in luminescence over the wild-type strains (Miyamoto *et al.*, 2000; Lupp *et al.*, 2003). These results are similar to what was observed in *V. harveyi*, and indicate that LuxO serves as an inhibitor of luminescence in both species.

The LuxO protein from MJ1 has been characterized in greater detail than that of ES114. The MJ1 protein shares 72% identity with LuxO from *V. harveyi*, and contains the conserved REC domain and putative site of phosphorylation, D47, in its N-terminus (Fig. 18.4A) (Miyamoto *et al.*, 2000). LuxO also contains a centrally located putative σ^{54}-interaction domain, with a predicted ATP-binding site, and a C-terminal putative DNA binding domain (Fig. 18.4A). These features are conserved in the ES114 protein, as the two *V. fischeri* proteins differ at only two residues (V/G114 and G/C177). Thus, like the *V. harveyi* protein, *V. fischeri* LuxO likely activates transcription at a σ^{54}-dependent promoter (Fig. 18.7) to control expression of genes involved in luminescence.

In support of that idea, Miyashiro *et al.* (2010) demonstrated that, as in other vibrios, LuxO promotes the transcription of an sRNA, designated *qrr1*. Unlike *V. harveyi*, which contains five *qrr* genes, and *V. cholerae*, which contains four *qrr* genes (Lenz *et al.*, 2004), *V. fischeri* contains only one *qrr1* gene (Miyashiro *et al.*, 2010). Based on the model generated in *V. harveyi* (Fig. 18.9A), Qrr1 from *V. fischeri* is predicted to bind to and destabilize the *litR* mRNA transcript, in conjunction with Hfq, resulting in little or no LitR protein expression. Thus, LitR would not be available to activate transcription of *luxR*, which is required to induce the *lux* operon and light production. The following data support this model: (1) in a *qrr1* mutant, *litR* transcript levels and luminescence were increased, similar to a *luxO* mutant (Miyamoto *et al.*, 2003; Miyashiro *et al.*, 2010), suggesting that Qrr1 negatively regulates luminescence through controlling *litR* transcript levels; (2) *qrr1* expression in wild-type cells was high at low cell density and decreased at high cell density (Miyashiro *et al.*, 2010), indicating quorum sensing control. Furthermore, *qrr1* expression remained low in the *luxO* mutant, indicating that LuxO activates *qrr1* expression at low cell density; (3) luminescence in a *luxO qrr1* double mutant was indistinguishable from either of the single mutants, indicating that these regulators function in the same pathway; (4) overexpression of *qrr1* in the *luxO qrr1* double mutant resulted in a decrease in *litR* transcript levels (Miyashiro *et al.*, 2010), suggesting that Qrr1 alone is sufficient to regulate *litR* mRNA transcript levels. Together, these data indicate that, like in *V. harveyi*, the *V. fischeri* LuxO protein is indirectly involved in regulating *litR* transcript levels through regulating the transcription of the sRNA Qrr1. Furthermore, these data demonstrate that LuxO connects the Lux phosphorelay to the LuxR/I circuit, since LitR is the direct transcriptional activator of *luxR* (Fig. 18.9C).

During squid colonization, a *luxO* mutant exhibits an initiation defect: when used to inoculate juvenile *E. scolopes*, the *luxO* mutant only reaches 37% of the colonization level of wild-type *V. fischeri* at 12 hours post-inoculation (Lupp and Ruby, 2005). This is an early time point, before the animals become fully colonized; thus, these data indicate that the *luxO* mutant has a defect in initiating colonization. However, at 72 hours post-inoculation, the colonization levels of the wild-type strain and the *luxO* mutant are comparable (Lupp and Ruby, 2005), suggesting that the *luxO* mutant is only defective at initiation, not persistence. This is perhaps not surprising, as at this late time point, luminescence is fully induced and LuxO should be inactivated. The *luxO* mutant is also defective for colonization when the wild-type strain is present: when juvenile squid were inoculated with a mixture of wild-type and *luxO* mutant strains, the *luxO* mutant exhibited a colonization disadvantage relative to the wild-type strain 24–48 hours post-inoculation (Lupp and Ruby, 2005; Hussa *et al.*, 2007; Miyashiro *et al.*, 2010). Whether the competition defect can be attributed to the initiation defect of the *luxO* mutant, or to the loss of control over a persistence factor not required during single strain colonization, remains to be determined.

What could account for the initiation defect of the *luxO* mutant? As mentioned above, bioluminescence is only required at later stages of colonization (i.e. persistence) (Visick *et al.*, 2000; Lupp and Ruby, 2005) and thus, light per se cannot account for the initiation defect of the *luxO* mutant. Array experiments revealed that mutations in *luxO* impacted expression of motility genes, and in support of this, *luxO* mutants exhibited decreased migration in motility assays (Lupp and Ruby, 2005; Hussa *et al.*, 2007). Previous studies have demonstrated that motility is essential for the initiation of colonization (e.g. Graf *et al.*, 1994), and thus the motility defect could be responsible for the initiation defect of the *luxO* mutant. Alternatively, other targets of LuxO activity identified in the array analysis, including regulatory proteins and genes involved in nutrient uptake and metabolism, could account for the initiation defect of the *luxO* mutant. Thus, the precise role of LuxO in initiation (Fig. 18.2C) remains unclear.

In conclusion, LuxO links the Lux phosphorelay to the LuxR/I circuit (Fig. 18.9C) by regulating transcription of a single sRNA, *qrr1*, that likely binds and destabilizes the *litR* mRNA transcript. Furthermore, LuxO is necessary for efficient colonization of the squid host, since a *luxO* mutant exhibits a defect in initiating colonization. Because its role in bioluminescence cannot account for the initiation defect, LuxO must control an additional gene(s) involved in this stage of colonization. Whether this additional gene(s) is controlled directly by LuxO or through downstream regulators (e.g. Qrr1 and/or LitR) remains unknown. Therefore, further studies of LuxO and LuxO-mediated activities are necessary to fully understand how LuxO impacts host colonization.

The autoinducer synthases – regulators of LuxO activity

In many two-component systems, the SK has been well characterized, while the signal(s) sensed by these components are unknown or less well characterized. However, in *V. fischeri* there have been limited studies of the hybrid SKs (AinR and LuxQ), while multiple studies have examined the importance of the signal synthase proteins AinS and LuxS, which produce the AI signal molecules C_8-HSL and an uncharacterized molecule termed AI-2, respectively (Fig. 18.9C). In the following sections, we will discuss how AinS and LuxS regulate LuxO activity and thus luminescence, as well as what is known about how AinS and LuxS influence the symbiosis between *V. fischeri* and its squid host.

AinS

AinS synthesizes the AI, *N*-octanoyl-homoserine lactone (C_8-HSL), proposed to signal through the hybrid SK AinR (Fig. 18.9C) (Gilson *et al.*, 1995). C_8-HSL also appears to serve as a second AI that directly activates LuxR-mediated transcription of the *lux* operon (Kuo *et al.*, 1994, 1996), but we will not focus on this aspect of regulation here. AinS was initially identified and characterized as a luminescence regulator in MJ1 (Kuo *et al.*, 1994, 1996; Gilson *et al.*, 1995), although recent studies have focused on AinS from the squid symbiont ES114. Thus, we will focus our discussion on studies involving AinS from this squid symbiont.

In *V. harveyi*, LuxM (a homologue of AinS) synthesizes an AI sensed by the hybrid SK LuxN, which functions through LuxU to control the activities of LuxO (Fig. 18.9B) (Ng and Bassler, 2009). Thus, although the AinS-produced signal is distinct from that produced by LuxM, it likely functions in a similar manner, signalling to the hybrid SK AinR to regulate LuxO activity in *V. fischeri* (Fig. 18.9C). Indeed, an *ainS* mutant of ES114 produced no detectable light in culture (Lupp *et al.*, 2003). Surprisingly, this mutant also exhibited a growth yield defect, reaching only 75% of the wild-type level (Lupp *et al.*, 2003). This growth yield defect could be suppressed by the addition of exogenous C_8-HSL, indicating that the loss of the signal molecule was indeed responsible for the phenotype. Thus, it appears that, like LuxO, the AinS-produced signal regulates more than just luminescence. In support of this idea, an array analysis that compared transcripts from wild-type and *ainS* mutant cells revealed differences in the control of genes involved in metabolism and nutrient uptake (Lupp and Ruby, 2005). Subsequently, Studer *et al.* (2008) demonstrated that the *ainS* mutant was unable to utilize the acetate that is normally secreted and assimilated by *V. fischeri* cells. As a result, the mutant cultures became acidified, which caused the loss of cell viability. Thus, AinS activity appears to regulate multiple cellular functions, including luminescence and acetate utilization.

The *V. harveyi* paradigm predicts that the AinS-produced signal, C_8-HSL, should function to control LuxO activity. If this were true, then the phenotypes of the *ainS* mutant should depend on LuxO function. Indeed, this was the case: a double *ainS luxO* mutant exhibited luminescence and growth yield phenotypes indistinguishable from the *luxO* single mutant (i.e. high luminescence and normal growth) (Lupp *et al.*, 2003). These results are consistent with the idea that AinS or, more specifically, C_8-HSL acts upstream of LuxO to control both luminescence and acetate utilization.

To understand how *ainS* itself was regulated, Lupp and Ruby (2004) evaluated levels of C_8-HSL and *ainS* transcription during growth of ES114. They found that over time the concentration of C_8-HSL increased over 2000-fold, concurrent with the increase in luminescence. Furthermore, *ainS* transcription was initially low, but increased exponentially until cells reached stationary phase, consistent with the increase in the amount of C_8-HSL produced during growth. These data suggested that *ainS* is autoregulated. The same pattern of *ainS* transcription occurred in *luxR* mutants, indicating that *ainS* was autoregulated independently of the LuxR/I circuit. However, autoregulation depended upon the Lux phosphorelay, as *ainS* transcription was decreased in an *ainS* mutant, but could be restored to wild-type levels by the addition of a mutation in *luxO* (Lupp and Ruby, 2004). Thus, the C_8-HSL signal not only controls luminescence and metabolism, but also its own production, in a manner that depends on LuxO.

Although the *ainS* mutant produced no light in culture (Lupp *et al.*, 2003), similar to a *luxA* mutant (Visick *et al.*, 2000), during colonization it produced detectable levels of light (between 10% and 40% of that observed for the wild-type strain) (Lupp *et al.*, 2003). These data indicate that other factors contribute to luminescence control during symbiosis. This mutant also exhibited an initiation defect: at an early time (12 hours) following inoculation, the *ainS* mutant reached only 45% of the wild-type colonization level (Lupp and Ruby, 2005). This defect occurred after the aggregation stage, as this mutant was not defective in the timing of aggregate production or the size of the aggregate. Finally, the *ainS luxO* double mutant exhibited the same initiation defect as the *luxO* single mutant (37% and 36% colonization, respectively), which wasn't much different from the single *ainS* mutant (45% colonization). These results support the idea that AinS and LuxO function in the same pathway, and suggest that the initiation defect of the *ainS* mutant may be due to some LuxO-regulated process.

In addition to its initiation defect, the *ainS* mutant exhibited a persistence defect: at 24 hours post-inoculation, the levels of colonization by the mutant were 75% that of wild-type, and further decreased to 20% at 72 hours post-inoculation (Lupp and Ruby, 2004). However, this persistence defect was not due to the defect in acetate utilization: an *acs* mutant, which is similarly defective in acetate utilization, achieved the same level of colonization as the wild-type strain at 48 hours post-inoculation (Studer *et al.*, 2008). Finally, disruption of *luxO* in the *ainS* mutant background restored colonization levels to that achieved by both the wild-type and single *luxO* mutant strains, indicating that the defect of the *ainS* mutant depends upon LuxO function (Lupp and Ruby, 2005).

In summary, the AinS-produced signal (C_8-HSL) controls multiple phenotypes via its control over LuxO activity, and plays roles in symbiosis during at least two stages of colonization. Given the differences in the environments at these stages, it seems reasonable to expect that different C_8-HSL-controlled processes are important in initiation versus persistence. However, further studies of AinS are necessary to determine which genes under C_8-HSL control participate at these different times, and what other regulators might contribute to dictate the appropriate cellular response.

LuxS

In *V. harveyi*, LuxS produces AI-2, a furanosyl borate diester (Chen *et al.*, 2002). *V. fischeri* also encodes a LuxS protein, and it appears to produce a similar autoinducer molecule: supernatants from wild-type *V. fischeri*, but not a *luxS* mutant contained AI-2 activity, as measured by the ability to activate luminescence of a *V. harveyi luxS* mutant (Lupp and Ruby, 2004). Furthermore, AI-2 from *V. fischeri* appears to signal through LuxP/Q as it does in *V. harveyi* (Fig. 18.9): wild-type supernatant from *V. fischeri* was unable to induce luminescence in a *V. harveyi luxQ* mutant. To date, however, there have been no studies of LuxP or LuxQ in *V. fischeri*, nor confirmation of the structure of the autoinducer molecule.

In *V. fischeri*, a *luxS* mutant reaches about 70% of the luminescence achieved by the wild-type strain (Lupp and Ruby, 2004), suggesting that AI-2 contributes to luminescence regulation in *V. fischeri*, although not to the same extent as C_8-HSL. Consistent with its mild effect on luminescence, *luxS* transcription was low and unchanged throughout growth of the cells in culture, and AI-2 levels varied only 4-fold. These data indicate that *luxS* is constitutively expressed, rather than autoregulated like *ainS*. Finally, as predicted from the *V. harveyi* model, AI-2 from *V. fischeri* functions upstream of LuxO: the luminescence output of a double *luxS luxO* mutant resembled the *luxO* single mutant (i.e. increased luminescence) (Fig. 18.9B and C) (Lupp and Ruby, 2004). Together, these data suggest that LuxS plays a role, albeit a minor one, in controlling LuxO activity and thus luminescence regulation in culture.

Consistent with this relatively small role in luminescence regulation in culture, the *luxS* single mutant had no observable impact on either symbiotic bioluminescence or squid colonization (Lupp and Ruby, 2004). However, combining the *luxS* and *ainS* mutations resulted in a synergistic defect in colonization: the double mutant reached colonization levels that were 50–75% that of the *ainS* single mutant at both 24 and 48 hours post-inoculation (Lupp and Ruby, 2004). In contrast, this synergistic effect did not appear to occur during colonization initiation: at 12 hours post-inoculation, the *luxS ainS* mutant was not different from that of an *ainS* single mutant (both were approximately 45% that of wild-type), indicating that either AinS plays distinct roles during initiation and persistence, or that AI-2 (produced by LuxS) plays a stronger role during persistence than in initiation. In either case, LuxS and AinS appear to function in the same pathway to regulate luminescence and LuxO activity, since an *ainS luxS luxO* triple mutant did not exhibit a persistence defect, but resembled the *luxO* single mutant.

In summary, the Lux phosphorelay plays multiple roles during the symbiosis between *V. fischeri* and *E. scolopes* (Fig. 18.2B and C). The *V. harveyi lux* phosphorelay provides a good model for understanding luminescence regulation in *V. fischeri*, but it is clear that the pathways of these organisms differ in a number of ways, including the number of *qrr* genes and the existence of the LuxR/I circuitry in the symbiont (Fig. 18.9). Many questions remain. For example, because luminescence itself cannot account for the various symbiotic defects of regulatory *lux* mutants, what are the downstream targets that are responsible? Do the hybrid SKs LuxQ and AinR function similarly to their *V. harveyi* homologues, and if so, what is the strength of each input in controlling luminescence? Does the Lux pathway in *V. fischeri* regulate biofilm formation or other colonization factors as it does in *V. harveyi* and *V. cholerae* (Lilley and Bassler, 2000; Miller *et al.*, 2002; Zhu *et al.*, 2002; Vance *et al.*, 2003)? The answers to these questions and many others will aid in our understanding of how bacteria use quorum sensing during host interactions.

ArcA

Lux is not the only pathway that controls bioluminescence in *V. fischeri*. The ArcA/ArcB TCS system is involved in repressing bioluminescence, as well as regulating a variety of other cellular functions (Bose *et al.*, 2007). In *E. coli*, the ArcA/ArcB two-component system is thought to respond to the redox state of the cell (Gennis and Stewart, 1996; Georgellis *et al.*, 2001). Under reducing conditions, the hybrid SK ArcB initiates a phosphorelay, autophosphorylating on a conserved histidine residue and subsequently serving as a phosphodonor to activate the RR ArcA (Georgellis *et al.*, 1997). Phosphorylated ArcA then serves to either positively or negatively regulate transcription from a variety of genes (Lynch and Lin, 1996; Salmon *et al.*, 2005).

In *V. fischeri*, *arcB* encodes a hybrid SK that is 57% identical to ArcB from *E. coli* and contains the conserved PAS, HisKA, and REC domains, as well as conserved residues involved in redox sensing (C180 and C241) and phosphotransfer (H292, D576, and H717) in *E. coli* (Georgellis *et al.*, 1997; Malpica *et al.*, 2004; Bose *et al.*, 2007). *arcA* encodes a RR that is 84% identical to ArcA from *E. coli* and contains the putative, conserved N-terminal REC domain, with the predicted, conserved site of phosphorylation (D54) and a C-terminal wHTH DNA binding domain (Fig. 18.4A) (Bose *et al.*, 2007). To determine whether the putative ArcA from *V. fischeri* had the same activity as the *E. coli* protein, Bose *et al.* (2007) introduced the *arcA* allele from *V. fischeri* into an *E. coli arcA* mutant; this *E. coli* mutant exhibits sensitivity to redox dyes, such as toluidine blue, due to changes in the electron flow in this mutant (Ruiz *et al.*, 2006). The *V. fischeri arcA* allele was able to restore toluidine blue resistance, demonstrating that ArcA from *V. fischeri* is functionally conserved (Bose *et al.*, 2007).

To further investigate how the role of ArcA in *V. fischeri* compares to *E. coli*, Bose *et al.* (2007) generated an *arcA* mutant and evaluated transcription of a putative ArcA-controlled gene. Specifically, control over the succinate dehydrogenase (*sdh*) gene, which is repressed by ArcA in *E. coli* (Ravcheev *et al.*, 2007), was assessed using a reporter construct with the *sdh* promoter fused to *lacZ*. Transcription from this reporter was increased in the *V. fischeri arcA* mutant (Bose *et al.*, 2007). Thus, it appears that *V. fischeri* ArcA functions like the *E. coli* protein.

The *V. fischeri arcA* mutant exhibited additional phenotypes. For example, it exhibited a slightly slower growth rate (Bose *et al.*, 2007, Hussa *et al.*, 2007) and a decrease in motility (Hussa *et al.*, 2007). This mutant also exhibited a substantial (approximately 500-fold) increase in bioluminescence (Bose *et al.*, 2007). This increased luminescence could be restored to wild-type levels when the mutant was complemented with the wild-type *arcA* allele, but not with an *arcA* allele in which the putative, conserved site of phosphorylation (D54) was changed to a glutamate (Bose *et al.*, 2007); in other RRs, this mutation results in a protein that is active but not phosphorylated (see, for example, Sanders *et al.*, 1989, 1992; Freeman and Bassler, 1999). These results suggest that ArcA represses luminescence, and requires the putative, conserved site of phosphorylation to do so.

One mechanism by which ArcA could control luminescence is to serve as a direct repressor by binding at the *lux* operon. In support of this hypothesis, the intergenic region between *luxR* and *luxI* contains a putative ArcA binding site that matches the core consensus sequence (5′-ATGTTAA-3′) recognized by *E. coli* ArcA (Liu and De Wulf, 2004; Bose *et al.*, 2007). More importantly, purified, phosphorylated ArcA from *E. coli* was able to bind to this promoter region in *in vitro* experiments (Bose *et al.*, 2007). Thus, ArcA serves as a direct repressor of luminescence in *V. fischeri*.

Given the known function of ArcA/ArcB in *E. coli*, it seemed probable that this system responds to the redox state of the cell in *V. fischeri* as well. Previous studies had demonstrated that activation of ArcA/ArcB from *E. coli* could be achieved by the addition of the reducing agent dithiothreitol (DTT) (Malpica *et al.*, 2004); thus, the same could be true for ArcA/ArcB from *V. fischeri*. Indeed, the addition of DTT to wild-type *V. fischeri* led to a 22-fold reduction in luminescence, but had no effect on light production by the *arcA* mutant (Bose *et al.*, 2007). Therefore, the ArcA/

ArcB two-component system in *V. fischeri* appears to function as a redox-responsive system, similar to the homologous system in *E. coli*.

Since ArcA affects luminescence and certain metabolic processes in *V. fischeri*, it seemed likely that ArcA might also play a role in symbiosis. Indeed, during mixed inoculation experiments using the *arcA* mutant and the wild-type strain, the wild-type strain out-competed the mutant by 4-fold at 48 hours post-inoculation (Bose *et al.*, 2007), although this defect was not observed at 24 hours post-inoculation (Hussa *et al.*, 2007). These results suggest that ArcA may be important later in host colonization (Fig. 18.2C); however, it is also possible that a 4-fold effect would not have been identified in the screen by Hussa *et al.* (2008). Surprisingly, the *arcA* mutant did not exhibit a colonization defect in single strain assays: there was no difference in the start of luminescence induction or number of bacteria per squid as compared with the wild-type strain (Bose *et al.*, 2007). These data suggest that ArcA might be inactivated during colonization (Fig. 18.2C). In support of this idea is the fact that the wild-type strain exhibits low specific luminescence (luminescence per cell) in culture, but increased specific luminescence in the squid, while the specific luminescence of the *arcA* mutant is similarly high in both conditions (Bose *et al.*, 2007). Thus, inactivation of ArcA may be important during colonization, although it would be necessary to examine the impact of constitutively active alleles of *arcA* and/or *arcB* during colonization to support this idea.

In summary, ArcA is a RR that exerts a strong impact on luminescence control in *V. fischeri*. It is likely that ArcA-mediated repression of luminescence in culture can account for the observation that the squid symbiont ES114 exhibits a 1000-fold decrease in specific luminescence in culture relative to that achieved during symbiosis (Boettcher and Ruby, 1990). It will be interesting to understand what other functions, if any, are controlled by ArcA that may contribute to colonization.

Global regulation and host development

During symbiotic colonization, *V. fischeri* promotes a series of morphogenic changes in the squid light organ (Fig. 18.10). For example, mucus secretion, which facilitates aggregate formation on the surface of the light organ, ceases in animals colonized by *V. fischeri* (Fig. 18.10B and C) (Nyholm *et al.*, 2002). In addition, the epithelial cells within the appendages on the surface of the light organ (Fig. 18.10) undergo apoptosis (Foster and McFall-Ngai, 1998). Subsequently, the appendages undergo regression and are lost over a 4-day period (Montgomery and McFall-Ngai, 1994; Doino and McFall-Ngai, 1995). These changes are a normal part of light organ maturation and considered hallmarks of successful colonization; likely, most if not all of these

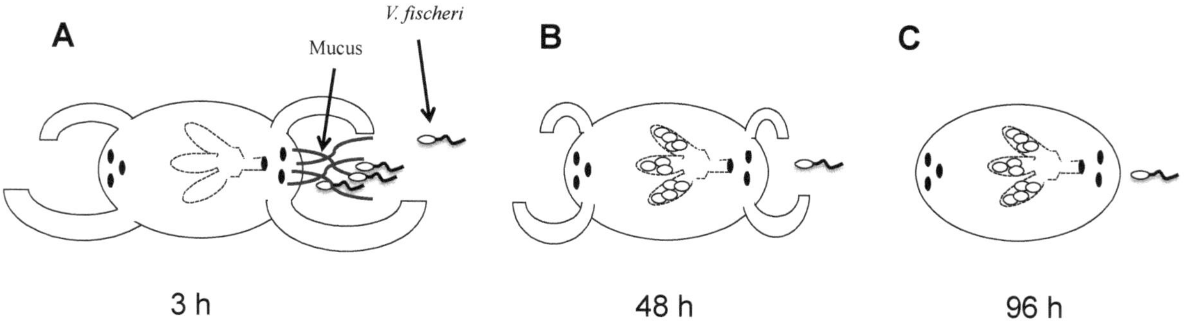

Figure 18.10 Host development induced by *V. fischeri*. (A) Shortly after hatching, *E. scolopes* juveniles secrete mucus from ciliated epithelial appendages present on the surface of the light organ. The mucus facilitates aggregation by *V. fischeri*. (B) Within 48 h, colonizing *V. fischeri* cells have signalled for mucus secretion to cease, and for the apoptosis and regression of the surface appendages. (C) Within 96 h, the appendages have fully regressed. The combinations of these developmental events restricts subsequent colonization by additional *V. fischeri* cells.

changes contribute to the specificity of the association and prevent subsequent 'superinfection' by additional *V. fischeri* cells. These developmental events are triggered by bacterial cell wall components released from colonizing bacteria, in particular LPS and a component of peptidoglycan (Foster *et al.*, 2000; Koropatnick *et al.*, 2004). Although these signals are known, much remains to be learned about the bacterial factors controlling host development. In this section, we focus on one RR, GacA, that appears to play a role in a variety of host developmental events.

GacA

The GacS/GacA two-component system (BarA/UvrY in *E. coli*) is used by a variety of bacteria to mediate bacteria–host interactions such as colonization, production of host- and tissue-specific effectors, and adaptation to the host environment (Heeb and Haas, 2001). This system has been well studied in bacteria involved in pathogenic associations (e.g. *Salmonella* spp., *V. cholerae*, and *Erwinia carotovora*), as well as in bacteria involved in benign (mutualistic) associations (e.g. *Pseudomonas* spp.–plant interactions) (Heeb and Haas, 2001). In *V. cholerae*, for example, the RR GacA, termed VarA, is involved in mediating pathogenesis (Wong *et al.*, 1998). The *V. fischeri* GacA protein is 85% identical to VarA, and, like other GacA proteins, contains a putative N-terminal REC domain with the predicted, conserved site of phosphorylation (D54) and a putative C-terminal helix–turn–helix DNA binding domain (Fig. 18.4A) (Whistler and Ruby, 2003).

The construction and analysis of a *gacA* mutant of *V. fischeri* revealed a significant role for this regulator in controlling a variety of cellular functions as well as symbiosis. For example, the *gacA* mutant exhibited altered colony morphology (smaller, less yellow, and translucent) when grown on nutrient agar and achieved a lower growth yield when cultured in a variety of complex liquid media (Whistler and Ruby, 2003). Furthermore, the *gacA* mutant was unable to grow on minimal media containing simple sugars as carbon sources (Whistler and Ruby, 2003). These results suggest that GacA regulates processes involved in growth-substrate utilization (Fig. 18.2B). The *gacA* mutant also exhibited decreased siderophore activity, and altered motility and chemotaxis (i.e. appeared to deplete serine from the media more rapidly than the wild-type strain) (Whistler and Ruby, 2003).

GacA also appears to regulate luminescence: the mutant produced no detectable light in culture, although this defect does not appear to be due to changes in the production or concentration of the AI signals (Whistler and Ruby, 2003). These data suggest that GacA may regulate luminescence via another mechanism, although this mechanism has yet to be determined. One possible explanation for how GacA regulates luminescence in *V. fischeri* comes from data in *V. cholerae*. VarA (the GacA homologue) controls the production of sRNAs that inhibit CsrA, a protein involved in post-transcriptional regulation of processes as diverse as biofilm formation and carbon metabolism (Romeo *et al.*, 1993; Romeo, 1998). CsrA also controls luminescence, apparently by affecting the activity of the RR LuxO (Lenz *et al.*, 2005). Thus, a similar mechanism may occur in *V. fischeri*, which contains a homologue of CsrA, as well as the inhibitor sRNAs (Kulkarni *et al.*, 2006).

During colonization, the *gacA* mutant exhibited an initiation defect: the mutant colonized only 51% of squid, and required longer incubation times (14–18 hours) and a higher inoculum to colonize 100% of squid (Whistler and Ruby, 2003). This defect was likely due, at least in part, to the inability of the mutant to aggregate outside of the light organ (Whistler *et al.*, 2007). Furthermore, animals that became colonized by the *gacA* mutant contained much lower bacterial loads than wild-type-colonized animals (approximately 1.5×10^3 vs. 1.2×10^5 bacteria per squid, respectively); thus, GacA appears critical for the accommodation stage as well (Fig. 18.2B and C) (Whistler and Ruby, 2003). Finally, GacA does not appear to be necessary for persistence, despite playing a role in regulating cellular functions important for persistence, such as luminescence and siderophore production (Graf and Ruby, 2000; Visick *et al.*, 2000): squid colonized by the *gacA* mutant maintained consistent (low) colonization levels up to 48 hours (Whistler and Ruby, 2003). Thus, GacA is an important regulator for both initiating symbiosis and achieving normal population densities within the light organ.

During the initial colonization experiments, an intriguing observation was made: some squid colonized by the *gacA* mutant were luminescent, while others were not (Whistler and Ruby, 2003). Because mutation of *gacA* impacted luminescence in culture, this observation warranted further investigation. Whistler and Ruby (2003) determined that (1) the luminescent squid contained higher population densities than non-luminescent squid, and (2) bacterial cells collected from *gacA*- and wild-type-colonized squid exhibited no difference in specific luminescence per cell (Whistler and Ruby, 2003). Together, these data indicated that as long as sufficient cell densities were reached, the cell density-dependent induction of luminescence could occur, permitting the squid to become luminescent. Thus, in contrast to what was observed in culture, GacA does not appear to be necessary for regulation of luminescence within the host.

Because the *gacA* mutant exhibited severe defects in colonization, Whistler and Ruby (2007) wondered whether the *gacA* mutant was defective in promoting morphogenesis of the light organ. Indeed, this mutant was unable to induce normal levels of apoptosis (Whistler *et al.*, 2007). This phenotype could be due to differences in the LPS profile of the *gacA* mutant: the LPS profile of the mutant, while resembling that of the wild-type strain, also contained lower molecular weight species not produced by the wild-type strain (Whistler *et al.*, 2007). Whether the defect in apoptosis is a result of the altered LPS profile of the mutant (and thus, potentially, the production of the wrong type of LPS, which could alter recognition between the bacterium and the host) or the decreased production of LPS (due to the lower colonization levels of the *gacA* mutant) or another explanation remains to be determined.

Squid colonized with the *gacA* mutant not only failed to undergo apoptosis but also continued to secrete mucus and permit bacterial aggregation at the light organ pores 96 hours post hatching, long after wild-type-colonized animals ceased mucus secretion (Whistler *et al.*, 2007). These data suggested that a *gacA*-mutant-colonized host might be susceptible to a secondary colonizer. Indeed, animals colonized with the *gacA* mutant were displaced from the light organ upon inoculation with the wild-type strain 60 hours after hatching and primary colonization (Whistler *et al.*, 2007). These results confirmed previous studies that suggested that cessation of mucus shedding limits subsequent colonization by other bacteria once a productive association is established (Fig. 18.10B and C) (Nyholm *et al.*, 2002). Furthermore, they suggested the possibility that the host can recognize defective symbionts, and respond by promoting colonization by other potential partners.

In conclusion, GacA appears to be a global regulator of a variety of processes in *V. fischeri*. Although it is difficult to pinpoint the cause of the numerous symbiotic defects that occur upon the loss of *gacA*, these studies have none-theless increased our understanding of the role the bacteria play in influencing host development. In particular, the *gacA* studies have provided insights into how colonized animals limit subsequent colonization by additional *V. fischeri* bacteria. The targeted disruption of members of the GacA regulon in the future will permit a dissection of the various phenotypes impacted by GacA and their importance in symbiosis.

Concluding remarks

V. fischeri encodes numerous TCS regulators, and a significant subset of them play key roles in the ability of this organism to form a productive association with its symbiotic partner, the squid *E. scolopes*. *V. fischeri* clearly experiences different environments as it transitions from seawater to the site of colonization (Figs. 18.1 and 18.2), and likely it detects different signals in each location. For the regulators involved in quorum sensing, the signals are known and are self-produced, but depend on appropriate conditions for their production (i.e. growth to high cell density). For some other TCS regulators, such as ArcA/ArcB, the signal (redox state) is suspected but as yet unverified. Finally, for a number of other regulators, such as RscS and Syp, the signals recognized are as-yet unknown. Thus, an important area of investigation lies in understanding the specific environments experienced at the stage of colonization that requires the activity of a particular TCS system.

Detailed studies of TCS regulators have revealed their involvement at each of the known stages of symbiosis: initiation, accommodation, and persistence (Fig. 18.2). For example, RscS, SypG, and SypE are involved in biofilm formation and aggregation at the initiation stage, and FlrC and CheY are necessary for motility (and likely chemotaxis) somewhere between the initiation and accommodation stages (Fig 18.2B and C). In contrast, LuxO is important at both the initiation and persistence stages, likely regulating non-luminescence processes during the initiation stage and luminescence and/or other factors during persistence (Fig. 18.2B and C). A better understanding of the regulons of each regulator, both in the presence and absence of its signal (if known), will provide testable hypotheses for determining which downstream targets are critical at specific stages of symbiosis.

Of the many TCS regulators encoded by *V. fischeri*, the most important subclass for symbiosis appears to be the σ^{54}-dependent RRs. Only six of the 40 RRs are in this class. However, four of these, LuxO, SypG, FlrC, and VpsR, have been shown to play important roles in symbiosis. NtrC (VF_0095), an additional member of that class, also appears to play a role in symbiosis, based on competition assays reported in the global survey of RRs (Hussa *et al.*, 2007); this regulator has yet to be characterized further. Given the importance of these regulators in symbiosis, it is perhaps not surprising that an *rpoN* mutant, defective for production of σ^{54}, is also defective for symbiotic colonization (Wolfe *et al.*, 2004). This type of regulation lends itself to tighter control: RNA polymerase carrying σ^{54} is unable to promote transcription in the absence of an activator protein, which provides the energy (via ATPase activity) necessary for transcription initiation (i.e. open complex formation) (Fig. 18.7); the result is a decrease in basal levels of transcription (Buck *et al.*, 2000). Additionally, these activator proteins typically require a signal to promote ATPase activity: in the absence of a signal (e.g. phosphorylation in the case of a RR) ATPase activity is off, while the presence of a signal promotes ATPase activity. Given this information, one could consider σ^{54}-dependent promoters as either 'on' or 'off', depending on the presence of environmental conditions and signals that dictate whether the regulatory proteins are activated. Thus, the use of σ^{54}-dependent RRs during symbiotic colonization may be an important mechanism by which certain processes are turned 'on' or 'off' at various times during colonization.

One observation made from this review is that inactivation of certain TCS regulators may be a common mechanism during symbiosis. For example, at least three regulators (SypE, LuxO, and ArcA) appear to be inactivated during colonization (Fig. 18.2C). This observation begs the question, how many regulators must become inactivated to promote symbiosis? The traditional approach of using gene disruptions to evaluate function may not be sufficient to determine whether a given TCS component plays a role in symbiosis, as disruption is equivalent to inactivation. For example, deletion of the RR *sypE* did not lead to noticeable defects in colonization, but the generation of a constitutively active (constitutively inhibitory) mutant led to a severe colonization defect (Morris *et al.*, 2011). Thus, it may become necessary to make point mutations in conserved residues to evaluate the roles of both activation and inactivation of a given regulator to fully test the necessity of a particular regulator in symbiotic colonization. Such point mutation studies, while difficult to accomplish, are becoming more feasible with the availability of the genome sequence and new approaches in mutant construction. Undoubtedly, such studies would provide invaluable insights into the regulatory mechanisms in place during colonization. The foundation is laid with a wealth of knowledge regarding traits and regulators critical to this symbiotic association. Future work should begin to fill in the details and provide a better understanding, in general, of bacteria–host interactions.

Acknowledgements

We thank Elizabeth Hussa and members of the Visick and Wolfe labs for critical reading of the manuscript. We also thank past members of the Visick lab for images of biofilm and non-biofilm forming strains of *V. fischeri*. Work in the Visick laboratory studying the role of TCS regulators in biofilm formation and symbiosis was supported by NIH grant GM59690 awarded to K.L.V.

References

Adler, E., Donella-Deana, A., Arigoni, F., Pinna, L.A., and Stragler, P. (1997). Structural relationship between a bacterial developmental protein and eukaryotic PP2C protein phosphatases. Mol. Microbiol. *23*, 57–62.

Baikalov, I., Schroder, I., Kaczor-Grzeskowiak, M., Cascio, D., Gunsalus, R.P., and Dickerson, R.E. (1998). NarL dimerization? Suggestive evidence from a new crystal form. Biochemistry *37*, 3665–3676.

Barrios, H., Valderrama, B., and Morett, E. (1999). Compilation and analysis of σ^{54}-dependent promoter sequences. Nucleic Acids Res. *27*, 4305–4313.

Bassis, C.M., and Visick, K.L. (2010). The cyclic-di-GMP phosphodiesterase BinA negatively regulates cellulose-containing biofilms in *Vibrio fischeri*. J. Bacteriol. *192*, 1269–1278.

Bassler, B.L., Wright, M., and Silverman, M.R. (1994). Sequence and function of LuxO, a negative regulator of luminescence in *Vibrio harveyi*. Mol. Microbiol. *12*, 403–412.

Boettcher, K.J., and Ruby, E.G. (1990). Depressed light emission by symbiotic *Vibrio fischeri* of the sepiolid squid *Euprymna scolopes*. J. Bacteriol. *172*, 3701–3706.

Boettcher, K.J., Ruby, E.G., and McFall-Ngai, M. (1996). Bioluminescence in the symbiotic squid *Euprymna scolopes* is controled by a daily biological rhythm. J. Comp. Physiol. A *179*, 65–73.

Bose, J.L., Kim, U., Bartkowski, W., Gunsalus, R.P., Overley, A.M., Lyell, N.L., Visick, K.L., and Stabb, E.V. (2007). Bioluminescence in *Vibrio fischeri* is controlled by the redox-responsive regulator ArcA. Mol. Microbiol. *65*, 538–553.

Bose, J.L., Rosenberg, C.S., and Stabb, E.V. (2008). Effects of *luxCDABEG* induction in *Vibrio fischeri*: enhancement of symbiotic colonization and conditional attenuation of growth in culture. Arch. Microbiol. *190*, 169–183.

Bourret, R.B., Hess, J.F., and Simon, M.I. (1990). Conserved aspartate residues and phosphorylation in signal transduction by the chemotaxis protein CheY. Proc. Natl. Acad. Sci. U.S.A. *87*, 41–45.

Boylan, M., Miyamoto, C., Wall, L., Graham, A., and Meighen, E. (1989). Lux C, D and E genes of the *Vibrio fischeri* luminescence operon code for the reductase, transferase, and synthetase enzymes involved in aldehyde biosynthesis. Photochem. Photobiol. *49*, 681–688.

Bren, A., and Eisenbach, M. (1998). The N terminus of the flagellar switch protein, FliM, is the binding domain for the chemotactic response regulator, CheY. J. Mol. Biol. *278*, 507–514.

Buck, M., Gallegos, M.T., Studholme, D.J., Guo, Y., and Gralla, J.D. (2000). The bacterial enhancer-dependent σ^{54} (σ^{N}) transcription factor. J. Bacteriol. *182*, 4129–4136.

Cao, J.G., and Meighen, E.A. (1989). Purification and structural identification of an autoinducer for the luminescence system of *Vibrio harveyi*. J. Biol. Chem. *264*, 21670–21676.

Chen, X., Schauder, S., Potier, N., Van Dorsselaer, A., Pelczer, I., Bassler, B.L., and Hughson, F.M. (2002). Structural identification of a bacterial quorum-sensing signal containing boron. Nature *415*, 545–549.

Correa, N.E., Lauriano, C.M., McGee, R., and Klose, K.E. (2000). Phosphorylation of the flagellar regulatory protein FlrC is necessary for *Vibrio cholerae* motility and enhanced colonization. Mol. Microbiol. *35*, 743–755.

Correa, N.E., Peng, F., and Klose, K.E. (2005). Roles of the regulatory proteins FlhF and FlhG in the *Vibrio cholerae* flagellar transcription hierarchy. J. Bacteriol. *187*, 6324–6332.

Crookes, W.J., Ding, L.L., Huang, Q.L., Kimbell, J.R., Horwitz, J., and McFall-Ngai, M.J. (2004). Reflectins: the unusual proteins of squid reflective tissues. Science *303*, 235–238.

Darnell, C.L., Hussa, E.A., and Visick, K.L. (2008). The putative hybrid sensor kinase SypF coordinates biofilm formation in *Vibrio fischeri* by acting upstream of two response regulators, SypG and VpsR. J. Bacteriol. *190*, 4941–4950.

Davidson, S.K., Koropatnick, T.A., Kossmehl, R., Sycuro, L., and McFall-Ngai, M.J. (2004). NO means 'yes' in the squid–vibrio symbiosis: nitric oxide (NO) during the initial stages of a beneficial association. Cell. Microbiol. *6*, 1139–1151.

DeLoney-Marino, C.R., Wolfe, A.J., and Visick, K.L. (2003). Chemoattraction of *Vibrio fischeri* to serine, nucleosides, and N-acetylneuraminic acid, a component of squid light-organ mucus. Appl. Environ. Microbiol. *69*, 7527–7530.

Doino, J., and McFall-Ngai, M. (1995). A transient exposure to symbiosis-competent bacteria induces light organ morphogenesis in the squid host. Biol. Bull. *189*, 347–355.

Dunlap, P.V., and Kuo, A. (1992). Cell density-dependent modulation of the *Vibrio fischeri* luminescence system in the absence of autoinducer and LuxR protein. J. Bacteriol. *174*, 2440–2448.

Dunn, A.K., Karr, E.A., Wang, Y., Batton, A.R., Ruby, E.G., and Stabb, E.V. (2010). The alternative oxidase (AOX) gene in *Vibrio fischeri* is controlled by NsrR and up-regulated in response to nitric oxide. Mol. Microbiol. *77*, 44–55.

Dutta, R., and Inouye, M. (2000). GHKL, an emergent ATPase/kinase superfamily. Trends Biochem. Sci. *25*, 24–28.

Engebrecht, J., and Silverman, M. (1984). Identification of genes and gene products necessary for bacterial bioluminescence. Proc. Natl. Acad. Sci. U.S.A. *81*, 4154–4158.

Engebrecht, J., Nealson, K., and Silverman, M. (1983). Bacterial bioluminescence: isolation and genetic analysis of functions from *Vibrio fischeri*. Cell *32*, 773–781.

Flemming, H.C., and Wingender, J. (2010). The biofilm matrix. Nat. Rev. Microbiol. *8*, 623–633.

Foster, J.S., and McFall-Ngai, M.J. (1998). Induction of apoptosis by cooperative bacteria in the morphogenesis of host epithelial tissues. Dev. Genes Evol. *208*, 295–303.

Foster, J.S., Apicella, M.A., and McFall-Ngai, M.J. (2000). *Vibrio fischeri* lipopolysaccharide induces developmental apoptosis, but not complete morphogenesis, of the *Euprymna scolopes* symbiotic light organ. Dev. Biol. *226*, 242–254.

Freeman, J.A., and Bassler, B.L. (1999). A genetic analysis of the function of LuxO, a two-component response regulator involved in quorum sensing in *Vibrio harveyi*. Mol. Microbiol. *31*, 665–677.

Gennis, R.B., and Stewart, V. (1996). Respiration. In *Escherichia coli* and *Salmonella typhimurium*: Cellular and Molecular Biology, Neidhardt, F.C., ed. (ASM press, Washington, DC), pp. 217–261.

Georgellis, D., Lynch, A.S., and Lin, E.C. (1997). *In vitro* phosphorylation study of the arc two-component signal transduction system of *Escherichia coli*. J. Bacteriol. *179*, 5429–5435.

Georgellis, D., Kwon, O., and Lin, E.C. (2001). Quinones as the redox signal for the arc two-component system of bacteria. Science *292*, 2314–2316.

Geszvain, K., and Visick, K.L. (2008a). Multiple factors contribute to keeping levels of the symbiosis regulator RscS low. FEMS Microbiol. Lett. *285*, 33–39.

Geszvain, K., and Visick, K.L. (2008b). The hybrid sensor kinase RscS integrates positive and negative signals to modulate biofilm formation in *Vibrio fischeri*. J. Bacteriol. *190*, 4437–4446.

Gilson, L., Kuo, A., and Dunlap, P.V. (1995). AinS and a new family of autoinducer synthesis proteins. J. Bacteriol. *177*, 6946–6951.

Graf, J., and Ruby, E.G. (1998). Host-derived amino acids support the proliferation of symbiotic bacteria. Proc. Natl. Acad. Sci. U.S.A. *95*, 1818–1822.

Graf, J., and Ruby, E.G. (2000). Novel effects of a transposon insertion in the *Vibrio fischeri glnD* gene: defects in iron uptake and symbiotic persistence in addition to nitrogen utilization. Mol. Microbiol. *37*, 168–179.

Graf, J., Dunlap, P.V., and Ruby, E.G. (1994). Effect of transposon-induced motility mutations on colonization of the host light organ by *Vibrio fischeri*. J. Bacteriol. *176*, 6986–6991.

Grau, B.L., Henk, M.C., Garrison, K.L., Olivier, B.J., Schulz, R.M., O'Reilly, K.L., and Pettis, G.S. (2008). Further characterization of *Vibrio vulnificus* rugose variants and identification of a capsular and rugose exopolysaccharide gene cluster. Infect. Immun. *76*, 1485–1497.

Gunsalus, R.P., Kalman, L.V., and Stewart, R.R. (1989). Nucleotide sequence of the *narL* gene that is involved in global regulation of nitrate controlled respiratory genes of *Escherichia coli*. Nucleic Acids Res. *17*, 1965–1975.

Heeb, S., and Haas, D. (2001). Regulatory roles of the GacS/GacA two-component system in plant-associated and other Gram-negative bacteria. Mol. Plant Microbe Interact. *14*, 1351–1363.

Hussa, E.A., O'Shea, T.M., Darnell, C.L., Ruby, E.G., and Visick, K.L. (2007). Two-component response regulators of *Vibrio fischeri*: identification, mutagenesis, and characterization. J. Bacteriol. *189*, 5825–5838.

Hussa, E.A., Darnell, C.L., and Visick, K.L. (2008). RscS functions upstream of SypG to control the *syp* locus and biofilm formation in *Vibrio fischeri*. J. Bacteriol. *190*, 4576–4583.

Jackson, M.D., Fjeld, C.C., and Denu, J.M. (2003). Probing the function of conserved residues in the serine/threonine phosphatase PP2Calpha. Biochemistry *42*, 8513–8521.

Jones, B.W., and Nishiguchi, M.K. (2004). Counterillumination in the Hawaiian bobtail squid, *Euprymna scolopes* Berry (Mullusca: Cephalopoda). Mar. Biol. *144*, 1151–1155.

Jones, B.W., and Nishiguchi, M.K. (2006). Differentially expressed genes reveal adaptations between free-living and symbiotic niches of *Vibrio fischeri* in a fully established mutualism. Can. J. Microbiol. *52*, 1218–1227.

Key, J., Hefti, M., Purcell, E.B., and Moffat, K. (2007). Structure of the redox sensor domain of *Azotobacter vinelandii* NifL at atomic resolution: signaling, dimerization, and mechanism. Biochemistry *46*, 3614–3623.

Kishii, R., Falzon, L., Yoshida, T., Kobayashi, H., and Inouye, M. (2007). Structural and functional studies of the HAMP domain of EnvZ, an osmosensing transmembrane histidine kinase in *Escherichia coli*. J. Biol. Chem. *282*, 26401–26408.

Koropatnick, T.A., Engle, J.T., Apicella, M.A., Stabb, E.V., Goldman, W.E., and McFall-Ngai, M.J. (2004). Microbial factor-mediated development in a host-bacterial mutualism. Science *306*, 1186–1188.

Kulkarni, P.R., Cui, X., Williams, J.W., Stevens, A.M., and Kulkarni, R.V. (2006). Prediction of CsrA-regulating small RNAs in bacteria and their experimental verification in *Vibrio fischeri*. Nucleic Acids Res. *34*, 3361–3369.

Kuo, A., Blough, N.V., and Dunlap, P.V. (1994). Multiple N-acyl-L-homoserine lactone autoinducers of luminescence in the marine symbiotic bacterium *Vibrio fischeri*. J. Bacteriol. *176*, 7558–7565.

Kuo, A., Callahan, S.M., and Dunlap, P.V. (1996). Modulation of luminescence operon expression by N-octanoyl-L-homoserine lactone in *ainS* mutants of *Vibrio fischeri*. J. Bacteriol. *178*, 971–976.

Lee, K.H., and Ruby, E.G. (1994). Effect of the squid host on the abundance and distribution of symbiotic *Vibrio fischeri* in nature. Appl. Environ. Microbiol. *60*, 1565–1571.

Lenz, D.H., Mok, K.C., Lilley, B.N., Kulkarni, R.V., Wingreen, N.S., and Bassler, B.L. (2004). The small RNA chaperone Hfq and multiple small RNAs control quorum sensing in *Vibrio harveyi* and *Vibrio cholerae*. Cell *118*, 69–82.

Lenz, D.H., Miller, M.B., Zhu, J., Kulkarni, R.V., and Bassler, B.L. (2005). CsrA and three redundant small RNAs regulate quorum sensing in *Vibrio cholerae*. Mol. Microbiol. *58*, 1186–1202.

Lilley, B.N., and Bassler, B.L. (2000). Regulation of quorum sensing in *Vibrio harveyi* by LuxO and σ^{54}. Mol. Microbiol. *36*, 940–954.

Liu, X., and De Wulf, P. (2004). Probing the ArcA-P modulon of *Escherichia coli* by whole genome transcriptional analysis and sequence recognition profiling. J. Biol. Chem. *279*, 12588–12597.

Lupp, C., and Ruby, E.G. (2004). *Vibrio fischeri* LuxS and AinS: comparative study of two signal synthases. J. Bacteriol. *186*, 3873–3881.

Lupp, C., and Ruby, E.G. (2005). *Vibrio fischeri* uses two quorum-sensing systems for the regulation of early and late colonization factors. J. Bacteriol. *187*, 3620–3629.

Lupp, C., Urbanowski, M., Greenberg, E.P., and Ruby, E.G. (2003). The *Vibrio fischeri* quorum-sensing systems *ain* and *lux* sequentially induce luminescence gene expression and are important for persistence in the squid host. Mol. Microbiol. *50*, 319–331.

Lyell, N.L., Dunn, A.K., Bose, J.L., and Stabb, E.V. (2010). Bright mutants of *Vibrio fischeri* ES114 reveal conditions and regulators that control bioluminescence and expression of the *lux* operon. J. Bacteriol. *192*, 5103–5114.

Lynch, A.S., and Lin, E.C. (1996). Transcriptional control mediated by the ArcA two-component response regulator protein of *Escherichia coli*: characterization of DNA binding at target promoters. J. Bacteriol. *178*, 6238–6249.

McCarter, L.L. (2001). Polar flagellar motility of the *Vibrionaceae*. Microbiol. Mol. Biol. Rev. *65*, 445–462.

McCarter, L.L. (2006). Regulation of flagella. Curr. Opin. Microbiol. *9*, 180–186.

McFall-Ngai, M., and Montgomery, M.K. (1990). The anatomy and morphology of the adult bacterial light organ of *Euprymna scolopes* Berry (Cephalopoda: Sepiolidae). Biol. Bull. *179*, 332–339.

Malpica, R., Franco, B., Rodriguez, C., Kwon, O., and Georgellis, D. (2004). Identification of a quinone-sensitive redox switch in the ArcB sensor kinase. Proc. Natl. Acad. Sci. U.S.A. *101*, 13318–13323.

Mandel, M.J., Wollenberg, M.S., Stabb, E.V., Visick, K.L., and Ruby, E.G. (2009). A single regulatory gene is sufficient to alter bacterial host range. Nature *458*, 215–218.

Martinez-Hackert, E., and Stock, A.M. (1997). The DNA-binding domain of OmpR: crystal structures of a winged helix transcription factor. Structure *5*, 109–124.

Mascher, T., Helmann, J.D., and Unden, G. (2006). Stimulus perception in bacterial signal-transducing histidine kinases. Microbiol. Mol. Biol. Rev. *70*, 910–938.

Meighen, E.A. (1991). Molecular biology of bacterial bioluminescence. Microbiol. Rev. *55*, 123–142.

Miller, M.B., Skorupski, K., Lenz, D.H., Taylor, R.K., and Bassler, B.L. (2002). Parallel quorum sensing systems converge to regulate virulence in *Vibrio cholerae*. Cell *110*, 303–314.

Millikan, D.S., and Ruby, E.G. (2002). Alterations in *Vibrio fischeri* motility correlate with a delay in symbiosis initiation and are associated with additional symbiotic colonization defects. Appl. Environ. Microbiol. *68*, 2519–2528.

Millikan, D.S., and Ruby, E.G. (2003). FlrA, a σ^{54}-dependent transcriptional activator in *Vibrio fischeri*, is required for motility and symbiotic light-organ colonization. J. Bacteriol. *185*, 3547–3557.

Millikan, D.S., and Ruby, E.G. (2004). *Vibrio fischeri* flagellin A is essential for normal motility and for symbiotic competence during initial squid light organ colonization. J. Bacteriol. *186*, 4315–4325.

Miyamoto, C.M., Lin, Y.H., and Meighen, E.A. (2000). Control of bioluminescence in *Vibrio fischeri* by the LuxO signal response regulator. Mol. Microbiol. *36*, 594–607.

Miyamoto, C.M., Dunlap, P.V., Ruby, E.G., and Meighen, E.A. (2003). LuxO controls *luxR* expression in *Vibrio harveyi*: evidence for a common regulatory mechanism in *Vibrio*. Mol. Microbiol. *48*, 537–548.

Miyashiro, T., Wollenberg, M.S., Cao, X., Oehlert, D., and Ruby, E.G. (2010). A single *qrr* gene is necessary and sufficient for LuxO-mediated regulation in *Vibrio fischeri*. Mol. Microbiol. *77*, 1556–1567.

Montgomery, M.K., and McFall-Ngai, M.J. (1992). The muscle-derived lens of a squid bioluminescent organ is biochemically convergent with the ocular lens. Evidence for recruitment of aldehyde dehydrogenase as a predominant structural protein. J. Biol. Chem. *267*, 20999–21003.

Montgomery, M.K., and McFall-Ngai, M. (1993). Embryonic development of the light organ of the Sepiolid squid *Euprymna scolopes* Berry. Biol. Bull. *184*, 296–308.

Montgomery, M.K., and McFall-Ngai, M. (1994). Bacterial symbionts induce host organ morphogenesis during early postembryonic development of the squid *Euprymna scolopes*. Development *120*, 1719–1729.

Morris, A.R., and Visick, K.L. (2010). Control of biofilm formation and colonization in *Vibrio fischeri*: a role for partner switching? Env. Microbiol. *12*, 2051–2059.

Morris, A.R., Darnell, C.L., and Visick, K.L. (2011). Inactivation of a novel response regulator is necessary for biolfilm formation and host colonization by *Vibrio fischeri*. Molec. Microbiol. *82*, 114–130.

Ng, W.L., and Bassler, B.L. (2009). Bacterial quorum-sensing network architectures. Annu. Rev. Genet. *43*, 197–222.

Nyholm, S.V., and McFall-Ngai, M.J. (1998). Sampling the light-organ microenvironment of *Euprymna scolopes*: description of a population of host cells in association with the bacterial symbiont *Vibrio fischeri*. Biol. Bull. *195*, 89–97.

Nyholm, S.V., and McFall-Ngai, M.J. (2004). The winnowing: establishing the squid–*Vibrio* symbiosis. Nat. Rev. Microbiol. *2*, 632–642.

Nyholm, S.V., Stabb, E.V., Ruby, E.G., and McFall-Ngai, M.J. (2000). Establishment of an animal–bacterial association: recruiting symbiotic vibrios from the environment. Proc. Natl. Acad. Sci. U.S.A. *97*, 10231–10235.

Nyholm, S.V., Deplancke, B., Gaskins, H.R., Apicella, M.A., and McFall-Ngai, M.J. (2002). Roles of *Vibrio fischeri* and nonsymbiotic bacteria in the dynamics of mucus secretion during symbiont colonization of the *Euprymna scolopes* light organ. Appl. Env. Microbiol. *68*, 5113–5122.

Nyholm, S.V., Stewart, J.J., Ruby, E.G., and McFall-Ngai, M.J. (2009). Recognition between symbiotic *Vibrio fischeri* and the haemocytes of *Euprymna scolopes*. Env. Microbiol. *11*, 483–493.

Popham, D.L., Szeto, D., Keener, J., and Kustu, S. (1989). Function of a bacterial activator protein that binds to transcriptional enhancers. Science *243*, 629–635.

Prouty, M.G., Correa, N.E., and Klose, K.E. (2001). The novel σ^{54}- and σ^{28}-dependent flagellar gene transcription hierarchy of *Vibrio cholerae*. Mol. Microbiol. *39*, 1595–1609.

Ravcheev, D.A., Gerasimova, A.V., Mironov, A.A., and Gelfand, M.S. (2007). Comparative genomic analysis of regulation of anaerobic respiration in ten genomes from three families of gamma-proteobacteria (*Enterobacteriaceae*, *Pasteurellaceae*, *Vibrionaceae*). BMC Genomics *8*, 54.

Romeo, T. (1998). Global regulation by the small RNA-binding protein CsrA and the non-coding RNA molecule CsrB. Mol. Microbiol. *29*, 1321–1330.

Romeo, T., Gong, M., Liu, M.Y., and Brun-Zinkernagel, A.M. (1993). Identification and molecular characterization of *csrA*, a pleiotropic gene from *Escherichia coli* that affects glycogen biosynthesis, gluconeogenesis, cell size, and surface properties. J. Bacteriol. *175*, 4744–4755.

Römling, U., Gomelsky, M., and Galperin, M.Y. (2005). c-di-GMP: the dawning of a novel bacterial signalling system. Mol. Microbiol. *57*, 629–639.

Ross, P., Weinhouse, H., Aloni, Y., Michaeli, D., Weinberger-Ohana, P., Mayer, R., Braun, S., de Vroom, E., van der Marel, G.A., van Boom, J.H., *et al.* (1987). Regulation of cellulose synthesis in *Acetobacter xylinum* by cyclic diguanylic acid. Nature *325*, 279–281.

Ruby, E.G. (1996). Lessons from a cooperative, bacterial–animal association: the *Vibrio fischeri-Euprymna scolopes* light organ symbiosis. Annu. Rev. Microbiol. *50*, 591–624.

Ruby, E.G., and Asato, L.M. (1993). Growth and flagellation of *Vibrio fischeri* during initiation of the sepiolid squid light organ symbiosis. Arch. Microbiol. *159*, 160–167.

Ruby, E.G., and Nealson, K.H. (1976). Symbiotic association of *Photobacterium fischeri* with the marine luminous fish *Monocentris japonica*; a model of symbiosis based on bacterial studies. Biol. Bull. *151*, 574–586.

Ruby, E.G. Urbanowski, M., Campbell, J., Dunn, A., Faini, M., Gunsalus, R., Lostroh, P., Lupp, C., McCann, J., Millikan, D., *et al.* (2005). Complete genome sequence of *Vibrio fischeri*: a symbiotic bacterium with pathogenic congeners. Proc. Natl. Acad. Sci. U.S.A. *102*, 3004–3009.

Ruiz, J.A., Fernandez, R.O., Nikel, P.I., Mendez, B.S., and Pettinari, M.J. (2006). dye (*arc*) Mutants: insights into an unexplained phenotype and its suppression by the synthesis of poly (3-hydroxybutyrate) in *Escherichia coli* recombinants. FEMS Microbiol. Lett. *258*, 55–60.

Salmon, K.A., Hung, S.P., Steffen, N.R., Krupp, R., Baldi, P., Hatfield, G.W., and Gunsalus, R.P. (2005). Global gene expression profiling in *Escherichia coli* K12: effects of oxygen availability and ArcA. J. Biol. Chem. *280*, 15084–15096.

Sanders, D.A., Gillece-Castro, B.L., Stock, A.M., Burlingame, A.L., and Koshland, D.E., Jr. (1989). Identification of the site of phosphorylation of the chemotaxis response regulator protein, CheY. J. Biol. Chem. *264*, 21770–21778.

Sanders, D.A., Gillece-Castro, B.L., Burlingame, A.L., and Koshland, D.E., Jr. (1992). Phosphorylation site of NtrC, a protein phosphatase whose covalent intermediate activates transcription. J. Bacteriol. *174*, 5117–5122.

Silversmith, R.E., and Bourret, R.B. (1999). Throwing the switch in bacterial chemotaxis. Trends Microbiol. *7*, 16–22.

Sitnikov, D.M., Schineller, J.B., and Baldwin, T.O. (1995). Transcriptional regulation of bioluminesence genes from *Vibrio fischeri*. Mol. Microbiol. *17*, 801–812.

Stabb, E. (2006). The *Vibrio fischeri-Euprymna scolopes* light organ symbiosis. In The Biology of *Vibrios*, Thompson, F.L., Austin, B., and Swings, J., eds (ASM Press, Washington, DC), pp. 204–218.

Stabb, E., Schaefer, A., Bose, J.L., and Ruby, E.G. (2008). Quorum signalling and symbiosis in the marine luminous bacterium *Vibrio fischeri*. In Chemical Communication Among Bacteria, Winans, S.C., and Bassler, B.L., eds (ASM Press, Washington, DC), pp. 233–250.

Stevens, A.M., Dolan, K.M., and Greenberg, E.P. (1994). Synergistic binding of the *Vibrio fischeri* LuxR transcriptional activator domain and RNA polymerase to the *lux* promoter region. Proc. Natl. Acad. Sci. U.S.A. *91*, 12619–12623.

Stock, A.M., Robinson, V.L., and Goudreau, P.N. (2000). Two-component signal transduction. Annu. Rev. Biochem. *69*, 183–215.

Studer, S.V., Mandel, M.J., and Ruby, E.G. (2008). AinS quorum sensing regulates the *Vibrio fischeri* acetate switch. J. Bacteriol. *190*, 5915–5923.

Taylor, B.L., and Zhulin, I.B. (1999). PAS domains: internal sensors of oxygen, redox potential, and light. Microbiol. Mol. Biol. Rev. *63*, 479–506.

Teather, R.M., and Wood, P.J. (1982). Use of Congo red–polysaccharide interactions in enumeration and characterization of cellulolytic bacteria from the bovine rumen. Appl. Env. Microbiol. *43*, 777–780.

Vance, R.E., Zhu, J., and Mekalanos, J.J. (2003). A constitutively active variant of the quorum-sensing regulator LuxO affects protease production and biofilm formation in *Vibrio cholerae*. Infect. Immun. *71*, 2571–2576.

Visick, K.L., and Skoufos, L.M. (2001). Two-component sensor required for normal symbiotic colonization of *Euprymna scolopes* by *Vibrio fischeri*. J. Bacteriol. *183*, 835–842.

Visick, K.L., Foster, J., Doino, J., McFall-Ngai, M., and Ruby, E.G. (2000). *Vibrio fischeri lux* genes play an important role in colonization and development of the host light organ. J. Bacteriol. *182*, 4578–4586.

Wang, Y., Dufour, Y.S., Carlson, H.K., Donohue, T.J., Marletta, M.A., and Ruby, E.G. (2010a). H-NOX-mediated nitric oxide sensing modulates symbiotic colonization by *Vibrio fischeri*. Proc. Natl. Acad. Sci. U.S.A. *107*, 8375–8380.

Wang, Y., Dunn, A.K., Wilneff, J., McFall-Ngai, M.J., Spiro, S., and Ruby, E.G. (2010b). *Vibrio fischeri* flavohaemoglobin protects against nitric oxide during initiation of the squid–Vibrio symbiosis. Mol. Microbiol. *78*, 903–915.

Weiss, D.S., Batut, J., Klose, K.E., Keener, J., and Kustu, S. (1991). The phosphorylated form of the enhancer-binding protein NTRC has an ATPase activity that is essential for activation of transcription. Cell *67*, 155–167.

West, A.H., and Stock, A.M. (2001). Histidine kinases and response regulator proteins in two-component signaling systems. Trends Biochem. Sci. *26*, 369–376.

Whistler, C.A., and Ruby, E.G. (2003). GacA regulates symbiotic colonization traits of *Vibrio fischeri* and facilitates a beneficial association with an animal host. J. Bacteriol. *185*, 7202–7212.

Whistler, C.A., Koropatnick, T.A., Pollack, A., McFall-Ngai, M.J., and Ruby, E.G. (2007). The GacA global regulator of *Vibrio fischeri* is required for normal host tissue responses that limit subsequent bacterial colonization. Cell. Microbiol. *9*, 766–778.

Wier, A.M., Nyholm, S.V., Mandel, M.J., Massengo-Tiasse, R.P., Schaefer, A.L., Koroleva, I., Splinter-Bondurant, S., Brown, B., Manzella, L., Snir, E., *et al.* (2010). Transcriptional patterns in both host and bacterium underlie a daily rhythm of anatomical and metabolic change in a beneficial symbiosis. Proc. Natl. Acad. Sci. U.S.A*107*, 2259–2264.

Wigneshweraraj, S., Bose, D., Burrows, P.C., Joly, N., Schumacher, J., Rappas, M., Pape, T., Zhang, X., Stockley, P., Severinov, K., *et al.* (2008). Modus operandi of the bacterial RNA polymerase containing the σ^{54} promoter-specificity factor. Mol. Microbiol. *68*, 538–546.

Wolfe, A.J., Millikan, D.S., Campbell, J.M., and Visick, K.L. (2004). *Vibrio fischeri* σ54 controls motility, biofilm formation, luminescence, and colonization. Appl. Env. Microbiol. *70*, 2520–2524.

Wong, S.M., Carroll, P.A., Rahme, L.G., Ausubel, F.M., and Calderwood, S.B. (1998). Modulation of expression of the ToxR regulon in *Vibrio cholerae* by a member of the two-component family of response regulators. Infect. Immun. *66*, 5854–5861.

Yildiz, F.H., and Visick, K.L. (2009). *Vibrio* biofilms: so much the same yet so different. Trends Microbiol. *17*, 109–118.

Yildiz, F.H., Dolganov, N.A., and Schoolnik, G.K. (2001). VpsR, a member of the response regulators of the two-component regulatory systems, is required for expression of *vps* biosynthesis genes and EPS(ETr)-associated phenotypes in *Vibrio cholerae* O1 El Tor. J. Bacteriol. *183*, 1716–1726.

Yip, E.S., Grublesky, B.T., Hussa, E.A., and Visick, K.L. (2005). A novel, conserved cluster of genes promotes symbiotic colonization and σ-dependent biofilm formation by *Vibrio fischeri*. Mol. Microbiol. *57*, 1485–1498.

Yip, E.S., Geszvain, K., DeLoney-Marino, C.R., and Visick, K.L. (2006). The symbiosis regulator *rscS* controls the *syp* gene locus, biofilm formation and symbiotic aggregation by *Vibrio fischeri*. Mol. Microbiol. *62*, 1586–1600.

Zenno, S., and Saigo, K. (1994). Identification of the genes encoding NAD(P)H-flavin oxidoreductases that are similar in sequence to *Escherichia coli* Fre in four species of luminous bacteria: *Photorhabdus luminescens*, *Vibrio fischeri*, *Vibrio harveyi*, and *Vibrio orientalis*. J. Bacteriol. *176*, 3544–3551.

Zhu, J., Miller, M.B., Vance, R.E., Dziejman, M., Bassler, B.L., and Mekalanos, J.J. (2002). Quorum-sensing regulators control virulence gene expression in *Vibrio cholerae*. Proc. Natl. Acad. Sci. U.S.A. *99*, 3129–3134.

Ziegler, M.M., and Baldwin, T.O. (1981). Biochemistry of bacterial bioluminescence. In Current Topics in Bioenergetics (Academic Press Chestnut Hill, MA), pp. 65–113.

Index

A

B

C

H

L

Q

R

S

T

U

V

Z

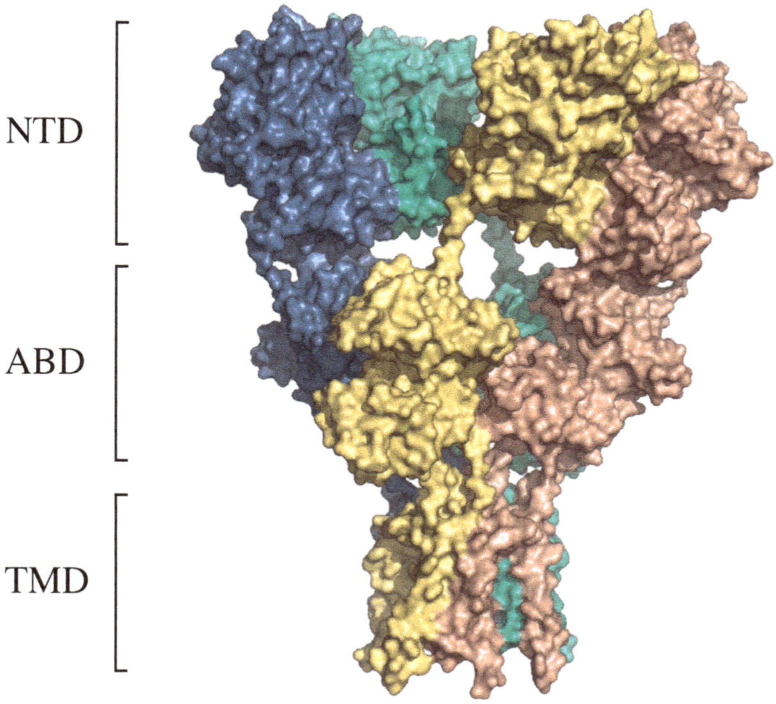

Figure 4.5 Space-filling model of an AMPA receptor (GluA2, pdb 3KG2). NTD, ADB and TMD represent the N-terminal domain (also known as ATD for amino-terminal domain), the agonist binding domain (aka LBD for ligand-binding domain) and the transmembrane domain (TMD), respectively. The NTD and ABD are both Venus flytrap domains (see text). GluA2 forms a homotetramer, and the four monomers are shown in different colours. The cavity of the ABD is empty and the ionic channel is closed in the structure presented here. The linker between the NTD and ABD (best seen in the yellow monomer) participates in signal transduction. Note that the structure displayed here is the only full-length structure available for this type of receptor. However, the system described in the text is that of the NMDA receptors, because more functional details are available for NMDA than for AMPA receptors.

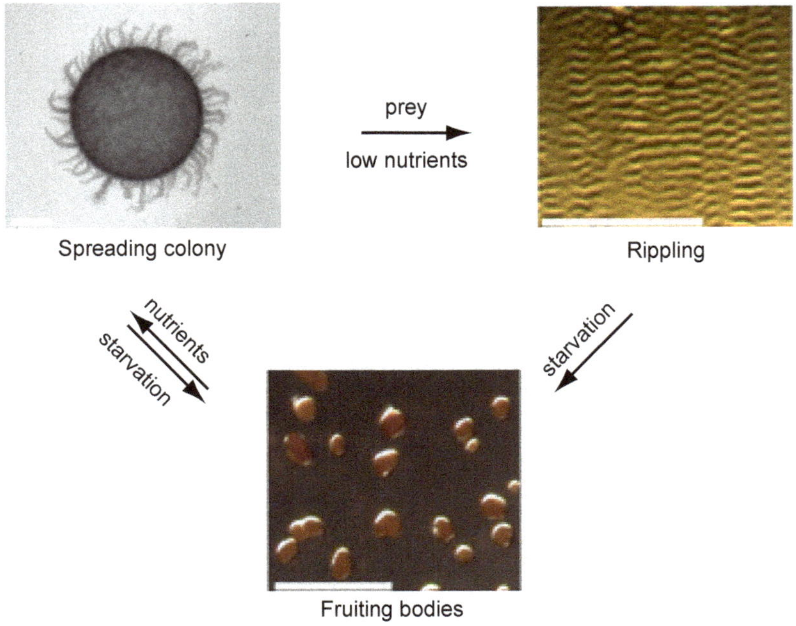

Figure 14.1 The three cellular patterns formed during the *M. xanthus* life cycle. *M. xanthus* cells organize into three types of patterns in response to nutrient levels. Scale bars: 1 mm.

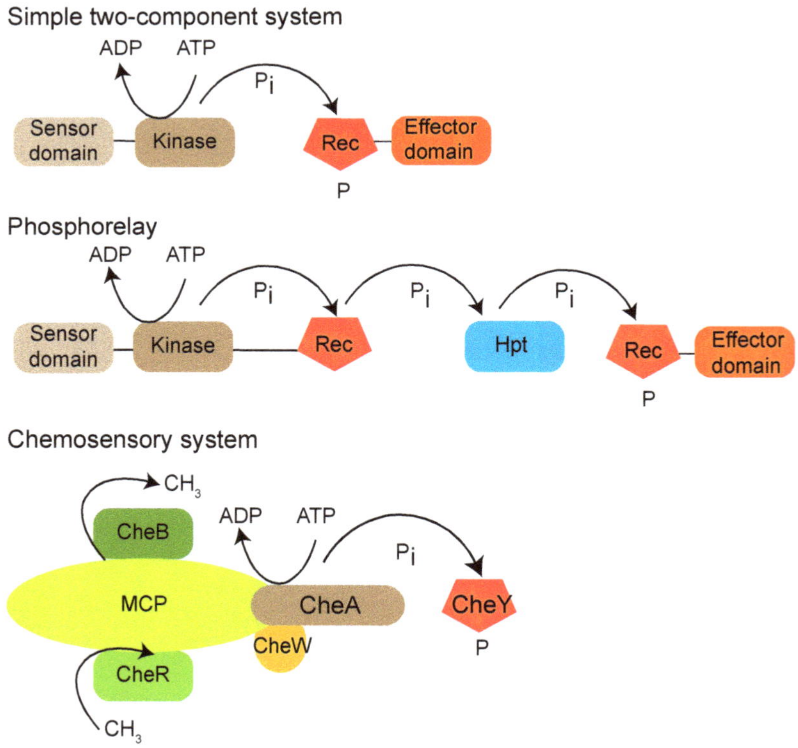

Figure 14.2 Structure of generic two-component signal transduction pathways. Schematics show structure and phosphotransfer reactions in a simple two-component system, in a phosphorelay, and in a chemosensory type of two-component system.

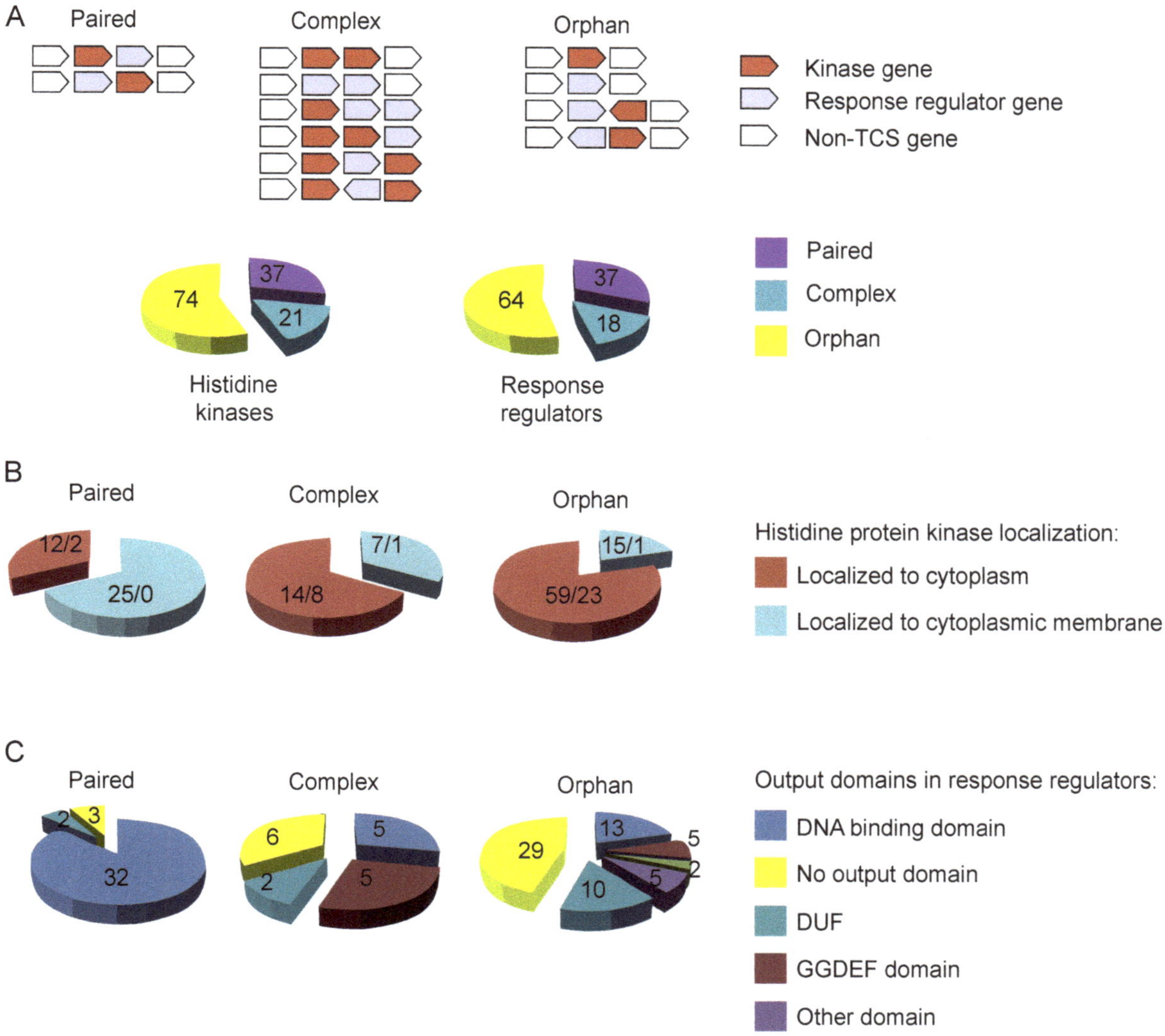

Figure 14.3 Two-component systems of *M. xanthus.* (A) Definition of genetic organization of genes of two-component systems. The pie-diagrams show the distribution of histidine kinase and response regulators encoded by genes in the different configurations. (B) Localization of histidine kinases according to genetic organization. The second number indicates the number of hybrid kinases. (C) Output domains in response regulators according to genetic organization. DUF, domains of unknown function; other domains indicate domains of known function.

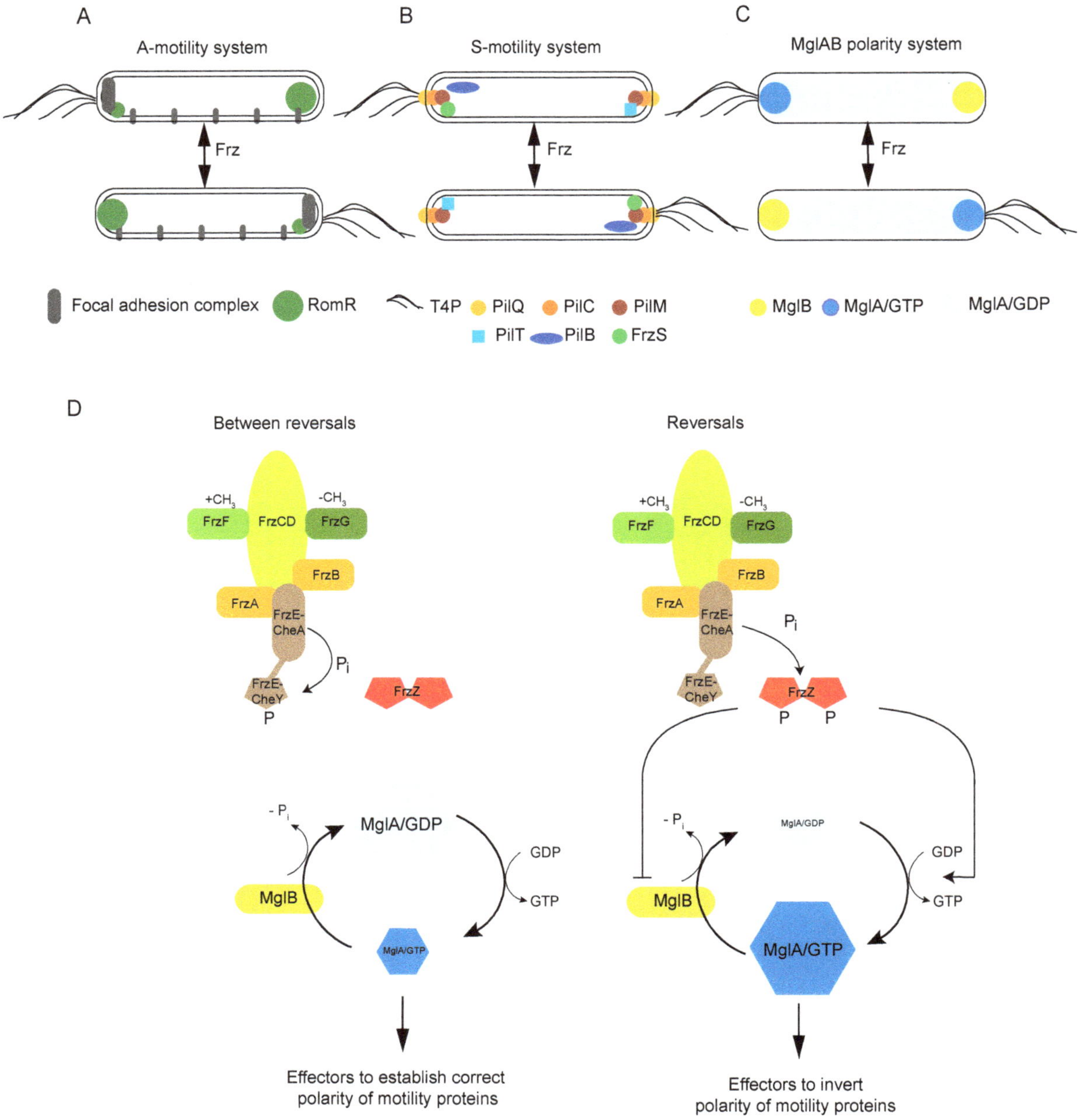

Figure 14.4 The chemosensory Frz system in combination with the MglA GTPase regulate the leading/lagging cell pole polarity axis. (A). Key components of the A-motility system. T4P indicate the leading cell pole. Focal adhesion complexes contain the proteins AglZ, AgmU, AglQ, AglR and AglS. These complexes are assembled at the leading pole and disassemble at the lagging pole. The polar clusters of RomR are indicated. The two cells demonstrate the localization of A-motility proteins before and after a reversal. (B) Key components of the S-motility system. T4P indicate the leading cell pole. The six T4P proteins that have been localized are shown in the indicated colour code. The two cells demonstrate the localization of T4P-proteins before and after a reversal. (C) Spatial regulation of MglA and MglB. Between reversals MglA/GTP binds at the leading pole and MglB at the lagging cell pole. The two proteins bind to opposite pole by a mechanism that involves mutual exclusion. MglA/GDP is diffusely localized to the cytoplasm. During a reversal and in a Frz-dependent manner, MglA/GTP is released from the leading cell pole and relocates to the lagging cell pole. Subsequently, MglB relocates to the new lagging cell pole. (D) Model of Frz chemosensory system. The two proposed alternative phosphate flows from the kinase domain of FrzE (FrzE-CheA) to either the CheY domain in FrzE (FrzE-CheY) or to the receiver-domains in FrzZ are shown. The signal that governs the direction of the phosphate flow is unknown. MglA is a nucleotide-dependent molecular switch and MglB its cognate GAP. Between reversals, MglB functions as a GAP. During reversals, FrzZ-PP is thought to either directly or indirectly act as a GEF or to inhibit GAP activity of MglB to stimulate accumulation of MglA/GTP.

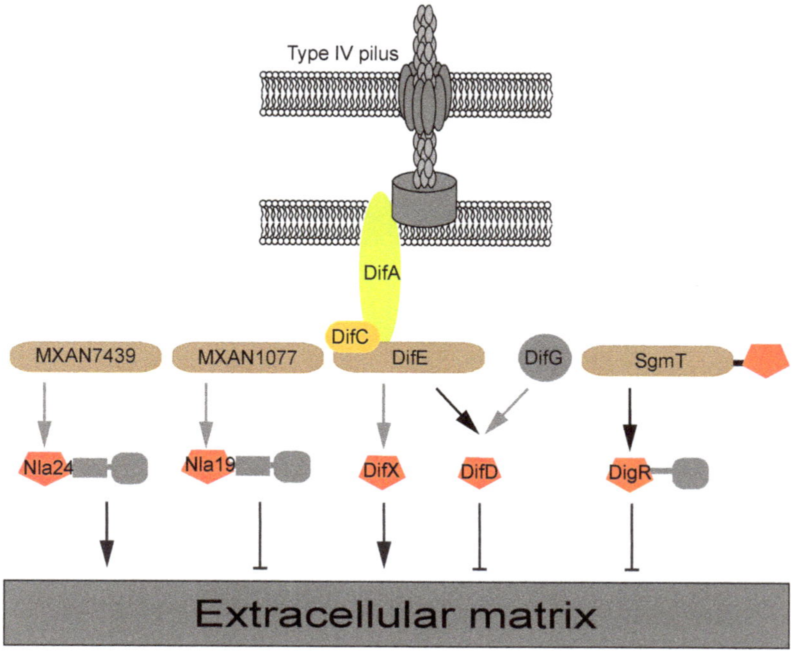

Figure 14.5 Regulation of extracellular matrix accumulation by proteins of two-component systems. A T4P is shown spanning the cell envelope and potentially interacting with DifACE proteins. From the DifE histidine protein kinase, the phosphoryl group is thought to be transferred to an unknown response regulator (DifX) to stimulate ECM accumulation or to DifD to inhibit ECM accumulation. Histidine protein kinase domains are indicated in light brown boxes; receiver domains are indicated by red pentagons. Nla24 and Nla19 are NtrC-like enhancer binding proteins. DigR is a DNA binding response regulator. Arrows or inhibitions indicate positive or negative effects on extracellular matrix accumulation. Grey arrows indicate that phosphotransfer has not been shown.

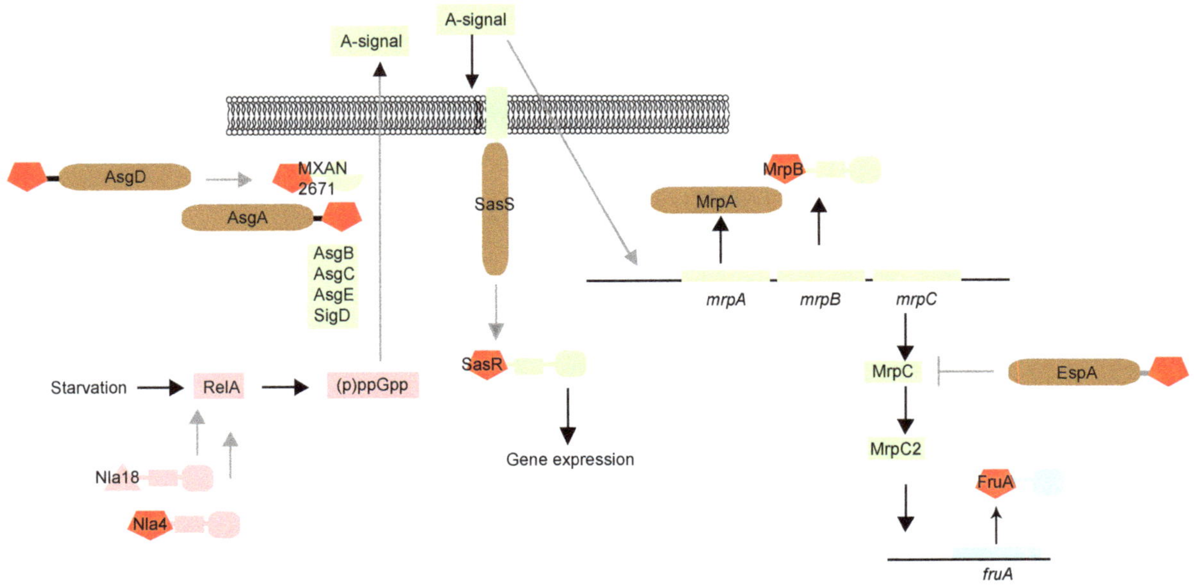

Figure 14.6 A-signal transduction pathway. A-signal synthesis is induced by the RelA-dependent stringent response with the accumulation of (p)ppGpp. By an unknown mechanism, this results in A-signal synthesis. The six proteins required for A-signal synthesis are indicated. Histidine protein kinase domains are indicated in light brown boxes; receiver domains are indicated by red pentagons; other domains are indicated in grey. Nla4, Nla18, SasR and MrpB are enhancer binding proteins; FruA is a DNA-binding response regulator. Light red indicates components of the stringent response; light green indicates components involved in synthesis and response to the A-signal; and, light blue indicates components in the C-signal transduction pathway. It is not known if AsgD acts in a pathway with the paired response regulator MXAN2671. Potential interactions are indicated by grey arrows.

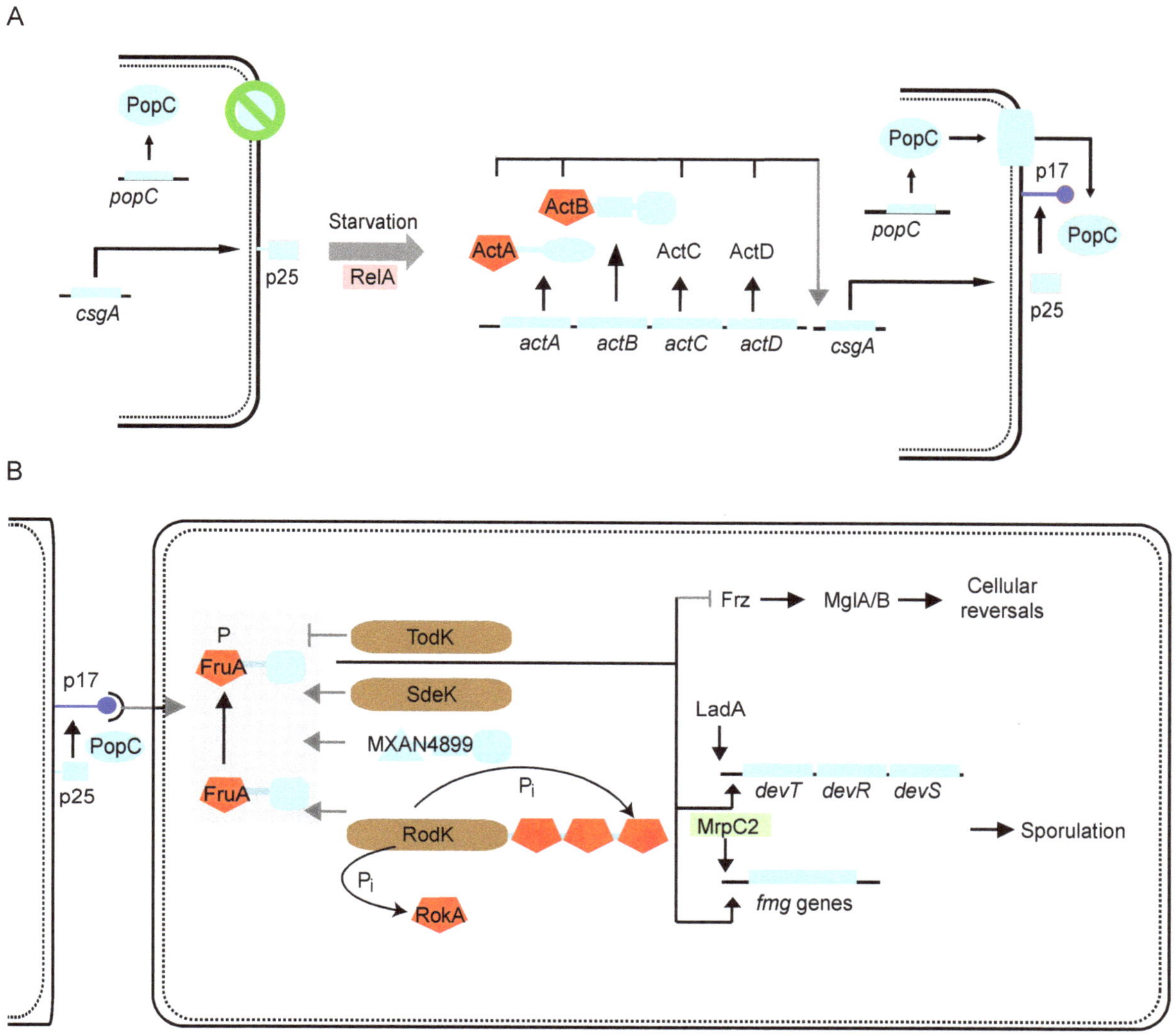

Figure 14.7 C-signal transduction pathway. (A) Synthesis of p17, the C-signal. The left-hand part of the schematic illustrates a vegetative cell and the right-hand part a starving cell. p25 and p17 are shown anchored in the outer membrane. The secretion system important for PopC secretion is not known. (B). The C-signal transduction pathway. The signalling event in the pathway is thought to be the interaction between p17 and a receptor on a neighbouring cell. This receptor remains to be identified. FruA is shown to be phosphorylated in response to C-signalling; genetic evidence supports this model, however, phosphorylation of FruA has not been demonstrated and a cognate kinase not identified. The grey box behind FruA and FruA-P indicates the level of the pathway where the signal transduction pathways defined by SdeK, TodK, RodK/RokA and MXAN4899 converge with the C-signal transduction pathway. Histidine protein kinase domains are indicated in light brown boxes; receiver domains are indicated by red pentagons. Potential interactions and indirect interactions are indicated by grey arrows. MrpC2 is marked on light green to indicate that it is induced by the A-signal pathway. MXAN4899 is an enhancer binding protein with an N-terminal FHA domain.

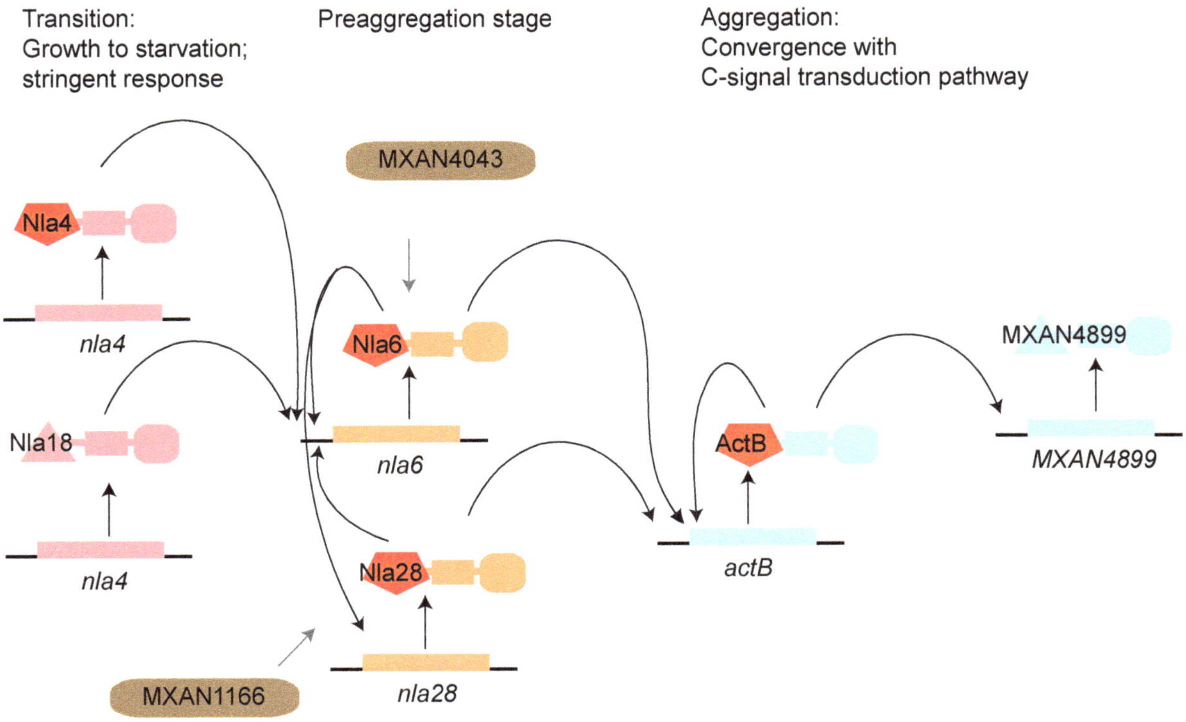

Figure 14.8 Enhancer binding protein cascade. Nla 4 and Nla18 are shown in light red to indicate their connection to the stringent response; Nla6 and Nla28 are shown in light orange and are not known to converge with other pathways than the one shown; ActB and MXAN4899 are shown in light blue to indicate that they connect to the C-signal transduction pathway. It is not known if MXAN1166 and MXAN4043 are the cognate kinases of Nla28 and Nla6, respectively. Histidine protein kinase domains are indicated in light brown boxes; receiver domains are indicated by red pentagons.

www.ingramcontent.com/pod-product-compliance
Lightning Source LLC
LaVergne TN
LVHW071345210926
848486LV00003B/230

* 9 7 8 1 9 0 8 2 3 0 0 8 9 *